HENDRIK ADRIAAN VAN REEDE TOT DRAKENSTEIN (1636-1691) AND HORTUS MALABARICUS

Figure 1. Hendrik Adriaan van Reede tot Drakenstein (1636-1691). Engraving about 1684, attributed to Pieter Stevensz. van Gunst, from Hortus Malabaricus. Muller 4479a.

HENDRIK ADRIAAN VAN REEDE TOT DRAKENSTEIN (1636-1691) AND HORTUS MALABARICUS
A CONTRIBUTION TO THE HISTORY OF DUTCH COLONIAL BOTANY

by
J. HENIGER
Biohistorical Institute, Utrecht

A.A.BALKEMA/ROTTERDAM/BOSTON/1986

Published with the financial support from
the Netherlands Organization for the Advancement of Pure Research (Z.W.O.).

ISBN 90 6191 681 X

© 1986 A.A.Balkema, P.O.Box 1675, 3000 BR Rotterdam, Netherlands

Distributed in USA & Canada by: A.A.Balkema Publishers, P.O.Box 230, Accord, MA 02018

Printed in the Netherlands

Contents

Introduction	VII
Summary	XI
Samenvatting	XIV

PART ONE: HENDRIK ADRIAAN VAN REEDE TOT DRAKENSTEIN

1. Van Reede in the Netherlands 1636-1650	3
2. Van Reede abroad 1650-1678	7
Cape of Good Hope 1657	7
Batavia 1657	10
The conquest of Malabar 1658-1663	13
Inspections of Malabar 1663-1667	18
First captain and sergeant-major of Ceylon 1667-1669	27
Commander of Malabar 1670-1677	31
Extraordinary Councillor of India 1677	51
3. Van Reede in the Netherlands 1678-1684	57
Arrival in the Netherlands	57
The Equestrian Order and the States of Utrecht	57
The Van Reede family	57
Hortus Malabaricus	59
Appointment as Commissioner-General of the Western Quarters	64
4. Van Reede abroad 1684-1691	69
Cape of Good Hope 1685	69
South Ceylon 1685	76
Bengal 1686-1687	77
Coromandel 1687-1689	78
North Ceylon 1689-1690	78
Tuticorin 1690-1691	78
Malabar 1691	80
Death and funeral	81
5. Epilogue	87
Francina van Reede	87
Elisabeth Antonia van Panhuizen	90

PART TWO: HORTUS MALABARICUS

Introduction	95
6. Bibliography	97

	Description	97
7.	Viridarium Orientale of Matthew of St. Joseph	105
	The Paris Codex of Viridarium Orientale	107
8.	Drawings of Hortus Malabaricus	125
	Description	125
9.	Renaissance botany	139
10.	Plant descriptions of Hortus Malabaricus	143
	General scheme of the descriptions	143
	Genesis of the descriptions	144
11.	Commentaries	153
	Arnold Syen	153
	Jan Commelin	159
12.	Insertion of Hortus Malabaricus in the botanical literature	171

PART THREE: ANNOTATED LIST OF PLANTS

Signs and abbreviations	178
Annotated list of plants	179
Volume 1	179
Volume 2	187
Volume 3	194
Volume 4	199
Volume 5	206
Volume 6	211
Volume 7	218
Volume 8	223
Volume 9	229
Volume 10	239
Volume 11	250
Volume 12	257
Appendices	265
Map of Malabar	273
Literature	275
Index of persons	281
Index of geographical names	285
Index of plant names	289
Index of animal names	293
Index of ship names	295

Introduction

In the 17th and 18th centuries the Dutch Republic played a prominent part in Europe in the study of botany. Representatives of all strata of Dutch society, at home and overseas, contributed to the flourishing of botanical science in the Netherlands. Not only professional botanists, professors of botany, and directors of botanical gardens, but also amateurs, soldiers and sailors, surgeons, chemists, physicians, merchants, administrators, politicians, and even the stadtholders of the Dutch Republic were engaged in collecting, growing, describing, and classifying plants. Draughtsmen, painters and engravers, translators and correctors collaborated closely with well-equipped printing offices in publishing splendidly illustrated works, which still constitute a climax in botanical world literature. Well-stocked gardens, museums of natural curiosities, and libraries in this country at that time furnished the botanist with such a wealth and variety of material for study that Paul Hermann (1646-1695), one of the then coryphaei, could justly speak of the *Paradisus Batavus*, the Dutch Paradise. Allured by the fame of Dutch botany, the young Linnaeus here spent some years, 1735-1738, to complete his schooling.

The flourishing of Dutch botany in the 17th and 18th centuries was largely due to the world-wide commercial empire of the East and West India Companies. Besides the commercial profits, by which botanical activities in the Netherlands could be financed, their ships brought countless unknown exotic plants from Asia, Africa, and America to the Netherlands. With unremitting zeal Company officals of all ranks and classes sent reports, descriptions, drawings, exsiccates, seeds, fruits, cuttings, and even complete plants to the mother-country. The study of exotic plants led to the pre-eminent Dutch specialism of exotic or colonial botany. Some Company officials had an opportunity to carry out thorough research on the floras of their stations. The names of Marcgraf and Piso, Bontius, Rumphius, Hermann, Van Reede, Cleyer and Kaempfer, Oldenland, and many others are linked indissolubly with the exploration of the flora of the exotic regions within the Dutch sphere of influence.

Dutch colonial botany was characterized by the making of herbaria of dried plants, the composition of codices of plant drawings, and the description of plants. The significance of colonial botany consisted above all in phytography. The Company official did not have at his disposal up-to-date scientific literature at his station by means of which to cudgel his brains about problems of botanical nomenclature and taxonomy. That is why Dutch botanists at home and their colleagues in other European countries were faced with the difficult task of classifying, at a great distance from their habitats, the luxurious abundance of forms of exotic plants and bring them into line with the European flora with which they were so familiar.

Against the background sketched above I have devoted this study to Hendrik Adriaan van Reede tot Drakenstein (1636-1691), a Utrecht nobleman, who as a commander of Malabar in the service of the United East India Company composed a flora of Malabar, the present federal State of Kerala in India, which he published in 1678-1693 under the title of *Hortus Indicus Malabaricus, Continens Regni Malabarici apud Indos celeberrimi omnis generis Plantas rariores*, below briefly referred to as Hortus Malabaricus.

The life and work of Van Reede form a fascinating chapter in Dutch colonial botany. They are eminently suited for sketching a picture of the botanical activities in the East and the interaction with Europe.

Hortus Malabaricus has become famous for various reasons in the history of botany. In the first place it is a climax in 17th-century botanical literature because of its ample size of twelve folio volumes, its detailed descriptions of plants, and its magnificently produced engravings in double folio size. In addition Hortus Malabaricus became known for the fascinating account of its genesis, which Van Reede himself has described in the preface to the third volume. Finally -and this is the most important point- Hortus Malabaricus was one of the main sources of Linnaeus for his knowledge of the tropical flora of Asia. After this, Van Reede's work was, and still is, consulted by taxonomists in their critical studies of Linnaean species.

Van Reede is ranked with his contemporaries and fellow-servants of the Company, the Ceylon botanist Paul Hermann and the Ambon naturalist Georg Everhard Rumphius. The three of them laid the main foundation for the Dutch knowledge of the tropical flora of Asia.

Van Reede was in the first place a servant of the Company, and only in the second place an amateur of botany.

During a career of nearly thirty years he rose from a simple midshipman to the all-powerful commissioner-general of the Western Quarters of Asia, admired by his friends and feared by his enemies. He showed himself to be a formidable warrior, a successful negotiator, an authoritarian administrator, and above all a prolix rapporteur on almost all facets of colonial policy and politics. In apparent contrast with this is Van Reede's love of Malabar, the country, the people, and the flora, as Kalff (1905) established. In my biography of Van Reede I have sought to refute this supposed contrast. By means of a detailed description of his career, particularly in Ceylon, Malabar, and Batavia in the years 1656-1677, I would show that his research about the Malabar plants is completely in line with the policy of the Company concerning the provision of medicaments and the practical search for useful plants. I have gone into less detail as regards Van Reede's activities as a commissioner-general in the years 1684-1691, because in that period he hardly occupied himself any more with the flora of Malabar. I did examine how he used this high office to stimulate botanical research at the Cape of Good Hope and in Ceylon.

I have given relatively much attention in the biography to persons who played a part in his life, career, and botanical activities. As organizer of Hortus Malabaricus Van Reede gathered about him a large circle of collaborators of all manner of nationalities and capacities. The most important minor biography within this Van Reede biography is that of Matthew of St. Joseph, the "conditor" or founder of Hortus Malabaricus.

In the analysis of Hortus Malabaricus the bibliographical, historical, and botanical aspects are discussed in order to ascertain the plan and the structure of the work. In the light of Matthew's earlier botanical writings, the *Viridarium Orientale*, the influence of Paul Hermann on the plant descriptions will be discussed. Through lack of sources I have not been able to study the actual contributions of the Malabar scholars to the description in more detail. The annotations of Arnold Syen and Jan Commelin to each species have ultimately determined the systematical arrangement of Hortus Malabaricus, and their notes on the preserved original drawings give some idea of their problems.

The reactions to Hortus Malabaricus and its incorporation in other botanical works have been studied by reference to the writings of fifteen botanists from the 17th and 18th centuries, pre-Linnaeans, Linnaeus, and Linnaeans. Since after Van Reede until the end of the Company's administration (1795) little further botanical research took place in Malabar, they mainly identified and interpreted Van Reede's plants with the aid of the Ceylon herbaria of Paul Hermann and Pieter Hertog and the Coromandel herbarium of Samuel Browne. I have not ventured on a new interpretation of the plants in Hortus Malabaricus, because I lack the taxonomic experience needed. Moreover, some years ago a renewed research was started by the Department of Botany of Calicut University, Kerala (Manilal et al. 1977). On the other hand I have added a list of all the plants with their native names described and illustrated in Hortus Malabaricus, with references to the original drawings and the quotations from the above-mentioned botanists who incorporated the material in their works. This list serves on the one hand as a survey of the botanical literature consulted by me, and on the other hand as a list of sources for the purpose of facilitating future historical and botanical research.

Sources and literature

In the literature the first essential biographical data about Van Reede are found in the writings of his contemporaries, the surgeon Schouten (1676), the soldier Herport (1669), the merchant Havart (1693), and the clergyman Valentijn (1724-1726). After a long silence, Veth (1887) provided some supplementary data by reference to material from the records. Entirely new data, based on examination of the records, were published recently by s'Jacob (1976), Fournier (1978, 1980), and Heniger (1980). Nevertheless, considerable gaps have remained, such as Van Reede's youth, his family relations, and his stay in Ceylon, Batavia, and Utrecht. By means of detailed examination of the records I have tried to fill these gaps further and to supplement the known data.

In spite of many investigations I have not succeeded in tracing personal records of Van Reede. That is why I had to base the biography of Van Reede primarily on the records of the Dutch East India Company in the General State Archive in The Hague. In this context use was made especially of the so-called "Overgekomen Brieven" (Letters Received), consisting of letters and reports, with numerous annexes, sent by the colonial administrators to Lords XVII and the Chamber of Amsterdam. In particular the letters from Malabar, Ceylon, Cape of Good Hope, Surat, and Batavia are rich sources for a reconstruction of the colonial career of Van Reede.

The share of Van Reede in the "Overgekomen Brieven" consists of a bulky mass of letters and reports produced by him as a commander of Malabar and as a commissioner-general of the Western Quarters. Their contents cover practically all the facets of colonial administration. Also in letters and reports of his superior Rijklof van Goens, governor of Ceylon, and of his predecessors and successors as commanders of Malabar, Van Reede's conduct is very often discussed. Only two long reports of Van Reede have been published in full: the Memorandum on Malabar of 1677, the most important document in the historiography of Dutch Malabar, by s'Jacob (1976), and the inspection report on Cape of Good Hope of 1685, with interesting data on Van Reede's view of applied botany, by Hulshof (1941).

Another part of the records of the East India Company that is important for Van Reede's career consists of the resolutions of Lords XVII, the resolutions of the Chamber of Amsterdam, and their correspondence with the High Government in Batavia. The State Archive of Utrecht contains material of a widely varied nature with reference to Van Reede. The Huydecoper Family Records contain the letter-books, with annexed diaries, of Joan Huydecoper van Maarsseveen, which on the one hand shed a vivid light on the factions within the East India Company, in which Van Reede played a part, and on the other hand furnish information on the botanical commerce between Amsterdam and the East in the days of Van Reede. The records of the Houses of Zuilen and of Amerongen supply detailed information about the Van Reede family in the 16th and 17th centuries. The records of the States of Utrecht contain the resolutions of the "Ridderschap" (Equestrian Order) and of the States, of which two administrative bodies Van Reede was a member in the years 1678-1684. With respect to the genesis of Hortus Malabaricus I have been able to make use of two important unpublished sources. In the British Library in London are to be found the original drawings underlying the engravings of the first ten volumes of Hortus Malabaricus. The drawings of volumes 11 and 12 still are not to be found. In the Musée d'Histoire Naturelle in Paris there is a manuscript of *Viridarium Orientale* of Matthew of St. Joseph, the descriptions and drawings of which give a good idea of the original plan of Hortus Malabaricus.

Acknowledgments

I am greatly indebted to the following persons and staffs of institutions for their assistance and criticisms: P. Brederoo (Leiden), Ms Ph.I. Edwards (London), Ms M. Fournier (Leiden), A. Govindankutty (Leiden), H.K. s'Jacob (Groningen), L. Krijnen (Vianen), Th. Laurentius (Voorschoten), P. Leenhouts (Leiden), H.L.Ph. Leeuwenberg (Vianen), K.S. Manilal (Calicut, Kerala), D.O. Wijnands (Wageningen); and the staffs of the General State Archive The Hague, the State Archive Utrecht, the Municipal Archive Utrecht, the Municipal Archive Amsterdam, the Department of Western Manuscripts of the British Library of London and the library of the Musée d'Histoire Naturelle of Paris. Sincere thanks are due to Ms C. Dikshoorn (Zeist) for the English translation of the manuscript, to Ms I.N. van Splunder (Utrecht), Ms E.M.M. Kool (Utrecht), and Ms T.W. de Vries-Koppert (Utrecht) for typing and preparing it for the printer, to Ms C.M.J. Hielkema (Utrecht) for bibliographical assistance, and to H. van der Veer (Utrecht) for assistance with photographs.

Abbreviations

ARA	Algemeen Rijksarchief, 's-Gravenhage; General State Archive, The Hague
BL	British Library, London
DTB	Doop-, trouw- en begraafregisters; Baptismal, marriage, and funeral registers
GAA	Gemeentearchief Amsterdam; Municipal Archive Amsterdam
GAU	Gemeentearchief Utrecht; Municipal Archive Utrecht
GAV	Gemeentearchief Vianen; Municipal Archive Vianen
MHN	Musée d'Histoire Naturelle, Paris
NNBW	Nieuw Nederlandsch Biografisch Woordenboek, see Literature
n.p.	no pagination
RAU	Rijksarchief Utrecht; State Archive Utrecht
VOC	Archief Verenigde Oost-Indische Compagnie; Archive Dutch East India Company

Reproduction of photographs by courtesy of
Atlas Stolk Rotterdam Fig. 19
British Library London Figs 34, 59-63, 65, 67, 68, 70, 72, 74, 76-80, 82
Iconographisch Bureau The Hague Fig. 4
General State Archive The Hague Fig. 13
Musée d'Histoire Naturelle Paris Figs 25, 27, 29, 38, 44, 46, 48, 50, 52, 56
Rijksmuseum Amsterdam Fig. 3
State Archive Utrecht Figs 2, 15, 16, 20
University Library Leiden Fig. 9
University Library Utrecht Figs 1, 5-8, 10-12, 14, 21-24, 26, 28, 30-33, 35-37, 39-43, 45, 47, 49, 51, 53-55, 57, 58, 64, 66, 69, 71, 73, 75, 81, 83-85

Summary

This study aims at a threefold purpose: (1) a description of the life of Hendrik Adriaan van Reede tot Drakenstein (1636-1691) as a servant of the Dutch East India Company and as organizer of the study of colonial botany; (2) an analysis of the previous history and genesis of his Hortus Malabaricus, published in the years 1678-1693, dealing with the flora of Malabar; (3) a survey of the process of insertion of Hortus Malabaricus in the botanical literature of the 17th and 18th centuries.

In Part One a detailed biography of Van Reede is given. In this Kalff's view is investigated that there is a contradiction between Van Reede's imperative behaviour as commissioner-general of the Western Quarters and his love of the land, the people, and the flora of Malabar. This investigation is hampered by the absence of personal records of Van Reede, so that an opinion about Kalff's view is mainly based on the records of the Company in the General State Archive at The Hague.

Hendrik Adriaan van Reede tot Drakenstein was born as a younger son from a Utrecht noble family that played a leading part in the political, administrative, and cultural life of this Dutch province. He grew up in a setting in which diplomacy, military and maritime affairs, reclamation and colonization, architecture and garden architecture, and above all honour and splendour predominated and in which a liberal religious attitude was not unknown. In this setting, however, there was little interest in the pursuit of the sciences. Van Reede accordingly has not had a scientific training, also owing to the untimely death of his parents and his departure from the Netherlands at the age of fourteen.

In 1656 he entered the service of the Company. On the way to the East he made friends with Joan Bax van Herentals and Isaac de Saint-Martin, who like him were to fill prominent posts in the Company's administration and were also to make contributions to the development of botany in the Dutch colonies. During the stays at the Cape of Good Hope and Batavia Van Reede received his first impressions of exotic nature. After this he took part as a soldier in the expeditions of Rijklof van Goens, which resulted in the definitive conquest of Ceylon and Malabar. Owing to his heroic behaviour during the conquest of Cochin (1663) he gained rapid promotion, thanks to Van Goens' patronage. As councillor, "regedore maior", captain, diplomat, and inspector in Malabar in 1663-1667 he ranged through that country, became thoroughly acquainted with the Malabar society, gained the respect of the native princes, notably the rāja of Cochin, Vīra Kērala Varma. Furthermore he was impressed by the luxuriant tropical nature and grew to appreciate the scientific knowledge of the Brahmins. In this period arose his wish to get to know the flora of Malabar, in particular the medicinal plants.

Van Reede continued his career as sergeant-major of Ceylon, 1667-1669. At that time he was mainly occupied by the Madurese war, which he brought to a good conclusion by his successful defence of Tuticorin. It was at that very time that Robert Padbrugge was exploring the natural resources of Ceylon, in connection with which the medical shop at Batavia, directed by Andries Cleyer, requested the government of Ceylon for assistance in the research for medicinal plants, with a view to improving the way in which the medicaments within the Company were supplied. This development was not only an inducement for the arrival of Paul Hermann, the founder of Ceylon botany, but also Van Reede's motive for compiling Hortus Malabaricus.

As commander of Malabar, 1670-1677, Van Reede at first had to occupy himself with the restoration of the declined authority of the Company in the commandment, for which he made use of his military and diplomatic talents and of his influence on the rāja of Cochin. By his victory over the Zamorin of Calicut (1671), his averting of a threatening French invasion (1672), and the establishment of the Union of Mouton (1674) he restored quiet. He was less successful in commercial policy. He contributed to the improved supply of medicaments by the establishment of a laboratory in Cochin, which also implied a first step on the road to independent botanical research in Malabar. The arrival of Matthew of St. Joseph in Cochin (1674) was the birth of Hortus Malabaricus. Van Reede himself indeed had already made an attempt to take in hand the investigation of the Malabar flora, but the pressure of his activities as commander formed a hindrance.

During his missionary work Matthew had compiled a voluminous illustrated manuscript about medicinal plants, *Viridarium Orientale*, which he wanted to be published by the Leiden professor of oriental languages Jacob van Gool. After an adventurous, but vain pursuit by the Company of the manuscript it ultimately ended up in Italy, where the Bolognese professor of botany Giacomo Zanoni was to publish parts of it. Notes and drawings similar to those of *Viridarium Orientale* served as the first version of Hortus Malabaricus. A visit of Paul Hermann to Cochin led to a second, definitive version.

In the preparations for Hortus Malabaricus Van Reede had the assistance of many collaborators, both Europeans and Malayali, whom he collected by virtue of his authority as commander and through his prestige among the natives.

Nevertheless the study of Malabar plants met with a premature end because in 1677 Van Reede felt obliged to leave Malabar. Through his authoritarian attitude in the council of Malabar he had made enemies, who exploited a long-smouldering conflict between him and his former patron Van Goens about the relationship of the commandment of Malabar to the government of Ceylon. The laboratory and the work on Hortus Malabaricus were severely criticized, and some collaborators, such as Matthew of St. Joseph, were put in an unfavourable light. Also thanks to the support of Joan Huydecoper van Maarsseveen, a director of the Company, Van Reede was given the office of extraordinary councillor of India in Batavia (1677).

In Batavia he continued the work on Hortus Malabaricus with the assistance of, among others, Willem ten Rhijne, but once again Van Reede was forced to fall back in the conflict with Van Goens, which was growing more and more fierce. When he had returned to the Netherlands (1678), Van Reede at first shunned Company's affairs. He settled in Utrecht, where he was admitted into the Equestrian Order of Utrecht and became a member of the States of Utrecht. In those functions he engaged in creditable activities for the reconstruction of the province of Utrecht, which had suffered heavily under the French occupation.

In the continuation of the work on Hortus Malabaricus Van Reede had to contend with a depression. The publication of the first two volumes, in 1678 and 1679, did not come up to his expectations. He was greatly affected by the death of his first editor, the Leiden professor of botany Arnold Syen, and of his publishers Johannes van Someren and Jan van Dyck. Moreover, from Batavia he was criticized by Van Goens, who had meanwhile become governor-general. Not until 1681 did Van Reede resume his botanical work by concluding a publishers' contract. With the assistance of the Utrecht professor of botany Johannes Munnicks as editor and the Amsterdam amateur of botany Jan Commelin as commentator he made some following volumes of Hortus Malabaricus ready for the press. In volume 3 (1682) in a long preface he rendered an account of the aim, the scope, and the genesis of the work, in which he also gave an enthusiastic description of the land, the people, and the flora of Malabar. These pages, the only ones by his hand which were printed besides the dedications to the volumes, also formed a defence against Van Goens.

Gradually Van Reede became involved again in Company's affairs, first as adviser, until in 1684 he was appointed commissioner-general of the Western Quarters, his terms of reference being the fight against corruption.

With great scrupulousness, relentless severity, and remarkable judgment he performed that task, which took him by way of the Cape of Good Hope, Ceylon, Bengal, and Coromandel to Malabar, during which he far exceeded the time allotted to him and the limits of his mandate and by which he definitively established his reputation as a merciless inspector and a builder of fortresses and towns squandering large sums of money. At the Cape his measures with respect to the Company's garden and his approval of an expedition to Namaqualand by Simon van der Stel contributed to a further development of Cape botany, which had already started under Bax.

From Ceylon, Bengal, and Coromandel he sent plants to the Amsterdam Municipal Garden, founded by Huydecoper and Commelin, which benefited the commentaries of the other volumes of Hortus Malabaricus. Finally, in 1691, he ordained an annual consignment of seeds from the Western Quarters to the Netherlands. In particular the government of Ceylon kept this up until the second half of the 18th century, to the advantage of the further development of Ceylon botany.

In the epilogue about the descendants of Van Reede a vain search for his personal records is reported.

In conclusion it may be said that Kalff's view is not correct. In the service of the Company Van Reede had a magnificent career, which owing to his special knowledge, his strong and imperious attitude, and his prestige among the natives brought him to the high office of commissioner-general. The enmity he incurred was due to his devotion to duty rather than to ambition. This same devotion to duty resulted in the compilation of Hortus Malabaricus.

Viewed in this light, there is no contradiction between his imperative behaviour as commissioner-general of the Western Quarters and his love of the land, the people, and the flora of Malabar. He used this office to create conditions for a greater flourishing of botany at the Cape of Good Hope and in Ceylon.

In Part Two the previous history and genesis of Hortus Malabaricus as well as its insertion in the botanical literature of the 17th and 18th centuries is described. This is preceded by a short bibliography of the original Latin edition of Hortus Malabaricus, so as to give some insight into its complicated structure.

The next chapter contains a description of the Paris codex of *Viridarium Orientale* of Matthew of St. Joseph. On the basis of a discussion of the drawings, plant names, and descriptions in this manuscript a picture is given of the first version of Hortus Malabaricus.

In the chapter on the drawings of Hortus Malabaricus a survey is given of the still existing drawings which have served as models for the engravings for the volumes 1 to 10. It is made plausible that these drawings are merely copies of the original drawings which Van Reede left behind with Ten Rhijne in 1678, when he left Batavia. It also describes to what extent the drawings were complete and which additions were not published in Hortus Malabaricus.

The chapter concerning Renaissance botany expounds the purpose and the methods and means by which the predecessors of Van Reede's commentators identified and classified newly found plants. The importance of Bauhin's *Pinax Theatri Botanici* for the commentaries is emphasized.

In the chapter on the plant descriptions first a survey of their structure is given. This is followed by an extensive discussion of the genesis of the descriptions and the contributions made by Matthew, Hermann, Malabar physicians and princes, translators, clerks, and editors.

Subsequently the achievements of the commentators Arnold Syen and Jan Commelin are discussed. For each of them a survey is given of the means, such as a botanical garden, herbarium, museum, and library, which were at their service in the dicussion, identification, and classification of Van Reede's plants. For Syen his Ceylonese collection, originating from Hermann, is found to play a central part in the identification. Syen's problems in classifying are elucidated by means of the numerals inscribed on the drawings of Hortus Malabaricus. They ultimately resulted, against Van Reede's original intention, in a Bauhinian arrangement of the plants in the volumes 1 and 2.

Commelin, more so than Syen, has been able to reap the fruits of the rapidly developing colonial botany. Thanks to the exertions of Huydecoper, Van Reede, and many others he could study many tropical plants in the Amsterdam Municipal Garden and other gardens for the purpose of the identification, in which context another Ceylon herbarium, originating from Hermann, rendered him good services. For the classification he followed at first Bauhin, but afterwards he switched over to the plant system of John Ray, which he applied as far as possible for the arrangement of the volumes 3 to 12.

The last chapter deals with the insertion of Hortus Malabaricus in the botanical literature by reference to a group of selected botanists. It is ascertained what means and methods they used and what results they arrived at. The most important authors were Ray, Johannes Burman, and Linnaeus. Ray fitted by far the larger part of Van Reede's plants in his plant system and by summaries of the plant descriptions made Hortus Malabaricus more accessible. Johannes Burman compared the Ceylonese flora on a large scale with that of Malabar, for which he made use of Commelin's Ceylon herbarium from Hermann and Pieter Hertog's Ceylon herbarium. Linnaeus followed Burman's method in his description of Hermann's own Ceylon herbarium and thus established, in modern taxonomic respects, the relationship between Hortus Malabaricus and the flora of Ceylon.

In Part Three an Annotated List of Plants is given. This list consists of a concordance of the engravings and drawings of Hortus Malabaricus as well as a survey of the references to the botanical literature discussed in chapter 12. The list further serves as a source of relevant identifications of Van Reede's plants before 1795.

Samenvatting

Deze studie beoogt een drieledig doel: (1) een beschrijving van het leven van Hendrik Adriaan van Reede tot Drakenstein (1636-1691) als dienaar van de Nederlandse Oost-Indische Compagnie en als organisator van de beoefening van koloniale plantkunde; (2) een analyse van de voorgeschiedenis en het ontstaan van zijn Hortus Malabaricus, gepubliceerd in de jaren 1678-1693, handelend over de flora van Malabar; (3) een overzicht van het proces van de incorporatie van de Hortus Malabaricus in de Europese botanische literatuur van de 17de en 18de eeuw.

In Deel Een wordt een gedetailleerde biografie van Van Reede gegeven. Daarin wordt Kalff's opvatting onderzocht dat er een tegenstelling zou bestaan tussen Van Reede's gebiedende gedrag als commissaris-generaal van de Westerkwartieren en zijn liefde voor het land, het volk en de flora van Malabar. Dit onderzoek wordt belemmerd door het ontbreken van een persoonlijk archief van Van Reede, zodat een oordeel over Kalff's opvatting hoofdzakelijk gebaseerd is op het archief van de Compagnie in het Algemeen Rijksarchief in 's-Gravenhage.

Hendrik Adriaan van Reede tot Drakenstein werd geboren als een jongere zoon van een Utrechtse adellijke familie die een leidende rol speelde in het politieke, bestuurlijke en culturele leven van deze provincie. Hij groeide op in een milieu, waarin diplomatie, leger en vloot, ontginning en kolonisatie, bouwkunst en tuinkunst, en bovenal eer en glorie de boventoon voerden, en waarin een vrijzinnige godsdienstige houding niet onbekend was. In dit milieu bestond echter weinig belangstelling voor wetenschapsbeoefening. Van Reede heeft dan ook, mede door de vroege dood van zijn ouders en zijn vertrek uit Nederland op veertienjarige leeftijd, geen wetenschappelijke opleiding genoten.

In 1656 trad hij in dienst van de Compagnie. Op weg naar de Oost sloot hij vriendschap met Joan Bax van Herentals en Isaac de Saint-Martin. Zij zouden, evenals hij, vooraanstaande ambten in het bestuur van de Compagnie gaan bekleden en eveneens bijdragen tot de ontwikkeling van de plantkunde in de Nederlandse koloniën. Tijdens de oponthouden aan Kaap de Goede Hoop en in Batavia ontving Van Reede zijn eerste indrukken van de exotische natuur. Daarna nam hij als soldaat deel aan de expedities van Rijklof van Goens, die leidden tot de definitieve verovering van Ceylon en Malabar. Door zijn heldhaftig gedrag bij de verovering van Cochin (1663) maakte Van Reede snel promotie dankzij Van Goens' bescherming. Als raad, "regedore maior", kapitein, diplomaat en inspecteur in Malabar doorkruiste hij in de jaren 1663-1667 dat land, leerde de Malabaarse maatschappij grondig kennen en verwierf zich de achting van de inheemse vorsten, met name van de rāja van Cochin, Vīra Kēreala Varma. Bovendien kwam hij onder de indruk van de weelderige tropische natuur en kreeg waardering voor de wetenschappelijke kennis van de Bramanen. In deze periode ontstond zijn wens om de flora van Malabar, met name de geneeskrachtige planten, te leren kennen.

In 1667-1669 vervolgde Van Reede zijn loopbaan als sergeant-majoor van Ceylon. Hij werd toen voornamelijk in beslag genomen door de Madurese oorlog, die hij door zijn succesvolle verdediging van Tuticorin tot een goed einde wist te brengen. Juist in die tijd onderzocht Robert Padbrugge de natuurlijke hulpbronnen van Ceylon. Naar aanleiding hiervan verzocht Andries Cleyer, het hoofd van de medicinale winkel te Batavia, aan het gouvernement van Ceylon om hulp bij het onderzoek naar geneeskrachtige planten, teneinde de medicamentenvoorziening binnen de Compagnie te verbeteren. Deze ontwikkeling was niet alleen een aanleiding tot de komst van Paul Hermann, de grondlegger van de plantkunde van Ceylon, maar ook Van Reede's motief om de Hortus Malabaricus samen te stellen.

Als commandeur van Malabar, 1670-1677, moest Van Reede zich aanvankelijk bezighouden met het herstel van het verminderde gezag van de Compagnie in het commandement. Hierbij maakte hij gebruik van zijn militaire en diplomatieke gaven en van zijn invloed op de rāja van Cochin. Met zijn overwinning op de Zamorin van Calicut (1671), zijn afweer van een dreigende Franse inval (1672) en de oprichting van de Unie van Mouton (1674) herstelde hij de rust. Minder succes had hij in de handelspolitiek. Door de stichting van een laboratorium te Cochin droeg hij bij tot een verbetering van de medicamentenvoorziening. Dit betekende tevens een eerste stap op weg naar zelfstandig botanisch onderzoek in Malabar. De komst van de Italiaanse arts, missionaris en botanicus Mattheus van St. Jozef naar Cochin (1674) was de geboorte van de Hortus Malabaricus. Weliswaar had Van Reede reeds een poging gedaan om zelf het onderzoek naar de Malabaarse flora ter hand te nemen, maar zijn drukke werkzaamheden als commandeur vormden een belemmering.

Mattheus had tijdens zijn missiewerk een omvangrijk,

Samenvatting

geïllustreerd manuscript over geneeskrachtige planten, het *Viridarium Orientale*, samengesteld, dat hij wilde laten uitgeven door de Leidse hoogleraar in Oosterse talen Jacob van Gool. Na een avontuurlijke, maar vergeefse jacht op het manuscript door de Compagnie belandde het uiteindelijk in Italië, waar de Bolognese hoogleraar in de plantkunde Giacomo Zanoni gedeelten ervan zou publiceren. Soortgelijke aantekeningen en tekeningen als die van het *Viridarium Orientale* dienden als eerste versie van de Hortus Malabaricus. Een bezoek van Paul Hermann aan Cochin leidde tot een tweede, definitieve versie.

Van Reede liet zich tijdens de voorbereidingen tot de Hortus Malabaricus bijstaan door vele medewerkers, zowel Europeanen als Malayali, die hij krachtens zijn autoriteit als commandeur en door zijn prestige bij de inheemse bevolking om zich heen verzamelde.

Niettemin kwam aan de studie van de Malabaarse planten een voortijdig einde, doordat Van Reede zich in 1677 gedwongen voelde Malabar te verlaten. Door zijn autoritaire houding in de raad van Malabar had hij zich vijanden gemaakt, die inspeelden op een reeds lang smeulend conflict tussen hem en zijn vroegere beschermheer Van Goens over de verhouding van het commandement Malabar tot het gouvernement Ceylon. Het laboratorium en de bezigheden aan de Hortus Malabaricus werden ernstig bekritiseerd en sommige medewerkers, zoals Mattheus van St. Jozef, in een kwaad daglicht gesteld. Mede dankzij de steun van Joan Huydecoper van Maarsseveen, een bewindhebber van de Compagnie, kreeg Van Reede het ambt van raad extra-ordinair te Batavia (1677).

Met de hulp van onder meer Willem ten Rhijne zette hij in Batavia het werk aan de Hortus Malabaricus voort, maar opnieuw moest Van Reede wijken in het steeds feller worden conflict met Van Goens. Teruggekeerd in Nederland (1678) hield hij zich aanvankelijk ver van Compagnie's aangelegenheden. Hij vestigde zich in Utrecht, waar hij werd opgenomen in de Ridderschap en lid van de Staten van Utrecht werd. In die ambten maakte hij zich verdienstelijk voor de wederopbouw van de provincie Utrecht, die zwaar geleden had onder de Franse bezetting.

Bij de voortzetting van het werk aan de Hortus Malabaricus had Van Reede met een inzinking te kampen. De publicatie van de eerste twee delen, in 1678 en 1679, beantwoordde niet aan zijn verwachtingen. De dood van zijn eerste redacteur, de Leidse hoogleraar in de plantkunde Arnold Syen, en van zijn uitgevers Johannes van Someren en Jan van Dyck troffen hem zwaar. Bovendien kreeg hij vanuit Batavia kritiek te verduren van Van Goens, die inmiddels gouverneur-generaal was geworden. Eerst in 1681 vatte Van Reede, met het sluiten van een uitgeverscontract, zijn botanische werkzaamheden weer op. Met behulp van de Utrechtse hoogleraar in de plantkunde Johannes Munnicks als redacteur en de Amsterdamse plantenliefhebber Jan Commelin als commentator maakte hij enkele volgende delen van de Hortus Malabaricus persklaar. In deel 3 (1682) legde hij in een lang voorwoord verantwoording af van het doel, de opzet en het ontstaan van het werk. Hierin gaf hij ook een geestdriftige beschrijving van het land, het volk en de flora van Malabar. Dit geschrift, het enige van zijn hand dat buiten de opdrachten van de delen gedrukt werd, was tevens een verweerschrift tegen Van Goens.

Geleidelijk aan raakte Van Reede weer betrokken bij aangelegenheden van de Compagnie, eerst als adviseur, totdat hij in 1684 benoemd werd tot commissaris-generaal van de Westerkwartieren met de opdracht de corruptie te bestrijden. Met grote nauwgezetheid, onverbiddelijke gestrengheid en opmerkelijk inzicht voerde hij die taak uit, die hem leidde lang Kaap de Goede Hoop, Ceylon, Bengalen, Coromandel en Malabar, waarbij hij de hem toegemeten tijd en de grenzen van zijn opdracht verre overschreed en waardoor hij zijn faam als genadeloze inspecteur en geldverspillende bouwer van forten en steden definitief vestigde. Aan de Kaap droegen zijn maatregelen met betrekking tot de Compagnie's tuin en zijn toestemming aan Simon van der Stel tot een expeditie naar Namaqualand bij tot een verdere ontwikkeling van de Kaapse plantkunde, die reeds onder gouverneur Bax van Herentals begonnen was.

Vanuit Ceylon, Bengalen en Coromandel stuurde hij planten naar de door Huydecoper en Commelin gestichte Amsterdamse Stadstuin. Deze planten kwamen ook de commentaren van de overige delen van de Hortus Malabaricus ten goede. Tenslotte ordineerde hij in 1691 een jaarlijkse verzending van plantenmateriaal uit de Westerkwartieren naar Nederland. Met name het gouvernement van Ceylon heeft zich tot in de tweede helft van de 18de eeuw daaraan gehouden tot voordeel van de verdere ontwikkeling van de Ceylonese plantkunde.

In het naschrift over de nakomelingen van Van Reede wordt verslag van een vergeefse speurtocht naar zijn persoonlijke archief gedaan.

Concluderend kan gezegd worden, dat Kalff's opvatting niet juist is. Van Reede had in dienst van de Compagnie een schitterende loopbaan, die hem door zijn kennis van zaken, zijn krachtige, gebiedende optreden en zijn prestige bij de inheemse bevolking tot het hoge ambt van commissaris-generaal bracht. De vijandschap die hij zich op de hals haalde, was eerder een gevolg van zijn plichtsbetrachting dan van heerszucht.

In dit licht gezien bestaat er geen tegenstelling tussen zijn gebiedende houding als commissaris-generaal van de Westerkwartieren en zijn liefde voor het land, het volk en de flora van Malabar. Hij heeft dit ambt gebruikt om voorwaarden te scheppen tot een grotere bloei van de plantkunde aan de Kaap de Goede Hoop en in Ceylon.

In Deel Twee worden de voorgeschiedenis en het ontstaan van de Hortus Malabaricus beschreven alsmede de incorporatie van dit werk in de botanische literatuur van de 17de en 18de eeuw. Dit wordt voorafgegaan door een korte bibliografie van de oorspronkelijke Latijnse uitgave van de Hortus Malabaricus, teneinde een inzicht te geven in de ingewikkelde structuur ervan.

Het volgende hoofdstuk bevat een beschrijving van de Parijse codex van het *Viridarium Orientale* van Mattheus van St. Jozef. Aan de hand van de tekeningen, plantenamen en beschrijvingen in dit manuscript wordt een beeld van de eerste versie van de Hortus Malabaricus gegeven.

In het hoofdstuk over de tekeningen van de Hortus Malabaricus wordt een overzicht gegeven van de nog bestaande tekeningen, die als voorbeelden voor de gravures van de delen 1-10 hebben gediend. Daarin wordt aannemelijk gemaakt dat deze tekeningen gecopieerd zijn naar de oorspronkelijke tekeningen, die Van Reede in 1678 bij zijn vertrek uit Batavia bij Ten Rhijne heeft achtergelaten. Eveneens wordt beschreven, in hoeverre de tekeningen voltooid waren en welke toevoegingen niet in de Hortus Malabaricus gepubliceerd werden.

Het hoofdstuk over de plantkunde tijdens de Renaissance geeft het doel en de methoden en middelen weer, waarmee de voorgangers van Van Reede's commentatoren pas ontdekte planten identificeerden en classificeerden. Hierbij wordt de betekenis van Bauhin's *Pinax Theatri Botanici* voor de commentaren naar voren gebracht.

In het hoofdstuk over de plantebeschrijvingen wordt eerst een overzicht van hun structuur gegeven. Daarna volgt een uitvoerige bespreking van het ontstaan van de beschrijvingen en de bijdragen die Mattheus, Hermann, Malabaarse artsen en vorsten, vertalers, klerken en redacteuren hebben geleverd.

Vervolgens worden de prestaties van de commentatoren, Arnold Syen en Jan Commelin, besproken. In een overzicht van de middelen die hen ten dienste stonden bij de discussie, identificatie en classificatie van Van Reede's planten, komen aan de orde de botanische tuin, herbaria, museum en bibliotheek. Bij Syen blijkt zijn Ceylonese collectie, die afkomstig was van Hermann, een centrale rol in de identificatie te spelen. De problemen waarvoor Syen zich bij de classificatie gesteld zag, worden verduidelijkt aan de hand van zijn aantekeningen op de tekeningen van de Hortus Malabaricus. Zij leidden uiteindelijk, tegen de oorspronkelijke bedoeling van Van Reede in, tot een Bauhiniaanse rangschikking van de planten in de delen 1 en 2.

Commelin heeft, meer dan Syen, de vruchten van de zich snel ontwikkelende koloniale plantkunde kunnen plukken. Dankzij de inspanningen van Huydecoper, Van Reede en vele anderen kon hij ten behoeve van de identificatie vele tropische planten in de Amsterdamse Stadstuin en andere tuinen bestuderen, waarbij een ander Ceylonees herbarium, dat eveneens van Hermann afkomstig was, hem goede diensten bewees. Bij de classificatie volgde hij aanvankelijk Bauhin, maar daarna ging hij over op het plantensysteem van John Ray, dat hij, voor zover dat mogelijk was, toepaste op de rangschikking van de delen 3-12.

Het laatste hoofdstuk handelt over de incorporatie van de Hortus Malabaricus in de botanische literatuur aan de hand van een groep geselecteerde botanici. Hierbij wordt nagegaan, welke middelen en methoden zij gebruikten en tot welke resultaten zij kwamen. De belangrijkste schrijvers waren Ray, Johannes Burman en Linnaeus. Ray paste het overgrote deel van Van Reede's planten in zijn plantensysteem in en maakte door samenvattingen van de plantebeschrijvingen de Hortus Malabaricus beter toegankelijk. Johannes Burman vergeleek op grote schaal de flora van Ceylon met die van Malabar, waarbij hij gebruik maakte van Commelin's Ceylonese herbarium van Hermann en van Hertog's herbarium. Linnaeus volgde Burman's methode bij zijn beschrijving van Hermann's eigen Ceylonese herbarium en vestigde zodoende in modern-taxonomisch opzicht het verband tussen de Hortus Malabaricus en de flora van Ceylon.

In Deel Drie wordt een geannoteerde lijst van planten gegeven. Deze lijst bestaat uit een concordantie van de gravures en tekeningen van de Hortus Malabaricus, alsmede uit een overzicht van de verwijzingen naar de botanische literatuur die in hoofdstuk 12 wordt besproken. De lijst dient tevens als bron van relevante identificaties van Van Reede's planten vóór 1795.

Part One
Hendrik Adriaan van Reede tot Drakenstein

1
Van Reede in the Netherlands 1636-1650

Not much is known about Van Reede's early days. Hendrik was born in 1636 as the youngest child of Ernst van Reede, lord of De Vuursche and Drakenstein, and Elisabeth Utenhove. His parents had him christened on Sunday, 13 April of that year, in the Old Church in Amsterdam[1].

Ernst van Reede at first fulfilled the function of forester of the Land of Utrecht, 1619-1626[2]. At that time he lived in the city of Utrecht at St. Janskerkhof[3]. During the last years of his life he was a member of the Council of the Admiralty in Amsterdam, 1633-1640[4].

Hendrik's mother died in February 1637, and his father followed her shortly afterwards, in October 1640, into the grave[5]. The couple had had twelve children in all, some of whom died at an early age[6]. The many orphans, and thus also Hendrik, were placed under the guardianship of an uncle, Godard van Reede van Nederhorst, a brother of Ernst van Reede[7].

But this uncle died already in 1648, when Hendrik was only twelve years old. Perhaps Hendrik's eldest brother Gerard, who had meanwhile come of age, then took over the guardianship[8]. The actual guardianship of Gerard cannot have been of long duration, for 'from the beginning of my fourteenth year I have done nothing but travel beyond the borders of my native country', Hendrik wrote about himself. From this it also follows that he will not have received much schooling; in his later career, too, he had no opportunity for this or, as he himself testified: 'I do not excel either in varied and profound learning or in accurate knowledge of botany, spices, and medicaments'[9].

Although we know next to nothing about Hendrik's boyhood, something can indeed be said about the setting from which he sprang and the extent to which this may have influenced his later behaviour and interests. Hendrik's later life was to be dominated by a complex of coherent factors. In his career we shall be concerned by turns with nobility, military affairs, administration, diplomacy, finance, reclamation, trade, colonization, religion, building affairs, and -last but not least- science. Although it cannot strictly be proved, it may be assumed that already in his youth Hendrik came into contact with the majority of these factors in his direct environment.

By birth Hendrik belonged to a very distinguished and influential noble family, which, divided into different branches, produced many military men, diplomats, and public servants from the 16th to the 18th century. The centre of the Van Reedes lay in the province of Utrecht, where they owned extensive estates. They owed their standing to the ownership of various manor-houses (privileged castles and country-houses), which gave them access to the Equestrian Order of Utrecht, and thus to prominent public offices. During the 17th century the Van Reedes held important posts in the States of Utrecht and in the States General[10].

Among the older relations of Hendrik we find as the best-known persons his uncle and guardian, Godard van Reede van Nederhorst, negotiator in the Treaty of Münster (1648), which put an end to the Eighty Years' War against Spain[11]. Another uncle was Johan van Reede van Renswoude, president of the States of Utrecht and ambassador in England and Brandenburg[12]. His famous cousin, Godard Adriaan van Reede van Amerongen, served as ambassador in Denmark, Sweden, and Brandenburg[13]. They helped to make possible the development of the Republic to a leading European power.

The honour and glory of the family was a delicate point with the Van Reedes. Amidst the predominantly middle-class society in the Republic the noble families formed a minority, which tended to decrease rather than increase. In Hendrik's branch of the family a distasteful affair had occurred. His grandfather, Gerard van Reede van Nederhorst, had married far beneath his high rank. By doing so, he endangered the full nobility of his descendants, and consequently also their access to the Equestrian Order. With difficulty he could possess himself of his patrimony, the parentage of his wife remained obscure[14]. Nevertheless Hendrik bore the fictitious arms of his grandmother quite undismayed and considered himself to belong to the high nobility of Utrecht[15]. His imperious attitude later on as commander of Malabar, and as commissioner-general, may partly be attributed to his sense of honour as a nobleman.

Through intermarriages the Van Reedes formed a well-knit clan[16] which played a dominant role in the Equestrian Order. Thus they were able to pave the way to a successful career also for relatives who were less richly blessed with worldly goods, and to raise the status of the family by this means. Thus, the offices of forester of the Land of Utrecht and of councillor of the Admiralty in Amsterdam were allotted to Hendrik's father, Ernst van Reede, who himself did not own a recognized manor-house.

Against this general background the young Hendrik

Figure 2. House Drakenstein near De Vuursche about 1650. Anonymous Indian ink drawing, 18.2 × 39.5 cm. RAU, Topographical Atlas 1119:185.

will have grown up. He was to make his own contribution to the honour and glory of the Van Reedes by dedicating several volumes of Hortus Malabaricus to his nearest relations[17].

Apart from their role in the States of Utrecht, the Van Reedes made many contributions to the cultural and economical flourishing of the province of Utrecht. In part the rise of the Utrecht country-seat is due to their delight in building, while large parts of the then still existing wilderness were reclaimed thanks to their exertions.

The Van Reedes devoted much attention to their manor-houses. They altered and embellished them. Sometimes on the site of these old-fashioned medieval castles they erected entirely new country-seats. A well-known example is the reconstruction of the house of Nederhorst, which was completed about 1635 by Hendrik's uncle and guardian, Godard van Reede van Nederhorst. Hendrik will undoubtedly have known this beautiful country-seat quite well in his youth[18]. The castle of Amerongen was embellished considerably by Godard Adriaan van Reede van Amerongen[19]. Gerard van Reede van Renswoude built the famous, still existing house of Renswoude[20]. Since 1640 Hendrik's own brother, Gerard van Reede van Drakenstein, built the remarkable octagonal house of Drakenstein, from which the two brothers were to derive the second part of their surname, Van Reede van (tot) Drakenstein (Fig.2)[21]. In the same period the house of Rijnhuizen was erected by their brother-in-law, René van Tuyll van Serooskerken[22].

In addition Hendrik's nearest relations were greatly interested in reclamation and colonization projects. His grandfather, Gerard van Reede van Nederhorst, and the latter's sons invested a good deal of money in draining Horstermeer[14]. In 1626 Hendrik's father bought the manor De Vuursche with vast waste lands and began to reclaim them and make them habitable. Shortly after 1640 Hendrik's brother Gerard erected the church and the village of De Vuursche and spent nearly the whole of his fortune on the completion of this project[23]. Hendrik's uncle and guardian Godard too was not behindhand in this respect. He took part in the damming-in of Heer-Hugowaard in North Holland, dug the Rhedervaart in the village Nederhorst-den-Berg, and after 1640 engaged in the foundation of the colony Nederhorst on Staten Island, New York[24]. In his colonial career Hendrik's passion for building and pioneering was to become a subject of fierce criticism

All these building and reclaiming activities weighed upon the finances of the Van Reedes. Hendrik's father for a long time had groaned under heavily burdened family estates[25]. In 1636 he even became involved in an embezzlement scandal, which reached such a height that the States General intervened in it[26]. Hendrik's brother Gerard finally went bankrupt[8]. The uncle and guardian Godard and his descendants, too, were hard put to it to keep their heads above water[27]. One is inclined to think of a family trait when later, in his Malabar period, Hendrik is reproached with constant impecuniosity and excessive financial demands.

In military matters Hendrik had an impressive example. His mother's father was the famous colonel Antoni Utenhove, who during the Eighty Years' War had become widely known because of his stubborn, but vain defence of the fortresses of Oostende (1604) and Emmerich (1606) against the Spanish general Spinola[28]. Much less known are Hendrik's uncles by marriage, Maurits Lodewijk de la Baye and Ernst van Abcoude van Meerten, who also served in the army of the States[29]. A fact of more direct influence on the young Hendrik will have been the fate of his elder brother Frederik, captain in the army of the Dutch West India Company in Brazil. Frederik was one of the numerous casualties in the disastrous battle of the Guarapes of 19 February 1649, which heralded the expulsion of the Dutch from Brazil by the Portuguese[30]. In Hendrik's family therefore there existed a military tradition, and a certain heroic aspect cannot even be denied to it. In this respect Hendrik was to consolidate that tradition by the heroic part he played in the conquest of Cochin in 1663.

A special aspect with which Hendrik was to be confronted in Malabar was the religious situation there. The Van Reedes in general appear to have been orthodox Protestants, or at least behaved as such; otherwise it would not have been very well possible for them to hold public offices in the Republic. There was, however, one striking exception among them: Hendrik's father, Ernst van Reede, was a convinced follower of the heterodox Arminians, who took rather a liberal attitude towards people of other persuasions. In 1626 Ernst became involved in Arminian riots in Utrecht, and later too, when he lived in Amsterdam, he continued to side with the Arminians[31]. Hendrik's indulgent attitude in Malabar towards the "heathen" and the Roman Catholics, and his friendships with the Discalced Carmelite Matthew of St. Joseph and the Spinozist clergyman of Cochin, Johannes Casearius, both of them co-founders of Hortus Malabaricus, might be attributed to his childhood.

In scientific respects, the Van Reedes did not have much demonstrable interest. Among the older generations only Frederik van Reede van Amerongen published a genealogical booklet (1595), in which he tried to enhance the lustre of his family by means of falsifications[32]. With the exception of Hendrik himself, the best-known member of the younger generations was Frederik Adriaan van Reede van Renswoude, who studied at the Faculty of Arts of Utrecht University and who afterwards was in correspondence with the famous Delft naturalist Antoni van Leeuwenhoek on the control of insects in orchards. But Hendrik was to become acquainted with Frederik Adriaan only after his return from the East.

An interest in overseas regions was unmistakable among Hendrik's nearest relations, especially in the West India Company. His father and his brother Gerard, as members of the Council of the Admiralty of Amsterdam, were concerned with the military equipment of that Company. His other brother Frederik served in the Brazilian army of the same Company, and his uncle and guardian Godard erected the settlement Nederhorst in New York, then still a colony of that Company.

Notes

1 The text of the registration of baptism of Hendrik reads: 'On Sunday 13 ditto [April 1636] we received for their covenant as follows: the child of "Joncheer" Ernst van Reden van de Vuijrs and of "Joffrouw" Elijsabet Uijtenhove, in the presence of Petronella Reuters, wife of Mr. Elbert Spiegel, and "Joncheer" Charles Uijtenhove, Sr. Pieter Martsz Hoefijser, tax collector of the Convoy. All these were witnesses to the baptism of Hendrick' (GAA, DTB 7, Register Baptisms Old Church, p.83). It is striking that the child did not receive both Christian names, Hendrik Adriaan, which he used himself later on and by which he is known in history.

2 In the existing literature there is sometimes confusion with Ernst's cousin Ernst van Reede tot Amerongen (1599-1635), the marshal of the Overkwartier and of Eemland (NNBW vol.3 (1914):1005-1006).

3 St.Janskerkhof in Utrecht was then still surrounded by the so-called claustral houses of the former Minster of St.Jan. Ernst probably lived in the claustral house no.16 (now Hotel des Pays-Bas), which he had on lease from 1619 to 1631. Diagonally opposite this house Ernst's brother Godard owned the claustral houses nos 12 and 13, situated on the Drift, at the corner of the Nobelstraat. Godard had the house no.12 on lease since 1624 and the house no.13 since 1632 (RAU, Records Chapter St.Jan 155, Accounts Minor Chamber 1624-1645).

4 It is not known on the ground of what title he fulfilled this office. It is true that the Equestrian Order of Utrecht had the gift of a seat in the Admiralty of Amsterdam, but Ernst was not a member of the Equestrian Order.

5 Elisabeth Utenhove died on 13 February 1637 (NNBW vol.3 (1914):1010). Her death was registered in the week of 20 February 1637 at the Orphans' Court of Utrecht: '"Joffr." Elisabeth van Wttenhoven, wife of "Joncheer" Eerst van Rhede, lord of De Vijers, Draeckesteijn, Councillor of the Admiralty of Amsterdam, leaving her husband with legitimate minor children; deceased in Amsterdam'. Although she died in Amsterdam, it appears from a note in the margin of this registration that she was buried in the Cathedral of Utrecht (GAU, DTB C 1 a 2, dated 20 February 1637). Her husband Ernst van Reede died on 17 October 1640 in Utrecht (NNBW vol.3 (1914):1010). Two days later he was also buried in the Cathedral: 'on the XIX ditto [October 1640] "Jor." Eernst van Reede van de Vuijrsch, Draeckesteijn, etc., Councillor of the Admiralty of Amsterdam, with great tolling of the bells for three hours and pauses, at the canon's half-price, since he is the brother of the Dean – XV fs' (RAU, Records Chapter Cathedral 702, vol.9, Account 1640-1641, under the heading 'Receipt concerning the tolling of the bells'). His death was registered as follows at the Orphans' Court of Utrecht in the week of 26 October 1640: '"Jor." Ernst van Reede, lord of De Vuyrs, Draeckesteyn etc., Councillor of the Admiralty of Amsterdam, leaving legitimate major and minor children, of whom the Lord of Nederhorst is an Uncle on the father's side, Cathedral, 3-0-0' (GAU, DTB C 1 a 2, dated 26 October 1640).

6 Charles Utenhove, lord of Rijnestein, in 1636, in the codicil to his last will mentioned the then living children of his sister Elisabeth, wife of Ernst van Reede, in the following order: Gerrit, Jan, Frederik, Anthonie, Karel, Hendrik, Agnes, Machteld, Margaretha, and Maria (GAU, Notarial Records U 012a015 (protocols W. van Galen):208-208v, dated 7 September 1636 OS). A previously born son Hendrik, christened on 29 December 1633, and a daughter Geertruid, christened on 14 December 1634, both in the Remonstrant Church in Amsterdam, apparently had al-

7   ready died at the moment this codicil was made (GAA, DTB 301:2,6).
7   When the death of Ernst was registered, the lord of Nederhorst was mentioned as the nearest relation of the children left by Ernst, see note 5. This entry was important for the Orphans' Court, which in general exercised the general superintendence of the orphans and therefore had an interest in knowing to whom the guardianship was due, in this case to the lord of Nederhorst.
8   Gerard van Reede van Drakenstein (?-1669) inherited from his father De Vuursche and Drakenstein. In 1641 he succeeded his father as a councillor of the Admiralty of Amsterdam (NNBW vol.3 (1914):1010-1011).
9   Hortus Malabaricus vol.3 (1682):(iii); English translation in Heniger 1980:41.
10 For a survey of the Utrecht nobility and the dominant role of the Van Reedes, see Wittert van Hoogland 1909 vol.1.
11 NNBW vol.3 (1914):1025-1026.
12 NNBW vol.3 (1914):1037-1038.
13 NNBW vol.3 (1914):1007-1009.
14 NNBW vol.3 (1914):1024-1025.
15 For the arms of Hendrik, see the discussion of his portrait in chapter 6.
16 The great interwovenness of the Van Reedes is difficult to describe clearly; see Appendix 1.
17 See chapter 3.
18 The house of Nederhorst in the municipality of Nederhorst-den-Berg is still in the original 17th-century condition; see the description and illustrations of it in Moes & Sluyterman 1914 vol.2:101-115.
19 The medieval castle of Amerongen no longer exists. In 1673 it was destroyed by the French (Mulder 1949:34, figs 3 and 4).
20 Moes & Sluyterman 1915 vol.3:42. In November 1985 Renswoude was partly destroyed by fire.
21 Schenk & Spaan 1967:17-18. The house of Drakenstein was the residence of Her Majesty Queen Beatrix, when she was still the Crown Princess of the Netherlands.
22 Van Gulick 1960:259.
23 On the seigniory of De Vuursche and the country-seat of more than 100 ha situated in it, see RAU, Records Chapter St.Jan 333, Register Long Leases vol.2:cxviii-cxx; ibid. 149, Accounts Major Chamber 1625-1628; ibid. 902, Documents about the seigniory of De Vuursche 1612-1613.
24 The company for the foundation of the colony Nederhorst was established by Godard van Reede van Nederhorst on 13 June 1640 OS (GAU, Notarial Records U 021a008 (protocols G. Vastert):38-38v).
25 RAU, Records House Zuylen 110, Portioning of 1 and 9 December 1615.
26 Ernst van Reede and other members of the Admiralty of Amsterdam are said to have embezzled in September 1636 over 14,000 guilders of the Dutch West India Company. On 17 June 1637 the sheriff of Amsterdam summoned them before the court. But meanwhile they had addressed themselves to the States General, which acquitted them (GAU, City Records II,3663).
27 Moes & Sluyterman 1914 vol.2:102.
28 Wittert van Hoogland 1912 vol.2:575; Nederland's Adelsboek vol.16 (1918):149.
29 In 1616 Ernst van Abcoude van Meerten married Lucia van Reede, a sister of Ernst. Maurits Lodewijk de la Baye in 1630 married Anna Maria van Reede, another sister of Ernst (NNBW vol.3 (1914):1024).
30 NNBW vol.3 (1914):1010; Boxer 1957:213-216.
31 At least in 1633 and 1634 Ernst had his children Hendrik and Geertruid christened in the Amsterdam Remonstrant Church, see note 6. For his participation in the Arminian riots in Utrecht, see Evers 1933.
32 NNBW vol.3 (1914):1006.

2
Van Reede abroad
1650-1678

From his fourteenth year, from about 1650, Van Reede according to his own statement was abroad[1]. It is not precisely known how he started his career. It is alleged that he ran away from home when he was fourteen and then enlisted in an East-Indiaman[2], but this allegation cannot be verified in the records of the Company[3].

His career with the Company only started a few years later, in September 1656[4]. He then became a soldier in the army of the Company. In that month of September 1656 Lords XVII, the directorate of the Company, met in Middelburg for their customary autumn session[5]. There they discussed the results of the fleet returning from Batavia, which had put into port in the past summer, and decided to fit out a new fleet of fifteen ships and to man it with a crew of 3,600, two-fifths of whom were to consist of soldiers. An important item in that meeting of Lords XVII was the discussion of the successful military operations of the Company against the Portuguese in Southern Ceylon. During that discussion the victorious conqueror himself, Rijklof van Goens, was also present. He then stood up for his views on a continuation of the operations with attacks on Northern Ceylon and on the coasts of India, the last bases of the Portuguese empire in Asia. Lords XVII agreed to these war projects and thus, retrospectively, decided the further career of a young soldier like Van Reede. It was obvious that at least a part of the departing soldiers would be used to execute the plans of Van Goens. In the next years Van Goens destroyed practically the whole of the Portuguese power, conquered Malabar on the western coast of India, and finally installed Van Reede, who had become his protégé, as commander there.

In the months of October-December 1656 the ships left the Dutch ports in small groups for the ultimate destination Batavia under the command of Van Goens as admiral. The route to be followed was the usual one in those years: first they revictualled at the Cape of Good Hope and next they sailed directly to Batavia. The ships of Van Goens successively called at the Cape from February to May 1657, and arrived in Batavia during May to August of that year[6].

It is tempting to assume that Van Reede served on Van Goens' flag-ship "Oranje" and that he thus got to know his chief and later opponent at close quarters from the first moment after he left the Netherlands. However, the data for this are lacking. It is not known in which ship Van Reede had been placed[7].

CAPE OF GOOD HOPE 1657

On his way to the East, in the spring of 1657, Van Reede called at the Cape of Good Hope. There, shortly before, in 1652, Jan van Riebeek had founded the Dutch settlement which served as revictualling station on the long road to the East Indies. The East-Indiamen usually called there for one or two weeks, sometimes more. As a rule the ship's crew was not allowed to go ashore at the Cape, with the exception of the sick and the dead. Yet Van Reede had the opportunity to view the recently built fortress and the Company's garden. In 1685, when he inspected the Cape extensively as commissioner-general, he observed:

'That same evening [19 April 1685] I viewed the Fortress, the suburb, and the Company's garden, but although it was evening and in the clear moonlight things stood out clearly enough, I found nothing that resembled this place, or showed features of it, as it was in the year 1656 (!), when I passed there, except only the unchangeably high mountains, so that I perceived such great improvement with amazement'[8].

At the Cape Van Reede must have had the first opportunity to become acquainted with the Dutch colonization methods and with the exotic flora and fauna. But unfortunately he has told us no more than the above about his first visit to the Cape.

If nevertheless we wish to get some idea of the impression which the Cape of Good Hope made about 1656 on observant passing travellers such as Van Reede, we must consult visitors such as the merchant Joan Nieuhof (1654), the chief surgeon Gijsbert Heeck (1655), and another surgeon, Wouter Schouten (1658)[9]. We shall meet Nieuhof and Schouten later again as Van Reede's brothers-in-arms during the conquest of Malabar. In their journals they praised the pleasant climate, were amazed about the plants and animals unknown to them, described in dismayed undertones the manners and customs of the Hottentots, and admired with chauvinist enthusiasm the Company's garden.

About the Company's garden Heeck wrote:

'To the West side of the Fort, along the flowing stream whence we drew water, there was now a fine enclosed Garden, where a Dutch Farmer of Amsterdam was set with his wife and children, to sow and cultivate the same, living there in a little house built of reeds, looking after the milch-cows, sheep, pigs and hens, and doing other such household tasks, providing the Administrator's table with fresh butter, milk, vegetables, fruits and such like that can be grown here. Carrots, cabbage, beetroots or carrot-salad, beets, onions, cress, sorrel and cornsalad grow here freely, as also radishes and waterlemons; but parsley, marjoram, sage, tarragon, artichokes, asparagus are meagre and few. Chervil will not grow here at all: the white cabbages do not stand up, and I never saw runner-beans or peas, and believe that these could not stand up to the terribly strong winds that come down over the Table Mountain. This Garden is being daily extended, and is surrounded and cut through by many channels leading from the stream, for the irrigation of the same'[10].

Nieuhof further added to the list of plants grown there: 'In the Garden grow olive, oranges, peaches, apricots, and other fruit-trees'[11].

As is well-known, the cultivated vegetables and fruits served to revictual the ships of the East India Company calling at the Cape, and especially to counteract the greatly feared illness of scurvy among the crew. In our view the Company's garden, which later was to become so famous as an acclimatization garden of useful plants and which was even to assume the character of a botanical garden, will probably have been no more than a large kitchen garden as yet at the time of Van Reede's visit[12]. Nevertheless, the creation of the Company's garden in its early days was already an impressive achievement. Indeed, for the first time in the history of the East India Company the Dutch had succeeded in permanently introducing a large group of European foodplants in the midst of an underdeveloped population, whose feeding habits were deeply repugnant to them, and under entirely different climatological circumstances.

The Company's garden at the Cape confronts us with an intriguing problem, which was discussed already at that time by fits and starts among the more scientifically interested servants of the East India Company. This concerned the question whether the medicaments should be supplied from the Netherlands or in the overseas settlements themselves. Usually at a central point -the Company chemist's shop in Amsterdam- the so-called surgeon's chests were filled with European medicaments and sent along with the East-Indiamen. But on the long voyage these medicaments were so greatly liable to decay that in the long run, on the way and certainly at the destination, they were no longer effective[13].

There were different possibilities for solving this problem. The first step had just been taken with the founding of the Company's garden at the Cape of Good Hope, where it was possible to restore the general health conditions of the crew, halfway the voyage to the East. The second step required much greater intellectual exertion, to wit making use of the local native medicinal knowledge, in order to make the sending of the surgeon's chests superfluous. This implied that in principle in every settlement a study had to be made of its natural history (minerals, animals, but mainly plants) so as to trace the medicinally interesting materials, and that the experiences of native doctors had to be laid down in writing. Already in 1601, on the eve of the foundation of the East India Company, Carolus Clusius, the then coryphaeus of European botany and superintendent of Leiden University Garden, pointed out the necessity of such studies[14]. Unfortunately, during the whole of its existence the Company was little inclined to issue coercive directives for natural-history research in its settlements. It is ultimately due to the individual interest of a motley row of officials of the Company that exotic botany was to flourish so much in the Netherlands and overseas. Van Reede himself was to look upon his Hortus Malabaricus as a contribution to the solution of the problem of the supply of medicaments. But it would be saying too much that the twenty-year-old soldier already became aware of this problem during his first visit to the Cape.

Moreover, at that time there was not much occasion for this, for about 1657 it was not yet possible to speak of a coherent knowledge of South African natural history. The interest in the native flora and fauna indeed was hesitantly growing in those days. Thus, from the beginning, commander Van Riebeek grew some native plants in the Company's garden and had the forests in the environs examined, but he was mainly interested in foodplants and useful timber and firewood[15]. The only published information on Cape plants then available consisted of twelve species which had been brought to Europe in the first twenty-five years of the 17th century and had been described and/or illustrated there by Clusius (1605), Lobelius (1605), Sweertius (1612), and Bodaeus a Stapel (1644). The best-known of these plants, which made a great stir, were the African Tulip (*Haemanthus coccineus* Jacq.), the Carrion-Flower (*Stapelia variegata* L.), and the Red-Hot Poker (*Kniphofia uvaria* (L.) Hook f.). The majority of these Cape plants will have been collected during incidental explorations in the hills[16].

Probably there was not yet any question of introduction of living Cape plants into Dutch gardens in those days. The first introductions from the Cape are found only in the private garden of Hieronymus Beverningk since 1663 and in Leiden University Garden in 1668[17].

This poor knowledge of the Cape flora is reflected in the not very detailed remarks about it by the visitors. Nieuhof was impressed by the 'very dense and lovely woods' on the slopes of the Table Mountain and drew attention to their usefulness as firewood and timber, but he was able to think of only a few trees with names, such as 'wild almond trees' (*Brabeium stellatifolium* L.) and 'wild pineapple trees' (*Podocarpus latifolius*?). He was even more at a loss in the case of other plants: 'The flat fields and valleys are overgrown with grass and sweet-smelling herbs and flowers'[18]. Heeck, who also made a trip to the Table Mountain, saw on his way 'milkwood trees', 'bushes', 'long grass', and 'a quantity of unknown shrubs'[19]. Schouten in turn climbed the Lion-Hill: 'we found it set with pleasant herbs, long grass, and many well-smelling flowers, but with few trees'[20].

- Nieuhof also made an attempt to gain some knowledge about the use of the native medical materials, but the result was disappointing:

'Those Hottentot doctors seem to have some knowledge, at least of how to sew up a wound; but the scars remain as if it were cauterised. They carry their charms and medicines with them as do our quacks. The herbs they keep in closed-up tortoise-shells, but the little roots, claws, teeth and small horns of animals (since in the use of these lies their art, and they also have some knowledge of their effects)'[21].

All in all, the situation at the Cape of Good Hope around 1657 was illustrative as regards Dutch exotic botany. As beginning colonizers, the Dutch were primarily interested in economic plants. It was only when the astonishment about the natural scenery surrounding them had somewhat abated that they started to look a little more closely at the plants in their new colony and in an unorganized way went in search of medicinal plants and the native knowledge of them, with in the background the vague wish to improve the supply of the Company with medicaments. During his first visit to the Cape Van Reede no doubt did not receive much more than this picture of exotic botany. But still it was a picture which was to dominate for a long time his further wanderings through Malabar and in Ceylon, in regions which still had to be conquered and put under control.

How different the situation was to be, nearly thirty years later when Van Reede inspected the Cape in 1685, when the study of the Cape flora was flourishing and he, then a recognized expert in exotic botany, came at the right moment! Then the world of science expected that after the success of Hortus Malabaricus he would tackle a Hortus Africanus[22].

Van Reede's visit to the Cape was a starting-point in another respect as well. On the long sea voyage from the Netherlands to the Cape he made friends for life with his fellow-soldiers Isaac de Saint-Martin (c.1629-1696) (Fig.3) and Joan Bax van Herentals (c.1637-1678), who like himself were of noble descent. For many years on end the three of them were to be thrown together in the Company's army. They fought in Malabar and served in Ceylon, until each of them went his own way. Saint-Martin was to become sergeant-major at Batavia, a high post of command; Joan Bax was to end his life as governor of the Cape of Good Hope, while Van Reede was to return to Malabar as commander of that district.

Another comrade of Van Reede, Karel van Tetterode, who later was to become a wood-cutters' boss at the Cape, related about this 'that with Mr. Hendrik Adriaan van Rhede, Commissioner-General, who had recently visited him, Mr. St. Martin (then Sergeant-Major), and the deceased Mr. Bacx, formerly governor here, when all of them were still soldiers, he had dined at one and the same mess as their fellow-soldier'[23].

Van Reede too, at the height of his career, was able to recall the care-free trio of former days: 'more than thirty

Figure 3. Isaac de Saint-Martin (c.1629-1696). Painting on canvas attributed to Jan de Baen. Rijksmuseum Amsterdam, Cat. no. A 4156.

years ago they were young, penniless, jobless and brave, and with muskets on their shoulders went aboard a ship for the Indies'[24].

In particular his friendship with Joan Bax was important to Van Reede for more than one reason. In later years Bax kept up a constant correspondence with his influential uncle by marriage Joan Huydecoper van Maarsseveen, the burgomaster of Amsterdam and also one of the directors of the East India Company. Bax kept his uncle continually informed of their vicissitudes in the East, especially when the three friends came into conflict with their chief, Rijklof van Goens. Once he had returned from the East, Van Reede was to become acquainted with the circle of burgomaster Huydecoper, to which also Jan Commelin, one of the editors of Hortus Malabaricus, belonged. It was to Huydecoper that Van Reede was to dedicate volume 3 of Hortus Malabaricus. It was also to Huydecoper that Van Reéde owed his second stay in the East, as commissioner-general. A special feature of the friendship with Joan Bax was to become their concurrent interest in natural history. In Ceylon, and especially later, as governor of the Cape of Good Hope, Bax was to take great pains to present Huydecoper with the most divergent natural curiosities, including a codex with illustrations of Cape plants, and thus contributed to the development of Cape botany[25].

Notes

1. See chapter 1, note 9.
2. NNBW vol.3 (1914):1011.
3. The name Van Reede does not occur in the resolutions of Lords XVII, nor in the few preserved ship's pay-books from the period 1650-1656.
4. This appears from a letter of Van Reede and the Council of Malabar to Lords XVII, of 28 October 1675, in which he tenders his resignation as commander of Malabar. He then mentions that in September next his tenure of office will end, and that he has then been in the service of the Company for twenty years (VOC 1308:595-596).
5. This autumn session lasted from 25 September to 13 October 1656; for the subjects discussed during the meetings, see VOC 150.
6. For the sailing dates of Van Goens' fleet, see VOC 3991:386 (Memorandum of arrival of the ships at the Cape of Good Hope in 1657), and De Hullu 1904:161,179,200,203,206, 225-226). Rijklof van Goens Sr. (1619-1682), one of the great men of the Company, left for the East in 1628; since 1653 he fought the Portuguese in Ceylon and India as admiral and general of the naval and land-forces in the Western Quarters; governor of Ceylon 1662-1663 and 1665-1675, director-general of India 1675-1678 and Governor-General 1678-1681. His son Rijklof van Goens Jr. (1642-1687) succeeded him as governor of Ceylon 1675-1679; commissioner-general of India 1678 and Councillor of India 1684 (NNBW vol.6 (1924):588-591; Wijnaendts van Resandt 1944:59-60,62-63).
7. A simple means for the reconstruction of the career of a servant of the Company is the ship's pay-book. In this book the pay received by the crew of an East-Indiaman putting out to sea is accurately posted up. But the ship's pay-books for the years 1656-1661 are lacking mainly in the records of the Company, also that of the ship in which Van Reede must have sailed in 1656.
8. Hulshof 1941:14. Here Van Reede mentioned 1656 as the year of his calling at the Cape. Apparently he made a mistake, because he entered the Company's service only in September 1656, see note 4.
9. Joan Nieuhof, merchant in the ship "Kalf", visited the Cape of Good Hope from 9 February to 12 March 1654; Gijsbert Heeck, on board the "Vereenigde Provinciën", called at the Cape from 2 to 15 April 1655; and Wouter Schouten, surgeon on board the "Nieuwpoort", stayed there from 25 July to 1 August 1658 (Raven-Hart 1971 vol.1:10,27,32,33,42,48,49,52).
10. Raven-Hart 1971 vol.1:38.
11. Raven-Hart 1971 vol.1:13.
12. Karsten 1951:12.
13. See in more detail about this problem, section A recommendation to examine plants.
14. Hunger 1927 vol.1:267; Heniger 1973:38.
15. Karsten 1951:60-66.
16. On the Cape plants from the pre-colonization period, see White & Sloane 1937 vol.3:1109-1113; Hutchinson 1946: 555; Reynolds 1950:72-74. For the recent habitats of these plants I consulted Adamson & Salter 1950.
17. Veendorp & Baas Becking 1938:78. For the Cape plants which were sent as exsiccates or as seeds and bulbs to Beverningk, see Breyne 1678:2,22-23,66-67,69,85,102-103, 130,136-137,139-140,171.
18. Raven-Hart 1971 vol.1:14,23.
19. Raven-Hart 1971 vol.1:40-41.
20. Raven-Hart 1971 vol.1:49.
21. Raven-Hart 1971 vol.1:22.
22. See chapter 4, section Cape of Good Hope 1685.
23. This anecdote is mentioned by Valentijn 1726 vol.5:135; see also Hulshof 1941:140-141, note 1. On Saint-Martin, see De Haan 1910 vol.1, pt.2:15-21; on Bax, see Boeree 1943: 488-489.
24. Van Reede related this in 1685 in a conversation with the Frenchman François-Timoléon de Choisy, a member of the French embassy to Siam, who called at the Cape of Good Hope on his way there (Raven-Hart 1971 vol.2:267).
25. See Huydecoper's letter-books and diaries, 1671-1678, in RAU, Records Huydecoper Family 55 and 56. Joan Huydecoper van Maarsseveen (1625-1704) was a member of the council of Amsterdam 1662-1704, thirteen times burgomaster in the period 1673-1693, and since 1666 director of the East India Company (Elias 1903 vol.1:518-520).

BATAVIA 1657

After the stay at the Cape of Good Hope the voyage continued to Batavia. As has been said, the ships of the fleet of Van Goens arrived there in the course of May-August 1657[1]. In those months therefore Van Reede and his friends Joan Bax and Isaac de Saint-Martin must also have arrived in Batavia.

I have not succeeded in documenting the following five years, 1657-1661, of Van Reede's life. This is due to the fact that the records of the East India Company from this period contain hardly any biographical data about soldiers below the rank of officers. It is only in 1661, when Van Reede has attained the rank of lieutenant in Malabar, that he appears on the stage again. But it would seem plausible to me that the first steps of Van Reede's career in Asia may be seen in connection with the military operations of Rijklof van Goens in Ceylon and Malabar since 1657, because already in 1662 Van Reede will be found to be greatly respected by this general (Fig.4)[2].

Van Goens himself arrived in Batavia on 1 July 1657. During his short stay in the Netherlands the conquest of the southern part of Ceylon had meanwhile been completed with the occupation of the Portuguese capital Colombo. In meetings with the High Government of India he discussed the consequences of the fall of Colombo and he planned the tactics and strategy now to be followed against the Portuguese. The plan was first to attack North Ceylon and then attempt to defeat the Portuguese in India decisively. In Batavia Van Goens gathered a war fleet and military forces, with which he was to sail to Ceylon on 5 September 1657[3].

During the preparations of Van Goens the soldiers of the expeditionary forces had an opportunity for some time to look about in and around Batavia. This town, the Queen of the East, as the Dutch liked to call the capital

Figure 4. Rijklof van Goens (1619-1682). Anonymous engraving after M. Balen. Muller 1879a.

1655 had a good opportunity to compare Batavia in the past and the present with each other. The following description of Batavia and its environs was inserted by Heeck in his journal on 4 August 1655.

The town had greatly changed since his last visit. Heeck admired the new residence of the Governor-General and the new town hall. The fortifications of Batavia had been extended considerably. The water-ways within the town had been improved so much that the ships could now sail upstream across the town, which greatly reminded him of the typical Dutch canals in the towns at home.

But outside the town walls Heeck was struck by the first results of Dutch colonization under the tropical sun. There, up to two or three hours' distance outside the town the gardens and plantations extended which the Dutch had laid out after they had pulled up the wood and drained the marshes. He was delighted by the 'very beautiful plantations of sugar cane, batatas, pumpkins, cucumbers, runner-beans, cabbage, carrots, lettuce, fockie fockies and all sorts of other pot-herbs, as well as by elegant gardens extremely fruitful in coconuts, betel, pinang, oranges, limes, grapefruits, papayas, mangoes, durians, bananas, pine-apples, and a variety of other fruits'. He also admired 'some beautiful pleasure-grounds, adorned with all manner of beautiful flowers and aromatic herbs, brought together from various regions, which serve to amuse and give pleasure to the owners and their parties, for they often visit them by going up the river in pleasure-prams. Thus, most citizens live among great, luxurious, and special pleasures, for around (the town of Batavia) there is now peace with the Javanese, so that it is safe enough outside the town, as I myself have seen and experienced several times with some friends on horseback when we travelled more than three hours' distance up-country'[5].

of their Asiatic empire, had been founded by them in 1619. It was the economic, political, and military power-centre from which, by sailing-vessels and hand-written letters, the High Government of India, consisting of the Governor-General and the Council of India, ruled numerous trading-stations and settlements in the immense expanse from South Africa to Japan and from Persia to Java. After a difficult initial period, in which Batavia and the Dutch had had to stand various heavy sieges by Javanese forces, the town developed in a short time into one of the most beautiful European cities in Asia[4].

As far as I could ascertain in his writings, Van Reede himself has not made any reference to his stay in Batavia in 1657. Just as in the case of the Cape of Good Hope, we therefore have to consult again a journal of a contemporary if we wish to imagine the picture which Batavia presented to Van Reede in those days.

Gijsbert Heeck, the upper surgeon, who had been in Batavia before in 1636 and in 1644, on his third visit in

Like the Cape of Good Hope, Batavia also played a part in supplying the ships of the East India Company with medicaments. In the castle of Batavia the surgeon's shop was accommodated[6]. From this shop the surgeon's chests, which had been sent from the Amsterdam chemist's shop of the Company, were forwarded on to their destination. The surgeon's shop also dealt with all the requests from the trading-stations for medicaments, and it supplied them in so far as they were in stock in Batavia. If the shop did not keep them in stock, the request was forwarded on to the chemist's shop in Amsterdam. The stocks of the surgeon's shop seemed to have consisted not only of European medicaments, but partly also of Asiatic medicaments. Thus in 1642 drugs were bought from Persia[7]. The management of the shop was entrusted to an upper surgeon, who in addition also ran a policlinic and trained surgeon's apprentices[8]. It is only incidentally known whether he also carried out research on tropical medicaments. Thus in 1643 a description of the medicinal properties of "rais de deos" was sent from Malacca to the Netherlands[9]. In the sixteen-sixties the surgeon's shop was to be managed by Andries Cleyer, who was to take great pains to promote research on medicinal plants in Asia and with whom Van Reede also was to come into contact indirectly[10].

Even though in the case of the surgeon's shop of Batavia around 1657 it is as yet hardly possible to speak of any scientific research about the natural history of Asia, still some important steps had already been taken outside the direct field of activity of the shop by Carolus Clusius (1526-1609) in Leiden and Jacobus Bontius (1592-1631) in Batavia. They were interested primarily in the Indian Archipelago, and Java in particular[11]. Already since the first voyage of the Dutch to Java (1596-1599) the sailors regularly brought back natural objects from the Archipelago to Amsterdam, where they were in great demand among collectors and scientists. Clusius, who at that time was supervisor of Leiden University Garden, spared no pains or expense to lay down the spoils of natural science in those years in descriptions and illustrations. He published these data in his *Exoticorum Libri Decem* (1605) and thus started the glorious Dutch tradition of investigation of East-Indian nature[12]. After the first voyages the research stagnated for almost a generation, owing partly to the fierce fight of the Dutch for hegemony in Java. Only in 1627 the physician Jacobus Bontius, who had taken his doctor's degree at Leiden University, arrived in Batavia, where by the side of his official activities he worked indefatigably on Java's medical and natural curiosities. Over ten years after his death, in 1642, his medical observations, *De Medicina Indorum*, appeared; thus Bontius, though posthumously, laid the foundation of tropical medicine. His observations on the flora and fauna of Java still were not available when Van Reede arrived in Batavia. Only in 1658 Willem Piso published them in *De Indiae utriusque res naturales et medicae*[13]. In the Netherlands, in the botanical gardens of Leiden, Amsterdam, Breda, and Utrecht, after the activities of Bontius, attempts were made to grow some species of plants from the Indian Archipelago, such as *Biophytum sensitivum* (L.) DC., the Touch-me-not, *Sorghum saccharatum* (L.) Moench., and *Bambusa arundinacea* (Retz.) Willd., the Indian reed or Bamboo[14]. In the "Ambulacrum", the Leiden Museum of Natural History, especially the bamboo sticks were regarded for generations as symbols of the relations between Leiden University Garden and Java[15].

Owing to lack of data there is little knowledge about the state of affairs in East-Indian natural research in the fifties of the 17th century. If we are to believe Sirks, after the work of Bontius the research in natural science in Java had stopped again[16]. It is indeed curious that in Batavia, after all the capital of the East India Company, no activity of any significance is to be found until after 1660. This was to change only with the arrival of the previously mentioned Andries Cleyer, about 1664, and Willem ten Rhijne, in 1673, who like Cleyer was a physician greatly interested in tropical natural history[17].

If nevertheless we wish to have some idea of the knowledge of the natural history of Java at the time of the arrival of the young Van Reede in Batavia, we have to turn back to the work of Bontius.

The keynote of the attitude of the European of those days was one of amazement at the wealth of the tropical nature of Java. Bontius gave utterance to that attitude at the beginning of his fifth book in the questioning words:

'Tell me what strange herbs Java produces on her fertile fields and what is spread on her rich bosom; and what are the fishes swimming in her neighbouring sea; and what are the birds, that fly in the free air; and which snakes crawl along with scaly coils. And what are the wild animals hidden in the deep shadows of the woods. Do you want to behold the wonderful garden of the Hesperids?'[18].

But after the amazement the wish arose to describe and illustrate this wealth of nature. His interest was first of all in medicinal plants and animals with a view to diseases such as beriberi, dysentery, cholera, jaundice, and malaria, which made many victims among the servants of the Company. In the second place Bontius paid much attention to foodplants, spices, and other natural products of some commercial value in the Indian Archipelago. It is striking how much value Bontius attached to the knowledge of the Javanese and other Asiatics in addition to his own experience[19]; he thus followed the advice of Clusius from 1602 to note down for the exotic natural products 'their names in their manner and for what purpose they can be used'[20].

But just as at the Cape of Good Hope, in Java the European could only study wild flora and fauna in a limited region. Bontius had to confine himself to the immediate environs of Batavia, because further inland he was threatened by hostile natives and savage animals. Even thirty years after Bontius, Gijsbert Heeck did not venture more than three hours' distance outside Batavia. All the greater was the gratitude of Bontius for the material and information brought to him by kindly disposed natives from more remote regions.

Although Bontius worked and wrote as a pioneer in Batavia, he was not altogether without literature about Asiatic natural history. He could regularly fall back on three older authors, the Portuguese physicians Garcia da Orta (c.1500-c.1568) and Cristobal Acosta (c.1525-c.1594) and the Dutch traveller Jan Huygen van Linschoten (1563-1611). Like Bontius, these three authors lived for a fairly long time in Asia, on the western coast of India. Da Orta and Van Linschoten lived in Goa, while Acosta spent some years in Cochin in Malabar, then still in the hands of the Portuguese. Since no other writings but those of these three were available, Bontius repeatedly compared his own observations from the environs of Batavia with their observations from Goa and Cochin. Bontius even devoted a separate writing, the first book of his *De Medicina Indorum*, to a criticism of twenty chapters of Da Orta's *Coloquios* (1563)[21]. In this way Bontius pointed out the general resemblances between the two floras and faunas.

Thus, many years before Van Reede began to occupy

himself with the flora of Malabar, Bontius already accidentally brought the natural treasures of the western coast of India within the commercial interests of his masters of the East India Company at Batavia.

Notes

1 See the preceding section, note 6.
2 See the following section.
3 De Hullu 1904:203.
4 Stapel 1939 vol.3:117-182.
5 ARA, Colonial Acquisitions 1903, XV, Journal of Gijsbert Heeck's third voyage to the East Indies, n.p.
6 Since 1667 the surgeon's shop was referred to as medical shop.
7 VOC 1135:734.
8 Schoute 1929:142-143.
9 VOC 1141:203-204.
10 Schoute 1929:161-163; Van Nuys 1978; De Haan 1903.
11 As a guide for the following I used Sirks 1915:4-12.
12 For the start of research on East-Indian nature, see more in detail Hunger 1927 vol.1:281-283, and Heniger 1973.
13 For Jacobus Bontius, see Van Andel 1931.
14 For these species, see under their pre-Linnaean names, *Herba viva Christoph. à Costa*, *Milium indicum sive Sorghum rubrum* and *Arundo indica*, respectively, in the garden catalogues of Leiden (Vorstius 1633), Amsterdam (Snippendael 1646), Breda (Brosterhuysen 1647), and Utrecht (Regius 1650).
15 For the Leiden bamboo sticks, see Heniger 1973:39,47.
16 Sirks 1915:12.
17 Cleyer and Ten Rhijne will be discussed below.
18 *Opuscula Selecta Neerlandicorum de Arte Medica* vol.10 (1931):211.
19 *Opuscula* vol.10 (1931):XXVI-XXVII,XXX.
20 Hunger 1927 vol.1:267: 'Memorandum for the Chemists and Surgeons who are to sail to East India on the fleet in the year 1602'.
21 See *Animadversiones in Garciae ab Orta*, in *Opuscula* vol.10 (1931):2-51.

THE CONQUEST OF MALABAR 1658-1663

As previously mentioned, on 5 September 1657 Rijklof van Goens sailed away from Batavia to assail the Portuguese in India. He first turned against the northern part of Ceylon, where before the end of 1657 he dislodged the Portuguese by the occupation of Jaffanapatnam[1]. Subsequently Van Goens crossed to the mainland of India and in the early part of 1658 conquered Tuticorin, the centre of the famous pearl-fishery, about which later, in 1668, Van Reede was to write a detailed inspection report[2]. In the following breathing space Van Goens encamped his troops in Colombo and prepared himself for the last battle against the Portuguese.

Van Goens did not venture to push through directly to Goa, the capital of the Portuguese empire in Asia. With the offensive means available to him he considered that town to be impregnable. For that reason he directed his attention to Malabar, which he rightly looked upon as the core of the Portuguese military power.

In the next five years, 1658-1663, the Company attacked Malabar with constantly growing force. It is in this war that Van Reede's career in Malabar started, and I will therefore go into this fight in somewhat greater detail, the more so as this struggle from the first led him through regions and places which we shall regularly come across afterwards in Hortus Malabaricus.

Malabar is the name which designates the western part of India, situated between Cape Comorin, the extreme southern tip of India, and the river Mangalīēr in the North, and between the mountains of the Ghats in the East and the Arabian Sea in the West. The area is very elongated; the coastline has a length of about 960 km as the crow flies, while the land is only 30 to 100 km broad.

The later Dutch commandment of Malabar, however, was larger than Malabar proper. It also comprised the realm of Canara, where the Company owned a trading-station at Bārssalūr, and the realm of Bijapūr, where the Dutch had a trading-station at Vengurla[3].

Malabar consists roughly of three landscapes: the mountain slopes of the western Ghats, followed by their spurs, the hilly country, and finally the level land along the coast towards the Arabian Sea. From the mountain slopes numerous rivers flow westwards to the Arabian Sea. In the level land they form a continuous chain of lagoons and marshes. The coast is difficult of access for heavy-draught ships, because it gradually slopes into the sea. There are only a few good natural harbours, such as those of Quilon, Cochin, and Calicut.

The tropical climate of Malabar, with temperatures above 30 degrees centigrade, is moist, especially during the wet monsoon from May to September. In that period the plain is flooded. The Europeans with their heavy war-material avoided undertaking campaigns during the wet monsoon as much as possible.

The vast majority of the inhabitants of Malabar in the 17th century were the Malayali, who had their own special culture. Van Reede estimated their number at more than three million fighting men[4]. Among them, the group of the Nāyars, the warriors of Malabar, was feared most by the Europeans. Although they fought mainly with swords and shields, owing to their great numbers and unusual ferocity they landed a European army more than once in difficulties.

Malabar was not a unitary state in the 17th century. The country was cut up into a confused mass of small kingdoms with greatly interwoven spheres of influence. The principal rulers were the Kolathiri of Cannanūr, the Zamorin of Calicut, the rāja of Cochin, and the rāja of Travancore. The most formidable ruler was undoubtedly the Zamorin of Calicut, who succeeded in maintaining his independence from the Europeans. The key position in the colonial domination, however, was held by the rāja of Cochin. The kingdom of Cochin was very attractive for the Europeans because of its central position in the pepper district and the good roadstead of the town of Cochin.

Ever since Antiquity, Malabar had played a great part in Eurasian trade. The country produced the best pepper of India, and further coconut, rice, areca, cardamom, ginger, bananas, teak and sandalwood, and Indian copal[5]. The Romans, and later the Arabs, transported these products to the Mediterranean area, where they gained almost legendary fame. In the hope of breaking through the Arabian spice monopoly, the Portuguese settled on the western coast of India since the end of the 15th century. They founded a colonial empire with Goa as capital and a chain of fortified places along the Indian coast. On the coast of Malabar they built a strong system of fortresses in Cannanūr, Cranganūr, Cochin, and Quilon, and by means of contracts with the rājas they secured the pepper monopoly. Especially Cochin developed, under the Portuguese rule, into a veritable European town with many churches and monasteries, and it was protected by strong defences. Next to Goa, Cochin was looked upon as the most important town of the Portuguese in Asia.

However, when the Portuguese command of the sea weakened, since the end of the 16th century other European powers, such as England and the Dutch Republic, succeeded in breaking through the Portuguese blockade of western India and establishing commercial relations with Malabar. The Dutch were well-informed about the exotic products and the Asiatic trade routes through the enthusiastic description of Jan Huygen van Linschoten, who had lived in Goa for a long time[6].

Rijklof van Goens had to organize altogether five expeditions against Malabar in order to subdue the Portuguese, and partly also the Malayali. The first three expeditions, in 1658, 1660, and 1661, were not much of a success[7]. A humble gain consisted in an alliance with the Cochinese claimant to the throne, Vīra Kērala Varma, for dethroning the ruling rāja of Cochin, Rāma Varma, who sided with the Portuguese. As a hostage, the claimant offered his brother of the same name, with whom Van Reede later became friends.

The decision in the fight for Malabar fell in the last two expeditions. In the fourth, from November 1661 to April 1662, Van Goens gathered a fleet of 29 ships and more than 4,000 men. The latter included lieutenant Van Reede, the naval chaplain Philippus Baldaeus, the surgeon Wouter Schouten, and Joan Nieuhof, who attended to the army train. All four were afterwards to immortalize the country, the people, and the nature of Malabar in writings. First of all the expeditionary forces landed near Quilon, which was occupied on 8 December 1661 after a battle against the Nāyars. A Dutch garrison remained behind, while the main body of the army sailed further to the north. There Cranganūr was besieged from 3 to 15 January 1662; it had to surrender. There again a garrison was left behind. Subsequently, on 3 February 1662, Van Goens laid siege to Cochin. The claimant to the throne, Vīra Kērala Varma, had joined him. The army was threatened on the flank by the royal palace of Cochin, where rāja Rāma Varma with his Nāyars stayed. A few days after the landing Van Goens attacked the palace. He then promised Van Reede the first captaincy becoming vacant if he could save rānī Gangādhara Laksmī, the old queen of Cochin, unharmed from the turmoil of battle. The fight for the palace ended in a massacre. The rāja and two of his brothers were killed, but Van Reede found the rānī on the roof of the neighbouring temple and put her in safety.

In Malabar relations Van Reede's action was very significant because, especially now that the rāja had been killed, the rānī had great influence on the designation of the successor to the throne, and thus on the future relations between Cochin and the Company. The contemporaries did not omit to mention Van Reede's action. Thus the surgeon Wouter Schouten wrote: 'only the old Queen of Malabar's Cochin was taken prisoner by the Ensign "Jonkheer" Henderik van Rhede, because she had favoured the Portuguese as much as she had been hostile to us. But the General kept her alive and she was treated well, the more so because of the intercession of the king, our Friend, whose Aunt she was. However, she was taken into custody, because she was not trusted rather than for her beauty; for she was an ugly, old woman, but adorned with Gold chains and trinkets, which stood out wonderfully against her black skin'[8].

In spite of Van Reede's success, Van Goens could not keep up the siege of Cochin. On 2 March 1662 he beat a retreat. In the conquered fortresses of Cranganūr and Quilon he left garrisons behind. Van Reede was posted as a provisional captain to the garrison of Cranganūr (Fig.5). With the main body of the army, Baldaeus, Nieuhof, and Schouten left the coast of Malabar. The latter alone was not to return there.

Van Goens went to Batavia for reinforcements. He again raised an army, in which the Swiss soldier-painter Albrecht Herport also served. The fleet left Batavia on 26 August 1662. A small vanguard under the command of Isbrand Godske sailed in support to the garrison of Cranganūr, where he arrived on 10 September. He marched with the garrison to Cochin in order to occupy, while awaiting the main body, a strategically situated fortress opposite the town (24 September). The main body appeared shortly afterwards before Cochin and on 28 October 1662 the decisive siege started.

As a captain, Van Reede sat on the council of war of the Company's army and thus took part in the consultations. His signature is therefore regularly to be found at the foot of the minutes of the council of war[9].

During the siege Van Goens summoned the Dutch claimant to the throne, Vīra Kērala Varma, who had fled to Mannar after the abortive first siege. The prince and his brother travelled to Quilon, where Nieuhof, who represented the Company there at that moment, received them in November 1662. The brothers, however, fell seriously ill, and on the way to Cochin the claimant to the throne died. Fortunately the rānī recognized the brother, who was also called Vīra Kērala Varma, as successor to the throne.

Van Reede took an active part in the war operations

Figure 5. View of Cranganūr. Etching by J. Kip from Schouten 1708.

during the siege[9]. In December 1662 he occupied "Papeneiland" near Cochin and the Bōlghatti island near Ērnakulam. Together with his friend Isaac de Saint-Martin and Nāyar auxiliaries he subsequently expelled the pro-Portuguese claimant Gōda Varma from Anchi Kaimal[10]. Van Reede had returned to the army camp before Cochin when on 5 January 1663 it was decided to assault the town. The next day the Dutch forced their way into Cochin amidst heavy fighting. On 7 January the town and the fortress capitulated.

With the surrender it was also decided that the entire Portuguese population, including the Roman Catholic clergy of European origin, was to leave the town. The execution of the capitulation treaty was entrusted, among others, to Van Reede[11].

One day after the fall of Cochin, Rijklof van Goens entered the conquered fortress like a prince. The Dutch general literally put the crown on the work when on 6 March 1663 he crowned Vīra Kērala Varma as rāja of Cochin with his own hands[12].

With the fall of Cochin the fight was not yet at an end. In February Van Goens sent strong armed forces, including also Van Reede, to Cannanūr (Fig.6), which after a short siege was conquered on the Portuguese[13]. Thus the Portuguese had definitively been expelled from Malabar.

The same forces also occupied Parūr at the end of February[14].

On 26 February 1663 Van Reede was to be found in the palace of the rāja of Cranganūr. Together with the second merchant Cornelis van Essem he there submitted personally the pact of friendship to the Zamorin of Calicut for signature and afterwards transferred to this prince and to the rāja, according to previously made agreements, the town of Cranganūr conquered early in 1662 and the artillery captured there[15].

In March 1663 Van Goens himself still directed the subjection of Purakkād, a vassal of the rāja of Cochin. This time, too, Van Reede, now together with Saint-Martin, submitted a peace treaty to the rāja of Purakkād for signature, on 14 March[16]. On 20 March 1663 Van Goens formally regulated the relationship between the Company and the Cochinese kingdom in a 'Contract and eternal alliance', in which 'the king of Cochin recognizes that he has been restored by the Honourable Company in his kingdom and that consequently he accepted the Company as his protector'. Together with the upper merchant Cornelis Valkenburg, Van Reede was present to witness the signature of this final document of the Malabar war between the Portuguese and the Dutch[17].

Van Goens now considered his conquest as completed.

Figure 6. View of Cannanūr. Anonymous engraving from Nieuhof 1682.

Two days later he left Malabar and left the country behind in the care of occupying forces. In his Instruction of 22 March 1663 he set forth the organization of the Dutch colonial administration, in which he allocated to Van Reede a prominent position as councillor of Malabar, president of the town council of Cochin, and envoy to leading potentates[18].

The very first impressions which the Dutch conquerors received of the land and the nature of Malabar can be found in the writings of Wouter Schouten and Philippus Baldaeus. During the military exploits they could pay attention only to the most remarkable things and wrote less in detail than Joan Nieuhof and later Van Reede, who were to spend a much longer time in Malabar. Schouten and Baldaeus confined themselves to the environs of Cochin in their descriptions.

The first sight of Malabar was worded by Schouten (Fig.7) as follows:

'Many wonderful Sea-coasts of Asia show in a very charming way far into the sea their pleasant appearance and agreeable position; but in my opinion the Coast of Malabar excels them all'[19].

And once they had landed, the Dutch marched 'along the pleasant Shore of the Kingdom of Cochin, which was overshadowed fairly, airily, and agreeably by a fine Grove of very good fruit-bearing Coconut trees'; and about the village of Aldea, near Cochin, Baldaeus wrote:

'which was very pleasantly adorned with fine Dwellings, in the fashion of the country, and above with Coconut and other Trees ... we saw here, inland, delightful fields, plains, ponds, and walks, which were separated from each other by all manner of fruit-bearing Trees, and planted on the outside very fairly and very pleasantly'[20].

Figure 7. Wouter Schouten (1638-1704). Etching and engraving by C. Hagen from Schouten 1708.

Figure 8. Philippus Baldaeus (1632-after 1672). Engraving by A. Blooteling after Sijdervelt from Baldaeus 1672.

Baldaeus (Fig.8) added to this description of a tropical culture landscape:

'Cochin is considered not to be as healthy as other places on these coasts, and this on account of the lowness of the land and the numerous marshes; otherwise there is plenty of all sorts of Fish and Flesh here, and great delight of the Rivers and inland waterways, enriched with several well-planted, very delightful small Islands, where the Portuguese usually had their Pleasure houses, as also on the main Land'[21].

About the flora and the fauna in the environs of Cochin, Baldaeus could not tell very much. He mentioned that it was a land rich in fish, and what he thought most remarkable was the Goa cod-fish, which was brought to the market in Cochin. For the rest he did not get beyond an enumeration of the natural products of the country: pepper, cardamom, ginger, "borborri", aloe, bezoar stones, saltpetre, wax and honey, amber, cloves, nuts, and mace[22]. He was impressed by the addiction of the population to opium:

'which the Inhabitants eat until they are not only foolish, but also half-mad', and said about the then fifty-year-old Zamorin of Calicut: 'he was already dozing on account of his immoderate eating of Amfiun or Opium'[23].

In Wouter Schouten's work, on the other hand, we already find a first beginning of a natural history of Malabar. It is true that owing to his short stay of a few months in Malabar he was only able to describe in some detail a handful of striking and curious plants. But nevertheless he paid attention to details such as the shape of the leaves, the colour and the smell of the flowers, and the medicinal and industrial use of these plants.

First of all he described the coconut palm, which by its majestic shape and frequency puts a stamp on the coastal landscape. He discussed at length the widely varied uses of the coconuts, the leaves, and the bark by the natives. He told his readers about the gigantic Wonder-tree or Root-tree with its curious 'offshoots', about the Arbor Triste, a beautiful specimen of which he had seen already previously in the front yard of the church of Cranganūr, about the Sensitive Trees around Cochin, whose snapping seeds had greatly scared him during the dramatic episode of the siege, about the Datura which renders men insane, the aromatic Cardamom, and the lac-tree. The other plants were enumerated by him only by their names: white and black rice, "Kitsery", katjang, pepper, onion, "Borreborry", ginger, batatas, bananas, mangoes, carambolas, pine-apples, tamarind and opium[24].

About the fauna around Cochin, Schouten gave details

of spectacular animals such as the camel, the jackal, the bat, and poisonous and strangling snakes. Further, like Baldaeus, he drew attention to the abundance of fish of the coastal land. He could refer to the birds of Malabar only by the names of species known in Europe: tame ducks and mallards, geese, herons, hens, peacocks, partridges, turtle doves, parakeets, parrots, and singing-birds[25].

Notes

1. Aalbers 1916.
2. See section First captain and sergeant-major of Ceylon, note 5.
3. Some of the following data about Malabar have been taken from Van Reede's Memorandum of Malabar of 1677, see s'Jacob 1976:192-196.
4. This number given by Van Reede in the Memorandum of 1677 may be exaggerated, but it does show that Malabar was densely populated according to European standards of those days.
5. Warmington 1974:181-217.
6. In his *Itinerario* (1596) Van Linschoten discussed in great detail a variety of plants, animals, and minerals from India which were important for trade; see Kern & Terpstra 1956:51-183.
7. The following survey of the conquest of Malabar has been taken from s'Jacob 1976:XL ff.
8. Schouten 1708:217-218. Here Schouten erroneously called Van Reede ensign instead of lieutenant. Less detailed reports about this event in Baldaeus 1672:116, and Nieuhof 1682:120.
9. See the resolutions of the council of war of 14 December 1662, 5 January and 25-31 January 1663 (VOC 1239: 1169-1172,1198-1199v,1206a-c).
10. Gōda Varma, who died in April 1692, ruled as rāja of Cochin in 1662 between the two sieges. The Company considered him as a Portuguese puppet, but Van Reede, whose political opponent he was to be later, gave a more favourable opinion about him in the Memorandum of Malabar of 1677: 'because all the evidence agrees that he has many good gifts and the greater part of all hearts and affections' (s'Jacob 1976:19,152).
11. For the capitulation treaty of Cochin, see Heeres 1931:232,No.CCLXVI.
12. Nieuhof 1682:124: 'This brother was the nearest to the crown, and was also ... crowned with a gold crown, in which the mark of the Company was cut'.
13. The conquest of Cannanūr was described by Albrecht Herport, who witnessed the siege (L'Honoré Naber 1930:108-113). The surrender of Cannanūr by the Portuguese was signed on 15 and 16 February 1663 (Heeres 1931:234-237,No.CCLXIX).
14. L'Honoré Naber 1930:113-114.
15. For the pact of friendship with the Zamorin and the transfer of Cranganūr, see Heeres 1931:237-239,No.CCLXX, dated 22/26 February 1663.
16. For the peace treaty with Purakkād, see Heeres 1931:240-242,No.CCLXXII; Valentijn 1726 vol.5, pt.2: 36-37; Van der Chijs 1891:184-185.
17. Heeres 1931:242-246,No.CCLXXIII. On the differences in interpretation of the patronage of the Company over Cochin, see s'Jacob 1976:XLIX-LI.
18. s'Jacob 1976:3-13.

19. Schouten 1708:261.
20. Schouten 1708:213.
21. Baldaeus 1672:115.
22. Baldaeus 1672:100.
23. Baldaeus 1672:110,104.
24. Schouten 1708:279-282.
25. Schouten 1708:283-286.

INSPECTIONS OF MALABAR 1663-1667

Councillor of Malabar

The first years after the conquest of Malabar by the Dutch were dominated by the consolidation of the position of the Company. Until 1670 the commandment of Malabar came under the jurisdiction of the government of Ceylon. Consequently, Rijklof van Goens as governor of Ceylon had great influence on the affairs in Malabar, frequently to the great annoyance of the commanders, who considered his activities as interference with their responsibilities. The general policy lines were laid down by Rijklof van Goens in his Instruction of 1663. They were meant to obtain a firm footing in the protectorate-kingdom of Cochin, to get trade monopolies from the Malabar princes by negotiations, and, if necessary, to use military force against reluctant princes[1]. It was the task of the commander and the council of Malabar to carry these policy lines into effect. As a councillor of Malabar, Van Reede actively co-operated in this. He did his work with so much success that he was ultimately considered as the pre-eminent Malabar authority by his superiors, particularly Rijklof van Goens, as well as by the native princes[2].

It is not known to me to what extent Van Reede as a councillor regularly attended the meetings of the council in Cochin. Nor are any data available to me about his activities as president of the town council of Cochin. Still, it seems to me that in the years 1663-1667 Van Reede will not have been in the town of Cochin very often. From the following sections it will become evident that he was frequently travelling about and stayed at the courts of different rājas.

Regedore maior of Cochin 1663-1665

Already in 1663 Van Reede received the curious honour of being appointed "regedore maior" of the kingdom of Cochin[3]. The recently crowned rāja of Cochin, Vīra Kērala Varma, was assisted in the government of his kingdom by a council of Nambūthiri or Malabar Brahmins. The daily administration was in the hands of a first councillor, in Portuguese called "regedore maior". Van Reede, who was the first Dutchman to hold this office, later gave the following account of this:

'The daily events are mostly administered by a man who might be called first councillor, if he did not moreover exercise another rule. But since he has the function of a stadtholder rather than one resembling that of a councillor, the Portuguese called him "regedore maior" or supreme administrator. For this function the king generally uses distinguished people, but then those who have more experience and knowledge than the personal power of nâyars, always looking to dignity of birth, because the Malabars attach particular importance to this. It is to him that the stewards and governors, sheriffs and tenants have to render account, while he attends to everything that concerns the king and the kingdom'[4].

The choice of Van Reede for this function, a lofty one in Malabar eyes, will be connected with the fact that it was difficult for the râja to hold his ground against his own courtiers, who were especially after the treasury. The râja originated from the highlands and was thus, properly speaking, a stranger in Cochin. The Company, as protector of the râja, was very anxious to procure him a solid position in his own kingdom. Van Reede's noble birth, too, will have contributed to the choice, if he was to be on a par with the high-born Nambûthiri, with whom he had to form the royal council. And finally his rescue of the râni Gangâdhara Laksmî from the massacre of February 1662 will have gained him much respect among the Malayâli. Van Reede took great pains and spent much time in restoring the king's authority and the ruined treasury. He did not even scruple to support the râja in internal conflicts with a body of troops. But after one year he had enough of it, because his exertions were hardly successful at all: 'but everything was fruitless, and the labour and worry were swept away by the wind'[5]. In spite of this, Van Reede presumably continued to hold the function of "regedore maior" until February or March 1665. After that time he was far away from Cochin[6].

Despite his disappointing experience Van Reede had had a unique opportunity at the royal court of Cochin to become better acquainted with some facets of the Malabar society. It is from this period that his friendship with the râja Vîra Kêrala Varma will date, at whose orders numerous plants were afterwards collected for Hortus Malabaricus[7]. The fact that personally he knew the râja very well appears from his letter to Lords XVII of 1676 talking about the king's eloquence, foolery, inconstancy, and credulity[8].

About the Nambûthiri, whom Van Reede knew as colleagues in the royal council, he wrote as follows:

'Nambûthiri or priests are Malabar Brahmins. Some of them have no other occupation but the temple service, and are exempt from all wordly cares, constantly occupied with heathen wisdom, star-gazing, and natural science. They are very good and sober people, living a godly and modest life; they never eat anything that is alive or gives life, and they drink no other beverage but water, honey, milk, and butter. Among them there are also many who serve at the royal courts, both as councillors and as envoys to their neighbours, either in peace or in war, and to whom the most important affairs of government and prosperity are entrusted, as well as the education of the children of all the principal kings and lords. These people are pleasant to associate with, at least for those who know them and are able to adapt themselves a little to their way of life, for their mastery and wisdom at the same time render them arrogant'[9].

Perhaps it was from the royal council of Nambûthiri that Van Reede got the idea of using later a similar board as a discussion forum in the preparations for Hortus Malabaricus.

In November 1663, together with his friend Joan Bax, he was in Vengurla, where they negotiated, among others, with the Discalced Carmelite fathers about the possible return to Malabar of their bishop Giuseppe di S. Maria. Van Reede will then hardly have suspected that one of those negotiators, Matthew of St. Joseph, was to become ten years later his chief support in the composition of Hortus Malabaricus.

Meanwhile Van Reede by no means neglected the interests of the Company. In January 1664 he conducted a punitive expedition to Alangâdu, to avenge the assassination of the prince of Parûr[10]. After this, in March 1664, he visited the archdeacon of the Christians of St. Thomas in the highlands of Malabar[11].

In the next year, 1665, in January-February, Van Reede went as an envoy to Goa. On his way there he also visited the Dutch factory in Vengurla[12]. In February-March he took part in the punitive expedition against the râja of Kârthi-kappalli. In the decisive battle Van Reede was in command of the right wing of the Dutch army, and his friend Joan Bax of the left wing. At that time the soldier Albrecht Herport, who took part in this expedition, made a general view of the battle against the Nâyars (Fig.9)[13].

Envoy 1665

As early as March 1665 Van Reede was in Quilon, in South Malabar, where he was engaged in planning the new Dutch fort. On 30 March he left Quilon for a long diplomatic mission along the south coast of India. We may assume that his office of "regedore maior" of Cochin was now definitively a thing of the past.

Van Reede reported on this journey to his superior Rijklof van Goens in two extensive reports[14]. It is in these reports that Van Reede wrote for the first time about Malabar and adjacent regions. We shall find that -in between all his diplomatic activities- he began to reflect with his observations about the landscape, the climate, and the people upon the Malabar society, into which as a soldier he had penetrated by force.

During his journey through the South of India Van Reede first of all visited, on 2 april, the râja of Travancore, 'who caused me to be welcomed for half a mile by one of his Regedores with 30 to 40 Nâyars with bells, further doing me all honour and friendship in their way'. In spite of this welcome he percieved that the Malayâli had the same aversion from Christianity as he had from their 'superstition'. Nevertheless, that same day Van Reede concluded a pepper contract and a defence alliance with the râja[15], and during the exchange of presents he was honoured by the king with 'a Gold bracelet and a silk garment'. The râja of Travancore must have been Râma

Figure 9. Inset: battle against the rāja of Kārthi-kappalli in 1665. Dutch main force (1) with auxiliary Nāyar forces (2) commanded by Van Reede on the right wing and Joan Bax on the left wing. Etching by Conrad Meijer after Wilhelm Stettler from Herport 1669.

Varma, who ruled from 1662 tot 1671[16]. About him Van Reede wrote: 'This rāja seems to be very energetic, ready, and even capable of stipulating, writing, and signing -an unusual thing- on one day a contract without his deputies being allowed to eat or drink. He is taller than most men and has a distinguished appearance, which commands respect'.

One day later, on 3 April 1665, Van Reede arrived in Tēngāpattanam, where the Company had a small factory. There he conversed with the christianized population and talked with Portuguese Jesuits still living there, so as to find a new relationship with the Company. On that occasion he procured extensive information on the geography and demography of the coastal region between Quilon and Cape Comorin[17]. But he could also describe this region by his own observation. The following passage from his report may be looked upon as a first sign of his interest in the natural history of Malabar:

'The shores between Quilon and Tēngāpattanam, where the Christians live, are mostly of loose sand, as around Cochin, and several rivers which, before they go inland, run close behind the sand of the dunes, in some places they are very narrow, so that they can easily be shot across with a musket. None of these rivers but is annually stopped by the sea, and when they are at all open, in their mouth there is little water, much swell and shedding, though some of them are fairly deep inland, like those of Āttingal and Tēngāpattanam, and go very far inland, having their source in the highlands, as is said by those who know this. From Tēngāpattanam to the Cape the land deteriorates and the people improve, being much more kindly disposed and amiable, but also poorer than in the land to the north, full of rocks and thorns and wholly without coconut trees, so that none but palmyra palms grow there, in wholly infertile soil; the shores, covered behind by high rocky mountains, are also much rougher and owing to the tremendous dashing of the sea cannot be reached by shallops or vessels: the inhabitants use Cattapanels or paleguas'[18].

Van Reede continued his journey along the coast of India. Via Cape Comorin he left Malabar, and on 11 April reached Tuticorin in the kingdom of Madurai. There he negotiated with the Neik and the Teuver, the princes of this country. The Teuver even welcomed him with a great retinue of elephants, camels, horses, and men and presented him with a written history of the rise of his power. Just as in Tĕngāpattanam, Van Reede here also procured thorough information about the geography and demography of the coastal region from Cape Comorin up to Madurai[19].

A less pleasant feature of the journey was the investigation into the behaviour of Joan Nieuhof, who at that time was the Dutch chief at Tuticorin. The negative report which Van Reede and his colleague Laurens Pijl drew up about him was to spell the end of Nieuhof's career in the Company[20]. On 26 June Van Reede received the order to arrest Nieuhof, who had meanwhile been transferred to Quilon, on a charge of malversation. At a rapid pace Van Reede travelled back to Malabar, 'taking an unusual route and across the mountains, being hindered not a little in the lands of Madurai by the awfully dry and hot gales and in the lands of the Malabars no less so by the heavy and hard rains, wet roads, strongly flowing rivers, and poor facility in crossing them'[21].

But in the meantime difficulties had arisen in Malabar. Before Van Reede was able to occupy himself with Nieuhof, in July and August 1665 he travelled about in Kāyamkulam to mediate in a war between the rājas of Travancore, Tekkumkūr, and Cochin. He again gave evidence of the thoroughness with which he tried to do his work, by submitting a detailed report of the 'Beginning and causes of the Malabar war such as this matter was told to me by Malabars'[22]. About 10 August 1665 Van Reede succeeded in bringing about peace between the warring rājas. That his exertions with respect to the internal affairs of Malabar were appreciated is evident from a present of the rāja of Travancore, for he 'presented me with a Gold Medal with a chain, on which his arms and name are cut, as well as his own sword'[23].

Still Van Reede was not very satisfied with the results attained. By way of an apology for the circumstantial statements in his report he observed:

'also in order to show how improper it is for the Honourable Company to deal with these false people, their disputes and quarrels, of which it is so difficult to know the cause and of which one might so easily judge wrongly, since we do not understand their language or their interest very well'[24].

At first sight his statement about 'these false people' appears to be a harsh judgment about the national character of the Malabar people, which may partly also have been caused by his disappointing experience as "regedore maior" of Cochin. But how favourably his self-criticism that he does not understand the language and the interests of the Malayali contrasts with oversimplified utterances of commanders such as Jacob Hustaert and Isbrand Godske, who wrote in the same period about the 'perfidy of the Malabars', 'the Malabars who are all of a villainous disposition', and 'perfidy, deceit, and infidelity is no disgrace among them'[25]. The opinion of Joan Nieuhof on the Malayali, however, comes closer to that of Van Reede. He sought the cause of the changeableness and inconstancy of the native population in their addiction to opium[26].

Chief of Quilon 1665-1667

After the end of the Malabar war, in August 1665, Van Reede went to Quilon in order to take measures against Nieuhof (Fig.10). He arrested him in September and sent him to Batavia for trial[27]. Van Reede himself undertook the management of the Quilon factory.

Nieuhof, however, was acquitted in Batavia from the charges brought against him, although for the time being the Company did not wish to make use of his services any longer[28]. The arrest of Nieuhof was the first controversy of Van Reede with a colleague within the Company. More conflicts with others were to follow, until he himself became the victim.

Figure 10. Joan Nieuhof (1618-1672). Anonymous engraving from Nieuhof 1682. Van Someren 3886b.

Apart from the correctness of the charge, the responsibility for the negative report about Nieuhof must not be laid on the shoulders of Van Reede alone. Indeed, Laurens Pijl, the later governor of Ceylon, to whom the Amsterdam Botanical Garden was to owe many a shipment of plants, also signed the report. It is, however, tragic that with Nieuhof a capable servant of the Company disappeared from the coast of Malabar. This is all the more tragic because, like Van Reede, he evinced a greater interest in the Malabar country and its natural history than his contemporaries[29]. A potential co-operation in this field between the two men had thus in advance become impossible.

After the arrest of Nieuhof, Van Reede returned from Quilon to Tuticorin in order to write there his report about his long journey through the south of India, which report he completed on 1 October[30]. After this he crossed to Ceylon and presented his report on 7 October to his superior, Rijklof van Goens, the governor of Ceylon.

Van Goens was greatly satisfied with the work performed by his subordinate, who had proved to be, not only a good soldier, but also a successful diplomat. As long as Van Reede was in Colombo, Van Goens made use of his fresh acquaintance with Malabar affairs during meetings of the council of Ceylon[31]. Towards the Governor-General in Batavia, too, Van Goens made no secret of his appreciation of Van Reede. In a long eulogy, in which he gave his arguments why he had appointed Van Reede as chief of Quilon, he wrote:

'Considering the objectionable behaviour of the merchant Nieuhof and because we had so few capable and faithful merchants, we have been obliged to appoint the above-mentioned captain Hendrik van Reede as chief of Quilon, Karunāgappalli, and Kāyamkulam, and to add thereto Tēngāpattanam and the shore of Carembaly or Morenbril as far as Pagodinso... We hope we have not made a mistake in making this choice, not only because the said Van Reede has already had good experience and is much liked by the Malabar princes, but also because through his vigilance Quilon and the whole shore will be preserved, if necessary, in time of war, since he is a constantly sober and seasoned soldier, and what he may still lack in style of trading, will no doubt be made good within a short time by his ability. Your Honours will see from our resolution, which we have presented to him with your consent, what marks of honour the Neik of Madurai and Travancore has given him'[32].

Before Van Reede could return to his station Quilon, Van Goens first sent him on a tour of inspection to the Dutch possessions in North Ceylon. The following has been taken from his report about this journey[33].

In November and December 1665 Van Reede visited the settlements of Jaffna and Mannar, especially for the purpose of inspecting the defensive strength of their forts. In Jaffna, supported by the chief Joriphaes Vosch[34], he tried to persuade the clergyman Baldaeus residing there to stay in Ceylon for another two years. Baldaeus did not give in to their insistence and soon after left for the Netherlands, where he was to start the publication of his book about Ceylon and Malabar[35].

During this journey, too, Van Reede evinced an interest in the nature surrounding him. Thus he wrote about the running-wild of the culture landscape near Mannar:

'Tellipoles and Miletis are neglected hedges of letter or milkwood that in the course of the time have become very dense woods, in which the game, both boars and deer, live and hide in great numbers, thus causing several lands lying close about them to remain uncultivated: if these wastes were cut down, the wood would also be suitable for lime-burning. The aforesaid game multiplies so plentyfully that the crops in the fields can hardly be preserved, and this multiplication takes place because no Inhabitant (as in the days of the Portuguese) is able to kill or catch any game'[36].

After this interlude in Ceylon Van Reede returned to Malabar to undertake the management of the Company factory in Quilon definitively. Quilon was the centre of the pepper, cinnamon, and opium trade in South Malabar. The Company had the monopoly of that trade there. The chief of the factory had to provide for the purchase of the pepper and cinnamom which were brought there by native merchants. Conversely, the Company imported opium. The chief was constantly engaged in concluding contracts, paying customs duties, and checking smuggling[37]. In reality the post of chief of Quilon was reserved to a qualified merchant. Van Goens therefore hesitated in the above-mentioned eulogy of Van Reede whether on this post a soldier was the right man. But probably Van Reede's military capacities were decisive, because in Van Goens' view Malabar, and in particular Quilon, was an important base for the island of Ceylon in case of war. Ultimately Van Goens hoped that his protégé, besides a successful diplomat, would also prove to be a good merchant.

Unfortunately, as far as is known, Van Reede has not left behind a report about his stay in Quilon. From the correspondence of his superiors Godske, the commander of Malabar, and Van Goens, the governor of Ceylon, in the years 1666-1667 it appears, however, that Van Reede did not like his work as chief of Quilon. His contacts with the native merchants were not very satisfactory[38]. He was required to drive away the Portuguese Jesuits with whom he had become acquainted in 1665[39]. He had to examine Nieuhof's shaky administration of Quilon[40]. The building of the Quilon fort, in which as a soldier he was greatly interested, was disappointing owing to all sorts of silly things[41]. When moreover through no fault of his own he got involved in a dispute of competence with the first merchant Jan van Almonde, that was the last straw for him. Van Almonde had arrived in Cochin round the turn of the years 1666-1667 and had subsequently been stationed in Quilon under the command of Van Reede. But this highly qualified merchant, with 28 years' trading experience, justly refused to be apprenticed to a captain 'who has never seen any

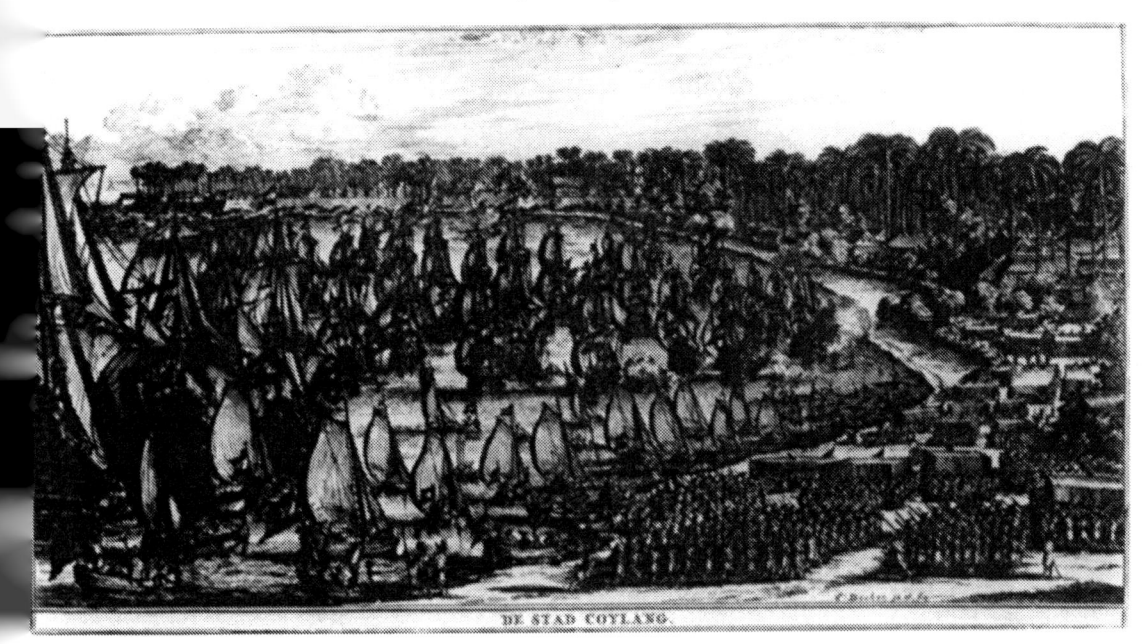

Figure 11. Landing of Van Goens' expeditionary forces at Quilon in 1661. Etching by Conrad Decker from Schouten 1708.

trading'[42]. Anyhow, Van Reede tendered his resignation from the Company. His friend Joan Bax, since 1663 councillor of Malabar, did the same. It is not known what reasons Bax adduced for his own resignation. Perhaps he did not agree with the proceedings in the council of Malabar, in which commander Godske exposed his violent conflict with Van Goens about the Malabar government policy[43]. Rijklof van Goens, who was not only governor of Ceylon, but also commander of the Company's army of the Western Quarters, became embarrassed by this threatening loss of two captains. In view of the Second Anglo-Dutch War raging in Europe he could not spare any seasoned soldiers for averting a possible English attack in India[44]. He appeased Van Reede and Bax with an appeal to their military honour:

'Van Reede and captain Bax have requested their dismission to the mother country, but we have advised them both not to soil their reputation by their departure as long as the war with England lasts'[45].

Both captains accepted the inevitability of their situation. Nevertheless, a good opportunity soon presented itself for eliminating with one blow the existing displeasure among the principal servants of the Company in Malabar. Commander Godske in turn tendered his resignation. Van Goens heartily endorsed the request of his troublesome opponent, and at the same time recommended his own protégé, Van Reede, to the High Government in Batavia as successor in the commandership of Malabar. But the High Government decided otherwise. On 22 August 1667 it appointed Lucas van der Dussen as commander, transferred Van Reede to Ceylon, and kept Bax for the time being in Cochin[46]. Van Goens indeed protested against this decision, because he preferred to see a capable soldier like Van Reede as commander, but finally he had to resign himself to it[47]. Thus Van Reede's stay of six years in Malabar came to an end.

As has been said, Van Reede himself has not written about Quilon. All the more detail has been given by his predecessor there, Joan Nieuhof, who was chief of Quilon from 1663 to 1665 and who became acquainted earlier than Van Reede with South Malabar during diplomatic journeys[48].

Nieuhof related that Quilon was the oldest fortification of the Portuguese in India. It used to be a small but beautiful town, not far from the coast, with a direct communication by water through the interior with Cochin, Cranganūr, and Kāyamkulam. After its conquest by the Dutch, on 8 December 1661, the area of the town was considerably reduced and nearly all the buildings were demolished (Fig.11). The only buildings that remained intact within the cutting-off, were the St. Thomas Church, some conventual buildings, and the finest residences of the Portuguese. The fort, in which formerly the Portuguese governor of Quilon resided, henceforth served as the residence of the Dutch chief and the servants of the Company. Nieuhof liked the stay in Quilon very much. He wrote:

'Behind these houses are beautiful gardens, planted with coconut and other Indian trees, which flourish there

Figure 12. Animals of Malabar. Anonymous etching from Nieuhof 1682.

plentifully ... The air is extremely healthy'.

About the environs he remarked:

'The land around Quilon, too, is very delightful and is considered the most fertile of all India. It is very densely planted with all manner of Indian fruit-trees, so that the roads alone are free. Much pepper grows there, which climbs high into the Indian trees with its tendrils and is relieved of its fruit in January and February. Everywhere many beautiful gardens are seen, planted with Mangoes and other Indian fruits'.

Writing about the nearby village of Quilon de China, he said:

'The houses there usually have beautiful gardens - especially those along the rivers- in which there grow all manner of trees, fruits, flowers, and herbs. Among other things one sees there very fine lemons, which do not grow on trees, but on bushes'.

Nieuhof was also fascinated by the evening quiet:

'Upstream one sees charming summer-houses behind the gardens, where it is very pleasant, particularly in the evening, when the greatest heat is over. The inhabitants are then fishing there with the angling-rod and sometimes catch a good deal of fish'[49].

In Quilon Nieuhof amply availed himself of the opportunity for a profound study of the land, the people, and the nature of Malabar (Fig.12)[50]. In many respects his observations agree with what Van Reede was later to write about this in Hortus Malabaricus[51]. Nieuhof dwelt at length on the tropical climate of Malabar, with its long and heavy rains in winter and the great heat in summer. He was greatly interested in Malabar architecture; he admired the bamboo houses of the common people and the stone palaces of the princes. He gave details about the eating habits of the Malayali and their addiction to opium. He devoted many pages to the Malabar castes. When speaking about the Brahmins and their erudition, he wrote:

'They also apply themselves to astronomy and other

sciences, and resemble our Philosophers and teachers. There are also some who practise medicine, pharmacy, and all other professions'.

He described the Nāyars as great warriors and distinguished them from opium eaters running amok. The common people were classified by him into "Moukois", fishermen and salt-makers, Tivas, treeclimbers, Parruas, pearl-fishers, and peasants not specified further.

If one compares Nieuhof's statements about the flora around Quilon with the quotations taken above from Van Reede's official reports about Travancore, Madurai, and Ceylon, one receives the impression that at that time Nieuhof had gone a little more thoroughly into the natural history of Malabar than Van Reede. But one should not overlook the fact that Van Reede's opportunities of writing about nature were limited. In fact, at that time he had to draw up reports about colonial administration, not about tropical nature. Still, there is an indication that Van Reede's interest in the nature of Malabar in the years 1661-1667 developed further than can be inferred from his official reports. In his preface to volume 3 of Hortus Malabaricus he related how on his numerous journeys through Malabar he was fascinated by the overwhelming wealth of the flora. The passage in question, which is probably the most impressive one ever written by a European about the Malabar vegetation, in my opinion is an echo from the years 1661-1667, when he was roaming through Malabar as a soldier, a diplomat, and an inspector. This passage precedes his narrative about his decisive meeting with the physician and botanist Matthew of St. Joseph in 1674. Van Reede wrote explicitly that he received his impressions when 'on account of my office [I] fairly often had to make journeys through very well-cultivated regions ... and dense forests'. Later, as a commander of Malabar in 1670-1677, Van Reede no longer made many journeys, namely only to Vaipin, Cranganūr, and Kāyamkulam, whereas during his first stay in Malabar he was constantly on his way. Moreover he cited memories from the highlands of Malabar, with which he became acquainted in March 1664, when on a visit to the arch-deacon of the Christians of St. Thomas, and in June-July 1665, when travelling from Tuticorin to Travancore:

'After I had been with the Indians for a considerable number of years and on account of my office fairly often had to make journeys through very well-cultivated regions, on the way I observed large, lofty, and dense forests. I saw that they were pleasing through the marvellous variety of the trees of the same kind in the same forest if one were to search for this. I have seen a great many trees encircled by and entangled with other plants, while the latter, fixed in the soil with their roots, had formed such a confused mass with the tree with which they had become entangled that it could hardly be distinguished. I rather frequently saw (to put it thus generally) many ivies of various kinds clinging to one tree and moreover shooting up in the very branches of the trees, and also various plants against the bare trunk, so that it was often very pleasant to behold, on one tree, leaves, flowers, and fruits of ten or twelve different kinds displayed. And yet they did not harm this tree in any way, so that the trunks of such trees were very close to each other and very thick, or at all events they lifted their heads in the air to an elegant height of as much as eighty feet (of the ordinary length), adorned all round with green herbs of all kinds, so that there was not left any part that was not covered. The branches also, wide-spreading and shady on all sides, finally rose to a very great and really extraordinary height. Nay, sometimes they bent down and, taking root again in the fertile soil, sprouted again and shot up. One might regard one such tree as a magnificent, elegant, and delightful palace, whose vaults were supported by as many columns as one could discern branches. And in such a palace a great many people could easily be sheltered, in order to shun the inclemency of the climate, for instance excessive showers and the heat of the sun; and thus these forests resembled a house of a very elegant structure rather than virgin forests. And I have observed that not only the fertile soil extending in the plains was thus adorned, but that even the rough rocks and the steeps of the mountains were equally full of luxuriant forests, nay, the most barren rocks abounded nevertheless with the splendour of a rich vegetation of trees and herbs; some of these plants had settled on the top, others had made cracks, into whose fissures they penetrated with their roots. Every land and field extending in the plains abounded so much with plants and trees of every kind (as I have said above of the forests) and radiated such fertility that indeed every piece seemed to have been cultivated by the careful hand of some gardener and planted in a very elegant order. Indeed, even the pools - and one may wonder about this!-, the marshes, nay, the borders of the rivers which carried salt water displayed several plants, with which they were almost completely covered. Did I speak of abundance? There was no piece, not even the smallest, even of the most barren soil which did not display some plants.

Since I perceived all this repeatedly, this had led me to believe that I judged not without reason that this part of India was truly and rightly the most fertile part of the whole world and that it was largely similar to the island of Taprobana (which nowadays is called Ceylon), especially to that part which is situated in the same climate as the Malabar region[52]. I was often seized by the desire to explore and examine the leaves, flowers, bark, and roots of those plants. And then I found that they frequently had a very sweet smell and a penetrating taste. And when I asked the natives who accompanied me on my journeys whether they knew anything about these plants, they not only disclosed the names, but also knew very well their curative virtues and use. I have often witnessed this on the way, when one of our companions suffered from some complaint, either an internal or an external one, although they -these Indians- were not endowed with either medical or botanical knowledge; indeed, they spend their whole life in the military service, in agriculture, or in some craft'[53].

Notes

1 See the Instructions of Rijklof van Goens of 22 March 1663, of Jacob Hustaert of 6 March 1664, of Rijklof van Goens of 5 April 1666, and of Isbrand Godske of 5 January 1668 in s'Jacob 1976:3-81.
2 See Van Goens' eulogy of Van Reede below.
3 s'Jacob 1976:LXX,note 292.
4 See the Memorandum of Van Reede of 1677 in s'Jacob 1976:118.
5 Thus Van Reede in a retrospect of this time in his Memorandum of 1677 (s'Jacob 1976:132). See also the statement of his superior, commander Jacob Hustaert, in his Instruction of 6 March 1664, about Van Reede as "regedore maior" (s'Jacob 1976:31-32).
6 His successor as "regedore maior" of Cochin was Olatche, whose appointment was communicated by the raja on 18 May 1666 to the commander of Malabar (Heeres 1931:336-337,No.CCCVII).
7 See the dedication of vol.3 of Hortus Malabaricus (1682) to Vīra Kērala Varma.
8 Letter of Van Reede and the council of Malabar to Lords XVII, of 9 December 1676 (VOC 1321:914-914v).
9 Van Reede's Memorandum of 1677 (s'Jacob 1976:98).
10 An account of the expedition to Alangādu was given by the Swiss soldier Albrecht Herport (L'Honoré Naber 1930:140-142).
11 See the Instruction of 6 March 1664 of Jacob Hustaert. No further information is available about this mission of Van Reede (s'Jacob 1976:25).
12 According to Albrecht Herport (L'Honoré Naber 1930:146), but he could not say anything about the purpose of Van Reede's mission to Goa.
13 L'Honoré Naber 1930:146-158, and engraving VIII. The piece treaty with the raja of Kārthikappalli was concluded on 27 February 1665 (Heeres 1931:315-317,No.CCXCVIII).
14 The one report is dated 25 June 1665, Tuticorin, the other 1 October 1665, Tuticorin. They were submitted to Rijklof van Goens on 7 October 1665 (VOC 1251:723-839).
15 This treaty was concluded on 25 April 1665 (Heeres 1931:323-326,No.CCCI).
16 s'Jacob 1976:123, note 290.
17 In his report (VOC 1251:737-739) Van Reede gave a tabulated survey of this region. He mentioned 92 villages in Travancore, 6,400 Moors, 41 vessels, 53 churches, 9 rivers, and 21,190 Christians.
18 VOC 1251:740. Palmyra palms, "jagerbomen" in the original Dutch text, are Carimpana HM 1:11-12:9 and Ampana HM 1:10:10, *Borassus flabellifer* L. Paleguas, "balcken" in the original text, are native rowing-boats as pictured by Van Linschoten in *Itinerario* (1596); see Kern & Terpstra 1956 pt.2:28 and the engraving facing p.25.
19 In another tabulated survey (VOC 1251:784-785) Van Reede counted in this region 22,640 Moors, 2 rivers, 32 churches, 266 vessels, and 28,396 Christians.
20 This report, 'Trade of the Merchant Joan Nieuhof', is a section of Van Reede's report on Travancore and Madurai of 1 October 1665 (VOC 1251:759-774). Laurens Pijl (?-1705?) was chief of Tuticorin 1665, commander of Jaffnapatnam 1673-1679, first commander 1679-1681 and governor of Ceylon 1681-1692, and Councillor of India in Batavia 1692-1705 (Wijnaendts van Resandt 1944:62,63).
21 VOC 1251:786-787.
22 VOC 1251:792-799.
23 VOC 1251:831.
24 VOC 1251:831.
25 See the Instruction of Jacob Hustaert of 6 March 1664 and the Memorandum of Isbrand Godske of 5 January 1668 (s'Jacob 1976:14,21,49). Other quotations of the Dutch colonial opinion on the Malabars are to be found in s'Jacob, p.LXXIX.
26 Nieuhof 1682:143.
27 Joan Nieuhof (1618-1672), envoy of the Company in China 1655-1657, chief of Tuticorin and afterwards of Quilon 1663-1665, lived in Batavia until 1670, and disappeared during an excursion on Madagascar in 1672 (Van der Aa vol.13 (1868):215-217).
28 s'Jacob 1976:30, note 170.
29 On Nieuhof's statements about the natural history of Malabar, see below.
30 Van Reede left Quilon on 4 September 1665, travelled across Travancore, and arrived in Tuticorin on 11 September (VOC 1251:838-839).
31 See the extensive letter, of 23 October 1665, of Van Goens and the council of Ceylon to the Governor-General and the Council of India, on the state of affairs in the government of Ceylon, under the competence of which Malabar also came, and a similar letter of 30 October 1665. Van Reede signed both these letters as councillor of Ceylon (VOC 1253:1608-1658,1667-1682).
32 Letter of 23 October 1665 of Van Goens and the council of Ceylon to the Governor-General and the Council of India (VOC 1253:1632-1633).
33 'Report of the journey from Colombo to Mannar, Jaffnapatnam and its subordinate guards and forts delivered ... by Hendrik van Reede, Capt. on 4 December 1665' (VOC 1251:931-944). The journey took from 4 November up to and including 4 December 1665.
34 Joriphaes Vosch later returned to the Netherlands. There he lived in the house Te Vliet in Lopikerkapel (near Utrecht), which he sold in 1699 to Antoni Carel van Panhuizen tot Voorn, the husband of Van Reede's adopted daughter Francina; see also chapter 5.
35 Philippus Baldaeus (1632-after 1671), since 1656 clergyman at Galle, served in Van Goens' army which conquered Mannar and Jaffnapatnam in 1658, and accompanied the Malabar expedition of 1661-1662, clergyman at Jaffnapatnam until 1665, returned to the Netherlands in 1666, since 1669 clergyman at Geervliet (NNBW vol.2 (1912):81-82).
36 VOC 1251:939.
37 See the survey on the trade in Malabar, and the Memorandum of Isbrand Godske of 5 January 1668 on Quilon (s'Jacob 1976:LVII-LXI,51-52,57).
38 Letter of Godske to Van Goens, of 31 July 1666 (VOC 1256:334v-335v).
39 Letters of Godske, Coulster, and the council of Malabar to Van Goens, of 12 September 1666 and 30 October 1666 (VOC 1256:391-392,400).
40 Letter of Godske to Van Goens, of 31 July 1666 (VOC 1256:340-342).
41 Letter of Godske to Van Goens, of 31 July 1666, and letters of Godske, Coulster, and the council of Malabar to Van Goens, of 2 October 1666 and 30 October 1666 (VOC 1256:340-342,365,400).
42 Letter of Jan van Almonde to the Chamber of Amsterdam, of 8 March 1667, Quilon (VOC 1255:1056-1064).
43 On the conflict between Godske and Van Goens, see s'Jacob 1976:LXVIII.
44 On the Anglo-Dutch relations in India, see s'Jacob 1976:LXI-LXII.
45 Letter of Van Goens and the council of Ceylon to Lords XVII, of 25 February 1667, Colombo (VOC 1255:923).

46 Letter of the Governor-General and the Council of India to Van Goens and the council of Ceylon, of 5 October 1667 (VOC 1261:80v).
47 Letters of Van Goens and the council of Ceylon to Lords XVII, of 8 November and 17 November 1667 (VOC 1265:915-916,787).
48 Nieuhof concluded in South Malabar the commercial treaties with the rāja of Karunāgappalli on 7 February 1664 and with the queen of Quilon on 29 February/2 March 1664 (Heeres 1931:261-263,No.CCLXXIX, and 270-271, No.CCLXXXII).
49 Nieuhof 1682:121-123.
50 Nieuhof 1682:141-161.
51 Except these observations Nieuhof's editor also included notes on some fifty Malabar plants. Nearly all of them were taken from volumes 1 (1678) and 2 (1679) of Hortus Malabaricus, see Nieuhof 1682:161-180.
52 Here Van Reede referred to the climatological resemblance between Malabar and the south-western part of Ceylon.
53 Hortus Malabaricus vol.3 (1682):(iv-vi); Heniger 1980: 42-44.

FIRST CAPTAIN AND SERGEANT-MAJOR OF CEYLON 1667-1669

The transfer of Van Reede to Ceylon took him back into the active life of a soldier, an envoy, and an inspector. His appointment as first captain of Ceylon implied that he was in command of all the troops of the Company in the government, including the garrisons of Malabar and the southern coast of India. In this high military rank he was directly subordinate to Van Goens, who was the commander of the army in the Western Quarters. On the strength of his rank Van Reede also sat on the council of Ceylon in Colombo, but since he was frequently travelling, he did not attend many meetings of the council.

His first action as first captain of Ceylon was the dispatch of a personal letter of thanks to the High Government for his appointment, on 26 January 1668, but it contained at the same time the request 'to favour him with the name of honour and Character of a Major, since there is no other difference between those two offices'[1]. As a matter of fact the commander of the Ceylon armed forces usually had the title of sergeant-major. Van Goens strongly supported this request of Van Reede:

'the aforesaid Van Reede is a man who may render the Honourable Company in this Island and under this government great services, as Your Honours may easily see from his acivities and writings. His special request is that you may honour him with the Character of Major, in our opinion a small matter indeed, for which it is not worth while to dismiss such a man from the service of the Honourable Company'[2].

The High Government took no risks in this affair of honour, and in its letter of 7 October 1668 conferred the desired title of sergeant-major on Van Reede. That it was not only a matter of honour appears from the raise of his salary, from 80 guilders monthly as a captain to 120 guilders as a sergeant-major, which was incidentally granted to him[3].

During the dispatch of his letter of thanks of January 1668 Van Reede was in Tuticorin. There he received from Van Goens the mandate to undertake a mission to the Neik of Madurai. Van Reede had already visited the Neik previously, in April 1665. This time he was intended to obtain from that prince the trade monopoly from Cape Comorin to Nāgappattinam, as well as the right to build a fort somewhere on the southern coast of India[4]. The mission of Van Reede, which took from 17 February to 16 May 1668 and on which he reported at length, was not successful. The attitude of the Neik was outright hostile, and it was only after he had patiently waited for a long time that Van Reede was received by him. He could not achieve much beyond handing the Neik the presents he had brought. To his indignation the Neik also pinched his own diamond ring, the value of which he himself estimated at least at 80 rixdollars.

On his return to Tuticorin, Van Reede just missed the famous pearl-fishing there, which was annually held and in which the Company had a share in the proceeds. Van Reede confined himself to verifying the accounts, and was able to report that the Company had made a profit of 125,000 guilders[5].

In July 1668 he was in Colombo to report on his activities in the south of India[6]. There he then spent some quiet months while awaiting his appointment as a sergeant-major. He even declined an offer of Van Goens to become desāve or chief of a district somewhere in Ceylon[7]. Only at the end of 1668, when his appointment became known, did he become available again for special assignments[8].

Van Goens had fallen out once more with the commander of Malabar. After Isbrand Godske, this time it was the turn of Lucas van der Dussen to be the butt of his displeasure. After a vain investigation by Isaac de Saint-Martin into the behaviour of Van der Dussen[9], on 1 February 1669 Van Reede in turn was charged with looking into the Malabar affairs. The choice of Van Reede for this mission was rather unfortunate, because Van der Dussen was aware that Van Goens doubted the latter's military capacities and would have preferred Van Reede as commander of Malabar[10]. Thus it happened that Van der Dussen remonstrated when on 8 February 1669 his opponent appeared on the roads of Quilon and subsequently waited upon him in Cochin. The commander doggedly refused to recognize the qualifications of Van Reede as commissioner of inquiry in Malabar. In a flaming row with him Van Reede could not undertake anything useful, and on 10 April 1669 he returned to Colombo, greatly offended[11].

One day later, on 11 April, Van Goens discussed the situation in the council of Ceylon. He was furious that his orders were not obeyed in Malabar. He saw to it that Van der Dussen was discharged as commander of Malabar, that is to say: subject to the approval of the High Government. At the same time Van Reede was ordered to arrest Van der Dussen and to send him to Batavia[12].

Thus after the affair with Joan Nieuhof, Van Reede was faced once more with the task of taking a firm line with a colleague. But this time things did not go as far as this. In fact, during his commission in Malabar the Neik of Madurai had taken a more and more menacing attitude towards the Dutch settlement in Tuticorin. There was a rumour that he was gathering a big army with a view to interfering with the approaching annual pearl-fishing. In the same meeting of 11 April 1669 in which the Van der Dussen affair had been dealt with, Van Goens decided to take vigorous action against the Neik. He sent Van Reede with a detachment of 60 soldiers to the threatened Tuticorin[12].

When he arrived there, Van Reede found that with his modest force he could not provide sufficient protection to ensure that the pearl-fishing could take place. The pearl-divers, who had already flocked together, fled and the inhabitants of Tuticorin, too, left the town to escape from the forthcoming struggle between the Neik and the Company. The Neik earnestly meant to settle accounts once for all with the Dutch intruders. His army grew to 15,000 foot and 500 horse, while with his artillery he tried to batter down Tuticorin. As Van Reede afterwards stated in his report of what he called the Madurese war: 'We then were entirely besieged and surrounded as it were by enemies'. With Laurens Pijl, the Dutch chief of Tuticorin, he organized the defence, made sallies to keep the enemy at bay, and gained time by negotiations to permit the arrival of reinforcements. In October 1669 indeed Van Goens in person took the core of the Ceylon army to the besieged fortress. A little more than 1,100 foot and 50 horse were now available to Van Reede for starting the decisive struggle against tenfold odds. On 31 October, in a large-scale sally, he conducted his army to victory[13].

Only when on 30 December 1669 the peace treaty with the Neik, which put an end to the Madurese war[14], had been concluded, did Van Reede leave Tuticorin to go to Malabar, not in order to arrest Van der Dussen, as had originally been intended, but to take over the administration of that commandment himself[15].

In an exultant letter Van Goens informed Lords XVII in Amsterdam of the defeat of Madurai and emphasized unambiguously that Van Reede had conducted the defence of Tuticorin. Van Goens was heartily sorry that at the orders of the High Government he was not allowed to follow up this success with a campaign in the south of India to make some further conquests, but he consoled himself with the thought that he would send the Ceylon army, which now was in the field after all, to Malabar 'in order also to settle the affair with the Zamorin'[16]. And indeed the greater part of the army, this time under the command of Joan Bax, was to follow its former commander to Malabar.

There can be no doubt but that, in his period as a sergeant-major of Ceylon, Van Reede definitively gained a reputation as a formidable commander-in-chief and a troublesome, if not dangerous, man for his fellow-servants of the Company. One can hardly fail to get the impression that in those respects he was beginning to resemble considerably his protector Rijklof van Goens.

A different question is that to what extent in that same period Van Reede could study the land, the people, and the nature of the tropics more thoroughly. It must be said beforehand that he did not have much time for this. The mission to Madurai, the inquiry in Malabar, and finally the prolonged siege of Tuticorin, absorbed almost his whole attention. His reports about these operations contain hardly any observations on the tropical world. All the same, there are some indications that Van Reede's interest in it continued.

A case that in itself was insignificant, though it was typical of Van Reede, was the attention he paid to the elephant that accompanied him on his mission to Madurai, in the early part of 1668. To his astonishment the animal travelled only by night. After a few days, however, the elephant fell ill. Van Reede did not know what to do, until to his relief the Pulle, a vassal of the Neik of Madurai, sent him a 'medicine master', who succeeded in putting the animal on its feet again with medicaments. Van Reede, who had already been confronted in Malabar with the knowledge of the Nambūthiri in the field of natural science, was impressed by the achievement of the Madurese physician. He gladly sacrificed a few days to enable the animal to recover. The fact that shortly afterwards, through an accident, the elephant broke its hip and presumably had to be killed detracted nothing from his appreciation of the 'medicine master'[17].

It is a pity that on two occasions Van Reede missed the pearl-fishing of Tuticorin. The first time, in 1668, the fishing had just ended when he arrived in Tuticorin after his mission to Madurai. The second time, in 1669, the fishing did not take place on account of the Madurese war. Still, he must have gained a good idea of the proceedings from the enthusiastic report given of it in 1668 by Laurens Pijl, the chief of Tuticorin, which was also signed by Van Reede. Pijl was greatly interested in the unpredictable behaviour of the oysters as a result of which the search for the annually shifting oyster-banks was quite an adventure[5].

During the Madurese war Van Reede was directly confronted with the problem of the supply of medicaments for the army. The siege of Tuticorin lasted so long a time that the stocks of the hospital in Colombo, from which the medicaments were obtained, got exhausted. Even after the decisive sally of 31 October 1669 against the Neik, when the peace negotiations had already started, the medical situation in Tuticorin remained critical. The army surgeon did not venture to search for medicinal herbs in the field, in enemy territory. The lack of medicaments became one of the reasons which induced Van Reede and his staff in the end to expedite the peace treaty[18].

His report on the Madurese war induced Van Reede to study the political history, partly in defence of Van Goens, who was accused of having started the war on his own initiative[19]. Before reporting on the war itself, Van Reede related the colonial history of Madurai since the arrival of the Portuguese and the development of the relations between the Neik and the Company since 1644[13]. Probably this forms the origin of his historical interest such as it was to be expressed in his Memorandum of 1677 about Malabar.

A recommendation to examine plants

In the course of 1669, when Van Reede was in the besieged Tuticorin fortress, the government of Ceylon received a letter from Batavia containing some very interesting passages on the problem of the supply of medicaments within the Company. In fact, on 24 April 1669 the High Government wrote that the doctor of the castle of Batavia, Andries Cleyer, had learned that in Ceylon there grew large quantities of some important medicinal plants, in particular colocynth apples in the Jaffnapatnam region and salsaparilla near Caleture in the neighbourhood of Colombo. The doctor, who superintended the medical shop in Batavia, was eager to receive preparations of these plants, in order to ascertain whether the Company could be relieved of the costly importation from the Netherlands. The High Government availed itself of the opportunity to urge a scientific investigation into medicinal plants in general:

'thus we shall be greatly pleased if you would recommend Doctor Robertus Padbrugge (whom we take to be also a good herbalist) as well as the master surgeons and others who have knowledge of medicinal herbs to speculate and satisfy their curiosity about these things in order to discover in Ceylon all that might be found to be a great comfort there'[20].

Similar letters had also been sent by the High Government at the request of the same Cleyer to the government of Coromandel[21] and to the directorate of Bengal[22].

The inducement for these letters was the rapidly increasing interest within the Company in information on natural products. In the first half of the 17th century the emphasis was still largely on the highly profitable trade in spices, but after 1650, when the Company extended its influence and power in Asia more and more, there was a growing realization that other branches of economic life in Asia might also yield tempting profits, such as mining, plantations, and the trade in drugs and medicaments. Above, in the discussion of Van Reede's first visits to Cape of Good Hope and Batavia in 1657, it was already pointed out that the supply of medicaments within the Company formed a problem, in which at that time there was a greater actual interest at the Cape than in Batavia. Another point that has already been brought up was the critical situation in which Van Reede came to be in Tuticorin, when the supply of medicaments stagnated there[23].

About 1664, however, Andries Cleyer settled in Batavia. In 1667 he became chief of the medical shop there, which under his direction developed in a short time into a central storehouse with a laboratory, which began to produce medicaments and supply them to the ships, outposts, and the town chemist's shop of Batavia. Cleyer, however, had to work frequently with raw materials brought from Amsterdam[24]. In order to evade the high costs of transport and the poor quality of raw materials from Europe, he sought to import them from the outposts to Batavia. Naturally, 'speculations' (statements) about these much less familiar materials and their botanical suppliers were also indispensable for Cleyer's work. In 1668 he had already achieved a first success by sending for drugs from the Cape[25].

In this light one has to see Cleyer's requests for preparations and information in the letters of 1669 to Ceylon and other districts. In the letter to Ceylon, Cleyer addressed in particular Dr Robert Padbrugge, whom he presumed to be a good herbalist. Padbrugge was a graduated physician and a cousin of Joan Huydecoper van Maarsseveen, a director of the Company[26]. In his official reports Padbrugge evinced great interest in natural history. Later, when he was governor of Ambon (1682-1687), he was to become known as protector of Rumphius, who was then studying the natural history of Ambon. In 1668, in the early part of his career, Padbrugge was the best physician available to the Ceylon governor Van Goens[27]. On 14 January Padbrugge submitted to him a short, but interesting report about the natural resources of Ceylon[28]. In this he drew attention to the favourable possibilities for iron-mining and extraction of saltpetre, and to the immediate vicinity of fastflowing waters and vast forests, which were suitable for timber and firewood and for the manufacture of wheels. In addition -and that is more interesting for us- he drew attention to the fact that in those forests there were numerous edible and medicinal plants, which 'easily feed inhabitants roaming in the forest and keep them in good health'. Of the medicinal plants observed by him he enumerated some thirty species which were known to him from classical medicine, and concluded from this that 'in case of illness there are so many resources that one could produce not only a whole list of the familiar, but moreover a whole book of the unfamiliar ones'.

It is this report of Padbrugge to which the High Government and Andries Cleyer referred in their letter of 24 April 1669 with the recommendation to continue the search for medicinal plants. Lords XVII, too, who got to see the report at a somewhat later moment, endorsed the request in their letter of 9 May 1669[29].

As long as Malabar was still dependent on the government of Ceylon, the recommendation of course also applied to the commander of Malabar. As a councillor of Ceylon, Van Reede, although he did not attend many meetings of the council, will no doubt have heard of the recommendation. It would seem plausible that at this moment, shortly before he entered upon his function as commander of Malabar, the idea of an examination of the flora of Malabar will have arisen.

It is also remarkable that the High Government did not entrust the Ceylon research to one person, but to a group of persons, namely a physician, surgeons, and 'others who have knowledge of medicinal herbs'. We shall find this idea of co-operation of experts in a joint project developed in Van Reede's Hortus Malabaricus into an admirable organization of co-operating princes, scholars, draughtsmen, and interpreters. It is tempting to suppose that the letter of 24 April 1669 was the formal starting-point of Hortus Malabaricus. But this does not seem to be true. When later on Van Reede was assailed on account of his scientific work in Malabar, he never referred to the respective passage from this letter either in his official writings or in the preface to volume 3 of Hortus Malabaricus.

Padbrugge's report and the recommendation by the High Government, endorsed by Lords XVII, has had a consequence for Ceylon itself as well: the arrival of the physician and botanist Paul Hermann, who was to found the floristics of Ceylon and thus, both directly and indirectly, was to exercise great influence on Hortus Malabaricus[30].

Padbrugge probably returned to the Netherlands shortly after the submission of his report, on 14 January 1668, without being able to continue his research on the natural resources of Ceylon[31]. Van Goens was frustrated by this, but with his customary energy he immediately asked Lords XVII to send him another qualified physician and a few chemists. In that same year 1668 Lords XVII sent two chemists with good botanical knowledge, Abraham Goetjens and Willem de Witte, to Ceylon, but they could not or would not yet comply with the request for a physician at that moment. Only after repeated insistence of Van Goens did they appoint, on 15 November 1671, Paul Hermann, who arrived in Ceylon in the course of 1672[32].

Strongly supported by Van Goens and later by his son of the same name and successor as governor, during his stay of nearly eight years in Ceylon (1672-1680), Hermann accumulated a large collection of natural curiosities. Especially his Ceylon herbarium, the so-called *Musaeum Zeylanicum*, and his animals in spirits, the *Museum Indicum*, caused a stir later on, when he settled in Leiden. Hermann mainly concentrated on the phytography and the taxonomy of the Ceylon flora. For this he made use of the medico-botanical knowledge of the native physicians and he thoroughly studied the meaning of Singalese plant names. Already during his stay in Ceylon he became widely known in the Netherlands because he regularly sent dried plant material with short descriptions to friendly botanists in Europe. Thus Arnold Syen, Leiden professor of botany, and Jan Commelin, Amsterdam amateur-botanist, received from him small, but representative collections of Ceylon plants, which they in turn were to use in their commentaries on Hortus Malabaricus[33].

Apart from this indirect influence, Paul Hermann also influenced the composition of Hortus Malabaricus in a more direct way. When Van Reede had been commander of Malabar for some time and had already started his study of the Malabar flora, in or about 1674 Hermann visited Cochin. On that occasion he gave some advice on the first of version Hortus Malabaricus which induced Van Reede to revise the original plan[34].

Hermann's work in Ceylon also played a part in Van Reede's life in a less agreeable way. Below we shall refer repeatedly to the genesis and the festering of the conflict between Van Reede and Van Goens on questions of policy in Malabar and Ceylon. In the increasingly fierce discussion, which was to be conducted with all possible means, Van Goens in the long run also ranked the results of Hermann's botanical studies in Ceylon far above Van Reede's passionate enthusiasm for the Malabar flora. It cannot be proved whether and to what extent Hermann agreed with the opinion of Van Goens, but it is quite possible that the relationship between Van Reede and Hermann was spoiled to such an extent by Van Goens' attitude that the two men, after their repatriation to the Netherlands, failed to collaborate closely in the publication of Hortus Malabaricus[35].

Notes

1 VOC 1268:1081.
2 VOC 1268:1193, letter of 4 July 1668 of Van Goens and the council of Ceylon to the Governor-General and the Council of India.
3 VOC 892:658, letter of 7 October 1668 of the Governor-General and the Council of India to Van Goens and the council of Ceylon.
4 VOC 1268:1054-1055, letter of 25 January 1668 of Van Goens and the council of Ceylon to Lords XVII.
5 VOC 1268:1151-1154v, 'Report of what happened during the fishing held the 5th time by the Honourable Company on the Coast of Madurai and the banks of Tuticorin, Pondecail, Cailpatnam as far as Manapar ... under the survey of and by Hendrik van Reede, first Captain of the Ceylon government', dated 27 June 1668 (Tuticorin). This fishing lasted from 15 March to 12 May 1668. 483 Vessels and 16,359 people took part in it.
6 VOC 1268:1159-1191, 'Report and daily notes of the events of the Journey and negotiations during the mission to the Neik of Madurai by the Captain Hendrik van Rheede', dated 31 June 1668 (Colombo) and handed to Van Goens in Colombo on 2 July 1668.
7 Van Reede was in any case in Colombo in July and August, for in that period he also signed the letters of Van Goens and the council of Ceylon of 13 July and 16 August 1668 (VOC 1268:1196-1199,1209-1210). About Van Reede's waiting attitude in these months, see the letter of 4 July 1668 of Van Goens and the council of Ceylon to the Governor-General and the Council of India (VOC 1268:1193).
8 In the letter of 7 October 1668 of the Governor-General and the Council of India to Van Goens and the council of Ceylon the title of sergeant-major of Ceylon was conferred on Van Reede (VOC 892:658). On 1 February 1669 Van Reede already bore that title (VOC 1273:1456-1457).
9 VOC 1268:1118-1118v, 'Report submitted by Saint-Martin on 10 May 1668 ... on account of his mission to Cochin'.
10 VOC 1273:1456-1462v contains three documents, of 1 February 1669, concerning the lines of conduct which Van Reede had to follow towards Van der Dussen.
11 About his action as commissioner of inquiry in Malabar, Van Reede wrote a 'Report and short account' (VOC 1273:1463-1471).
12 VOC 1273:1538-1552, resolution of the council of Ceylon of 11 April 1669.
13 The report of the Madurese war and the siege of Tuticorin in VOC 1270:150-194, 'Reasons and causes of the begin-

ning and continuation of the Madurese war against the State of the Honourable Company ... by Hendrik van Reede tot Drakenstein, sergeant-major of the militia of the Island and the Government of Ceylon ... and Laurens Pijl, first merchant and chief of the sea-ports of Madurai, at the Tuticorin Office', dated 19 December 1669 (Tuticorin).
14 The peace treaty with the Neik of Madurai in VOC 1270:528-529.
15 VOC 1277:1668-1669, letter of 16 January 1670 of Van Reede and the council of Malabar to the Governor-General and the Council of India.
16 VOC 1270:1ff, letter of 9 January 1670 of Van Goens and the council of Ceylon to Lords XVII.
17 VOC 1268:1159-1160.
18 VOC 1270:521-524, resolution of 23 December 1669 taken in Tuticorin by Van Reede, Joan Bax, Laurens Pijl, and the others for the adoption of the peace with the Neik.
19 VOC 1270:14, letter of 9 January 1670 of Van Goens and the council of Ceylon to Lords XVII.
20 VOC 893:297-298, letter of 24 April 1669 of the Governor-General and the Council of India to Van Goens and the council of Ceylon.
21 VOC 893:335-336, letter of 24 April 1669 of the Governor-General and the Council of India to the governor Antoni Paviljoen and the council of Coromandel, in which Cleyer asked for 'colocynth apples, vomit nuts, and dried myrabolans' to be sent, and 'in which Mr. Joachim Fijbeecq (also being a good herbalist) is to be recommended to add such medicinal herbs, which grow on the coast'.
22 VOC 893:470-471, letter of 11 June 1669 of the Governor-General and the Council of India to the directorate of Bengal: 'since Mr. Jacob Frederik Strick Berts, first surgeon in Bengal (because he has long resided in the land of the Mogul and has roamed through it), must necessarily have obtained great experience and knowledge of many medicinal drugs and herbs in those lands, Your Honour will be well-advised to recommend and animate him to compose a list of all such things as he will think that they might profitably have been gathered in the said countries and sent hither so as to reduce the demands made on the home country'.
23 See above.
24 Van Nuys 1978:14.
25 VOC 893:135-137,895-899, letters of 31 January and 15 November 1669 of the Governor-General and the Council of India to commander Jacob Borghorst and the council of Cape of Good Hope.
26 Robert Padbrugge (1637/8-1703) studied medicine in Leiden, where he took his degree in 1663. He was twice in the East. It is not known when he went the first time; he worked in Ceylon and repatriated from there presumably in 1668. In 1670 he sailed to Ceylon the second time, now as first merchant; after this he was governor of Ternate (1676-1682) and of Ambon (1682-1687). Finally (1687-1688), he was Extraordinary Councillor of India in Batavia. In 1688 he returned to the Netherlands as admiral of the returning fleet. He then settled in Amersfoort, where he also died (NNBW vol.8 (1930):1249-1250). Although Huydecoper repeatedly reminded his cousin Padbrugge in his letters of their family relationship, I have not been able to ascertain in what way exactly they were related (RAU, Records Huydecoper Family 55,57,58, letters of 1672, 1676, 1681, and 1683).
27 VOC 1261:286v, letter of 25 January 1668 of Van Goens and the council of Ceylon to Lords XVII.
28 VOC 1261:356-356v, 'Report of the second merchant Robertus Padbrugge, containing his experience in the Island of Ceylon', dated 14 January 1668.
29 VOC 319, n.p., letter of 9 May 1669 of Lords XVII to the Governor-General and the Council of India.
30 Paul Hermann (1646-1695) studied medicine in Wittenberg, Leipzig, and Padua, where he took his degree in 1670. After his stay in Ceylon, 1672-1680, he became a professor of botany and director of the botanical garden of Leiden University, 1680-1695. (Botter-Weissleder 1949; Heniger 1969:527-560; Smit 1969:69-88).
31 I have not been able to ascertain exactly when Padbrugge left Ceylon. In 1670 in any case he was in the Netherlands, when Lords XVII engaged him again on 3 September (VOC 152, Resolutions of Lords XVII, n.p.). But from the course of events concerning the re-allocation of his station in Ceylon, which is described hereinafter, one might infer that he repatriated already in 1668.
32 See the letters of January 1670 and 19 December 1670 of Van Goens and the council of Ceylon to Lords XVII (VOC 1270:690v-691; VOC 1274:95). For Hermann's appointment, see VOC 152, resolution of Lords XVII of 8 August 1671, and VOC 239, resolution of the Chamber of Amsterdam of 15 November 1671.
33 See chapter 11.
34 See chapter 10.
35 See chapter 3.

COMMANDER OF MALABAR 1670-1677

Although Van Reede had been appointed commander of Malabar already by a letter of 26 September 1669 of the High Government (see Appendix 2), he could not enter upon the administration of Malabar at once. At the moment of his appointment he was in Tuticorin to resolve the conflict with the Neik of Madurai. When Van Goens received the news of Van Reede's appointment, he considered it wiser that Van Reede should first carry the current peace negotiations with the Neik to a successful conclusion before leaving for Malabar. For that reason Van Goens sent captain Salomon Silvester in advance to Cochin to regulate the transfer of the commandment.

On 30 December 1669 the parting commander Van der Dussen transferred the administration to Silvester and to Gelmer Vosburg, first merchant and second-in-command in Cochin. On the same day he left the town to go to his new station in Persia. After the peace with the Neik had been signed, Van Reede travelled to Cochin, where he arrived on 11 January 1670. On that same day he was inaugurated as commander of Malabar[1]. Van Reede now entered upon the period of his life, until March 1677, which was to cause him many frustrations in his function, but lasting scientific fame.

Considering the circumstances, the High Government had no doubt made a good choice by appointing Van Reede as commander of Malabar. Indeed, Van Reede was a man of the first hour, he had already shown that he was endowed with various talents and that he knew the country very well. As a soldier, Van Reede had had a considerable share in the conquest and the pacification of Malabar. As an administrator he had served in Cochin and Quilon, while during his inspections of Vengurla, Quilon, and Cochin he had proved to be a keen fighter of corruption. Van Reede's diplomatic talents had been demonstrated in Purakkād, Tekkumkūr, Goa, and Travancore. Moreover he had special relations with the Cochin court as a result of his deliverance of the rāni and his activities as "regedore maior". Finally, through his service in Ceylon Van Reede could be deemed to be acquainted with the strategic position of Malabar in relation to Ceylon. All things considered, Van Reede, who when entering upon his commandership was only 33 years of age, had an excellent record of service. It had not therefore been an exaggeration on the part of Van Goens when he had recommended his subordinate to the High Government in the words: 'endowed with extraordinary talents ... to be able to manage those people [the Malayali], who are a difficult people'[2]. All the more grievous was the surprise for Van Goens when he learned that with the appointment of the new commander at the same time the commandment of Malabar was detached completely from his government of Ceylon. Van Reede was accountable only to the Governor-General and the Council of India. The High Government thus hoped it had put an end to the many years of wrangling between Malabar and Ceylon, the more so as it was known to them that Van Goens and Van Reede were on good terms with each other[2]. In spite of this, Van Goens was to resist the secession of Malabar in every possible way during Van Reede's administration of Malabar, but he was to find his former subordinate a strong opponent, who managed to repay him in kind with a good deal of verbal force. From 1670 onwards the struggle against Van Goens and his followers was to govern a considerable part of Van Reede's career. In that struggle, Hortus Malabaricus was to be involved in the long run.

As commander of Malabar, Van Reede was in control generally of affairs of administration, trade, militia, justice, and church. More in particular he was chairman of the so-called Political Council of Malabar, for short the Council, chairman of the Council of Justice, and commander-in-chief of army and navy, including the supervision of the defences and the shipbuilding yard. The principal subordinate with whom Van Reede had to do in the exercise of his function was the first merchant Gelmer Vosburg, who was entrusted with the trade and the management of the finances of the commandment and

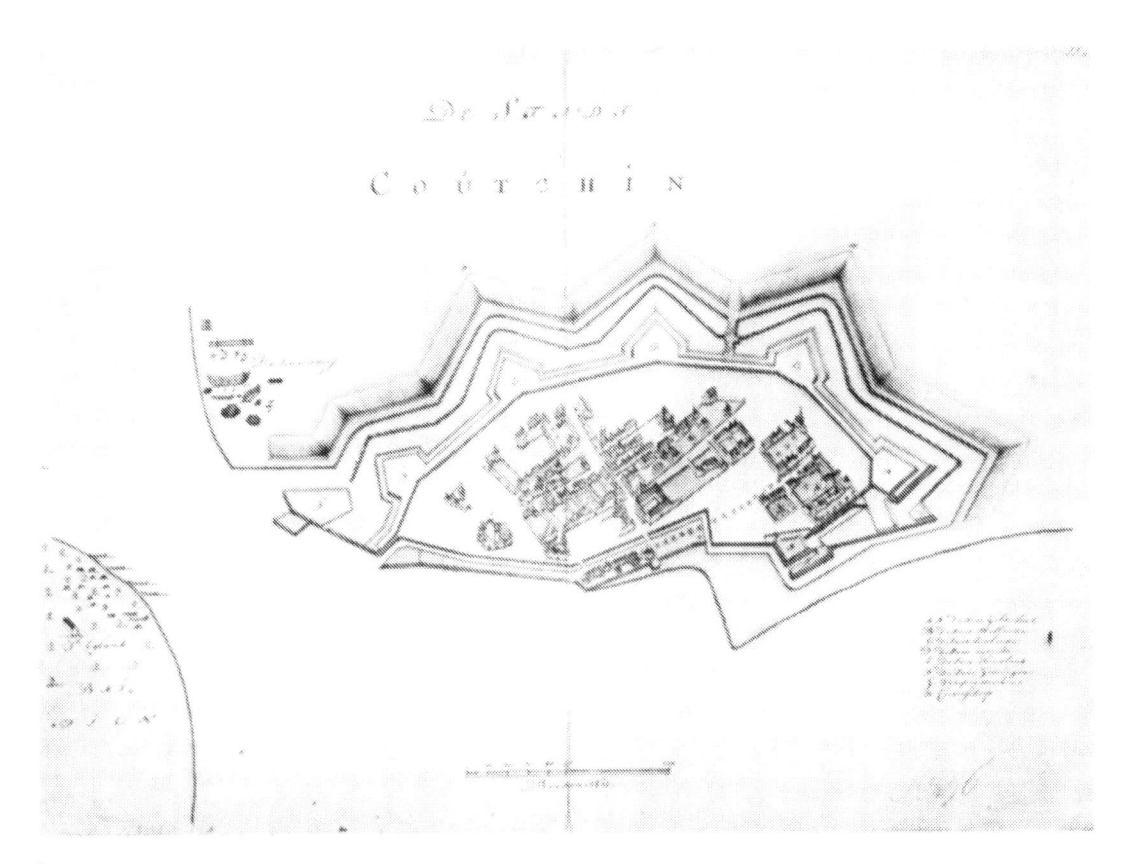

Figure 13. Plan of the town Cochin in 1677. Anonymous coloured Indian ink drawing, 53 × 74 cm. ARA, Collection Leupen 896.

who also acted as Van Reede's second-in-command or substitute[3].

Van Reede had his residence in Cochin, the capital of Malabar (Fig.13). He lived in the commander's house on the embankment, overlooking the mouth of the Cochin river. There was also a garden attached to the house[4]. The Dutch Cochin was only a fragment of the former Portuguese Cochin. During the siege of 1663 Cochin was a fairly large town with more than 12,000 inhabitants. But after the conquest by the Dutch, Van Goens had ordered a drastic reduction of the area of the town to an oval-shaped fortress, which inside its walls had an area of about 22 hectares. The Portuguese town was razed to the ground and turned into marshes wherever possible[2]. During Van Reede's adminstration the Dutch Cochin numbered only 165 habitable houses[5], so that the population was at most 1,000 souls, including the garrison.

On the landside this little town was protected by a system of bastions, counterscarps, curtains, and moats, which had been designed in 1663 and which Van Reede, too, had to strengthen a good deal.

Of the numerous churches which once formed the pride of the Portuguese Cochin, the Dutch spared only three. The church of Nossa Senhora Boa Viagens had been adapted for Protestant service, the cathedral S. Cruz had been degraded to the Company's warehouse, and only in the monastic church of S. Francisco was mass allowed to be celebrated for the few remaining Catholic Tupasses[6].

Outside the fortifications lay the shipbuilding yard, where small ships were built for the use of the Malabar commandment[7].

From this modest settlement Van Reede had to promote the Dutch commercial interests in an area which was four to five times that of the Dutch Republic in Europe. His commandment embraced not only the coast of Malabar itself, but also that of the adjacent Canara. Later Van Reede in addition had to administer the factory in Vengurla, to the north of Goa[8].

Malabar had been divided into three districts by the Company. Under the jurisdiction of Cochin were the forts of Cranganūr and Pallippuram and the factory in Purakkād. The Quilon fort was the principal place of the southern factories in Kāyamkulam and Tengāpattanam. The Cannanūr fort, on the extreme northern border of Malabar, was administered by its own chief. These districts comprised a great many military outposts, a few of which were situated inland, but most of them on the shore. Some outposts were regular little forts. The forts and outposts were mainly concentrated in the region between Cranganūr and Quilon, which was evidently the centre of the Dutch power in Malabar[9].

For the protection of the interests of the company Van Reede had at his disposal a military force of only 300 to 600 Europeans, the majority of them garrisoned in Cochin. Further, for rounds and patrols he made use of contingents of Tupas mercenaries. During the military expeditions, moreover, auxiliary troops of Nāyars could be appealed to[10].

The great interest of the East India Company in this region was due to the pepper of Malabar. This climbing-plant grew on the hills and against the mountain slopes and was common there, but it was found especially in the territory of the rāja of Tekkumkūr. Further Malabar yielded cinnamon, areca, several kinds of timber, cardamom, and products of the coconut palm. Opium, on the other hand, was imported into Malabar from Persia and Surat and exchanged for pepper[11].

When Van Reede arrived as commander, the Company had just taken the first step in the commercial policy[12]. In the period from 1663 to 1667 it had acquired practically all the pepper monopolies from the rājas on the Malabar coast. It also had received important cinnamon and opium monopolies. It was for Van Reede to put these monopolies into effect, the Company counting on a yield of 2,000 tons of pepper a year. The Dutch trading centres were mainly to be found in Cochin, Quilon, and Cannanūr. From there the Company maintained commercial relations with the surrounding kingdoms.

The commercial monopolies of the Company contained a weak link in the person of the Zamorin. This prince had granted only small pepper and opium monopolies, while he had reserved a free trade for himself. Thus, a considerable part of the trade movement of Malabar escaped from the control of the Dutch.

The politics of the Company in Malabar also ran parallel to the commercial relations. It supported in the first place its protégé, the rāja of Cochin, who was highly respected traditionally in Malabar, in consequence of which the Company could also count on the assistance of the rājas subordinate to Cochin. In addition the Company also highly valued the friendship of the princess of Quilon.

However, from of old, Malabar was divided between two parties fighting each other: southern Malabar under the leadership of the rāja of Cochin and northern Malabar under the command of the Zamorin. After the Portuguese, the Dutch also got involved in this eternal conflict, and since 1666 a state of war existed between the rāja of Cochin and the Zamorin about the possession of Cranganūr, in which the former was able to hold his own thanks to the patronage of the Company. In 1670 the war had not yet been decided by a long chalk. In reality Van Reede was chosen as the new commander because Van Goens expected that this soldier and diplomat would solve the Zamorin problem.

A particularly complicated matter was formed by the religious and linguistic problems in Malabar. The Malayali were divided into Hindus, Moslims, and Christians of St. Thomas. Especially the latter were the subject of the Roman Catholic mission during the Portuguese government, so that some of them were under the authority of Rome, but the original groups adhered to the Orthodox or Syrian Church. During the rule of the Protestant Dutch the Roman Catholic mission among the Christians of St. Thomas was continued, which was a thorn in the flesh of the Company. Moreover, there were two more groups of Roman Catholics, namely converted Hindus called the Lascarins, and the descendants of Portuguese-Malabar marriages, the Tupasses.

Of the European Roman Catholics, after the banishment of the Portuguese population to Goa in 1663, only the representatives of some monastic orders had been left. In Cochin there were still Franciscans for ministering to the Tupasses there. In the rural districts the Jesuits and the Discalced Carmelites contested each other the control of the native population.

An important incidental aspect of this variety of religions was the phenomenon that it was greatly dependent on the strata of society. Thus the Brahmin caste of the Hindus comprised the principal rajas, the Moslims included the most powerful merchants, the Lascarins the soldiers, just as the Tupasses, who were also manual workers and small traders.

If the Company was to keep on good terms with all these groups, which were important for its colonial status in Malabar, it had to act in a tolerant way in the religious policy. As a rule the Dutch, who in their native country were well acquainted with the co-existence of divergent religions, found no difficulty in doing so. They were actuated much less than the Portuguese by the desire to convert the heathen. The Dutch attitude towards the Roman Catholic priests in Malabar was a different matter, for they constantly suspected them of attempts at infiltration of Rome and Portugal. Still, among the Dutch, too, there was a movement of more orthodox Protestants which, under the leadership of the classis of Amsterdam, was greatly in favour of converting the population of Malabar to Protestantism. In Cochin there were two Reformed clergymen, Marcus Mazius and Johannes Casearius, who not only had to minister to the needs of the Europeans there, but also had to promote the conversion of the natives to Protestantism[13]. They were not very successful in this, partly owing to the language barrier. The clergymen were required to know a fair number of languages. They had to know some Latin and Greek in order to understand the Syrian liturgy of the Christians of St. Thomas. The Portuguese language served for preaching to the Tupasses. Probably Mazius and Casearius had been able to pick up the elements of Portuguese in the missionary centre of Jaffnapatnam in Ceylon, which functioned in this respect as an intermediate station for posting to Malabar. Once they had arrived in Cochin, the clergymen were expected to get a sufficient command of those languages in Malabar by practice and self-study.

The linguistic problem was not only involved in the conversion of the natives to Protestantism. The Dutch administrative machinery, too, urgently needed reliable interpreters. Thus Van Reede did not know Malayalam or Sanskrit. At first he was almost wholly dependent on Tupasses and Brahmins as interpreters, such as Vinaique Pandito and Emanuel Carneiro, for talking with the natives[14]. A special circumstance was that shortly before 1668 the soldier Herman Hasencamp had been sent to Kōttayam, in the kingdom of Tekkumkūr, to teach the Christians of St. Thomas some Latin. Hasencamp's presence in Kōttayam soon gave rise to a small school, where not only Latin, but also Dutch, Malayalam, and even Sanskrit were taught. Since the beginning of 1668 Dutch boys also went to school there to receive a training for interpreter. The raja of Tekkumkūr was personally very much interested in the school. He sent to it prominent Brahmins as teachers for the young people, and to the asthonishment of the Dutch he himself also taught them. During his commandership Van Reede paid much attention to the school of Kōttayam and could also reap the first fruits of it. In 1674 the first Dutch boy finished his studies to the satisfaction of his Malabar teachers and was apprenticed by Van Reede at the secretary's office of Cochin[15]. Unfortunately the name of this pupil is not known, but perhaps he was the same person as Christiaan Herman van Donep, the secretary of the town of Cochin in 1675, who then did translations for Van Reede's Hortus Malabaricus.

As regards the discharge of the general administration of Malabar, from the very first Van Reede was faced with serious problems. In the first place he found an understaffed council, which he brought to full strength as best he could in that small Dutch community by appointing new members[16]. In this he was not always very fortunate, for on second thoughts some of them were found to be unequal to their task. Secondly -and this was much more serious- from the first Van Reede could not get on with his substitute Gelmer Vosburg[17], although the latter was quite competent. This controversy largely paralysed the alertness of the council of Malabar.

In order to bypass both these obstacles, Van Reede in the long run assumed an authoritarian attitude, which indeed will not have been alien to him as a soldier and nobleman, but which was generally not accepted gratefully, either by the Dutch in Malabar or in Batavia. During the last years of his commandership Van Reede himself had the obligatory reports to Batavia and Amsterdam drawn up, without previous discussion with his council, and pressed its members to sign those documents. But thus we are indeed certain that Van Reede's views, also in more cultural and scientific matters, have come down to us practically unobstructed and that his initiatives in these things were carried out almost without any difficulty. Unfortunately the discord in the council of Malabar also permeated the lower administrative bodies, so that there, too, Van Reede was faced with many unpleasantnesses.

The administration of Malabar was adapted as much as possible to the succession of the seasons. The most important affairs had to be dispatched as much as possible during the dry season, because in the rainy season, from May to September, traffic on land and by sea was greatly impeded. There was then not much sense in carrying out military operations, undertaking tours of inspection, hunting the smugglers, or working on the defences. Moreover, the months of August to November were the season of the rice harvest and the second planting, so that in those months hardly any native workers were available. The consequence was that the months of April to November formed a slack season for the Dutch civil servants, in which arrears of work could be made up and in which they had relatively much leisure, like Van Reede and his collaborators, for engaging in scientific research.

The first years of Van Reede's commandership were taken up by military and diplomatic offensives for restoring the authority of the Company in Malabar, which had declined under the previous commanders. A complication in this respect was formed by the threatening attitude of a French naval squadron.

First of all Van Reede had to settle accounts with the Zamorin of Calicut, who had taken hostile action since 1666 against the rāja of Cochin. The issue of the struggle was the possession of the strategic fort of Cranganūr, which in that year the Dutch had seized in the name of the rāja[18]. For want of military forces the rāja and the Company had hardly been able to ward off the advancing Zamorin. After the victory in the Madurese war and the subsequent peace of Tuticorin, Van Goens sent the Ceylon expeditionary army under the command of Joan Bax[19] to Malabar, where the latter arrived shortly after Van Reede. Although this expedition had been planned long before, Van Reede at once seized the opportunity to emphasize his first action as commander with a great display of power. The Dutch army, numbering 900 soldiers on this occasion, marched into the territory of the Zamorin and defeated him on 23 March 1670. In the battle the king lost not only 300 casualties in killed and wounded, but his own wife, too, was mortally wounded, while a young son of his was wounded in the leg. The Dutch retreated with small losses to Cranganūr. Because of the approach of the rainy season Van Reede could not undertake any further action, and moreover the expeditionary army had to return in good time to its base in Ceylon[20].

The Zamorin by no means intended yet to give up the struggle. Having been warned by a Dutch defector that the Ceylon militia had meanwhile left Malabar again, he made an attack on the fort of Cranganūr with 700 to 800 men on 1 June 1670, in the rainy season, of all times. The Dutch garrison, however, repelled the attack[21].

Van Reede now had to wait for the end of the rainy season before he could force a decision. He gathered a formidable army, consisting of 950 men of the Dutch militia, 200 Tupasses, 200 Lascarins, and 2,500 Cochinese Nāyars, the latter under the command of Rāma Varma, the third prince of Cochin, and of the rāja of Cranganūr, to capture the kingdom of Cranganūr. From 20 October to 28 December 1670 the allies besieged the so-called tower of Cranganūr, which was defended in vain by the Zamorin with 11,000 men and a battery of 11 guns. After the fall of the tower, on 24 January 1671, Van Reede returned triumphantly to Cochin[22].

However, the subsequent negotiations with the kingdom of Calicut to arrive at an acceptable peace took a very slow course. Towards the end of the year 1671 this matter became urgent, because Van Reede was then warned by the government in Batavia that a war with France was impending and that in that case the Company was apprehensive for the safety of Malabar and Ceylon[23]. Van Reede actually carried through the peace negotiations and brought about an armistice between Cochin and Calicut, which was signed on 6 February 1672 by the Zamorin in his palace in Bagueur. A few days later, on 11 February, Van Reede also signed the treaty, in behalf of the Company, in Noordwijk (Mandurty), the northernmost outpost which the Dutch had in the kingdom of Cranganūr[24]. In spite of the bloody warfare the Zamorin was to contribute to Hortus Malabaricus by sending plants, such as Kathou-theka-maravara HM 12:49:25, to Van Reede.

Hardly had Van Reede returned to Cochin when he had to discover of how little significance was the armistice with the Zamorin. Behind his back the latter had granted to the French East India Company the use of a factory in Aicotta (Azhikkōdu), at a short distance from the Dutch Cranganūr fort. On 15 February 1672 a formidable French fleet of 13 sail appeared off Ponnāni, and in the evening it lay in front of the Cranganūr river to enforce the French claims. The next day, on 16 February, the French landed, occupied Aicotta, claimed the kingdom of Cranganūr for the French crown and the French Company, and conquered the Cranganūr fort. As suddenly as they had come they disappeared again. On 17 February their fleet lay off Pallippuram and Van Reede feared an invasion of the island of Vaipin. Post-haste he ordered the native militia to march to the threatened spot, while he himself with the Dutch troops rushed forward on the inland waters. Fortunately it did not come to a fight with the strong French forces, for the French commander François Caron sent word that he only wanted to claim Aicotta for the French. After this, the fleet sailed slowly in the direction of Cochin, with Van Reede cautiously following along the coast of Vaipin. Finally the French disappeared towards Tēngāpattanam. A week later, on 25 February 1672, the French merchant De Flacourt appeared, to open the factory at Aicotta, but Van Reede managed to put him off[25].

Although the French invasion on the coast of Malabar had finally blown over, the Dutch had been greatly frightened. Van Reede now had to exert himself to put Malabar in a state of defence. He did not himself have a war fleet at his disposal to prevent a landing. For defence at sea he was dependent on Van Goens, who as admiral was in command of the Company's fleet in the waters of India and Ceylon. Van Reede on his part worked hard on the strengthening of the Dutch forts in Malabar. He mainly concentrated on the forts of Cochin, Cannanūr, and Quilon. In the years of the French threat, from November 1671 to the end of 1674, he spent over 600,000 guilders on their improvements[26].

The pressing necessity of this work is evident from the fact that Van Reede made a tour of inspection to southern Malabar in the middle of the rainy season, in August 1672.

Amidst the problems with the Zamorin and the

French, nearer home Van Reede also had many worries about the rāja of Cochin, Vīra Kērala Varma. Personally, Van Reede got on well with the rāja, as appears from his dedication of volume 3 (1682) of Hortus Malabaricus to this king, but politically the situation was quite different. According to the views of Van Reede the political problems were partly due to the character of the rāja. He had been born about 1630 in the highlands and had been educated entirely in a religious and scholarly atmosphere. In his youth he does not seem to have occupied himself with politics. However, when in 1663 his elder brother, the claimant to the throne of Cochin, died unexpectedly, according to the Malabar law of succession the crown dropped into his lap. Vīra Kērala Varma never disguised from Van Reede that he disliked the government business. He preferred to remain aloof and devote himself to meditation and study. Once he had come to the throne, in spite of himself, he was confronted with the particularly precarious situation of Cochin. This kingdom was no match by itself for the Zamorin of Calicut, a fact which was aggravated by the discord among the Cochin princes and by the slight loyalty of the vassal states. Moreover, the rāja felt the pressure of the alliance with the Company, which he could not do without if he was not to be ruined altogether. Vīra Kērala Varma decidedly was not the right person to drag the Cochin kingdom out of its problems. On the contrary, in the opinion of the Dutch, and also of Van Reede, he did a good deal to make an even greater mess of things[28]. In the course of the years the rāja turned out to be a wilful prince, who was found to be very envious of the respect shown to other people. His interest in religious matters was revealed in an almost theocratic government, which was accompanied by great donations of money to temples. The exchequer, which was not very flourishing as it was, thus became involved in great difficulties, the more so because he pawned many crown-lands in order to get money. Symptomatic of his attitude in financial and religious affairs was the conversation between him and Van Reede early in 1673, when the rāja came to say goodbye before departing for Trichur in order to sacrifice 10,000 fanums, 'on which occasion we sought to dissuade him from it, adding that if he omitted to do so, he would become rich; but it was no use: he professed that it was a greater honour for the King of Cochin to be a supporter of the pagodas or temples than to be rich in money'. Equally typical was the complaint of Van Reede upon this: 'If one should tell such rulers that the world cannot be governed with a paternoster in the hand, they would no doubt change their lives'[29]. It was also a serious thing that Vīra Kērala Varma was not very lucky in the choice of his counsellors. The rāja left practically the whole of the adminstration to the Canara merchant Perimbala and his followers, greatly to the displeasure of the princes and the vassals of Cochin, who thus lost much power and financial advantages. The prestige of the rāja, an important factor in Malabar society, had been impaired so much that highwaymen did not scruple to loot the immediate neighbourhood of his palace[30]. The Company, too, threatened to lose its influence on the rāja. It feared, not unjustly, that it might be swept away from Malabar if the confused situation continued.

In this situation Van Reede had to try to improve and reinforce the prestige of the Company. The means for this were limited. The small contingent of Dutchmen, indeed, was unable to compel a solution by force. Large-scale military aid from Ceylon or elsewhere, too, was out of the question, because the Company did not wish to make new conquests, which would entail insuperable expense. Van Reede could choose between regular small punitive expeditions and a replacement of the ruling rāja on the one hand, as advocated by Van Goens[31], and a more diplomatic action on the other hand. The first choice was financially, organizationally, and politically difficult to execute. For that reason Van Reede sought from the beginning of his administration to create intensive diplomatic relations with the various strata of Malabar society, with a view to establishing a kind of balance of power, in which the Company would be able to hold its own and effectuate its trade monopolies[32].

The first step taken by Van Reede was the appointment of captain Burghart Uytter as controller of the royal finances, approximately the same position as Van Reede had formerly occupied at the Cochin court. Uytter grew to be a capable negotiator, who travelled about the country during the whole period of Van Reede's administration, as an envoy to the courts of the rāja's and other princes[33]. Van Reede himself only very rarely made diplomatic journeys. It was not in agreement with the views of the princes that the commander should visit them personally to discuss affairs of state. Thus Van Reede confined himself to receptions in or near the town of Cochin.

At first Uytter appeared to make directly for success. He became acquainted fairly soon with the complicated Cochinese relationships. He succeeded in bringing about a reduction of the burden of debt of Vīra Kērala Varma[34] and in entering into relations with the discontented Cochinese princes[35]. Uytter even managed to induce the rāja, who led rather an ambulatory life, to come personally to Cochin. The first meeting with Van Reede took place on 28 April 1671. The commander tried to reconcile Vīra Kērala Varma with the princes and some other rulers who were also present, but it was a half-hearted affair. When on 13 October of that year the rāja appeared in Cochin the second time, he agreed to the institution of a council of state of six of the principal lords[36].

The point seemed to have been carried, but in the following years 1671-1673 the rāja could or would not break away from the influence of his favourite Perimbala. The council of state was usually ignored, the rāja continued his roaming existence, stayed in temples rather than at the court, antagonized the Nāyars of Pārur and Curūrnādu, and set the lords of Mouton at loggerheads. Perimbala was assuming more and more the position of a viceroy, and it was even rumoured that he was seeking to drive the Company out of the country[37]. Van Reede could undertake little to check these developments. At the end of 1673, however, events took a different turn. The discord in the Cochin kingdom had become so explosive that the opponents of Perimbala finally convened a national assembly in Mouton. On 3 December 1673 they concluded an alliance, the so-called Mouton Union, against internal and external enemies and for the defence of their privileges and freedoms. Van Reede sent a representative to this assembly in the person of captain Uytter, who joined the alliance in behalf of the Company. The commander now seized the chance. In February 1674 he summoned the principal allies and the rāja to Cochin. Together with Van Goens, who at that time lay with the Company's fleet on the Cochin roads, he forced Vīra Kērala Varma to his knees in the confirmatory treaty of 23 February 1674. The rāja joined the Mouton Union, dropped Perimbala, and promised to respect the council of state. One of the seats on the council of state was assigned to Burghart Uytter, to observe the interests of the Company[38].

Now that unity had been restored in the kingdom, the Dutch administration of Malabar ran in more quiet channels, and Van Reede could devote himself to commercial, cultural, and scientific matters.

During the growing disturbances in 1673 the trade of the Company in Malabar had all but stopped[39]. Until May 1674 Van Reede had not been able to purchase pepper, the principal product of Malabar[40]. It was not until 1675/76 that the pepper trade got under way again[41]. Van Reede did not succeed, partly also owing to the difficult political situation, in putting the pepper monopoly completely into effect. The European competition and the smuggling defeated this object. Gradually he also discovered that Cochin as the chief trading-town had had its day[42].

Another failure in the trade policy had to be accepted by Van Reede in the areca trade. The nuts and leaves of the areca palm, *Areca catechu* L., were used in the very popular betel-chewing[43]. Despite Van Reede's violent resistance -he feared great difficulties with the Malabar areca-growers- he had to obey an order of the High Government to check this trade as much as possible in favour of the profitable areca trade of Ceylon[44]. On the other hand Van Reede tried to start an areca culture of his own. In the island of Vaipin, across the Cochin river, within view of the commander's residence, he caused a plot of sandy soil with an area of 5 hectares to be reclaimed, and planted there an orchard of 50,000 areca palms, interspersed with coconut trees. In or near this orchard he built, between February and June 1675, a brick house with a labyrinth and a pond. The purpose of this plantation, popularly known as the Company's garden, was threefold. Van Reede thus hoped to have in the long run a marketable lot of areca. In addition he used the garden as a location for official receptions of rājas and other grandees who were not very keen on entering the Cochin fort. For the Dutch population of Cochin the garden was a favourite pleasure resort, and even the only possibility of recreation without an escort, since the natives did not generally care for the Dutch to make trips in the neighbourhood[45]. The foundation of the Company's garden by Van Reede recalls the reclamation activities of his father, his brother, and his uncle in the Netherlands during his youth[46].

This plantation has not done Van Reede much credit. His second-in-command Vosburg violently criticized the great manpower -800 to 1,200 coolies and 50 to 80 bricklayers- and the great expense (50,000 to 60,000 guilders) that were said to have been squandered. He accused the commander of having eliminated the money under the item of fortifications[47].

Van Reede's successor Lobs, too, worked on the embellishment of the Company's garden. Early in 1677 he ordered the construction of another brick house. But in the rainy season the structure, which had got on to the first floor, collapsed. Then, in November and December 1677, Lobs ordered the ruin to be demolished down to the foundations[48].

The orchard did not yield the commercial success which Van Reede had anticipated. The soil was found to be too high, too sandy, and too poor. The areca palms had to be watered constantly, and they wasted away when in 1681 the watering had stopped. The coconut trees could hardly hold their own. Consequently the charges exceeded the profit[49].

Still, the Company's garden was maintained. As late as 1698 Zwaardekroon stated that the garden was used for receptions of prominent Malayali[50].

After 1674 Van Reede also occupied himself with the problem of the supply of medicaments, a problem which formed one of the motives for the composition of Hortus Malabaricus.

As already stated, in 1669, when Van Reede was still a sergeant-major of Ceylon, the government there received a recommendation from the High Government to conduct a research on plants. This was connected with the endeavour of Andries Cleyer to organize the production of Asiatic medicaments and the search for the botanical suppliers in the medical shop and the adjoining laboratory in Batavia. Although in my view this recommendation must not be considered as the formal starting-point of Hortus Malabaricus, I have pointed out that the idea to proceed to study the flora of Malabar must then indeed have occurred to Van Reede. A step in the direction of independent botanical research in Malabar was taken by Van Reede when, at his own initiative, in the commander's residence in Cochin he fitted up a laboratory, in which the chemist Paulus Meysner was set to work. He there got a furnace and a distillation apparatus and was ordered to 'extract from famous medicinal herbs, fruits, and roots ... their waters, oils, and salts, and to examine in what respect they are equal to or excel the European ones, in order to provide the Company's medical shop therewith and to avoid the annual expense of many ineffective leaves, roots, seeds, and ointments from Batavia to Ceylon'. The most important result of the work of

Meysner was, in Van Reede's opinion, the distillation of oil from roots of wild cinnamon. The oil was used with favourable results in the Cochin hospital[51]. I do not know the exact data of the foundation of the laboratory and whether Van Reede consulted with Cleyer about the fittings of the laboratory and the planning of the research. In the correspondence between Van Reede and Batavia no indications about this are to be found. But it is indeed possible that another scholar, Matthew of St. Joseph, who will be referred to presently, gave him advice on this subject. In any case the High Government had not been consulted in advance about the fitting-up of the laboratory in Cochin.

Van Reede, therefore, accepted the entire responsibility when his opponents subjected his laboratory, his 'puppet' as they called it with a sneer, to criticism. The consequence was that in its letter of 22 October 1675 the High Government prohibited the continuation of the research. It was thought that the knowledge of the distillation of oil from cinnamon would be detrimental to the market in Europe if the results of the research became generally known[52]. Van Reede defended himself against the prohibition in his letter of 10 February 1676. He gave an account of the planning and the purpose of the laboratory, pointed out the success of the oil from cinnamon in the hospital, and recalled that the method of distilling cinnamon oil had already been known for years among the Singalese in Ceylon, who used the oil in betel-chewing[51]. Van Reede did not obey the prohibition. The work in the laboratory continued, also under his successor Lobs, and was stopped only in 1678[53].

A special moment in the life of Van Reede was the renewal of his acquaintance with the Italian physician and missionary Matthew of St. Joseph, with whom at one time, in 1663 in Vengurla, he had negotiated about a possible return of the Discalced Carmelites to Malabar. Early in 1674 Van Reede sent for him to Cochin, for consultation at the sick-bed of a few servants of the Company[54]. Matthew then turned out to be an excellent authority on Malabar society, a smart diplomat, a skilful physician, an experienced botanist, and a pleasant talking partner, with whom Van Reede contracted a cordial friendship. Their cooperation was to become the basis of Hortus Malabaricus.

Matthew of St. Joseph

Matthew, whose own name was Pietro Foglia, was born about 1617 in Marcianise, a small town between Naples and Capua. He was a son of Scipio Foglia and Vittoria Cortesi. He studied medicine at Naples University, where he was also said to have taken his doctor's degree. In 1639 he entered the order of Discalced Carmelites in Naples and then assumed the monastic name of Matthew St. Joseph. In 1644 he was sent to the East as a missionary by his order. As fas as is known, he never returned to Europe. He died in 1691 in India[55].

At first Matthew worked in the 'Mission in Syria', and he stayed for four years in the monastery of Mount Carmel in Palestine, where Coelestinus of St. Liduina (Pieter van Gool, 1597-1672) was his superior. Coelestinus was a brother of the renowned Leiden professor of oriental languages Jacob van Gool (1596-1667), who was greatly interested in botany[56]. At the request of the latter, Matthew studied the flora of the Lebanon, of which he made rough plant sketches.

In 1648 he was transferred to the 'Mission to Persia and the Indies'. He worked for some time in Bassora, the present-day Basra in Iraq. There he learned Arabic, Turkish, and Persian, but his missionary work among the Moslims was not very successful. He became convinced that the missionaries had better concentrate on medical care, because there were so few physicians in the East. In Basra he wrote a book about medicine, which he wanted the Jesuits in India to print. It was also in Basra that Matthew copied illustrations of plants from the work of Saladin Artafa[57].

In 1651 Matthew left Basra and settled in the Carmelite residence in Diu, a Portuguese possession on the coast of Gujarat. From there, about 1655, he undertook a voyage to Mozambique, where he did not fail to take the opportunity of studying the flora of East Africa[58].

In his further travels in the East his love of botany never left him. Wherever his missionary work took him, he made drawings of useful plants and notes about their properties. His collection of notes and drawings became a large one, which was known among his contemporaries as *Viridarium Orientale*, a small part of which was published by Zanoni (1675) and Monti (1742)[59].

This is not the right place to disclose the numerous wanderings of Matthew. I will confine myself to his journeys in India. In 1656 Matthew was ordered to leave for Malabar, where the Discalced Carmelites had an important missionary province, which was known as the Church of the Serra or the Church of Angimal (Ankamali). In Malabar, in the years 1658-1663, during his tours he availed himself of the opportunity to study the flora of this area and the medicinal virtues of the plants.

The transfer of Matthew to Malabar, to the Church of the Serra, was connected with the wish of Pope Alexander VII to reconcile the Roman Church with the Malabar Christians of St. Thomas and to nullify the influence of the Syrian patriarch. A commission under the leadership of an apostolic commissioner, Giuseppe di S. Maria, who, like Matthew, was a Discalced Carmelite[60], left Italy on 22 February 1656. The secretary of the commission, Vincenzo Maria di S. Caterina da Siena, wrote a very detailed report of this voyage, which he published in 1672 under the title of *Il Viaggio All' Indie Orientali*. At the end of 1656 the commission on board of a Dutch ship arrived in Surat. There Matthew joined the company. On 19 November the commission left Surat for Malabar. Via Goa and Vengurla, where a meeting with the 'captain' of the Dutch factory took place, the commission arrived in Cannanūr about January 1657. After a good deal of squabbling in Malabar the apostolic commissioner succeeded in asserting his authority, and on 22 July 1657 the

Catholic Christians of St. Thomas, during a solemn mass, took the oath of allegiance to the Roman Church[61].

During the journey in India the commission also had an open eye for the geography and ethnography and the natural history of the regions passed through. In his report the secretary Vincenzo Maria devoted great attention to cultivated plants, medicinal plants, ornamental plants, quadrupeds, birds, and snakes[62]. It is highly probable that Matthew assisted him in drawing up this part of the report, for in writing about the medicinal properties of the plants Vincenzo Maria stated: "La compagnia di personatà pratica delle loro virtù, mi diede commodità d'osseruare di molte, quali aggiungerè alle descritioni particulari"[63]. He mentioned only very incidentally the habitats in Malabar. In Calicut and Canara the Carmelites admired the coconut palms. In Cannanūr and its neighbourhood they studied the "Zenzaro", the "Cardamomo", and "il fico grande". In Mouton they saw "Sabsanta" and a snake called "Aianda Polagen". In Porka they met a huge crocodile; in this connection Vincenzo Maria gave some particulars about the cult of the crocodile in the neighbourhood of Cochin. Finally, in a river near Cochin, they caught a giant snake, "Perimpambo", which they dried on the Piazza delle Dogona in Cochin. Vincenzo Maria also mentioned Brahmin physicians, in particular their art of curing ulcers.

When after the completion of their task, on 7 January 1658, the apostolic commission embarked in Cochin to return to Italy, Matthew of St. Joseph remained behind in Vaipin to act temporarily as substitute of the apostolic commissioner. Shortly afterwards, on 10 March 1658, from Goa there arrived Giacinto di S. Vincenzo, who took over Matthew's function of substitute commissioner[64].

It is not certain whether Matthew then stayed in Malabar, for about December 1660 he was in Surat, to wait there again for Giuseppe di S. Maria, who undertook for the second time a journey of inspection to Malabar, now as apostolic vicar[65]. Giuseppe arrived on 14 May 1661 in Cochin, where he was proclaimed bishop of the Church of the Serra. It is not clear what position Matthew then occupied, but Giuseppe expressly mentioned that Matthew was his companion[66].

The position of the Carmelites in Malabar soon became critical. The Dutch laid the first siege to Cochin, November 1661-March 1662. In vain the terrified bishop, who was in the besieged town, tried to obtain protection from "Il Rickloff Generale", Rijklof van Goens. Great was the joy among the Carmelites when the feared commander had to give up the siege[67].

But during the second siege, when Cochin capitulated on 7 January 1663, bishop Giuseppe and other priests fell into the hands of the Dutch. It is not probable that Matthew, too, was in the conquered town, for the bishop later related that only the Carmelites Marcello, the substitute apostolic commissioner at that moment, and Gottifredo di S. Andrea were with him then[68].

According to the capitulation treaty all the Roman Catholic priests of European origin had to leave Malabar. Bishop Giuseppe protested with Van Goens against his banishment, referring to the papal commission letter, from which according to him it appeared that his diocese comprised only Tekkumkūr and Vadakkumkūr, and not Cochin. But Van Goens remained adamant and stated that the Malabar Catholics 'are amply provided with their own priests and are quite content with them'[69].

The bishop resigned himself for the time being and took the odious insinuation of the Dutch conqueror about the native clergy to heart. On 1 February 1663 he consecrated the Malabar Alexander da Campo as bishop of the Church of the Serra and a few days later, on 4 February, suffered himself to be transported to Goa in the company of two Carmelites[70].

When in November 1663 the peace between Portugal and the Netherlands became known in Goa, Giuseppe tried once again to come to an understanding with the Dutch. He sent Matthew of St. Joseph and Gottifredo di S. Andrea to Vengurla to negotiate there with the "Capitane e Consiglieri Van Ree, e Bax" about his return to the Serra. But the Dutch negotiators made it plain to the Carmelites that they preferred Alexander da Campo as bishop[71].

Ultimately Giuseppe, realizing the impossibility of a return to Malabar, decided to return to Rome. On 22 January 1664 he left Goa in the direction of Surat. On the way he called in Vengurla to say goodbye to the two Carmelites Matthew and Gottifredo, who were still there[72]. Matthew will have availed himself of the opportunity to present the bishop with the manuscript and the drawings of his *Viridarium Orientale*, or part of it, with a view to having it printed in Europe. I will refer to this presently.

The bishop continued his journey to Surat, where on 10 April 1664 he went on board the ship "Carbares", which took him to Basra. From here he travelled to Rome, where he arrived on 6 May 1665[73].

Matthew of St. Joseph remained in the East. Since Malabar had become inaccessible to him, he returned to the mission in Persia. About 1666 he was in Bender Abbas, and in 1667 and 1668 in Isfahan. Early in 1668, however, Matthew planned to return to the Serra again. Via Goa he arrived in Malabar by the end of 1668 or the beginning of 1669[74].

During this Persian period, 1666-1668, Matthew must have managed to send other parts of his *Viridarium Orientale* to Europe, for in 1669 the Carmelite Michael di S. Eliseo in Milan received from him a small book with drawings of plants. Another Carmelite, Valerius of St. Joseph, who like Matthew worked in the 'Mission to Persia and the Indies' in 1671, took another booklet to Rome. Both booklets got into the hands of Zanoni, who published from them in his *Istoria Botanica* (1675)[75].

The reason for Matthew's return to Malabar was later intimated by Van Reede in his letter of 15 May 1674 to the Dutch Governor-General in Batavia. The Pope, or the

Apostolic College, had sent the Carmelite to Malabar as coadjutor to the Malabar bishop, who had difficulty in holding his own against the Orthodox and Portuguese priests. 'This man [Matthew] has now observed this duty for five years on land in this service, without ever coming within the Company's limits', Van Reede stated[76].

Bishop Alexander da Campo, who had remained loyal to Rome, indeed governed the churches in the 'highlands nearest to the shore', in the spurs of the Malabar mountains. A less correct statement is Van Reede's allegation that before 1674 Matthew never operated within the Company's territory. One year before, in 1673, the Dutch commander had permitted him to build two churches, one in Chathiath, on the bank of the Cochin inland waterway near Ernākulam, the other a little more to the north, in Warapoly[77]. Van Reede wisely concealed this goodwill towards Matthew, for the building of Catholic churches decidedly was not in keeping with the Company's policy.

As has been said, early in 1674 Matthew was sent for to Cochin for medical consultation. On that occasion the Carmelites turned out to be a shrewd diplomat. He offered to put himself, together with the bishop, under the Company's protection in exchange for their loyalty to the Company. The idea rather appealed to Van Reede, because the attempts to convert the Christians of St. Thomas to Protestantism were altogether unsuccessful, and he had to seek for other methods to increase his hold on the native population. He transmitted Matthew's offer to the government in Batavia and suggested the admission of two or three more coadjutors, Romans or Neapolitans, to Malabar, the more so because the Carmelites were 'elderly men, who were conversant with medicine'. Pending the ultimate decision of the High Government Matthew stayed in Cochin. Presumably he then gave Van Reede information on the complicated religious relations in Malabar, on which the commander reported at length in his letter of 15 May 1674[78].

The meetings between the two men in 1673 and 1674 must also have led to the plan of composing together a work on the Malabar flora, for already in 1674 Van Reede asked the Brahmins Ranga Botto, Vinaique Pandito, and Apu Botto to come to Cochin to collaborate on Hortus Malabaricus[81].

As to his attempt to reach an agreement with Matthew on the position of the Church of the Serra in Malabar, Van Reede trod on extremely dangerous ground. On the part of the Carmelites his attitude was acclaimed, for 'Making use of his studies in botany, he [Matthew] was able to gain the favour and liking of Van Reede, the Dutch governor of Cochin, a fanatic heretic, converted in time by Fr. Matthew into a friend and benefactor of the Carmelites and their missions'[8]. But within the Company his friendship with Matthew was to be used as a vicious weapon against him.

The Company and Matthew's Viridarium Orientale

In the preface to volume 3 of Hortus Malabaricus Van Reede straight away recognized Matthew of St. Joseph as the "conditor", the founder of the work. He exposed therein what Matthew's contributions had been. It is, however, curious that in doing so he did not mention that already previously, during his wanderings in the Near and Middle East, Matthew had composed a work similar to Hortus Malabaricus, namely *Viridarium Orientale*. It appears improbable that Matthew should not have told him about his earlier botanical studies, the more so because he originally intended to publish *Viridarium Orientale* in the Netherlands.

About 1662, shortly before the fall of Cochin, from India Matthew had addressed to the previously mentioned Leiden orientalist Jacob van Gool the request to publish his notes and drawings of oriental plants. The work consisted of 'eight books and more than one thousand drawings and figures'. Van Gool agreed to this and persuaded Lords XVII on 22 August 1663 to provide for the transport of Matthew's manuscripts to the Netherlands[81].

Thus a hunt for those manuscripts started, since in consequence of the war between Portugal and the Netherlands it was not clear where the author, who was very ambulant, and his manuscripts were to be found. On 25 August 1663 Lords XVII sent letters to the administrators of the Dutch settlements in Ceylon, Coromandel, Persia, and Bengal, requesting them to trace the abode of Matthew[82]. The Company nearly succeeded in laying hands on the precious manuscripts. When in April 1664 Matthew's bishop Giuseppe di S. Maria, on the way to Rome, called at Surat after the fall of Cochin, he talked there about his vicissitudes with some Franciscan monks and told them that he had manuscripts of Matthew with him, which he was to take to Rome, to have them published there. As stated above, the bishop left Surat for Basra on 10 April. Only one day later, on 11 April, there arrived in Surat one of the circulating letters of Lords XVII, addressed to the director of the Dutch factory there, containing the order to trace Matthew's manuscripts and send them to the Netherlands. The director at once inquired about the matter, but had to learn from the Franciscan monks in Surat that the bird, with his precious luggage, had just flown[83]!

It will no doubt always remain an open question whether Matthew even heard of this amazing coincidence. But he must certainly have learned that his *Viridarium Orientale* arrived safe and sound in Italy, for in April 1673 already he wrote from Warapoly to Zanoni, who was then engaged in Bologna in publishing his manuscripts[84]. This was the end of the exertions of Jacob van Gool and the Company with Matthew's fruits of his botanical research in India in the years 1656-1663.

From the foregoing it may be inferred that, in view of the war conditions in Malabar, Matthew gave up his original plan to have his manuscripts published by Van Gool in the Netherlands, and then, making a virtue of necessity, sent them to Italy to have them printed there. The later fortunes of *Viridarium Orientale* of Matthew of St. Joseph and their botanical contents will be discussed further in chapter 7. But here it can already be mentioned that Matthew made illustrations of plants in a rather sketchy way. In quality they are often inferior to the drawings that were later to be made for Hortus Malabaricus. In the corresponding texts in *Viridarium Orientale* Matthew confined himself to summary descriptions of the plants with remarks about their medicinal properties and applications. *Viridarium Orientale* contained plants from all the regions in which Matthew roamed since 1644, when he went to the monastery of Mount Carmel. Unfortunately he mentioned only few habitats, so that generally it cannot be ascertained exactly where he has botanized and drawn. With respect to Malabar it appears from several examples that he also worked on *Viridarium Orientale* during his travels in this country. Although *Viridarium Orientale* is not the basis or the starting point of Hortus Malabaricus, still the drawings and descriptions give an idea of the capacities of Matthew as a botanist at the moment when he and Van Reede met in Cochin and decided to collaborate.

As has been said, it remains curious that in his preface to volume 3 of Hortus Malabaricus Van Reede referred neither to the contents nor to the adventurous history of *Viridarium Orientale*. A possible explanation of this may be that out of prudence Van Reede wished not to give too much prominence to Matthew's role in the Catholic mission, in order not to give offence to critics in the predominantly Protestant Netherlands. For the floristics of Malabar the coming of Matthew to Cochin may be called an extremely fortunate circumstance. He and Van Reede had completely different backgrounds, culturally and religiously as well as politically, but they apparently found each other in their similar interest in exotic plants and their medical use. As I have previously set forth already, the Company sought in Asia for a solution of the problem of the supply of medicaments. Van Reede in the first instance contributed to this by establishing a laboratory in the commander's residence in Cochin. But an even better opportunity was offered by the presence of an experienced, widely travelled botanist like Matthew. It is natural to presume in this situation that Van Reede saw his chance to establish for the use of the Company a much grander project than a laboratory research alone. With Matthew at hand, he was able to extend the research to descriptions and illustrations of the medicinal plants of Malabar. Or, as Van Reede himself wrote in his preface to volume 3 about one of his motives for the composition of Hortus Malabaricus:

'it would involve great profit for the Illustrious East India Company, which indeed would be able to save those expenses which it spends on transporting medicaments to that place (India). Indeed, it would be possible to use, at less expense and with greater profit, Indian medicaments, either the same or at any rate with the same, if not superior virtues'[85].

For Matthew on his part it must have been attractive, after *Viridarium Orientale*, which was a practical survey of medicinal plants in the Near and Middle East rather than a thorough study of them, to continue his botanical research, to concentrate it on Malabar plants, and to develop it on a large scale. As we shall see presently, the means which the Dutch commander could get at his disposal for the performance of the work were much more extensive than Matthew, as a single individual, would ever have obtained. The latter seems to have found the question to what extent his status as a Catholic missionary would be injured by his alliance with a heretic commander less important, since in Basra he had become convinced that in the East medical care carried more weight than conversion.

Collaborators in Malabar

As appears from the prefaces to volume 1 of Hortus Malabaricus, the core of the work was established in the years 1674 and 1675[86]. The year 1676, when Van Reede had resolved to leave Malabar, will have been employed for making additions and corrections. In March 1677 Van Reede left Malabar.

In this short period of less than three years, hundreds of Malabar plants were collected, drawn, described, and discussed. It cannot be imagined but Van Reede called in the assistance of numerous other collaborators besides Matthew of St. Joseph. A motley row of Europeans and Malabars collaborated as plant collectors, draughtsmen, informants, translators, writers, etc. It is fascinating to see how Van Reede availed himself of knowledge and talents which were present in Malabar at that time.

Among his own subordinates, the servants of the Company, he found some capable people. First of all there was the clergyman of Cochin, Johannes Casearius (c.1640-1677), who assisted Matthew in making the Latin version of the manuscript of Hortus Malabaricus. It has been mentioned before that the clergymen of the Company in Cochin were expected to have a fairly broad knowledge of languages, to wit of Latin, Greek, Portuguese, and Malayalam.

Casearius was born about 1640 in Amsterdam. He studied at Leiden University, first philosophy from 1659 and subsequently theology from 1661[87]. During his student days in Leiden Casearius was a housemate of the famous philosopher Baruch de Spinoza in Rijnsburg, who taught him Cartesian philosophy and tried to show him the way in unorthodox free-thinking[88]. Casearius completed his studies in 1665 at Utrecht University, where Henricus Regius, also a Cartesian, taught medicine and botany[89]. As was customary at that time for students of theology, he did not take his degree, but on 5 October 1665 took his examination as a clergyman before the classis of Amsterdam. Although he passed, he was admonished by the examiners to apply himself more to the Bible and orthodox writings. His Spinozistic background will have been one of the causes why he could not get a start as a clergyman in the Netherlands. In 1666 he vainly tried to get a living in Smyrna, Turkey. In 1667, in Amsterdam, he married Isabella Brent. When at the end of that year the Company informed the classis of Amsterdam about the shortage of clergymen in the East, Casearius was one of the candidates; he was examined, and appointed on 2 January 1668[90]. On the "Sparendam" he sailed via Batavia to Colombo[91].

He may have been able to follow a short course in Ceylon at the Jaffna school to learn some Portuguese and a few elements of the Indian languages. The church council of Ceylon detailed him in Cochin as a colleague of Marcus Mazius. In the course of 1669 Casearius must have arrived in his station[92]. In Cochin he made himself conversant further with Portuguese and Malayalam. On the ground of a good testimony of Van Goens the High Government in Batavia cherished high expectations of his linguistic knowledge, with a view to the missionary work in Malabar[93]. Casearius was therefore the right man indeed for helping Van Reede and Matthew with philological problems concerning Hortus Malabaricus.

More personally, too, Casearius had a relationship with Van Reede. In fact he was the catechist of Francina, the adopted daughter of the commander. The other clergyman of Cochin, Marcus Mazius[94], was not involved in the scientific work. Van Reede could not get on with him. Although he recognized that Mazius was a good preacher, he considered that this man had a disagreeable temper, and he reproached him with a want of Christian charity. In April 1675 a smouldering conflict came to a head when Francina van Reede was to be admitted as a member of the Dutch Reformed Church. Mazius opposed this, for he held that her church-going 'had been somewhat negligent'. Casearius, as president of the Cochin church council, tried to appease this 'affair full of bitterness and passion'. But Van Reede took the matter very ill, accused Mazius of stirring up trouble in Cochin, and caused the church council to send the reluctant clergyman instantly away to Batavia[95].

Among the soldiers of the Cochin garrison Van Reede found the draughtsman Antoni Jacobsz. Goetkint. He signed the first figure of volume 1 of Hortus Malabaricus. He belonged presumably to the Antwerp artists' family of that name[96]. I do not know when he entered the service of the Company and how he came to be in Cochin. During Van Reede's commandership he held the rank of a sergeant. After the death of captain Burghart Uytter, in 1677, he was provisionally promoted an ensign. He was to distinguish himself in 1678, when he saved Matthew of St. Joseph from an attempt at kidnapping[97]. The other draughtsman known by name, Marcelis Splinter, who signed figure 39 of volume 6, stemmed from an old family of painters and sculptors who worked on the Cathedral of the city of Utrecht. Marcelis himself was born in Amsterdam in 1634[98]. About his stay in Malabar no data are available. Still, he must have been an employee of the Company there, because in the period of Van Reede's commandership there were no Dutch free-burghers in Malabar as yet. The identity of the other draughtsmen -by his own account Van Reede made use of three or four draughtsmen[99]- is not known. In the absence of contemporary muster-rolls of the Company in Malabar it cannot be guessed either who they could have been.

Matthew and Casearius were assisted in their linguistic problems by Christiaan Herman van Donep, who was employed as a secretary in Cochin. He translated the statement of 20 April 1675 of the Malabar physician Itti Achudem from Portuguese into Latin[100]. Perhaps Van Donep is identical with the pupil of Kōttayam who completed his studies there with distinction in 1674[15]. The town council, which concerned itself mainly with the administration of justice in minor civil cases, consisted of three Company's servants and four Tupasses. Van Donep therefore may be presumed to have had some juridical in addition to linguistic knowledge.

Van Reede submitted the botanical and medical questions concerning Hortus Malabaricus to an advisory board of fifteen to sixteen scholars, physicians and botanists, whom he had partly chosen from his own staff:

'The plants ... were subjected, if this was worthwhile, to the examination of skilful physicians and botanists, whom I had convoked for that purpose, both from my staff and from among people of some reputation. Naturally I took care that the pictures of the plants were shown to this board, which sometimes consisted of fifteen or sixteen scholars, and that by means of an interpreter they were asked whether they knew those plants and their names and curative virtues'[101].

In this case again Van Reede did not mention any names, but we may attempt to ascertain from which group of Company's servants those staff members may have come. In Cochin there were a hospital and a medical shop, and it is natural to seek the collaborators among the medical staff. The Company did not have a qualified physician in Malabar. In the hospital there were seven surgeons, among whom the chief surgeon Simon Kadensky and Johan Cero were the most prominent[102]. The most likely one is Kadensky, who during Van Reede's commandership married a sister of Andries Cleyer, the director of the medical shop in Batavia. I do not know whether Kadensky and Cleyer had scientific contacts with each other via correspondence. But Kadensky's interest in Malabar plants is known from a list of medicinal plants and other materials drawn up by him in 1687[103]. The ordinary surgeons, too, must not be underestimated. In spite of Schoute's rather negative criticism of the capacities of the Company's surgeons in general, they also included 'good herbalists' such as Strick in Bengal and Fijbeecq in Coromandel.

The medical shop was directed by the previously mentioned chemist Paulus Meysner, who also worked on the distillation of cinnamon oil in Van Reede's laboratory. Meysner is the only member of the medical staff whose contributions to Hortus Malabaricus are recorded, for in the preface to volume 1 Casearius mentioned the insertion of some results of experiments in distillation in the plant descriptions[104].

Van Reede tried to overcome the barriers within the board by means of an interpreter. There was no lack of interpreters in Malabar. Since the Kōttayam school turned out its first pupil in 1674 he will not have had difficulty in finding a suitable person for this. Perhaps his interpreter was the previously mentioned Van Donep. One may also think of the preceptor of the school himself, Wisdorpius, who had succeeded as such the deceased Hasencamp[15]. Even the garrison bookkeeper Pieter Minnes may be thought of, because he was regularly used as a translator from Portuguese into Dutch[105]. It is certain that Manuel Carneiro acted as a translator on the board. According to the statement of Itti Achudem in volume 1 of Hortus Malabaricus he translated the latter's pronouncements on medicinal virtues and botanical properties from Malayalam into Portuguese. But, since his name had a Portuguese ring, Carneiro will not have been a Dutchman or a German. Presumably he was a Tupas of Portuguese-Malabar birth.

All in all, the intellectual climate among the Company's servants in Malabar was fairly favourable when, after the arrival of Matthew of St. Joseph, Van Reede began to look for collaborators for Hortus Malabaricus. Among the clergymen he found a university graduate who was conversant with Latin, Portuguese, and Malayalam; he could dispose of at least two draughtsmen originating from well-known artists' families; in scientific problems he could rely on the experience of the interested chemist and possibly on one or more surgeons; and finally he could facilitate the discussion by means of some experienced linguists. It is striking that the small community of Company's servants in Malabar could yield so many people who had a knowledge of and interest in cultural matters.

The other group of the Malabar intelligentsia was formed by the Malayali. Here, too, Van Reede found people who were suited and willing to assist him. First of all Van Reede appealed to scholarly Brahmins. According to their own statement in volume 1 of Hortus Malabaricus, Ranga Botto, Vinaique Pandito, and Apu Botto, three Brahmins from the Cochin kingdom, collaborated since 1674. In the Company records no further data about them have been found. Probably they belonged to the Nambūthiri, the priests among the Brahmins, about whom Van Reede wrote elsewhere:

'Some of them have no other occupation but the temple service, and are exempt from all worldly cares, being constantly occupied with gentile wisdom, star-gazing and natural science'[106].

Of the Brahmins, Vinaique Pandito acted as interpreter from Sanskrit, or Konkani, into Portuguese. From another section of Malabar society came Itti Achudem, a physician from Karapurram, the coastal district of Mouton, to the south of Cochin; he lived in the house Coladda[107]. He belonged to the Chogāns, a caste occupying itself with the profitable coconut industry. On account of their skill in climbing trees the Dutch also called them the 'tree climbers'[108]. Among the Chogāns we must also reckon the native collaborators whom Van Reede described as 'experts in plants, to whose care it was entrusted to collect for us finally from everywhere the plants with the leaves, flowers, and fruits, for which they even climbed the highest tops of the trees'[109].

Van Reede took great pleasure in attending meetings of his advisory board. In particular the knowledge and the behaviour of the Brahmins greatly fascinated him. In the preface to volume 3 of Hortus Malabaricus he wrote about this:

'However, this board had been brought together from various parts of Malabar, so that, because they did not know each other, they were moved by ambition, a sense of honour, and suspicion -of course, in order to win favour- if they ignored something, to cloak and conceal it. And they did all they could not to disgrace themselves if they dishonestly attributed either superior or indeed inferior virtues to the plants.

I often attended a most delightful entertainment, for instance when these Brahmins or gentile philosophers disagreed and disputed with each other by weight of arguments, which they took from maxims, rules, verses from Antiquity, and books of their ancestors who where renowned for their learning. Indeed, they disputed and strongly defended their own opinions, but with incredible modesty, such as you might even miss in the most distinguished philosophers of the world, without any acerbity, mental disturbance, or neglect to respect each other's opinion. They honour Antiquity and the first inventors of their sciences with the most pious reverence, and by them they judge their own views and also their own experiences, and they subject them to *their* authority. And as regards medicine and botany, the knowledge of these sciences is preserved in verses, the first line of which begins with the proper name of the plant, whose species, properties, accidents, forms, parts, location, season, curative virtues, use, and the like they then describe highly accurately. And they did this so skilfully that, if anyone mentioned the proper name of some plant, any Brahmin will at once answer you, stating whatever has been and can be said about it. Although, however, this method of teaching, which requires a tenacious memory, seems to be rather difficult, still they impress these verses with playful ease on the memory of the young, which they say is then strongest. Later the docile minds of adolescence and manhood retain them faithfully. The first invention of these disciplines (namely, medicine and botany) is considered to be so old that they show books by authors of whom all affirm by constant asseveration that they lived four thousand years ago'[110].

Criticism of Van Reede and his collaborators

The scientific work of Van Reede and his collaborators did not remain exempt from criticism. Like the areca orchard, the country house, and the laboratory, Hortus Malabaricus also was exposed to attacks from within the Company.

The criticism came in the first instance, again, from the Malabar second-in-command Gelmer Vosburg and was directed at the persons of Matthew of St. Joseph and Johannes Casearius. In his report of 7 June 1675 about the administration of Van Reede in Malabar, Vosburg wrote:

'which Italian or Neapolitan has managed to worm himself so much into the good graces of the Commander Van Reede that he is almost considered a saint'.

And after a resounding diatribe against Roman priests in general, he continued about 'this crafty Italian':

'He is said to be an expert doctor of medicine and a good herbalist; that is why he has all the more opportunity to ransack every hole, to call on the sick, and visit the houses ... now in Cochin they are engaged in making with this Carmelite a herbarium of all sorts of herbs, roots, plants, etc., in which the clergyman Casearius is also a great helper'.

In Vosburg's opinion Casearius would have done better to apply himself to the study of Portuguese, so as to undertake with more success the conversion among the Catholics[111]. It is striking how in this criticism religious intolerance and proselytism are opposed to medical aid and scientific work.

In discussing this report in the Council of India, Van Goens, who had meanwhile become director-general of India in Batavia, added that Matthew now evidently had supreme power over all the Christians in Cochin and that Van Reede had even had a house, just outside the town, fitted up for him for the celebration of mass[112]. The Council of India was intimidated by this antipapism, and the governor-general Maatsuiker, himself a Roman Catholic by origin, was forced in the letter of 22 October 1675 to order Van Reede to banish Matthew from the kingdom of Cochin:

'At the same time for good reasons we also order you herewith to expel, at sight of the present letter, the Carmelite Father Matthew without any delay not only from the town, but from the whole Cochin province, and as much further to the north as possible, since this is a serious matter to us'.

Furthermore, Casearius was given a stern reprimand:

'with our further request to exhort the Reverend Casearius to do his duty and consult somewhat more the books of his profession than others outside it'.

He was admonished to apply himself to native and Dutch Christianity[113].

One cannot but get the impression that the clergyman Mazius sent to Batavia also put in a word in the attacks on the positions of Matthew and Casearius. In Batavia the church council had meanwhile justified Mazius as to his attitude in the conflict round Francina van Reede. Van Goens strongly supported him in the Council of India so as to procure him a new appointment[114]. The information he could provide about Malabar and religious affairs will no doubt have been opportune for Van Goens.

With the orders of the High Government concerning Matthew and Casearius the position of Hortus Malabaricus was shaken. Casearius broke down as a result of the pressure and in February 1676 tended in his resignation[115]. It was even more serious that also the leader of the research, Van Reede, already in October 1675, independently of the orders of Batavia, had intimated to Lords XVII in the Netherlands that he wished to repatriate. At that time already Van Reede was greatly disappointed by the repeated attacks on his policy as commander of Malabar[116].

Pending the official agreement with the two resignations, the work on Hortus Malabaricus continued, but presumably in a mood of little optimism. In June 1677 Van Reede and Casearius were to leave together for Batavia.

Van Reede was obliged to use the remaining time mainly in submitting defences to Lords XVII in Amsterdam and to the High Government in Batavia. He will not therefore have had much opportunity to occupy himself intensively with Hortus Malabaricus.

In this situation at the end of 1676 Van Reede was instructed by the High Government to collect, on behalf of Stadtholder William III of Orange, birds, plants, bulbs, and seeds in Malabar[117]. In September 1675 the Stadtholder had intimated to the Company in Amsterdam that he wished to receive annually from the East 'all sorts of animals, birds, tissues, cabinets, and other curiosities'[118]. Lords XVII had agreed to this[119]. Not only Malabar, but also Cape of Good Hope[120] and Ceylon received instructions to collect natural curiosities. Ceylon, where Rijklof van Goens Jr. was then governor, received the flattering mandate to send a couple of living elephants, birds, and other tame animals, cinnamon trees and pepper trees, and other rare plants to the Netherlands[121]. Whilst Van Goens Jr. had time and opportunity to carry out the order and in 1679 sent among other things, elephants to the Netherlands[122], Van Reede was not able to fill his part. He will no doubt have lacked sufficient time to undertake anything substantial before his approaching departure from Malabar. As far as is known, his successors as commanders, too, did not send any natural curiosities to the Stadtholder.

Let us revert to the difficult position in which Matthew found himself. Van Reede made a serious effort to keep him still in Cochin for the time being. In February 1676 he tried to convince the government in Batavia of the importance of the priest's presence in Cochin for the religious policy of the Company[123]. But Van Reede's position as a commander at that time was already so shaky that the Council of India remained adamant[124]. Moreover, to some extent it was put in the right as to its standpoint when in that same year 1676 a commission of eight Carmelites arrived in Malabar. They had received a mandate from the Pope and a letter of safe conduct from Lords XVII to put affairs in order among the Christians of St. Thomas of Malabar in so far as the claims of Rome were concerned. The essential issues they were to deal with was the person of Matthew, who, according to their charge 'for some years past has not behaved as religiously as was seemly and has refused his superiors the respect and humility he owed to them, and has recently changed at his own discretion without having authority, nay, against the order of the prelates, relying solely on the power of the commander, and made his own dispositions'. Their intention was to compel Matthew to leave Malabar and to appoint in his stead another coadjutor of bishop Alexander. Van Reede did his best to protect his friend from that commission. He therefore obstructed the election of the coadjutor and expelled the commission from the territory of the Company. Of course the Council of India will not have failed to see that Van Reede's argument of Matthew's indispensability in the Malabar religious policy was rebutted by the fact that the Pope himself had dropped the priest. In spite of this precarious affair Van Reede persisted in his view about Matthew. When he left Malabar, he commended him emphatically to the protection of his successor Jacob Lobs. According to Van Reede, Matthew was 'a special friend of the Dutch nation, a merry and kind companion'. In his opinion it was due to the animosity of the Portuguese clergy that the inquisition had been set at the priest. Van Reede introduced him to Lobs as the best advisor in native religious affairs that he could wish for[125]. With the departure of Van Reede, Matthew's collaboration in Hortus Malabaricus also terminated.

Lobs complied in so far with Van Reede's wish that he kept Matthew on the pay-roll of the Company[126]. This will also have been furthered by the fact that Van Reede, who himself sat on the Council of India in 1677, could still protect him for some time from Batavia.

Early in 1678 the commission of the Carmelites suddenly reappeared in Cochin, evidently with the intention to try to pilot Matthew as yet from Malabar under the new commander. The fathers burst in upon him in his church in Pallurutti, to the south of Cochin, and carried him off to Warapoly. Commander Lobs wrote about this as follows:

'when our old Father Friar Matthew officiated at night in his small church in Pallurutti, an hour's distance outside the town, they burst in upon the old man, dragging him like a criminal into their vessel, and went on to Warapoly, intending to take him from there further on, down the Cranganore River, in a vessel lying in readiness there for the purpose, to Goa in perennial imprisonment, thus thinking that they had carried the day here'.

Lobs got wind of the kidnapping and sent the ensign Antoni Goetkint, one of the former draughtsmen of Hortus Malabaricus, to liberate Matthew[127]. The commission of the Carmelites complained to the High Government[128]. Van Reede had meanwhile left Batavia and consequently could no longer do anything to spare Matthew. The High Government was fed up with the affair and expressly instructed the following commander, Martin Huisman, to settle the position of Matthew in Malabar definitively[129]. The High Government was willing to allow him to stay on, provided that he retired from the order of the Carmelites[130]. In connection with this, in 1679 Matthew travelled to Persia and in August 1680 returned to Cochin again, with the request to be allowed to continue his mission under the orders of his superiors and 'to end the little of his lifetime that was left to him here'.

However, Huisman told him that he could only be admitted if he left his order. Then Matthew travelled empty-handed to Surat, 'quite worn with age'[131]. Thus Matthew disappeared from the history of Malabar. He died in 1691[55].

The question as to what happened with the other European collaborators of Van Reede after 1677 can only be answered in a fragmentary way. It has been mentioned before that Casearius went to Batavia with him in 1677. The translator Christiaan Herman van Donep also appears to have left for Batavia and still to have worked there for Van Reede. The further fortunes of the draughtsman Marcelis Splinter and the translator Manuel Carneiro are not known. In any case they no longer appear on the muster-roll of Malabar in 1688, so that they must have either left or died before that time.

We know a little more about the other draughtsman, Antoni Goetkint. After the decease of captain Uytter, at the end of 1677, he had been promoted provisionally by commander Lobs from a sergeant to an ensign[127]. But the Council of India did not approve of the promotion, because it wanted to economize on the Malabar staff. Goetkint did not accept the put-back and in April 1679 left to Batavia with his family[132].

It is sad indeed that we have to find that the small group of Europeans, which since 1674 collaborated in Malabar with such great success in getting Hortus Malabaricus started, was driven from Malabar within a few years through envy, intolerance, and stinginess, or foundered in obscurity.

Of the Malayali among Van Reede's collaborators we only know the further career of Vīra Kĕrala Varma, the rāja of Cochin. After Van Reede had left Malabar, the king relapsed into the old error. He began to rely again on his old favourite Perimbala, who was given the management of the royal finances. The exchequer was ransacked so much that the rāja was obliged to pawn even his crown jewels[133]. Moreover, he was induced by Perimbala to start a new war against the Zamorin of Calicut, which earned him the displeasure of the Company. Once again voices were heard demanding his deposition for his slackness, which was prevented by a new constitution of May 1681 in Chennanmangalam. It was no good, for in 1684 a revolt broke out in Mouton, which could only be crushed with the aid of Company's troops. In 1685 the rāja fell seriously ill, and he died in 1687[134]. Van Reede later, during his long tour as commissioner-general of the Western Quarters, could no longer meet the rāja, for he did not arrive in Malabar until 1691. It cannot be ascertained whether Vīra Kĕrala Varma has ever seen a volume of Hortus Malabaricus, in which so much botanical and medical knowledge of his people was stored.

Departure from Malabar 1676-1677

In his letter of 28 October 1675 to Lords XVII Van Reede officially tendered his resignation as commander of

Malabar. He informed them that in September next (1676) his contract with the Company would terminate. He would then have served for twenty years, and therefore wished to repatriate[116]. He showed Lords XVII unmistakably on 9 December 1675 how greatly frustrated he felt about the developments round his position as commander:

'therefore the Commander Hendrik Van Reede (after the dispatch of the present letter) requests the High Government most urgently for his relief from here, while offering his services to the Honourable Company somewhere else in accordance with his qualifications, or to go home again ... since he cannot be of advantage to Your Honours in Malabar or acquit himself with honour or receive thanks from the High Government of India'[135].

He sent a copy of this letter to Batavia. Pending the decisions of his superiors, in February 1676 he meanwhile remitted a part of his money to Batavia[136].

The reactions of the High Government and Lords XVII to Van Reede's resignation were different, as will appear presently.

On 7 August 1676 the Counsil of India resolved 'to relieve Van Reede at his earnest request from the commandment of Malabar', without taking any further decision about his subsequent career. A few days later, on 17 August, Jacob Lobs[137], who at that moment was in Ceylon, was appointed as Van Reede's successor. The Council also restored Gelmer Vosburg to his earlier function of second-in-command of Malabar[138]!

There can be no doubt that in the re-appointment of Vosburg, who had been sent away from Malabar by Van Reede himself and who had slandered his commander in Batavia, one has to see the hand of the most powerful member of the Council of India, Rijklof van Goens.

At the head of the squadron of four ships on 26 September 1676 Gelmer Vosburg left Batavia for Ceylon. His luggage contained the documents concerning the resignation of his former superior Van Reede[139]. The notice of dismissal of Van Reede did not contain a single word of thanks for his twenty years' loyal service to the Company[140]. The retirement of Van Reede as commander of Malabar could not have been arranged more unworthily by Van Goens and his associates: no clear future any more within the Company, no thanks for services rendered, and a bitter enemy restored.

Vosburg first called for the new commander Lobs in Ceylon. On the way from Colombo to Cochin, to the annoyance of Lobs, he never ceased to paint the situation in Malabar in the blackest possible colours[141]. After their arrival in Malabar on 9 December 1676[142] Vosburg did not fail to make it clear to Van Reede that the tables were now being turned. He spread the story that the retiring commander would have to answer for his 'knavery' in Batavia[143]. The thoughtless attitude of Vosburg in a few months poisoned the atmosphere in the Malabar administrative machinery of the Company to such an extent that Lobs deemed it wiser to send him back again to Batavia as soon as possible. He fully endorsed Van Reede's opinion 'that the spite, haughtiness, and quarrelsomeness of my second-in-command alone is the cause that things went awry'[144].

Meanwhile Van Reede took all the necessary measures for his departure. He posted up his successor in detail about the state of affairs in Malabar. Later Lobs praised him on account of the good condition in which he left behind the police, the judicature, and the finances and excused him for the backlog of maintenance of the Company's buildings in Cochin, which were the responsibility of Vosburg[145]. Subsequently Van Reede wrote letters to all the rajas, princes, and lords with whom he had worked in good understanding, informing them of the impending transfer of the administration, to which they responded by confirming their friendship towards the Company[146].

On 14 March 1677 Van Reede officially transferred the commandment of Malabar to Lobs[147]. It only remained for him to remit the remainder of his money to Batavia[148] and to put the finishing touches to the Memorandum of transfer, which he signed on the day he left, on 17 March, and by which he bade farewell to Malabar in a worthy way[149].

I refrain from discussing Van Reede's Memorandum in detail here. I would refer to the full text, notes, and discussion in s'Jacob's thesis *Nederlanders in Kerala*, where this author shows that the document must still be looked upon as an important state paper in the whole history of Malabar. In an impressive way Van Reede exposed in it his comprehensive knowledge of the land and the people of Malabar. He described in detail the history, the political, social, and religious relations. The essence of the Memorandum is formed by a description of the interests and possessions of the Company as well as of the commercial products of Malabar. Finally Van Reede gave many recommendations to his successor on the way in which according to him the policy in Malabar should be executed. For a long time after his death the attention of every prospective commander of Malabar was to be drawn to the great importance of the Memorandum as a vademecum for the complicated relations in this tropical country.

I have already indicated above that a part of the Memorandum, in particular concerning the religious situation, might be based on information from Matthew of St. Joseph. Unfortunately Van Reede did not write in the Memorandum about the study of science in Malabar. He did not refer at all to the desirability of continuing the botanical research which he had started on a grand scale in his Hortus Malabaricus. Perhaps he realized, considering the criticism of his scientific work during the past years, that after his departure the circumstances in Malabar would be too unfavourable to insist on this.

On 18 March 1677, seen off with a salute from the Cochinese garrison, Van Reede set sail on board the "Pouleron". In heart-rending contrast with the dignity of his Memorandum, the decaying wooden wheels and mountings under the booming guns fell to pieces[150].

On board the "Pouleron" were also the ex-clergyman and collaborator of Hortus Malabaricus, Johannes

Casearius, and his family[151]. It is highly probable that Van Reede also took along with him his daugther Francina, for later she was, in the Netherlands, in the company of her father.

The passage to Batavia did not take place wholly without incidents. Shortly after the departure from Cochin the "Pouleron" was overtaken by the squadron of Willem Volger, who was going from his former station Surat to Batavia. On board the "Silversteyn", one of Volger's ships, was Van Reede's opponent Gelmer Vosburg! Jointly they sailed for Batavia, where a following round in the conflict about Malabar was to be fought out[152].

On the way the squadron encountered heavy weather. The "Pouleron" was struck by 'a terrific thunderbolt'; the mainmast and topmast were washed away, but the ship was saved. With a delay of one day after Volger's ships Van Reede arrived on the roads of Batavia, on 13 May 1677[153].

Notes

1   Letter of Lucas van der Dussen to the Governor-General and the Council of India, of 6 January 1670 (VOC 1277:1667-1667v); letter of Van Reede and the council of Malabar to the same, of 16 January 1670 (VOC 1277:1668-1669).
2   VOC 893:692, letter of 26 September 1669 of the Governor-General and the Council of India to Van Goens and the council of Ceylon.
3   The competencies of the commander, his substitute, and lower servants were described by Van Goens in his instructions of 22 March 1663 for the chiefs and the council of Malabar; see s'Jacob 1976:6-11.
4   ARA, Leupen Collection 896. The garden, which is referred to in Van Reede's Memorandum of 1677, can be recognized on the map by the row of trees behind the buildings (s'Jacob 1976:232).
5   VOC 1274:144, letter of 15 August 1670 of Van Reede and the council of Malabar to the Governor-General and the Council of India.
6   VOC 1308:611, letter of 15 May 1674 of Van Reede and the council of Malabar to the Governor-General and the Council of India; s'Jacob 1976:233.
7   VOC 1274:155, letter of 15 August 1670 of Van Reede and the council of Malabar to the Governor-General and the Council of India.
8   See Van Reede's Memorandum of 1677 in s'Jacob 1976:84. Later, in the preface to volume 3 of Hortus Malabaricus, Van Reede estimated the coast-line at 120 miles, 888 km, and the depth of the country at 10 to 30 miles, 75 to over 200 km (Heniger 1980:58, note 11).
9   s'Jacob 1976:LI-LII.
10 VOC 1274:150, letter of 15 August 1670 of Van Reede and the council of Malabar to the Governor-General and the Council of India.
11 s'Jacob 1976:XXI.
12 Unless otherwise stated, the following survey of the trade, politics, religion, and language in Malabar has been taken from s'Jacob 1976:XXXVII-XXXVIII,LI-LXV.
13 Marcus Mazius was sent as a clergyman to Cochin by Van Goens in November 1666 (VOC 1256:177v, letter of 12 November 1666 of Van Goens and the council of Ceylon to the Governor-General and the Council of India; VOC 1256:290v, letter of 23 November 1666 of Van Goens to Godske and the council of Malabar). After a conflict with Van Reede Mazius had to leave Malabar in 1675. Johannes Casearius is already mentioned in 1669 as a clergyman in Cochin, see note 96. In March 1677 Casearius, along with Van Reede, left Malabar for Batavia.
14 For the role of the two said interpreters in the realization of Hortus Malabaricus, see chapter 10.
15 In 1666 Herman Hasencamp, who was born in Cologne, Germany, and died on 2 November 1670 in Tekkumkūr, was in Colombo. Van Goens doubted his religious orthodoxy and placed him for the time being in the secretary's office there. In 1668 he already gave lessons in Kōttayam. After his death he was succeeded by Johannes Wisdorpius. The rāja of Tekkumkūr, who himself taught Sanskrit, died in 1674 (VOC 1256:164, letter of 12 November 1666 of Van Goens and the council of Ceylon to the Governor-General and the Council of India; VOC 1262:831v-832, letter of 12 March 1668 of Van der Dussen and the council of Malabar to Lords XVII; VOC 1274:162-162v, letter of 15 August 1670 of Van Reede and the council of Malabar to the Governor-General and the Council of India; VOC 1284:2099, letter of 20 April 1671 of Van Reede and the council of Malabar to the Governor-General and the Council of India; VOC 1299:380, letter of 15 December 1674 of Van Reede and the council of Malabar to Lords XVII).
16 VOC 1274:147,148v, letter of 15 August 1670 of Van Reede and the council of Malabar to the Governor-General and the Council of India.
17 Gelmer Vosburg arrived in the East in 1657. In 1665 he was chief in Cannanūr, then stayed in Ceylon, and worked with some interruptions in Malabar in 1669-1687. After this he served as director of Surat, 1687-1692, and governor of Malacca, 1692-1697, where he died on 19 February 1697 (s'Jacob 1976:135, note 345; Wijnaendts van Resandt 1944:184-185,209,285-286).
18 A summary of the war between Cochin and the Zamorin and its backgrounds is to be found in Van Reede's Memorandum of 1677 (s'Jacob 1976:140-146).
19 In the same letter of 26 September 1669 in which Van Reede had been appointed commander of Malabar by the High Government Joan Bax had been appointed his successor as sergeant-major of Ceylon (VOC 893:677).
20 The campaign of March 1670 is described in the letter of Van Reede and the council of Malabar to the Governor-General and the Council of India, of 27 March 1670 (VOC 1277:1671v-1673v). In the battle Van Reede lost 9 dead and 43 wounded.
21 Letter of Van Reede and the council of Malabar to the Governor-General and the Council of India, of 14 August 1670 (VOC 1277:1694).
22 The siege of the palace of the Zamorin is described in the letter of Van Reede and the council of Malabar to the Governor-General and the Council of India, of 8 March 1671 (VOC 1284:2057-2068v,2072) and in the letters of 10 March and 29 November 1671 to Lords XVII (VOC 1279:1004-1004v and 523v-540 respectively).
23 Letter of the Governor-General and the Council of India to Van Reede and the council of Malabar, of 22 October 1671 (VOC 895:754-755).
24 Letter of Van Reede and the council of Malabar to Lords XVII, of 3 March 1672 (VOC 1279:1002-1002v). According to a map of the neighbourhood of Cranganūr of 1678 (ARA, Leupen Collection 892) Noordwijk was situated at

a distance of 1,400 roods (about 5,200 m) to the north of Cranganūr on the river.

25 For the French actions on the coast of Malabar, see the letter of Van Reede and the council of Malabar to Lords XVII, of 3 March 1672 (VOC 1279:1002v-1003) and the correspondence of Van Reede with Caron, of 17 and 18 February 1672, and with De Flacourt, of 25 February (VOC 1279:1005-1006 and 1005v-1007 respectively).

26 A specified survey of the cost of the fortifications of Cannanūr, Cochin, Quilon, Cranganūr, and Pallippuram, in the years 1663-1674, is to be found in the letter of Van Reede and the council of Malabar to the Governor-General and the Council of India, of 27 March 1676 (VOC 1321:939). The great increase of the building activities can be inferred from the more than 700,000 guilders spent in the years 1663 to 1670 inclusive as against the more than 600,000 guilders in the next four years.

27 VOC 1295:287, letter of 22 April 1673 of Van Reede and the council of Malabar to the Governor-General and the Council of India.

28 VOC 1321:914-914v, letter of 9 December 1676 of Van Reede and the council of Malabar to Lords XVII.

29 Letter of Van Reede and the council of Malabar to the Governor-General and the Council of India, of 22 April 1673 (VOC 1295:282v).

30 Letter of Van Reede and the council of Malabar to Lords XVII, of 16 February 1670 (VOC 1277:1679).

31 Letter of Van Reede and the council of Malabar to the Governor-General and the Council of India, of 3 October 1671 (VOC 1284:2110-2133), about Van Goens' proposal to depose the raja of Cochin.

32 Thus Van Reede and the council of Malabar wrote on 29 November 1671 to Lords XVII that they did not like to interfere too forcibly in Malabar matters 'in order to get rid of our bad repute among the Malabars (who hate us) that we lord it everywhere with violence and severity, although this ill-natured people does not consider that it is not prudent to provoke the powerful or to pay back tyranny when one has been put in the wrong when it is not possible to achieve this by means of justice and fairness' (VOC 1280:590v-591).

33 See the letter of Van Reede and the council of Malabar to the Governor-General and the Council of India, of 22 April 1673 (VOC 1295:280), in which it was argued in favour of Uytter's raise of salary that he 'has always carried out all the Company's matters outside at the Cochinese court as well as in the country about the Malabar princes with all zeal and diligence'. Uytter died in 1679 in Cochin (s'Jacob 1976:179, note 493).

34 See the letter of Van Reede and the council of Malabar to the Governor-General and the Council of India, of 14 and 15 August 1670 (VOC 1274:121v-122) and of the same to Lords XVII, of 29 November 1671 (VOC 1280:553v).

35 See Uytter's reports of 28 March 1671 about his meeting with Rāma Varma, Gōda Varma, and the "regedore maior" in Manike Magalan, and of 3 April 1671 about his meeting with Erorma in Irinjālakkuda (VOC 1284:2101-2105v).

36 Letter of Van Reede and the council of Malabar to Lords XVII, of 29 November 1671 (VOC 1280:569-570,627v-628).

37 See the letters of Van Reede and the council of Malabar to Lords XVII, of 29 November 1671 (VOC 1280:530-531), to the Governor-General and the Council of India, of 10 April 1672 (VOC 1288:574), of 21 July 1672 (VOC 1288:592v,601-601v), of 22 April 1673 (VOC 1295:282-282v) and to Lords XVII, of 10 November 1673 (VOC 1291:590), and of 15 December 1674 (VOC 1299:364).

38 Van Reede and the council of Malabar described the events round the Union of Mouton and the ratification treaty of Cochin in detail in their letter of 15 December 1674 to Lords XVII (VOC 1299:363-373).

39 Letter of Van Reede and the council of Malabar to the Governor-General and the Council of India, of 22 April 1673 (VOC 1295:290v).

40 Letter of Van Reede and the council of Malabar to Lords XVII, of 15 December 1674 (VOC 1299:383-383v); of the same to the Governor-General and the Council of India, of 27 March 1676 (VOC 1321:932v).

41 Letter of the Governor-General and the Council of India to Lords XVII, of 30 September 1676 (VOC 1315:67v).

42 In 1680 commander Huisman fully recognized that the great native shipping to Cochin during the Portuguese administration had come to nothing under the Dutch (VOC 1361:536). On the increasing competition of the English in Malabar, see s'Jacob 1976:LXI-LXIV.

43 Betel-chewing is a stimulant (Lewin 1889:12-13,16-17).

44 Letter of the Governor-General and the Council of India to Van Reede and the council of Malabar, of 26 September 1676 (VOC 900:430-431). The areca trade of Malabar and Ceylon was an important item in the controversy between Van Reede and Van Goens, which ran through Van Reede's correspondence like a continuous thread.

45 See the report of Gelmer Vosburg, of 7 June 1675 (VOC 1308:805v); the journal of Rijklof van Goens Jr. during his inspection of Malabar (VOC 1307:694), in which his visit to the Company's garden on 20 February 1675 is mentioned; letter of the Governor-General and the Council of India to Van Reede and the council of Malabar, of 26 September 1676 (VOC 900:431); 'Note and Explanations' of Huisman and the council of Malabar to Lords XVII, of 18 April 1681 (VOC 1370:2227-2228).

46 See chapter 1.

47 The report of Gelmer Vosburg, of 7 June 1675 (VOC 1308:805v).

48 Letter of the Governor-General and the Council of India to Huisman and the council of Malabar, of 5 November 1678 (VOC 902:1377); letter of Huisman and the council of Malabar to the Governor-General and the Council of India, of 21 April 1679 (VOC 1349:1551v-1552); the report of Huisman as commissioner of Malabar, of 17 July 1679 (VOC 1349:1574v).

49 'Note and Explanations' of Huisman and the council of Malabar to Lords XVII, of 18 April 1681 (VOC 1370:2227-2228).

50 See the Memorandum by Hendrik Zwaardekroon, of 31 May 1698, in s'Jacob 1976:294.

51 Letter of Van Reede and the council of Malabar to the Governor-General and the Council of India, of 10 February 1676 (VOC 1321:895).

52 Letter of the Governor-General and the Council of India to Van Reede and the council of Malabar, of 22 October 1675 (VOC 899:442-443).

53 Letter of Huisman and the council of Malabar to the Governor-General and the Council of India, of 17 April 1679 (VOC 1349:1550).

54 Letter of Van Reede and the council of Malabar to the Governor-General and the Council of India, of 15 May 1674 (VOC 1308:620).

55 Unless otherwise stated, the biographical data of Matthew have been derived from Monti 1742:(xxvi)-(xxxv) and Chick 1939 vol.2:960-963. According to Monti, Matthew died in Tatta, in the province of Sindi; Chick, on the other

hand, states that he died in Cochin and was buried in Warapoly.
56 NNBW vol.10 (1937):287-289.
57 The identity of Saladin Artafa is unknown to me. A number of Matthew's copies of Saladin Artafa were later published by Zanoni 1675 and Monti 1742.
58 Zanoni 1675:23; Hermann 1698:75-76; Monti 1742:25-26.
59 See chapter 7.
60 Giuseppe di S. Maria (1623-1689) made two expeditions to Malabar, in 1656-1659 and in 1660-1665. He was successively bishop of Hieropolis, Bisignano, and Città di Castello (Chick 1939 vol.2:943-946, and Giuseppe's biography by Eustachio di S. Maria 1719, which contains his portrait).
61 Eustachio di S. Maria 1719:99.
62 Vincenzo Maria di S. Caterina da Siena 1672:332-409.
63 Vincenzo Maria di S. Caterina da Siena 1672:333. Monti 1742:(xxxv), also thought that Vincenzo Maria's survey of Indian plants has been derived from Matthew's commentaries.
64 Vincenzo Maria di S. Caterina da Siena 1672:412; Giuseppe di S. Maria 1672:53; Eustachio di S. Maria 1719:128-129.
65 The report of the second expedition was published in 1672 by Giuseppe di S. Maria himself: *Seconda Speditione All'Indie Orientali*. Eustachio di S. Maria 1719:161-330, gives supplementary information.
66 Giuseppe di S. Maria 1672:36,52; Eustachio di S. Maria 1719:192,203.
67 Eustachio di S. Maria 1719:243-245.
68 Giuseppe di S. Maria 1672:150.
69 Eustachio di S. Maria 1719:255-262; VOC 1239:1696-1696v, letter of 19 February 1663 of Van Goens to Lords XVII.
70 Giuseppe di S. Maria 1672:150; Eustachio di S. Maria 1719:272-276. Alexander da Campo or Chandy, kattanār of Kuravilanādu, in 1684 was still bishop of the autochtonous church of the Christians of St. Thomas (s'Jacob 1976:42,165,168,170,212).
71 Giuseppe di S. Maria 1672:191-194.
72 Giuseppe di S. Maria 1672:202.
73 Giuseppe di S. Maria 1672:203,211-212; Eustachio di S. Maria 1719:330.
74 Chick 1939 vol.2:962.
75 Monti 1742:(xxvi).
76 VOC 1308:620, letter of 15 May 1674 of Van Reede and the council of Malabar to the Governor-General and the Council of India.
77 Kindly communicated by Dr. K.S. Manilal, Calicut.
78 VOC 1308:618v-621v, letter of 15 May 1674 of Van Reede and the council of Malabar to the Governor-General and the Council of India.
79 Hortus Malabaricus vol.1 (1678):(xi)-(xii).
80 Thus Florencio del Niño Jesus in *La Orden de Sta Teresa* (1923), quoted by Chick 1939 vol.2:962.
81 VOC 105:426-427, Resolutions of Lords XVII, meeting of 22 August 1663.
82 VOC 318:656-658.
83 VOC 1242:1100-1101v, letter of 29 April 1664 of Dirk van Adrichem, director, and the council of Surat to the Governor-General and the Council of India; Van der Chijs 1893:371, meeting of 23 September 1664 of the High Government.
84 Zanoni 1675, preface.
85 Hortus Malabaricus vol.3 (1682):(vii); Heniger 1980:45.
86 Thus the statements of the Malabar collaborators in the preface of Hortus Malabaricus vol.1 (1678):(iii)-(xii).
87 *Album Studiosorum Academiae Lugduno Batavae* 1875:474,489. In both registrations, of 1659 and 1661, Casearius is mentioned as being 20 years of age. Perhaps he is identical with Johannes, the son of Cornelis Casier and Susanna Jans, who was christened on 4 January 1643 in Amsterdam (GAA, DTB).
88 Meinsma 1896:180-182. It appears from a correspondence of February 1663 between Spinoza and Simon de Vries that Casearius then had already lived for some time with the former. Spinoza considered him to be 'still too childish and volatile, and moreover bent on novelty rather than truth', but he hoped 'that after a few years he will remedy these faults of his youth; nay, in so far as I dare to judge his character, I am certain of this; that is therefore why his disposition induces me to like him' (Drs. H.L.Ph. Leeuwenberg, Vianen, drew my attention to the relationship between Casearius and Spinoza).
89 De Vrijer 1917; *Album Studiosorum Academiae Rheno-Traiectinae* 1886:59.
90 Meinsma 1896:185-186.
91 See the letters of the deputees of the Classis Amsterdam to the church council of Batavia, of 27 March and 19 November 1668, and of the same to the church council of Galle in Ceylon, of 3 April 1668 (GAA, Private Records 379, no.41:27,33,39).
92 According to the letter of the church council of Ceylon to the deputees of the Classis Amsterdam, of 22 January 1669, Mazius was then only in Cochin as a clergyman (GAA, Private Records 379, no.60:82). It appears from the letter of the Governor-General and the Council of India to Van Goens and the council of Ceylon, of 26 September 1669, that Casearius was already in Malabar on that date (VOC 893:663).
93 See the letters of the Governor-General and the Council of India to Van Goens and the council of Ceylon, of 26 September 1669, 31 July 1670, and 17 October 1670 (VOC 893:663; VOC 894:455,665) and the letter of Van Reede and the council of Malabar to the Governor-General and the Council of India, of 15 August 1670 (VOC 1274:142).
94 Mazius, who was born in "Abbenhoesen" (?), was already in Cochin in 1667 (GAA, Private Records 379, no.41:24, letter of the deputees of the Classis Amsterdam to the church council of Batavia, of 16 December 1667, Amsterdam; VOC 1469:341, General Rolls of the Government of Ceylon, of December 1690).
95 For the conflict concerning the confirmation of Francina van Reede, see the resolutions of the church council of Cochin, of 5, 8, and 10 April 1675, and the letter of that church council to the church council of Batavia, of 22 April 1675 (VOC 1308:677-678v,683v-685v). Mazius, with his family, left Cochin on or shortly after 22 April 1675 (Letter of Van Reede and the council of Malabar to the Governor-General and the Council of India, of 22 April 1675, in VOC 1308:654).
96 Pieter Goetkint (who died in 1583) and his sons Pieter (who died in 1625) and Antoni (who died in 1644) (Von Wurzbach 1906 vol.1:594; Thieme & Becker 1921 vol.14:317-318) worked in Antwerp. Perhaps Antoni Jacobsz. was a nephew or a grandson of the latter. Antoni Goetkint later became a captain in Ceylon, where he died before 5 July 1691 (De Vos 1903:420).
97 Letters of Lobs and the council of Malabar to the Governor-General and the Council of India, of 5 January and 24 April 1678 (VOC 1340:1475v,1488-1489v).
98 The Splinter family is already found in Utrecht from 1501. The eldest known representative is Gerrit (I) Splinter, who since 1505 was a foreman of the works of the Cathedral.

Marcelis (I) Splinter, who was buried in Utrecht on 24 January 1619, was a painter in the service of the Cathedral and also worked in the St. Nicholas Church. His son Gerrit (III) was born about 1606; later he moved to Amsterdam, where in 1631 he married Margaretha Willemsdr. van Emden, widow of Cornelis van de Bloock. On 30 March 1634 in the Old Church in Amsterdam from their marriage, Marcelis (II), the draughtsman of Van Reede, was christened (Swillens 1925:58-59,63-64; Maandblad van Oud-Utrecht vol.16 (1941):20; RAU, Records Cathedral Chapter 702, vol.5; GAU, City Records II, 3243, Conveyances 1619, vol.2:104-106, and Conveyances 1622, vol.2:56-58; GAU, DTB B Id 1:88; GAA, DTB 6:402; DTB 7:58,232; DTB 437:62).

99 'Three or four painters, who stayed with me in a convenient place, at once accurately depicted the living plants' (Van Reede in his preface to volume 3 of Hortus Malabaricus, p.(ix); Heniger 1980:47).
100 Hortus Malabaricus vol.1 (1678):(ix)-(x).
101 Hortus Malabaricus vol.3 (1682):(x); Heniger 1980:47-48.
102 VOC 1274:160v, letter of 15 August 1670 of Van Reede and the council of Malabar to the Governor-General and the Council of India.
103 In November 1669 in Batavia, 31 years old, the sister of Andries Cleyer, Anna Elisabeth Cleyer, born in Kassel, married Gottschalk Trap, of Lübeck, assistant of the Company. As a widow she remarried, at an unknown date, Simon Kadensky. Their daughter Barbara Margaretha Kadensky, born on 11 August 1678 in Cochin, who died on 31 March 1702, in Ceylon, married Adam van der Duyn, commander of Malabar 1708-1709 (De Vos 1903:695-696; Wijnaendts van Resandt 1944:150,187-188).
104 Hortus Malabaricus vol.1 (1678):(iv).
105 VOC 1274:160v, letter of 15 August 1670 of Van Reede and the council of Malabar to the Governor-General and the Council of India.
106 s'Jacob 1976:98.
107 In the Latin version of his statement of 20 April 1675, Itti Achudem called himself: "Doctor Malabaricus Natione Chego, gentilis & naturalis in Carrapuràm, seu terre dicta Codda Carapalli, habitator aedium dictarum Coladda" (Hortus Malabaricus vol.1 (1678):(xi)). The translation from the original Malayalam reads: 'the hereditary Malabar physician born at Collada house of Coddacarappalli village of Carrappuram and residing therein' (Manilal 1980:115).
108 See Van Reede's Memorandum of 1677 in s'Jacob 1976:100: 'The first among them (in this case the Sūdrās) are the tree-climbers, otherwise called silgos, who are also bound to war and arms. These people usually serve to teach the nayros in the fencing school; further their occupation is to tap the coconut trees and to make suri, arrack, and sugar therefrom, so that usually they have plenty of pelf'.
109 Hortus Malabaricus vol.3 (1682):(ix); Heniger 1980:47.
110 Hortus Malabaricus vol.3 (1682):(x); Heniger 1980:48.
111 VOC 1308:818,819.
112 VOC 1313:403v, account of Van Goens about Malabar, of 13 August 1675.
113 VOC 899:440,442-443,446, letter of 22 October 1675 of the Governor-General and the Council of India to Van Reede and the council of Malabar; see also s'Jacob 1976:LXXI-LXXII.
114 VOC 1313:404, account of Van Goens about Malabar, of 13 August 1675; VOC 899:446, letter of 22 October 1675 of the Governor-General and the Council of India to Van Reede and the council of Malabar. Mazius was afterwards sent to Ceylon, where he arrived in November 1676; he was stationed in Negombo. He was still there in 1678, but was later transferred to Colombo, where he still officiated in 1690 (GAA, Private Records 379, no.60:122,128, letters of the church council of Ceylon to the Classis Amsterdam, of 7 January 1677 and 8 December 1678; VOC 1469:341, General Rolls of the Government of Ceylon, of December 1690).
115 VOC 1321:902v, letter of 10 February 1676 of Van Reede and the council of Malabar to the Governor-General and the Council of India.
116 VOC 1308:595-596, letter of 28 October 1675 of Van Reede and the council of Malabar to Lords XVII.
117 VOC 900:525, letter of 19 October 1676 of the Governor-General and the Council of India to Van Reede and the council of Malabar.
118 VOC 240, Resolutions of the Chamber of Amsterdam, n.p., meetings of 2 and 9 September 1675.
119 VOC 320, Letter-book of Lords XVII 1673-1681, n.p., letter of 28 September 1675.
120 About the Cape present to Stadtholder William III, see chapter 4, section Cape of Good Hope 1685.
121 VOC 900:424,516-517, letters of 26 September 1676 and 18 October 1676 of the Governor-General and the Council of India to Van Goens Jr. and the council of Ceylon.
122 Coolhaas 1971 vol.4:295, General missive of 13 February 1679.
123 VOC 1321:894v, letter of 10 February 1676 of Van Reede and the council of Malabar to the Governor-General and the Council of India.
124 VOC 900:444, letter of 26 September 1676 of the Governor-General and the Council of India to Van Reede and the council of Malabar: 'we cannot imagine that the expulsion of the Carmelite Father would have such a harmful effect as the commander Van Reede seems to declare'.
125 The account of the action of the commission of the Carmelites in 1676 was given by Van Reede in his Memorandum of 1677; see s'Jacob 1976:170-172.
126 VOC 1352:236v, Indication of the principal items, drawn up by commissioner Huisman: 'The subsistence money that has been granted to a Neapolitan Carmelite friar Matthew from the Company's pay-desk at the order of the Commander Lobs'.
127 VOC 1340:1488v-1489v, letter of 24 April 1678 of Lobs and the council of Malabar to the Governor-General and the Council of India. The attempt to kidnap Matthew must have taken place early in 1678. In the letter of 5 January 1678 (VOC 1340:1456v-1476) Lobs did not yet say anything about this affair, and moreover it was only then that he recommended Goetkint, who was still a sergeant at that moment, to be promoted an ensign.
128 See the letters of this commission to the High Government in De Haan 1907:246-253.
129 VOC 902:1054, Instruction of the Governor-General and the Council of India to Huisman as commander of Malabar.
130 VOC 903:1044, letter of 21 September 1679 of the Governor-General and the Council of India to Huisman and the council of Malabar.
131 s'Jacob 1976:171; VOC 1360:1729v, letter of 28 April 1680; VOC 1361:519, letter of 19 June 1680; VOC 1361:552v, letter of 31 August 1680, all of them by Huisman and the council of Malabar to the Governor-General and the Council of India.
132 VOC 1349:1510v-1511, letter of 4 April 1679 of Huisman,

133 VOC 1349:1442v-1443, letter of 17 August 1678 of Lobs and the council of Malabar to the Governor-General and the Council of India. See also note 96.
133 VOC 1349:1442v-1443, letter of 17 August 1678 of Lobs and the council of Malabar to the Governor-General and the Council of India.
134 VOC 1410:552v-553, letter of 28 November 1685 of Vosburg and the council of Malabar to Lords XVII; s'Jacob 1976:LXXXVI.
135 VOC 1308:735, original letter of 9 December 1675 of Van Reede and the council of Malabar to Lords XVII.
136 VOC 1321:902v, letter of 10 February 1676 of Van Reede and the council of Malabar to the Governor-General and the Council of India. On this occasion Van Reede remitted 2,000 rixdollars at ½% a month to be paid into the bank at deposit rate.
137 Jacob Lobs, who died on 2 November 1688 in Batavia, commander of Malabar, 1677-1678, and governor of Ternate, 1682-1686 (Wijnaendts van Resandt 1944:183-184).
138 Van der Chijs 1903:190,196.
139 Van der Chijs 1903:233.
140 VOC 900:428-448, letter of 26 September 1676, of the Governor-General and the Council of India to Van Reede and the council of Malabar.
141 VOC 1329:1389, letter of 17 March 1677 of Lobs to the Governor-General and the Council of India.
142 VOC 1329:1330v, letter of 17 March 1677 of Lobs, Van Reede, and the council of Malabar to the Governor-General and the Council of India.
143 VOC 1329:1386, letter of 14 March 1677 of Lobs to the Governor-General and the Council of India.
144 VOC 1329:1391, letter of 17 March 1677 of Lobs to the Governor-General and the Council of India.
145 VOC 1329:1416, letter of 22 May 1677 of Lobs and the council of Malabar to the Governor-General and the Council of India.
146 VOC 1329:1333, letter of 17 March 1677 of Van Reede, Lobs, and the council of Malabar to the Governor-General and the Council of India.
147 VOC 1329:1385v, letter of 14 March 1677 of Lobs to the Governor-General and the Council of India.
148 VOC 1329:1335v, letter of 17 March 1677 of Van Reede, Lobs, and the council of Malabar to the Governor-General and the Council of India. Van Reede then remitted 2,864 3/8 rixdollars.
149 For English readers reference may be made to *Memoir written in the year 1677 A.D. by Hendrik Adriaan van Rheede* (Madras, 1911).
150 VOC 1329:1405v-1406, letter of 22 May 1677 of Lobs and the council of Malabar to the Governor-General and the Council of India.
151 VOC 1329:1335,1335v, letter of 17 March 1677 of Van Reede, Lobs, and the council of Malabar to the Governor-General and the Council of India. Casearius took 3,000 rixdollars with him from Malabar.
152 Willem Volger was director of Surat from 1672 to 1676 and after this of Bengal from 1677 to 1678, where he died on 6 January 1679 in Ougli (Wijnaendts van Resandt 1944:31-32). About Volger's stay in Cochin and the deposition of Vosburg as second-in-command of Malabar, see VOC 1329:1394,1397v-1400, letter of 2 April 1677, and fol.1403v, letter of 22 May 1677, both of Lobs and the council of Malabar to the Governor-General and the Council of India. The passage of Vosburg to Batavia on the "Silversteyn" was mentioned on 9 May 1677 (Van der Chijs 1904:125).
153 Van der Chijs 1904:125,134.

EXTRAORDINARY COUNCILLOR OF INDIA 1677

During the first weeks of his stay in Batavia Van Reede probably had nothing to do. His friend Isaac de Saint-Martin, who served since 1672 as a captain in the garrison of Batavia, was absent during the whole of the year 1677 on account of the war against the Mataram kingdom in East Java[1], so that he could not help him.

Possibly the Council of India did not know very well what to do with Van Reede and was waiting for orders from the Netherlands. On 4 June 1677 he was given a temporary job as an assistant in an appeal case before the Council of Justice[2]. But a few days later[3] the ship the "Blauwe Hulk" brought the important letter of 21 October 1676 from Lords XVII to the High Government, in Batavia, which was to have far-reaching consequences for some prominent servants of the Company.

For this we have to dwell for a moment upon the meeting of Lords XVII in Amsterdam of 16 October 1676, in which, among other things, the resignation of Van Reede, contained in his letter of 9 December 1675, was discussed. The reaction of the Lords to that letter was significant: 'The letter written to us by the Commander Van Reede and the Council ... is a cause of concern for us'[4]. In fact, at that moment Lords XVII were contending with a much wider problem than the Malabar conflict alone. The ill-balanced composition of the Council of India gave much cause for concern. The old and ailing governor-general Joan Maatsuiker was no match for the forceful director-general Rijklof van Goens Sr., who quite contrary to the wishes of Lords XVII persistently advocated a total conquest of Ceylon. His opponent was Cornelis Speelman, a councillor who had won his spurs as a capable administrator and a successful commander-in-chief in the Eastern Quarters and who, if possible, was an even more forceful personality than Van Goens[5]. The political game within the Council was played so passionately that this threatened the effectiveness of this supreme administrative body of the Company in Asia.

On the advice of the Chamber of Amsterdam Lords XVII resolved to make the political relations somewhat more balanced. The guide-line chosen was the appointment of 'capable and experienced ministers'. Since Van Goens decidedly could not be reproached with a lack of capability or experience with a view to the very great age of Maatsuiker and the latter's expected decease, he was promised the function of governor-general, while Speelman was then to take over the directorate-general. Further, Lords XVII supplemented the Council with four trusty authorities, including Willem Volger, then still director of Surat, and Van Reede[6]! In their letter of 21 October 1676, in which Lords XVII informed the High Government of these resolutions, they wrote about Van Reede:

'whom by the present we appoint for that purpose to the same quality of Extraordinary Councillor, adding the salary of an ordinary councillor of two hundred guilders a month standing for it, which shall take effect when he has entered upon that function and has taken the oath'[4].

On 15 June 1677 Van Reede took the oath in Batavia and took his seat in the Council of India. Instead of being called to account, as Vosburg had suggested, Van Reede was now expected to co-direct with Van Goens the empire of the Company. From 15 June to 22 November 1677 inclusive, the day before he left for the Netherlands, Van Reede indeed attended all the meetings of the Council, as appears from his signature to the resolutions taken[7].

The councillors were expected to make their contributions to the administration with the specialized knowledge they had acquired in previous functions. Naturally, sooner or later Van Reede's advice about Malabar and Ceylon would be asked, and thus the duel with Van Goens might again come to a head. In this situation it was unfortunate for Van Reede that he could not seek the support of Speelman, the principal opponent of Van Goens in the Council. During the whole of the year 1677 Speelman was not present in Batavia, because in that period, as commander-in-chief, he was conducting the war against the Mataram kingdom, in which Saint-Martin also participated[1].

Not until August 1677 did the Company's policy in the Western Quarters come up for discussion. In the meeting of 26 August there were long discussions about Ceylon, which resulted in a series of resolutions and in an express request addressed to Van Reede 'to serve them in writing with the knowledge and experience of the interests of the Honourable Company in the said Island of Ceylon, where, having served many Years in different qualities, he has also witnessed personally many important affairs and events'[8]. Van Reede accepted this mandate, and was able to employ the talent for which he was feared, dreaded, or admired: the writing of a detailed, sound report.

Two days later, on 28 August, the situation in Malabar was discussed, in particular Van Reede's transfer Memorandum. It will have given him satisfaction that this Memorandum was adopted as the basis of the Malabar policy[9]. In the letter of 7 September 1677 to the Malabar commander Lobs the Council informed the latter that the Memorandum was considered 'very illuminating'[10]. It might be possible that the Council's favourable opinion about Van Reede's Memorandum was partly also affected by an attempt of Van Goens to mollify him in his report about Ceylon. Nevertheless Van Reede composed a report that was devastating for Van Goens' ideas about Ceylon. Van Reede must have been conscious of the dangerous position into which he would get, for he did not submit his report to the Council of India; at least, Van Goens did not set eyes on it, although the latter did learn something about its contents. Later, when Van Reede had returned to the Netherlands, the game that he played became clear: his report was not meant for the Council of India, in which he would undoubtedly be exposed to the revenge of Van Goens, but it was intended for Lords XVII! Joan Huydecoper van Maarsseveen, one of the Lords XVII who opposed Van Goens, wrote about this to Van Reede's friend Joan Bax, then governor of the Cape:

'Mr. Reede van Drakenstein has submitted to the meeting of Lords XVII a circumstantial report, which (if possible) I will send to you, and about which general Van Goens, in a missive written to Mr. Valckenier and Mr. Van Dam, requests that the report which the said Van Reede has sinisterly withheld from him in Batavia, be sent to him, in order that he may punish his infamous and shameless lies (which he has been informed indirectly are contained therein), and to reveal his wickedness and hypocrisy'[11].

It needs no argument that in this atmosphere of distrust, hatred, and dissension within the Council of India there could be no question of meaningful co-operation. On 8 October 1677 therefore Van Reede resigned from the Council, in order to repatriate with the next return fleet. In the respective minutes the Council caused the following statement to be recorded:

'although we should have welcomed it if he should have an opportunity and inclination to continue here in India for some more years and attend this meeting'[12].

One is inclined to take this expression of regret addressed to Van Reede with a grain of salt and as being prompted by crocodile tears of Van Goens. But five days later, on 13 October, another councillor, Constantin Ranst, also resigned 'in very serious terms' from the Council. With express reference to Van Reede the same expression of regret was addressed to Ranst[13]. Ranst had owed his appointment as a councillor, in 1675, to his distant cousin Huydecoper van Maarsseveen and may therefore be looked upon as his partisan[14]. The double resignation of Van Reede and Ranst from the Council points to the fact that the conflict had come to a head. The expressions of regret may then be interpreted as worried statements of remaining councillors, who now had to do without the support of these two against Van Goens.

Still, Batavia in 1677 also offered Van Reede something else besides a vehement political conflict. At that time there were in the town Andries Cleyer, Herman Grimm, Willem ten Rhijne, and Johannes Casearius, the latter having come with Van Reede from Malabar. These were people all of whom were concerned with natural history in one way or another. It may be assumed that Van Reede contacted them in order to exchange ideas about his Hortus Malabaricus project.

In the preface to volume 3, Van Reede related that in Batavia he continued the work on Hortus Malabaricus, which he had been obliged to interrupt through his departure from Malabar. He found in Batavia a few plants of which in Malabar he had been able to have only incomplete illustrations made, but which he could now complement with the aid of painters. However, he did not mention their names. Perhaps Van Reede could rely on Hendrik Claudius, who was later to make many watercolours of Cape plants and animals by order of Cleyer, but who was then still in Batavia[15].

A very unfortunate factor for the good progress of the work, however, was the death of Johannes Casearius, who until that time had provided the Latin text of the descriptions in Hortus Malabaricus. This former clergyman of Cochin had fallen ill during the passage from Malabar to Batavia. On 13 July 1677, on account of his indisposition, he obtained permission to repatriate[16]. Shortly afterwards, about September, before he was able to leave, he died. Clergymen and elders of Batavia wrote in their letter of 29 November 1677 to Lords XVII:

'The Reverend Johannes Casearius ... who, having arrived here while ill, requested and obtained from Your Honours permission to repatriate, about two months ago departed this life and thus travelled to his eternal home'[17].

Van Reede profoundly regretted the premature death of his collaborator:

'Still I intended to complete the work I had begun both on the way and in Batavia. But oh, what a heavy loss! While I was pondering about this, our faithful Achates, that associate in the work, that Apollo, the irrigator of this Hortus, that most learned and reverend man, whom we loved so much, Casearius, was snatched away from us in the prime of life by bitter death and departed this life'[18].

His illness will not have enabled Casearius to contribute much to Hortus Malabaricus in Batavia. Van Reede had to look for another linguistic collaborator. He found such a man in the person of Christiaan Herman van Donep, 'who had a fluent pen and was sufficiently conversant with the Latin language, who arranged with me the adversaria and the collectanea insofar as it could be done, for which it was very helpful that most of those plants were still fresh in my memory'[18]. Van Reede here implies that Van Donep, who had already done some translations for him previously, in Cochin, was also in Batavia in 1677. I have not succeeded in finding any documents confirming his stay in this town.

Since Van Reede had been obliged to leave Matthew of St. Joseph behind in Malabar, he sought in Batavia for a botanist who would be able to assist him further. The situation now differed greatly from that of more than twenty years ago, the last time he was in Batavia. At that time not a scholar or writer of any reputation who occupied himself with natural history was present in the town. Now, twenty years later, it is striking to see that Van Reede had a fairly wide choice.

The function of chief of the medical shop in Batavia was then still held by Andries Cleyer, whom we came across already in 1669 when he was trying to centralize the research on medicinal plants in Batavia. Cleyer belonged to the circle of intimate friends of the councillor Cornelis Speelman[19]. He had just (in 1676) concluded a very profitable contract with the Company about the supply of medicaments, by which he earned no end of money[20] and which enabled him to engage as a private person painters of plants and animals, such as Claudius[21] and Cornelis Abramsen[22], and an experienced gardener, George Meister[23].

However, Van Reede states nowhere whether he discussed problems concerned with Hortus Malabaricus with Cleyer. Still, there is an indication that the two men

Figure 14. Malayali using a palm-leaf as a sunshade. The second man from the left suffers from elephantiasis. Detail of the engraving of Codda-pana, Hortus Malabaricus vol.3 (1682), Tab.2.

had contacts with each other about scientific subjects. In 1685 Christian Mentzel published a long passage from a letter of Cleyer to him, which dealt with the horrible disease of elephantiasis among the Malabar Christians of St. Thomas[24]. In this letter Cleyer mentioned that the Brahmin physicians in Malabar did not know any cure for this disease. It is quite possible that he got his information from Van Reede (see Fig.14). On the other hand, however, Simon Kadensky, the upper surgeon of Cochin, was a brother-in-law of Cleyer so that the latter may have obtained his data about elephantiasis in Malabar straight from Kadensky.

Perhaps Van Reede might have found most assistance for Hortus Malabaricus from the Danish physician Herman Grimm. Grimm had worked for some years in Ceylon in the vicinity of Paul Hermann, whom he praised for his great knowledge of the Ceylon flora[25]. In Batavia Grimm supervised the surgeons, while he 'had also undertaken to prepare all such medicinals and chemicals as can be made in this country'[26]. Apparently he worked along the guide-lines propagated by Cleyer and evidently to the satisfaction of the High Government, which in its letter of 13 February 1679 to Lords XVII testified about him:

'Here a laboratory has also been founded, in which by ... Grimm many medicaments are prepared and of which many are found in the forests and the gardens in Batavia itself'[27].

In connection with this activity Grimm published in 1677, in the year when Van Reede was in Batavia, an interesting piece of writing, the *Laboratorium Chymicum, Gehouden op het voortreffelijcke Eylandt Ceylon*. This booklet, which is probably the first work of natural science that came from the Company's printing office in Batavia[28], contained a survey of all the medicaments and their preparation, of animal, vegetable, and mineral origin, with which Grimm had daily come into contact during his stay in Ceylon[29].

Van Reede therefore could not have wished for a more

suitable collaborator than Grimm in Batavia: an experienced physician who was acquainted with medicinal plants in Ceylon. However, a few obscure hints from the dedication of the *Laboratorium Chymicum* make us suspect that Grimm was involved in the political strife in the Council of India. In the first sentences already he used expressions such as 'Ingratitude is the greatest Sin' and 'From the House of the Ungrateful Misfortune will never budge', phrases which recall similar statements of Rijklof van Goens Sr. about Van Reede a few years previously. The cat is let out of the bag in the letter of 20 November 1677, in which Grimm presented his *Laboratorium Chymicum* to Lords XVII[30]. In this he requested them, in accordance with Van Goens' ideas about Ceylon, 'to consent to the foundation in Ceylon of a general Laboratory for the whole of India, since [I] can testify that I have never beheld any more suitable place for this on the face of the earth'. In this same letter he also intimated that 'for that purpose (with the support of the Honourable Director-General R:v:G) I have published a draft of such medicaments'. In other words: Grimm had the *Laboratorium Chymicum* printed at the instigation of Van Goens, and the director-general apparently used this physician as a pawn on the political chess-board of Ceylon. Notwithstanding the scientific qualities which Grimm undeniably possessed[31], Van Reede will not have considered it wise to associate with this protégé of Van Goens[32].

Finally he approached a more neutral person, the Dutch physician Willem ten Rhijne, the well-known Japan-traveller[33]. Ten Rhijne had studied medicine in Franeker, in 1664-1666, and in Leiden in 1668, where Florens Schuyl was a professor of botany at that time. In 1673 he was chosen by Lords XVII to go to Japan by the order of the Shogun to send a physician with botanical and chemical experience to Decima[34]. On his way there, during a victualling period of nearly four weeks, Ten Rhijne botanized in the Cape. He stayed in Japan for more than two years, 1674-1676, and took part in two journeys to the court of Jedo, in order to offer the requested services to the Shogun. In spite of this, the Japanese government took hardly any notice of his presence in Jedo, a fact which greatly frustrated Ten Rhijne[35]. During his forced inaction he applied himself to Japanese medicine and to the flora of Japan. Without any further explanation he was absolved from his obligations by the Japanese, so that he had to leave Decima without anything done[36]. On 13 December 1676 the disappointed Ten Rhijne arrived in Batavia. Here he found it difficult to get a start, because he was presumably crossed by the influential Andries Cleyer, with whom he was later to have a flaming row. Only on 27 July 1677 did he get a function as governor of the Leper House in Batavia[37].

Although Van Reede praised Ten Rhijne as 'a man of rich experience and great learning', one may wonder whether as a Japan botanist he was able to make important contributions to the knowledge of Malabar plants. According to Van Reede 'he added thereto in Latin the curative virtues and use of plants which I had noted down in certain adversaria in order that, if an unfortunate event should disturb our work, at least something might still be left of it'[18]. From this it appears that Van Reede called in his assistance as a physician rather than as a botanist. Still there is one passage in Hortus Malabaricus from which it may be inferred that Ten Rhijne also concerned himself with the plant descriptions proper. The description of Todda-panna HM 3:9-14 contains the statement that, besides in Malabar, this tree also occurred on the Japanese islands Roukiou and Lequas, and that from this tree the Japanese prepare the "Sagou"[38]. Although Ten Rhijne is not expressly mentioned as informant in this passage, we may assume that it is due to him.

The collaboration between Van Reede and Ten Rhijne was only of short duration, at the utmost six months. In that period Van Reede must have gained great confidence in the integrity of the physician, for when he began to realize that his position in the Council of India was untenable and his repatriation was imminent, he confided to Ten Rhijne a complete copied set of the drawings and the descriptions of Hortus Malabaricus:

'Scarcely had I started to review my work when very valid reasons urged me to leave the Indies altogether. And I therefore took care that the pictures of the plants were depicted anew and that the descriptions which had been made up to that moment were copied. I left this copy behind with the distinguished Doctor Ten Rhijne, in order that, if either shipwreck or the necessity to throw things overboard upset our voyage, he might again send another copy of the Hortus Malabaricus to Europe. He might repeat this until at last they should have reached a suitable port. But it pleased Almighty God to restore us safe and sound to our paternal home again'[18].

It is an unanswered question whether the set of drawings and descriptions which remained behind with Ten Rhijne after Van Reede's departure still exists. After his death, in 1700, Ten Rhijne's scientific manuscripts vanished without leaving a trace. Possibly Van Reede's precautions with respect to the transport of drawings and descriptions to Europe later on inspired the governor-general Camphuijs, who was secretary of the Council of India in 1677, to do the same with the manuscripts and drawings of Rumphius' *Herbarium Amboinense*. What Van Reede had been spared on his way to Europe, did happen to Rumphius in 1692. The ship that transported the first six books of the *Herbarium* was sunk by the French. Thanks to Camphuijs' copy another set could be sent to Europe[39].

Departure from Batavia

As said above, on 8 and 13 October 1677 respectively Van Reede and his colleague Ranst resigned from the Council of India. A return fleet of six ships was just being put in the roads at that time. Ranst, who as ordinary councillor was one grade superior to Van Reede, was appointed admiral, sailing on the "Azië", and Van Reede vice-admiral, sailing on the "Cortgene". It is also characteristic of the uncertain political situation that the merchant Joan van Oosterwijck, a cousin by marriage of Huydecoper van Maarsseveen, was also leaving Batavia, as rear-admiral of the same fleet on board the "Vrije Zee"[40].

On 17 November governor-general Maatsuiker offered a farewell dinner to the departing councillors[41]. On 22 November both of them took part for the last time in a meeting of the Council of India, as did Maatsuiker, who was mortally ill and since then no longer presided over this body, leaving the chairmanship to the triumphant director-general Van Goens[42].

One day later, on 23 November, Ranst and Van Reede took leave of the Council of India, and with their family and train boarded the fleet that was lying in readiness. They were seen off by the governor-general, the Council, 'and a not inconsiderable crowd of male and female friends, as spectators'. On 24 November the return fleet left the roads of Batavia[43].

No calling at the Cape

The voyage to the Netherlands proceeded differently from what was usual. On account of the French Naval War (1672-1678) the return fleet did not call at the Cape, but sailed on without a stop to Pernambuco in Brazil, where victualling took place. The fleet reached Texel on 30 June 1678[44]. Thus Van Reede missed a meeting with his old friend Joan Bax van Herentals, who was governor of South Africa since 1676[45]. This would have been the last opportunity, for Bax died on 29 June 1678 of a chest disease[46]. The activities of Bax in the field of Cape natural history will be discussed in chapter 4, in the section on Van Reede's inspection of the Cape in 1685.

Notes

1. Encyclopaedie van Nederlandsch-Indië vol.6 (1932):350. For the Mataram war, 1676-1678, see Stapel 1939 vol.3: 398-405.
2. VOC 692:69-70, Resolutions of the Governor-General and the Council of India, 1677-1678, meeting of 4 June 1677.
3. The "Blauwe Hulck" arrived in Batavia on 15 June 1677.
4. VOC 320, Outgoing Letters of Lords XVII, 1673-1681, general missive of 21 October 1676.
5. NNBW vol.6 (1924):1254-1257; Stapel 1936.
6. VOC 240, Minutes of the Chamber of Amsterdam, 1675-1678, n.p., meeting of 16 October 1676. The two other newly appointed members of the Council of India were Balthasar Bort, governor of Malacca, and Antoni Hurt, governor of Ambon.
7. Van der Chijs 1904:78 ff.
8. VOC 692, Resolutions of the Governor-General and the Council of India, 1677-1678, pp.137-150.
9. VOC 692, ibid., pp.153-156.
10. VOC 901, Outgoing Letters of the Governor-General and the Council of India, p.552, letter of 7 September 1677.
11. RAU, Records Huydecoper Family 56, Letter-book 1675-1678, letter of 19 November 1678.
12. VOC 692, Resolutions of the Governor-General and the Council of India, 1677-1678, p.173.
13. VOC 692, ibid., p.178.
14. RAU, Records Huydecoper Family 56, Letter-book 1675-1678, letter of 2 September 1676 of Huydecoper to Constantin Ranst. Constantin Ranst (1635-1714) was successively chief of Tonkin, 1665-1667, of Japan, 1667-1668, and director of Bengal, 1669-1673. Moreover he was extraordinary councillor (1668) and ordinary councillor (1675) of India (Wijnaendts van Resandt 1944:29-30,146,302; Elias 1903 vol.1:542-543).
15. See chapter 4, section Cape of Good Hope 1685.
16. Van der Chijs 1904:222.
17. VOC 1323:231.
18. Hortus Malabaricus vol.3 (1682):(xv); Heniger 1980:53.
19. Cleyer was one of the executors of the will of Speelman in 1684; Cleyer's wife Catharina van Rensen and Aletta Hinlopen, the widow of Van Reede's friend Joan Bax van Herentals, saw to the winding-up of the movables and the condolences after the funeral; and Cornelis, the little son of Andries Cleyer born in 1677, was presumably a godchild of Speelman, to whom he bequeathed a considerable legacy (Stapel 1936:191-197,203).
20. Van Nuys 1978:14-15.
21. In 1681 Claudius concluded a contract with Cleyer to collect and draw, at his expense, plants at the Cape, as well as send animals and medicinal stones to Batavia (Van Nuys 1978:23-25). In 1685 Van Reede, during his inspection of the Cape, was to use the drawings of Claudius as a botanical guide.
22. Abramsen at first worked for Rumphius, but later went into the service of Cleyer. In the years 1690-1692 in Batavia he copied the coloured illustrations from the first six books of Rumphius' *Herbarium Amboinense*, before the originals were to be sent to the Netherlands (Lotsy 1902:57).
23. Muntschick 1984.
24. Published in *Miscellanea Curiosa sive Ephemeridum Medico-Physicarum Caesario-Leopoldinae Academiae Naturae Curiosorum*, Decade II, Anno III (1685), *Observatio* XIII, pp.52-53, Fig.9: *De S. Thomae Christianis Indiae Or. pedibus strumosis*.
25. Grimm 1677:87.
26. VOC 693, Resolutions of the Governor-General and the Council of India, 1677-1678, pp.9-10, meeting of 22 February 1678, in which, on the strength of this work, which he had already been doing for some time, he was given a salary raise from 60 to 90 guilders a month.
27. Coolhaas 1971 vol.4:295.
28. The printing office of the Company in Batavia was founded in 1668 (Encyclopaedie van Nederlandsch-Indië vol.1 (1917):642).
29. See also the preface of Peeters to his facsimile edition (1982) of Grimm's *Laboratorium Chymicum*.
30. VOC 1323:236-238, containing the original letter of Grimm with his perfectly preserved wax seal, accompanied by a catalogue of medicaments, materials, and instruments annually required for Ceylon.
31. In the coming years Grimm was to publish some more important writings in the field of tropical pharmacy.
32. In 1681 Grimm repatriated as a ship's surgeon on board the "Land van Schouwen", the flag-ship of Rijklof van Goens Sr., who then returned to the Netherlands as admiral of the return fleet (De Haan 1919:763).
33. Willem ten Rhijne (1647-1700), see NNBW vol.9 (1933): 861-863; Dictionary of Scientific Biography vol.13 (1976): 282-283.
34. The order of the Shogun to send a physician dated from

the court journey to Jedo of 1667 (ARA, Records of the Dutch Factory in Japan 81, Journal 1667-1668, dated 6 November 1667). Only on 9 May 1672 did Lords XVII take steps to look for a qualified physician. From five physicians applying for the job, on 6 February 1673 Willem ten Rhijne was chosen. His competitors were Pieter Boddens, Daniel Godtke, Samuel Manteau, and Adriaan van der Poel (VOC 152, Resolutions of Lords XVII, 1668-1673, n.p., meeting of 9 May 1672; VOC 239, Resolutions of the Chamber of Amsterdam, 1670-1674, n.p., meetings of 13 October 1672, 12 January, 23 January, 6 February, and 10 April 1673).

35 A circumstantial report on his grievances against the Japanese was written by Ten Rhijne in Jedo on 20 April 1676 to the Dutch chief Joan Camphuijs, the later governor-general (ARA, Records of the Dutch Factory in Japan 89, Journal 1675-1676, pp.65-72).

36 ARA, Records of the Dutch Factory in Japan 89, Journal 1675-1676, p.139, dated 9 September 1676.

37 VOC 692, Resolutions of the Governor-General and the Council of India, 1677-1678, p.114.

38 The reference is to the Ryu-Kyu islands (Loochoo islands) to the southwest of Kyūshū.

39 Lotsy 1902:48-49.

40 Van der Chijs 1904:395, meeting of 16 November 1677. On 13 March 1674, in Ougli, Bengal, Joan van Oosterwijck (c.1640-1699) married Cornelia Isabella Hinlopen (1656-1719), a sister's daughter of Joan Huydecoper van Maarsseveen; thus he became a brother-in-law of Joan Bax van Herentals. When returned in the Netherlands, Van Oosterwijck later became a member of the city council of Amsterdam, director of the Society of Surinam and director of the East India Company (Elias 1905 vol.2:616; RAU, Records Huydecoper Family 56, Letter-book 1675-1678, letter of 23 January 1677 of Huydecoper to Van Oosterwijck; letter of 19 November 1678 of the same to Joan Bax.

41 Van der Chijs 1904:395.

42 Joan Maatsuiker died six weeks later, on 4 January 1678.

43 Van der Chijs 1904:399,400.

44 VOC 5057, Journal of Paulus Sleeswijk, second mate on the "Vrije Zee", 1677-1678; VOC 5056, Journal of Jan Neck van Monnikendam, third watch on the "Azië".

45 On 2 November 1674 Joan Bax was promoted by Lords XVII and through the kind offices of Huydecoper, whose niece Aletta Hinlopen he had married in 1669, from commander of Galle, Ceylon, to governor of the Cape (RAU, Records Huydecoper Family 55, Letter-book 1671-1674, letter of 2 November 1674 of Huydecoper to Bax). He arrived in the Cape on 3 January 1676 and was inaugurated on 14 March next (VOC 4012, Journal of the Cape 1676, fol.171v,204v-205).

46 VOC 4014:287v-288, letter of 30 June 1678 of Hendrik Crudop and the council of the Cape to the Governor-General and the Council of India.

3
Van Reede in the Netherlands 1678-1684

ARRIVAL IN THE NETHERLANDS

The homeward bound fleet with which Van Reede returned to the Netherlands put into the roads of Texel on 30 June 1678[1]. In the next session of Lords XVII admiral Ranst reported on the Company's affairs in the meeting of 29 August in Amsterdam[2]. It was only in the following session, of 18 October to 1 November, that Van Reede's report on Ceylon, which he had written by order of the Council of India, was discussed by Lords XVII. Huydecoper, who as a member of Lords XVII learned with satisfaction how Van Goens Sr. was criticized, could not, however, prevent the appointment of the Ceylon governor Rijklof van Goens Jr. to the high post of commissioner-general of India, with the mandate to inspect all the factories of the Company. He did manage to obtain that his nephew Joan Bax van Herentals was to take over the function of Van Goens Jr. in Ceylon, with the rank of extraordinary councillor[3].

After Van Reede had received permission, on 28 October, to cash his bills of exchange worth nearly 11,500 guilders, he was able to put a period to his Indian career[4]. From the journals of Huydecoper it may be inferred that in the next years, until 1683, Van Reede no longer occupied himself actively with affairs of the Company[5].

THE EQUESTRIAN ORDER AND THE STATES OF UTRECHT

Van Reede settled in the province of Utrecht. In the first days of December 1679 he bought the manor-house of Mijdrecht[6]. On 7 December the Stadtholder William III ordered that he should be admitted into the Equestrian Order of Utrecht. On 11 December he was admitted into this body[7].

The ownership of a manor-house was one of the principal conditions for admission to the Equestrian Order. The manor in Mijdrecht was a small estate of about 4 hectares, situated close to the church of the village of the same name. Presumably the house was only a square dwelling, as may be seen in a drawing of 1650 (Fig.15). Although Van Reede called himself lord of Mijdrecht since 1679, it is not probable that he ever lived on the estate. Indeed, he owned a house in the city of Utrecht, where he had his home[8].

When Van Reede returned to Utrecht, this province was passing through a difficult period. The country had suffered severely during the French occupation in 1672 and 1673. A war levy had all but exhausted the finances, while the standing of the province had been seriously injured by its not very active attitude during the French offensive.

During his stay in the Netherlands Van Reede diligently worked on the reconstruction of Utrecht. He missed hardly any meeting of the Equestrian Order. He mainly occupied himself with the restoration of the shaken finances and the administration of the property of the Equestrian Order. After some minor mandates, on 12 November 1680 he took a seat in the Audit Office[9]. On 23 August 1684 he exchanged this function for a seat as a deputy of the States of Utrecht. On 6 October of that year he also assumed the function of "watergraaf" (director) of the Nieuwe Vaart on the Steenstraat[10]. There was no indication that before the end of 1684 he was to leave the Netherlands for good.

THE VAN REEDE FAMILY

That Van Reede chose Utrecht for his place of residence will partly also have been prompted by the fact that a great many of his relatives lived there. He made or renewed the acquaintance of younger and older relatives, in consequence of which he entered the closely knit clan of the titled and influential Van Reedes. The Van Reede family was then at the zenith of its power and glory in the province of Utrecht[11]. Several representatives, like Van Reede himself, were members of the States of Utrecht and inhabited castles or country-seats with magnificently laid-out gardens. In his turn, the 'Indian cousin', as Van Reede was called by his relatives[12], was to contribute to the glory of his family by dedicating six of the twelve volumes of Hortus Malabaricus to them.

Of his brothers and sisters, it was Machteld, Agnes, and Margaretha who were still alive. With his sister Agnes he must have had a special relationship, for even in his lifetime Van Reede annually allowed 200 guilders to her, presumably to support her none too flourishing finances[13]. Moreover later, in his last will of 1691, he was to bequeath a legacy for life to her[14].

At that time Agnes was the widow of Reinoud van Tuyll van Serooskerken. She lived in the castle of Zuilen,

Figure 15. House of Mijdrecht about 1650. Anonymous Indian ink drawing, 18.9 × 29.2 cm. RAU, Topographical Atlas 1119:161.

close to Utrecht, with her son Hendrik van Tuyll van Serooskerken (1642-1692), lord of Zuilen[15]. In 1665 the latter had married his cousin and stepsister Anna Elisabeth Van Reede van Nederhorst, who brought into marriage the castle of Zuilen. With this manor-house he was admitted into the Equestrian Order of Utrecht in 1667. Since 1669 he represented the States of Utrecht in the States General. Thanks to a safeguard, the castle of Zuilen escaped the pillage of the French in 1672-1673. Near the castle was a small, but beautiful garden[16].

Van Reede got on good terms with the lord of Zuilen, who in 1682 nominated him, together with Paul de la Baye, a guardian to his children[17]. When in 1684 Van Reede went to Asia as commissioner-general, he put his house in Utrecht at the disposal of the lord of Zuilen for his occupation[18]. Van Reede dedicated volume 4 of Hortus Malabaricus, of 1683, to him, recalling their friendship.

Paul de la Baye was a brother-in-law of the lord of Zuilen, whose daugther Trajectina boarded with him[19]. De la Baye served as a colonel in the infantry of the States. His son Maurits Cesar was to accompany Van Reede on his expedition to Asia and was later to marry the latter's daughter Francina[20]. Furthermore the lord of Zuilen maintained financial relations with Willem van Nassau, lord of Zuilestein. To the latter, Van Reede was to dedicate volume 6, of 1686. In 1674 he was admitted with the manor-house of Zuilestein into the Equestrian Order of Utrecht. He was on very friendly terms with his cousin, the Stadtholder William III, whom he often accompanied as a soldier in his campaigns. Since 1687 the lord of Zuilestein played a considerable role in the preparation and execution of the Glorious Revolution. The estate of Zuilestein, situated not far from Amerongen, originally formed a part of the big property of the Van Reedes, but after a family quarrel it was sold in 1630 to the Stadtholder Frederik Hendrik, who had the house and the gardens embellished. In 1640 he bestowed them on his bastard son Frederik van Nassau, the father of Willem. In the second half of the 17th century the gardens were greatly extended, while the original plan was maintained. The gardens survived the French occupation, such in contrast with the neighbouring Amerongen[21].

Presumably Godard Willem van Tuyll van Serooskerken (1647-1708), lord of Welland, will also have moved in the environment of the lord of Zuilen. Although they were only distant cousins, Godard Willem was to act, after Hendrik Jacob's death, as the guardian of the latter's surviving children[17, 18]. Van Reede was to dedicate volume 9, of 1689, to Welland. In a somewhat isolated position is baron Diederik van Baar (?-1683), a Guelder nobleman, who had been married to Van Reede's sister

Maria. Apart from the vague generalities which Van Reede wrote about this brother-in-law in the dedication of volume 7, of 1688, the nature of their relationship is not known. Van Baar was admitted in 1649 into the Equestrian Order of the Veluwe. From 1675 to 1683 he represented Gelderland in the States General, in which he was the chairman of the important commission of military affairs. It is curious that he had died a few years before the volume of Hortus Malabaricus dedicated to him appeared[22].

Beyond this close family-circle Van Reede also met some more distant cousins. From his mother's side he may have known from his youth his cousin Hendrik van Utenhove (1630-1715), lord of Amelisweerd. To him, Van Reede dedicated volume 10, of 1690. Van Utenhove was admitted in 1674 into the Equestrian Order of Utrecht with the manor-house of Amelisweerd. He was a colonel and later lieutenant-general in the infantry of the States and functioned as military governor of Hasselt and 's-Hertogenbosch successively[23].

With his distant cousin Frederik Adriaan van Reede van Renswoude (1659-1738) our Van Reede as the only one of his family seems to have shared an interest in scientific research. In later years the lord of Renswoude was on friendly terms with the famous Delft naturalist Antoni van Leeuwenhoek, with whom he carried on a correspondence about the control of insects in orchards. In 1681 he inherited the estate of the branch of Van Reede van Renswoude and was admitted in 1684 with the manor-house of Renswoude into the Equestrian Order of Utrecht, whose chairman he became later. As a member of the States General he acted as Dutch ambassador at the peace negotiations in Aachen (1712) and in Utrecht (1713). Through him the manor Renswoude, built in 1654, was given a beautiful garden lay-out in the French style[24]. Van Reede dedicated volume 8, of 1688, to him.

Already in 1679 Van Reede came into contact with the principal representative of his family, Godard Adriaan van Reede van Amerongen (1621-1691). On 22 October 1679 Van Reede settled with his nearest relations a matter concerning the estate of his brother Karel, who had died in Berlin in December 1672. They then charged Godard Adriaan, who was about to leave for Berlin as ambassador, with winding up the inheritance. In the only autograph letter of Van Reede known to me, that of 17 May 1680, he thanked his cousin for his good offices (Fig.16)[25].

Godard Adriaan had inherited from his father a rich estate, including the manor-house of Amerongen[26]. With this he was admitted in 1642 into the Equestrian Order of Utrecht. Since 1652 he represented his province in the States General. He gained much fame as ambassador of the States General in Scandinavia, Poland, Germany, and Spain. Out of gratitude for his merit the Danish king bestowed on him the Order of the Elephant (1659) and later also raised him to the Danish peerage, with the title of baron (1671). The States of Utrecht raised Amerongen on his behalf to a free seigniory (1676). Godard Adriaan and his wife Margaretha Turnor took a great deal of trouble to embellish the castle of Amerongen, dating from the Middle Ages, and the gardens belonging to it. During the French invasion of the Republic, in 1672, Louis XIV and the Duke of Orleans took up their residence there. But in 1673 the French destroyed the castle out of revenge for Godard Adriaan's diplomatic activities in Berlin, where he then resided. Thereafter, during the period from 1674 to 1680, under the direction of his wife one of the most magnificent Dutch countryhouses was erected on the mediaeval foundations. It is not known exactly when the famous gardens to this house were laid out, and whether our Van Reede ever knew them. He dedicated volume 5, of 1685, to the lord of Amerongen. The dedication was dated Utrecht, 1 July 1684.

A relative of Van Reede belonging to the circle of the lord of Amerongen rather than that of Zuilen was Godard Willem van Tuyll van Serooskerken, lord of Welland, mentioned already previously. As a young orphan he had been educated by his uncle Amerongen, whose protégé he was to remain his whole lifetime. Welland had been a canon of Oudmunster since 1671 and as such was a member of the States of Utrecht. In 1672 he conducted the delegation of Utrecht for the negotiations on the surrender with the conquering king Louis XIV. In consequence of this he fell into disgrace with the Stadtholder William III and was deposed in 1674 as a member of the States. Since that time he held a minor position as dikereeve of Lekdijk Bovendams. After the death of William III, in 1702, he became a member of the States General. In 1695 he bought the manor-house of Nederhorst from the estate of Hendrik Jacob van Tuyll van Serooskerken, to whose children he was guardian. Near this house, in 1700, he had a beautiful garden lay-out designed by the landscape gardener Jan van Staden[27].

HORTUS MALABARICUS

During his six years' residence in Utrecht Van Reede occupied himself intensively with the edition of Hortus Malabaricus. The manuscripts he had brought away from Malabar had indeed been edited further in Batavia by Ten Rhijne and Van Donep, but many notes had not yet been recast into complete plant descriptions. Judging by the extant originals in the British Library in London, a number of drawings had not yet been completed and had sometimes remained in the stage of pencil sketches[28]. The extensive task that awaited him was made even more difficult by some other factors, such as a disturbance of the general plan of the work, the rapid succession of the deaths of some collaborators, the poor results of the sale of the first published volumes, and the fierce criticism of the High Government from Batavia. This caused a depression and a lethargy, from which Van Reede was roused only in 1681.

In consequence of the absence of personal correspondence of Van Reede with his collaborators, in particular Arnold Syen and Jan Commelin, it is not very well possi-

tweelen gegeven zijn, voornamentlik Bruiningen, Van
Swelm, en van alsolen het onleden van 't Lam ooe
Blijnd

Loue Eedel Gebomer Heer en Heere

Vleede Ed Ye J8 ootmoedigen dienaer

H v Reede

utrecht 17/7 / may
a.º 1681

ble to give an accurate account of the successive stages of the edition of Hortus Malabaricus. On the other hand Van Reede himself has given a survey of the problems confronting him, in the preface to volume 3. When this survey is supplemented with a few official documents, this provides a picture of the way in which he set about it.

When in June 1678 Van Reede arrived in the Netherlands, shortly before, in April or May, the first volume of Hortus Malabaricus had been brought out by the Amsterdam publishers Johannes van Someren and Jan van Dyck. Arnold Syen, the Leiden professor of botany, who had written the commentaries and also edited this volume, presented a copy to the Curators of his university on 8 May[29].

The first volume had been dedicated to the governor-general Joan Maatsuiker. Lords XVII also received a few copies, which they sent to Batavia. In their letter of 31 August 1678 to the High Government they drew attention to the importance of the book with a view to the supply of medicaments in Asia:

'Lately there was printed and published in this country a certain book, entitled Hortus Malabaricus, well-known to you and a few copies of which are being sent to you herewith. We have been informed that the second volume will follow, in which the curative power of the plants and herbs occurring on the Coast of Malabar is described, which suggested to us the idea whether that Coast might not supply many of those herbs which are needed successively for the medicinal shop, in order to relieve the home country the more, and since for bringing this about someone would be expressly required, whether for this we could not successfully employ Dr. Hermans, who had previously been sent to Ceylon and is still residing there, a person whom we trust to have the requisite qualities for it. But these are mere speculations and we shall be pleased to do whatever may be thought out further or better by you for the achievement of our purpose'[30].

In this diplomatically formulated passage Lords XVII therefore suggested to the High Government to send Paul Hermann from Ceylon to Malabar with a view to occupying himself there with the supply of medicaments. Probably this passage was prompted by Van Reede, who had met Hermann in Cochin.

Meanwhile Maatsuiker had died, and Van Goens had succeeded him as governor-general. The reaction of Van Goens to the first volume of Hortus Malabaricus was furious. In the general missive of 13 February 1679 the High Government replied with the following sharp diatribe:

'With medicaments our people in Ceylon will not only be able to provide us ourselves, but also in due time provide us with them plentifully ... Great diligence in this respect is displayed by the doctor and herbalist Hermansz, who (as reported to us) has already described in Ceylon more than 10,000 unusual plants, shrubs, and herbs, among them a great many which have never been known to any authors. The island of Ceylon is so blessed a land of all sorts of valuable and splendid medicines as any land on the earth can be, all this so plentifully that before long only very few medicaments need be sent from the home country. The book you have sent us, called Hortus Malabaricus, does not contain anything that is known and plentifully to be obtained in Ceylon, so that one need not to go to Malabar for this at all. Those who know both sustain that the herbs and medicines in Ceylon are as much more potent as Ceylon cinnamon exceeds wild Malabar cinnamon'[31].

Apart from the ridiculous tenfold exaggeration that Paul Hermann had described more than ten thousand Ceylon species[32], this criticism reveals that above all Van Goens wanted to maintain the superiority of his island to Van Reede's Malabar. On the other hand Van Goens was indeed right in saying that the first volume of Hortus Malabaricus contained many species also occurring in Ceylon. But this argument was merely based on allegation and not on a publication about Ceylon plants of the same quality as Van Reede's first volume.

Van Goens' rejection of Hortus Malabaricus had no direct consequences for the opinion of Lords XVII. Long before the general missive of 13 February 1679 had reached the Netherlands, about March of that year the second volume appeared. Van Reede dedicated this second volume to 'his protectors and friends' Gillis Valckenier, Joan Hudde, Joan Huydecoper, and Pieter van Dam. This dedication was undoubtedly a political move of Van Reede. Valckenier was then the supreme leader of the Amsterdam government. Hudde and Huydecoper were his influential partisans. Moreover, Valckenier and Huydecoper played a prominent part in Lords XVII, of whom for many years Van Dam, as secretary, was the pivot[33]. Lords XVII then bought twelve copies from the publishers Van Someren and Van Dyck so as to bear part of the heavy printing expense. They sent six copies to Batavia and distributed the remaining six among the various Chambers of the Company in the Netherlands[34].

With this the conflict between Van Reede and Van Goens about Hortus Malabaricus and the medico-botanical resources of Ceylon and Malabar was not yet at an end. The coming of Van Goens to the Netherlands in 1682 was to induce Van Reede to make a countermove, which I will discuss presently.

While the criticism of Van Goens about the first volume was already disagreeable for Van Reede, he was affected even more seriously by the loss of his commentator and

Figure 16. Handwriting and signature of Hendrik Adriaan van Reede tot Drakenstein at the end of his letter of 7/17 May 1680, Utrecht, to Godard Adriaan van Reede van Amerongen. RAU, Records House Amerongen II 141.

publishers. Arnold Syen died on 21 October 1678; the publisher Jan van Dyck died in the same month and the latter's colleague Jan van Someren a month later, in November[35].

Van Reede soon found the Amsterdam amateur-botanist Jan Commelin willing to complete Syen's commentaries of the second volume, so that this volume could appear already in 1679. But the publishing-house of Van Someren and Van Dyck objected to the continuation of the project on account of threatening losses. The grand lay-out of the work with double-folio engravings involved great expense. In 1721 the commercial value of the copper plates and the stocks of Hortus Malabaricus amounted to 2,963 guilders[36]. Discouraged, Van Reede observed:

'There were very few purchasers, so that one might believe that so important a work was little appreciated. It appeared that each of the remaining laborious and cumbersome volumes which were still being prepared by me would have the same fate as the earlier ones. I assure you that all this caused me the greatest disgust at incurring such great expense and spending our money for the pleasure of others'[37].

Finally it was painful for Van Reede to find that, unknown to him, the general plan of Hortus Malabaricus had been interfered with. Van Reede's original plan had been to publish three substantial volumes: a first volume with the trees, a second with the shrubs, and a third with the herbs. Already in 1677, or earlier, he had sent a batch of drawings and descriptions to the Netherlands[38]. At that time a misunderstanding as to the total scope of Hortus Malabaricus arose with the printer or Arnold Syen and they probably thought that the whole work consisted only of this material. The result was that the printer indeed maintained the original plan of the division into trees, shrubs and herbs, but that he added some shrubs to the volume with the trees so as to give more body to the volume, and for the same reason added some herbs to the volume with the shrubs[39].

This interference is the cause of the fact that the systematic sequence of Hortus Malabaricus, taken as a whole, makes a confused impression[40]. All the above-mentioned disappointing experiences caused Van Reede's final lamentation:

'I therefore saw the *Hortus*, which I had laid aside as a memento of our former exertions, through utterly different eyes than I did before, not caring much about fame or libel and gossip. And thus our *Hortus*, which had been composed little by little, was dropped'[41].

Van Reede's depression lasted until 1681. At the instance of 'well-wishers of botany' and the publishing-house of Van Someren and Boom, a new firm, he then resumed his work on Hortus Malabaricus.

Van Reede first tried to solve the problem of the disturbed sequence on his own. When he did not succeed in doing so, he enlisted the assistance of Jan Commelin, who proposed a satisfactory solution. From the existing notes Van Reede himself composed Dutch descriptions of plants, which he caused subsequently to be translated into Latin. Unfortunately these translations were not correct. Then he found the Utrecht professor Johannes Munnicks willing to attend to the descriptions as well as the Latin translations[42].

On 17 January 1681 Van Reede, Munnicks, and Commelin concluded a contract with the publishers Pieter van Someren, Hendrik Boom, and the widow of Dirk Boom to publish the remaining volumes of Hortus Malabaricus. In the first instance they agreed to publish the third and the fourth volume. These volumes were to appear within one year after the descriptions and drawings had been submitted. Subsequently the parties agreed in more general terms that the whole work was to be submitted and published from year to year until it should be completed. The publishers took the whole expense of the publication upon themselves and moreover promised the three writers a royalty. Van Reede was to receive ten unbound copies of each volume, Munnicks twelve, and Commelin six[43].

With this contract the division of labour for the further production of Hortus Malabaricus had been fixed once for all. From the volumes appearing since then we can infer that until his death in 1692 Commelin provided for the systematic sequence and the commentaries, while Munnicks took upon him to make the descriptions and the translations. The role of Van Reede in the project is not quite clear. Presumably, as the 'main author' he had the supervision and confined himself to writing the dedications.

In recruiting Commelin and Munnicks as collaborators, Van Reede undoubtedly made a good choice, but certainly not the best possible one. At that time the Netherlands numbered other eminent botanists besides these, such as Abraham Munting, Pieter Hotton, and Paul Hermann.

Abraham Munting (1626-1683) had been a professor of botany and director of the botanical garden of Groningen University since 1658. As a botanist he was an orthodox scientist, who was not much interested in new developments in taxonomy. On the other hand he was quite famous as a grower of exotic plants. Thanks to his contacts with America, Asia, and Africa he succeeded in introducing in the Netherlands, in his well-functioning greenhouse, remarkable novelties such as the banana tree. In his publications, however, he emphasized the horticulture rather than the phytography of the plants he grew[44]. A decisive reason for Van Reede for not involving him in Hortus Malabaricus may have been the great distance of Groningen from Utrecht and Amsterdam, which would have rendered regular contact with Van Reede and the publishers difficult.

Pieter Hotton (1648-1709) was a physician in Amsterdam. After the death of Arnold Syen, under whom he had studied botany, for a short time (1678-1680) he filled the Leiden chair of botany and the directorship of the botanical garden, in anticipation of Paul Hermann's arrival from Ceylon. In 1692 Hotton succeeded Commelin as botanist of the Amsterdam Municipal Garden. When Hermann died in 1695, Hotton succeeded him as the Leiden professor of botany[45]. In his later life Hotton showed himself to be an erudite scholar, a bibliophile, an excellent Latinist, and a sound botanist with a great interest in the history of botany and in contemporary systematics. However, he published little. Apart from his inaugural address on the history of botany (1695), the only work of his that appeared, in 1702, was a treatise on the Ceylon medicinal plant Acmella, *Spilanthes paniculata* Wall. ex DC. From his extant correspondence it appears that he has contributed a good deal to the propagation of John Ray's systematic views by editing a new, amended edition of the latter's *Methodus Plantarum* (1703)[46]. However, in 1681, when the publishers' contract about Hortus Malabaricus was concluded, Hotton returned again, after his temporary professorship in Leiden, to his medical practice in Amsterdam. From that period no contacts of him with botanists are known.

Johannes Munnicks (1652-1711) was a professor of botany and anatomy and director of the botanical garden of Utrecht University since 1680. He became widely known for publications in the field of surgery. Like Hotton, he was an excellent Latinist, and in addition Munnicks engaged in translating medical works. Apart from his collaboration on Hortus Malabaricus, he has not published any botanical works at all. Nor did he ever have a reputation as a botanist, although Van Reede praised him for his 'unique interest in botany'. The then Utrecht botanical garden was not suited for growing exotic plants. There were no greenhouses, and the elevated garden was highly sensitive to the Dutch climate. That in spite of this Van Reede recruited him as a collaborator will have been due to the simple fact that Munnicks also lived in Utrecht, so that regular oral consultations about the descriptions and the translations were guaranteed. Moreover, Van Reede honestly admitted that the academic status of Munnicks as a professor also played a part[47]. As long as Van Reede was still in Utrecht, Munnicks collaborated on the volumes 3 (1682), 4 (1683), and 5 (1685). After this, he passed his task on to his pupil Theodorus Janssonius van Almeloveen (1657-1712), who accounted for volume 6 (1686). Ultimately the six remaining volumes were attended to by the Amsterdam physician Abraham van Poot (1638-1707)[48].

Jan Commelin (1629-1692) was a wealthy drugs merchant in Amsterdam. In 1672, when the Stadtholder William III came into power and a part of the Amsterdam government was deposed, Commelin was given a seat in the city council. Like Munting, he was a capable horticulturist. In his greenhouse, heated by stoves, in his country-seat Zuiderhout near Haarlem he very successfully grew citrus fruit, about which he published in his *Nederlandtze Hesperides* (1676). Although Commelin had not had a university training, in the course of the years he developed into a good amateur-botanist. Because of his special attention to habitats and localities of indigenous plants he may be considered as the first genuine Dutch florist[49]. His botanical rambles along the river Vecht regularly ended in the country-seat Goudestein, in Maarsseveen, owned by his political colleague in the Amsterdam government, Joan Huydecoper[50]. Commelin also maintained contacts with prominent botanists such as Jacob Breyne[51] and Paul Hermann. The latter sent him from Ceylon a small, but interesting herbarium of Ceylon plants, which at present is to be found in the Institut de France in Paris[52]. To a certain extent therefore Commelin was quite qualified as a collaborator on Hortus Malabaricus. Moreover Van Reede could hardly ignore him, because to his satisfaction Commelin had provided the commentaries of volume 2.

The question, however, is why Van Reede did not also involve Paul Hermann in Hortus Malabaricus, since the latter had botanized for some time in Malabar and his unique knowledge of the flora of Ceylon had to be considered very advantageous for the commentaries. After the death of Arnold Syen Paul Hermann had been appointed his successor by the Curators of Leiden University. He arrived in 1680 from Ceylon in Leiden, where he taught until his death in 1695[53]. At that time Hermann was indisputably the best botanist in the Netherlands. In his lectures, which attracted many students, also from abroad, he was the first among Dutch professors to devote great attention to taxonomy, in particular to the systems of Morison and Ray. Among his foreign correspondents he reckoned Jacob Bobart, of Oxford, Jacob Breyne, and the French taxonomist Tournefort. By building tropical greenhouses he elevated the Leiden botanical garden to a centre of colonial botany in Europe. His publication program comprised such comprehensive projects as the natural history of Ceylon, the flora of South Africa, and a survey of exotic horticulture in the Netherlands. Hermann was a very busy man, who for that reason alone already will have had little time for any collaboration on Hortus Malabaricus. On the other hand, it is plausible that Van Reede will have had objections to inviting Hermann, who appears from Van Goens' outburst against Hortus Malabaricus to have been a pawn of the governor-general. Nevertheless it appears from Commelin's commentaries that Hermann now and then assisted him. Although in the period when the publishers' contract was concluded Paul Hermann would have been by far a better choice than Jan Commelin as a commentator, no one could then foresee that Commelin was shortly to become the founder of the Amsterdam branch of colonial botany.

In November 1682 the Amsterdam city council founded the Municipal Garden at Nieuwe Plantage and entrusted the administration to Huydecoper and Commelin as commissioners. Huydecoper made use in a well-considered way of his position as director of the Company to order exotic plants to Amsterdam. He strongly urged the colonial administrators to undertake botanical research in the colonies and stimulated the production of drawings of plants on the spot. His letters to Van der Stel (Cape of Good Hope), Lamotius (Mauritius), Pijl (Ceylon), Cleyer and Ten Rhijne (Batavia), and others, which will be discussed later, testify to his unremitting zeal to elevate the Amsterdam Municipal Garden in a short time to the level of the Leiden botanical garden. Presumably it was also his suggestion that in 1683 the painter Jan Moninckx started the impressive series of water-colours of the most notable plants that grew and bloomed in the Amsterdam garden[49]. Thus, Commelin, who had the day-to-day direction, had plenty of opportunity to check his commentaries on Hortus Malabaricus by the plant material that came pouring from the East. It may be stated without exaggeration that the foundation of the Municipal Garden of Amsterdam, at a crucial moment in the history of Hortus Malabaricus, contributed considerably to the success of Van Reede's project.

The third volume of Hortus Malabaricus appeared in 1682. Van Reede dedicated it to Vīra Kērala Varma, the rāja of Cochin. His revived interest was expressed in particular in the long preface to this volume. It contains a frank report of the genesis of Hortus Malabaricus and an enthusiastic description of the land, the people, and the flora of Malabar, of which a frequent use has already been made in this study.

In an earlier publication I have pointed out that this preface should nevertheless be treated with some caution[54]. It was more than a justification by Van Reede of the object, the plan, and the execution of Hortus Malabaricus. In contrast with the accuracy we have learned from his official letters and reports, in the preface he was sometimes remarkably vague in the presentation of actual events. Moreover he reacted sharply to critics who called in question his status as a botanical author. And finally he alluded to a conspiracy which had forced him to leave Malabar and Asia. In my opinion the preface is not only an introduction to a botanical work, but also a political defence against Van Goens.

In 1679 and 1680 Van Goens had expressed the wish to resign as governor-general. One of the reasons was that he could not hold his own against the opposition of Cornelis Speelman and company in the High Government[55]. In their letter of 29 October 1680 Lords XVII dismissed him honourably and on account of his great merit towards the Company offered his son a seat in the High Government[56]. In December 1681 Van Goens left Batavia[57], and the next year he arrived in the Netherlands. Van Goens, who still had a great reputation as the conqueror of Ceylon and Malabar, was welcomed like a prince in Amsterdam.

There can be no doubt that Van Reede greatly dreaded the repatriation of the ex-governor-general. Viewed in this light, the political aspect of the preface to volume 3 of Hortus Malabaricus will have been a move of Van Reede against the threat formed by the personal presence of his arch-enemy in the headquarters of the Company. Van Goens, however, died shortly after his return in November 1682. The great aversion to Van Goens in the circles in which Van Reede moved appears from Huydecoper's violent, almost pathetic outburst of hatred. As one of the burgomasters he prevented the funeral in Amsterdam and sent the funeral procession to The Hague[58].

APPOINTMENT AS COMMISSIONER-GENERAL OF THE WESTERN QUARTERS

After having lived for some years comparatively quietly in a provincial town and in the family circle, from 1683 onwards Van Reede was gradually drawn into the affairs of the Company again.

In a letter of 24 April of that year to Laurens Pijl, Huydecoper, who congratulated him on his appointment as governor of Ceylon as successor of Rijklof van Goens Jr., enunciated that the policy with respect to that island, certainly now that Van Goens Sr. had died, would have to be altered, 'as has been very intelligently suggested by the lord of Drakenstein in his exposition and as a consultation is to be held about it shortly'[59]. In June Lords XVII -Huydecoper was then president- requested Van Reede to give his advice on the reorganization of the administration of Malabar[60]. Van Reede could not resist the temptation, and from that moment he became involved in a steadily developing plan for an energetic fight against the corruption and private trade of the Company in Asia. This is not the place for going at length into the numberless meetings devoted to this subject by Lords XVII. It will be quite evident that Van Reede, with his great knowledge about Malabar and Ceylon, made important contributions to the decision-making.

In October 1683 his friend Isaac de Saint-Martin repatriated and came to stay with him in his home in Utrecht[61]. After the termination of the Mataram war[62] Saint-Martin had been promoted to the rank of sergeant-major of Batavia, the highest military rank of the Company. He contributed in 1680 to the downfall of the independent sultanate of Ternate, but his superiors were not satisfied with his partly miscarried intervention in the Bantam war in 1682[63]. Perhaps this criticism was one of the reasons for Saint-Martin to repatriate. But once he had returned to the Netherlands, Lords XVII succeeded in persuading this capable commander again in June 1684 to fill his post once more and to accept a seat in the Council of India[64]. In anticipation of his departure Saint-Martin was frequently in the company of Van Reede, and together they regularly conferred with Huydecoper[65]. Evidently the conversations dealt with the Company's affairs in Asia and with an appointment of Van Reede as commissioner-general for the fight against corruption and private trade, for later Huydecoper stated: 'I contributed a great deal to the advancement of the Lord of Mijdrecht'[66]. Finally in the meeting of 20 October 1684 Lords XVII proposed to request Van Reede to undertake the leadership of a commission of inquiry to be sent to Bengal, Coromandel, Ceylon, and other places. A deputation from Lords XVII under the leadership of Huydecoper went to Utrecht to discuss the proposal with him.

A few days later, on 25 October, the deputies reported on their negotiations. The next day, on 26 October, Lords XVII discussed the report and took the following decision. Van Reede was to represent Lords XVII as commissioner-general in a rank above the governors and directors in Asia. He received a royal salary of 1,000 guilders a month and board free. Finally, upon his return he could count on a remuneration for the services rendered and the prospect on a half-pay of 3,100 guilders a year if after his return he should not be able to get at once a new function, as a compensation for the loss of his seat as a deputy of the States of Utrecht. Van Reede, who had already gone to Amsterdam on 23 October, was summoned to appear in person in the meeting after this decision. He then accepted the leadership of the commission[67].

The following weeks passed in feverish activity, to make it possible to leave with the outward bound fleet before the end of the year. Lords XVII had to complete the commission and Van Reede had to settle his private affairs in Utrecht. It was the intention that Van Reede was to depart with the outward bound fleet of 18 ships with a crew of 3,000. One half of the ships and their crew was to be paid by the Chamber of Amsterdam. From the Amsterdam ships Van Reede selected the "Bantam" to sail first to the Cape of Good Hope. He meant to change there to the "Purmer", which would have to take him to Bengal[68].

The commission was completed with Isaac Solman as the second and Johannes Bacherus as the third member. As a substitute of Van Reede, if necessary, the previously mentioned Laurens Pijl, governor of Ceylon, was designated. The secretary of the commission was to be the naval cadet Hendrik Zwaardekroon, who thus started a brilliant career, in the course of which he was finally to become governor-general[69].

The instruction for Van Reede as commissioner-general of the Western Quarters was established definitively on 29 November[70]. In the instruction nothing was formally fixed about measures to be taken with respect to natural history, medico-botanical research, or the collecting of plants. Still, it will appear from the following discussion of Van Reede's expedition that before his departure several persons requested him to provide for the forwarding of plant material from the Company's settlements which he was going to inspect. Those persons included the Stadtholder William III, the Grand Pensionary Caspar Fagel, the Leiden University curator Hieronimus Beverningk, Paul Hermann, and naturally also Huydecoper and Commelin.

In Utrecht Van Reede sold all his furniture, discharged his servants, and relinquished his house for occupation to his cousin Hendrik Jacob van Tuyll van Serooskerken, lord of Zuilen[71]. I do not know to whom he entrusted the management of his manor-house of Mijdrecht. Van Reede's own equipment for the expedition was not inconsiderable. Later he was criticized for this, and not quite unjustly so. People censured especially the pomp and circumstance with which he then surrounded himself and the crowd of all sorts of persons, in particular young Utrecht noblemen, who wanted to take part in the journey. Indeed, Van Reede did not scruple to gather round his person a retinue of some twenty persons[72]. First of all he took his daughter Francina with him. And already on 27 October 1684, one day after his acceptance of the commission, he engaged Sandrina Reets as lady-companion of his daughter[73]. Further Van Reede's young nephew Maurits Cesar de la Baye joined the company as a naval cadet[74]. To his horror Huydecoper learned that his youngest son, the fifteen-year-old Jan Elias, who did not care much to be a student, was also quite prepared to make a trip to the East in the retinue. The father sharply rebuked his son, saying that such a thing had never occurred with a son of a ruling burgomaster of Amsterdam and that upon his return (Van Reede was at first to stay away for three years) he would be no more than a naval cadet or a common soldier or a student[74]. Another person boarding the "Bantam" was Isaac de Saint-Martin, who was to sail on with this ship to Batavia to resume his function as sergeant-major there and to occupy the chair of extraordinary councillor of India[72].

In the meeting of 24 November Van Reede took his leave of the Utrecht Equestrian Order. One day later this body learned from a missive of the Stadtholder William III that he was to be maintained in all the prerogatives of the Equestrian Order[76]. Finally Van Reede also took his leave of Lords XVII on 11 December and sailed in the yacht the "Amsterdam" to Texel, where the outward bound fleet lay in readiness and where he arrived on 13 December. On 24 December 1684 the fleet weighed anchor[72].

Notes

1 Bruijn et al. 1979 vol.3:106-107.
2 VOC 108, Resolutions of Lords XVII, 1675-1680, n.p., meeting of 29 August 1678.
3 VOC 108, ibid., meeting of 25 October 1678; see also Huydecoper's letter of 19 November 1678 to Joan Bax (RAU, Records Huydecoper Family 56, Letter-book 1675-1678).
4 VOC 108, ibid., meeting of 28 October 1678.
5 At the back of his Letter-book Huydecoper kept journals in telegraphese in which he noted accurately all visits of and to family members, friends, acquaintances, and politicians.
6 The purchase of the house in Mijdrecht must have taken place between 30 November and 7 December 1679. The seller, Isabella Uijttenbogaert, the widow of Herman Schade, charged Jacob Servaes on 30 November to convey the house to Van Reede. It appears from the letter of 7 December of Prince William III (see following note) that at the moment Van Reede was already in possession of the house. It was not until 22 June 1680 that he was enfeoffed with the house by the dean of St. Jan in Utrecht, to whom it was in fee (RAU, Records House Zuylen 94, vol.1, dated 22 June 1680).
7 RAU, Records States of Utrecht 731, vol.6, n.p., Resolutions of 1670-1681, meeting of the Equestrian Order of 1 December 1679 OS, in which there is a copy of the letter of William III of 7 December 1679. Since 1674 the Stadtholder was entitled to appoint personally members of the Equestrian Order.
8 RAU, Records House Zuylen 94, vol.1, documents of the lawsuit of the Van Tuylls van Serooskerken against Antoni Carel van Panhuizen and Francina van Reede, dated 2 October 1695. It is not known to me where Van Reede lived in Utrecht. According to the registers of conveyances he

has not bought a house there. In any case he was already in Utrecht on 6 July 1678, when he had a power of attorney drawn up (GAU, Notarial Records U 069a001 (protocols G. van Toll):223, dated 6 July 1678).
9 The Audit Office of Utrecht consisted since 1657 of three persons. No resolutions of the Audit Office in the period 1640-1764 have been preserved, so that we have no notion of Van Reede's work in this office (Muller et al. 1915: 154-155).
10 RAU, Records States of Utrecht 731, Resolutions of the Equestrian Order, vol.6, n.p., meetings of the Equestrian Order of 12 December 1679, 6 and 12 January, 5 April, 12 November 1680; ibid.731, vol.7, n.p., meeting of 6 February 1682; ibid.736, Brief Minutes of the Equestrian Order, n.p., meetings of 23 August and 6 October 1684.
11 See chapter 1.
12 RAU, Records House Amerongen I 386, letter of 16 February 1679 of Godard van Reede, lord of Ginkel.
13 RAU, Records House Zuylen 125, Receipted bills of Gerard van Reede.
14 VOC 1505:420-422, last will of Hendrik Adriaan van Reede tot Drakenstein of 17 September 1691 in Cochin; see also Appendix 5.
15 For the occupation of the castle of Zuilen and the garden, see RAU, Records House Zuylen 94, vol.2, Accounts ... of the goods of the minor children Van Tuyll van Serooskerken, transmitted on 25 April 1695.
16 On Hendrik Jacob van Tuyll van Serooskerken, christened on 10 April in Utrecht and deceased on 24 July 1692, see Moes & Sluyterman 1912 vol.1:204-205; vol.2:103-104; NNBW vol.3 (1914):1025; Wittert van Hoogland 1909 vol.1:448-449,548-549; RAU, Records House Zuylen 88-94.
17 GAU, Notarial Records U 069a001 (protocols G. van Toll):237, dated 1 May 1682. On account of the long absence abroad of Van Reede, in 1691 and 1692 the lord of Zuilen transferred the guardianship of his children to Godard Willem van Tuyll van Serooskerken (ibid.: 253-253v,263-263v, dated 8 June 1691 and 26 July 1692).
18 RAU, Records House Zuylen 94, vol.1, Documents of the lawsuit of Godard Willem van Tuyll van Serooskerken, lord of Welland, as guardian of Reinoud Gerard van Tuyll van Serooskerken.
19 RAU, Records House Zuylen 94, vol.2:63v,64v.
20 Paul de la Baye, deceased on 1 September 1699, styled himself baron of Theil, lord of Noudepont sur Varne, Schavan(?), and Malay. In 1662 he married Elizabeth (deceased on 29 September 1689), a sister of the lord of Zuilen. Moreover, De la Baye and Van Reede were cousins. In the St. Nicholas church in Utrecht he had a beautiful, still extant sepulchral monument erected. For him, see RAU, Records House Zuylen 38, letter of B.M. de Jonge van Ellemeet to F.C.C. van Tuyll van Serooskerken, dated 25 May 1944; GAU, City Records II, 3243, Conveyances 1684, vol.2:167-170, dated 7 November 1684; GAU, Library 1255, E. van Engelen c.1730: *Grafs en wapen der kerken van Uytrecht*, vol.1:545-546 (Ms); De Jonge van Ellemeet 1939:98.
21 On Willem van Nassau, born at Zuilestein, christened on 7 October 1649 in The Hague and deceased on 2 July 1708 at Zuilestein, see NNBW vol.1 (1911):1358-1359,1367-1368; Moes & Sluyterman 1914 vol.2:74-79,86; The Dictionary of National Biography vol.21 (1921-1922):1341-1343; Wittert van Hoogland 1909 vol.1:172-177.
22 Diederik van Baar died on 16 July 1683. In 1655 he married Maria van Reede, who died on 10 August 1662 and was buried in Arnhem (d'Ablaing van Giessenburg 1859:296, 340,381; Japikse 1937 vol.3:6).
23 Wittert van Hoogland 1912 vol.2:78-79.
24 On Frederik Adriaan van Reede van Renswoude, who was born on 22 February 1659 in Utrecht and died on 12 December 1738 there, see Wittert van Hoogland 1909 vol.1:469; NNBW vol.3 (1914):1034-1036; Moes & Sluyterman 1915 vol.3:28,43-45; Van Gulick 1960:241-243; Taets van Amerongen 1914:24-34.
25 RAU, Records House Amerongen II, 141-142, Documents concerning the inheritance of Karel van Reede tot Drakenstein, 1672-1684. Karel van Reede was governor of Oranienburg, a property of Frederik Willem, Elector of Brandenburg.
26 On Godard Adriaan van Reede van Amerongen, who was born on 6 January 1621 in Amerongen and died on 9 October 1691 in Copenhagen, see Moes & Sluyterman 1912 vol.1:60-67; NNBW vol.3 (1914):1006-1009; Mulder & Slothouwer 1949:32-95.
27 On Godard Willem van Tuyll van Serooskerken, who was christened on 14 November 1647 in Utrecht and died on 19 February 1708 in The Hague, see Wittert van Hoogland 1912 vol.2:368-369; Moes & Sluyterman 1914 vol.2:101-115; Van Vliet 1961:647-648; Van de Bunt 1949:6 ff., Van der Bijl 1975:135-199; RAU, Records Chapter Oudmunster 50 and 26,vol.1.
28 The original drawings of Hortus Malabaricus are discussed extensively in chapter 8.
29 Fournier 1980:9, note 15; Heniger 1980:66, note 50.
30 VOC 320, Letter-book of Lords XVII, 1673-1681, n.p., letter of 31 August 1678 of Lords XVII to the Governor-General and the Council of India.
31 Coolhaas 1971 vol.4:294-295.
32 Judging from the extant Ceylon herbaria of Hermann in London, Paris, and Leiden, Hermann knew at most one thousand Ceylon species.
33 Gillis Valckenier (1623-1680), a member of the city council and burgomaster of Amsterdam, director of the Company, was the political leader of the Amsterdam government since 1667; especially after 1678 he had supreme power (Elias 1903 vol.1:CXV-CXXX, 478-479). Joan Hudde (1628-1704), the well-known mathematician, was a member of the city council and burgomaster of Amsterdam, since 1679 director of the Company, and since 1680 councillor of the Admiralty of Amsterdam. He was a cousin of Valckenier, whom he succeeded in 1680 in the political leadership of the city. Caspar Commelin dedicated to him his *Flora Malabarica* (1696) (Elias 1903 vol.1:528-529; NNBW vol.1 (1911):1172-1176). Pieter van Dam (1621-1706), advocate for the Courts of Holland and Utrecht, since 1652 was a director and secretary of the Company (Elias 1903 vol.1:206; NNBW vol.1 (1911):681-682).
34 VOC 108, Resolutions of Lords XVII, 1673-1680, n.p., meetings of 17 and 18 March 1679.
35 Van Eeghen 1965 vol.3:110-111; ibid., 1967 vol.4:128-131. Heniger 1980:66-67, notes 53,54.
36 Heniger 1980:66, note 51.
37 Hortus Malabaricus vol.3 (1682):(xvi)-(xvii); Heniger 1980:54-55.
38 See chapter 11, section Arnold Syen.
39 Hortus Malabaricus vol.3 (1682):(xv)-(xvi); Heniger 1980:53-54.
40 The systematic sequence in Hortus Malabaricus will be discussed in detail in chapter 11.
41 Hortus Malabaricus vol.3 (1682):(xvii); Heniger 1980:55.

42 Hortus Malabaricus vol.3 (1682):(xvii)-(xviii); Heniger 1980:55-56.
43 GAU, Notarial Records U 080a006 (protocols W. Zwaardekroon):241-241v, dated 7 January 1681 OS (=17 January 1681). For the complete text, see Appendix 3. The way in which the work was to be submitted and published is formulated somewhat vaguely in the text of the contract. Fournier 1980:15, thinks that the authors and publishers "agreed to prepare and to publish two volumes each year". If Fournier's interpretation is correct, the two parties have not kept to the contract, for only in the year 1688 two volumes, 7 and 8, appeared simultaneously.
44 Andreas 1953:52-79.
45 Veendorp & Baas Becking 1938:83,95-96; Wijnands 1983:10.
46 Den Tonkelaar 1983:10-64.
47 NNBW vol.10 (1937):656-657.
48 Theodorus Janssonius van Almeloveen, physician at Amsterdam and Gouda (1687); since 1697 professor of Greek, history, and eloquence, and since 1701 professor of medicine at Harderwijk University (NNBW vol.6 (1924):31-32). Abraham van Poot, a chemist at Amsterdam, studied medicine and took his medical degree at Utrecht University in 1666 (Van der Aa 1875 (vol.15):418-419; Album Studiosorum Academiae Rheno-Traiectinae 1886:60; Ketner 1936:22; GAA, DTB 481:504, DTB 495:434, DTB 1086:101).
49 Hunger 1925; Wijnands 1983:6-9.
50 Huydecoper mentioned a number of visits to Commelin in his journals. About their political relationship he observed in a letter of 17 February 1684:'burgomaster Hudde (the patron of both of us)' (RAU, Records Huydecoper Family 58, Letterbook 1683-1686).
51 Breyne sent, via Commelin, a copy of his *Prodromus Fasciculi Rariorum* (1680) to Huydecoper in 1681 (RAU, Records Huydecoper Family 57, Journal 1681, dated 4 June 1681: 'presented by Mr. Breyne via Commelin with his treaties').
52 About this, see more in detail chapter 11, section Jan Commelin.
53 For biographical data on Paul Hermann, see chapter 2, section First captain and sergeant-major of Ceylon, note 30.
54 Heniger 1980:35-41.
55 See chapter 2, section Extraordinary Councillor of India.
56 VOC 320, Letter-book of Lords XVII, 1673-1681, n.p., letter of 29 October 1680 of Lords XVII to the Governor-General and the Council of India.
57 De Haan 1919:763, meeting of 18 December 1681: Van Goens Sr. left Batavia on 17 December 1681 as admiral of the return fleet on board the "Land van Schouwen". On board the same ship Herman Grimm, the author of *Laboratorium Chymicum*, also returned to Europe. Another ship of this fleet took Hendrik Claudius, the draughtsman of the Cape natural history who later became known, to the Cape of Good Hope.
58 RAU, Records Huydecoper Family 57, Letter-book 1679-1682, letter of 24 December 1682 of Huydecoper to the governor-general Speelman. In his letter of 15 December 1682 to Aletta Hinlopen, the widow of Joan Bax van Herentals, Huydecoper reported on Van Goens' funeral, which had taken place on 21 November.
59 RAU, Records Huydecoper Family 58, Letter-book 1683-1686, letter of 24 April 1684 of Huydecoper to Laurens Pijl. Laurens Pijl arrived in India in 1662. He served as first merchant in Canara (1662), and after this as chief of Tuticorin (1665), commander of Jaffnapattanam in Ceylon 1673-1679. He governed Ceylon since 1679 with the title of first commander, then as governor 1681-1692. Finally he was a member of the Council of India 1687-1705 (took his seat in 1692) (Wijnaendts van Resandt 1944:63).
60 VOC 109, Resolutions of Lords XVII, 1681-1685, n.p., meeting of 21 June 1683.
61 Saint-Martin left India with the return fleet which left Batavia on 25 February 1683. After a stop from 11 June to 6 July at the Cape, the fleet arrived at Texel on 26 October 1683 (Bruijn et al. 1979 vol.3:114-117). RAU, Records Huydecoper Family 58, Letter-book 1683-1686, letter of 29 December 1683 of Huydecoper to Aletta Hinlopen, in which Huydecoper showed himself to be somewhat irritated because Saint-Martin had not yet paid him a courtesy visit, 'while he was staying in Utrecht with Mr. Reede van Drakenstein'.
62 See chapter 2, section Extraordinary Councillor of India, note 1.
63 Stapel 1939 vol.3:413-422; RAU, Records Huydecoper Family 58, Letter-book 1683-1686, letter of 25 April 1683 of Huydecoper to Aletta Hinlopen on Saint-Martin's share in the Bantam war.
64 VOC 109, Resolutions of Lords XVII, 1681-1685, n.p., meetings of 24 March and 24 June 1684.
65 See Huydecoper's Journal of 1684 in RAU, Records Huydecoper Family 58, Letterbook 1683-1686, dated 2 May, 26 June, 7 and 8 July, 9 August, 2 and 29 September.
66 RAU, Records Huydecoper Family 58, Letter-book 1683-1686, letter of 7 October 1685 of Huydecoper to Simon van der Stel, commander of the Cape of Good Hope.
67 VOC 109, Resolutions of Lords XVII, 1681-1685, n.p., meetings of 19,20,24, and 26 October 1684.
68 ARA, VOC 109, ibid., meetings of 24 October and 23 November 1684.
69 Isaac Solman and Laurens Pijl were appointed in the meeting of 28 November 1684. Johannes Bacherus was appointed in the meeting of 25 November 1684; Bacherus had formerly been merchant in Surat, but he was now promoted to the rank of first merchant on a salary of 100 guilders a month, 1,000 guilders for his equipment, and the promise of a remuneration after services rendered (VOC 109, Resolutions of Lords XVII, 1681-1685, n.p.). Hendrik Zwaardekroon (1667-1728) was, after his secretaryship of Van Reede's commission, commander of Jaffnapatnam 1694-1696, commissioner of Malabar 1696-1698, director of Surat 1699-1701, secretary of the High Government in Batavia 1703-1704, a member of the Council of India 1704-1718, and finally governor-general 1718-1725 (NNBW vol.7 (1927):1352-1354; Wijnaendts van Resandt 1944:186-288).
70 VOC 321:243v-255, Instruction for H.A. van Reede tot Drakenstein as commissioner-general of the Western Quarters of December 1684.
71 In order to be able to defray his preparations, he took an advance of 8,000 guilders on his salary (VOC 109, Resolutions of Lords XVII, 1681-1685, n.p., meeting of 23 November 1684).
72 The retinue of Van Reede, consisting of naval cadets and a bodyguard of soldiers, can be reconstructed from the men who changed in May 1685 at the Cape from the "Bantam" to the "Purmer" (VOC 5328, Ship's paybook of the "Bantam" 1684; VOC 5325, Ship's paybook of the "Purmer").
73 Sandrina Reets, then aged 16 or 17, was a daughter of the

Utrecht surveyor Gabriel Reets. The contract was concluded by the father with Van Reede. Van Reede then took the guardianship upon him and was given the right to give the girl in marriage in the East (K ... 1956:101-102).

74 For him, see the chapters 4 and 5.
75 RAU, Records Huydecoper Family 58, Letter-book 1683-1686, letter of 6 November 1684 of Huydecoper to his son Jan Elias. Jan Elias Huydecoper van Maarsseveen (1669-1744) later became a councillor and burgomaster of Amsterdam (Elias 1905 vol.2:700-701).
76 RAU, Records States of Utrecht 736, n.p., Brief Minutes of the Equestrian Order.

4
Van Reede abroad
1684-1691

From the last phase of Van Reede's life an overwhelming amount of information has come down to us. Van Reede performed his task as a commissioner-general of the Western Quarters particularly accurately. Together with his secretary Zwaardekroon, he produced since 1684 more than seven thousand folios of letters and reports about his activities at the Cape of Good Hope, in Ceylon, in India, and in Bengal. It goes beyond the scope of the present study to follow him step by step on his mission. I will confine myself to the main lines of his inspection expedition, for which I base myself on the studies of Hulshof (1941) about the Cape, of Van Goor (1978) and Gommans (1984) about Ceylon, and of s'Jacob (1976) about Malabar. On the other hand I shall go at length into the developments in botany which had taken place in particular at the Cape and in Ceylon until the arrival of Van Reede, into the way in which he promoted those developments, and into the consequences of his relevant reports and instructions.

CAPE OF GOOD HOPE 1685

The voyage from Texel to the Cape of Good Hope took nearly four months[1]. On 19 April 1685 Van Reede set foot ashore there.

At that moment Rijklof van Goens Jr., the son of his former opponent, had been stationed since October 1684 as commissioner in the Cape Colony in the name of the High Government[2]. At first Van Reede did not intend at all to inspect the colony already on the outward voyage. But such bad news about Van Goens' behaviour at the Cape reached his ears, that he decided to take action at once. In particular Simon van der Stel, the commander of the Cape, complained about the dictatorial behaviour of the commissioner, in consequence of which he had lost all authority in the colony. After a disagreeable conversation on 28 April with Van Goens, who had the impudence to behave in a rude way towards the commissioner-general as the representative of Lords XVII, Van Reede sent him back to Batavia then and there.

Van Reede could not refrain from observing haughtily in his report of this meeting, which procured him the decisive victory in the fifteen-year-old conflict with the Van Goenses:

'every increase of the prestige and the means of his father and himself was due to favour and the services of the Honourable Company alone, because the origins of these persons, who had entered so obscurely and unrecognizable from outside, were miserable rather than remarkable, as was sufficiently recognizable to everyone'[3].

The inspection of the Cape lasted from 21 April to 16 July 1685. From the first moment Van Reede did not leave any doubt about the fact that he had come to restore the authority of Lords XVII, to trace abuses, and to punish the culprits severely. He put forth a feverish activity, to which the journal published by Hulshof, testifies in an impressive way. He thoroughly inspected the castle of the Cape, took measures to improve the fortifications, and was regularly to be found among the workers. He made amendments in the administration and the finances, removed incompetent officials, and appointed more capable people in their stead. He gave directions for the extension of the trade in Cape wines. He regulated the supervision of the hospital, took the initiative for the building of new churches, and accommodated the educational apparatus in a special school building. With a view to improving the relations with the Hottentots he advised the colonists to study the life and the thought of the natives and to learn their language, because 'as regards knowledge, brains, fairness, and reasonableness, in so far as required in their housekeeping and civil manner of government, they are not second to any other peoples'. Van Reede pitied in particular the slaves and the serfs of the Company. He advocated a Christian treatment of the poor wretches and tried to give them greater legal security.

Van Reede did not confine himself only to the settlement at Table Bay. In the company of Isaac de Saint-Martin, Johannes Bacherus, and commander Van der Stel he made a tour of fifteen days, 16-31 July, in an open vehicle drawn by four lean mules in the hinterland of the Cape Colony[4]. The tour took them to Rondebosje, De Schuur, Hottentots Holland, False Bay, and Stellenbosch. In Stellenbosch Van Reede was confronted with the colonization policy of the Company in its acute shortage of pasture-lands and corn-fields. He ordained on the spot the promotion of Stellenbosch to a separate colony under its own "landdrost" (bailiff). From Stellenbosch the tour took them back to Klapmuts and Rondebosje.

On an excursion to the wood "Het Paradijs" Van Reede and Saint-Martin met their old friend from the very first phase of their colonial career, Karel van Tetterode, who was woodcutter's boss there[5]. Van Reede offered him his aid, but the old chap 'asked him to allow him to remain what he was'[6].

Via the Steenberg and after a visit to the copper-mine in the Koperberg the company returned to the castle of the Cape.

During this excursion Van Reede concerned himself especially with the Cape agriculture, forestry, and horticulture, stock-breeding, hunting, and mining. He was impressed by the South-African landscape, the prevailing climate, and the flora and fauna. In his journal he made numerous remarks about this, which will be discussed presently.

Since the previous stay of Van Reede at the Cape, in 1656, a great many changes had taken place in the field of the research on the Cape flora and fauna. Whereas at that time the study of nature was still in its infancy, commander Jan van Riebeek grew only a few indigenous plants in the Company's garden, and the callers could make no more than vague remarks about the Cape plants and their applications[7], now, thirty years later, the Cape botany had reached maturity and the Cape plants roused a vivid interest among experts and amateurs in Europe and Asia. Van Reede's inspection of the Cape in 1685 promoted this development. At the time of its founder, Jan van Riebeek, the Company's garden at the Cape was intended in the first instance for the cultivation of vegetables and fruit for the use of the ships calling there. Under Van Riebeek's successors the garden was enlarged with a medical section, which began to play an important part in the supply of medicaments for the Company. At the Cape, European medicinal plants were grown, which were not only used in the Cape Colony itself and sent along with the ships calling there; they were also exported to Batavia and Colombo.

Again it seems to have been Andries Cleyer, the director of the medical shop in Batavia, who started the share of the Cape in the supply of medicaments. It appears that in 1669, when he urged the administrators of Ceylon, Coromandel, and Bengal to undertake a research of medicinal plants[8], he had already been in contact for some time with the Cape commander Borghorst, from whom he obtained garden seeds and artichoke plants for the Batavian gardens and bought drugs for the medical shop[9]. In 1675 governor Godske promised Cleyer 'that he would see to it that henceforth there would not be so many stems among the said Herbs and that they would be packed more carefully'[10]. The second-in-command Crudop also dealt with requests from Batavia, such as in 1678:

'Meanwhile there have been shipped for the medical shop in Batavia eight bags with various drugs, specified more fully in the enclosed Catalogue, while the Herbs not yet included (when they have matured sufficiently) will follow by the earliest opportunity'[11].

With Colombo the Cape maintained an even more frequent connection with respect to medicinal plants. In 1677 governor Bax sent 'as many Garden seeds as the enclosed reports indicate' and 'plants and seeds such as the enclosed report indicate' to Ceylon[12], like Crudop in 1678:

'The medicinal Herbs lately ordered from here are being sent to you herewith: they consist of 5 bags, all of them with the Company's mark, and the authorities are recommended to preserve them from wetness'[13].

The Cape governors also experimented with the growing of Asiatic medicinal plants. Crudop asked in 1678 from Batavia and Ceylon for medicinal plants in exchange for those which had been sent there[11,13]. Van der Stel continued the exchange of medicinal plants. In 1680 he wrote:

'The medicinal drugs ordered for the chemist's shop in Batavia were caused by us to be gathered as much as was possible in the present season of the year; they have also been properly purified and the quantity is in accordance with the enclosed report'.

In exchange Van der Stel asked them to send 'all sorts of seeds and plants as well as pips and grains' to the Cape with a view to growing them there[14].

The search for indigenous medicinal plants, too, was stimulated from the Company's garden. Thus, in 1677 Bax urged the sergeant of Hottentots Holland, Laurens Visser, 'to shift as best you can with indigenous Medicinal herbs'[15].

From the above it is evident that in the seventies of the 17th century the Company's garden was beginning to play a role as acclimatization garden for medicinal plants from Europe, Asia, and also from the Cape itself.

However, an even more interesting fact is the rise of the scientific research at the Cape. This was due to the cooperation of calling botanists, Cape governors and gardeners with botanists and amateurs in the Netherlands. The most prominent calling botanists were Paul Hermann and Willem ten Rhijne. Hermann stayed at the Cape in April 1672 on his way to Ceylon and in March 1680 on the way back to the Netherlands[16]. He then accumulated a voluminous herbarium and a collection of drawings, which were intended for a flora of the Cape, *Prodromus Plantarum Africanarum*. Owing to his early death no such publication ever saw the light[17]. Ten Rhijne called at the Cape shortly after Hermann in October-November 1673 and also made a herbarium and drawings. He sent part of his Cape collections to Jacob Breyne in Danzig, who published about this in *Exoticarum Plantarum Centuria Prima* (1678)[18].

Of even greater importance was the stimulus emanating from the repeatedly mentioned director of the Company, Joan Huydecoper van Maarsseveen[19], who founded the botanical relations between the Netherlands and the Cape definitively in a many years' correspondence with his nephews, the Cape governors Joan Bax and Simon van der Stel. It was also through the influence of Huydecoper that in 1674 Bax was promoted by Lords XVII from the rank of commander of Galle, Ceylon, to that of governor of the Cape, where he assumed his function in 1676[20].

Already during his stay in Ceylon Bax had presented his uncle annually with all sorts of natural curiosities[21]. Once he had arrived at the Cape, Bax continued to send these presents as a service in return for Huydecoper's good offices in connection with his appointment. He showered on his uncle gifts of live canaries, jackdaws, siskins, Cape geese, parakeets from Madagascar, skins of wild cats, and a chalice made out of a rhinoceros horn. In addition he also sent chests of seeds and bulbs of Cape plants, and even a herbarium consisting of '2 boxes of dried Cape flowers'. At the Cape Bax had even found a draughtsman, who started to make a series of watercolours of plants, of which he sent Huydecoper successively a few booklets[22].

Huydecoper, in turn, became interested in the vicissitudes of the Cape Colony. In 1676 on his own initiative he sent to the Cape one of his own villagers, the gardener Jan Hendriksz. de Beet, from Maarssen, who was entrusted by Bax with the direction of the Company's outer garden Rustenburg in Rondebosje[23]. Huydecoper provided personally for the consignment of seeds of Dutch hazelnut trees, oaks, and beeches, thus contributing to the further development of the outer garden[24].

Bax also carried out the preparation of the Cape share in the Company's present to the Stadtholder William III[25]. In May 1676 he sent bulbs, plants, and wild almond trees[26], and in March 1677 there followed another collection of Cape plants and live animals, including monkeys from Madagascar and a young, tamed rhinoceros[27]. The shipments were transferred, also through Huydecoper, to Daniel Desmarets, the French court chaplain of William III, who housed the plants and animals in the gardens and the menagerie of the prince's palace of Honselaarsdijk[28].

Meanwhile Huydecoper had exerted himself to help Bax to rise in the hierarchy of the Company. On 25 October 1678 Lords XVII nominated him governor of Ceylon and extraordinary councillor of India as the successor of Rijklof van Goens Jr[29]. Huydecoper recommended Bax, when leaving for Ceylon, to take along the draughtsman of the Cape plants, in order to illustrate flowers, trees, snakes, etc. there. He also requested him to transfer the book with drawings of Cape plants to his successor at the Cape[30]. Apparently the notice that Bax had died already on 29 June 1678 had not yet reached the Netherlands.

As has been said, during the administration of Bax, Van Reede passed the Cape without calling[31]. He could not therefore acquaint himself with the developments in Cape botany which were then taking place through the activities of his friend.

Huydecoper now used his influence to have his other nephew, Simon van der Stel, appointed commander of the Cape in 1679[32]. Huydecoper requested him to send bulbs and seeds from the Cape, and birds and flowers from Madagascar and Mauritius. He sent a memorandum to instruct Isaac Lamotius, the administrator of Mauritius, how to make a herbarium. Huydecoper also wished to receive illustrations of flowering plants from the Cape and Mauritius[33]. In 1681 and 1682 Van der Stel sent chests with bulbs, Cape geese, a parrot from Mauritius, and booklets with drawings of plants[34]. However, Huydecoper did not keep most of the presents for himself. Near his country-house Goudestein, in Maarsseveen on the Vecht, he had no botanical garden or greenhouse, but he liked to distribute natural curiosities to friends and acquaintances in the Netherlands to promote the good relations. Thus he presented Cape seeds and bulbs to Agnes Block, of the country-house Vijverhof near Loenen, and to the previously mentioned Daniel Desmarets[35].

The private wishes of Huydecoper, who after all was a director of the Company, appear to have been in contradiction with the policy of Lords XVII to banish the botanical hobby as much as possible from the shipping traffic[36]. Huydecoper was aware of this and urged Van der Stel to fulfil his wishes 'without wronging the Company' and to confine himself to modest shipments of Cape plants[37]. The situation, however, changed at once when in 1682 the Amsterdam Municipal Garden or Hortus Medicus was founded, and Huydecoper and Jan Commelin were entrusted with its direction. From that moment Huydecoper, who was also burgomaster of Amsterdam, did his utmost to induce Van der Stel to ship as many Cape plants as possible for the new garden. Jan Commelin and Hieronymus Beverningk drew up a special list of the required species. Moreover Huydecoper insisted on a vigorous continuation of the series of drawings of Cape plants[38].

From this moment the Cape botany developed very greatly. Many species which were sent to the Amsterdam garden in the next years by Simon van der Stel, and later by the latter's son Willem Adriaan, were published by Jan and Caspar Commelin in *Horti Medici Amstelodamensis Rariorum Plantarum* (1697, 1701)[39].

Huydecoper also contributed in another way to the dissemination of the knowledge of the Cape flora. In 1683 he caused those of his drawings of Cape plants which Bax and Van der Stel had sent to him to be copied for the use of the aforesaid Jacob Breyne in Danzig. Breyne's set of copies constitutes the nucleus of the codex *Flora Capensis*, now owned by the Brenthurst Library in Johannesburg. According to Gunn & Du Plessis a great many of Breyne's copies served as models for engravings which were published in 1739 by Johann Philipp Breyne, Jacob's son, in *Prodromi Fasciculi Rariorum Plantarum*[40].

Van der Stel now also applied himself to acclimatizing rare plants in the Company's garden before shipping them to the Netherlands: cinnamon trees for the Amsterdam garden, and at the request of the Dutch Grand Pensionary Caspar Fagel, who owned a beautiful botanical garden at his country-house Leeuwenhorst near Noordwijk, cinnamon, clove, and camphor trees[41]. By way of exception, in 1684 Fagel even obtained permission from the Company to send a collector to the Cape at his own expense[42].

It may be assumed that, in his exertions for Huydecoper, Van der Stel availed himself of the services of the chemist Hendrik Claudius. Claudius reached the Cape from Batavia in the early part of 1682. He had been charged by Andries Cleyer to draw and paint Cape plants, to collect drugs and minerals, and to start a herbarium at the latter's own expense. In 1683 Van der Stel sent Claudius along as a secretary and draughtsman on an expedition to Namaqualand[43].

From the above survey of the development of Cape botany since 1669 the disappointment of Van der Stel can be understood when in 1684-1685 Rijklof van Goens Jr. inspected the Cape as a commissioner during seven months[2]. Van Goens seems to have had little or no interest in the achievements of Bax and Van der Stel so far in the Company's garden and its dependences and with respect to Cape botany. Van der Stel could not prevent that Van Goens gave away a part of the best soil of the Company's garden to a favourite[44]. Nor could he stop the commissioner extirpating the acclimatizing cinnamon trees which were destined for the Amsterdam Municipal Garden[45]. It is possible that the rare trees for the Grand Pensionary Fagel were then also destroyed.

Van der Stel was greatly relieved by the coming of Van Reede. The latter delivered him from Van Goens, but what is more, in Van Reede he found an inspector with whom he could amply exchange views about the cultivation and the examination of plants and the exploration of the new territories. One of the first objects of inspection was the Company's garden and its neighbourhood[46]. With his characteristic thoroughness Van Reede looked at the situation of the garden and the physical character of the soil.

He observed that the garden was not situated on level ground, but on a fairly steep slope[47], and was covered with a layer of fertile black earth one and a half foot thick. He attributed the formation of that layer to the washing-away of the surrounding highlands during heavy rains. He pointed out that underneath the black earth there was yellow and white clay, which, when exposed to sun and air, was transformed into ferruginous clay or iron stone, such as he had also observed in Java, Ceylon, and Malabar. At a greater depth he found blue and chalk-white ferruginous clay, without any admixture of sand, which was transformed into slush or soft mud when mixed with water. The process of the washing-away was studied by him in a more elevated territory, where the cover of vegetation 'grass, heather, and other rough growth', had disappeared, in consequence of which the uppermost layer was no longer retained by the roots, and the rain-water in a short time could produce big holes in the soil.

The Company's garden was irrigated by a system of ditches and dams, by means of which the water from the brook, coming from the Leeuwenberg, was conducted all over the garden without any difficulty. But when, upon the setting-in of the wet monsoon, in June, heavy downpours swept the garden, the system could not absorb the sudden huge water influx, as a result of which large parts of the flower-beds were washed away. Van Reede thereupon ordered a partition to be placed in the inlet of the garden[48].

On the very evening of his arrival already, 19 April, Van Reede was greatly pleased by the sight of the garden:

'But nothing can be compared with the pleasant and nice aspect of the Company's garden, the mere sight of which is able to comfort and refresh a man coming from that wild, raw, and merciless sea in a land without trees and bare highland. For the foot-paths, which extend further than the eye can reach, are planted on either side with high walls of pleasant green trees, where one can be safe in the heaviest south-easterly winds; the whole garden is divided into a great many square sectors or beds, planted with all sorts of fruit-trees and vegetables which, protected against evil winds, grow and flower plentifully, so that benefit and pleasure are combined here'.

Two days after this lyrical effusion, on 21 April, Van Reede was somewhat more matter-of-fact when inspecting the garden. The garden itself had an area of about 40 acres at that time[49]. The ground was divided by three broad walks and eight or nine cross-paths into 16 or 18 sectors or fields. Van Reede described the garden as follows:

'These fields, which make for the pleasant aspect of a garden and plantation, are planted with Cape shrubs or with alders or bay-trees, and both species are trimmed to the same height and thickness, 20 to 22 feet high and 6 to 8 feet thick. This height and thickness is very necessary, and without this nothing would be able to grow there because of the violent and hard south-easterly winds, which blow extremely hard here for some two-thirds of the year. Between these high and thick hedges the fields are enclosed. Several of them were planted with European fruit-trees such as apples, pears, cherries, peaches, as well as oranges, lemons, and pomegranates in the form of orchards, also along the paths and broad walks, and they are very luxuriant and thriving. The others are sown and planted with all sorts of greens, vegetables and roots ... In front at the entrance there is a brick wall, about 60 roods long and 10 feet high. Further it is surrounded by a green hedge and a ditch'.

Unfortunately Van Reede has not had a catalogue of the garden made, which might have given us a more detailed picture of the plants contained in it. According to Van Reede the Company's garden was tended by 54 male and female slaves under the direction of a Dutch head gardener[50]. He inspected the quarters of the gardening staff. Both slaves and white men were accommodated in a heavy, solid, square brick building, situated in the garden. That building also contained the living-rooms of the head gardener and the drying-rooms for storing the seeds, roots, and fruits.

Less attention was devoted by Van Reede to the outer garden Rustenburg in Rondebosje, where De Beet was head gardener[23]. The garden was 'partly planted with orchards of European fruit-trees and also divided, as at the Cape in the big garden, by hedges of green, some of the beds being filled with vines, others with cabbage and all sorts of vegetables, as well as lime and lemon trees'[51].

Van Reede was more interested in the viniculture of Rustenburg, which had been enlarged under Van der Stel. He suggested to the commander an improvement of the quality of the Cape wine, which he thought a little too sour, by removing the skins, pips, and stems from the grapes before proceeding to press them. He proposed to Lords XVII to promote and monopolize the viniculture in the Cape Colony considerably, 'for since we find that these regions of Africa near the headland are as suitable for growing vine as any countries in Europe, the Honourable Company might be able to provide from here the whole of the Indies with wine, brandy, and vinegar'.

Van Reede showed an interest in the full-grown European oaks of Rustenburg:

'on one side of the orchard there was a row of oaks, about 20 years old, some of which were as thick as a man's waist, but they had no tall trunks, being as it were polled, spread their branches very far from each other, and bore much and good fruit, clear evidence that the earth and the air might produce such trees'.

Whilst the Company's gardens were praised by him, Van Reede considered the condition of the woods in the Cape Colony very precarious in connection with the need of fire-wood and timber. Already during the administration of Van Riebeek the area of the indigenous woods was seriously declining[52]. Van Reede found an almost entirely deforested Table Valley[53]. Only at the foot of the Devil's Peak he discovered a last remnant of a thin underwood, which could not by any means cover the annual need of fire-wood, which he estimated at 1,000 to 1,200 cart-loads. Upon a closer study of the trees he found that they were not suitable for large-scale cultivation and were moreover susceptible to heathland fires. Still, he considered that, provided that the felling were carried out with discretion and the wild shoots were spared sufficiently, the underwood might still be of some importance.

Van Reede wondered where the Company and the freeburghers got the necessary wood. On his expedition upcountry he made the appalling discovery that the natural woods were being extirpated in an unjustifiable way. A little to the south of Rondebosje he saw a vast tract with tall felled trees, which had apparently been found useless as timber. The young trees had also been cut down. Van Reede greatly disapproved of such wastefulness in a country with so few trees:

'those who have neglected it so much have provided only for themselves and not at all for the future'.

On his expedition in the direction of the Houtbaai, too, he met with the same disconsolate picture: big plains full of stumps and stubs of felled trees. To Van Reede's astonishment Van der Stel thought that 'the woods would grow again'. He then explained to him that 'the felled wood grew again neither from the roots nor from the trunk when once it had been stripped of branches, but from the seed of remaining trees, and that here he was shown in an hour's walk that at first sight not one tree, neither old nor young, had been spared, but they had all been destroyed'. Van der Stel could not but admit that 'in the same way the woods were treated which were preserved and cut down for the Honourable Company; whence was that wood then to be fetched in the future?'.

Mindful of the attempts of Huydecoper during the rule of governor Bax to introduce Dutch timber into the Cape[54], Van Reede went in search of the results of the introductions. In Rustenburg he had already seen that European timber could thrive here very well. In the wood "Het Paradijs", in the cleft between the Table Mountain and the Devil's Peak, which had suffered less from felling, he asked for information from the woodcutter's boss Van Tetterode[56]. He found there two (!) European alders indeed, but for the rest only Cape timber, which, to his bewilderment, had been given by Scandinavian woodcutters the names of European trees, 'or what is usually made from it, which latter could be looked upon as being unknown to us, but the others having no resemblance at all either in leaves, fruit, bark, or wood, it is to be wondered at how they have been able to give names of known trees to strange growths, which do not in the least resemble them, and one is misled if one hears them mention elders, currant-bushes, ash-trees, alders, beeches, peartrees, and the like trees which do not resemble them in any respect'.

Here therefore Van Reede realized the importance of botanical research on Cape trees for forestry.

Van der Stel, however, swore to it that he had personally sown the imported seeds, but he admitted that he had not looked after them since. Van Reede did not like to accuse him of bad faith or insincerity, but did point out to him his negligence in supervising the foresters. He charged the commander to plan, wherever the woods had been destroyed, plots of 10 to 12 square roods with European timber and to provide for a quarterly supervision. With respect to the woods still existing he ordered that during the felling circular or square sectors of 6 to 7 roods should be spared, so as to make natural restoration possible.

During his expedition through the hinterland Van Reede had plenty of opportunity to admire and study the Cape nature.

When he took measures to regulate the hunting of lions, wolves, and leopards, which caused great damage in kraals and stables, it was pointed out to him that packs of 'wild dogs' made great havoc among the game. A few weeks later four of those animals were brought to him. Although they made the same noise as ordinary dogs, Van Reede observed that they had big, bare, and hairless ears. In his report he wrote about this:

'it is to be noted that no wild beasts or tame animals are seen here which are similar to those of Europe, except for oxen. For the sheep without wool and long hair are goats rather than sheep. And the same applies to other wild animals: it seems as if nature here, having made a combination of members of various species, had wanted to produce none but monsters. For thus one here sees the head of an elk, a billy-goat, a cow, a deer, and a roe combined with other bodies, and this makes us doubt how to call them, after the head, or after the body. Thus, this part of Africa, as regards the earth, plants, and trees, animals, both walking and creeping, as well as birds, has much that is extraordinary and does not resemble any other countries and regions'[55].

He observed the divergent character of the Cape fauna also for the flora. When he returned from his expedition, on 30 May he summarized his impressions about the Cape flora as follows:

'It should only be added to this that this protruding corner of the countries of Africa is also notable with regard to plants and trees, for wherever one goes, on low and moist, high and dry lands, everywhere one finds great plenty of fragrant flowers of the most brilliant and beautiful colours that one might imagine, most of which have bulbs such as they are not found in any other countries of the world. Nay, there is no mountain or rocks and stones so barren, bearing any green foliage, but there are found the most beautiful flowers, fragrant plants, even various kinds of heather which are fragrant in their leaves or flowers, so that doubtless that soil brings forth many strong herbs serving man's health if they were known and used in the right way, which will also be profitable for the Honourable Company to be examined'[56].

When a few days after this eulogy of the Cape flora the wet monsoon set in, Van Reede recorded:

'It was daily getting colder and the snow began to appear on the high mountain range towards the north-east, but the wind remained westerly, which prevented the speed in sending off the ships. One very clearly saw here the different nature of the plants and trees, for those from Europe began to lose their leaves, while on the other hand those from these regions, in consequence of the rain that had fallen, sprouted young shoots and leaves again and produced their bloom and flowers'[57].

Van Reede did not confine himself to making general comments on the Cape nature. During his inspection of the Cape he consulted the drawings and the herbarium of Hendrik Claudius. The French Jesuit-missionary Gui Tachard, who was at the Cape in June 1685 and met Van Reede several times, related:

'He [Claudius] had already completed two thick volumes of various plants, painted from nature, and has collected specimens of all kinds which he had pasted into another volume. Doubtless M. Van-Rhêden, who always keeps these books in his own apartment and showed them to us, intends publishing soon a *Hortus Africanus* after his *Hortus Malabaricus*'[58].

From Tachard's statement it may be inferred that Van Reede scrupulously observed the Cape plants. But it cannot as yet be proved that he actually intended to compose a *Hortus Africanus*, as Tachard suggested. The 'two thick volumes of various plants, painted from nature', and the volume of the herbarium with 'specimens of all kinds' of Claudius up to the present have not been traced, so that they cannot provide further information about Van Reede's preparations for a *Hortus Africanus*[59]. This does not alter the fact that in his eulogy of the Cape flora quoted above Van Reede gave the important advice that the Cape plants should be studied for the benefit of the Company. With this advice he followed up the current developments in Cape botany such as I have described them. For Van der Stel this must have been a stimulus, after his disappointing experiences with Van Goens Jr., to continue his explorations in the field of natural sciences.

During the inspection of the copper-mine at the foot of the Table Mountain near the Houtbaai Van der Stel stressed the mining problems with which he was faced. In view of his own theoretical knowledge of chemistry he doubted that this mine could even be rendered workable. He proposed to Van Reede that he himself should be allowed to make an expedition to Namaqualand, whence highly promising information about the presence of silver- and copper-mines had been coming in. Van Reede agreed to this and at the same time decided that a report should also be drawn up about 'what else there might be as to that nature as well as nations and people in that region'[60]. The expedition to Namaqualand took place from 25 August 1685 to 26 January 1686, after Van Reede had left the Cape. On that expedition Van der Stel took Claudius with him. The latter then made the water-colours of plants and animals of Namaqualand, which became famous and were copied many times after this. A number of the water-colours of plants were published by Plukenet in his *Phytographia* (1691)[61].

If we summarize what Van Reede achieved in a short time during his inspection of the Cape with respect to the exploration of the Cape nature, we can establish the following.

He took measures for the improvement of forestry, advocated the importance of the introduction of European timber, and drew attention to the poor knowledge of Cape timber. In addition, he gave suggestions for the improvement of viniculture and ordered the further exploration of Namaqualand.

For us it is more important that he drew the attention of Lords XVII emphatically to the extraordinary character of the Cape fauna and flora. He praised the Company's garden not only for its usefulness, but also for the 'pleasure' it gave, and pointed out the desirability of a search for useful plants. By dismissing commissioner Van Goens Jr., he eliminated an imminent stagnation in the search that had already started. He stimulated the progress of the search by his interest in the work of Claudius and in a report about the nature and the people of Namaqualand.

In general it may be said that all these measures, recommendations, and remarks of Van Reede implied a confirmation of Van der Stel's policy in the exploration of the Cape nature since 1679. But in the conclusion of his inspection report Van Reede also gave an opinion about the person of Van der Stel, in which he gave a scarcely veiled warning to the Cape commander. Van Reede was convinced that Van der Stel was an honest, faithful, and diligent servant of the Company. But he had not failed to note that the commander showed the greatest diligence in matters in which he took pleasure. 'But', Van Reede pursued, 'since the Commandment consists of many things that must be done, maintained, and continued because of their purpose and usefulness, they do not depend on our temperament and approval, but on our duty. Therefore in that sense many reasons have occurred to me for showing him the difference between our own pleasure in certain parts of our service and what we have to do in general, even against our sentiment and pleasure, because we have to serve our masters, and not ourselves'[62].

In other words, Van Reede decidedly appreciated the enthousiasm displayed by Van der Stel, but he also warned him not to be guided by his personal preferences, but by his duty. This passage recalls Van Reede's own experiences in Malabar. At that time he had to bear a good deal of criticism, which was partly directed against his work on Hortus Malabaricus, which his opponents regarded as an irrelevant hobby. I cannot avoid the impression that Van Reede wanted to caution the commander against this danger, criticism of 'our own pleasure'. It will be discussed hereafter to what extent these were prophetic words addressed to Van der Stel and his subordinates.

I would now make a few observations about the consequences which Van Reede's inspection has had for the further exploration. His report fell on fertile soil among Lords XVII, or, as Huydecoper expressed it: 'his work there was greatly appreciated'[63].

Van der Stel, too, seems to have been very pleased with Van Reede's inspection. Long after the commissioner-general had left the Cape, Van der Stel immortalized his visit with the foundation of the settlement Drakenstein on the Bergrivier, on 16 October 1688[64].

The expedition to Namaqualand was very successful for Van der Stel. By reference to the journal of this expedition illustrated by Claudius, Huydecoper asked for the consignment of plants and animals from that region[65].

Of course Van der Stel continued to send Cape bulbs and seeds to the Amsterdam Municipal Garden[66]. He took pains to enrich the Company's garden with exotic plants. In 1687 he wrote to Lords XVII:

'as also the cultivation of all sorts of Asiatic and American root crops and tree-fruit, the Honourable Commander sparing no pains to get hold of those of the outward bound and home-bound ships, both Dutch and foreign'[67].

A wonderful acquisition was, a year later, in 1688, the present of Andries Cleyer from Batavia. The latter's gardener George Meister then brought a large collection of Asiatic plants to the Cape, including the camphor-tree and the tea-shrub from Japan, as a result of which the damage formerly caused by Van Goens Jr. was remedied again. Van der Stel availed himself of the opportunity to entrust Meister with a consignment of seventeen cases with trees, flowering plants, and garden plants, which were destined for the Stadtholder William III, the Grand Pensionary Fagel, and the Amsterdam Municipal Garden[68].

For Claudius the expedition to Namaqualand had an unfortunate consequence. When the previously mentioned Jesuit-missionary Tachard in March 1686, on the way back from Siam to France, called for the second time at the Cape, Claudius handed him a report, a map, and some drawings of plants and animals of that expedition. When he was back in Paris, Tachard published all this material in his *Voyage de Siam* (1686)[69]. Apparently Claudius did not realize that with his communicativeness towards the Jesuit he was acting against the interests of the Company and that he had given priority to his 'own pleasure' over his duty to the Company, against which Van Reede had warned.

Even before Van der Stel got hold of Tachard's book, he already seemed to have got wind of Claudius' liberality to the Jesuit and he no longer trusted the draughtsman. The commander indeed had reason for his suspicion against Jesuits, for he had discovered that one of his gardeners, the Frenchman Pierre Couchetez (Pieter van der Coste), acted at the Cape as a contact of the "Missions étrangères" in St. Germain (Paris)[70]. In this light it becomes intelligible that in his letters of April or June 1686 Van der Stel requested Huydecoper to send him 'a boy from the fatherland who can make water-colours' and a 'stray botanist'[71].

When in June 1687 Tachard, for the second time on his way to Siam, again visited the Cape, Van der Stel set eyes on a copy of the *Voyage de Siam*. Van der Stel considered Claudius' indiscretions about Namaqualand as treason and banished him from the Cape. This expulsion must have taken place between June 1687 and April 1688[72].

Huydecoper, jointly with Jan Commelin, exerted himself to have the place of Claudius, which had thus become vacant, filled again[73]. On 30 December 1687 Hendrik Bernard Oldenland (c.1663-1697), a pupil of the Leiden professor of botany Paul Hermann, sailed in the ship "Den Helder" from Texel to the Cape, where he went ashore on 25 June 1688[74]. The arrival of Oldenland at the Cape of Good Hope heralded a new period in the Cape botany. As head gardener of the Company's garden he brought its scientific aspect to great prosperity. He introduced at the Cape Hermann's taxonomic and nomenclatural views in botany, composed a voluminous herbarium, and applied himself to the cultivation of medicinal and other plants pouring in from Europe, Asia and America[75].

We may regard the banishment of Claudius and the arrival of Oldenland as consequences, though indirectly, of Van Reede's inspection of the Cape in 1685. One of the main criticisms of Van Reede with respect to Van der Stel was the poor condition of the Cape forestry. Huydecoper took great pains to be of use to his nephew to change this situation. In his letters of 1686, 1687, and 1694 he gave

Van der Stel detailed advice on the cultivation of a variety of European woods: abeles, pines and spruce-firs, poplars, beeches, elms, lime-trees, oaks, olive-trees, chestnuts, walnut-trees, horse-chestnuts, plane-trees, etc.[76]. As a commissioner of the Amsterdam Municipal Garden, Huydecoper will no doubt have received this information from his fellow-commissioner Jan Commelin.

Despite of Huydecoper's support the arboricultures did not take root very well. Lords XVII in 1690 urged Van der Stel to display somewhat greater diligence in the planting of woods, in order that the Cape Colony should not get into trouble on account of a shortage of firewood and timber. The cultivation of olive-trees, too, did not get off to a good start according to Lords XVII[77]. In 1690 the Company sent to the Cape Jan Hertog (c.1663-1722), a brother of Willem Hertog, the hortulanus of the Leiden University garden, who was recommended to Van der Stel because of his 'knowledge ... of a variety of native and exotic plants, especially of their nomenclature and cultivation'[78].

The assistance of Jan Hertog was not much use. The critical remarks of Lords XVII about Van der Stel's policy in forestry increased more and more. To this were added serious complaints about his expensive colonization projects of Stellenbosch and Drakenstein, to which Van der Stel devoted much attention. He did not succeed in refuting the objections to his policy, in spite of favourable reports of Oldenland and De Beet about the Company's garden and the outer garden Rustenburg[79]. By a resolution of 6 September 1696 Van der Stel was finally dismissed from the service of the Company![80]. Not only Claudius, but also Van der Stel appeared ultimately to have become a victim of their 'own pleasure'. In spite of all his enthusiasm and his merit for the exploration of the Cape nature still Van der Stel disregarded Van Reede's warning to concern himself primarily with the affairs of the Company, 'which must be done, maintained, and continued'.

Van Reede left the Cape on 16 July 1685 on board the "Purmer" and sailed to Ceylon in the company of the flute "Adrichem"[81]. His friend Isaac de Saint-Martin had already left a month before, on 13 June, by the "Bantam" for Batavia, whence in the coming years he was to send seeds of Javanese plants to the Amsterdam Municipal Garden[82].

SOUTH CEYLON 1685

On 13 October 1685 Van Reede arrived in Colombo, where he carried out his inspection for nearly two months, until 6 December of that year, with as much enthusiasm and application as at the Cape of Good Hope[83].

He soon came to be on good terms with the governor of Ceylon, Laurens Pijl[84]. Among the other members of the staff, however, he caused confusion by discharging some servants on account of incapacity or disobedience.

As a result of a shift in the hierarchy Thomas van Rhee thus became commander of Galle[85]. Van Reede took measures against private trade. He intensively inspected the fortifications of Colombo, Galle, and Negombo. He also tried to meet complaints of the Singalese population.

After Paul Hermann had left the island in 1680[86], to the great disappointment of the Ceylon government, the natural-history research had relapsed to the level of the consignment of medicinal plants and drugs to the medical shop in Batavia. Lords XVII tried to fill the vacancy of Hermann. In 1680 they sent to Ceylon the chemist Johannes van Sanen, who was said to be an expert in the search for and the preparation of native herbs. A grand botanical expedition of Van Sanen, however, ended in a complete failure because he had left the plants to decay[87]. But as soon as Van Reede had arrived in Ceylon the botanical activities began to flourish again. Apparently he clearly explained the wishes of his botanical friends in the home country, for since 1685 the Company's servants in Ceylon exerted themselves to oblige the commissioner-general.

Van Reede himself set the example by consigning in January 1686 to the Amsterdam Municipal Garden a collection of Ceylon roots, seeds, and trees, including cinnamon-trees, but it is not certain whether the cinnamon-trees ever arrived in Amsterdam. In a suspicious mood Huydecoper wrote to Van Reede:

'they will probably have been carried off by the commander Van der Stel'[88].

In December 1686 N. Elsevier, a Company's servant in Maturé, sent a small chest of bulbs and seeds[89]. With the same shipment the plants of Magnus Wichelman will also have been sent off[90]. In January 1688 from Colombo a collection of 145 Ceylon seeds was dispatched to Paul Hermann[91], while in January 1689 three chests with roots and seeds, destined for Hieronymus Beverningk, and three more cases with seeds and bulbs, destined for Huydecoper, were shipped[92].

That Van Reede indeed was the driving force behind these consignments of plants appears from the instruction drawn up by him later, on 5 March 1691, in Cochin. Although this instruction, strictly speaking, was not issued during Van Reede's inspection of South Ceylon, I will discuss it here because of its great importance for the future development of Ceylon botany during the Dutch rule[93].

The instruction of 1691 was destined for all the administrators of the Dutch colonies belonging to the Western Quarters. In this document Van Reede declared that when he left Europe he had been charged by prominent persons to cause seeds, bulbs, and roots to be sent annually from the East to the Netherlands. He stated emphatically that the plants were to be collected not only for the pleasure of scholars and amateurs, but also for the benefit of medical botany. Meanwhile he had learned that plants had been successfully grown from the seeds which had already been shipped during his presence in the

Western Quarters. He enlisted the aid of all the administrators to collect seeds and provide them with registers of their names. The seeds, with the exception of those from the Cape of Good Hope, were to be sent to Ceylon in order to be shipped from there to the Netherlands. He emphatically pointed out that this instruction did not refer to seeds of familiar garden plants, 'but to foreign, wild, and unnoticed seeds, such as they occur indifferently and can be found in the woods'. In order to avoid the risks of the voyage, the seeds and the registers were to be distributed over two parcels and to be sent by different ships. He gave the advice to put the seeds not in small bags or cloths, but in boxes, as he himself had formerly done successfully from Bengal. Finally he ordered that one parcel of seeds was destined for King William III and the other for Jan Commelin, of the Amsterdam Municipal Garden.

With this instruction Van Reede gave a blueprint of the way in which in his opinion colonial botany was to be studied within the Company. It is remarkable that he apparently did not expect that professional botanists would settle in the East to make investigations on the spot. What he had in mind was a system of collecting and transporting plants with Ceylon as centre. Just as in the report of his inspection of the Cape, he emphasized the importance of the wild flora for medical botany. If we recall his remark about the usefulness and the pleasure of the Cape garden and his advice to Van der Stel to give priority to duty over pleasure, we find that he now admitted, with the distribution of the seeds between Jan Commelin and King William III, that in practice usefulness and pleasure were the two pillars of botany. It would appear contradictory that Van Reede, considering his passionate interest in the Malabar flora and after his long conflict with Van Goens about Ceylon and Malabar, ultimately designated Ceylon as the plant-collecting centre. The solution of this question is simple. Ceylon, specifically Colombo and Galle, was the staple and the centre of trade of the Company in the Western Quarters. All the goods from the western part of Asia which were destined for Europe were accumulated in Ceylon and transported annually from there by the Ceylon squadron direct by way of the Cape to the Netherlands. It was therefore for a practical reason that Van Reede chose Ceylon as the plant-collecting centre. At that time he could not suppose that in the course of the eighteenth century Ceylon was to be not only the centre, but even the only Asiatic colony in the Western Quarters which was to ship botanical material to the Netherlands continually. In so far as I have learned hitherto from the records of the Company, only the government of Ceylon carried out the 1691 instruction for nearly a hundred years.

The botanical material that was collected in the last decade of the seventeenth century, during the administration of the Ceylon governor Laurens Pijl and his successor Thomas van Rhee, found its way to Jan and Caspar Commelin and the Amsterdam Municipal Garden, and was partly published in their *Horti Medici Amstelodamensis Rariorum Plantarum* (1697, 1701)[94]. The Company's surgeons in Colombo made up reports about medicinal plants, in which they enumerated the names, short descriptions, and medicinal properties[95]. A remarkable incident was the discovery of the curative effect of Acmella against renal calculi, to which Pieter Hotton devoted an article[96]. During the administration of Pijl or Van Rhee probably the two-volume codex of water-colours of Ceylon plants with short descriptions was also composed, which later came into the hands of the Leiden professor of botany Hermann Boerhaave and which is now in the University Library in Leiden[97]. From 1699 the surgeons of Colombo annually provided for the shipment to the Netherlands of seeds of 150 plant species on an average with their register. After the death of King William III (1702) his share was henceforth sent to the Leiden botanical garden. As far as is known, until 1779 the surgeons provided the botanical gardens of Amsterdam and Leiden with Ceylon seeds, sometimes complemented with herbarium material[98].

From the above we may conclude that, after the stay of Paul Hermann, Van Reede's visit to South Ceylon and his instruction of 1691 were the decisive impulse for the further development of the Dutch knowledge of the flora of Ceylon.

BENGAL 1686-1687

On 6 December 1685 Van Reede left South Ceylon and sailed to Bengal, where he arrived at the end of January and was to remain until February 1687. In Bengal Van Reede made a clean sweep of the Company's servants in order to suppress the notorious corruption there. He stayed mainly in Ougli on the Ganges, the main settlement of the Company there[99].

The rigorous action of Van Reede during the inspection of Bengal struck terror within the Company. Lords XVII, however, on the whole showed great satisfaction with his activities. Huydecoper observed that Van Reede was 'like an Angel sent to us from heaven', but he had also learned that his victims were bent on revenge[100]. At the background of this was probably the strong suspicion that some members of the High Government in Batavia had a hand in the Bengal corruption. In 1687 Lords XVII, under the chairmanship of Huydecoper, engaged in a violent discussion about the reorganization of the High Government, during which Van Reede was pushed to the fore as a candidate for the office of governor-general, but some of the Lords preferred that he should first return to the Netherlands to report[101].

In Bengal, too, Van Reede concerned himself with natural history. He presented the Bengal king Chahestachan with a tusked elephant, which he had specially sent for from Ceylon. In Bengal he bought falcons for the Ceylonese king Radja Sinha, including an exceptional specimen called "Basij"[102].

As mentioned before, from Bengal Van Reede also sent

seeds to the Netherlands. Jan Commelin described two species which were successfully grown, after arrival in 1688, in the Amsterdam Municipal Garden: *Alcea Bengalensis* and *Ficus bengalensis*[103]. Paul Hermann probably also received seeds from Van Reede, including *Vicia Benghalensis*[104].

COROMANDEL 1687-1689

After the inspection of Bengal Van Reede first sailed back to Colombo, which he reached in June 1687. In Colombo his nephew Maurits Cesar de la Baye was transferred to the service of the garrison of that city while his secretary Zwaardekroon boarded the "Geldria". Van Reede himself remained on the "Purmer"[105].

Since July 1687 Van Reede continued his inspection in Coromandel, which was to last into June 1689. He spent the first year in Nāgappattinam and the second year in Paliacatte. He left the inspection of more remote factories of Coromandel to his substitute Johannes Bacherus, who was not inferior to his principal as to severity in the fight against corruption. Van Reede used his stay in Nāgappattinam mainly to start the building of the fortress, which was to develop into one of the biggest Dutch fortifications in Asia, and the improvement of the harbour. By doing this he aimed at the elevation of this city to the capital of the Western Quarters. This object was never attained, but after his departure from Coromandel the head-quarters of the Company in this region were transferred from Paliacatte to Nāgappattinam[106].

In Nāgappattinam Van Reede had the opportunity to study the indigo industry and the dyeing of linen. In a long report he wrote as follows about the indigo plant:

'the small trees from which the Indigo (also called aniil) originates are 1½, 2, or 2½ feet high, with many sprays and small, roundish, and very bitter leaves, not unlike the Palm; they bring forth many purple, red, and bluish flowers, which are followed by siliques, filled with many small beans. They grow on Sand and sandy fields, of a saltpetre-like nature, clay-ground not being so good, are sown in January in ploughed earth, and are cut off at four fingers above the earth'.

And after a description of the harvesting of the leaves he continued with a report of the process of preparation of the dye, which he had caused to be carried out 'here, and in my room'[107].

It was a curious coincidence that during Van Reede's presence in Coromandel the German botanist Engelbert Kaempfer (1651-1716) was twice in Nāgappattinam for a few days. In the years 1684 to 1688 Kaempfer as a surgeon was in the service of the Company's factory in Gamron (Bender Abbas) in Persia. During his residence in Persia he studied the Persian language, geography, archaeology, and history as well as the flora and fauna of Persia. At that time he also wrote a monograph on the Persian date-palm[108]. In June 1688 Kaempfer left Persia on board the "Capelle" with Batavia as his ultimate destination, whence he was to undertake his famous voyage to Japan[109]. In the waters of Ceylon and southern India Kaempfer's ship got involved in the complicated traffic between the numerous Dutch settlements. Kaempfer moved about successively between Tuticorin, Nāgappattinam, Colombo, Cochin, and Nāgappattinam again[110].

His first stay in Nāgappattinam lasted from 23 September to 4 October 1688; he then studied the training of the cobra[111], by which he was fascinated. In June 1689 Kaempfer was again in Nāgappattinam, where Van Reede was about to leave for Tuticorin[112]. Kaempfer then availed himself of the opportunity to submit to the commissioner-general, in a letter of 10 June 1689, the financial problems with which he was faced. Apparently the fame of Van Reede as the author of Hortus Malabaricus had also reached Kaempfer, for along with the above-mentioned letter he sent him a copy of his monograph on the Persian date-palm[113]. However, it is not known, though it is rather plausible, that Van Reede and Kaempfer then also met in Nāgappattinam. At all events Kaempfer alleged in a letter to Father Raphael du Mans, who lived in Persia, that Van Reede had offered him a post as upper surgeon in Coromandel[114].

NORTH CEYLON 1689-1690

The next region to be inspected by Van Reede was the possession of the Company in the northern part of Ceylon. He stayed there from 23 June 1689 to 2 March 1690. His principal station was Jaffnapatnam[115]. In North Ceylon Van Reede devoted much attention to trade. He stimulated the elephant trade and put new life into the declined pearl-fishery by the discovery of new pearling-grounds. As elsewhere, he discharged incapable servants, took measures for the improvement of the fortifications, and made reforms in the finances. One of the most interesting things brought about by Van Reede was the foundation of the seminary of Jaffna in February 1690. This school was intended for the training of Singalese as clergymen, teachers, catechists, clerks, and interpreters, and a good deal of scope had been made for the teaching of Protestant theology and the native languages[116]. The teaching of the native languages, Singalese and Tamil, will have influenced the fairly correct rendering of the native names of plants which the eighteenth-century surgeons were to use[98].

Meanwhile Lords XVII were beginning to show some anxiety about Van Reede's long absence, and moreover they were losing sight of his activities[117].

TUTICORIN 1690-1691

After North Ceylon Van Reede inspected Tuticorin from April 1690 to January 1691. On that occasion he also visited Alur in Travancore. Meanwhile he had exchanged the ship the "Purmer" for the "Dregterland".

Figure 17. Mausoleum of Hendrik Adriaan van Reede tot Drakenstein at Surat. Photograph by L.C. Rookmaker.

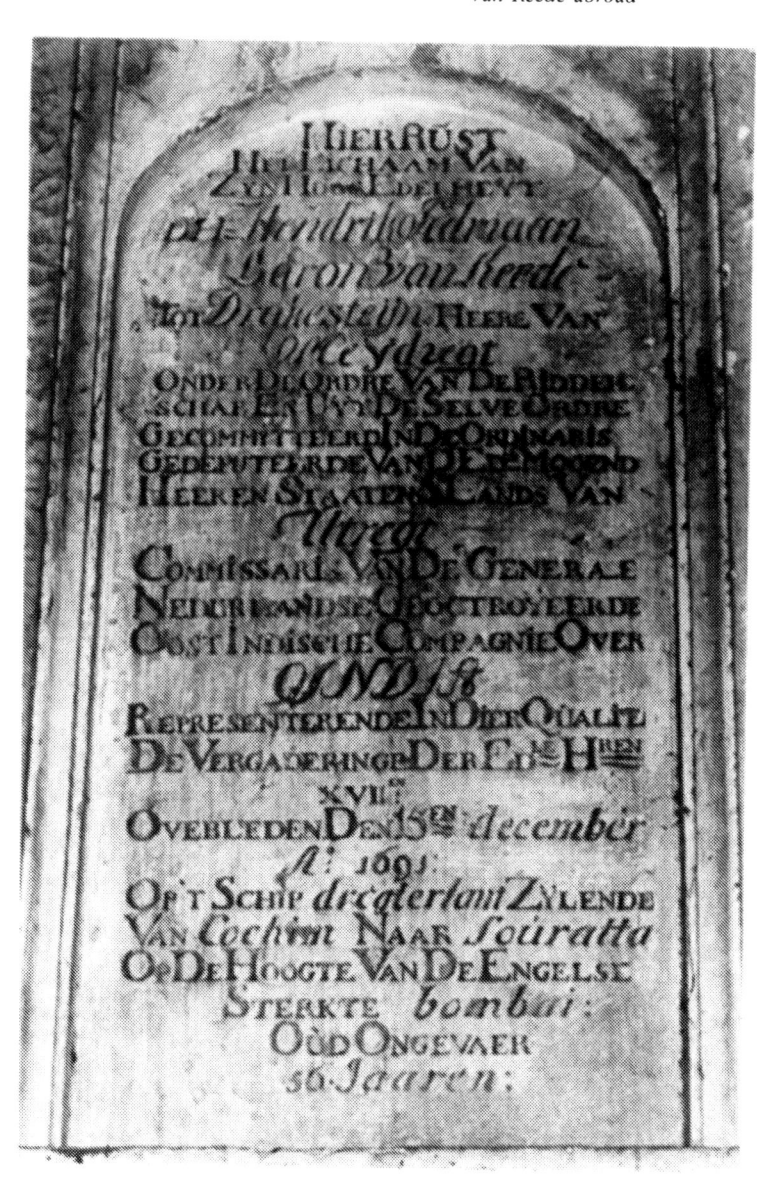

Figure 18. Epitaph of Hendrik Adriaan van Reede tot Drakenstein in his mausoleum at Surat. Photograph by L.C. Rookmaker.

MALABAR 1691

From Tuticorin Van Reede went to Malabar. He appeared there on the roads of Cochin on 10 February 1691[119], and landed on 18 February[120]. A good deal had changed in Malabar since Van Reede had left this commandment in 1677. His friend, rāja Vīra Kĕrala Varma, had died in 1687[121]. His successor Rāma Varma, rāja from 1687 to 1693, had been involved in difficulties since 1690, both with the Company and with Van Reede's former opponent, the Vettatu prince Gōda Varma. This so-called Vettattu revolt raged in the northern part of Malabar. Van Reede spent five months of his stay in Malabar in suppressing the revolt by force. In this period he issued the previously discussed instruction of 5 March 1691 on the collection of plants in the Western Quarters[122].

During the war, however, Van Reede fell ill, and from 14 October 1691 the commander of Malabar, Isaac van Dielen, had to take over the command of the campaign from him[123]. He suffered from an intestinal complaint[124] and did not have at hand, as Van Dielen had two years previously, the medical assistance of an experienced medical practitioner such as Kaempfer[125].

Van Reede probably anticipated that he would no

Figure 19. Funeral procession of Hendrik Adriaan van Reede tot Drakenstein at Surat on 3 January 1692. Etching by Jan Luyken from Havart 1693. Atlas van Stolk 2873.

longer be able to complete his mandate as commissioner-general. At all events he wanted to be on the safe side, and already on 17 September 1691, in Cochin, he made his secretary Zwaardekroon draw up his last will, in which he nominated his daughter Francina his sole heir[126].

The illness seems to have put him in an inconstant disposition. On 23 November he composed an instruction for commander Van Dielen, in which he regretted that owing to want of time and to illness he had not succeeded in dispatching many urgent affairs in Malabar. The same evening he still had a long conversation with the commander, without informing him that his ship "Dregterland" was ready to sail. The next day Van Reede sailed off from Cochin for Surat, to the great astonishment of Van Dielen[127].

DEATH AND FUNERAL

The voyage from Cochin to Surat was dominated by the change for the worse of Van Reede's illness and his rather sudden death off Bombay on 15 December 1691. His last weeks were described as follows in a letter of 27 April 1692 of the council of Surat to the Governor-General and the Council of India[128]:

'after His Honour, the commissioner Van Reede, of blessed memory, had reduced most of the rebellious Malabar princes to obedient submission, he left, while ill, on 24 October of the past Year by the ship Dregterland from Cochin to Surat, and day by day his illness became worse during the voyage, until at last, on 15 December next, almighty God called him to his eternal home off the English fortress of Bombay from the midst of his important project; before the hour of his decease he had not been able to suspect that his death was so near. He had not made any dispositions'.

On 24 December, nine days after Van Reede's death, the "Dregterland" appeared on the roads of Surat. The secretary of the commission, Hendrik Zwaardekroon, sent a letter to the Company's factory, in which he communicated the decease of the commissioner-general[129]. The body of Van Reede was put ashore and embalmed here[130]. In the meeting of 26 December the council of Surat decided, 'at the earnest request' of Van Reede's daughter Francina, to provide for a pompous funeral according to 'this glorious national character' and the important office of the deceased. At the same time it was decided to erect a tomb on his grave (Figs.17-18). Pending the ultimate decision from Batavia they took no responsibility for the expenses of the funeral and the tomb, but they were willing to advance them to Francina van Reede[131].

The preparations for the funeral took a great deal of

time. Thus it was only on 31 December that it was decided to render military honours to the deceased commissioner-general in the following to his grave[132]. The funeral took place on 3 January 1692 in the presence of numerous representatives from Dutch, Indian, Jewish, Armenian, and English circles, as also the ambassador of Abessinia (Fig.19)[133]

Notes

1. The journal of the voyage, from 11 December 1684 to 19 April 1685, is to be found in VOC 1421:545-579v. The squadron consisted of the big ship "Bantam", the yacht "Purmer", and the flutes "Adrichem" and "Eemnes".
2. Raven-Hart 1971 vol.2:254. After his governorship of Ceylon, 1675-1679, Van Goens Jr. for a short time had been commissioner-general of India, 1679-1680, but in the exercise of that office he had got into the bad books of Lords XVII. In 1681 he returned to the Netherlands. After the death of his father, in 1682, he returned to the East as a member of the Council of India (Wijnaendts van Resandt 1944:62; VOC 320, n.p., letters of 20 June and 29 October 1680 of Lords XVII to the Governor-General and the Council of India).
3. About Van Reede's action against Van Goens Jr., see more in detail Hulshof 1941:9-10,23-88. On 7 May 1685 Van Goens embarked on the "Ridderschap", which put to sea on 10 May (ibid.:74-75,86).
4. Hulshof 1941:103-149.
5. See chapter 2, section Cape of Good Hope 1657.
6. Hulshof 1941:140-141; Valentijn 1726 vol.5:135.
7. See chapter 2, section Cape of Good Hope 1657.
8. See chapter 2, section First captain and sergeant-major of Ceylon 1667-1669.
9. Letters of 31 January and 15 November 1669 of the Governor-General and the Council of India to Borghorst and the council of the Cape of Good Hope (VOC 893:135-136,895-899). Jacob Borghorst was commander of the Cape from 1668 to 1670.
10. VOC 4011:80, letter of 7 January 1675 of Godske and the council of the Cape of Good Hope to the Governor-General and the Council of India. We have already met Isbrand Godske as commander of Malabar, 1665-1667. After his directorate of Persia, 1667-1679, he had returned by way of Batavia to the Netherlands, upon which he was appointed governor of the Cape, 1672-1676 (Wijnaendts van Resandt 1944:180-181,245).
11. VOC 4014:297-297v, letter of 28 August 1678 of Crudop and the council of the Cape of Good Hope to the Governor-General and the Council of India. Hendrik Crudop was provisional governor of the Cape after the death of Joan Bax van Herentals (29 June 1678) until the arrival of Simon van der Stel (14 October 1679) (VOC 4014:287v-288; VOC 4015:290-290v).
12. VOC 4013:176v,206, letter of 12 July 1677 of Bax and the council of the Cape of Good Hope to the Governor-General and the Council of India.
13. VOC 4014:309, letter of 31 October 1678 of Crudop and the council of the Cape of Good Hope to the Governor-General and the Council of India.
14. VOC 4015:188v-189,192, letter of 31 May 1680 of Van der Stel and the council of the Cape of Good Hope to the Governor-General and the Council of India.
15. VOC 4013:190-190v, letter of 17 April 1677 of Bax to Laurens Visser.
16. Curiously enough, the exact dates of Hermann's visits to the Cape have never been established. On 15 November 1671 he was engaged by the Chamber of Amsterdam to be employed as physician in Ceylon (VOC 239, n.p.). It is not known by what ship he sailed to Ceylon. Two big ships with destination Ceylon, the "Gouda" and the "Amersfoort", had left long before Hermann's appointment. Perhaps he sailed in a small flute, the "Ipensteyn", which put to sea on 11 December 1671, bound for Ceylon. This ship arrived on the roads of the Cape of Good Hope on 19 April 1672 and at Tuticorin on 28 July 1672 (Raven-Hart 1971 vol.1:159,165; Bruijn et al. 1979 vol.2:174-175). Perhaps an accurate study of Hermann's Cape herbarium of 1672, which is in the British Museum for Natural History, Sloane Herbarium 75, can show whether Hermann really collected plants in the month of April 1672 at the Cape. After a stay of almost eight years in Ceylon Hermann left that island by the ship "Sumatra" on 3 January 1680 (VOC 1343:297-297v,393-393v, letter of 3 January 1680 of the governor and the council of Ceylon to Lords XVII). The "Sumatra" arrived at the Cape in March 1680 and left the roads again on 29 March (Raven-Hart 1971 vol.2:225). Consequently, Hermann's second visit to the Cape must have taken place in March 1680.
17. For more detailed discussions of Hermann's Cape collections, see Gunn & Codd 1981:184-185, and Heniger 1969: 530-534.
18. Gunn & Codd 1981:342.
19. On Huydecoper, see chapter 3, section Appointment as Commissioner-General of the Western Quarters.
20. Joan Bax was appointed on 2 November 1674 by Lords XVII as governor of the Cape, see Huydecoper's letter to him of even date (RAU, Records Huydecoper Family 55, Letter-book 1671-1674). Bax did not arrive at the Cape until 3 January 1676 and was installed on 14 March next (Raven-Hart 1971 vol.1:179).
21. RAU, Records Huydecoper Family 55, Letter-book 1671-1674, letters of 3 September 1671, 26 September 1672, and 27 September 1673 of Huydecoper to Bax. From Ceylon Bax shipped pots with Mango and Acacia, and coconuts, "taloets", oil from the fruit of the cinnamon-tree, and living civet-cats, parakeets, and Negombo devils.
22. RAU, Records Huydecoper Family 56, Letter-book 1675-1678, letters of 30 November 1676, 7 November 1677, and 19 November 1678 of Huydecoper to Bax. The name of the draughtsman of the water-colours was not mentioned by Huydecoper. The Cape herbarium which Bax sent to Huydecoper in two boxes is presumably the same as the two-volume *Herbarium Vivum*, which was sold in 1704 at the public sale of Huydecoper's son Joan (1656-1703) for 34 guilders; the catalogue of the sale contained the note: 'The intending Buyers should know that all these are Herbs which have grown and been dried with their Flowers at the Cape of Good Hope'. Apparently Hans Sloane was the buyer of this herbarium, for in the British Museum for Natural History the Sloane Herbarium contains nos 77 and 78: "Plants gathered at the Cape of Good Hope which belonged to Mr. Meerseveen and were bought at the auction of his books in Holland"; about the shipments of Bax to Huydecoper, see in more detail Rouweler 1982:17-19,27-29.
23. Jan Hendriksz. de Beet was christened on 14 December 1645 in Maarssen, the same parish to which Huydecoper's seigniory Maarsseveen also belonged (RAU, DTB Maarssen 31:55). In his letter of 1 May 1676 Huydecoper introduced him to Bax as an experienced kitchen-gardener

and drew attention to his knowledge of the laying-out of the parks and gardens of the country-seats along the Vecht (RAU, Records Huydecoper Family 56, Letter-book 1675-1678). De Beet worked until 1697 as gardeners' boss of Rustenburg (VOC 4038:854v, Muster-roll of officers, soldiers, and sailors, of 1 July 1697).

24 RAU, Records Huydecoper Family 56, Letter-Book 1675-1678, letter of 30 November 1676 of Huydecoper to Bax. In 1695 De Beet gave a survey of the impressive size of the Rustenburg plantations (VOC 4034:250-253, inventories of Rustenburg and Steenberg by Jan Hendriksz. de Beet, dated 5 May 1695).

25 The wish of the prince to be presented annually with a variety of natural curiosities by the Company was already mentioned in the section about Van Reede's commandership of Malabar.

26 RAU, Records Huydecoper Family 56, Letter-book 1675-1678, letter of 25 September 1676 of Huydecoper to Daniel Desmarets, and of 30 November 1676 to Bax.

27 VOC 4012:33, letter of 14 March 1677 of Bax and the council of the Cape of Good Hope to Lords XVII; VOC 4013:785. The rhinoceros died during the voyage. The preserved skin was presented by the Company to Leiden University and exhibited in the "Ambulacrum" in the botanical garden (VOC 240, Resolutions of the Chamber of Amsterdam 1675-1678, n.p., meeting of 29 July 1677; Rookmaker 1976:88-89).

28 RAU, Records Huydecoper Family 56, Letter-book 1675-1678, letter of 25 September 1676 of Huydecoper to Desmarets; VOC 240, Resolutions of the Chamber of Amsterdam, n.p., meeting of 13 July 1677.

29 VOC 108, Resolutions of Lords XVII 1675-1680, n.p., meeting of 25 October 1678.

30 RAU, Records Huydecoper Family 56, Letter-book 1675-1678, letter of 19 November 1678 of Huydecoper to Bax.

31 See chapter 2, section No calling at the Cape 1678.

32 VOC 108, Resolutions of Lords XVII 1675-1680, n.p., meeting of 18 March 1679. When Bax was appointed, the Cape Colony had already been reduced from a government to a commandment. Van der Stel arrived on 12 October 1679 by the ship "De Vrije Zee" at the Cape and was installed two days later (VOC 4015:289v-290v, Journal of the Cape of 1679).

33 RAU, Records Huydecoper Family 57, Letter-book 1679-1682, letters of 22 May and 20 October 1680 of Huydecoper to Van der Stel.

34 RAU, Records Huydecoper Family 57, Letter-book 1679-1682, letter of 8 November 1681; ibid 58, Letter-book 1683-1686, letter of 15 January 1683, both of Huydecoper to Van der Stel.

35 RAU, Records Huydecoper Family 57, Letter-book 1679-1682, letter of 29 July 1681 of Huydecoper to Mrs De Flines (Agnes Block) and letter of 1 August 1681 to Desmarets; ibid., Diary, dated 2 and 3 August 1681.

36 In their letter of 18 October 1677 to the High Government Lords XVII wrote that the previous summer they had found the homeward-bound fleet 'covered and obstructed in such a way with boxes, and that in such great numbers, as if they were whole gardens resulting in so great a weakening and damaging of the ship by all the weight on top that we were obliged to write off and prohibit herewith the sending of all those cuttings, trees, and plants, as well as the making of those gardens, and this now and then for the private use of those Chiefs'. In their letter of 29 October 1680 Lords XVII sharpened once more their prohibition of excessive transport of private goods (VOC 458, Letter-book of Lords XVII 1673-1681, n.p.).

37 RAU, Records Huydecoper Family 57, Letter-book 1679-1682, letters of 22 May 1680 and 8 November 1681 of Huydecoper to Van der Stel.

38 RAU, Records Huydecoper Family 58, Letter-book 1683-1686, letters of 15 January and 13 December 1683 of Huydecoper to Van der Stel, and of 27 May 1683 to Jacob Breyne. From the letter of 13 December it appears that the first consignment of Cape bulbs and seeds for the use of the Amsterdam Municipal Garden arrived in 1683.

39 Wijnands 1983:213.

40 In their annotated facsimile edition of the *Flora Capensis* Gunn & Du Plessis (1978) were not aware of the possibility that this codex might be a copy. Oliver (1980a,b) showed that this *Flora Capensis* is a copy of the florilegium in the Botanical Research Institute (BRI) in Pretoria. Rouweler (1982), on the basis of letters of Huydecoper to Jacob Breyne, came to the conclusion that the *Flora Capensis* largely consists of the set of copies which Jacob Breyne received from Huydecoper (see also Kuijlen 1982:117) and that the BRI florilegium is possibly the original Cape codex of Huydecoper.

41 RAU, Records Huydecoper Family 58, Letter-book 1683-1686, letter of 7 October 1685, and ibid. 59, Letter-book 1686-1687, letter of 15 December 1686, both of Huydecoper to Van der Stel.

42 VOC 28, Resolutions of Lords XVII, n.p., meeting of 28 November 1684. The name of Fagel's collector is not known to me. Considering the date of consent of Lords XVII, it is plausible that this person sailed to the Cape by the same ships which also took Van Reede to the Cape.

43 Hendrik Claudius (c.1655-before 1697), whose career in the service of the Company is known only fragmentarily, was contracted in November 1681 in Batavia by Cleyer. Claudius sailed by the "Africa" to the Cape, where he arrived on 16 February 1682. In the history of the Cape natural history he became famous for his water-colours of plants and animals which he made during Van der Stel's expedition to Namaqualand in 1685 (Gunn & Codd 1981:117-119; Van Nuys 1978:23-24; Raven-Hart 1971 vol.2: 235).

44 For Van Reede's report of this event, see Hulshof 1941:28-29.

45 RAU, Records Huydecoper Family 59, Letter-book 1686-1687, letter of 15 December 1686 of Huydecoper to Van der Stel, in which the former reacted furiously: 'not being able to understand that Mr. Van Goens should have ordered severely that they would have to be extirpated, while the request of the Lords of Amsterdam and Governors about such a trifle ought to have counterbalanced his own, the more so as the request has taken place by the order of the meeting of the XVII'.

46 Hulshof 1941:14,24-30.

47 On the basis of several dimensions given in his report on the Company's garden it can be calculated that the average gradient of the garden was nearly 20%.

48 Hulshof 1941:169-170.

49 Van Reede did mention the length of the garden, 'about 200 roods', but not its width. Owing to the somewhat irregular plan the estimates of the area by the visitors vary, but they fluctuate round 40 acres (Raven-Hart 1971 vol.2:502).

50 The head gardener of the Company's garden at that time was Cornelis Heermans, of Haarlem, who according to the lists of the board money was the manager for a short time

only, 1684-1685 (VOC 4021:825; VOC 4022:539; Hulshof 1942:352-353).
51 The inspection of Rustenburg took place on 16 May 1685 (Hulshof 1941:101-112).
52 Karsten 1951:64-66.
53 Unless otherwise stated, the following considerations of Van Reede about deforestation of the Cape have been taken from his reports of 5, 21, and 25-28 May 1685 (Hulshof 1941:72-73,129-130,139-146).
54 See above, notes 23 and 24.
55 Hulshof 1941:93-94,137-139,233-234. Van Reede seems to have been the first visitor of the Cape to have empathically drawn attention to the barking and the ears of the 'wild dog', *Lycaon pictus*; see Raven-Hart 1971 vol.2:499, Wild Dogs.
56 Hulshof 1941:151.
57 Hulshof 1941:167.
58 Raven-Hart 1971 vol.2:181, translated from Tachard's *Voyage de Siam* (1686).
59 The research on the drawings of Claudius is still proceeding and comes up against many problems, which can in part only be solved by a thorough comparison of all the known codices of Cape plant drawings from the second half of the 17th century.
60 Hulshof 1941:89-93.
61 For a complete survey of Claudius' water-colours of Namaqualand, see Gunn & Codd 1981:36-38,118-119.
62 Hulshof 1941:226.
63 RAU, Records Huydecoper Family 59, Letter-book 1686-1687, letters of 15 December 1686, 12 April and 26 November 1687 of Huydecoper to Van der Stel.
64 VOC 4025:342-343v, Journal 1688.
65 RAU, Records Huydecoper Family 59, Letter-book 1686-1687, letters of 15 December 1686, 12 April and 26 November 1687 of Huydecoper to Van der Stel.
66 In his *Horti Medici Amstelodamensis Rariorum Plantarum* (1697) Jan Commelin published the following dated Cape plants. He received from Van der Stel in 1687 *Zantedeschia aethiopica* (L.) Spreng. and *Knowltonia vesicatoria* (L.f.) Sims, in 1689 *Trachyandra divaricata* (Jacq.) Kunth, and in 1691 in the Amsterdam Municipal Garden *Myrsine africana* L. blossomed (Wijnands 1983:46-47,140,154,179).
67 VOC 4025:426v-427, letter of 18 April 1687 of Van der Stel and the council of the Cape of Good Hope to Lords XVII.
68 Georg Meister left Batavia on 11 December 1687 by the "Waalstroom". He stayed at the Cape from 29 March to 30 April 1688, and arrived in the Netherlands on 12 August 1688. In Batavia he was a gardener in the private service of Andries Cleyer, with whom he made two voyages to Japan, in 1682/3 and 1685/6. In 1685 Huydecoper requested Cleyer to send plants for the Amsterdam Municipal Garden (Karsten 1951:167; Raven-Hart 1971 vol.2:340-352; Gunn & Codd 1981:46,250; Muntschick 1984; RAU, Records Huydecoper Family 58, Letter-book 1683-1686, letter of 28 November 1685 of Huydecoper to Cleyer).
69 For an English translation of Tachard's information about Van der Stel's expedition to Namaqualand, see Raven-Hart 1971 vol.2:281-294.
70 See the undated French letter to the "Supérieures et Directeurs des Missions étrangères" which was intercepted by Van der Stel (VOC 4025:570).
71 RAU, Records Huydecoper Family 59, Letter-book 1686-1687, letter of 15 December 1686 of Huydecoper to Van der Stel.
72 VOC 4025:4v-7v, letter of 26 April 1688 of Van der Stel and the council of the Cape of Good Hope to Lords XVII, in which he reports on the visit of the French squadron on its way to Siam, which called at the Cape from 9 to 28 June 1687, and on the banishments of Claudius and Cauchetez because of their understanding with the Jesuits. See also Gunn & Codd 1981:118, who suggest, however, that Van der Stel was envious of Claudius' discoveries.
73 RAU, Records Huydecoper Family 59, Letter-book 1686-1687, letter of 26 November 1687 of Huydecoper to Van der Stel: 'Mr. Commelin and I are engaged in sending you everything that is in your list, and also a botanist and a dauber'.
74 VOC 5359:108, Ledger and journal of the "Den Helder", 1687-1688: list of receipts and expenditure of "Hendrik Barent(!) Oldelant", naval cadet.
75 For Oldenland's importance for the Cape botany, see Karsten 1951:72-90; Gunn & Codd 1981:46-47,265-266.
76 RAU, Records Huydecoper Family 59, Letter-book 1686-1687, letters of 15 and 23 December 1686 and 26 November 1687, and ibid. 60, Letter-book 1694, letter of 11 December 1694 of Huydecoper to Van der Stel.
77 VOC 323, Letter-book of Lords XVII, n.p., letter of 27 August 1692 of Lords XVII to Van der Stel and the council of the Cape of Good Hope.
78 Karsten 1951:73, note 54. Jan Hertog, born in Aachen, Germany, left the Netherlands on 20 December 1690 as naval cadet on the "Pampus" and arrived at the Cape on 29 May 1691. With a short interruption, from 20 March 1708 to 31 August 1709, he remained in the service of the Company at the Cape until August 1715. At first he served as second gardener under Oldenland, but he succeeded him in 1697 as head gardener. From 1717 to 1722 he stayed as plant collector for Herman Boerhaave, the then Leiden professor of botany, in Surinam, where he died. Jan Hertog is not to be confused with Willem's son, Pieter Hertog (1695-1728), who composed a herbarium in Ceylon, which partly served as a basis for Burman's *Thesaurus Zeylanicus* (1737) (VOC 5378:98, Ledger and journal of the "Pampus", 1690-1691: List of receipts and expenditure of Jan Hertogh; Gunn & Codd 1981:178-179; Van Beek 1983).
79 Within the context of this study this is not the right place to go into all the complaints of Lords XVII against Van der Stel. He defended his policy at length in his letter of 9 May 1695 with many enclosures, including the reports of Oldenland and De Beet (VOC 4043:1-67,250-256).
80 VOC 112, Resolutions of Lords XVII, n.p., meeting of 6 September 1696. In spite of his discharge Van der Stel had to remain in office until the arrival of a new governor. In 1697 his son Willem Adriaan van der Stel, ex-alderman of Amsterdam, was appointed as such; it was only at the beginning of 1699 that he took over the administration from his father (VOC 323, Letter-book of Lords XVII, n.p., letters of 15 November 1696, 27 December 1697, and 31 July 1698 of Lords XVII to Simon van der Stel and the council of the Cape of Good Hope; Raven-Hart 1971 vol.2:445).
81 Hulshof 1941:241-242.
82 Hulshof 1941: 171-172; RAU, Records Huydecoper Family 59, Letter-book 1686-1687, letters of 15 December 1686 and 12 December 1687 of Huydecoper to Saint-Martin and Ten Rhijne respectively.
83 For a discussion of Van Reede's inspection of South Ceylon, see Gommans 1984:35-41.
84 Van Reede co-operated with him in Quilon (1665) and Tuticorin (1668-1669), see chapter 2.
85 Thomas van Rhee (1634-1701) came to the East in 1659; he was second-in-command in Nāgappattinam from 1674 to

1678, and since 1678 chief of Tuticorin; from 1681 he acted as head administrator in Colombo. After the retirement of Laurens Pijl he succeeded the latter as governor of Ceylon, 1692-1697 (Wijnaendts van Resandt 1944:63-64).
86 The Ceylon governor Rijklof van Goens Jr. and his successor Laurens Pijl repeatedly formulated their disappointment about the departure of Paul Hermann: 'in order that the Honourable Company may be served to their advantage for a long time yet with his plentiful knowledge of the herbs' (VOC 1343:31v, letter of 17 April 1679), 'we shall thus be deprived here of an expert in the knowledge of the native herbs and be obliged to order many drugs from elsewhere, which for the rest are available here in plenty' (VOC 1343:379, letter of 31 October 1679) and 'who through his experience and knowledge of the herbs has rendered the Honourable Company a great service through his presence here' (VOC 1343:393v, letter of 3 January 1680).
87 Den Tonkelaar 1983:25-26.
88 RAU, Records Huydecoper Family 59, Letter-book 1686-1687, letter of 13 December 1686 of Huydecoper to Van Reede in reply to the latter's (lost) letter of 23 November 1685, and letter of 15 December 1686 of Huydecoper to Van der Stel. Van Reede's consignment was shipped by the "Eenhoorn" (ibid., letter of 7 August 1686 of Huydecoper to Alexander de Munck). This ship left Ceylon on 22 January 1686, called at the Cape from 2 to 12 April 1686, and arrived in the Netherlands simultaneously with the return fleet from Batavia on 26 July 1686 (Bruijn et al., 1979 vol.3:122-123).
89 RAU, Records Huydecoper Family 59, Letter-book 1686-1687, letter of 9 December 1687 of Huydecoper to N. Elsevier in reply to the latter's letter of 4 December 1686 from Maturé.
90 RAU, Records Huydecoper Family 59, Letter-book 1686-1687, letter of 6 December 1687 of Huydecoper to Wichelman. Magnus Wichelman (1647-1705) arrived in the East in 1669; since 1674 he worked in Ceylon; thereafter he was commander of Malabar from 1697 to 1701 and director of Persia from 1701 to 1705 (Wijnaendts van Resandt 1944:186; s'Jacob 1976:283, note 1).
91 Bodleian Library (Oxford), MS Sherard 174:51-51v, "Semina à Ceylon a Dno Hermanno. Colombo 6 January 1688".
92 VOC 1447:990,1001v, bills of loading of the "Waterland" and the "Maas", dated 23 January 1689 (Galle).
93 VOC 1478: 1382-1385. For the complete text, see Appendix 4.
94 Wijnands 1983.
95 VOC 6990:1-13v, 'Short Description of the seeds and their properties, as well as their use', of 1697; VOC 6990:993-994v, 'Description of use and effect of the following roots and herbs', of 13 January 1699, by the surgeon Pieter de Hoey.
96 The discovery was made by a German soldier, Dirk Barentsz., during his many years' imprisonment in the interior of Ceylon. In the years 1690-1699 the surgeons Jan Pietersz. Maas, Joan Jacob Bruynink, Antoni Hoepels, and Pieter de Hoey wrote lengthy reports about Acmella and shipped herbarium material of the plant (Den Tonkelaar 1983:25-28,71-93,104-111,116-117).
97 Leiden University Library, BPL 126 D, *Icones mediocriter delineatae et coloribus pictae Plantarum malabaricum, cum descriptione hollandica virum earundem*, 2 vols. In spite of the title the codex does not contain species from Malabar, but from Ceylon. According to a note to it of 22 March 1977 of A.J. Kostermans the codex was made in 1670-1700.

98 The numerous catalogues of the Ceylon seeds are to be found in the Letters Received from Ceylon; see also Van Beek 1983.
99 Veth 1887:139-142; Gommans 1984:33-34; Lequin 1982. Van Reede's first letter from Ougli dated from 7 February 1686 (VOC 1429:921v). Many documents of Van Reede's stay in Ougli are to be found in VOC 1421, 1429, and 1435. When he left Ougli, he drew up an instruction for the director of Bengal on 21 February 1687 (VOC 1435:128-210).
100 RAU, Records Huydecoper Family 59, Letter-book 1686-1687, letter of 2 November 1687 of Huydecoper to Aletta Hinlopen.
101 RAU, Records Huydecoper Family 59, Letter-book 1686-1687, letters of 22 November and 2 December 1687 of Huydecoper to Van Reede and Saint-Martin respectively.
102 VOC 1421:140v-141, letter of 9 December 1686 of Van Reede to Lords XVII. The falcon "Basij" died a few days after arrival in Galle; two other falcons were presented by Laurens Pijl with a solemn letter of 13 February 1687 to the Ceylonese king (VOC 2167:668-670).
103 Commelin 1697:35-36,119-120; Wijnands 1983:143-144,152-153.
104 Hermann 1687:623-624,625 tab.
105 These changes can be inferred from the ship's paybooks of the "Purmer" and the "Bantam": VOC 5325, no.148 ("Hendrick Swaardecroon"), and VOC 5328, no.3. ("Joncker Hendrick Adriaan van Rheede tot Drakesteijn, etc.") and no.226 ("Maurits Cesar du Thel").
106 Veth 1887:142-153. Documents about Van Reede's stay in Coromandel are to be found in VOC 1435, 1438, 1439, 1477 (on fol.418-697 the instruction of 29 November 1689), and 1478.
107 VOC 1449:57-79v, letter of 1 December 1688 of Van Reede to Lords XVII.
108 Dictionary of Scientific Biography vol.7 (1973):204-206.
109 Meier-Lemgo 1968:144-145.
110 Meier-Lemgo 1968:151-157.
111 Kaempfer 1712:565-573; Meier-Lemgo 1933:114-118.
112 Van Reede wrote letters from Nāgappattinam on 15, 16 and 17 June 1689 (VOC 1478:490,500; VOC 1477:246,282).
113 Meier-Lemgo 1965:287.
114 Meier-Lemgo 1965:288-289.
115 For Van Reede's inspection of North Ceylon, see Gommans 1984:44-57 and VOC 1477-1478.
116 Van Goor 1978; Gommans 1984:46. Van Reede appointed as first governor of this seminary Johannes Frederik Stomphius, a brother-in-law of Paul Hermann (VOC 1479:494-498,512-527).
117 Gommans 1984:48.
118 The visit to Tuticorin and Alur is not mentioned by Veth 1887. The documents relating to this are to be found in VOC 1478. On 13 October 1690 Van Reede was on board the "Dregterland" on the roads of Jaffna (VOC 1478:980), during a short trip to Nalur in North Ceylon (VOC 1478:501,549,737,768,784) in September of that year.
119 VOC 1478:1353.
120 s'Jacob 1976:LXXV,221, note 2.
121 Vīra Kērala Varma died on 10 February 1687 at the age of about 73 years (VOC 1431:81v, letter of 30 April of Vosburg and the council of Malabar to the Governor-General and the Council of India).
122 See section South Ceylon 1685.
123 About the Vettattu revolt and Van Reede's campaign against Cranganūr, Parūr, Alangādu, the Ainikkur Nambidi and the Anchi Kaimals in the period of 18 March to 26

October 1691, see s'Jacob 1976:221-222,246-247,258-259. Isaac van Dielen (1652-1693) came to the East in 1668, since 1669 he served in Ceylon, and he was commander of Malabar, 1688-1693 (Wijnaendts van Resandt 1944:185; s'Jacob 1976:221).

124 When the body of Van Reede was balsamed in Surat, an incipient inflammation of the large intestine was diagnosed (Veth 1887:155).

125 When Kaempfer stayed in Malabar from December 1688 to May 1689, the private household of Van Dielen was ravaged by an infectious disease. Kaempfer succeeded in setting the commander on his legs again (VOC 1448:373v-374; VOC 1474:201; Meier-Lemgo 1952:187).

126 VOC 1505:420-422.

127 s'Jacob 1976:LXXV,221-251 (the instruction for Van Dielen, dated 'In the ship Dregterland, lying ready to sail to Surat on 23 November 1691'); and VOC 1478:1586.

128 VOC 1520:12-12v.

129 VOC 1529:397v.

130 VOC 1520:12v.

131 VOC 1529:92, resolution of the directors and the council of Surat, of 26 December 1691. The bond of Francina is dated 16 February 1692 in Surat. The funeral ran into a cost of 3,073 9/32 silver rupees, while the building of the tomb was estimated at 6,000 ± 600 to 700 silver rupees (VOC 1505:772-772v). For a description of Van Reede's tomb, see Greshoff 1906.

132 VOC 1529:402v, Journal of Surat, dated 31 December 1691, in which permission was given for some members of the militia and 60 to 70 native soldiers to accompany the funeral procession.

133 The report of the funeral and the disposition of the funeral procession is to be found in VOC 1529:404v-408v, Journal of Surat, dated 3 January 1692. For the complete text, see Appendix 6.

5
Epilogue

In the introduction I have already referred to the fact that apart from the drawings of Hortus Malabaricus none of Van Reede's personal possessions have been preserved. The following survey of the lives of his daughter and grand-daughter reflects the vain search for his inheritance.

FRANCINA VAN REEDE

In his will of 17 September 1691 Van Reede nominated Francina van Reede his sole heiress[1]. The true identity of this lady is rather obscure. The opinions of the biographers are divided: she is said to have been a bastard daughter, or an adoptive daughter, or even a legitimate daughter of Van Reede. In his last will Van Reede stated that he was a bachelor. This implies that Francina cannot have been his legitimate daughter[2].

In this same will Van Reede stated further about Francina:

'whose "gevader" he, testator, was, whom he educated, and whom he had always considered, and still considers, as his adopted daughter'.

The term "gevader" may mean that Van Reede was her godfather and that he had presented her for baptism. At all events Francina had been adopted by Van Reede.

In 1675, too, when a commotion arose in the church council of Cochin about the confirmation of Francina, it was submitted that Van Reede had 'assumed all the best and the well-being of this lady as if she were his daughter' and 'as being a Nonje in his house adopted as his daughter and dear to him'. Moreover, she was then also referred to as "Nonje Francina van Rheede" and "Juffr Francina van Rheede, alias Sipkens"[3]. Constantijn Huygens Jr. as well as Joan Huydecoper later called her the "nuna" or "nunia" of Van Reede. In the original meaning of the word they may have meant by this a descendant of the Portuguese who lived under the rule of the East India Company. But the same Huygens knew the story that Francina's mother was a married Indian woman, and that three men, including Van Reede, could be considered to be the father[4].

The solution of this story consists in Francina's "alias" Sipkens, a family name which nowadays still occurs in the province of Groningen. In my opinion Francina may have been born from the marriage of a Company's servant Sipkens, unknown to me, with a Malabar woman. Unfortunately the baptismal registers of Malabar from this period are missing, so that no further date of Francina's birth can be given. Her adoption by Van Reede may be explained by the fact that as a commander, in a letter of 1670, he wrote that the number of orphans in Malabar was greatly increasing and that the Protestant church in Cochin could not afford an orphanage, so that the Company had to provide for many children[5]. Possibly the disconsolate situation resulting for the half-castes induced Van Reede to adopt the girl Francina. Later this parentage, which was apparently obscure in the opinion of outsiders, played Francina foul tricks as heiress of Van Reede.

In so far as can be ascertained, Francina always accompanied her father on his voyages. Thus, she also stayed in Utrecht in the years 1678-1684, when her father sat on the States of Utrecht. It has been mentioned already previously that one day after having undertaken the commission to India, on 27 October 1684, Van Reede engaged Sandrina Reets as her companion[6]. On 11 December, the day on which Van Reede took leave of Lords XVII, Huydecoper previously paid a special visit to her[7].

Probably she accompanied her father during the whole of the inspection expedition along the Company's factories in India[8]. During his stay in Coromandel, from July 1687 to June 1689, the curious affair took place which led to the erection of the memorial in Trincomalee in November 1687[9]. She also took part in Van Reede's last voyage, from Cochin to Surat, and witnessed his death on 15 December 1691 off Bombay. It was Francina's 'earnest request' that her father should be buried in Surat 'with the fitting honours and reputation' and that a tomb should be erected on his grave 'according to the manner of this country and the quality of His Honour'. As said before, she undertook, as the sole heiress of her father, to pay the expense of his funeral and his tomb. The deed was drawn up on 16 February 1692 in the Company's garden on the outskirts of Surat[10].

It is questionable what happened in Surat with the possessions and the papers left by Van Reede, in so far as they were present on the ship the "Dregterland". Apparently the commission papers were entrusted to the secretary Hendrik Zwaardekroon. On 30 April 1692 he left with these papers on board the "Maas" for Tuticorin[11]. Since June of that year, he stayed in Colombo, pending further orders of Lords XVII[12]. Presumably he forwarded the commission papers to Batavia, for there, until 1747, letters and reports of Van Reede concerning his inspections of Jaffnapatnam and Malabar were deposited[13]. We must assume that Van Reede's private property came into the hands of Francina. At all events the will of her father was in her possession[14].

Meanwhile in Surat the rumour got about that the Malabar princes had revolted again. For that reason on 26 February 1692 the "Dregterland" was sent to Malabar. On board this ship were Francina van Reede and some members of the commission, bound for Ceylon[15]. Presumably Francina arrived in Ceylon in the course of April 1692. One of the first things she did there was to request the government of Ceylon to confirm the last will of her father. The governor and the council informed her in their resolution of 2 May 1692 that they could not do so, because the goods of the deceased were found to lie not in Asia, but in the Netherlands. She repeated her request a few days later, on 8 May, but the government of Ceylon again refused to comply with her request. On the other hand the government presented to her the bill of the credits of her father and of the expense of his funeral and tomb, amounting together to over 28,000 guilders. But Francina had already paid back 8,500 rupees into the Company's purse of Colombo and Galle, and for the rest she declared herself 'unable' to pay. She proposed to settle the final account in the Netherlands, and this was done[16].

In all probability in the course of 1692 Francina married in Colombo her cousin Maurits Cesar de la Baye, who was captain of the Ceylon militia at that time. On 15 December 1692 they made a mutual last will. The marriage lasted only a short time, for he died there already on 14 February 1693[17].

After the death of her father and her husband there was no longer any reason for Francina to stay in Asia, and even before 16 May 1693 she permanently left this part of the world by the "Purmer"[16], on her way to an uncertain future in the Netherlands. The ship the "Purmer" joined the return fleet, which put into port at the Cape of Good Hope on 12 June 1693.

Francina will probably have arrived in the Netherlands in the autumn of 1693. Shortly after her arrival she went to Utrecht, where she settled the inheritance of her deceased husband together with her father-in-law Paul de la Baye. She showed him their last will and an estate inventory, with which in a notarial deed of 14 November 1693 OS he declared himself satisfied[17].

In Amsterdam the settlement with the Company followed. On 4 February 1694 she received 5,379 guilders of her husband's pay[19]. The final account of her father was somewhat more complicated. At the time of his death Van Reede was entitled to a salary of 83,700 guilders. In Colombo he had taken out 13,814 guilders, while his funeral and tomb together had cost 14,697 guilders. Moreover, on his departure in 1684 Van Reede had taken out 8,000 guilders. Consequently, on 19 March 1694 she was paid the amount of 47,188 guilders[20]. Apparently at this moment the Company had not yet taken a decision whether it would make a contribution to the expense of Van Reede's funeral tomb. Fortunately the matter was settled soon afterwards. Lords XVII offered Francina 5,000 guilders as a compensation. She accepted this in the form of 'a piece of silver ... in which the merits of the deceased are to be engraved as a permanent memorial', as Huydecoper wrote in a letter of 6 May 1694 to Saint-Martin[21]. However sad the death of her father and her husband may have been, Francina had inherited from them a considerable capital and the manorhouse of Mijdrecht which made her a wealthy woman.

Already in December 1693 it became known that she intended to marry again[4]. The prospective husband was her Utrecht cousin Antoni Carel van Panhuizen tot Voorn, a sister's son of Van Reede, who was captain in the States' army. On 11 January 1694 OS they drew up their marriage contract[22] and on 28 January they were married in the French Church in Utrecht[23].

The couple settled in Utrecht, where in the years 1694 to 1698 three of their children were born[24]. The family presumably lived on the Nieuwe Gracht, near St. Pieters Kerkhof, where Van Panhuizen owned several houses[25]. In these years Francina became involved in a sordid affair, which threatened to ruin her existence. In his last will her father had destined, among others, a legacy of 2,000 guilders for his sister Agnes, the mother of his nephew and friend Hendrik Jacob van Tuyll van Serooskerken. Mother and son, however, died soon after Van Reede himself, even before this testamentary disposition could be carried into effect[26]. The children of Hendrik Jacob at that time were under the guardianship of Godard Willem van Tuyll van Serooskerken, lord of Welland, to whom Van Reede had dedicated formerly, in 1689, volume 9 of Hortus Malabaricus. Now the estate left by Hendrik Jacob was in a deplorable condition, so much so that his children could only accept his inheritance under benefice of inventory. The guardian, the lord of Welland, made a point of settling the finances of the children entrusted to him as favourably as possible. He was involved in a series of proceedings to ward off creditors and to collect outstanding debts[27]. In this situation the guardian at first accepted the legacy of Van Reede. But on second thoughts he refused it[28] and, what was more, he challenged the validity of Van Reede's will. In fact, it was extremely unusual in the Dutch law of succession of those days for an adoptive child of a testator to be nominated the sole heir while heirs-at-law were still living. In addition the rumour which already at the end of 1693 got about that Francina was really a bastard daughter of Van Reede, induced the lord of Welland to decide on 2 September 1695 to institute proceedings against Francina in order to secure the whole inheritance of Van Reede. On 11 November 1696 the Court of Utrecht, the supreme court of justice of that province, decided to refer the case to a commission. On the basis of the report of this commission the court gave judge-

Figure 20. House Te Vliet at Lopikerkapel in 1701. Etching by Daniel Stoopendaal after a drawing by C. Specht. RAU, Topographical Atlas 1120:18*.

ment on 27 March 1697. The manor-house of Mijdrecht was awarded to the children of Hendrik Jacob, but for the rest their claim was rejected[29].

A few years later, on 27 April 1700, Reinoud Gerard van Tuyll van Serooskerken was enfeoffed with the manor-house. That for him and his guardian the issue in the proceedings was the financial profit rather than justice or the political importance of a manor-house appears from the fact that on the same day Mijdrecht was transferred to Everard Booth[30]. Francina and her husband probably accepted the court's judgement with mixed feelings. On the one hand the loss of Mijdrecht prevented a registration of Van Panhuizen in the Equestrian Order of Utrecht, and thus a political career, but on the other hand the house had already been unfit for habitation for many years. When some time after the proceedings the opportunity to buy another manor-house in the province of Utrecht presented itself, they took it with both hands. Early in 1699 the heavily mortgaged property of Joriphaes Vosch, situated in Lopik, Lopikerkapel, and Zevenhoven, was put up for auction. From this, Van Panhuizen bought the manor-house Te Vliet at Lopikerkapel, which was knocked down to him on 27 February 1699[31]. On 21 October 1699 he was enfeoffed with it by the States of Utrecht[30]. Soon after the purchase Francina moved into the house with her family, for already on 14 May she acted as a sponsor in the nearby church and on 9 July her daughter Machteld Louise was baptized there[32].

The house Te Vliet was, and still is, an imposing building. An engraving from 1701 gives a good idea of it (Fig.20). The house was situated in the out-of-the-way hamlet of Lopikerkapel in the midst of the Utrecht polder-land and at a short distance from the river Lek. In this somewhat isolated environment Francina spent the remainder of her life. The family entered into relations with the local intelligentsia and became friends with the clergyman Daniel Broussard and the French schoolmaster Johannes Fransz. van Weert[33]. A special moment in this retired life must have been the visit of Hendrik Zwaardekroon, the former commission-secretary of Van Reede, in 1702. After Francina had left Ceylon, Zwaardekroon had continued his career in the service of the Company. He had first become commander of Jaffna, 1694-1696, and subsequently, as commissioner of Malabar, 1696-1698, he got the opportunity to complete her father's inspection of that region, which had remained incomplete. In the long report of 31 May 1698 about his findings there he made no secret of his admiration for the deceased commissioner-general. After this he concluded his Asiatic career for the time being with the post of director of Surat, 1699-1701[34]. Zwaardekroon's stay at the house Te Vliet was dominated by the birth of Francina's youngest child, Elisabeth Antonia, who was baptized on 22 October 1702 in the presence of this old friend[35].

After this, Francina was confronted with a series of domestic misfortunes. Her husband died in 1714, her only son Bartholomeus Cornelis in 1716[36]. Her daughters Henriette Adriana and Machteld Louise, too, appear to have died at an early age, for on 27 May 1718 Francina's youngest daughter Elisabeth Antonia was enfeoffed with the house Te Vliet as heiress of her brother[37]. Mother and daughter continued to live in the house, where Francina finally died on 6 June 1731. A week later she was buried in the church of Lopikerkapel[38].

ELISABETH ANTONIA VAN PANHUIZEN

The last years of Francina's life cannot have been simple in financial respects. The property of the Van Panhuizens was considerably burdened. This will no doubt have induced Elisabeth Antonia to get rid of several immovables during the first years after her mother's death. She sold the house Te Vliet on 2 July 1732 to Jan Louis van Hardenbroek, while the houses on the Nieuwe Gracht in Utrecht were disposed of in 1734 and 1735[39].

Elisabeth Antonia van Panhuizen had meanwhile settled in 1732 in the neighbouring little town of Vianen[40]. She became friends with the Walloon clergyman of Vianen, Guillaume Jalabert, a widower with young children[41]. Their affair developed into a scandal. In 1737 Jalabert was suspended as clergyman. Although the couple married on 27 March 1737[42], the suspension of Jalabert was maintained until September 1739[43]. After this time he was able to fulfil his function in Vianen undisturbed until his death in 1764[44].

An important factor for the future inquiry into the inheritance of Van Reede is the last will of 5 February 1737, which Elisabth Antonia drew up immediately before her marriage. She nominated her prospective bridegroom as her sole heir, 'with the exclusion of all and any persons who might be related to her, testatrix'[45]. This exclusion implied that neither the Van Reedes nor the Van Panhuizens could lay any claim to her inheritance.

The marriage of Elisabeth Antonia with Guillaume Jalabart remained childless. It is not known exactly when she died[46]. In 1747 Jalabert as her widower and heir sold a family-vault in the Cathedral in Utrecht[47].

In 1749 Jalabert married for the third time, with Gillette Jeanne Le Roux, a daughter of the burgomaster of Vianen, Alexander Le Roux[48]. From their marriage only one child was born, Alexander Le Roux Jalabert (1751-1803), a wealthy man, who played a role in the magistracy of Vianen and died a bachelor[49]. In the documents concerning the settlement of his inheritance no indications about Van Reede's inheritance have been found[50].

On the basis of the above survey of the lives of Francina van Reede and Elisabeth Antonia van Panhuizen it may be presumed that owing to their deteriorating financial situation they may have sold the estate of Van Reede privately around 1731.

Notes

1. VOC 1505:420-422.
2. The story of Van Reede's marriage with one Johanna Schade was launched by Wittert van Hoogland 1912 vol.2:505. Kalff 1905:255 imitated him in this. In NNBW vol.3 (1914):1011, Regt pointed out that Johanna Schade was married to a brother's son of our Van Reede, Frederik Hendrik van Reede.
3. See the meetings of the church council of Cochin on 5 and 10 April 1675 (VOC 1308:677,683v-684).
4. In a letter of 6 May 1694 Huydecoper wrote to Saint-Martin: 'The Nunia of the lord of Mijdrecht has married the son of the lord of Voorn' (RAU, Records Huydecoper Family 60, Letter-book 1694). Huygens in his journal mentioned on 27 December 1693 after an evening of card-playing: 'It was said there that the nuna or daughter of Mr. Van Rheede, who died in India, was to marry the eldest son of Mr. Van Voorn; that the said nuna was a daughter, born of an Indian woman, who had a husband, and that Van Rheede had gone to bed with that woman as well as two others, and that some people said that after they had diced for it, she had fallen to his, Van Rheede's share' (quoted by Kalff 1905:254). It is remarkable that, in NNBW vol.3 (1914):1012; Regt simply takes the story of this gambling for granted.
5. VOC 1274:146v-147, letter of 15 August 1670 of Van Reede and the council of Malabar to the Governor-General and the Council of India.
6. See chapter 3, section Appointment as Commissioner-General of the Western Quarters.
7. The journal of Huydecoper mentions on that day: 'In the morning I called on the daughter of the lord of Mijdrecht' (RAU, Records Huydecoper Family 58).
8. Trajectina van Tuyll van Serooskerken on 2 October 1695 in the proceedings against Francina van Reede testified about her opponent: 'and who remained with the deceased Lord of Meijert [Mijdrecht] during his arduous voyage and difficult affairs until his decease' (RAU, Records House Zuylen 94, vol.1, Hearing by the judge Everard Becker).
9. Jurriaanse 1942.
10. VOC 1505:772-772v; see also the preceding section Death and funeral.
11. The departure of Zwaardekroon from Surat was announced in a letter of 27 April 1692 by the directors and the council of Surat to the Governor-General and the Council of India: 'and by the same ship, at his earnest request, the merchant and secretary of the general commission, Hendrik Swaardecroon, also left with the commission papers deposited with him' (VOC 1520:22v). The day of Zwaardekroon's departure is mentioned in the Journal

of Surat of 21 December 1691 to 16 December 1692 (VOC 1529:445v-446).
12 VOC 1524:41, letter of 16 May 1693 of Van Rhee and the council of Ceylon to the Governor-General and the Council of India.
13 These papers arrived in Amsterdam in 1748 and were inserted in the Company's records only in 1754. Van Reede's papers about Tuticorin, 1690-1691, were collected in 1747 by Lords XVII in one file, VOC 1478. This may account for the fact that Valentijn 1726 vol.5, in his description of Ceylon and Malabar, could not make any statement about Van Reede's inspections of Jaffnapatnam, Tuticorin, and Malabar, no more than later authors, who were guided by Valentijn.
14 VOC 1505:419,422v-423. Thanks to these requests the last will of Van Reede has come down to us, though in a copy, for Francina had to submit it with her requests.
15 VOC 1520:13, letter of 27 April 1692 of the directors and the council of Surat to the Governor-General and the Council of India; VOC 1529:423v, Journal of Surat.
16 VOC 1524:90-91, letter of 16 May 1693 of Van Rhee and the council of Ceylon to the Governor-General and the Council of India. The amount of the sum she still owed was 28,534 guilders.
17 GAU, Notarial Records U110a004 (protocols H. van Hees):no.72, dated 14 November 1693 OS. In this deed Francina declared that Maurits Cesar de la Baye had died on 4/14 February 1693 in Colombo. Their last will of 15 December 1692 in Colombo is reproduced in a copy there.
18 Raven-Hart 1971 vol.2:399.
19 VOC 5328, Ship's pay-book of the "Bantam", no.226.
20 VOC 5328, ibid., no.3. Thus the settlement had not yet been completed, for Van Reede was also entitled to 1,000 guilders a month as salary and Francina still had 8,500 rupees put out at interest in Ceylon.
21 RAU, Records Huydecoper Family 60, Letter-book 1694.
22 GAU, Notarial Records U110a004 (protocols H. van Hees):no.87, dated 11 January 1694 OS.
23 GAU, DTB BIIa11: they had the banns read on 14 January 1694.
24 Their first child, Henriette Adriana, was baptized on 2 December 1694 in the Cathedral; she was buried on 28 December 1697. Their second child was Bartholomeus Cornelis, who was born in 1696, but of whom no certificate of birth has been found in Utrecht. Their third child, also called Henrietta Adriana, was baptized on 12 January 1698 in the Cathedral.
25 About his houses on the Nieuwe Gracht, see the conveyances of 7 September 1734, 28 January 1735, and 19 April 1735. Another house on the same canal had been inherited by him from Elisabeth Catharina van Panhuizen van Voorn (GAU, City Records II, 3254, Conveyances 1734 and 1735).
26 Agnes van Reede died on 26 May 1692 and her son on 24 July 1692, while Francina van Reede arrived in the Netherlands only in the autumn of that year.
27 The judgements of the Court of Utrecht concerning the proceedings instituted by Godard Willem van Tuyll van Serooskerken, as guardian of the children of Hendrik Jacob, are to be found in RAU, Judicial Records 188. Thus in the year 1693 four of his actions ended with a judgement. In November 1695 the children of Hendrik Jacob were obliged to sell the seigniory Vreeland to Pieter Reaal, lord of Nigtevecht (GAU, Notarial Records U078a005 (protocols H. Vyandt), dated 12 and 20 November 1695).
28 RAU, Records House Zuylen 94, vol.2, Account, Certificat (etc.) of the goods of the minor children of Van Tuyll van Serooskerken, transmitted on 25 April 1695, fol.39v-40: 'Received on the fifteenth of November from the Legacies of the late lord of Mijdrecht the sum of 2,000.-.-'. This item has been deleted with the note: 'Since the acceptors were never willing to acquiesce in the Last Will of the lord of Mijdrecht, this item should be cancelled'.
29 RAU, Judicial Records 188, n.p., Registers of civil judgements, vol.28, dated 11 November 1696 and 27 March 1697. Documents of the proceedings are to be found in RAU, Judicial Records 252 (Documents from the house of the deceased Everard Becker) and RAU, Records House Zuylen 94, vol.1.
30 Wittert van Hoogland 1912 vol.2:505-506. A copy of this enfeoffment is to be found in RAU, Records House Zuylen 794.
31 The announcement of the auction was published on 18 February 1699 at the townhall of Utrecht. The estate of Joriphaes Vosch that was posted up comprised the manorhouse Te Vliet with several lands in Lopik and Zevenhoven, and the seigniory of Zevenhoven. This estate was encumbered with a mortgage of 12,931 guilders. It appears from the printed publication of the auction that the farm Verwershoef of Vosch, with 40 acres of land in Lopikerkapel, was also sold (RAU, Records Chapter St. Mary 528).
32 RAU, DTB 289, Register of the Baptized in the Reformed Church in Lopikerkapel, 1621-1811.
33 Some children of Broussard were presented for baptism by the Van Panhuizens: on 14 May 1699 Fransina Margriet by Francina van Reede, on 20 April 1700 Carel Hendrick by Antoni Carel van Panhuizen, and on 28 July 1709 Cornelia Lydia by Bartholomeus Cornelis van Panhuizen. On 21 September 1710 the latter was also sponsor of Cornelis, the son of the schoolmaster (RAU, DTB 289).
34 For Zwaardekroon's career from 1694 to 1701, see NNBW vol.7 (1917):1352-1354; Wijnaendts van Resandt 1944:186, 288; s'Jacob 1976:LXXVII-LXXVIII, where on pp.283-359 Zwaardekroon's report on Malabar was also printed in full.
35 RAU, DTB 289: 'Lopikercapel 1702. On 22 October a child was baptized of Mr. Antoni Carel van Panhuis, Lord of Vliedt, and Mrs. Fransina van Rheede, his wife, in the presence of Mr. Pieter van Panhuis and Hendrik Swaerdekroon and miss Cornelia Agnes van Panhuis tot Voorn ... It was a daughter. The Name of the child is Elisabeth Antonia'.
36 Wittert van Hoogland 1912 vol.2:593.
37 RAU, Judicial Records 1869, Repertory of the feudal register 1576-1743, fol.240.
38 RAU, DTB 299:147, Register of the Reformed Church in Lopikerkapel of the receipts of duties for the opening of the graves (etc.) 1730-1811: 'In the year 1731, on the 6th of June Francina van Reede, lady of Vliet, rested in the Lord and was interred on the Wednesday following, the 13th, in the church of Lopiker Capel & Sevenhoven, to wit in her vault'.
39 The selling price of the house Te Vliet was 16,000 guilders. The valuation price for the taxes was a little higher, 16,500 guilders. The house was burdened since 3 February 1731 with a mortgage of 4,500 guilders at 3½% a year (GAU, Notarial Records U166a009 (protocols E. Vlaer), dated 2 July 1732; ibid. U186a002 (protocols G. van den Doorslagh):17-18, dated 13 February 1736 and 25 May 1736, fol.37-38v, dated 18 May 1736; RAU, DTB 301, Lists of conveyances and mortgages of Lopik and Lopikerkapel 1707-1762, dated 9 May 1731 and 28 November 1734; RAU, Judicial Records 1042, Protocols of Conveyances etc. of

Zevenhoven, vol.3, dated 3 February 1731). The houses on the Nieuwe Gracht were encumbered with a total mortgage of 18,000 guilders (see note 25).

40 At the end of December 1732 she was confirmed as a member of the Dutch Reformed Church in Vianen (GAV, Records of the Dutch Reformed Church 65:19).

41 ARA, Judicial Records of South Holland before 1811, Vianen 20:247-249v, dated 23 February 1737, in which Guillaume Jalabert, widower of Susanne Vauquet, declared that he had five children, namely Magdalena, aged 10, Emilia, aged 9, Margarita, aged 8, Susanna, aged 7, and Abraham André, aged 6.

42 ARA, DTB Vianen 5:121.

43 Horden Jz. 1952:168.

44 GAV 67:2v, Accounts of the Mourning garbs 1764: funeral of Guillaume Jalabert on 18 April 1764.

45 ARA, Notarial Records of South Holland before 1842, 7854 (protocols G.J. Hammius, Vianen), dated 5 February 1737.

46 The accounts of the mourning garbs of Vianen begin only in 1750.

47 ARA, Notarial Records of South-Holland before 1842, 7856 (protocols G.J. Hammius, Vianen), dated 6 December 1747.

48 ARA, DTB Vianen 9:2, dated 26 May 1749.

49 Alexander Le Roux Jalabert since 1778 was clerk of the Chamber of Justice and of the Feudal Court and dike-reeve of Vianen (ARA, Countship Audit Office Registers 810:137v-140v, dated 3 March 1778). He was buried in Vianen on 8 February 1803 (ARA, DTB Vianen 7).

50 ARA, Notarial Records before 1842, 7901 (protocols W. Pernis, Vianen), dated 18 February, 28 March, 18 and 22 November 1803; ARA, Judicial Records of South Holland before 1811, Vianen 28:178-183, dated 7 December 1805.

Part Two
Hortus Malabaricus

Introduction

Hortus Malabaricus may be considered as the first publication on the flora of a definite district in Asia. It describes and illustrates 740 plants of 17th-century Malabar, from Cape Comorin to Calicut, from the littoral and the lowland to the hills and mountains, covering a region of roughly 30 x 400 km. Although Hortus Malabaricus was in the first instance intended and planned as a contribution to the knowledge of tropical, medicinal plants for the use of the Dutch East India Company, it exceeds its initial scope by including many non-medicinal plants, such as aromatic plants, timber, food, and ornamental plants, and even wild plants without any obvious use. The life-size engravings and often extensive descriptions supply, certainly for that time, sufficient and usually accurate information on all parts of the plants as well as native names, flowering and fruit-bearing periods, and habitat, locality, and distribution. Commentaries on the plants, discussing earlier literature and attempting identification and scientific nomenclature, greatly add to the usefulness of the work.

Nevertheless, Hortus Malabaricus is not the first publication of some coherence on Asiatic plants. In the preceding part of this study reference has already been made to the earlier writings of Garcia da Orta, Cristobal Acosta, and Jacobus Bontius dealing with medicinal and other useful plants in the neighbourhood of Goa, Cochin, and Batavia. Although these works constitute the basis of Asian botany, they supply only general, mainly practical information without pretending to form thorough and orderly plant descriptions. The only 17th-century work which can signally bear comparison on all points with Hortus Malabaricus is Rumphius' *Herbarium Amboinense*, which contains a thousand plants, covering the Indian Archipelago. But this work, or a part of it, was completed only in 1692, about the time when the printing of Hortus Malabaricus was finished.

The scope, the detailed character, and the complexity of many divergent topics, such as phytography, taxonomy, medicine, geography, philology, art history, and bibliography, call for some limitations in the analysis of Hortus Malabaricus. In order to avoid a continuous enumeration of generalities, I have opted for a more detailed discussion of various aspects which throw light on the previous history and the genesis of Hortus Malabaricus and its insertion in European botany. In this context I will not concern myself with the medical aspects and the modern taxonomy of Hortus Malabaricus.

6
Bibliography

Of Hortus Malabaricus three editions are known. The original Latin edition, in folio, consisting of twelve volumes, was published in Amsterdam in 1678-1693; a title-page reissue of volume 1 appeared in Amsterdam in 1686. A Dutch edition, in folio, of the volumes 1 and 2, *Malabaarse Kruidhof*, translated by Abraham van Poot, was published in Amsterdam in 1689; a title-page reissue of this work appeared in The Hague in 1720. Finally, a Latin edition, in quarto, of volume 1, *Horti Malabarici Pars Prima*, edited and annotated by John Hill, was published in London in 1774[1]. In the history of the insertion of Hortus Malabaricus in European botany up to the end of the 18th century Van Poot's translation and Hill's edition hardly played any role. Consequently, only a short bibliographical description of the original Latin edition is given below.

Hortus Malabaricus contains three types of illustrations: Van Reede's portrait, the frontispiece of several volumes, and the plant engravings. Not all the sets of Hortus Malabaricus contain the portrait[2]. Moreover, it does not always appear in the same volume[3]. The portrait also occurs in the first volume of *Malabaarse Kruidhof* (both issues, of 1689 and 1720). This goes to show that the portrait was for sale separately and could be bound at will in one of the volumes. That is why the engraving of the portrait is described separately.

Hortus Malabaricus counts only one type of frontispiece. It occurs, with different inscriptions and legends, in the volumes 1 (both issues, of 1678 and 1686) and 3 (1682), and also in volume 1 of *Malabaarse Kruidhof* (both issues). The original plate will be described under volume 1 of 1678, while the differences in other volumes will be mentioned.

Hortus Malabaricus contains 791 uncoloured plant engravings in all, of which 712 are in double-folio and 79 in folio, depicting 729 kinds of plants. The plates have been pasted on separate slips, which in turn have been bound. The engravings have varying dimensions, measured along the plate-marks. The double-folio engravings generally measure (33.5-35.5) x (41.5-44.5) cm and the folio engravings (33.0-33.5) x (20.5-21.0) cm. The engravings have been carefully copied from the original drawings, though in reverse and with additions and omissions of leaves, flowers, fruits, and seeds[4]. Only two engravings have been signed by engravers, to wit Fig.1 in volume 1, representing Tenga, by Bastiaan Stoopendael (Fig.21), and Tab 39 in volume 6, representing Hina paretti, by Gonsalez Appelman (Fig.22). The great stylistic uniformity of the engravings of the plants does not allow of a detailed analysis of their individual contributions to the engraving work of Hortus Malabaricus, although there are slight differences in the execution of the Roman characters on the plates.

DESCRIPTION

Portrait

The unsigned engraving of 33.5 x 20.5 cms (Fig.1) has been divided into the portrait proper and a balustrade. Van Reede, shown en face and at three-quarter length, stands behind the balustrade. He is attired in armour, with a knotted neckerchief, bordered with broad strips of lace, and he wears a long, curling wig. On his left side the hilt of a sword is visible. The face of Van Reede is characterized by a wart on the root of the nose, a few light wrinkles in the forehead and the two cheeks, and a very small moustache. The left arm is slightly lifted, with the palm of the hand turned upwards. The right arm is bent at right angles, with in the hand a baton, resting on a table-top. On the table-top lies a draped cloth.

The background of the portrait is partly formed by a draped curtain on the left. On the right the background consists of a wooded landscape under a clouded sky, and intersected by water; in the water there is an approximately quadrangular island, with along the banks a structure suggesting fortifications (Cochin?).

On the balustrade there is a ten-line inscription reading:

"*Illustris-ac Genero: sissimus Dominus* HENRICUS ADRIANUS VAN RHEEDE, *Tot Draakestein, Toparcha in Meydrecht, Mala: barici regni quondam Gubernator, et supremi Consessus Belgici, qui est Bataviae Indorum, Senator extraordi: narius, nunc Equestris dignitatis Ordinibus Illustrium ac Praepotentium Ultrajectinae dioeceseos adscriptus, et hoc tempore Legatus ad res societatis Belgicae in India extra et intra Gangem inspiciendas et componendas.*"

Over this inscription is to be found the crowned Van Reede coat of arms of two indented fesses supported by two griffins. To the left and the right of the inscription on either side there are two vertical rows of four coats of arms with the names REEDE, GOOR, NIENRODE,

Figure 21. Tenga, Hortus Malabaricus vol.1 (1678), Fig.1. Engraving by Bastiaan Stoopendael after a drawing by Antoni Jacobsz. Goetkint.

RENESSE, DIEST, BRANDENBURG, VEELEN, SOLMS, UTENHOVE, DE KALE, KETEL, MULART, RENESSE, BORSSELEN, SCHIMMELPENŃ, and GAESBEECK.

Although the engraving is unsigned, it is ascribed by Regt to the Amsterdam artist Pieter van Gunst[5]. According to the inscription it represents Van Reede as commissioner-general of the Western Quarters. Since he left the Netherlands forever in December 1684, the portrait must have been drawn or painted in or shortly before 1684. The engraving may have been executed after his departure.

The sixteen coats of arms on the engraving, representing four generations of Van Reede's ancestors, reach back to the end of the Middle Ages. Apparently, they serve to emphasize his noble birth as required for the admission to the Equestrian Order of Utrecht. Among them there are distinguished Utrecht noble families such as Van Nijenrode, Van Renesse, and Van Gaasbeek[6]. But his ancestors belonging to the families Van Diest, Van Brandenburg, Van Veelen, De Kale, etc. can hardly be traced, either in the records of the Houses of Zuilen and Amerongen, or in the current genealogical literature, let alone that their noble status can be established. Possibly Van Reede derived his knowledge of his obscure coats of arms from scutcheons of deceased relatives[7] and from the genealogical booklet published by Frederik van Reede van Amerongen (1595)[8].

Volume 1 (1678)

Frontispiece. The unsigned engraving of 32.5 x 20.5 cm (Fig.23) represents a garden stuffed with palms and potted plants, surrounded by an arched pergola. In the centre of the garden stands an ornamental summer-house bearing a tablet reading "HORTUS INDICUS MALABARICUS". In front of it, beneath one of the arches, an Indian woman is sitting, holding a rake, with

Figure 22. Hina paretti, Hortus Malabaricus vol.6 (1686), Tab.39. Engraving by Gonsalez Appelman after a drawing by Marcelis Splinter.

a pruning-knife at her feet, and apparently symbolizing Indian botany. Four kneeling Malayali children, with elongated ear-lobes, are tendering potted plants to her, including a small tied-up tree. The legend of the engraving reads "AMSTELODAMI, Sumptibus JOANNIS VAN SOMEREN et JOANNIS VAN DYCK. Anno MDCLXXVIII.".

The same frontispiece, but with changed texts, is also found in the 1686 issue of volume 1 and in volume 3.

Title-page (Fig.24). HORTUS INDICUS MALABARICUS, *Continens* Regni Malabarici apud Indos celeberrimi omnis generis Plantas rariores, *Latinis, Malabaricis, Arabicis, & Bramanum Characteribus nominibusque expressas,* Unà cum Floribus, Fructibus & seminibus, naturali magnitudine à peretissimis pictoribus delineatas, & ad vivum exhibitas. *Addita insuper accuratâ earundem descriptione, quâ colores, odores, sapores, facultates, & praecipuae in Medicinâ vires exactissimè demonstrantur. ADORNATUS* Per HENRICUM van RHEEDE, van DRAAKENSTEIN, Nuperrimè Malabarici Regni Gubernatorem, nunc supremi Consessus apud Indos Belgas Senatorem Extraordinarium, & primum successorum loco ordinario destinatum. *ET* JOHANNEM CASEARIUM, Ecclesiast. in Cochin. *Notis adauxit, & Commentariis illustravit* ARNOLDUS SYEN, Medicinae & Botanices in Academia Lugduno-Batava Professor. [printer's mark] *AMSTELODAMI,* Sumptibus JOANNIS van SOMEREN, ET JOANNIS van DYCK. Anno MDCLXXVIII.

The printer's mark represents two farmers with a barn, a tree, and several ants; the legend reads: "NON AESTAS EST LAETA DIU, COMPONITE NIDOS.". The same mark is to be found on the title-pages of the following volumes 2-12.

Pagination. (xiv),110,(1) pp. The page numbers 75 to 78 actually cover eight pages.

Plates. 57 double-folio engravings, numbered Fig. or F. 1-57, depicting 50 kinds of plants. Fig.1 is signed "Anto-

Figure 23. Frontispiece of volume 1 (1678) of Hortus Malabaricus. Anonymous etching.

Figure 24. Title-page of volume 1 (1678) of Hortus Malabaricus.

ni Jacobi Goedkint. B.Stoopendael fecit."[9]. Fig.15 depicts Ambapaja as well as Papajamaram.

Text. Pp.(i)-(ii) dedication to Joan Maatsuiker, governor-general of the Dutch East Indies, by Van Reede and Joannes Casearius, clergyman of Cochin; pp.(iii)-(iv) preface by the same Casearius; pp.(v)-(vi) preface by Matthaeus à S. Joseph, Discalced Carmelite of the Italian Congregation, dated Cochin, 20 April 1675; pp.(vii)-(viii) statement in Malayalam by Emanuel Carneiro, sworn interpreter of the East India Company, dated Cochin, 20 April 1675, followed by the Latin translation by Christiaan Herman van Donep, secretary of the town of Cochin; pp.(ix)-(x) statement in Malayalam by the Malabar physician Itti Achudem, dated Cochin, 20 April 1675, followed by the Latin translation by Van Donep, based upon a Portuguese translation from Malayalam by Carneiro; pp.(xi)-(xii) statement in Konkani by the Brahmin physicians Ranga Botto, Vinaique Pandito, and Apu Botto, dated Cochin, 20 April 1675, followed by the Latin translation by Van Donep, followed by a Portuguese translation from Konkani by Vinaique Pandito; pp.(xiii)-(xiv) preface by Arnold Syen; pp.1-110 text of 52 chapters of plant descriptions, of which two have not been illustrated; p.(111) index, advice to the bookbinder, and errata.

Volume 1 (reissue 1686)

Frontispiece. The engraving is the same one as in the 1678 issue of volume 1, but the legend now reads "T'AMSTELAEDAMI, Sumptibus Viduae JOANNIS VAN SOMEREN, Haeredum JOANNIS van DYCK, HENRICI & Viduae THEODORI BOOM, Ao. MDCLXXXVI.".

Title-page. HORTI MALABARICI PARS PRIMA, DE VARII GENERIS ARBORIBUS ET FRUTICIBUS

SILIQUOSIS *Latinis, Malabaricis, Arabicis, Brachmanum characteribus nominibusque expressis,* Adjecta Florum, Fructuum, Seminumque nativae magnitudinis vera delineatione, colorum viriumque accurata descriptione, *ADORNATA Per Nobilissimum ac Generosissimum D.D.* HENRICUM van RHEDE tot DRAAKESTEIN, Toparcham in Mydrecht, quondam Malabarici Regni Gubernatorem supremi Consessus apud Indos Belgas Senatorem Extraordinarium, nunc vero Equestris Ordinis nomine Illustribus ac Praepotentibus Provinciae Ultrajectinae Proceribus adscriptum, *ET THEODORUM JANSON.* ab ALMELOVEEN M.D. *Notis adauxit, & Commentariis illustravit* JOANNES COMMELINUS. [printer's mark] *AMSTELAEDAMI.* Sumptibus Viduae JOANNIS van SOMEREN, Haeredum JOANNIS van DYCK, HENRICI & Viduae THEODORI BOOM. Anno MDCLXXXVI.
Pagination. (xii),110 pp.
Plates. As in the 1678 issue.
Text. As in the 1678 issue, however, pp.(xiii)-(xiv), with Syen's preface, and p.(111), with the index (etc.), have been omitted.

Volume 2 (1679)

Title-page. HORTI INDICI MALABARICI PARS SECUNDA DE FRUTICIBUS Regni Malabarici apud Indos celeberrimi, *Latinis, Malabaricis, Arabicis, & Bramanum Characteribus nominibusque expressis,* Unà cum Floribus, Fructibus & Seminibus, naturali magnitudine à peritissimis pictoribus delineatis, & ad vivum exhibitis. *Addita insuper accuratâ earundem descriptione, quâ colores, odores, sapores, facultates, & praecipuae in Medicina vires exactissimè demonstrantur.* ADORNATA Per HENRICUM VAN REEDE, TOT DRAAKESTEIN, Nuperrimè Malabarici Regni Gubernatorem, nunc supremi Consessus apud Indos Belgas Senatorem Extraordinarium, & primum successorum loco ordinario destinatum, *ET JOHANNEM CASEARIUM,* Ecclesiast. in Cochin. *Notis adauxit, & Commentariis* illustravit JOANNES COMMELINUS. [printer's mark] *AMSTELODAMI,* Sumptibus Viduae JOANNIS van SOMEREN, ET Haeredum JOANNIS van Dyck. Anno MDCLXXIX.
Pagination. (vi),110,(1) pp.
Plates. 56 double-folio engravings, numbered F. 1-57, depicting 52 kinds of plants.
Text. Pp.(i)-(ii) dedication to Gillis Valckenier, burgomaster and councillor of Amsterdam and director of the East India Company, Johannes Hudde, lord of Waveren, burgomaster and councillor of Amsterdam, Joan Huydecoper, lord of Maarsseveen and Neerdijk, late burgomaster and councillor of Amsterdam and director of the East India Company, and Pieter van Dam, advocate and general councillor of the East India Company, by Van Reede; pp.(iii)-(iv) panegyric on Van Reede and Casearius by Dr. Willem ten Rhijne, dated Batavia, 15 March 1678; pp.(v)-(vi) preface by Jan Commelin; pp.1-110 text of 55 chapters of plant descriptions, of which three have not been illustrated; p.(111) index and advice to the bookbinder.

Volume 3 (1682)

Frontispiece. The engraving is the same one as in the 1678 issue of volume 1, but the tablet on the summerhouse now reads "HORTI INDICI MALABARICI Pars tertia de ARBORIBUS.", and the legend "AMSTELODAMI, Sumptibus JOANNIS VAN SOMEREN et JOANNIS VAN DYCK. HENRICI et Viduae THEODORI BOOM. Ao. MDCLXXXII.".
Title-page. HORTI INDICI MALABARICI PARS TERTIA DE ARBORIBUS REGNI MALABARICI, *Latinis, Malabaricis, Arabicis, & Brachmanum Characteribus nominibusque expressis.* Unà cum Floribus, Fructibus & Seminibus, naturali magnitudine à peritissimis pictoribus delineatis, & ad vivum exhibitis. *Addita insuper accuratâ earundem descriptione, quâ colores, odores, sapores, facultates, & praecipuae in Medicina vires exactissimè exhibentur. ADORNATA Per Nobilissimum ac Generosissimum D.D.* HENRICUM VAN REEDE, TOT DRAAKESTEIN, Toparcham in Meydrecht, quondam Malabarici Regni Gubernatorem, supremi Consessus apud Indos Belgas Senatorem Extraordinarium; nunc vero Equestris ordinis nomine Illustribus ac Praepotentibus Provinciae Ultrajectinae Proceribus adscriptum, *ET JOHANNEM MUNNICKS,* M.D. In eademque facultate Anatomes ac Botanices in Academia Ultrajectina Professorem, & Reip. Ultraj. Poliatrum. *Notis adauxit, & Commentariis illustravit* JOHANNES COMMELINUS. [printer's mark] *AMSTELODAMI,* Sumptibus Viduae JOANNIS van SOMEREN, Haeredum JOANNIS van DYCK, & HENRICI & Viduae THEODORI BOOM. Anno MDCLXXXII.
Pagination. (xxii),87,(1) pp.
Plates. 64 double-folio engravings, numbered Tab. 1-64, depicting 33 kinds of plants.
Text. Pp.(i)-(ii) dedication to Noetavile-Virola, king of Cochin (etc.), by Van Reede; pp.(iii)-(xviii) preface by Van Reede; pp.(xix)-(xx) preface by Johannes Munnicks; pp.(xxi)-(xxii) preface by Jan Commelin; pp.1-87 text of 34 chapters of plant descriptions, of which one has not been illustrated; p.(88) index.

Volume 4 (1683)

Title-page. HORTI INDICI MALABARICI PARS QUARTA DE ARBORIBUS REGNI MALABARICI, *Latinis, Malabaricis, Arabicis, & Brachmanum Characteribus nominibusque expressis,* Unà cum Floribus, Fructibus & Seminibus, naturali magnitudine à peritissimis pictoribus delineatis, & ad vivum exhibitis. *Addita insuper accuratâ earundem descriptione, quâ colores, odores, sapores, facultates, & praecipuae in Medicina vires exactissimè exhibentur. ADORNATA Per Nobilissimum ac Generosissimum D.D.* HENRICUM VAN REEDE, TOT DRAAKESTEIN, Toparcham in

Meydrecht, quondam Malabarici Regni Gubernatorem, supremi Consessus apud Indos Belgas Senatorem Extraordinarium; nunc vero Equestris ordinis nomine Illustribus ac Praepotentibus Provinciae Ultrajectinae Proceribus adscriptum, *ET JOHANNEM MUNNICKS, M.D.* In eademque facultate Anatomes ac Botanices in Academia Ultrajectina Professorem, & Reip. Ultraj. Poliatrum. *Notis adauxit, & Commentariis illustravit* JOANNES COMELINUS. [printer's mark] AMSTELODAMI, Sumptibus Viduae JOANNIS van SOMEREN, Haeredum JOANNIS van DYCK, HENRICI & Viduae THEODORI BOOM. Anno MDCLXXX III

A variant of this title-page has "FRUCTIFERIS" after "DE ARBORIBUS" and "Anno MDCLXXIII." instead of "Anno MDCLXXXIII".

Pagination. (ii),125,(2) pp.

Plates. 61 double-folio engravings, numbered Pars 4. Tab. 1-61, depicting 57 kinds of plants.

Text. Pp.(i)-(ii) dedication to Hendrik Jacob van Tuyll van Serooskerken, baron of Vreeland, lord of Zuylen (etc.) and deputy of the States of Utrecht, by Van Reede; pp.1-125 text of 58 chapters of plant descriptions, of which one has not been illustrated; p.(127) index.

Volume 5 (1685)

Title-page. HORTI INDICI MALABARICI PARS QUINTA DE ARBORIBUS ET FRUTICIBUS BACCIFERIS. REGNI MALABARICI, *Latinis, Malabaricis, Arabicis, Brachmanum Characteribus Nominibusque expressis,* Unà cum Floribus, Fructibus & Seminibus, naturali magnitudine à peritissimis pictoribus delineatis, & ad vivum exhibitis. *Addita insuper accuratâ earundem descriptione, quâ colores, odores, sapores, facultates, & praecipuae in Medicina vires exactissimè exhibentur. ADORNATA Per Nobilissimum ac Generosissimum D.D.* HENRICUM VAN RHEDE, TOT DRAAKESTEIN, Toparcham in Mydrecht, quondam Malabarici Regni Gubernatorem, supremi Concessus apud Indos Belgas Senatorem Extraordinarium; nunc vero Equestris ordinis nomine Illustribus ac Praepotentibus Provinciae Ultrajectinae Proceribus adscriptum, *ET JOHANNEM MUNNICKS, M.D.* In eadem facultate Anatomes ac Botanices in Academia Ultrajectina Professorem, & Reip. Ultraj. Poliatrum. *Notis adauxit, & Commentariis illustravit* JOANNES COMMELINUS. [printer's mark] *AMSTELODAMI,* Sumptibus Viduae JOANNIS van SOMEREN, Haeredum JOANNIS van DYCK, HENRICI & Viduae THEODORI BOOM. Anno MDCLXXXV.

Pagination. (vi),120,(1) pp. The page number 47 is erroneously printed 37.

Plates. 60 double-folio engravings, numbered Pars 5. Tab 1-60 depicting 61 kinds of plants. Tab 33 depicts Karetta amelpodi as well as Katou belluta amelpodi.

Text. Pp.(i)-(ii) dedication to Godard Adriaan van Reede, lord of Amerongen (etc.), member of the States General and extraordinary ambassador to Brandenburg and Saxony, by Van Reede, dated Utrecht, 31 July 1684; pp.(iii)-(iv) preface by Johannes Munnicks; pp.(v)-(vi) preface by Jan Commelin; pp.1-120 text of 62 chapters of plant descriptions, of which one has not been illustrated; p.(121) index and errata.

Volume 6 (1686)

Title-page. HORTI MALABARICI PARS SEXTA, DE VARII GENERIS ARBORIBUS ET FRUTICIBUS SILIQUOSIS *Latinis, Malabaricis, Arabicis, Brachmanum characteribus nominibusque expressis,* Adjecta Florum, Fructuum, Seminumque nativae magnitudinis vera delineatione, colorum viriumque accurata descriptione, *ADORNATA Per Nobilissimum ac Generosissimum D.D.* HENRICUM VAN RHEDE TOT DRAAKESTEIN, Toparcham in Mydrecht, quondam Malabarici Regni Gubernatorem supremi Consessus apud Indos Belgas Senatorem Extraordinarium, nunc vero Equestris Ordinis nomine Illustribus ac Praepotentibus Provinciae Ultrajectinae Proceribus adscriptum, ET THEODORUM JANSON. AB ALMELOVEEN, M.D. *Notis adauxit, & Commentariis illustravit* JOANNES COMMELINUS. [printer's mark] *AMSTELAEDAMI.* Sumptibus Viduae JOANNIS van SOMEREN, Haeredum JOANNIS van DYCK, HENRICI & Viduae THEODORI BOOM, Anno MDCLXXXVI.

Pagination. (vi),109,(2) pp. The page numbers 101, 105, and 107 are erroneously printed 105, 109, and 111 respectively.

Plates. 61 engravings, 59 double-folio and 2 folio, numbered Pars 6. Tab 1-61, depicting 54 kinds of plants. Tab 39 is signed "Marcelis Splijnter in. fecit G.Appelman sculpsit"[10].

Text. Pp.(i)-(ii) dedication to Willem van Nassau, baron of Zuilestein and Leersum, member of the States of Utrecht (etc.), by Van Reede; pp.(iii)-(v) preface by Theodorus Janssonius van Almeloveen; p.(vi) panegyric on Van Reede and Janssonius by Henricus Christianus Henninius; pp.1-109 text of 54 chapters of plant descriptions; p.(111) index and errata.

Volume 7 (1688)

Title-page. HORTI MALABARICI PARS SEPTIMA, DE VARII GENERIS FRUTICIBUS SCANDENTIBUS; *Latinis, Malabaricis, Arabicis, Belgicis, Bramannum characteribus nominibusque expressis,* Adjecta Florum, Fructuum, Seminumque nativae magnitudinis vera delineatione, colorum viriumque accurata descriptione, *ADORNATA Per Nobilissimum ac Generosissimum D.D.* HENRICUM VAN RHEDE TOT DRAAKESTEIN, Toparcham in Mydrecht, quondam Malabarici Regni Gubernatorem, supremi Consessus apud Indos Belgas Senatorem Extraordinarium, nunc vero Equestris Ordinis nomine Illustribus ac Praepotentibus Provinciae Ultrajectinae Proceribus adscriptum, *Notis adauxit, & Commentariis illustravit* JOANNES COMMELINUS *In ordinem redegit, & latinitate donavit* ABRAHAMUS à POOT M.D. [printer's mark] *AMSTELAEDAMI.*

Sumptibus Viduae JOANNIS van SOMEREN, Haeredum JOANNIS van DYCK, HENRICI & Viduae THEODORI BOOM, Anno MDCLXXXVIII.

Pagination. (ii),111,(2) pp. The page number 65 is erroneously printed 63 and the page number 71 has been used twice on two successive leaves.

Plates. 59 double-folio engravings, numbered Pars 7 Tab: 1-59, depicting 57 kinds of plants.

Text. P.(i) dedication to Diederik baron of Baer, member of the States General (etc.), by Van Reede; p.(ii) preface by Abraham van Poot; pp.1-111 text of 57 chapters of plant descriptions; p.(113) index.

Volume 8 (1688)

Title-page. HORTI MALABARICI PARS OCTAVA, DE VARII GENERIS HERBIS POMIFERIS & LEGUMINOSIS; *Latinis, Malabaricis, Arabicis, Belgicis, Bramannum characteribus nominibusque expressis.* Adjecta Florum, Fructuum, Seminumque nativae magnitudinis vera delineatione, colorum viriumque accurata descriptione, *ADORNATA Per Nobilissimum ac Generosissimum D.D.* HENRICUM VAN RHEDE TOT DRAKESTEIN, Toparcham in Mydrecht, quondam Malabarici Regni Gubernatorem, supremi Consessus apud Indos Belgas Senatorem Extraordinarium, nunc vero Equestris Ordinis nomine Illustribus ac Praepotentibus Provinciae Ultrajectinae Proceribus adscriptum, *Notis adauxit, & Commentariis illustravit* JOHANNES COMMELINUS *In ordinem redegit, & latinitate donavit* ABRAHAMUS à POOT M.D. [printer's mark] *AMSTELAEDAMI.* Sumptibus Viduae JOANNIS van SOMEREN, Haeredum JOANNIS van DYCK, HENRICI & Viduae THEODORI BOOM, Anno MDCLXXXVIII.

Pagination. (ii),97,(1) pp. The page numbers 40 and 93 are erroneously printed 46 and 39 respectively; the page number 51 has been used on two successive leaves.

Plates. 51 double-folio engravings, numbered Pars 8 Tab. 1-51, depicting 47 kinds of plants.

Text. P.(i) dedication to Frederik Adriaan baron of Reede, lord of Renswoude (etc.), member of the Equestrian Order of Utrecht, by Van Reede; p.(ii) preface by Abraham van Poot; pp.1-97 text of 49 chapters of plant descriptions, of which two have not been illustrated; p.(98) index.

Volume 9 (1689)

Title-page. HORTI MALABARICI PARS NONA, DE HERBIS ET DIVERSIS ILLARUM SPECIEBUS *Latinis, Malabaricis, Arabicis, Brachmannum characteribus nominibusque expressis,* Adjecta Florum, Fructuum, Seminumque nativae magnitudinis vera delineatione, colorum viriumque accurata descriptione, *ADORNATA Per Nobilissimum ac Generosissimum D.D.* HENRICUM VAN RHEDE TOT DRAKESTEIN, Toparcham in Mydrecht, quondam Malabarici Regni Gubernatorem supremi Concessus apud Indos Belgas Senatorem Extraordinarium, nunc vero Equestris Ordinis nomine Illustribus ac Praepotentibus Provinciae Ultrajectinae Proceribus adscriptum, *In ordinem redegit & latinitate donavit* ABRAHAMUS A POOT M.D. *Notis adauxit, & Commentariis illustravit* JOANNES COMMELINUS. [printer's mark] *AMSTELAEDAMI.* Sumptibus Viduae JOANNIS van SOMEREN, Haeredum JOANNIS van DYCK, HENRICI & Viduae THEODORI BOOM, Anno MDCLXXXIX.

Pagination. (vi),170,(1) pp. The pages 97-98 are lacking in the successive numbering; the page number 155 is erroneously printed 153.

Plates. 87 engravings, 57 double-folio and 30 folio, numbered Pars 9 Tab. 1-87, depicting 85 kinds of plants.

Text. Pp.(i)-(ii) dedication to Godard Willem van Tuyll van Serooskerken, lord of Welland (etc.), by Van Reede; pp.(iii)-(iv) panegyric on Abraham van Poot by Nicolaus Kis; pp.(v)-(vi) panegyric on Van Poot by Nicolaus Apati; pp.1-170 text of 85 chapters of plant descriptions; p.(171) index.

Volume 10 (1690)

Title-page. HORTI MALABARICI PARS DECIMA, DE HERBIS ET DIVERSIS ILLARUM SPECIEBUS *Latinis, Malabaricis, Arabicis, Brachmannum characteribus nominibusque expressis,* Adjecta Florum, Fructuum, Seminumqne nativae magnitudinis vera delineatione, colorum viriumque accurata descriptione, *ADORNATA Per Nobilissimum ac Generosissimum D.D.* HENRICUM VAN RHEDE TOT DRAKESTEIN, Toparcham in Mydrecht, quondam Malabarici Regni Gubernatorem, supremi Concessus apud Indos Belgas Senatorem Extraordinarium, nunc vero Equestris Ordinis nomine Illustribus ac Praepotentibus Provinciae Ultrajectinae Proceribus adscriptum, *In ordinem redegit & latinitate donavit* ABRAHAMUS A POOT M.D. *Notis adauxit, & Commentariis illustravit* JOANNES COMMELINUS. *Reipublicae Amstelodamensis Senator, ac Horti Medici praefectus.* [printer's mark] *AMSTELAEDAMI.* Sumptibus Viduae JOANNIS van SOMEREN, Haeredum JOANNIS van DYCK, HENRICI & Viduae THEODORI BOOM. Anno MDCXC.

Seminumqne is a misprint for Seminumque.

Pagination. (ii),187,(3) pp.

Plates. 94 engravings, 65 double-folio and 29 folio, numbered Pars 10. Tab 1-94, depicting 94 kinds of plants.

Text. Pp.(i)-(ii) dedication to Hendrik van Utenhove, lord of Amelisweerd, member of the Equestrian Order of Utrecht (etc.), by Van Reede; pp.1-187 text of 94 chapters of plant descriptions; pp.(189)-(190) index.

Volume 11 (1692)

Title-page. HORTI MALABARICI PARS UNDECIMA, DE HERBIS ET DIVERSIS ILLARUM SPECIEBUS *Latinis, Malabaricis, Arabicis, Brachmannum characteribus nominibusque expressis.* Adjecta Florum, Fructuum, Seminumque nativae magnitudinis vera delineatione, colorum viriumque accurata descriptione,

ADORNATA Per Nobilissimum ac Generosissimum D.D. HENRICUM VAN RHEDE TOT DRAKESTEIN, Toparcham in Mydrecht, quondam Malabarici Regni Gubernatorem supremi Concessus apud Indos Belgas Senatorem Extraordinarium, nunc vero Equestris Ordinis nomine Illustribus ac Praepotentibus Provinciae Ultrajectinae Proceribus adscriptum, *In ordinem redegit & latinitate donavit* ABRAHAMUS A POOT M.D. *Notis adauxit, & Commentariis illustravit* JOANNES COMMELINUS. *Republicae Amstelodamensis Senator, ac Horti Medici praefectus.* [printer's mark] *AMSTELAEDAMI.* Sumptibus Viduae JOANNIS van SOMEREN, Haeredum JOANNIS van DYCK, HENRICI & Viduae THEODORI BOOM, Anno MDCXCII.

Pagination. 133,(1) pp. The page number 23 is erroneously printed 21; the pages 89-90 are lacking in the successive numbering.

Plates. 64 engravings, 62 double-folio and 2 folio, numbered Pars 11. Tab. 1-65 (the numbers 16 and 17 on one plate), depicting 62 kinds of plants.

Text. Pp.1-133 text of 63 chapters of plant descriptions, of which one has not been illustrated; p.(134) index.

Volume 12 (1693)

Title-page. HORTI MALABARICI PARS DUODECIMA, & ULTIMA, DE HERBIS ET DIVERSIS ILLARUM SPECIEBUS *Latinis, Malabaricis, Arabicis, Brachmannum characteribus nominibusque expressis.* Adjecta Florum, Fructuum, Seminumque nativae magnitudinis vera delineatione, colorum viriumque accurata descriptione, *ADORNATA Per piae memoriae Nobilissimum ac Generosissimum D.D.* HENRICUM VAN RHEDE TOT DRAKESTEIN, Toparcham in Mydrecht, Malabarici Regni Gubernatorem supremi Concessus apud Indos Belgas Senatorem Extraordinarium, & Equestris Ordinis nomine Illustribus ac Praepotentibus Provinciae Ultrajectinae Proceribus adscriptum. *In ordinem redegit & latinitate donavit,* ABRAHAMUS A POOT M.D. *Notis ex parte adauxit, & Commentariis illustravit* JOANNES COMMELINUS, *Reipublicae Amstelodamensis (dum viveret) Senator, ac Horti Medici Praefectus.* Accedit, praeter hujus partis specialem, generalis totius Operis Index. [printer's mark] *AMSTELAEDAMI.* Sumptibus Viduae JOANNIS van SOMEREN, Haeredum JOANNIS van DYCK, HENRICI & Viduae THEODORI BOOM, Anno MDCCIII.

The year of publication MDCCIII is a misprint for MDCXCIII.

Pagination. 151,(9) pp. The page number 79 is erroneously printed 97.

Plates. 77 engravings, 61 double-folio and 16 folio, numbered Pars XII. Tab. 1-79 (the numbers 36, 37, and 38 on one plate), depicting 77 kinds of plants.

Text. Pp.1-151 text of 77 chapters of plant descriptions; p.(152) index to volume 12; pp.(153)-(160) general index to all volumes.

Notes

1 For John Hill (1716-1775), English chemist and naturalist, and his edition of the first volume of Hortus Malabaricus, see Stafleu & Cowan 1979 vol.2:198-205, no.2777.

2 The copy of the Rijksherbarium, Leiden, does not contain Van Reede's portrait.

3 Stafleu & Cowan 1983 vol.4:751 cite the portrait bound in volume 1. Fournier 1980:22 mentions the portrait in volume 3. The copy in the Institute of Plant Systematics, Utrecht, shows a specimen bound in volume 9.

4 See chapter 8.

5 Pieter Stevensz. van Gunst (c.1659-after 1731), draughtsman, engraver, and etcher in Amsterdam; he also engraved a portrait of the Amsterdam burgomaster and amateur of botany Nicolaas Witsen (1641-1717) (NNBW vol.3 (1914):1013; Thieme & Becker vol.15 (1922):345-346; Von Wurzbach 1906 vol.1:622-623; Waller & Juynboll 1938:124).

6 For Van Reede's ancestors, see Ernst van Nijenrode (who died in 1558), lord of Zuilestein, in Wittert van Hoogland 1909 vol.1:169 and Maris 1956:104-105; Johan van Renesse van Baar Sr. (who died in 1562) and his son of the same name (who died in 1598), lords of Rijnestein, in Wittert van Hoogland 1912 vol.2:574, NNBW vol.3(1914):1063, and Rientjes & Böcker 1947:119-121; and Johanna van Gaasbeek van Abcoude (who died in 1578) in Van der Aa 1852 vol.1:247-248 and Wittert van Hoogland 1909 vol.1:544, 1912 vol.2:574.

7 The scutcheons of Gerard van Reede (1624-1670), lord of Nederhorst, one of Van Reede's cousins, in the Dutch Reformed Church of Nederhorst-ten-Berg, show the coats of arms of Diest, Veelen, Brandenburg, and Solms (Bloys van Treslong Prins & Belonje 1930 vol.4:185; see also NNBW vol.3 (1914):1024).

8 See chapter 1.

9 Bastiaan Stoopendael (1637-1693), draughtsman, engraver, and etcher in Amsterdam; he also engraved the plan of the Amsterdam Municipal Garden (1685) (Von Wurzbach 1910 vol.2:667; Waller & Juynboll 1938:315; Harmsen 1978: 70-73,83,135).

10 Gonsalez Appelman (?-?), engraver in Antwerp, Leiden, Amsterdam (since 1674), and Köln (GAA, DTB 693:77, DTB 1249:309, DTB 1250:82; Waller & Juynboll 1938:7-8).

7
Viridarium Orientale of Matthew of St. Joseph

In his preface to volume 3 of Hortus Malabaricus Van Reede gave all credit to the Discalced Carmelite Matthew of St. Joseph as the "conditor" or founder of the work:
'Indeed, we owe this man, who was leading a very honest life of true piety, the greatest possible gratitude, because he assisted us so honestly, faithfully, and with such untiring exertion. Nay, we rather mention this excellent man, in order to show that we venerate his memory and shall always do so, and renew the pleasant recollection of his friendship and goodwill towards us. I frankly declare that this first founder and originator of this *Hortus* has also subsequently been a sedulous cultivator'[1].

The contributions of Matthew of St. Joseph to Hortus Malabaricus were described as follows by Van Reede:
'When summoned, this venerable man did not hesitate to come to me in the town of Cochin and, when he had understood my purpose, viz., that a list and description of the plants of Malabar had to be made, for which I promised this venerable father my assistance, he very kindly and promptly offered me his services in turn. At the time he showed me some rough sketches of plants which he had formerly drawn as an aid to his memory with pen and ink, viz., one leaf, one flower, and one fruit of each plant, of which he had noted down in particular the curative virtues, with the aid of which this pious old man had helped, on his journeys, not only himself but also many others.

Thus the first beginnings of this *Hortus* were that from the rough sketches of plants, copied on other paper, arose a more accurate delineation of them through the care of the Rev. Father Matthew, who, relying solely on memory, turned out the natural forms of the plants in a wonderful way from the rough sketches. But then, because he delineated everything with a pen, and that only in outline, being inexperienced in painting with a brush, the pictures were not very much like the living plants. Moreover, the minor parts of the plants could not be shown accurately, or at least not distinctly, so that the true forms of the living plants could not be recognized at all therefrom, especially because in those pictures made with a pen neither light nor shadow could be shown somewhat accurately. Of course the excellent old man had not imagined that it would be necessary to take greater and more accurate pains in depicting and describing those plants, because he himself with the aid of his memory was indisputably acquainted thoroughly with the plants presenting themselves to him and their curative virtues. Moreover, these rough sketches had nearly always been made on his journeys, so that fairly often now the fruit of one plant, now the flowers of another plant were missing, and sometimes both. Nevertheless I corrected this defect as well as I could by seeking precisely for those plants and supplying the deficiencies. But still many plants have not been traced, either because he had not noted down the names accurately, or the venerable old man had received them from men who did not know the common names of the plants by which the botanists of Malabar designate each plant, or because they were plants from another climate. These approximately were the obstacles which disturbed the beginnings of the *Hortus Malabaricus*. Among them the greatest was that if you compared the drawings with the plants themselves, you would hardly have said that the pictures had been made from them, so poorly did they accord therewith. As to the description, this often cohered so little through perfunctory and fleeting attention that it might have referred to several plants, even if they were quite different'[2].

In short, although Van Reede fully recognized that he owed the first draft of Hortus Malabaricus to Matthew of St. Joseph, he was also aware of the insufficiency of the drawings and descriptions produced by the Carmelite. Influenced by the criticism, expressed by Paul Hermann during his visit to Malabar, on the first draft, too, Van Reede changed the plan of the project and started a new version with the aid of all the intelligentsia, collectors, and artists then available in Malabar. Although Van Reede did not explicitly state this, it is likely that Matthew henceforth mainly restricted himself to general assistance in the work.

Unfortunately, the manuscripts of Hortus Malabaricus have not been recovered, so that from them we cannot form a more detailed idea of Matthew's contributions and of Van Reede's first draft. From the above-mentioned quotations it is clear that the first draft was based on rough sketches and descriptive notes made by Matthew before 1674, the year of his arrival in Cochin under Dutch rule. It is obvious that there must be a relationship between these sketches and notes and *Viridarium Orientale*, that bulky botanical work, illustrated with more than a thousand pictures which, through the inter-

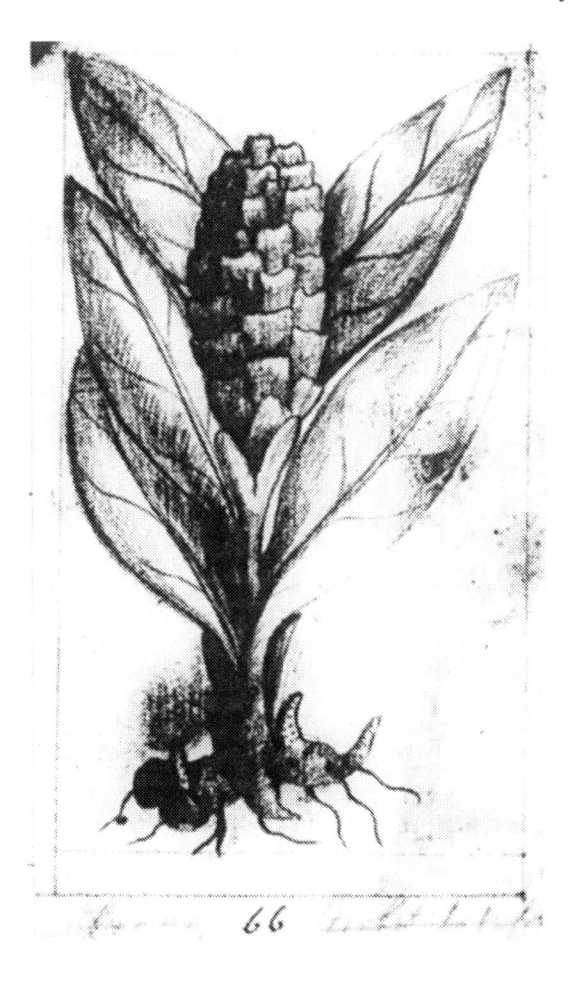

Figure 26. *Curcuma radice longa.* Engraving from Monti 1742, Tab.59.

Figure 25. Crocus Aladar. Drawing from *Viridarium Orientale*, MHN 1764, lib.1, fig.XX (Barrelier fig.66).

mediary of Giuseppe di S. Maria, Matthew had sent to Italy in 1663 to have it printed there[3]. It is not strictly necessary to assume that the sketches and notes which he initially showed to Van Reede were copies of *Viridarium Orientale*, since Matthew, after dispatching his manuscripts, might have started again to draw and describe. However that may be, Matthew's *Viridarium* may serve as a model of the first draft of Hortus Malabaricus.

The *Viridarium Orientale* has never been published as a separate work. Moreover, the manuscripts do not even appear to have survived in their entirety. As early as the 17th century they broke up into several parts, a circumstance by which a general survey is hampered.

The greater part of *Viridarium* brought to Italy by Giuseppe came into the hands of Coelestinus of St. Liduina, formerly Matthew's prior of Mount Carmel, who, in 1660, had exchanged this monastery for a professorship of oriental languages in Rome. Coelestinus' share in *Viridarium* is said to have consisted of about six hundred drawings with corresponding descriptions. He prepared an edition of his share, but owing to his premature death, in 1672, it was never published.

Another, much smaller, part came into the possession of Michael di S. Eliseo in Milan in 1669. Michael in turn passed on his share to Giacomo Zanoni[4], who was preparing a book on exotic plants written in Italian and published as *Istoria Botanica* in 1675. As late as 1671 Michael sent him a third, additional part of *Viridarium* through Valerius of St. Joseph. From the two collections he had at his disposal Zanoni inserted fifty-six of Matthew's plants into *Istoria Botanica*. Through this publication, three years before the appearance of the first volume of Hortus Malabaricus (1678), Matthew's name was thus already known to European botanists, and his plants, however moderate the published number of them as compared with the whole extent of *Viridarium* may be, found their way into the botanical literature.

In 1742 Gaetano Lorenzo Monti[5] published a latinized edition of Zanoni's book under the title of *Rariorum Stirpium Historia*. In this work, besides a bi-

Figure 28. Zambal. Engraving from Monti 1742, Tab.159.

Figure 27. Ringhini. Drawing from *Viridarium Orientale*, MHN 1764, lib.1, fig.CIII (Barrelier fig.108).

ography of Matthew, he also included some additions derived from Zanoni's manuscripts and drawings of *Viridarium*. At the same time he annotated Matthew's plants in so far as they had been published by Zanoni and partly renamed them in a more modern, though pre-Linnaean nomenclature.

While Zanoni and Monti give some idea of Matthew's drawings and descriptions, the Paris codex, based upon Coelestinus' collection, is still more interesting.

THE PARIS CODEX OF VIRIDARIUM ORIENTALE

This codex is accommodated in the library of the Musée d'Histoire Naturelle, MS 1764, in Paris. About 1667 or 1668 Coelestinus of St. Liduina submitted his version of *Viridarium Orientale* to the French Dominican friar Jacques Barrelier[6] for comment and advice in printing. Barrelier prepared a new edition. He copied all the plant descriptions from Coelestinus' version, had the drawings reduced to a smaller size, and attempted to rearrange the plants in a more suitable systematic order.

The Paris codex consists of two parts. Part 1 contains, in 57 folios, Barrelier's copy of Coelestinus' version of *Viridarium* with 232 chapters of plant descriptions referring to 256 figures, and 6 folios with an index to this, followed by 30 folios with Barrelier's own preparatory notes. Part 2 contains 111 folios with 211 reduced drawings. Barrelier's set of the drawings is not complete; at least some forty are missing. On the other hand, some drawings in this set are not discussed in the plant descriptions. The margins of the manuscript have been damaged by a fire, which had some effect on the closely written pages of the text, but the drawings are completely intact.

It must be said that the codex contains neither descriptions nor drawings by Matthew himself. But from the arguments following hereafter it may appear that the codex is very suitable to be used as a source for picturing Matthew's botanical activities. As regards the genuineness of the descriptions, Coelestinus, in a letter of 9 February 1669 to Barrelier inserted at the beginning of the codex, testified to his problems in translating Matthew's Italian

Figure 29. Giasson. Drawing from *Viridarium Orientale*, MHN 1764, lib.2, fig.XCI.

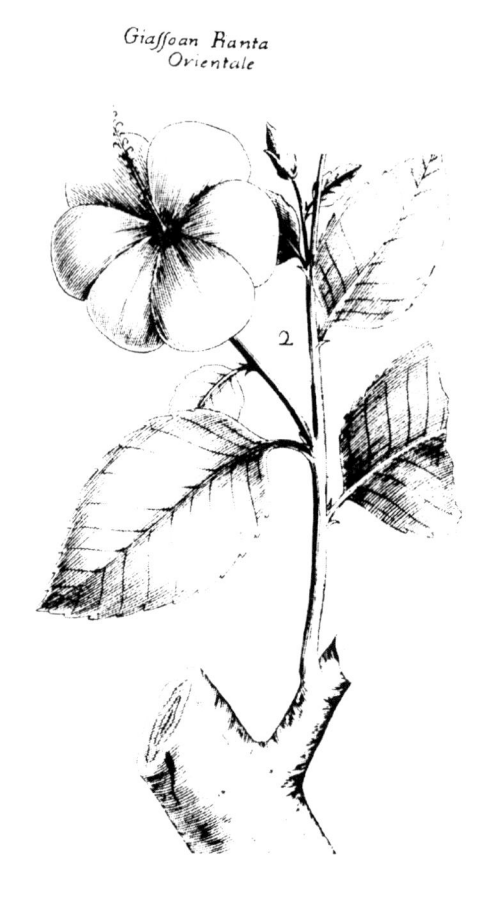

Figure 30. Giassoan. Engraving from Monti 1742, Tab.99. fig.2.

descriptions into Latin. Apparently Coelestinus did not paraphrase or summarize Matthew's texts. Indeed, his translations very closely resemble those published by Zanoni and Monti. As regards the drawings reduced by Barrelier, some of them show great similarity to the engravings published by Zanoni and Monti, for instance the drawing of Crocus Aladar in the codex (Fig.25) and the engraving of Curcuma radice longa (Fig.26)[7]. Other drawings in the codex show less similarity to the engravings, but in style and composition they are closely akin; see the drawing of Ringhini (Fig.27) and the engraving of Zambal (Fig.28)[8], and the drawing of Giasson (Fig.29) and the engraving of Giassoan (Fig.30)[9].

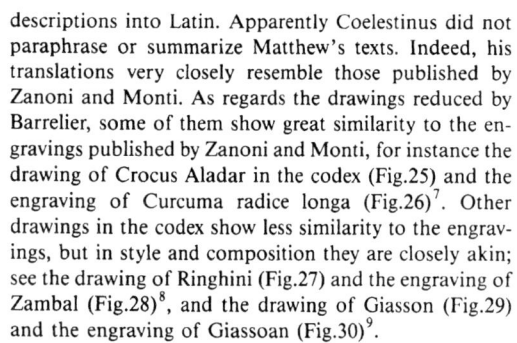

If we add the fifty-six figures and descriptions from Zanoni and Monti to the Paris codex, we have at our disposal 267 figures and 288 chapters of descriptions originating from Matthew's *Viridarium Orientale*. On the strength of them we can make the following remarks.

Drawings

Both Coelestinus and Zanoni confirm Van Reede's statement that Matthew made his drawings himself 'with pen and ink'. To the description of *Cassia fistula* in the codex Coelestinus added:

'the Reverend father Matthew, formerly my beloved colleague in the Mission to the East, who, from the Indies, sent me these figures delineated with his own hand and elucidated briefly'[10].

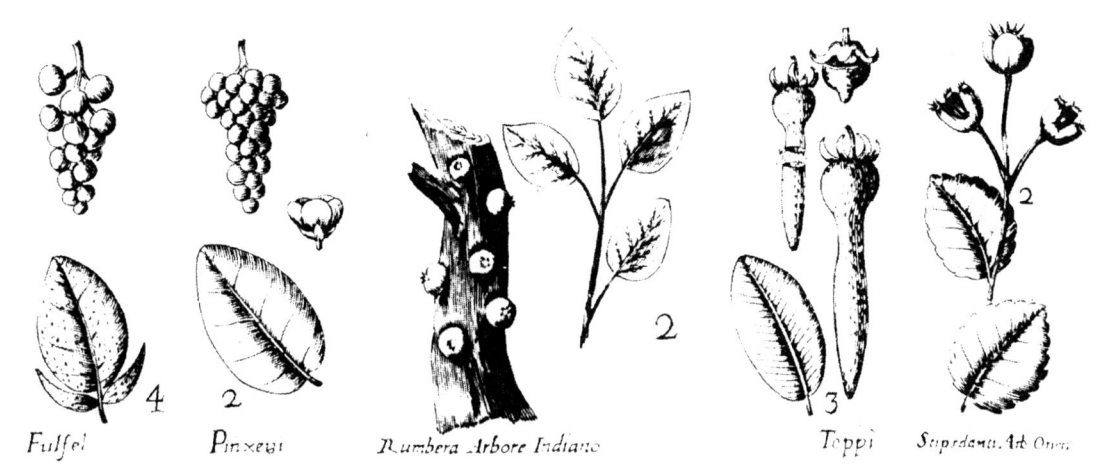

Figure 31. Fulfel, Pinxevi, Rambora, Toppi, and Stipsdanti. Details of engravings from Monti 1742, Tab.154. fig.4, Tab.165. fig.2, Tab.183. fig.2, Tab.165. fig.3, and Tab.75. fig.2.

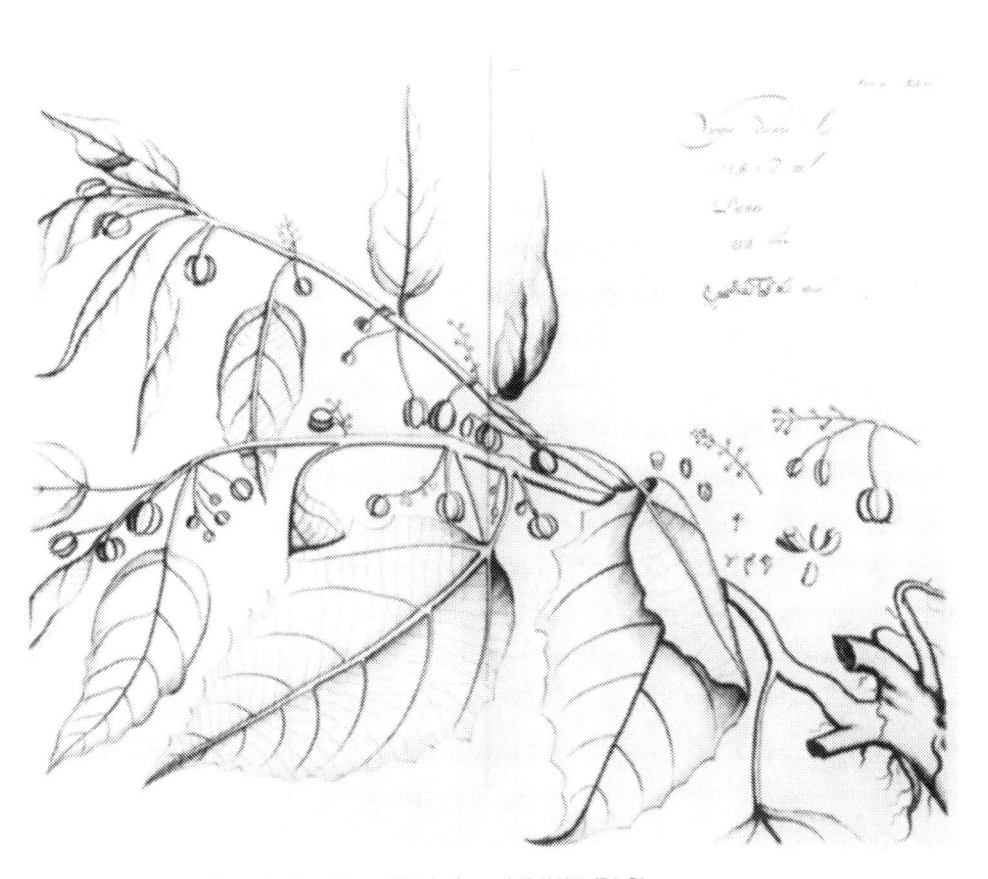

Figure 32. Naga-dante. Engraving from Hortus Malabaricus vol.10 (1690), Tab.76.

Figure 33. Edible bird's nests. Engraving from Monti 1742, Tab.185.

Figure 35. Bakeli. Engraving from Monti 1742, Tab.28.

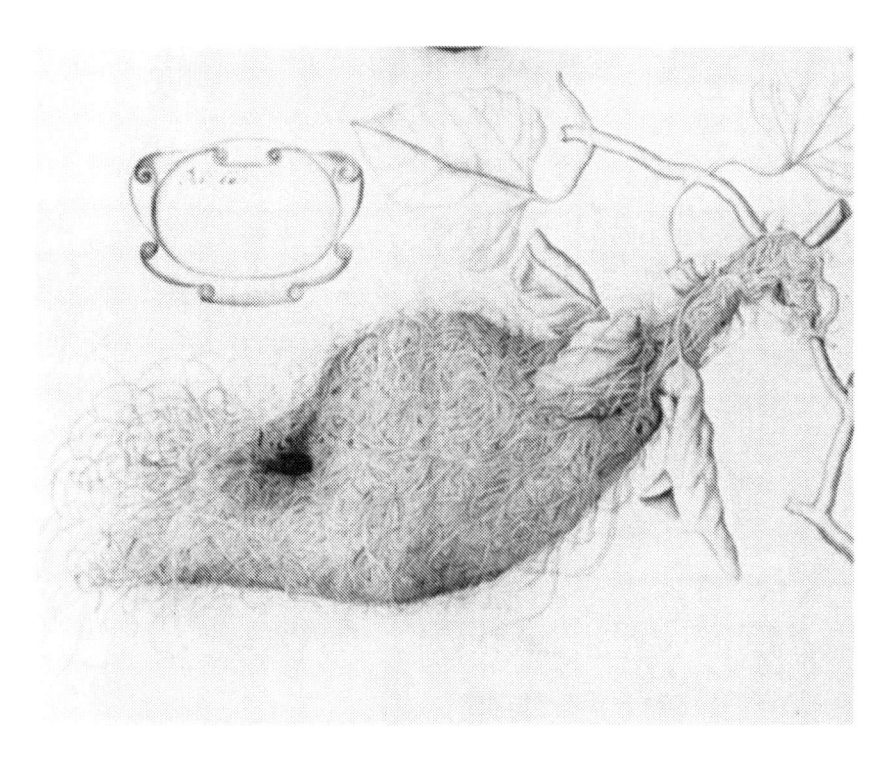

Figure 34. Paki-kudi, edible bird's nests. Drawing from *Horti Malabarici Icones*, BL Add MS 5030:124-125.

Figure 36. Totta Vari. Engraving from Monti 1742, Tab.131. fig.2.
Figure 37. Todda-vaddi. Engraving from Hortus Malabaricus vol.9 (1689), Tab.19.

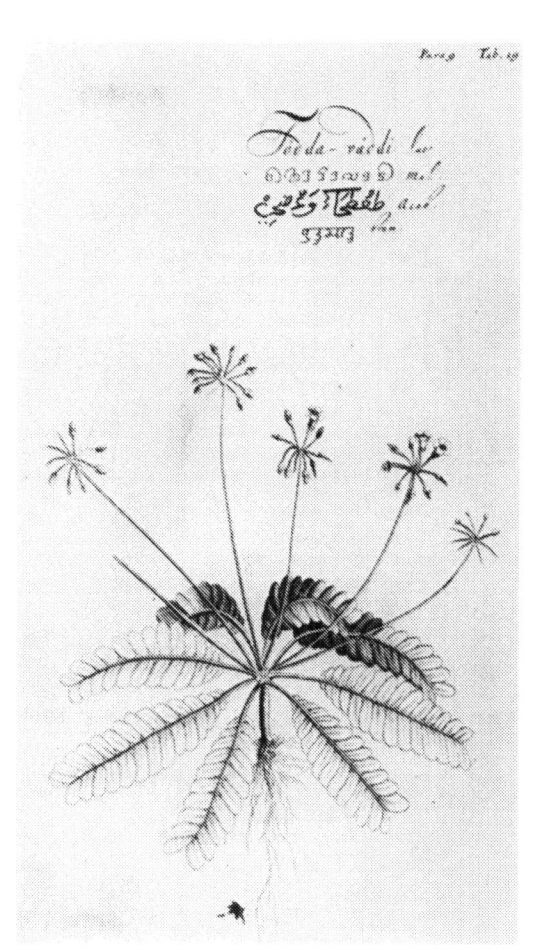

In the case of *Archam Indiae Orientalis arbor* Monti remarked:
'Matthew of St. Joseph ..., who spent his life in those remote regions, has drawn this picture with a writing-pen'[11].

In the case of *Mandragora*, observed by Matthew in Lebanon Mountain in his early days, Coelestinus wrote:
'the author of these books, the Reverend father Matthew of St. Joseph, who has drawn the singular plants from life with his own hand'[12].

Van Reede's remark that Matthew depicted only 'one leaf, one flower, and one fruit of each plant' agrees only partly with the available figures. Generally *Viridarium* contains three types of drawings. Indeed, Zanoni and Monti met with some twelve figures in their collection showing hardly recognizable plants with no more than separate flowers, fruits, and leaves, or an inflorescence: Fulfel, Pinxevi, Rambora, Toppi, and Stipsdanti (Fig.31)[13]. Only in the case of Stipsdanti did Monti venture to refer to botanical literature, namely Van Reede's Naga-dante HM 10:151:76 (Fig.32). On the other hand not a single plant of this type figures in the codex.

Another type is represented by some very stiffly drawn plants. Matthew copied them from the book of Saladin Artafa, at Basra, where he stayed in 1648-1651. In the codex no figure of this type occurs, although the descriptions sometimes include notes derived from Artafa's book.

For the greater part, however, Matthew's pictures are rather well-executed drawings of plants, consisting of herbs with roots and cut-off trunks and branches of trees and shrubs, usually showing flowers and fruits, sometimes with separate details of them. But the flowers and fruits have been executed in such a sketchy manner that they hardly offer any starting-point for determination. Before the leaves were drawn, they were pressed, in order to get accurate shapes, but as a result the plant depicted made a distorted impression. Figures of this type recall

Figure 38. Muri Kuti. Drawing from *Viridarium Orientale*. MHN 1764, lib.2, fig.LXXVII (Barrelier fig.82).

Figure 39 (opposite page). Muriguti. Engraving from Hortus Malabaricus vol.10 (1690), Tab.32.

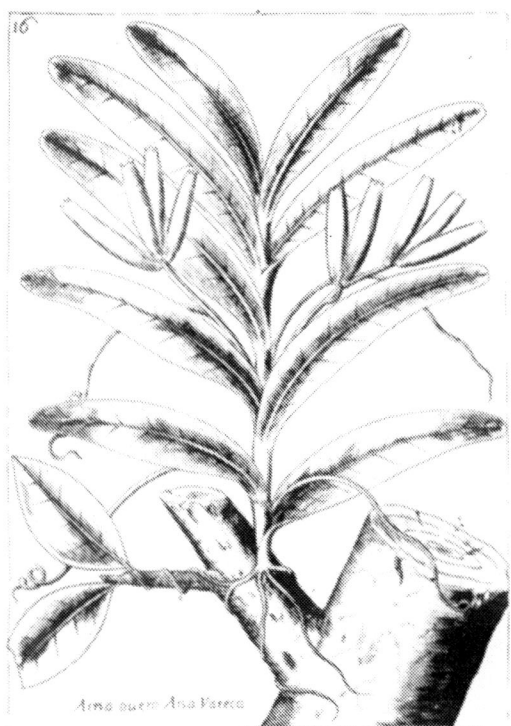

Figure 40. Arna Vareca. Engraving from Monti 1742, Tab.9.

Figure 41 (opposite page). Thalia maravara. Engraving from Hortus Malabaricus vol.12 (1693), Tab.4.

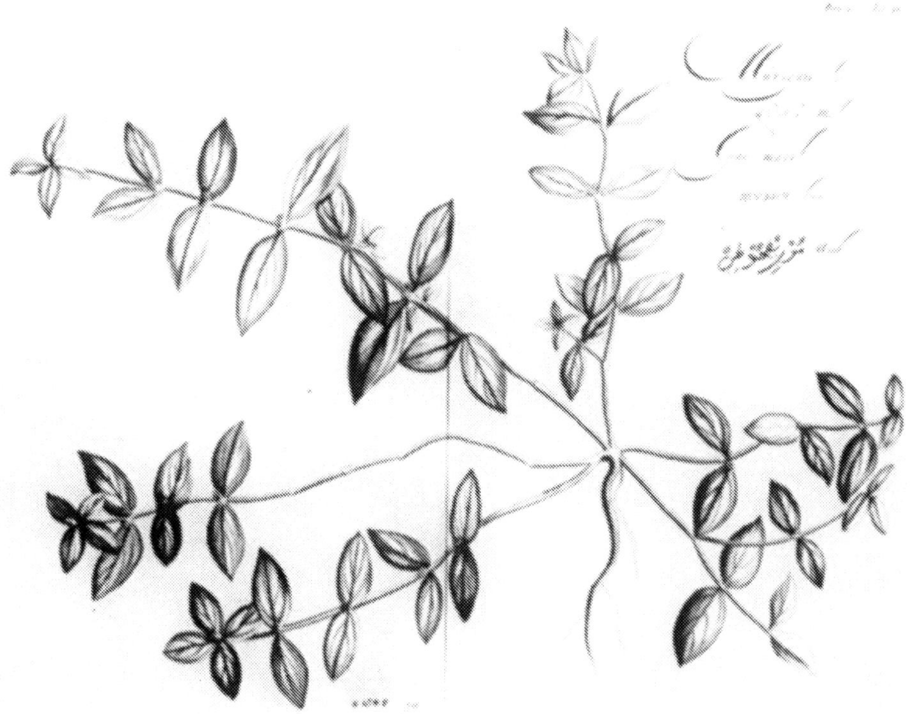

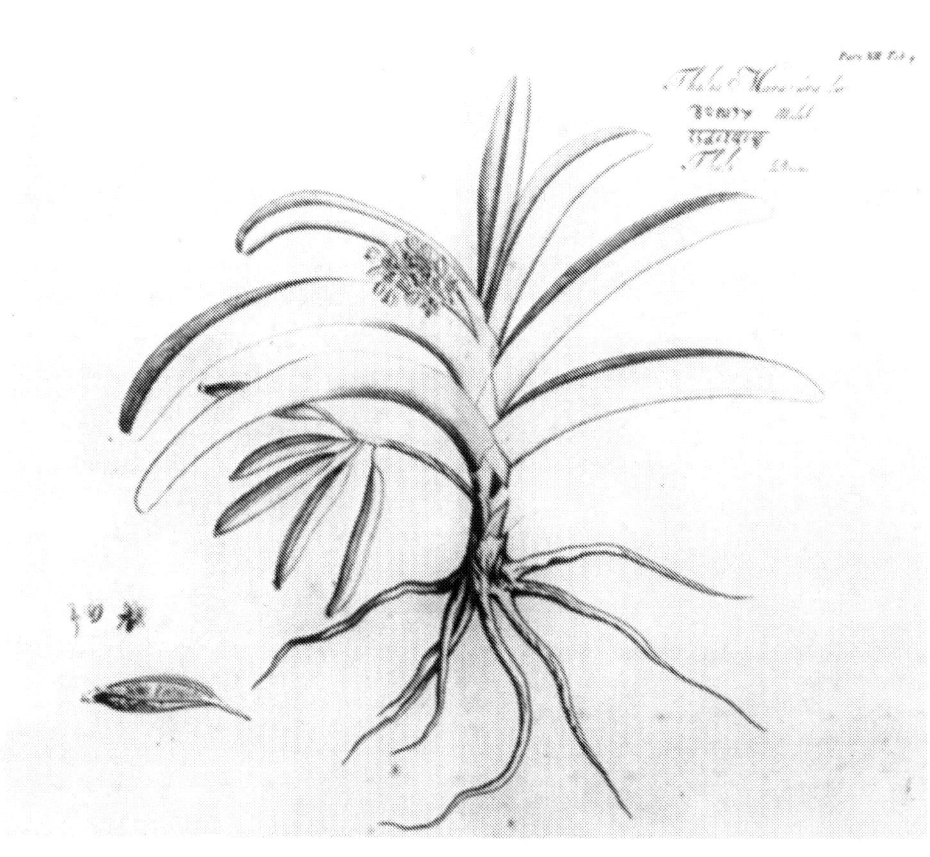

Figure 42. Arkasond. Engraving from Monti 1742, Tab.9.

Figure 43 (opposite page). Odallam. Engraving from Hortus Malabaricus vol.1 (1678), Fig.39.

Figure 44. Igera muri. Drawing from *Viridarium Orientale*, MHN 1764, lib.1, fig.LVI (Barrelier fig.62).

Figure 45 (opposite page). Isora-murri. Engraving from Hortus Malabaricus vol.6 (1686), Tab.30.

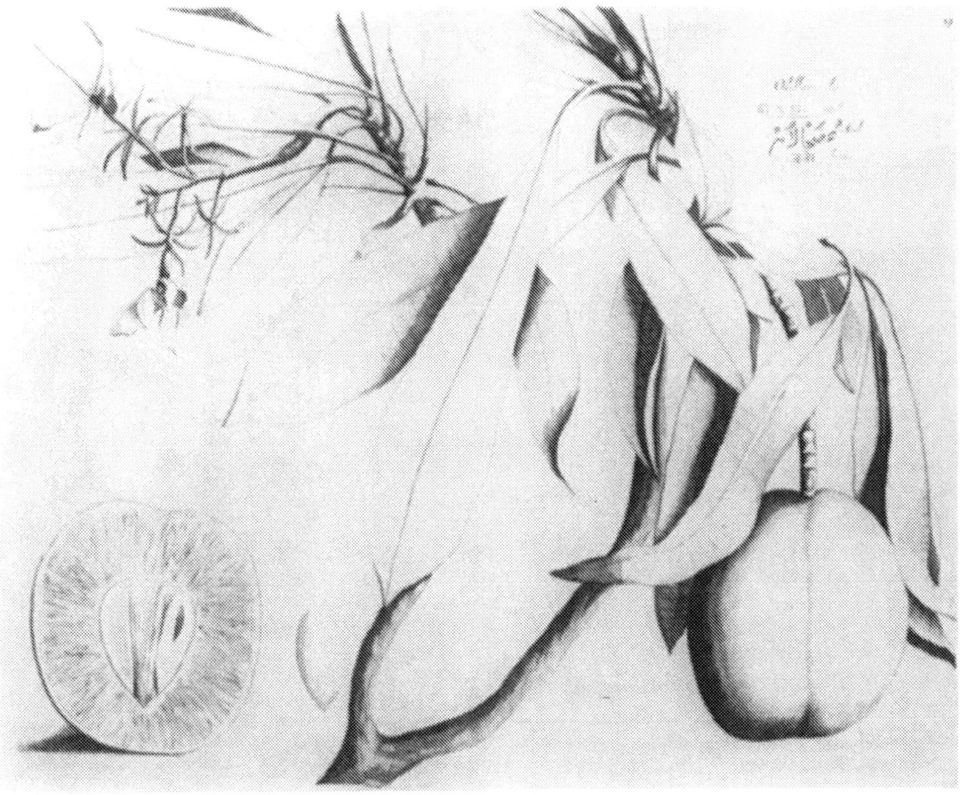

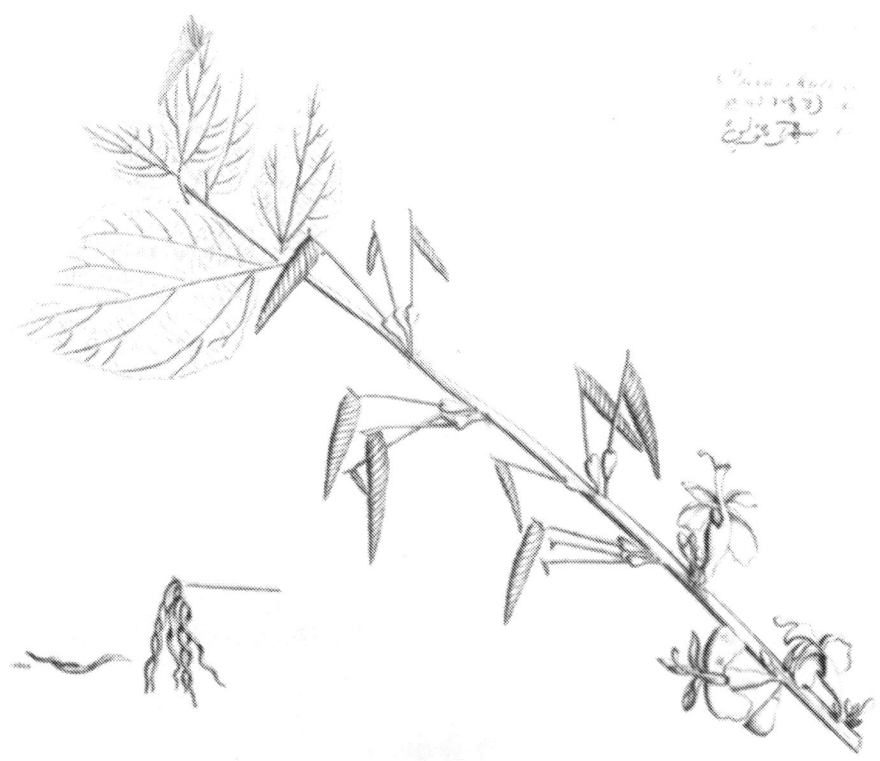

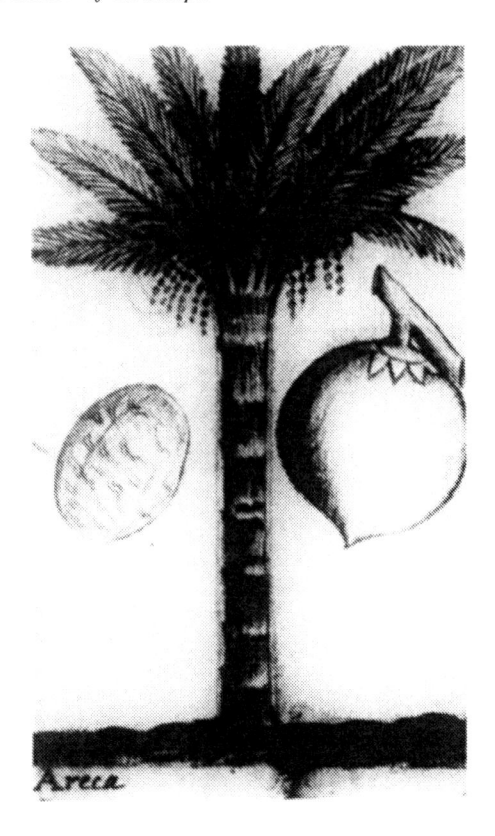

Figure 46. Areca. Drawing from *Viridarium Orientale*, MHN 1764, lib.1, fig.XXXIX (Barrelier fig.45).

Figure 47 (opposite page). Caunga. Engraving from Hortus Malabaricus vol.1 (1678), Fig.5.

Figure 48. Zingiber. Drawing from *Viridarium Orientale*, MHN 1764, lib.1, fig.LVII (Barrelier fig.63).

Figure 49 (opposite page). Inschi. Engraving from Hortus Malabaricus vol.11 (1692), Tab.12.

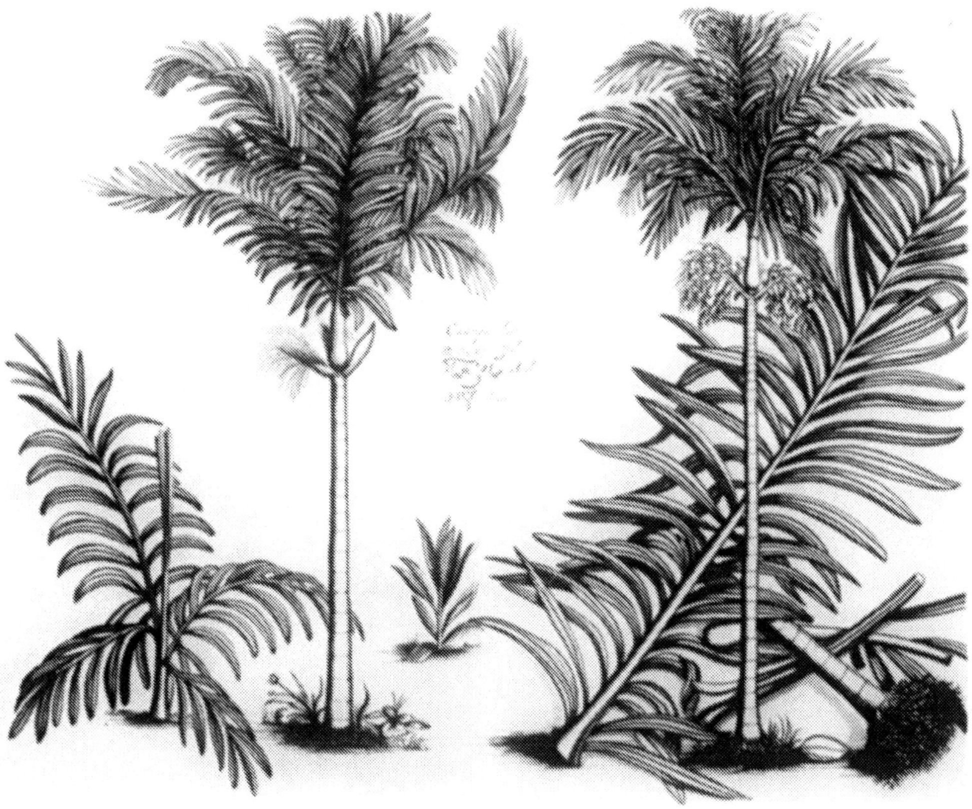

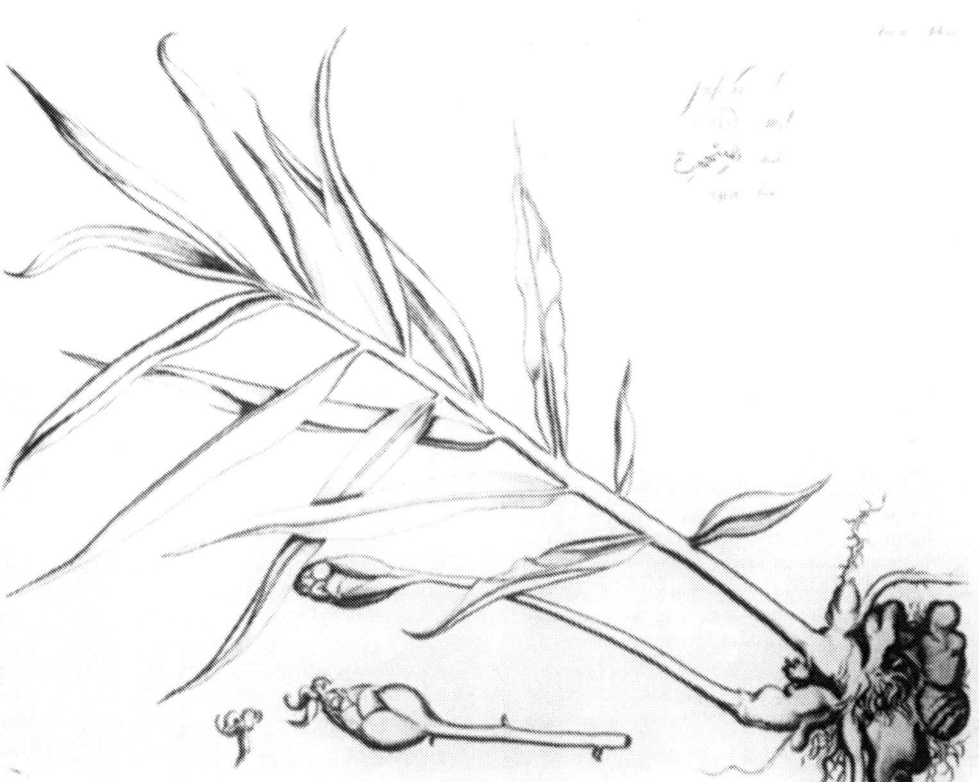

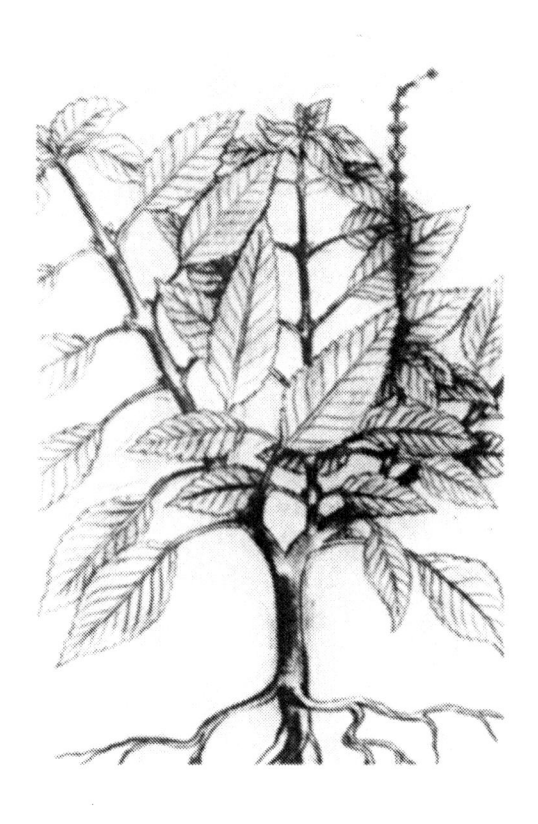

Figure 50. Kata Tartavè. Drawing from *Viridarium Orientale*, MHN 1764, lib.1, fig.LXXVII (Barrelier fig.83).

Van Reede's remark about Matthew's more elaborate sketches for the first draft of Hortus Malabaricus, namely 'that the true forms of the living plants could not be recognized at all therefrom'.

In general *Viridarium* reflects the route of Matthew's travels in the East in the years 1644 to 1663. In many of his descriptions he mentioned localities, regions, and countries where he observed his plants or from which he borrowed native names, and mentioned medicinal virtues and uses. The Paris codex has descriptions and drawings of plants growing in Cyprus, Palestine, Lebanon, Syria, Persia, Mozambique, India, and Ceylon. Zanoni and Monti published only plants from the four last-mentioned countries.

In connection with Hortus Malabaricus, Matthew's Indian plants are the most interesting ones. From Diu, his residence from 1651 to 1656, he discussed Turbith and Nai[14]. In his extensive discussion of palms he referred to Kaginri in the hills of Surat and Damao[15]. And from Goa he had Danti and some Aloes[16]. From Coromandel, whence there is no biographical evidence that he ever visited that region, he drew edible bird's nests (Fig.33)[17], and from the neighbourhood of Nagappattinam he pictured the Bakeli tree with curious birdlike fruits (Fig.35)[18]. Finally, some twenty-five plants were attributed to Malabar by Matthew. He mainly confined himself to geographical indications such as 'in the province of Malabar' or 'in the region of the Malabars'; he rarely mentioned a settlement or town, and only once a river. Of a considerable number of plants he exclusively mentioned Malabar native names, without making it clear whether he described and pictured those plants from specimens occurring in Malabar.

One of the very first plants observed by Matthew in the province of Malabar was Tota Vari. During the mission of the Carmelites, led by Giuseppe di S. Maria in Malabar in 1657-1658, he drew this species (Fig.36) and noted down:

'This wonderful herb is to be found in the region of the Malabars and is named Tota vari by the natives: there I set eyes on it on the octave of St. John the Baptist [1 July] in the year 1657'[19].

Monti identified this presumably first historical recording of a published figure of a Malabar plant with Van Reede's Todda-vaddi HM 9:33-34:19 (Fig.37), *Biophytum sensitivum* (L.) DC. Of all the other Malabar plants Matthew did not mention any dates of observation.

Other examples of Malabar plants are Muri Kuti, 'a

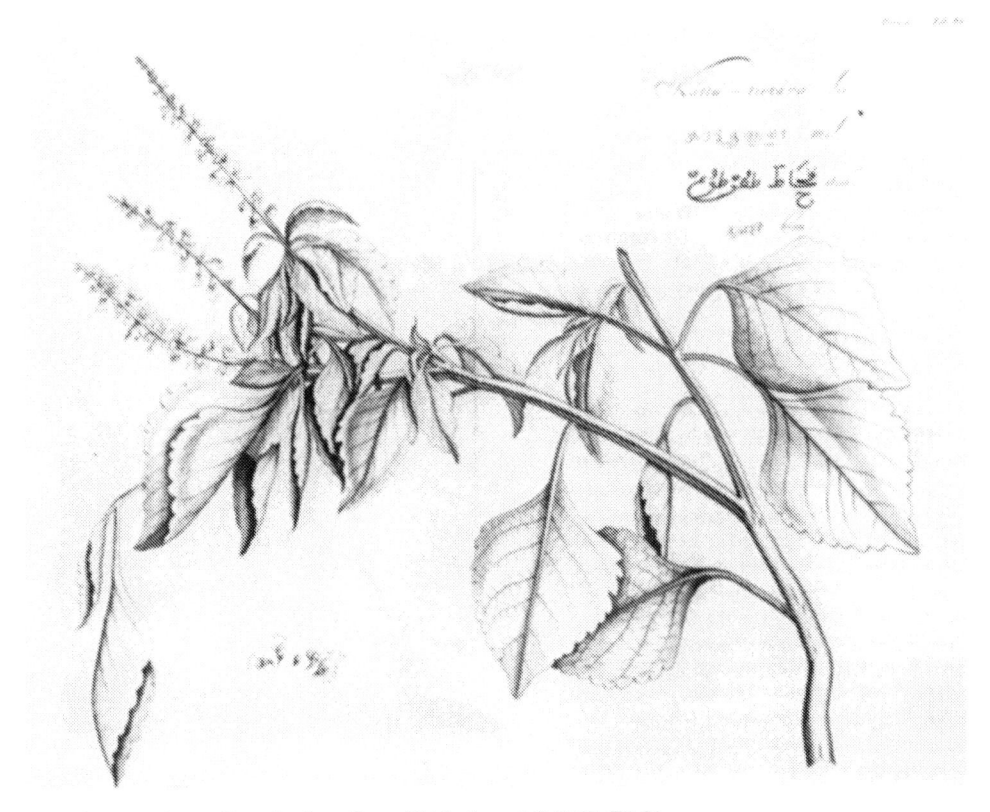

Figure 51. Kattu tirtava. Engraving from Hortus Malabaricus vol.10 (1690), Tab.86.

herb found among the Malabars' in the codex (Fig.38)[20], corresponding to Van Reede's Muriguti HM 10:63:32 (Fig.39), *Hedyotis Auricularia* L., and the epiphyte Arna Vareca in Zanoni (Fig.40), identified by Monti[21] with Van Reede's Thalia maravara HM 12:9:4 (Fig.41), *Epidendrum furvum* L. A habitat mentioned more accurately by Matthew is that of Arkasond (Fig.42), 'in a small brook in Mangati'[22], by which will be meant a small river in Ālangādu, in the centre of the missionary province of the Church of the Serra. This species resembles, according to Monti, Van Reede's Odallam HM 1:71:33 (Fig.43), *Cerbera Manghas* L., the Mangas Tree.

From Cochin, Matthew discussed the root of Igera muri in the codex (Fig.44)[23]. This species also figures in Zanoni's list of Matthew's Malabar plants transmitted by Valerius of St. Joseph, which Jan Commelin connected with Van Reede's Isora-murri HM 6:55-56:30 (Fig.45), *Helicteres isora* L. From Cochin, and also from Canara, Matthew mentioned Areca in the codex (Fig.46)[24], corresponding to Caunga HM 1:9-10:5-8 (Fig.47), *Areca Cathecu* L., the Areca or Betel-nut Palm. From Malabar, especially from Calicut, Matthew praised the high quality of the roots of ginger in his chapter on Zingiber in the codex (Fig.48), corresponding to Van Reede's Inschi HM 11:21-23:12 (Fig.49), *Amomum Zingiber* L.

Some of the plants mentioned by Matthew only by their Malabar native names are also to be found in Hortus Malabaricus. In the codex he described Kata Tartavè (Fig.50)[26], corresponding to Kattu tirtava HM 10:171:86 (Fig.51), *Ocimum gratissimum* L. On *Artocarpus integrifolia* L., the Jack Fruit Tree, originally a native from the forests of the Western Ghats, but widely cultivated for its enormous edible fruits, Matthew wrote in the codex (Fig.52):

'This is a large tree such as laurus or quercus, named by the Malabars giacca or Jaka; it has different, large and thick leaves. By the natives, however, it is named Canara and Guzerat pana or phanas'[27].

He depicted this tree several times, for Zanoni published two other drawings (Figs 53 and 54). The curious addition of the artillerist with the firing gun placed on a cut-off trunk, apparently served to emphasize the huge size of the tree and its fruits[28]. Monti identified it with Van Reede's Tsjaka-maram HM 3:17-20:26-28 (Fig.55).

One of the flowers in Malabar most beloved among the Carmelites was the Mountain Ebony, also known as the St. Thomas Flower, thus named after the apostle who brought Christianity to India. Although Matthew did not explicitly state that he observed this plant in Malabar, one may assume that his drawings were based upon Malabar specimens. In the codex, under Mandaru (Fig.56), he described the plant in only a few words[29]. Monti, whose father received an exsiccate from Batavia many years previously, explained that the sanguine splashes on the purple or pink petals are reminiscent of the martyrdom of St. Thomas (Fig.57)[30]. He identified the engraving of Matthew's drawing with Van Reede's Chovanna-mandaru HM 1:57-58:32 (Fig.58), *Bauhinia variegata* L.

Plant names

From both the codex and Zanoni it appears that Matthew did not attempt to connect his plants with pre-Linnaean nomenclature in the botanical literature then current. Probably he did not have relevant works at his disposal during his travels. The few quotations from Da Orta, Matthioli, and others in the codex seem to have been inserted by Coelestinus or Barrelier. Matthew confined himself to the native names then current as he learned to know them during his travels. His great interest in oriental languages may have contributed to his special attention to native names from Arabic, Sanskrit, Malayalam, Tamil, and Sinhala. However, the small number of Malabar plants in *Viridarium Orientale* cannot show decisively whether Matthew was actually master of Malayalam. Anyhow, he did have some knowledge of this language, for now and then he gave an explanation or a translation of a native name. We can compare some of his Malabar native names with those in Malayalam script on Van Reede's engravings, a main source of 17th-century Malayalam in this field.

Matthew's plant name Mandaru is a reasonable Roman transcription of the Malayalam name mantāram, and in Igera muri one can recognize isāramūri without much difficulty. Caida ciaka is apparently Matthew's transcription of kaita (or kayita) cakka. His Tota vari is not far removed from tottā vati. Sometimes his spelling reveals his Italian origin, as in giacca and ciaka, transcribed from the Malayalam cakka, which sounds tsjakka or tsjaka in Dutch transcriptions in Hortus Malabaricus and jack in English transcriptions[31].

Classification

The absence of pre-Linnaean nomenclature in *Viridarium Orientale* raises the question whether Matthew applied any systematic arrangement, a problem we shall also encounter in the editing of Hortus Malabaricus. Neither Zanoni nor Monti gives any indication, but in the Paris codex, in spite of its complicated structure, one can distinguish three different arrangements. The drawings, in Part 2, have been arranged in three "libri" or books: book 1 on trees, book 2 on shrubs, and book 3 on herbs. Each book is preceded by its own title-page.

The plant descriptions, however, have been arranged not in three, but in two books. This arrangement is the

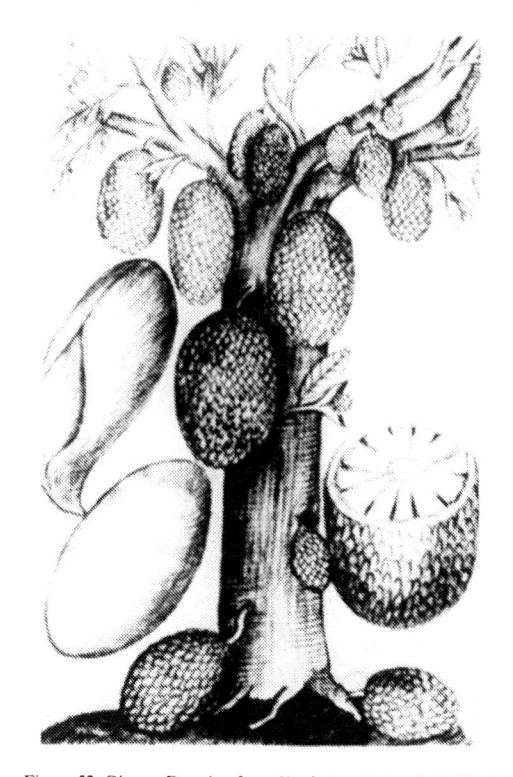

Figure 52. Giacca. Drawing from *Viridarium Orientale*, MHN 1764, lib.1, fig.XXV.

Figure 53. Jaca. Engraving from Monti 1742, Tab.90.

Figure 54. Jaca. Engraving from Monti 1742, Tab.91.

Figure 55 (below). Tsjaka-maram. Engraving from Hortus Malabaricus vol.3 (1682), Tab.26.

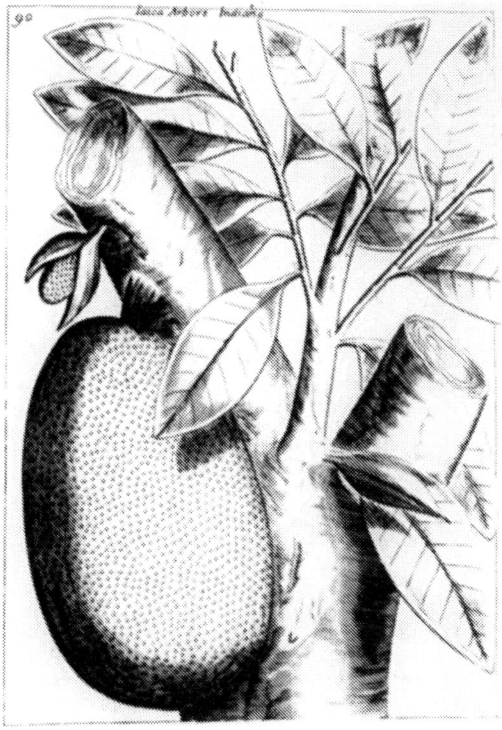
90

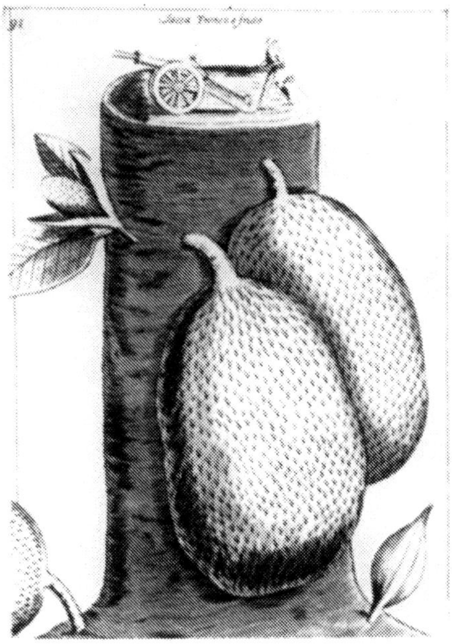
91

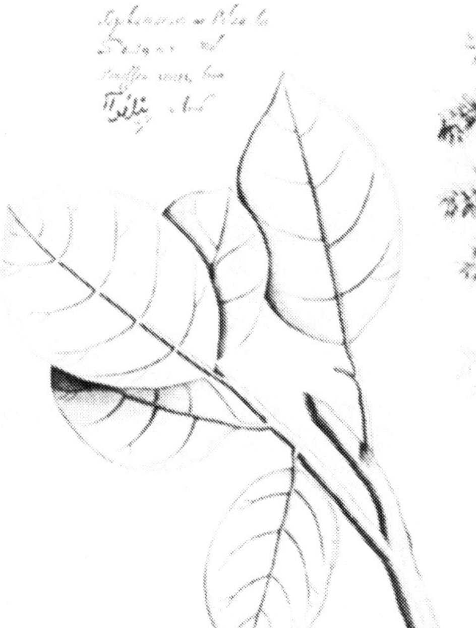

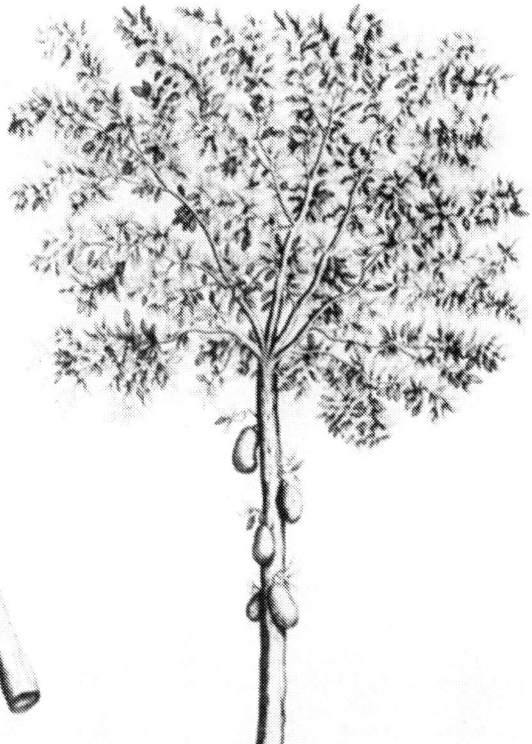

Figure 56. Mandaru. Drawing from *Viridarium Orientale*, MHN 1764, lib.1, fig.LXXXI (Barrelier fig.87).

Figure 57. Assitra. Engraving from Monti 1742, Tab.20.

result of Coelestinus' adaptation of Matthew's original manuscripts. The sequence of the figures to which Coelestinus refers in his version is completely different from that of the drawings. From numerous pencil notes it appears that Barrelier took great pains to restore the connection between the drawings and the descriptions. He also composed an extensive index of the plant names in Coelestinus' version, with many cross-references to synonymous native names and a concordance to the arrangement of the drawings in three books. Finally, Barrelier attempted to adapt Coelestinus' arrangement of the descriptions to the Bauhinian plant system. Thus he regrouped some ninety plants into taxa such as *Siliquosae, Pomiferae, Coniferae, Spinosae,* and *Asymbolae.*

From the above it may be concluded that the arrangement of the drawings in three books, together with Barrelier's concordance, reveals Matthew's original arrangement of *Viridarium Orientale* in trees, shrubs, and herbs.

A more detailed classification within one book is hardly noticeable. This is partly due to the absence of a number of drawings corresponding to the descriptions, and partly to the indeterminability of another number of them, such as the copies from Saladin Artafa. In general Matthew combined important fruit trees such as palms, areca, jambo, and carambola; among the shrubs, he discussed some mimosacea in sequence, and combined a number of convolvulaceous herbs. In his descriptions he occasionally used terms such as "genus" and "species" in a pre-Linnaean pragmatic sense. His "genus sarmentosum" comprised both woody and herbaceous climbing-plants, of which he placed the "genus stipitum et sarmentosum", or woody climbers, in book 2 on shrubs. Among the herbaceous climbers, mainly convolvulacea, Matthew distinguished a special "species cucumeris", or cucumbers, and a "genus vitilaginosum", or tendrilous plants. In book 3 on herbs the climbers are followed by "herbae serpentes super terram", or creepers. Among the edible herbs he distinguished a "genus olerum", or cabbages, and a "genus obsoniorum", or condiments. From his notes it appears that Matthew, as a physician, did not use any classification according to medical virtues and uses, or diseases.

From these fragments of classification one may infer that Matthew had some notion of plant affinities, but felt satisfied with arranging his plants according to simple features of general forms and practical uses.

Plant descriptions

A more definite system is to be found in Matthew's

Figure 58. Chovanna-mandaru. Engraving from Hortus Malabaricus vol.1 (1678), Fig.32.

descriptive method. As a rule he opened a description with the general determination of tree, shrub, or herb, sometimes more precisely classified by a "genus" or "species". Then he mentioned the habitat, locality, and finding place, the name of the plant, and its synonyms in other languages. He noted sizes in fingers, palms, feet, and yards, or simply mentioned them as large, medium, or small. Further he characterized leaves, flowers, fruits, and seeds by a few botanical terms, or otherwise he applied the "similis" method, referring mainly to Italian plants. He also observed colour, odour, and taste of the different parts. Matthew did not restrict himself to the description of only one specimen from a single locality. Usually he accumulated different information from different regions. Occasionally he discussed important vegetable products in detail. Finally, he paid much attention to medical virtues and uses, as well as preparations of medicaments.

In fact, *Viridarium Orientale* was no more than Matthew's medico-botanical notebook for the consultation of relevant information on medicinal plants during his travels, or in Van Reede's words:

'he had noted down in particular the curative virtues with the aid of which this pious old man had helped, on his journeys, not only himself, but also many others'.

Notes

1 Hortus Malabaricus vol.3 (1682):(ix); Heniger 1980:47.
2 Hortus Malabaricus vol.3 (1682):(vii)-(viii); Heniger 1980: 45-46.
3 For more detailed information on Matthew and the adventures of *Viridarium Orientale* in Malabar and Italy, see also chapter 2, section Commander of Malabar 1670-1677.
4 Giacomo Zanoni (1615-1682), professor of botany and director of the botanical garden of Bologna University (Monti 1742:(vii)-(xxi); Saccardo 1895 vol.1:176; 1901 vol.2:116).
5 Gaetano Lorenzo (or Cajetanus) Monti (1712-1797), professor of botany at Bologna University (Saccardo 1895 vol.1:112).
6 Jacques Barrelier (1606-1673), French botanist, travelled in France, Spain, and Italy. His main work, *Plantae per Galliam, Hispaniam et Italiam observatae* (Paris, 1714), was posthumously published by Antoine de Jussieu (Dictionnaire de Biographie française vol.5 (1951):579).
7 MHN, MS 1764,lib.1,fig.XX; Zanoni 1675:16,fig.LXXX; Monti 1742:86,Tab.59.

8   MHN, MS 1764,lib.1,fig.CIII; Zanoni 1675:208; Monti 1742:208,Tab.159.
9   MHN, MS 1764,lib.2,fig.XCI; Monti 1742:136,Tab.99.fig.2.
10 MHN, MS 1764,lib.1,fig.LXXXIV.
11 Monti 1742:20.
12 MHN, MS 1764,lib.2,fig.LVIII.
13 Zanoni 1675:98,fig.LXXV; 160,fig.LXXIV; 166,fig.LXXIX; 198-199,fig.LXXIV; 190,fig.XXXII; Monti 1742:117, Tab.154,fig.4; 183,Tab.165.fig.2; 190,Tab.183.fig.2; 223,Tab.165.fig.3; 191,Tab.75.fig.2.
14 MHN, MS 1764,lib.2,fig.XV: Turbith, no drawing present; lib.1, fig.XLII: Nai.
15 MHN, MS 1764,lib.1,fig.XXVI: Kaginri, no drawing present.
16 MHN, MS 1764,lib.1,fig.LXI: Danti, and fig.CIX: Aloe.
17 MHN, MS 1764,lib.2,fig.III, no drawing present; Zanoni 1675:21,fig.VI; Monti 1742:239-240,Tab.185.
18 Monti 1742:43,Tab.28.
19 Monti 1742:221-223,Tab.131.fig.2; Zanoni 1675:199-200,fig.LXI.
20 MHN, MS 1764,lib.2,fig.LXXVII.
21 Zanoni 1675:29,fig.IX; Monti 1742:23-24,Tab.16.
22 Zanoni 1675:24,fig.VIII; Monti 1742:12-13,Tab.9.
23 MHN, MS 1764,lib.1,fig.LVI.
24 MHN, MS 1764,lib.1,fig.XXXIX.
25 MHN, MS 1764,lib.1,fig.LVII.
26 MHN, MS 1764,lib.1,fig.LXXVII.
27 MHN, MS 1764,lib.1,fig.XXV.
28 Monti 1742:127,Tab.90-91.
29 MHN, MS 1764,lib.1,fig.LXXXI.
30 Zanoni 1675:26,fig.XV; Monti 1742:32-33,Tab.20. Monti's father was Giuseppe Monti (1682-1760), professor of botany and director of the botanical garden of Bologna University (Saccardo 1895 vol.1:112; 1901 vol.2:74).
31 Govindankutty 1983:244,247,252-254.

8
Drawings of Hortus Malabaricus

According to the hand-written catalogue of accessions to the Department of Manuscripts of the British Library, London, the British Museum purchased in Holland, about 1771:
"Five volumes, in folio, containing the original drawings in indian ink, of the plates in the work entitled 'Hortus Malabaricus', published in Amsterdam in twelve parts, folio, 1678-1703. The artist from the preface of the work, appears to have been the Carmelite friar named Matthaeus à S. Joseph".

Today this codex forms Additional Manuscripts 5028-5032.

The volumes were bought, in fact, at the auction of the collections of Bernard Siegfried Albinus in Amsterdam on 16 October 1771. The Albinus sale catalogue reads:

'The original drawings of the Hortus Malabaricus gathered by J. Commelin in Amsterdam 1686 being the drawings of the first volumes after which the plates of the preceding work have been made. 10 volumes, 5 ribs'[1].

The drawings correspond as to number and order to the engravings in volumes 1-10 of Hortus Malabaricus. As the sale catalogue states, this collection does not contain the drawings for the last two volumes. However, Jan Commelin, the principal commentator of the published work, wrote in his own hand on the title-page of the first volume of the codex that he had gathered 'all the original drawings' in 1686. This means that he must have made up a sixth volume with drawings corresponding to volumes 11 and 12 of the published work. Consequently, this sixth volume must have been detached from the codex between 1686 and 1771. An explanation of this dissociation would be that after Commelin's death, in 1692, when volume 11 of Hortus Malabaricus was published, the sixth volume was handed over to Abraham van Poot, the last editor, who was to prepare the publication of the 12th and final volume of the work. The first five volumes of the codex might have been left in the possession of Commelin's heirs, from which Albinus acquired them on a later, unknown date. The further vicissitudes of the sixth volume after Van Poot's death, in 1711, are not known. Up to the present the drawings for volumes 11 and 12 have not been traced.

DESCRIPTION

The drawings are bound in vellum and lettered on the spine "Horti Malabarici Icones Mus Brit Iuxe Emptionis". Each of the volumes measures 37.5 x 25.4 cm.

The codex contains 651 drawings (590 in double-folio and 61 in folio) and 3 separate text-leaves, which were numbered with pencil in 1979. Each drawing is followed by an unnumbered interleaf.

All the corresponding drawings and engravings exactly reflect each other. For a complete survey of their concordance, see Part Three.

BL Add MS 5028

The volume consists of 2 unnumbered folios, 227 numbered folios with unnumbered interleaves, and 3 unnumbered folios.

Folio 1 (text-leaf) reads: "Het Eerste Deel van alle de Originele Tekeningen vanden Hortus Malabaricus Bijeen vergadert door Jan Commelijn in Amsteldam A° 1686" (The first volume of all the original drawings of the Hortus Malabaricus gathered by Jan Commelin in Amsterdam in the year 1686) (Fig.59).

Folios 2-115 include all the 57 drawings in double-folio as published in volume 1 of Hortus Malabaricus. On folios 2-3, the drawing of Tenga, bottom left: "Antoni jacobus goedkint fecit"; on folios 4-5, another drawing of Tenga, top right: "A.J.G.f.".

Folio 116 (text-leaf) reads: "Originele Tekeningen vanden Hortus Malabaricus Het tweede Deel" (Original drawings of the Hortus Malabaricus, the second volume).

Folios 117-227 include all the 56 drawings (55 in double-folio and 1 in folio) as published in volume 2 of Hortus Malabaricus. On folio 227v: "Pars Secunda".

BL Add MS 5029

The volume consists of 3 unnumbered folios, 250 numbered folios with unnumbered interleaves, and 2 unnumbered folios.

Folios 1-126 include 63 drawings in double-folio, all published in volume 3 of Hortus Malabaricus, but the drawing for Neli-pouli HM 3, Tab.48, is missing. Folio 1 reads: "Hortus Malabaricus Pars 3" and folio 3 reads: "Hort: malab. Pars. 3 qui continet Arbori: Fructifer: houdende 63 figuren. sijnde 32 boomen 57 Figuren 26 bomen. 37 struike" (Hortus Malabaricus volume 3, which contains fruit-bearing trees, comprising 63 figures, being 32 trees, 57 figures, 26 trees, 37 shrubs).

Folios 127-250 include 62 drawings in double-folio, of

Figure 59. Title-page of volume 1 of the drawings of *Horti Malabarici Icones*, BL Add MS 5028:1.

Figure 61. Performance of a fakir. Drawing (unpublished) by Marcelis Splinter from *Horti Malabarici Icones*, BL Add MS 5030:122-123.

Figure 60. Unfinished drawing (not published) from *Horti Malabarici Icones*, BL Add MS 5029:247-248.

Figure 62. Watermark of the crowned Strasbourg lily in BL Add MS 5027.

Figure 63. Tinda-parua. Drawing from *Horti Malabarici Icones*, BL Add MS 5028:96-97.

which 61 were published in volume 4 of Hortus Malabaricus, and one is an unpublished drawing, on folios 247-248. Folio 127 reads: "Het 4e deel 56 fig: Nucifer:" (The 4th volume, 56 figures, nut-bearing plants).

BL Add MS 5030
The volume consists of 2 unnumbered folios, 245 numbered folios with unnumbered interleaves, and 3 unnumbered folios.

Folio 1 (text-leaf) reads: "Hortus Malabaricus Het 5e & Seste Deel Bij een vergadert door Jan Commelijn A° 1686" (Hortus Malabaricus, the 5th and sixth volumes, gathered by Jan Commelin in the year 1686).

Folios 2-121 include 60 drawings in double-folio, all published in volume 5 of Hortus Malabaricus. On folio 44 the original inscription, although pasted over, is still readable: "Arboris & Frutices bacciferis 't samen 53 figuren segge 53 N° 5 Het 5de deel 53 fig Arbor et Frutices Baccifer" (Berry-bearing trees and shrubs, together 53 figures, say 53, N° 5, the 5th volume, 53 figures, berry-bearing trees and shrubs).

Folios 122-123 is an unpublished drawing in double-folio of the performance of a fakir; folio 122 reads: "Scandentes siliquatae Leguminos et Apocyni species alia 53 55 stuke Het 9de Deel. 55 fig: darde boek 3 dardeel 3 n° 2" (Climbing pod-bearing leguminosae and other species of Apocynum 53, 55 pieces, the 9th volume, 55 figures, third book 3 third part n°2); folio 122v reads: "Marcelis Splynter".

Folios 124-125 contain an unpublished drawing in double-folio of a Paki-kudi.

Folios 126-245 include all the 61 drawings (59 in double-folio and 2 in folio) as published in volume 6 of Hortus Malabaricus; folio 126 reads: "Het 6de Deel 74 fig: Arbores siliquosae Florifera et alia" (The 6th volume, 74 figures, pod- and flower-bearing trees and others), folio 200 reads: "Arbor. var. gener: Arboris Floriferas & varia genere 42 fig:" (Various genera of trees. Flower-bearing trees and various genera, 42 figures), and folio 202 reads: "marcelis splijnter".

BL Add MS 5031
The volume consists of 3 unnumbered folios, 220 numbered folios with unnumbered interleaves, and 3 unnumbered folios.

Folios 1-118 include all the 59 drawings in double-folio as published in volume 7 of Hortus Malabaricus; folio 1 reads: "Frutices Scandetes 65 fig. Het 7de deel: Frutices et plantis Scandentes et alia:" (Climbing shrubs 65 figures. The 7th volume: climbing shrubs and plants and others).

Folios 119-220 include all the 51 drawings in double-folio as published in volume 8 of Hortus Malabaricus; folio 119 reads: "Plantis Scandens pomiferae & alia gener Scandentes Pomiferae, et Convolvul et Trifoliae & alia siliquata Het 8ste Deel" (Climbing fruit-bearing plants and other genera; fruit-bearing climbers, and convolvuli and trifoliates and other podded plants; the 8th volume).

BL Add MS 5032
The volume consists of 3 unnumbered folios, 304 numbered folios with unnumbered interleaves, and 2 unnumbered folios.

Folios 1-144 include 86 drawings (57 in double-folio

Figure 64. Tinda-parua. Engraving from Hortus Malabaricus vol.1 (1678), Fig.48.

and 29 in folio), all published in volume 9 of Hortus Malabaricus; folio 73 reads: "Het bekomt tot het 9de Deel 53 Fig: De plantis verticillat: Anomalis siliquosis" (It belongs to the 9th volume, 53 figures, on verticillate plants, pod-bearing anomalies).

Folios 145-304 include 94 drawings (66 in double-folio and 28 in folio), all published in volume 10 of Hortus Malabaricus; folio 145 reads: "Het 10de Deel 57 Fig: Miscellanaea" (The 10th volume, 57 figures, miscellaneous).

The drawings in the codex undoubtedly served as models for the engravings of Hortus Malabaricus. Besides the title and other inscriptions mentioned above the drawings, especially the backs, are covered with numerous notes, numbers, and scribbles, which clearly show the difficulties met by Van Reede's collaborators in arranging the great bulk systematically.

The handwriting on the title-pages of the different volumes is that of Jan Commelin, the principal commentator of Hortus Malabaricus, as can be easily established with the aid of his letters to Caspar Sibelius van Goor[2]. The notes with regard to the final sequence of the illustrations must have been written by the editors or commentators. Some drawings have directions for the engraver. On the back of the drawing of Isora-murri BL Add MS 5030:184-185 Commelin noted:

"van figura 31 tot 61 inclus syn om the snyden gezonden den 28/18 juni 1684" (figures 31 up to and including 61 have been sent for engraving on 28/18 June 1684).

The drawing of Katu-conna BL Add MS 5030:148-149 reads:

"memorie voor den plaatsnyder sodanich als het blomken, geteken a. afgebeeld is, soodanig moeten oock perfect gemaackt worden all de blommekens, aan de tacken b.b. behorende dewijle de zelve aan deese tekening gebreecken" (memorandum for the engraver: such as the small flowers marked a. has been depicted, all the small flower belonging to the branches b.b. must also be completed).

Finally, the codex contains a few drawings not published in Hortus Malabaricus. In BL Add MS 5029: 247-248 an unfinished drawing is to be found (Fig.60). In BL Add MS 5030:122-123 there is a drawing of the performance of a fakir signed by Marcelis Splinter, one of Van Reede's draughtsmen (Fig.61). On BL Add MS 5030:124-125, a drawing of a Paki-kudi, an edible bird's nest (Fig.34), Commelin wrote:

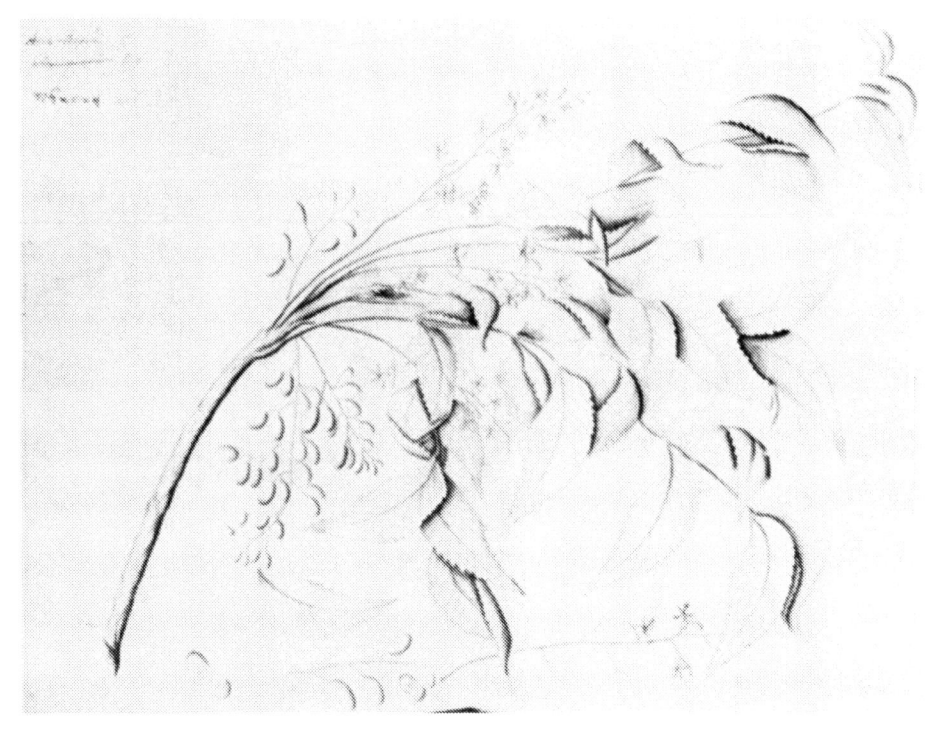

Figure 65. Aria-bepou. Drawing from *Horti Malabarici Icones*, BL Add MS 5029:229-230.

"dese plant moet voor een ander deel werden bewaart om dat het geen gewas of plant en is, maer alleen en nesje van Clijne vogelkens aen de tacken van Bomen hangende" (this plant must be kept for another volume because it is not a growth or plant, but only a small nest of small birds, hanging on the branches of trees).

Many drawings show the consequences of the engraving process: the outlines of the drawings have been scratched in with a sharp object and this is also visible on the backs of the drawings. The backs have been blackened, and marks of attaching the paper are visible at the corners. Kooiman & Venema (1942) explained this process on the basis of the production of the engravings for Jan Commelin's *Horti Medici Amstelodamensis Rariorum Plantarum* (1697). The engraver covered the back of the drawing all over with black or red chalk. He pasted the paper on a copperplate and pressed through the outlines of the figure with a burin in order to produce a rough sketch of the plant in black or red on the plate. In consequence, after the plate had been engraved and printed, the engraving turned out to be a reflection of the original drawing. Indeed, all the corresponding engravings of Hortus Malabaricus are reflections of the drawings in the codex.

Although it is certain that the drawings of this codex are the models for the engravings of Hortus Malabaricus, this does not imply that they were made, at the time, in Malabar or even in Asia. In the past it was a regular practice of multiplying drawings and paintings for different purposes. Van Reede, too, had his own set of drawings from Malabar copied as a safety measure in Batavia in 1677. The question is: does the codex contain Van Reede's own Malabar set, or the Batavia copy, or perhaps another copy?

When we study the watermarks and countermarks in the codex, they show great resemblance in all the paper used. The leading watermark is the crowned Strasbourg lily occurring in the drawings, text-leaves, unnumbered folios, and interleaves (see Fig.62). An exception is the coat of arms with a bend, crowned by a lily, in the unnumbered folios of BL Add MS 5031. The countermarks corresponding to the Strasbourg lily show some diversity: IHS, IHS and ET, and IHS and LM. Separate countermarks not found in connection with a watermark are IHS and PB, IHS and MC, MD, IV, CDG, and IVILLIDARY. All these marks point to Dutch paper manufactured in the second half of the 17th century[3].

The paper used by the Company's administration in Malabar and Batavia in 1674-1677, when the Malabar set and the Batavia copy were made, show several different watermarks. A considerable amount of paper then used

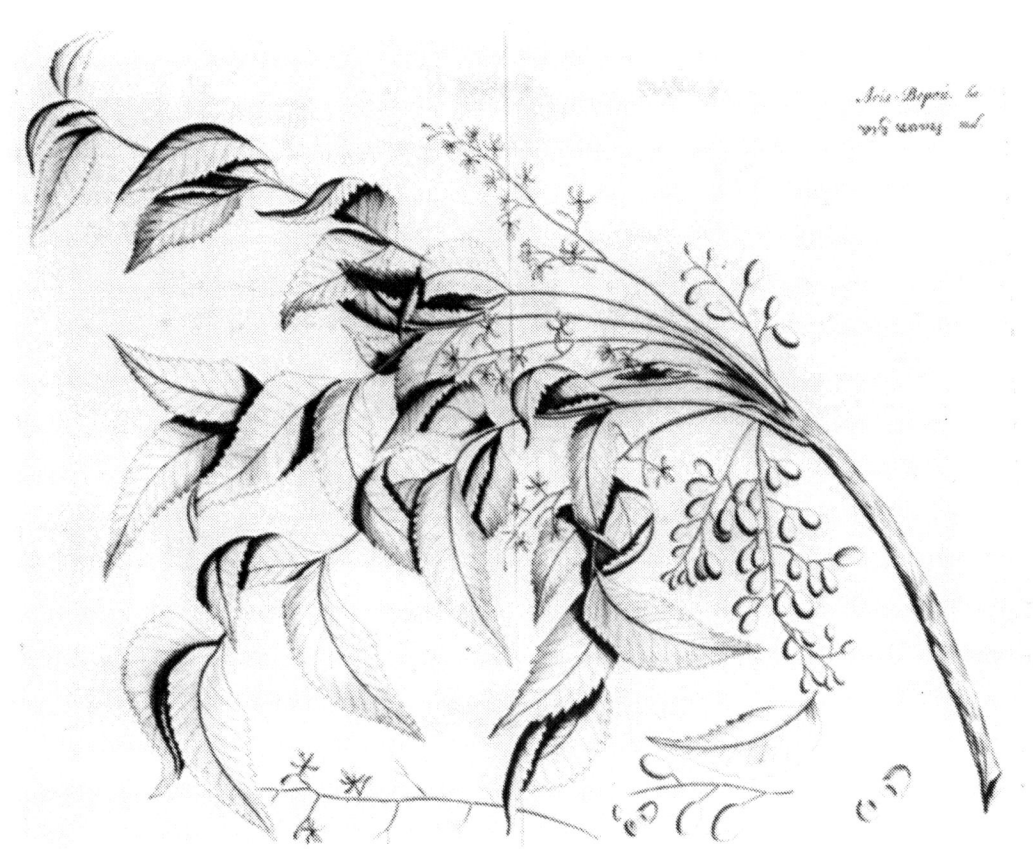

Figure 66. Aria-bepou. Engraving from Hortus Malabaricus vol.4 (1683), Tab.52.

Figure 67. Kareta-tsjori-valli. Drawing from *Horti Malabarici Icones*, BL Add MS 5031:89-90.

132 *Drawings of Hortus Malabaricus*

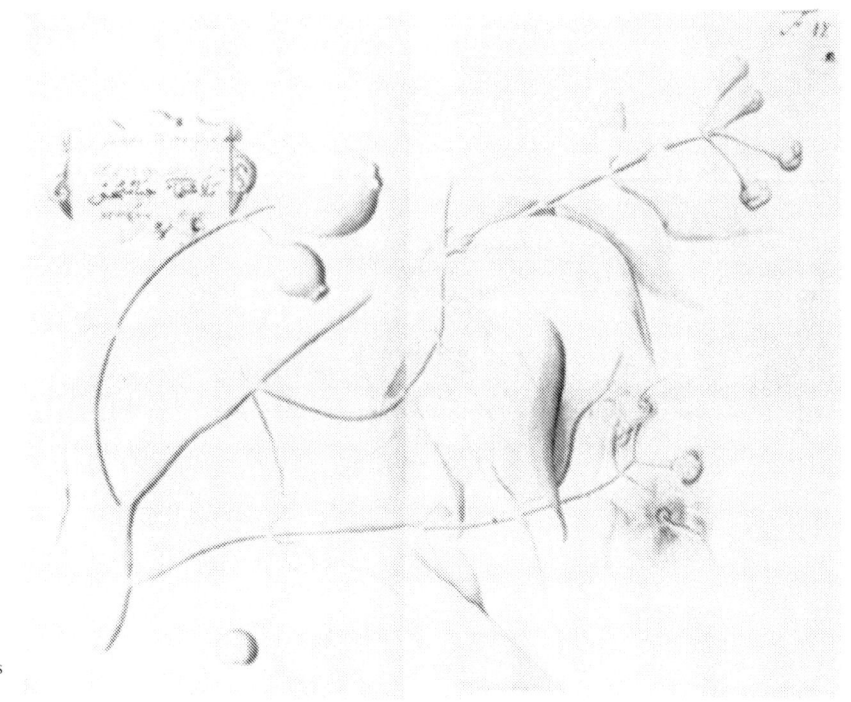

Figure 68. Malacca-schambu. Drawing from *Horti Malabarici Icones*, BL Add MS 5028:34-35.

Figure 69. Malacca schambu. Engraving from Hortus Malabaricus vol.1 (1678), Fig.17.

Figure 70. Kasjavo maram. Drawing from *Horti Malabarici Icones*, BL Add MS 5030:38-39.

Figure 71. Kasjavo maram. Engraving from Hortus Malabaricus vol.5 (1685), Tab.19.

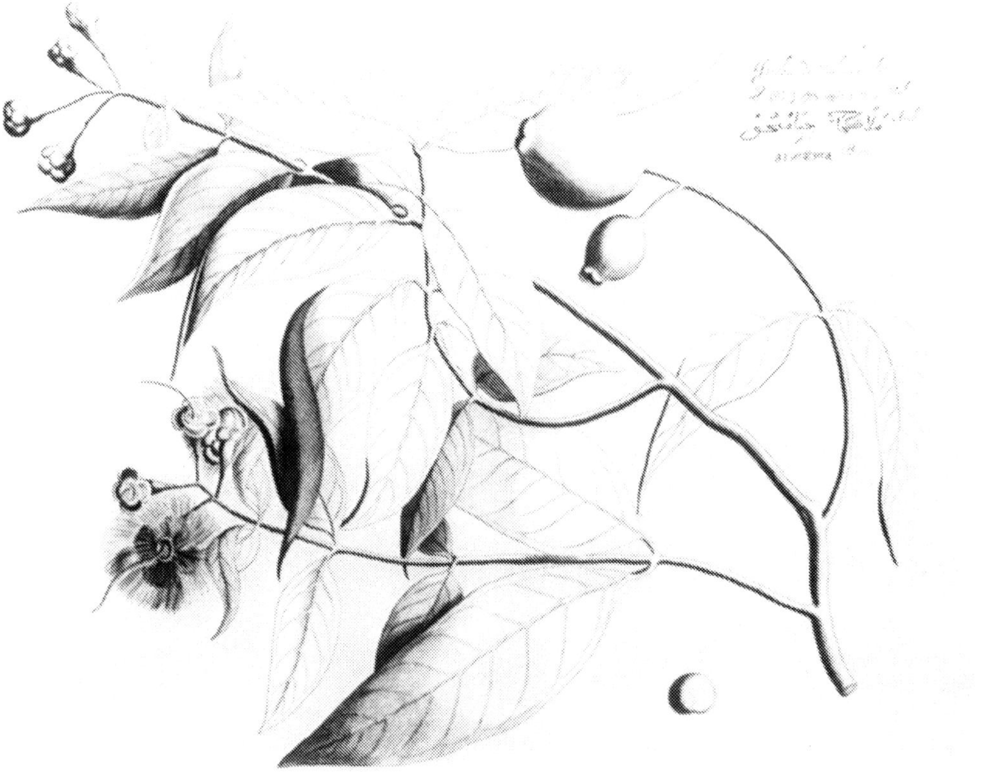

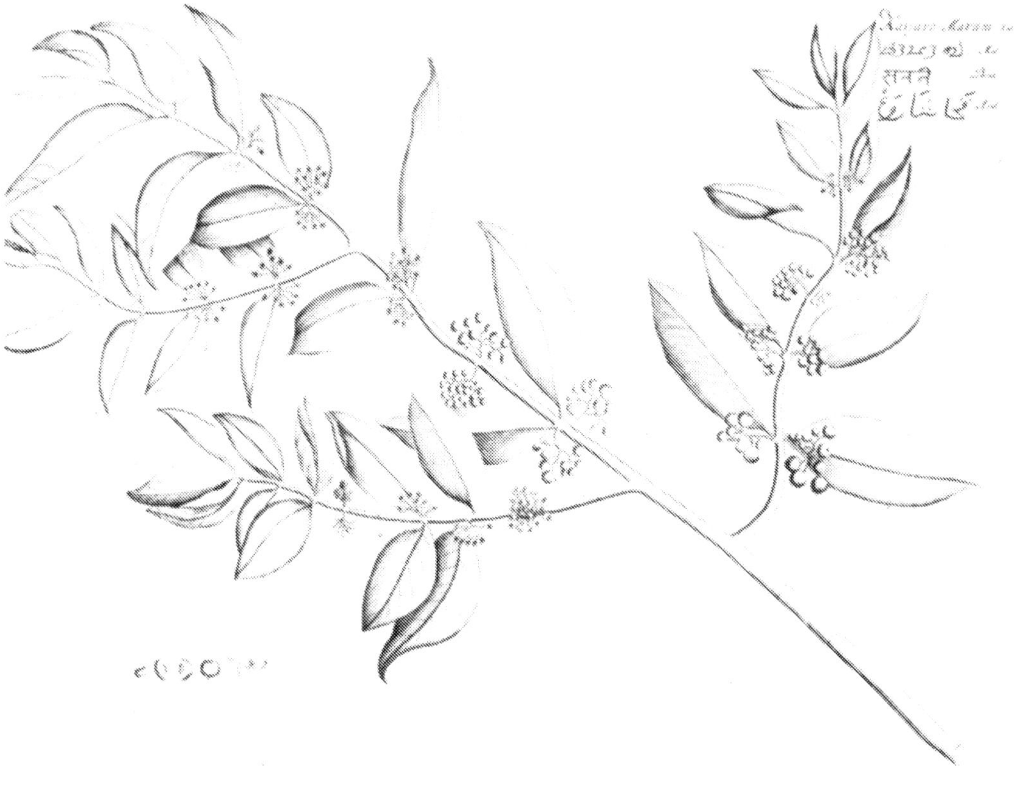

Figure 72. Oepata. Drawing from *Horti Malabarici Icones*, BL Add MS 5029:215-216.

in Cochin can be divided into two main categories: that with the fool's cap and that with the coat of arms of the city of Amsterdam. The fool's cap occurs in the paper of Van Reede's original letters to the High Government, of 15 May 1674, 22 April 1675, and 28 August 1676[4]. The Amsterdam coat of arms is to be found in the paper of his original letters of 28 October 1675 and 20 December 1676[5]. A third, but rare category is that with the crowned anchor cross as can be observed in the paper of Cochin copies of letters of some Malabar princes to Van Reede in 1676[6]. As far as I am aware, paper with the Strasbourg lily was not used in Cochin in those years in which Van Reede's own drawings for Hortus Malabaricus were made there!

However, in Batavia in 1677, when Van Reede stayed there, a copy of his Memorandum of 14 March 1677 to his successor, commander Lobs, was written on paper with the crowned Strasbourg lily and the countermarks IHS and PB[7]. This copy accompanied the General Missive of 24 November 1677 of the High Government to Lords XVII, which shows the Amsterdam coat of arms[8].

From the above we may conclude that in the years 1674-1677 paper with the crowned Strasbourg lily was not yet, or no more, in use in Malabar, and that it is not very likely that the drawings in the codex of the British Library mainly showing this watermark were made in Malabar. On the other hand, paper with this watermark was in use in Batavia, where Van Reede had his own set copied in 1677. Apparently Van Reede, on leaving Batavia for the Netherlands, entrusted Willem ten Rhijne with the care of his own set and took the Batavia copy with him to the Netherlands. I therefore suppose that the codex of the British Library is identical with the Batavia copy.

One point of discussion still remains. Volume 1 of Hortus Malabaricus was published in 1678, and on 8 May of that year Arnold Syen, its editor, presented the Curators of Leiden University with a copy. Since Van Reede did not arrive in the Netherlands until June 1678, he must have shipped the drawings for volume 1 on a previous date. This means that he must have sent the first instalment of copies of the drawings with ships returning to the Netherlands in the beginning of 1677, before leaving Batavia with the rest of the copied drawings.

All the drawings are in a good condition except the drawing of Moul-elavou BL Add MS 5029:101-102, which was pasted on a new sheet afterwards, and the drawings of Samstravadi and Katou-tsjeroe BL Add MS 5029:137-138 and 143-144, which have turned a pale green at the top of the sheets.

Many drawings clearly show that they were cut off after the engraving had been made. Thus a lot of notes and numbers on the drawings were unrecognizably mutilated, as a result of which a reliable interpretation of them is

Figure 73. Oepata. Engraving from Hortus Malabaricus vol.4 (1683), Tab.45.

rather hazardous. When the volumes of the codex were bound, the drawings were not cut off again, for they show different sizes. The double-folios vary from 33.7 x 44.4 cm to 38 x 50.6 cm, the folios from 35 x 23 cm to 37.5 x 26 cm.

In artistic respects, as compared with the engravings of Hortus Malabaricus, the drawings are somewhat disappointing. The sharply cut lines and the brilliance of the engravings are not found again in the drawings. When one goes through the codex, one can easily observe that the hundreds of pictures were not nearly so well finished off when they were handed over to the engravers. The unfinished state of the collection recalls Van Reede's account of the difficult conditions in which Hortus Malabaricus was accomplished in Malabar and Batavia.

His final plan of illustrating the work may be inferred from fully elaborated drawings. These have been neatly inked and washed, and provided with a borderline. Within the borderline the names of the plant in several languages are given in an inset which is often nicely decorated. A few of the most beautifully executed drawings even have landscapes.

But many drawings did not reach this state of completeness. They broke down in a state of a pencil draft and sketch, sometimes only partly inked. Other drawings have no inset or names within.

On the other hand, the unfinished appearance of the codex gives a good idea of the subsequent steps in producing the illustrations. At first the plant picture was drafted without many details in leaves and flowers. A good example of this is Tinda-parua BL Add MS 5028:96-97 (Figs 63-64). Then the details were filled in. After that the completed sketch was inked. Often the ink lines do not follow the pencil lines exactly, as can be seen in the drawing of Aria-bepou BL Add MS 5029:229-230 (Figs 65-66). Finally, the inked drawing was washed. A beautiful specimen of successful washing is Kareta-tsjorivalli BL Add MS 5031:89-90 (Fig.67), but a failure is Paeru BL Add MS 5031:199-200. Sometimes the washing effaced a detail as in Katou-naregam BL Add MS 5029:151-152. The next step was that of carrying the drawings, finished or not, to the copperplate. As said above, the engraver blackened the back of the drawing, pasted it on a plate, and tried to trace the outlines of the

Figure 74. Schunda-pana. Drawing from *Horti Malabarici Icones*, BL Add MS 5028:22-23.

pictures accurately. But in some cases, as in Malacca schambu BL Add MS 5028:34-35 (Figs 68-69), he got off the track and followed his own course several millimeters beyond the inked lines. The engraver had quite a task in adding omitted details to the pictures and in reconstructing unfinished drafts and sketches. In the engravings of Carim-curini and Bem-curini HM 2:31-34:20-21 he had to add several small flowers to the ears. In the engravings of Hummatu, Nila-hummatu, and Mudela-nila-hummatu HM 2:47-52:28-30 he invented new cross-sections of Datura fruits, which he possibly based upon specimens cultivated in the Netherlands[9]. In other engravings he mainly reconstructed leaves and flowers from examples elsewhere in the drawing. Sometimes he forgot, or was not ordered, to make reconstructions, as can be seen in the cut-off leaves in both the drawing and the engraving of Kasjavo maram BL Add MS 5030:38-39 (Figs 70-71). On the other hand, the engraver omitted many details. He obviously did so by order of the editor or commentator. The drawing of Oepata BL Add MS 5029:215-216 (Figs 72-73) shows a separate inflorescence crossed out, with beside it the note "gelt niet" (this does not count). In a few cases of minor importance the engraver changed the position of separate details.

The borderlines seem to have been drawn after the drafts, sketches, and insets had been made. In some sketches borderlines cut through drafted lines of the drawings; that is why marginal parts of these drawings were neither inked nor engraved. Some insets have also been cut through by borderlines. Moreover, in some cases the position of insets obstructed the full elaboration of important details, such as leaves and inflorescences (see Schunda-pana BL Add MS 5028:22-23 (Figs 74-75)). However, in the corresponding engraving the engraver ignored the inset and reconstructed the omitted details. It is most likely that first of all the insets were drawn and thereafter the plants.

One hundred of the insets have been embellished with human figures and with images derived from nature. The decorations have few or no associations with Malabar. They are more related to the decorative art of European baroque. From several humoristic scenes on the insets it is obvious that the draughtsman enjoyed the decorating work. About one-third of the insets have been embellished with fruits, flowers, and garlands, as might be expected in a collection of botanical drawings. But in the greater majority the draughtsman liked to add human figures, animals, real and fancy, and even a church with crosses (Fig.76) and a dramatic mask. Among the human figures there are men, bearded, bald, hatted, laughing, shouting, and women, gracious and wreathed, even a fighting couple, of which the wife is striking the underlying man with her slipper (Fig.77). A less numerous feature is formed by the heads (Fig.78), winged or not, of children, boys, and angels. Some insets bear snakes, birds (an owl and a bird of prey catching a snake), goats, and

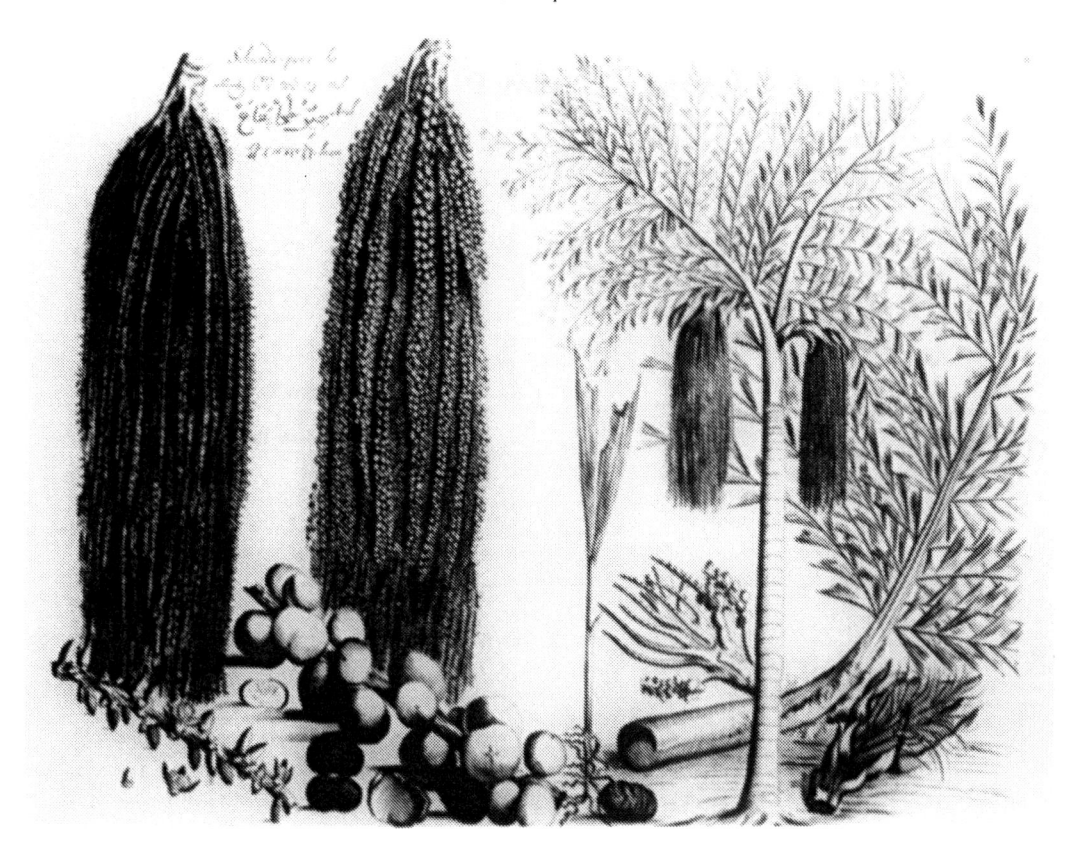

Figure 75. Schunda-pana. Engraving from Hortus Malabaricus vol.1 (1678), Fig.11.

Figure 76. Church with crosses. Inset of the drawing of Anachorigenam from *Horti Malabarici Icones*, BL Add MS 5028:196-197.

Figure 77. Fighting couple. Inset of the drawing of Pee-cupameni from *Horti Malabarici Icones*, BL Add MS 5032:283-284.

Figure 78. Heads with long noses. Inset of the drawing of Schanganam-pullu from *Horti Malabarici Icones*, BL Add MS 5032:180-181.

Figure 79. Merman. Inset of the drawing of Schageri-cottam from *Horti Malabarici Icones*, BL Add MS 5028:112-113.

Figure 80. Winged Dragon. Inset of the drawing of Panel from *Horti Malabarici Icones*, BL Add MS 5028:132-133.

lions. The fancy animals consist of snakes, birds, mermen (Fig.79), hermaphrodites, satyrs, winged dragons (Fig.80), and other monsters. But none of these lovely decorations were used in the engravings.

Notes

1. The Dutch text in *Pars Bibliothecae sive Catalogus Librorum Medicorum ... Bernardus Siegfriedus Albinus* (1771) reads: "De origineele Teekenningen van de Hortus Malabaricus bij een vergadert door J. Commelin in Amst. 1686 zynde de Teekenningen der Eerste deelen waar van de plaaten van 't voorige werk na gemaakt zyn. 10 deelen 5 ribbe". Bernard Siegfried Albinus (1697-1770) was professor of anatomy and surgery in Leiden University (NNBW vol.4 (1918):22-24).
2. BL, Sloane Manuscripts 2729:164,165,172,180,191, letters of 6 June 1684, 13 July 1684, 10 November 1684, February 1685, and 22 August 1686 of Jan Commelin to his cousin Caspar Sibelius van Goor, physician at Deventer. For a photograph of the third letter, see Wijnands 1983:7.
3. Churchill 1967:83-85, figs 400-437.
4. VOC 1308:597,642-669; VOC 1316:601-654.
5. VOC 1308:595-596; VOC 1316:600.
6. VOC 1308:797.
7. VOC 1329:1336v-1385.
8. VOC 1323:1-225v.
9. According to their garden catalogues, Paul Hermann (1687:583,584) and Jan Commelin (1689:338) cultivated these three kinds of Datura in the botanical gardens of Leiden and Amsterdam.

9
Renaissance botany

Modern European botany was born in the 15th and 16th centuries, when critical text editions of the works of the classical authors about botany, such as Theophrastos, Dioscorides, and Pliny, were printed. A highly interested host of scholars turned eagerly to Theophrastos' *Historia Plantarum* and *De Causis Plantarum,* to Pliny's *Historia Naturalis*, and to *De Materia Medica* of Dioscorides. Especially the physicians interested in botany devoted a large part of their energy to writing commentaries on these works. In the centre of interest were Dioscorides' sections on medical botany; this author was to remain the absolute authority in the field of medicinal plants till the 17th century[1].

The commentaries were concerned in the first place with the interpretation and identification of the plants of Dioscorides. The problems with which the early botanists were confronted were not trifling. In retrospect, Dioscorides' aphoristic plant descriptions, the absence of illustrations, and his deficient geographical data have formed almost insurmountable obstacles[2]. In spite of this, the general desire for identification of that time gave the impulse for the development of methods, criteria, and aids which in the long run were to deliver botany from the trammels of medical science and to elevate it into an independent discipline.

Dioscorides, who lived in Asia Minor (Turkey) and afterwards probably also in Italy, treated mainly plants from the eastern Mediterranean regions. As long as those regions were controlled by the hostile Ottoman sultanate, only very few Western physicians and botanists ventured on a most dangerous adventure of observing and collecting plants in the Near East. The stay-at-homes, consoled by the presumption then generally current that the classical plants might also grow in their immediate, familiar neighbourhood, thus devoted themselves to the study of regional floras in Europe, hoping that they would be able to compare their results successfully with the plants of Dioscorides.

Botanists of all nationalities, Italians, Spaniards, Frenchmen, Swiss, Germans, Englishmen, Belgians, and Dutchmen collected, cultivated, described, and pictured their native plants and, aided by the recently invented techniques of printing and engraving, published the results of their studies in sumptuous herbals, critical tracts, scientific correspondences, plant catalogues, itineraries, etc.

The rise of 16th-century botany owes a good deal to the development of two important aids: the herbarium and the botanical garden. About 1540 Luigi Anguillara[3] invented the technique of the conservation of plants pasted on paper. Henceforth besides the drawing, painting, or engraving of a plant the original dried specimen was to be used increasingly as an everlasting source of information and control. In 1544 the same Anguillara founded the first botanical garden in the University of Pisa, which was soon followed by Padova (1545)[4] and other Italian universities. Even before the end of the 16th century the principal European centres of university training possessed such institutional gardens, in which there were cultivated not only the current European medicinal plants, for the use of education in medical botany, but also tender exotic plants, acquired at often great expense and cherished in greenhouses or conservatories. The botanical garden of Pisa was the first in 1591[5] to found a cabinet or museum of natural history where all kinds of natural products could be stored. The living plants, continually cultivated and increased in number, together with the botanical objects in the cabinet, were inestimable sources for botanical studies in the following centuries.

During the discussion of the newly found plants soon the need for a definite method of plant description made itself felt. In the course of the 16th century a generally accepted method crystallized, which consisted of a morphological description from the bottom to the top of the essential parts of the plant: root, stem, leaf, flower, fruit, and seed. To the description there were added the locality, the seasons of sowing, flowering, and fructification, the medical virtues and uses, and other applications, a discussion of nomenclature and synonymy, and -the essence of 16th-century botany- a comparison with the relevant classical species. The botanical terminology, however, showed great diversity and depended on the verbal power of expression of the individual botanist in his native tongue or in Latin.

Already in the 16th century two aspects, nomenclature and classification, presented themselves which were to change the character of botany in the 17th and 18th centuries from vivid narrative descriptions fascinating the general public to barren ordered enumerations of plant names which were interesting only for a small public of specialists, from Dioscorides study to Linnaean botany. The growing importance of well-thought-out systems of nomenclature and classification found its origins in the pragmatic necessity of inventing instruments of communication on and presentation of an ever increasing number of plants which even before the end of the 16th century exceeded the classical plants tenfold.

The problem of the nomenclature was at first solved in a plausible way. Starting from a classical plant, scholars maintained its original Latin or Greek name and simply derived from it the nomenclature of the supposedly identical or related species collected somewhere in Europe. A complication was formed, however, by the continual disagreement among the 16th-century botanists about the identity of a given classical plant. In consequence of this the nomenclature of identical European plants often widely varied and gave rise to a disastrous confusion in scientific communication.

In classification Dioscorides' system was simply adopted, or a special system was derived from it which was revised on the strength of pragmatic criteria analogous to those of Dioscorides. The botanists indeed sometimes succeeded in establishing small taxa of related species by their intuition, but as long as the identity of the classical plants was disputed, there was little progress in the development of the classification systems according to natural grouping on higher levels.

In the 16th century as yet the Dioscorides study found an important rival in the exotic botany of the newly discovered worlds. The discovery of the sea-routes to Asia and America caused a great influx of unknown plants and botanical products into Europe. Only some medicinal and commercial plants could be connected with the writings of Dioscorides, Theophrastos, and Pliny, in so far as they were brought from neighbouring Asian regions such as India and Ceylon, which at one time, in Antiquity, had maintained commercial relations with the Mediterranean world. But the majority of these plants from America and remote Asian regions hardly had any affinity with or similarity to the classical plants or familiar European plants. The interpretation of totally new plant forms of an amazing diversity was a task far exceeding the experience of the Dioscorides study then accumulated. Methods, criteria, and resources of 16th-century botany were hardly sufficient for the adequate incorporation of all the new information into the existing apparatus of botanical knowledge.

Towards the end of the 16th century botany passed through a crisis. Attempts were then made to improve, adapt, and change the foundations of the discipline in different ways and with varying success. Carolus Clusius rejected the Dioscorides study as unprofitable[6] and turned to the plant description without allowing himself to be restricted by medico-botanical presuppositions. Roving through Europe, he devoted a great part of his long life to the observation and description of all the plants of local floras. His great proficiency in Latin and his skill in making summaries of the essentials from other people's descriptions enabled him to compose concise and pregnant descriptions of any humble plant. We shall find a worthy successor as to this skill in John Ray (1628-1705), who was to condense many descriptions in Hortus Malabaricus into short, but very useful surveys, of which Linnaeus was to make effective use.

Another botanist who disowned the Dioscorides study was Andrea Caesalpino (1519-1603), who, reverting to Aristotle's and Theophrastos' investigations of the essential parts of living beings, thoroughly studied flower, fruit, and seed, and as a result constructed a revolutionary system of plant classification, based on morphological characters of fruit and seed. But his principal divergence from the established tradition of Dioscorides' classification and its derivations will have been the cause that at first his system had little influence[7]. It was Ray again who accepted Caesalpino's classificatory criteria and built a far more successful system, which greatly affected Jan Commelin's arrangements of the later volumes of Hortus Malabaricus.

Contrasted with the innovations which Clusius and Caesalpino brought about in 16th-century botany was the work of the Swiss brothers Jean and Caspar Bauhin (1541-1613 and 1560-1624)[8], who recapitulated and extended the trends in the Dioscorides study to masterly surveys of all the botanical knowledge then known from the earliest days.

Jean Bauhin applied himself mainly to the description and illustration of thousands of plants, consciously following the generally accepted method, from root to seed, with additional biological, medical, and pharmaceutical information. He did not live to see his studies published, but his son-in-law Jean Cherler edited them into *Historia Plantarum Universalis* (1650-1651). This work, a treasure-house of 16th-century phytography, served as a model for the plant descriptions in Hortus Malabaricus, and in this descriptive respect Van Reede's flora of Malabar is firmly rooted in the best traditions of the Dioscorides study.

In close correspondence with his brother's writings, Caspar Bauhin specialized in the confusing problems of nomenclature, synonymy, and classification. In his main work, *Pinax Theatri Botanici* (1623), Bauhin was the first to introduce a consistent classification into "genera" and "species". True to the objectives of the Dioscorides study, he based his "genera" on the classical plants of Dioscorides, Theophrastos, and Pliny, and included all those described, but mainly European plants which in his opinion were identical with or related to them. Each "species" was defined by him by means of a diagnosis in which the name of the "genus" was followed by one or more striking characteristics of the plant. His diagnoses called into being the phrase names which were to dominate the nomenclature of plants until the binary names in Linnaeus' *Species Plantarum*. After each diagnosis Bauhin listed all the synonyms known to him, carefully mentioning names and authors, thus furnishing an unsurpassed, still valuable key to early botanical literature. In the establishment of his sections, or taxa of higher level, he succeeded in recognizing many natural groups of "genera", but in his classification of the sections he applied a somewhat obscure notion about the gradual complexity of plant forms, advancing from the supposedly simple forms such as grasses and bulbous plants, via dicotyledonous plants, ferns, mosses, and fungi to the supposedly perfect shrubs and trees.

Caspar Bauhin's fidelity to Dioscorides and his associates, however, prevented him from finding solutions for the nomenclature and classification of newly discovered exotic plants, which owing to their initially deficient descriptions and their deviating forms did not fit in with his concepts of "genera". For that reason, as an appendix to the trees, he ended the *Pinax* with a chaotic survey of exotic arbores, frutices, herbae, plantae, nuces, etc.

Both Arnold Syen's and Jan Commelin's nomenclature and arrangements in the first volumes of Hortus Mabalaricus hark back to Bauhin's *Pinax*. They tried to relate Van Reede's plants to Bauhin's "genera" and "species" and adopted the latter's phrase names or invented new ones derived from them. In the arrangements they followed Bauhin's sections as far as possible, but they did not hesitate to make changes therein at their own discretion.

From the above survey of the purpose and means of 16th-century botany we may conclude that Hortus Malabaricus is largely based on the issues of the Dioscorides studies. Aids such as herbarium and botanical garden, and methods in description, nomenclature, identification, and classification were born and developed at that time. Not only the results of the Dioscoridean tradition, but also views opposed to these issues were expressed in Hortus Malabaricus. These aspects form the basis for the now following analysis of the plant descriptions and commentaries in Hortus Malabaricus, which are intended to give the exertions of Van Reede and his many collaborators a place in the long, wearisome road from Renaissance botany to Linnaean botany.

Notes

1. For general literature on this chapter, see Sprengel 1817 vol.2:249-378; Winckler 1854:67-97; Meyer 1857 vol.4:207-399; Sachs 1875:3-79; Möbius 1937:4-38.
2. Berendes 1902; Dictionary of Scientific Biography vol.4 (1971):119-123.
3. Dictionary of Scientific Biography vol.1 (1970):167.
4. Gola 1947:11.
5. Martinoli 1963:5.
6. Hunger 1942 vol.2:221.
7. Dictionary of Scientific Biography vol.15, suppl.1 (1978): 80-81.
8. Dictionary of Scientific Biography vol.1 (1970):522-527.

10
Plant descriptions of Hortus Malabaricus

GENERAL SCHEME OF THE DESCRIPTIONS

In his preface to volume 1 Johannes Casearius explained the general scheme of the plant descriptions in Hortus Malabaricus. In these there were to be treated the form and appearance of the plants, their principal parts, such as roots, trunks or stems, leaves, flowers, fruits, and seeds, as well as the characteristics of colour and the native plant names in Mayalam and "Brahmin", written in their own scripts. He emphasized the care bestowed on the description of related plants in order to prevent confusion. Moreover, he drew attention to the contributions of Matthew of St. Joseph and Malabar physicians concerning the medical use and to the notes on the virtues of occasionally extracted oils. Casearius admitted that the descriptions were not always complete, but he did not doubt the life-size illustrations would contribute to a clear and distinct knowledge of the plants[1].

Van Reede and his collaborators indeed followed a fixed scheme of description. This scheme can readily be recognized by the separate paragraphs into which it is subdivided, or in the absence of paragraphs by italicized words in the consecutive text.

The first paragraph is devoted to the nomenclature, relationship, habitus, and habitat of the plant. The nomenclature consists of the Malayalam name, to which, as a rule, is added its synonym in "Brahmin" or Konkani. These foreign names are not printed in their own scripts, as Casearius suggested, but in Roman script. In addition, though less frequently, synonymous names in Portuguese and in Dutch also occur, and now and then, in the volumes 3 and 4, also in Malay and Japanese.

Occasionally the relationship to one or more similar plants is mentioned with references to related Malayalam names, such as, for instance, Atty-alu, Itty-alu, Arealu, and Peralu in HM 1:43-49:25-28.

The habitus is described as tree, shrub, herb, or creeper; the height of the plants is expressed in relation to human measures. Sometimes some general characteristics of the trunk or stem, wood, bark, position of the branches, foliage of trees, etc. are also added.

The habitat is referred to by terms indicating mountains, riversides, forests and groves, gardens and orchards, shadowy and sunny places, and sandy, rocky, humid, marshy, watery, and muddy soils. In the volumes 3 to 7, however, the habitat is a part of the locality or distribution of the plant at the end of the description.

In the next paragraphs the parts of the plants are dealt with. Dependent on the different editors of the volumes, the style of writing as well as the terminology vary from concise in the volumes 3 to 5, to fluent in the other volumes. The number and the length of the paragraphs vary according to the detailed or deficient character of the available information. The first two volumes contain many and detailed paragraphs, while starting with volume 9 the number and the length greatly decreases, to a few lines in volume 12. The consequence is that in the detailed descriptions the essential characteristics are often deluged by numerous minute details, while the short or fragmentary descriptions give only little relevant information. For practical reasons the sequence of the paragraphs is sometimes also slightly changed, but in general the scheme of root, trunk or stem, leaf, flower, fruit, and seed as announced by Casearius is followed. Within this scheme further subdivisions are to be distinguished, according to the respective anatomical structures. Attention is devoted to form and structure of the bark or rind, wood, medulla of roots and trunks, and leaves as well as of flower structures such as calyx and petals, sepals, stamens and anthers, pistils, ovaries and stigmata, and of secondary structures such as hairs, tendrils, thorns, and spines. Size, length, width, and diameter of the parts and their structures are mentioned both in qualitative terms such as short, small, medium, large, and thick, and in quantitative terms of human or standard measures. In addition such secondary characteristics as odour, colour, and taste are mentioned.

The paragraph on locality and distributions only occurs in the volumes 3 to 8; to this is added the habitat of the first paragraph. The statements range from villages, towns, and other places to regions and principalities in Malabar. Sometimes the distribution outside Malabar is also mentioned, specifically in the case of cultivated plants which have been introduced from elsewhere into Asia.

In the next paragraph the months and seasons, as well as the annual frequency of florescence and fructification are mentioned. To this are added notes on the ripening of fruit and seed and on the cultivation of the plant.

The paragraph on the use deals with the importance of the plants for commerce, industry, agriculture, and horticulture, notably of aromatic plants, timber, food plants, and ceremonial, cosmetic, and ornamental plants.

A special paragraph, often typographically separated

from the preceding ones, is devoted to the medicinal virtues, with general information on the preparation and administration of the medicaments. If a plant is not being used in native medicine, this is usually stated explicitly.

The last paragraph, typographically clearly separated from the main text, contains a commentary on the plant described. The commentaries of Hortus Malabaricus will be discussed in chapter 11.

GENESIS OF THE DESCRIPTIONS

The way in which the plant descriptions came into being was discussed in detail by Van Reede in the preface to volume 3. In spite of the fullness of detail, it is not very well possible to give a correct picture of the process, the more so because the manuscripts which formed the basis of the ultimate plant descriptions have not been recovered.

Van Reede related that during his travels in Malabar 'I was often seized by the desire to explore and examine the leaves, flowers, barks, and roots of those plants. And then I found that they frequently had a very sweet smell and a penetrating taste. And when I asked the natives who accompanied me on my journeys whether they knew anything about these plants, they not only disclosed the names, but also knew very well their curative virtues and use'[2] (Fig.81).

At first Van Reede did not succeed in laying down adequately the information he thus obtained, 'For no one goes to Malabar to apply himself to this, but all those who have gone there under the auspices of the Illustrious East India Company are compelled so much to perform their office accurately that they have no leisure to undertake this, even if they wished to'[3].

In the biographical part I have already brought out that Van Reede, especially in the years 1663-1667, when he traversed Malabar in many directions, had full opportunity to get a first notion of a description of the Malabar flora. In the first period of his commandership, in the years 1670-1674, however, owing to great pressure of work he was unable for the time being to carry this into effect. The arrival of Matthew of St. Joseph in Cochin, in 1674, was an opportunity to be seized by Van Reede for pushing on the long-fostered project. He proposed to the Carmelite 'that a list and description of the plants of Malabar had to be made, for which I promised this venerable father my assistance; he very kindly and promptly offered me his services in turn'. As regards the descriptions, Matthew apparently showed him a notebook in which 'he had noted down in particular the curative virtues'[3].

It soon became apparent that Matthew could indeed derive much information about plants from his memory, 'but still many plants have not been traced, either because he had not noted down the names accurately, or the venerable old man had received them from men who did not know the common names of the plants by which the botanists of Malabar designated each plant, or because they were plants from another climate'[4].

Figure 81. Conversation between a Malabar warrior and a European (Van Reede?). Detail of the engraving of Codda-pana Hortus Malabaricus vol.1 (1678), Fig.1.

In chapter 7, on Matthew's *Viridarium Orientale*, I have discussed his method of describing plants in more detail. Here suffice it to remark that he could furnish Van Reede with notes on the name, habitus, distribution, and general characteristics, but above all on the medical use of the plants. But in spite of all Matthew's good intentions Van Reede was bound to state that 'As to the description, this often cohered so little through perfunctory and fleeting attention that it might have referred to several plants, even if they were quite different'[4].

Up to this point Van Reede's survey of the genesis of the plant descriptions can readily be followed. But immediately afterwards he continued with a report about the visit of Paul Hermann to Malabar, which involves some problems as regards both the contents and the chronology. Van Reede reported as follows on Hermann's visit:

'there arrived in Malabar from the island of Ceylon the famous Paul Hermann, doctor of medicine, who had just been appointed professor of botany at the Illustrious University of Leiden. When he had seen that the beginnings of our *Hortus* were impeded by so many obstacles and by the rather inaccurate nature of the descriptions as well as the pictures, with modest kindness he saw fit to give us to understand that in this way it would hardly become a book of any importance. Nevertheless, the shortness of his stay with me was the cause that I could not discuss my collectanea of the *Hortus Malabaricus* with him, neither could he assist me with wise advice in this very complicated work. Circumstances being such, I wanted to begin the whole enterprise anew, and I devoted myself to describing the plants as well as I could and working at this *Hortus* in another way'[4].

From the preceding words it may be inferred that Hermann criticized the then state of the descriptions because of the inaccuracy, but that he had no time to explain how this could be remedied, and even less had an opportunity to study Van Reede's "collectanea". It would appear somewhat improbable that Hermann should not have wrapped up his criticism in hints for the improvement of the plant descriptions. In fact, there is a considerable difference between Matthew's own descriptions and the final descriptions such as they were published already in the volumes 1 and 2 of Hortus Malabaricus. The most important difference is that Matthew's paragraph on general characteristics has been replaced by a full description of all parts of the plant, from root, stem, and leaf to flower, fruit and seed. The question is who was responsible for this difference, or in other words: who has suggested to Van Reede a new scheme for the descriptions? In chapter 8, on the drawings of Hortus Malabaricus, I supposed that Van Reede sent the drawings and descriptions of volume 1 to Arnold Syen in the beginning of 1677. This implies that the final version of the descriptions in this volume must have come into being in Asia, either in Cochin or in Batavia.

Willem ten Rhijne, with whom Van Reede co-operated on Hortus Malabaricus in Batavia for some months in 1677, can hardly be considered as the designer of the new scheme, because in that case the original plan of the descriptions would first have had to be completely reorganized and extended before the drawings and descriptions of volume 1 could be sent to the Netherlands. It is implausible that they can then have reached Syen in Leiden in 1677.

A possible solution seems to be that Paul Hermann, during his visit to Malabar, suggested a new scheme for the plant descriptions to Van Reede. In this light it also becomes clear why Van Reede regretted that he could not discuss his "collectanea" with Hermann. These "collectanea" must be understood as notes and remarks on the plants observed. Presumably Van Reede hoped that Hermann could advise him to what extent he could use his notes and remarks in the reorganization of the new scheme. Anyhow, it appears plausible that with regard to the plant descriptions we may look upon Paul Hermann as one of the co-founders of Hortus Malabaricus.

Nevertheless, it remains curious that Van Reede, who elsewhere in the preface to volume 3 had nothing but praise for his collaborators, put such emphasis on the negative aspects of Hermann's contribution to the genesis of Hortus Malabaricus. A possible explanation might be that, in view of the political aspect of this preface, Van Reede did not care to thank a former subordinate of his enemy Van Goens too much, and certainly not to praise him too much, for his contribution to Hortus Malabaricus.

It is difficult to ascertain exactly when Hermann was in Malabar. Van Reede's statement that Hermann had shortly before been appointed professor of botany in Leiden is not correct. That appointment took place only in 1679[5], when Van Reede had already returned to the Netherlands. Hermann himself alleged in his *Paradisus Batavus* that he stayed in Cochin in 1677. On that occasion he met Matthew of St. Joseph and they discussed a dried leaf of *Arum bifolium Arabicum maculatum*[6].

As is known, on 18 March 1677 Van Reede left Cochin with destination Batavia, so that in the first months of 1677 Hermann may have been in Malabar. But this date can hardly be reconciled with the fact that it was only after Hermann's visit that Van Reede started the new scheme of the plant descriptions. In that case he must have sent the drawings and manuscripts of volume 1 to the Netherlands at an even later date. Apparently Hermann erroneously mentioned the year 1677, or else it was not in that year, but earlier that he discussed the problems with Hortus Malabaricus with Van Reede.

Unfortunately the records of the Company with reference to Malabar contain few detailed data on visits to Malabar of Company's servants who were stationed in Ceylon during Van Reede's commandership. Originally I thought that Hermann had been in Malabar in 1675. In February-March of that year his superior, the Ceylon governor Rijklof van Goens Jr., in the company of Van Reede, undertook a tour of inspection in Malabar. Although the available records of Van Goens' journey do not contain any decisive statement about this, it is just possible that Hermann accompanied his governor at that time in his capacity of physician[7].

On second thoughts, however, the year 1674 is more likely. From the sequel it will become apparent that, after his conversation with Hermann, Van Reede appealed to some Malabar physicians to assist him in the new plan of Hortus Malabaricus. On 20 April 1675 three of these physicians, Ranga Botto, Vinaique Pandito, and Apu Botto, stated that they had co-operated for two years from the Brahmin year "Palivanapaco" (1597) or 1674 according to the Western era, until 10 April of the Brahmin year "Requecao", or 1675[8]. This means that already in 1674 Van Reede started the new plan of Hortus Malabaricus. But since Matthew of St. Joseph had arrived in Cochin in that same year 1674 and had since worked on the first plan, it must be concluded from this that Hermann also spoke with Van Reede in 1674.

After this problematical passage about Hermann's visit, Van Reede continued in the preface to volume 3 his account of the genesis of Hortus Malabaricus with a detailed report about the execution of the new plan. At first he wanted to undertake the description of the plants himself, but he did not find time enough for it. Matthew of St. Joseph, too, according to him was too much engaged on his missionary work to co-operate completely.

For that reason Van Reede requested Johannes Casearius, one of the Dutch ministers of Cochin, to take over part of the work of Matthew, specifically the plant descriptions, although Casearius had never devoted himself to botany and 'for that reason he feared that his style, to which the terminology of botanists and physicians was unknown, would not be up to the dignity of so great a work'[9]. Van Reede was reticent about the task of Casearius. But from Casearius' own preface to volume 1 of Hortus Malabaricus it may be inferred that he occupied himself with arranging the relevant data which were put at his disposal by Matthew and the Malabar physicians, from which he subsequently composed a Latin version of the plant description[1].

Van Reede realized that Matthew's knowledge of the flora of Malabar did not go much beyond what the Carmelite needed in the daily practice as a priest and a physician, 'with the aid of which this pious old man had helped, on his journeys, not only himself but also very many others'[3]. Van Reede solved this problem by invoking the assistance of Malabar scholars and princes, in order to collect as many plants and as much botanical, medical, and other useful information as possible. This idea will have been suggested to him through the numerous and usually good contacts which he had so far maintained with several strata of Malabar society and through his appreciation of the knowledge of Indian scholars and scientists, of which he had given evidence in official reports. Moreover, his conversation with Hermann may have contributed to this, because the latter in his study of the flora of Ceylon frequently made use of medical and linguistic information of native origin.

In the preface to volume 3 Van Reede told the following about the method of collecting the plants:

'some physicians, both Brahmans and others, at my orders made in their own language lists of the best known and most frequently occurring plants, on the basis of which others again divided the plants according to the season in which they attracted notice because they bore either leaves or flowers or fruit. This catalogue according to the time of the year was then given to certain men who were experts in plants, who were entrusted with collecting for us finally from everywhere the plants with the leaves, flowers, and fruit, for which they even climbed the highest tops of trees. Having generally divided them into groups of three, I sent them to some forest. Three or four painters, who stayed with me in a convenient place, at once accurately depicted the living plants readily brought by the collectors. To these pictures a description was added nearly always in my presence'[10] (Fig.82).

The compilation of 'lists of the best known and most frequently occurring plants' was no uncommon occupation within the Company. It was connected with the constantly increasing interest in medicine supply, of which the Company's records of nearly all factories contain numerous examples. For Malabar the oldest known list dates from 1687; it was drawn up by Simon Kadensky, who was already chief surgeon in Cochin during Van Reede's commandership. Although the lists made up by the Malabar physicians were intended in the first instance for the plant collectors, they also contained information about the periods of flowering and fructification, which we can find regularly at the end of the plant descriptions in Hortus Malabaricus.

Figure 82. Draughtsman. Detail of the drawing of Codda-pana from *Horti Malabarici Icones*, BL Add MS 5029:3-4.

Some of the physicians who were responsible for the compilation of the lists are known by name. First of all they were the Brahmins Ranga Botto, Vinaique Pandito, and Apu Botto, coming from the district of Cochin. In their statement of 20 April 1675 they reported that they had caused trees, plants, herbs, and twiners with their flowers, fruit, and seeds to be taken to the town of Cochin by their servants who were acquainted therewith, in order to be drawn and described there[8].

Another physician was Itti Achudem, residing at Collada House in the coastal region of Mouton, who, according to his statement of the same date, communicated the medical virtues and properties of the trees, plants, herbs, and twiners from his book[11]. Although Itti Achudem did not write this explicitly, it is very likely that he also contributed to the listing and collecting of the plants, since he belonged to the caste of the Chōgans or 'tree-climbers'.

From Van Reede's words about his botanical excursions one gets the impression that the descriptions were composed on the site, at least in a rough form. But according to the statement of the three Brahmin physicians the collected plants were also taken to Cochin in order to be described there.

The lists of wanted plants were also sent by Van Reede to princes further distant:

'Moreover, I had courteously asked all the princes of Malabar kindly to inform us of plants which were rather well known by their form and their use. Thus, it happened that from time to time plants were brought to us from a distance of fifty or sixty leagues and that there was not one among those kings and princes but had cultivated some flowers in our *Hortus*'[9].

The distance of 'fifty or sixty leagues' (120 to 132 km) was in conformity with the distribution of the localities of the plants mentioned in the volumes 3 to 8 of Hortus Malabaricus. From this it can be inferred that the rāja of Cochin and his vassal princes of Parūr, Ālangādu, Cranganūr, and Vadakkumkūr will have collaborated, as will also the rāja of Tekkumkūr, who was greatly interested in the Dutch school of Kōttayam. In particular the rāja of Cochin, Vīra Kērala Varma, to whom Van Reede dedicated volume 3, according to him deserved the title of honour of Maecenas on account of his aid and assistance in the collection of numerous trees, shrubs, and plants[12].

The assistance given to him by the surrounding princes shows that Van Reede's botanical excursions did not extend throughout Malabar, but remained confined to the neighbourhood of Cochin and the neighbouring interior. This is in conformity with the fact that since 1674 Van Reede only very rarely undertook journeys in Malabar, and left the maintainance of diplomatic relations to captain Burghart Uytter. At that time Van Reede mainly confined himself to meetings in or near Cochin. Further on in the preface to volume 3 Van Reede described his botanical excursions in greater detail:

'Still, meanwhile the description of the plants which grow around the city of Cochin and whose curative virtues were proclaimed by the indigenous physicians as having been famous from extreme antiquity was rapidly approaching its end. And yet quite often there were plants for which we had to wait a whole year, and sometimes even longer, until we could secure one of their flowers or fruits. These plants were noted down as it were at random, in order that the deficiencies might be supplied properly when an occasion presented itself. Sometimes, if the nature of the office or other weighty causes required it, accompanied by all those whose work consisted in cultivating this *Hortus* and by a good many others, I travelled by water to the adjacent interior in order to rebuke on behalf of my superiors some Malabar prince or other, which was fairly often necessary in the public interest. When then on my journey, as it were to while away the time, I observed the luxurious forests and the delightful plains, I ordered my whole party to leave the ship. Then they wandered in groups through the forests, some of them in order to find booty, some of them because I delighted very much in it, many of them in order to satisfy their curiosity about new things, and others because they worked at our *Hortus*, ravaging all sorts of plants, especially those which they saw adorned with fruits and flowers. When this party, which often consisted of two hundred men, had collected plants there for some hours, they returned, laden with the plants they had collected from here, there, and everywhere, to our warships, which the Malabars call *Mansjous*. Here we examined and looked at our spoils and the painters roughly depicted the plants we had not yet seen, to which was added a rather short and rough description, these descriptions being later perfected in a more convenient place and time'.

In this way numerous plants were brought together, the descriptions of which gave Casearius work for many months. During these botanical stops even the princes came to the tents of the Dutch to observe their diligent activities[13].

A certain remarkable instrument used by Van Reede in the collection of data was the advisory board of scholars, which recalls the royal council of Cochin, consisting of Nambūthiri and of which in 1663-1665 he had also formed part as "regedore maior". The possible members of Van Reede's advisory board, both Malayali and Company's servants, have already been discussed by me before.

The principal subjects that were discussed were the native names, relationships, properties, forms and parts of the plants, their distribution, periods of flowering and fructification, and medical and other uses. Van Reede did not speak about the contributions of the Europeans to the discussions, but he took great delight in those of the Malayali. The Malabar scholars cited verses of antiquity and consulted books of their ancestors. Van Reede did not quote the titles of these books, but since their authors 'lived four thousand years ago' he probably referred to the great Indian medical collections of Caraka, Susruta, and Vagbhata[14]. Itti Achudem described his reference book more accurately as containing the names, medicinal virtues, and properties of the trees, plants, herbs, and twiners, while Ranga Botto, Vinaque Pandito, and Apu Botto stated that they derived their knowledge from the book *Manhaningattnàm* dealing with the names of the plants and their medicinal virtues and uses[15].

The answers of the board were recorded in a "commentarium", presumably a notebook; elsewhere there is a reference to "adversaria" or collections of remarks and observations[16]. A special problem was the translation of the answers. During the discussions of the board and, as we may assume, also during the botanical excursions, Van Reede made use of an interpreter, but he did not mention in what languages the desired information was given. This is of importance with respect to the native plant names, which were to be used in different languages and in different scripts in the descriptions and on the engravings in Hortus Malabaricus. Since the title-pages of Hortus Malabaricus read "Latinis, Malabaricis, Arabicis, & Bra(ch)manum Characteribus nominibusque expressis", it is generally thought that the Malabar scholars informed Van Reede in Malayalam, Arabic, and Sanskrit. But thanks to the careful studies by Manilal (1980) and Govindankutty (1983) we know that the actual situation was more complicated.

Malayalam

The Chogân physician Itti Achudem used not only Malayalam, but also Portuguese when commenting on the plants under discussion. He dictated his opinions in both languages to Emanuel Carneiro, one of the interpreters of the Company. Presumably Itti Achudem did not know sufficient Portuguese, for Carneiro translated his certified statement from Malayalam into Portuguese. It is not clear whether Carneiro in his turn knew Dutch. In any case he did not translate his own Malayalam statement into Dutch, but into Portuguese.

The statement of Itti Achudem has been engraved in Malayalam in the Kolezuthu script (Fig.83). This script, in which already for hundreds of years the ancient Malayalam texts had been written, was becoming extinct in Van Reede's time. Carneiro, on the contrary, wrote his own statement in the newer Aryaezuthu script (Fig.84). According to Manilal these engravings are the first, and possibly the only, simultaneously printed illustrations of both scripts. More important, however, is the question in what script the Malayalam plant names on the drawings in the British Library and on the engravings in Hortus Malabaricus are written. One might expect that for this the Kolezuthu script should have been used, because this was used by Itti Achudem, the only known learned, Malayalam-speaking informant of Van Reede, but both Manilal and Govindankutty have pointed out that the plant names are actually written in the Aryaezuthu script.

Moreover, Manilal suggested that Carneiro, who was acquainted with this script, might have written these names on the drawings or sketches. On the ground of his study of the orthography of the Malayalam plant names in the Aryaezuthu script Govindankutty concluded that these names are written not in one, but in several Malayalam dialects. This indicates that not Itti Achudem alone, but also other Malayali who have remained unknown formed part of Van Reede's board. This is in conformity with the latter's remark that 'this board had been brought together from various parts of Malabar'[17].

Plant taxonomists, however, do not make use of the Malayalam plant names in the Aryaezuthu script. They cite exclusively the Malayalam names such as they are transcribed into Roman characters, both on the engravings and in the headings of the plant descriptions. Govindankutty pointed out that there are sometimes great differences between the spelling in the Aryaezuthu script and that in the Roman script. According to him this is due to European transcribers, who partly misheard the spoken Malayalam plant names and partly were in trouble in rendering the unusual sounds in graphemes of

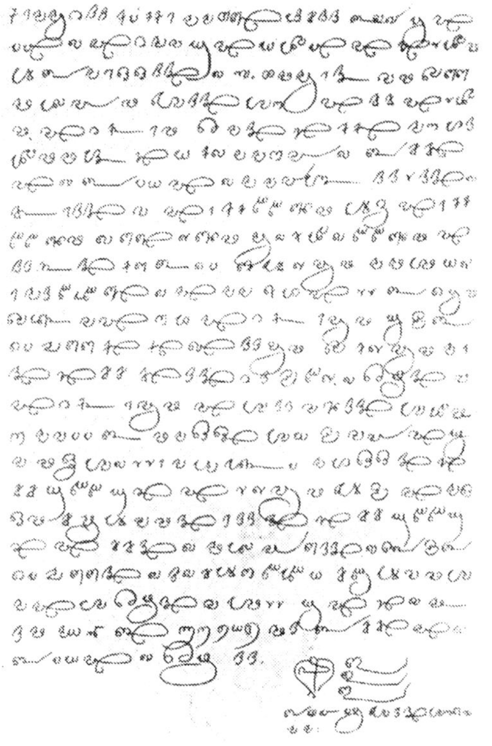

Figure 83. Statement of Itti Achudem in Malayalam in Kolezuthu script. Engraving from Hortus Malabaricus vol.1 (1678):(ix).

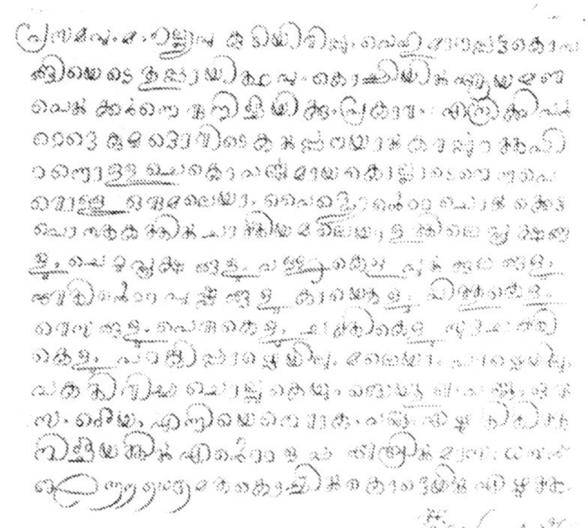

Figure 84. Statement of Emanuel Carneiro in Malayalam in Aryaezuthu script. Engraving from Hortus Malabaricus vol.1 (1678):(vii).

their own languages. Thus he recognized Dutch, German, and Portuguese-speaking transcribers among Van Reede's collaborators.

Arabi-Malayalam

On the engravings of the plants as a rule there are also plant names in Arabic script. One is inclined to assume that these are synonymous names in the Arabic language. However, in his preface to volume 3 Van Reede nowhere mentioned that he had consulted informants speaking Arabic. Nor do the statements of the Malabar physicians and the prefaces by Casearius and Matthew in volume 1 contain any reference to Arabic plant names. Again, no such names in Roman script are mentioned in the plant descriptions. Both Manilal and Govindankutty concluded on closer inspection that these names, despite Arabic script, are written in Malayalam and that they are usually identical with the Malayalam names in the Aryaezuthu script!

According to Govindankutty the use of the Arabic script may perhaps be accounted for by the fact that Van Reede wanted to make a connection with the Persian-Arabic herb commerce with Europe[18]. I agree with Govindankutty that it is quite possible that Matthew executed the Arabic transcription on the illustrations, since he had studied the Arabic language during his stay in Basra, in 1648-1651.

Konkani

The statement of the Brahmin physicians Ranga Botto, Vinaique Pandito, and Apu Botto is engraved in the Nagari script, the usual script of the Sanskrit language (Fig. 85). We find the same script on the drawings and on the engravings of the plants in Hortus Malabaricus with the addition of "Bram." or "Bramin.". In the plant descriptions the Brahmin plant names are transliterated into Roman characters. The Company's servants of that time assumed that the Brahmins of Malabar, writing Nagari script, also made use of the Sanskrit language. Thus in 1670 Van Reede reported that the Brahmins sent by the rāja of Tekkumkūr to the Dutch school of Kōttayam had started the instruction of 'the principal language called Samscortam'[19].

However, contrary to the opinion of most Western botanists in later times, neither the statement nor the Brahmin plant names are in Sanskrit, but in the current Konkani language, which differs considerably from Sanskrit. Konkani-speaking people, who originated from Konkan (Goa and neighbouring regions), were expelled from their land by Portuguese persecutions in the 16th century, and settled in Malabar, and also in Cochin[20]. A study of the Konkani plant names in Hortus Malabaricus is still wanting.

Portuguese

In the discussions of the board the Portuguese language is found to have occupied an important place as a vehicle, or at least as a transitional language. The action of Carneiro as an interpreter from Malayalam into Portuguese

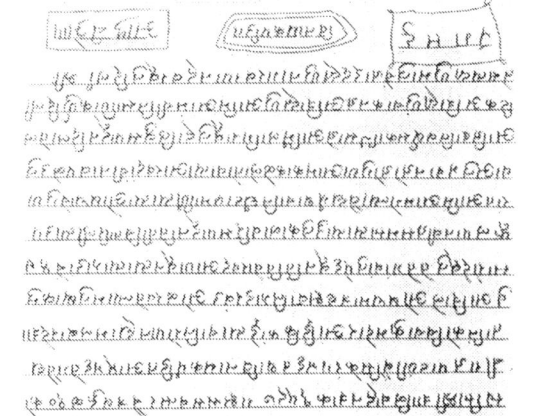

Figure 85. Statement of Ranga Botto, Vinaique Pandito, and Apu Botto in Konkani in Nagari script. Engraving from Hortus Malabaricus vol.1 (1678):(xi).

has already been discussed. The Brahmin physicians do not seem to have needed the assistance of an interpreter, for one of them, Vinaique Pandito, himself translated their statement from Konkani into Portuguese.

On the Netherlands side the town secretary of Cochin, Christiaan Herman van Donep, acted as interpreter. He knew not only Portuguese, but also Latin, as witness his translations of the statements from Portuguese into Latin. The interpreters will have been assisted in their activities by other bilingual or polyglot Europeans. Matthew of St. Joseph, who lived and worked for many years under Portuguese rule, was undoubtedly well-versed in the Portuguese language. Likewise Johannes Casearius and Pieter Minnes, another Dutch interpreter of Portuguese, will have contributed to a ready transmission of information.

The role played by the Portuguese is emphasized by the Portuguese plant names which figure as synonyms in the plant descriptions of Hortus Malabaricus.

The complicated activities of Van Reede's advisory board make it understandable that the transformation of the numerous, widely divergent data into a plant description in fluent Latin was a time-consuming affair. Round about his departure from Malabar, in 1677, only some hundred descriptions were completed to the extent that they could be sent to Europe to be printed.

In Batavia Van Reede worked, as best he could, at the other descriptions, hundreds of them. The illness and death of Casearius, who had travelled with him, deprived him in September 1677 of his experienced Latinist. The latter's task was taken over by Christiaan Herman van Donep, and in addition he received assistance from Willem ten Rhijne.

In Batavia Van Reede was faced with the task of arranging his "collectanea" and "adversaria", while Ten

Rhijne took upon himself to latinize the medical "adversaria". However, in the absence of botanists who by their own observation were acquainted with the flora of Malabar Van Reede had to rely partly on his own memory 'to review the whole work, correct the errors, supply many deficiencies, and from time to time complete the descriptions'. As has been said, upon his departure from Batavia, in November 1677, Van Reede left a copy of the descriptions, in so far as they were then completed, behind with Ten Rhijne.

When in June 1678 Van Reede had arrived in the Netherlands, he was confronted with problems connected with the publication of the first two volumes of Hortus Malabaricus. Apart from the fact that the arrangement of the two volumes was contrary to his own intentions, he had to endure 'the mockery and censure of critics' on the plant descriptions[21]. It is not clear to whom this criticism was due. In his preface to volume 3 Van Reede did not mention any author or publication by name. The sharp outburst contained in Van Goens' general missive of 1679 was directed only in general terms against Hortus Malabaricus, while the reviews, one of which had appeared since then in the *Journal des Scavans*, vol.7 (1680), did not contain any criticism of the plant descriptions as regards the contents. Perhaps Van Reede was referring to privately communicated criticisms, of which nothing is known.

Nevertheless, the criticism induced him to undertake the descriptions of the third and subsequent volumes of Hortus Malabaricus himself. With the aid of his "adversaria", "collectanea", and other notes, and relying on his memory, he composed the descriptions in Dutch and had them translated by one or more Latinists who had been put at his disposal by the printer. The quality of the translations, however, was so poor that he soon gave up this working method. In this situation Van Reede applied to the Utrecht professor of anatomy and botany, Johannes Munnicks, who was willing to undertake not only the latinization of the descriptions, but also 'the care of all the descriptions and everything relating thereto'[22]. This shows that Van Reede reverted again to his approved method of cooperation with Latinists such as Casearius and Van Donep, who had written Latin descriptions from arrangements of the "collectanea" and "adversaria". Although Munnicks' botanical knowledge, experience, and means must not be overrated, Van Reede considered him the most important participant in this stage of the final editing of Hortus Malabaricus, for he allowed him the highest author's fee, of twelve copies per volume, in the publishers' contract of 1681.

Here the question arises as to whether, and to what extent, Munnicks has been able to respond to the criticism of Casearius' descriptions in the first two volumes of Hortus Malabaricus. In this context it must be taken into account that we do not know the starting points of the two, to wit Van Reede's "collectanea" and "adversaria", and that we can only judge by the ultimate result.

On the whole Munnicks has not meddled with the general scheme of descriptions. As in the case of Casearius, there is a fixed sequence of the paragraphs on nomenclature, relationships, habitus, parts of the plants, use, and medical properties. Munnicks added a special paragraph on the distribution of the plant, which is lacking in Casearius' descriptions and in which he gave more details about the florescence and fructification periods than Casearius did. However, this difference may be due to lack of information rather than to negligence on the part of Casearius, because the last four volumes of Hortus Malabaricus do not have notes on distribution either. A more important aspect consists in the differences in the style of writing and in the botanical terminology. Casearius used fluent Latin, as a rule, with attention to an orderly syntax; he made up for his insufficient knowledge of terminology, of which Van Reede was aware, by specifications of relevant characteristics. Munnicks, on the other hand, concerned himself much less about a smooth style; he cluttered his sentences with long rows of characteristics in the then current terminology. As a result of this, Casearius' descriptions are indeed longer and more cumbrous, but they were more readable, while Munnicks' descriptions are shorter and more condensed, but rather tedious. In his own preface to volume 3 Munnicks remarked that in this way he followed Clusius, Jean Bauhin, Alpini, and other phytographers[23].

The frequency of publication of Munnicks' volumes, three in four years, is not in conformity with the agreements in the publishers' contract. In the latter it was agreed that the authors, Van Reede, Munnicks, and Commelin, were to deliver two volumes every year to the publishers, who in their turn were to publish them within one year after delivery. Van Reede, as to his share, seems to have kept to this agreement, for besides the preparatory work for the volumes 3, 4, and 5, in 1684, upon his departure for Asia, he had also three other volumes in readiness[24]. Munnicks, however, could not keep up with him. In his preface to volume 5 he complained 'that I can hardly hope to complete two volumes every year, I can even hardly release one volume'. He imputed this to the pressure of his work, both at the university and in his medical practice, as well as to delay on the part of the printers. He announced that Theodorus Janssonius van Almeloveen, his former pupil and now his sincere friend, was to take over his task and to edit the remaining volumes of Hortus Malabaricus[25].

Janssonius, who was a medical practitioner in Amsterdam, was fortunate in that for the compilation of the plant descriptions he could make use of the tropical collection of the recently founded Amsterdam Hortus Medicus and of the immediate presence of its commissioner Jan Commelin[26]. But in spite of Munnicks' announcement, Janssonius' cooperation on Hortus Malabaricus was short-lived, for he only edited the plant descriptions of volume 6 (1686). The reason of this will have been that he moved to Gouda, by which his necessary regular contact with the Amsterdam garden and Jan Commelin was rendered difficult. Janssonius' plant descriptions are different from those of his former teacher. He abandoned Munnicks' concise style and returned to Casearius' fluent Latin.

The last Latinist to edit Hortus Malabaricus was the Amsterdam physician Abraham van Poot. In contrast to his predecessors he was not hampered by lack of time or other restraining circumstances, for he finished his task of composing plant descriptions for six volumes in five years. Van Poot made light of his task to such an extent that he wrote only prefaces, to the volumes 7 and 8, consisting of a few lines from which we learn no more than that he was the Latin translator. Nevertheless, his share in Hortus Malabaricus should not be underrated. He completed the volumes 7 and 8 in 1688, for which he could still make use of the descriptions prepared by Van Reede himself. But Van Poot had to compose the descriptions for the remaining four volumes from the rough material of Van Reede's "collectanea" and "adversaria", which he had left behind in the Netherlands. Moreover he undertook, at the request of the publishers and with the approval of Van Reede, to translate the whole of the work into Dutch. According to his preface of 1 September 1688 the translation of the first two volumes of the *Malabaarse Kruidhof* (1689) was already finished in the same year. Unfortunately this translation was not a great success, for as late as 1719 sufficient copies of it were still on hand to make possible the title-page issue of 1720. This small success may account for the fact that Van Poot did not continue the translation of the other volumes.

The quality of Van Poot's plant descriptions is considerably inferior to that of his predecessors. Although he followed the fluent style of writing of Casearius and Janssonius, his descriptions are as a rule considerably shorter and contain relatively less information. His descriptions frequently also lack some paragraphs of the general scheme, such as nomenclature, habitus, roots, medical use, and from volume 9 all references to distribution. This inferiority is due to Van Reede's absence, as a result of which Van Poot could not consult him, especially in the case of the non-edited "collectanea" and "adversaria".

Summarizing, we may say that the plant descriptions in Hortus Malabaricus consist of several elements of highly different origin. Although Matthew of St. Joseph can be held responsible for the original scheme of the descriptions, the final scheme seems to have been planned at the advice of Paul Hermann during his visit to Malabar, presumably in 1674. Nevertheless, some elements from Matthew's scheme have been taken over in it, such as nomenclature, habitus, distribution, and medical and other uses. Van Reede at first derived his specific information from Matthew's notebook and memory. In a later stage he supplemented it on a large scale with information supplied to him by Malabar physicians and other Malayali, either by oral tradition or from medical treatises. In collecting the plants, Van Reede was assisted by Malabar physicians and several rājas, among them mainly the rāja of Cochin, while Van Reede himself confined himself to excursions in the neighbourhood of Cochin.

The rough descriptions were made partly on the spot, partly in Cochin. The plants and the gathered information were discussed in a polyglot advisory board of Malayali and Europeans, assisted by one or more interpreters. As a result of the discussions, native names in Malayalam and Konkani in Aryaezuthu and Nagari script respectively were added to the illustrations. On the basis of the notes, or "collectanea" and "adversaria" brought together in this way, Johannes Casearius in Malabar, Christiaan Herman van Donep in Batavia, and Johannes Munnicks, Theodorus Janssonius van Almeloveen, and Abraham van Poot in the Netherlands composed latinized descriptions. The differences in the descriptions are due to differences in the style of writing and the use of botanical terminology of the various authors as well as to differences in the information available in the "collectanea" and "adversaria".

Notes

1 Hortus Malabaricus vol.1 (1678):(iv).
2 Hortus Malabaricus vol.3 (1682):(vi); Heniger 1980:44.
3 Hortus Malabaricus vol.3 (1682):(vii); Heniger 1980:45.
4 Hortus Malabaricus vol.3 (1682):(viii); Heniger 1980:46.
5 Veendorp & Baas Becking 1938:83.
6 Hermann 1698:76 "Nam cum anno millesimo sexcentesimo septuagesimo Coutschinae haereremus". An illustration of this kind of Arum had been copied formerly by Matthew in Basra from the book of Saladin Artafa (Monti 1742:25-26,Tab.23.fig.3).
7 Heniger 1980:57, note 7;59, note 16.
8 Hortus Malabaricus vol.1 (1678):(xi)-(xii).
9 Hortus Malabaricus vol.3 (1682):(ix); Heniger 1980:47.
10 Hortus Malabaricus vol.3 (1682):(viii); Heniger 1980:47. From this passage one may conclude that Van Reede never made a herbarium of Malabar plants, see the discussion of the 12-volume herbarium, "Plantae Malabaricae", at Göttingen, by Johnston 1970:655.
11 Hortus Malabaricus vol.1 (1678):(ix)-(x); Manilal 1980:115-116.
12 Hortus Malabaricus vol.3. (1682):(i)-(ii).
13 Hortus Malabaricus vol.3 (1682):(xiii)-(xiv); Heniger 1980: 51-52.
14 Hortus Malabaricus vol.3 (1682):(x); Heniger 1980:48 and 62, note 29.
15 Heniger 1980:62, note 27. I did not succeed in tracing the identity of these books.
16 Hortus Malabaricus vol.3 (1682):(x),(xv); Heniger 1980:48, 53, and notes 26 and 45.
17 Hortus Malabaricus vol.3 (1682):(x); Heniger 1980:48.
18 Govindankutty 1983:265, note 9; Hortus Malabaricus vol.3 (1682):(vi); Heniger 1980:44.
19 VOC 1274:162-162v, letter of 15 August 1670 of Van Reede and the council of Malabar to the Governor-General and the Council of India.
20 Manilal 1980:113; Govindankutty 1983:244, note 11; personal communication of K.S. Manilal.
21 Hortus Malabaricus vol.3 (1682):(xvii); Heniger 1980:55.
22 Hortus Malabaricus vol.3 (1682):(xviii); Heniger 1980:56.
23 Hortus Malabaricus vol.3 (1682):(xx).
24 See Janssonius' preface to Hortus Malabaricus vol.6 (1686):(iii).
25 Hortus Malabaricus vol.5 (1685):(iii)-(iv).
26 Hortus Malabaricus vol.6 (1686):(iv).

11
Commentaries

ARNOLD SYEN

Arnold Syen (1640-1678) is the first, but also the least known commentator of Hortus Malabaricus. His life and work is known to us only from short biographical sketches by Boerhaave and Veendorp & Baas Becking[1]. These sketches are based upon Syen's own statements in Hortus Malabaricus, upon remarks of his botanical friends Morison, Breyne, and Hermann, and upon certain documents from the records of the Curators of Leiden University. None of Syen's scientific inheritance -manuscripts, herbaria, correspondence, etc.- has so far been traced. Of his publications, apart from his contributions to Hortus Malabaricus, we know only his medical dissertation *De Hydrope* (1659).

Syen studied and took his medical degree at Leiden University. After this he had a medical practice in Gouda, where he also owned a botanical garden, but it is not known what plants he cultivated there.

From 1670 to his untimely death, in October 1678, Syen was professor of botany in Leiden and director of the botanical garden as the successor of Florens Schuyl (1619-1669)[2]. In spring and in summer he demonstrated plants in the garden; in autumn and in winter he lectured on medical botany[3]. On his travels in England, France, and Germany he became well acquainted with the European flora and according to Boerhaave he was specialized in it. Syen did not consider himself as an expert in tropical botany, which appears from a remark in his letter presenting volume 1 of Hortus Malabaricus to the Curators of Leiden University on 8 May 1678:
'For a long time I have been in doubt whether I should dare to offer these rare plants from the Malabar region, which region occupies a vast part of India, to you, for these plants were not the object of my own special studies'[4].

When on his return to the Netherlands Van Reede first saw the recently published volume, he stated, possibly with some disappointment, that 'It contained a commentary, illustration, and comparison with the European plants'[5]. Syen's own modesty and the remark of Van Reede might suggest that he did not get beyond a comparison of tropical plants with the European flora known at that time, a comparison which according to modern taxonomic conceptions would be foredoomed to failure. From the following discussion of Syen's botanical resources and taxonomic conceptions it will appear that his attempt to furnish relevant commentaries on Malabar plants was a first step in the further development of tropical botany in the Netherlands. It is true that Syen's resources were limited and that we are insufficiently informed about their qualitative value, but still such fundamental elements of botanical science as the botanical garden, the herbarium, the botanical library, drawings, etc., already play a part in his work to bring about some systematic order in his contributions to Hortus Malabaricus.

Resources

The Leiden botanical garden
During Syen's directorship the Leiden botanical garden possessed the most comprehensive and most differentiated collection of living plants in the Netherlands. From the garden catalogue of his predecessor Schuyl (1668) it appears that this collection of more than 1,800 kinds of plants gave a fair picture of the Dutch indigenous flora as well as of the European exotic flora. For the determination of his garden plants Schuyl had mainly relied upon the botanical works of the three founders of Dutch botany, Dodonaeus, Lobelius, and Clusius.

The non-European flora -in particular the tropical plants- was then still represented by few species. Shortly before, during Schuyl's directorship, the first importation of South African plants took place[6]. Under Syen's directorship the Cape collection was enlarged further. Paul Hermann, his later successor, who botanized at the Cape in 1672, on his way from Texel to Ceylon, sent him an important lot of seeds, fruits, roots, and descriptions of plants[7]. In 1676 Syen received numerous botanical rarities from the Cape[8], which presumably came from a consignment of natural curiosities of the Cape governor Joan Bax to the Stadtholder William III[9]. The Cape collection could be kept alive without much difficulty in the moderately heated greenhouses of the Leiden garden.

It was, however, much more difficult for Syen to cultivate tropical plants as long as no hothouses had yet been built in Leiden. The first hothouses were not built until 1680-1687, under Hermann's directorship[10]. One of the tropical plants received by Syen was a Papaya-tree from the West Indies, which he was to compare later on with Van Reede's Papajamaram HM 1:24. It is of greater interest that since 1675, thanks to Hermann, Syen could start a -very modest- Ceylonese collection of living plants, which could be of service to him for the discussion of Malabar plants and which will presently be described in some detail.

The garden herbarium

It is not known with certainty whether a garden herbarium of the Leiden garden was available to Syen. Moreover, in practice 17th-century and 18th-century Dutch professors of botany always found it difficult to make a strict separation between their own private herbaria and the public garden herbaria. However, in 1661 and 1669-1671 Syen's townsman, the Leiden chemist Antoni Gaymans (1634-1680) made representative herbaria from the Leiden garden, and he provided his exsiccates with a botanical nomenclature which in all probability he derived from Schuyl's garden catalogue of 1668[11]. It is quite possible that Gaymans lent his herbaria to Syen for consultation, but unfortunately Syen does not refer in any way in his commentaries to a garden herbarium which might confirm this assumption.

The museum

The so-called "Ambulacrum", the oldest greenhouse in the Leiden botanical garden, also contained a museum of natural curiosities. It appeared from the catalogues of 1659 and 1670 that the museum mainly contained stuffed animals, natural products, and some ethnographical curiosities from Brazil and the East Indies[12]. In two cases Syen explicitly referred to material present in the "Ambulacrum". In his commentary on Ily HM 1:26 he mentioned two bamboos, 28 and 26 feet long respectively, which he thought had formerly been presented by the well-known Brazil naturalist Willem Piso. In reality the bamboos had been collected in 1600 in Bantam by Nicolaas Coolmans[13]. In the other case, in his dicussion of Caniram HM 1:68, Syen referred to a coloured illustration of the snake Naja & Naghaja, which he had received from Hermann from Ceylon and which hung in the "Ambulacrum".

The library

Syen will undoubtedly have possessed a private library of botanical literature, but information about this is lacking. Until far into the eighteenth century the Leiden botanical garden did not possess its own library, as did later the Amsterdam botanical garden. On the other hand the Leiden University Library was the most important library in the Netherlands.

In the first volume of Hortus Malabaricus Syen referred to some twenty publications, which gives us an idea of his working method and his thought. In broad outline he used two geographical groups of botanical descriptions: Asia and America. For Asiatic plants he used in the first place Acosta's treatise on aromatic and medicinal plants of Portuguese India (1578), however in the abbreviated Latin translation by Clusius (1601). No less important works for Syen were Da Orta's *Coloquios* (1563) on the plants of Goa and its environs, at least in the edition of Clusius (1601), and Van Linschoten's *Itinerario* with the commentaries of Paludanus (1596). In addition Syen consulted Bontius' descriptions of plants in the neighbourhood of Batavia (1658). Occasionally he also made use of De Flacourt's work on the natural history of Madagascar (1661). For American plants Syen regularly referred to Marcgrav's and Piso's work on the flora of Brazil (1658). He also used Recchus' edition (1651) of Hernandez' book on the natural history of Mexico and De Rochefort's description of the natural history of the Antilles (1658).

From the above it appears that Syen had at his disposal the most important works of his time on tropical plants, even though these works were essentially regionally orientated. But for his commentaries this limitation was no obstacle, because in the first volume of Hortus Malabaricus only generally known, widely cultivated tropical plants were described. On the other hand it is indeed important to find that for comparison with Malabar plants Syen referred not only to tropical plants from other Asiatic regions, such as Goa and Batavia, but also to American tropical plants. He thus laid the basis for a trend in earlier tropical botany which his successor as commentator, Jan Commelin, was to follow and develop further. This trend can be followed up to and inclusive of the *Species Plantarum* of Linnaeus[14]. The origin of Syen's comparison with plants from the New World consists in the supposition which was already popular in his time that under similar climatological conditions, wherever in the world, identical plants might be expected to occur.

Syen's Ceylonese collection

It has already been stated that Syen had Ceylonese material for study at his disposal: living plants in the Leiden botanical garden and objects in the museum. In addition he also possessed Asiatic herbaria. In his discussion of Marotti HM 1:66 he mentioned 'my dried Javanese, Ceylonese and other plants'. He did not give any further information about the origin and the identity of his Javanese exsiccates, but he may have received them from Andries Cleyer, of Batavia, who had extensive contacts with European botanists. For the rest, apart from the case of Marotti, Syen could not make any use of his Javanese herbarium. On the other hand Syen regularly discussed his specimens from Ceylon. It is not always quite clear, however, whether he had in mind exsiccates, living plants, museum objects, or merely manuscript notes. That is why we here speak only about his Ceylonese collection, without going too strictly into its exact nature.

To my knowledge Syen's Ceylonese collection is the first of this kind from that island which played a part in the history of botany. In the course of the present study the importance of Ceylonese collections for the identification of Van Reede's plants will become more and more prominent.

In one-third of his commentaries in volume 1 of Hortus Malabaricus Syen introduced, with a view to comparison with Van Reede's plants, all sorts of seeds, fruits, leaves, flowers, short notes, and descriptions, all originating from Ceylon. Usually Syen wrote explicitly that Paul Hermann had sent him this material from Colombo. He discussed a specimen in the private garden of Hieronymus Beverningk, one of the Curators of Leiden University, which also came from Hermann. But in other cases Syen did not mention the donor.

As regards Syen's anonymous material from Ceylon, this may have been a consignment of 1677 to the Stadtholder William III, consisting of an unspecified 'lot of trees, plants, and animals'[15]. This consignment was an answer of Rijklof van Goens Jr., governor of Ceylon, to the formal request of Lords XVII to all the factories of the Company to honour the Prince, who had succeeded shortly before in expelling the French armies from the Republic, with natural curiosities[16]. It is quite likely that Paul Hermann, at that time the most prominent naturalist in Ceylon, had actually provided for this consignment. It is also likely that Syen may have got his anonymous Ceylonese material from this consignment. Thus it may be assumed that Syen's Ceylonese collection originated either directly of indirectly from Paul Hermann.

In so far as can be reconstructed from stray data in Hortus Malabaricus, Syen received his Ceylonese material successively in the years 1675, 1676, and 1677. A crucial point in this chronology is the interpretation of the term "hoc anno" (in this year) in the mention of reception of botanical objects from Ceylon. Syen did not mean by this the year of publication of volume 1 of Hortus Malabaricus, 1678, but the preceding year 1677. The Asiatic return fleet, including also the squadron from Ceylon, always arrived in the Netherlands in summer. Since Syen presented volume 1 to the Curators of Leiden University already on 8 May 1678, he cannot have discussed in this volume Ceylonese material which might have arrived in the summer of 1678. Accordingly, Syen's "hoc anno" must be 1677, "anno praeterito" (last year) 1676, and "ante biennium" (two years ago) 1675. This interpretation is not contrary to the years 1675 and 1676 which Breyne, Hermann, and in one case Syen himself mentioned with respect to the arrival of some Ceylonese plants. Syen's "praecedentis annis" (in preceding years) allows of no other interpretation but 1676, 1675, or even earlier. There are, however, no indications that Syen already received material from Hermann in 1673 and 1674, although this is possible, because Hermann already arrived in Ceylon in 1672.

The following survey of Syen's Ceylonese collection has been arranged according to the donor and the year of arrival.

From Hermann's consignment of 1675 we know a dried twig with the inscription *Arbor Sancti Thomae* (etc.), which Syen compared with Chovanna-mandaru HM 1:57-58:32, *Bauhinia variegata* L. According to Breyne in that year Syen cultivated in the garden *Mimosa non spinosa major Zeylanica Domino Hermans* (*Aeschynomene aspera* L.), which died, however, in 1676, while he did not have the opportunity to dry this species or have it depicted[17]. Syen was more successful with the Asiatic variety of Sweet flag, *Acorus Calamus* L. var. *verus* L., which was still living in 1687[18] and which was used by Jan Commelin for a comparison with Van Reede's Va embu HM 11:99:48.

In Hermann's consignment of 1676 Syen found a dried part of a bamboo with leaves and flowers, with the inscription *Arundo Indica arborea maxima, cortice spinosa* and the Ceylonese name Nuayhas, evidently a slip of the pen for the more correct Unaghas, which he dicussed at Ily HM 1:25-26:16. This consignment also contained leaves and flowers of Ghoraka and of the related species Kanna Ghoraka, which he associated with Coddam-pulli HM 1:41-42:24, *Cambogia Gutta* L.

In Hermann's consignment of 1677 he found a dried branch with immature fruits, and besides seeds, with the inscription *Iambos Sylvatica, fructu cerasi magnitudine* and the Ceylonese name Walgambu, together with a short note by Hermann on the leaf of *Iambos major*. Syen adopted the first-mentioned inscription for Natischambu HM 1:29-30:18, *Eugenia malaccensis* L. The flowers with the Ceylonese inscription Aehaela, which Syen received in this year and which he discussed at Conna HM 1:37-38:22, *Cassia fistula* L., presumably were also sent by Hermann. Another plant arriving in this consignment was a living specimen addressed by Hermann to Beverningk, which Syen identified as *Mangas fructu venenato C.Bauhin* and which he compared with Odallam HM 1:71-72:39, *Cerbera Manghas* L. Finally this consignment also included a specimen of Indigo from Hermann, with the inscription *Polygala Indica frutescens, ex cujus foliis Anil sive Indigo conficitur* (etc.) and the Ceylonese name Awari, which he identified with Ameri HM 1:101-102:54, *Indigofera tinctoria* L.

For four species from Hermann no year of arrival was given. Syen received a nut with numerous seeds, with the Ceylonese name Telabo, which he tried in vain to cultivate for a comparison with Cavalam HM 1:89-90:49, *Sterculia Balangas* L. On the other hand, according to Breyne he succeeded in keeping alive an *Amaranthus spinosus* L.[19] and also an *Aeschynomene* (?) sp.[20]. A very interesting import was the cinnamon-tree, the economically most important cultivated plant of Ceylon, in the form of a small living tree, with seeds and roots, and with a description by the name of *Canella ex qua Cinnamomum* and the Ceylonese name Kurudu, which Syen discussed at Carua HM 1:107-110:57. At the same time Beverningk also received such a small tree. Unfortunately both specimens held out only two or three years in the cold of the Dutch winter[21]. Hermann also sent camphoric distillation products of cinnamon to Syen and Beverningk[22].

Among the anonymous consignments which may be ascribed to Hermann were, in 1677, resin from a palm-tree dealt with by Syen at Tenga HM 1:1-8:1-4, and a living banana-tree discussed at Bala HM 1:17-20:12-14. Undated consignments contained very large pods associated by Syen with Palega-pajaneli HM 1:77-78(1):43, *Bignonia indica* L.; next a fruit with the Ceylonese name

Kiridiwael, mentioned at Curuta-pala HM 1:83-84:46, *Tabernaemontana alternifolia* L., and finally a dried plant and seeds compared with Agaty HM 1:95-96:51, *Aeschynomene grandiflora* L.

It is tempting to suppose that Syen's Ceylonese collection, or at least the exsiccates, might be identical with one of the Ceylonese herbaria of Paul Hermann still existing in Leiden, London, or Paris. But Hermann's inscriptions quoted by Syen do not agree with those in the said herbaria. Apparently, Syen's specimens belonged to another herbarium volume, which seems to have been lost.

From the above surveys of literature and botanical objects consulted by Syen one may conclude that Van Reede's statement on the former's leaning towards European plants is only partly correct. Syen actually brought all the literature on tropical plants, as far as available to him, into the discussion. Moreover, he conscientiously made use of herbaria and other materials, recently received from Ceylon and Java, which contributed to his general knowledge of tropical plant forms. In this respect Syen owed very much to Paul Hermann, whose generosity in sending exsiccates, descriptions, and notes formed the basis of a first comparison of Van Reede's Malabar plants with the Ceylonese flora.

Interpretation, nomenclature, and classification

Before a discussion of Syen's interpretation, nomenclature, and classification it must be asked what drawings and descriptions were actually available to him.

Syen at first assumed that the material which Van Reede had sent to him and which was to be published as volumes 1 and 2 of Hortus Malabaricus constituted the whole work, and not only a part of it. From the prefaces to volume 1, written by Van Reede, Casearius, and Matthew of St. Joseph, it was not possible for Syen to infer that ultimately the work was to become much larger than what he had received from Asia for the purpose of editing and commentary. Nor does Syen seem to have been informed of Van Reede's intention to divide the whole work into three main divisions, part 1 on trees, part 2 on shrubs, and part 3 on herbs, and subsequently to split up each main division into several volumes in a certain systematic order. The consequence of these misunderstandings was that Syen looked upon the material received by him as a unity, in which he was to create some semblance of order, in so far as this was possible without his having a definite indication of Van Reede's intentions. It was an editorial problem in this context that the number of trees available to Syen was much larger than that of the shrubs and herbs. A division into three volumes, each devoted to one group, was unattractive, because in that case the volume with the trees would become much thicker than the two other volumes with shrubs and herbs. At the advice of the publisher, who preferred two volumes of the same size, Syen decided to include in volume 1 most of the trees and in volume 2 the remaining trees, supplemented with shrubs and herbs, until he reached the same size as

that of volume 1[23]. Thus he thwarted, quite unconsciously, Van Reede's intentions.

After his return to the Netherlands, in June 1678, Van Reede, upon inspecting volume 1 of Hortus Malabaricus, which had just been published by Syen, discovered that his original plan had fallen through. In fact, Syen and the publisher had completely blocked the classification of the trees, as a result of which the various groups of trees established by Syen could no longer be enlarged with new, related species brought from Asia by Van Reede. Thus, Syen opened volume 1 with five palm-trees illustrated with eleven engravings, immediately followed by the banana-tree and the papaya. Van Reede, however, also brought detailed descriptions and twenty-five drawings of three other palm-trees or palm-like trees, which obviously should have followed the palm-trees already published, by which a splendid survey of this group would have been given.

Van Reede realized that in the next volumes, starting with volume 3, he had to begin once more with a new, second classification of his remaining material, in consequence of which the unity of the whole work, however, would be broken. He foresaw that the reader might be frustrated by having to look up related plants in different volumes.

Interpretation and nomenclature

For the interpretation and nomenclature of a Malabar plant Syen used the following method, in which Bauhin's *Pinax* was an important guide.

In the first place he tried to associate the plant with the existing botanical and medico-botanical literature. For commonly known plants, such as the coconut tree, the Betel-nut Palm, the Cassia, and the Indigo, he referred to already existing and consequently published descriptions. Where these descriptions were quoted by Bauhin, Syen adopted his phrase names. Thus, the coconut tree, Tenga HM 1:1-8:1-4, was named *Palma Indica coccifera angulosa, Iansiat Indi* C.B.P.; the Betel-nut Palm, Caunga HM 1:9-10:5-8, received Bauhin's name *Palma cujus fructus sessilis Faufel dicitur*, and the banana tree, Bala HM 1:17-20:12-14, was styled *Palma humilis longis latisque foliis* C.B.P.

If Syen was not convinced of the identity of a Malabar plant with a species that had been described, he discussed quotations on related species from the literature. In the commentary on Nilicamaram HM 1:69-70:38, he quoted Aldini's *Hortus Farnesianus* (1625), Bauhin's *Pinax*, and Vallot's catalogue of the "Jardin Royal" in Paris (1665), but finally he opted for the new name *Arbor Acaciae foliis Malabarica, fructu rotundo, semine triangulo*. In other cases he put forward some of Hermann's material from Ceylon and consequently adopted Hermann's manuscript name, such as *Iambos Sylvatica, fructu cerasi magnitudine* for Natu-schambu HM 1:29-30:18. In a number of cases Syen could not trace sufficient relevant literature. He then classed the plant in a Bauhinian genus and derived from this a phrase name by adding one or more characteristic elements. In the Pariti group, HM

1:51-56:29-31, he was aware of its relationship with the European Alceas and he therefore invented the name *Alcea Malabarensis*. He separated the pentafoliate Cudupariti by the name *Alcea Malabarensis Pentaphylla* (etc.) from the two other species Bupariti and Pariti as *Alcea Malabarensis, Abutili folio*. The difference between these two species was characterized by him on the ground of the size and colour of their flowers, as *flore majore, ex albo flavescente* and *flore minore, ex albo flavescente, exterius subaspero* respectively. Another example is the Alu group, HM 1:43-50:25-28, which he could not find in the literature, but which he recognized as a species of *Ficus*; he therefore invented the name *Ficus Malabarensis* and distinguished the various plants by their leaves and fruits.

If Syen was not very sure of the identity of a plant with a Bauhinian genus, he emphasized the affinity or similarity by adding *similis* or *affinis* to the generic name. In his opinion Odallam HM 1:71-72:39 with its peach-like fruits had some similarity to *Persica* and he therefore named it *Persicae similis angusti folia* (etc.). He compared Agaty HM 1:95-96:51 to Sesban described by Alpini and named *Galega Aegyptiaca siliquis articulatis* by Bauhin, from which he derived *Galegae affinis Malabarica arborescens, siliquis majoribus articulatis*, thus maintaining the similar characteristic of articulated pods.

In many cases, however, Syen did not succeed in relating a Malabar plant to any known genus, though he was able to define its characteristics. In such a situation he did not invent a new generic name -a practice applied some decennia later on- but fell back on a simple, though archaic method of nomenclature of *Arbor Malabarica* or *Nux Malabarica*. Nevertheless he sought to include a systematic feature in such a name. The pod-bearing trees of the Mandaru group HM 1:57-63:32-35 were classed in *Arbor siliquosa Malabarica*, while the distinction between the kinds was defined on the ground of the bifid leaves and the colour of the flowers.

In only one case, Cat-ambalam HM 1:93, did Syen have no comment at all, because the drawing was missing.

Classification

For a more exact appreciation of Syen's share in the arrangement of Hortus Malabaricus the numbers on the drawings of the London codex contain some indications.

The drawings for the first two volumes bear four types of numbering. The most frequent type is the "F" series, which continues, with a short interruption, from F 1 to F 113, covering all 113 drawings published in both volumes. In volume 2 the numbers of the "F" series were changed twice, in contrast with volume 1. Surprisingly enough, we also meet with the "F" series on drawings destined for volumes 7 to 10. The numbers link on with those of volumes 1 and 2, and continue until 212, though with several omissions, which are to be attributed to the cutting-off of the margins of the drawings.

The second type is the "Fig." series, which at the beginning of volume 2 fills the short interruptions in the "F" series from Fig.58 to Fig.68, covering, among others, the Kaida group of Malabar plants.

The third type is the "f" series, also in volume 2, which runs parallel to the "Fig." series from f 1 to f 8.

Finally, the fourth type, which has no preceding letter, consists of two series of numerals alone: 1 to 51 on the drawings for volume 1, and 5 to 50 on the drawings for volume 2. On looking more closely, it is found that the numbers of the fourth type do not form a numbering of the drawings, but of the corresponding chapters of descriptions.

The first three types occur on the front of the drawings, but the fourth type occurs exclusively on the back.

The "f" series has been written in pale ink and shows clear traces that the drawings have been washed. We may therefore assume that the "f" series was written by Van Reede or one of his clerks before these drawings were sent to Europe. The "Fig." series on the drawings for volume 2 is written in Jan Commelin's careless hand, which we shall encounter numerous times in other types, not discussed here, on the drawings for the next volumes. From this it might be concluded that the long "F" series occurring on the drawings for volumes 1 and 2 as well as on those for volumes 7 to 10 has been written by Syen, but owing to lack of sufficient material for comparison, such as letters written by Syen, this cannot be established with certainty. The fourth type of numbers, also occurring in both volumes 1 and 2, are very similar to those of the "F" series, so that this type, too, may be attributed to Syen.

With some reservation we may draw the following conclusions from these numbers about Syen's share in arranging Hortus Malabaricus. There can be no doubt that volume 1 was arranged by Syen. From the unchanged numerals of the "F" series of its drawings, and also from the unchanged numerals of the fourth type listing the corresponding chapters, it is clear that he had no problems in definitely arranging the sequence of the plants. But it is also evident that he has still been able to occupy himself with the arrangement of volume 2 before death surprised him in October 1678. The twofold changes of the numerals of the "F" series indicate that Syen had more difficulty with the arrangement of this second volume, specifically in the sequence of the shrubs and the herbs. It would seem that the Kaida group, which bears no traces of the "F" series, could not be inserted by him in a suitable place. At first Syen wanted to start volume 2 with the plants of the Panel group, which bear the crossed-out numbers F 58 and F 59, and as appears from the unchanged numerals of the second series of the fourth type, continue with the plants in the sequence as actually published. Apparently, Commelin, who took over Syen's work, decided on the ultimate arrangement by opening volume 2 with the Kaida group.

The surprising feature, however, is that Syen's "F" series also occurs on drawings for volumes 7 to 10. This might mean that he already possessed these drawings with corresponding descriptions before Van Reede had returned from Asia. But this is contradicted by the fact that in preparing the publication of volume 1 Syen had failed to find the drawings and descriptions of two plants, Caicotten-pala and Ana-parua, although Van Reede had expressly stated in the text that Caicotten-pala should succeed Codaga-pala HM 1:85-86:47 and Ana-parua should succeed Tinda-parua HM 1:87-88:48. In effect Ana-parua occurs among the drawings for volume 7 under Syen's number F 178 and Caicotten-pala among the drawings for volume 10 under Syen's number F 142. It is improbable that Syen would have omitted the publication of these two drawings in volume 1 if they had already been available to him at that time, in 1677/8. In other words, Syen did not set eyes on the drawings for volumes 7 to 10 until after the publication of volume 1 in May 1678. Apparently, after his return to the Netherlands in June 1678, Van Reede handed his other material for Hortus Malabaricus, or at least the material for volumes 7 to 10, to Syen for editing.

From the numbering of the "F" series, which continues until F 212, it is evident that a few months before his death Syen was still able to study some hundred drawings of this new material. Summarizing, we may state that Syen has done more than merely edit volume 1. He also largely arranged volume 2, and after Van Reede's return he also started arranging the new material.

As has been said, when editing volumes 1 and 2 Syen had no insight into Van Reede's systematic intentions. Judging from the numbers on the drawings, it is not even likely that he or his collaborators in Asia were occupied in detail with classification problems. Only the washed numbers of the small "f" series on the drawings of the Kaida group indicate that he intended to keep this group together. Still, the presence of the "f" series may have been a point of reference for Syen. In the discussion of the contents of Hortus Malabaricus it has been pointed out that Van Reede and his collaborators placed small groups of two, three, four, or more plants with similar native names together. This is confirmed by Van Reede's washed numbers on the drawings of the Kaida group, consisting of Kaida, Kaida-taddi, Perin-kaida-taddi, and Kaida tsjerria. Presumably Syen used these small groups among the material at his disposal as nuclei of his arrangement, around which he could group the single plants. In volume 1 such small groups were the Panas (2), Schambus (2), Pullis (2), Alus (4), Paritis (3), Manadarus (3), Palas (3), and in volume 2, besides the Kaidas (4), also the Panels (2), Nosis (2), Schettis (3), Curinis (2), Hummatus (3), Avanacus (4), Schundas (3), Schorigenams (3), Callis (3), Schullis (4), Carambus (3), and Tageras (2). Syen has not been able to maintain this manner of grouping consistently. In fact, he had previously divided the material into trees, shrubs, and herbs. In consequence related plants became distributed over these three main divisions, although Syen was conscious of their relationship. An example is formed by the representatives of the modern family *Malvaceae*, which appear in three different places in the first two volumes. The tree-like Paritis, such as Bupariti (*Hibiscus populneus* L.), Pariti seu Tali-pariti (*Hibiscus tiliaceus* L.), and Cudupariti (*Gossypium arboreum* L.), named by Syen *Alcea Malabarensis* on account of their similarity to the pre-Linnaean Alceas from Europe, were placed by him among the trees in volume 1. The shrubby Schem-pariti (*Hibiscus Rosa sinensis* L.), however, as appears from its native name, being related to the tree-like Paritis and expressly referred to in Van Reede's text as a fourth species of the Pariti group, was placed by Syen among the shrubs in volume 2, although he must have been conscious of the fact that this plant also belonged to the pre-Linnaean Alceas. In the same way a herb with a non-related native name such as Cattu-gasturi (*Hibiscus Abelmoschus* L.), also a pre-Linnaean Alcea, strayed among the herbs at the end of volume 2. This example may explain the apparent confusion in classification with which not only a modern botanist, but also a historian of botany consulting Hortus Malabaricus, is frequently confronted.

Syen's "F" series on the drawings for volumes 7 to 10 has too many omissions to draw far-reaching conclusions therefrom. The numbers still existing only allow of the supposition that, after the herbs of volume 2, Syen wanted to continue with forty or more herbs from volumes 9 and 10 as published afterwards by Commelin, to be followed by a somewhat confused selection of herbs and shrubs from volumes 7 and 8. But this will not have been more than a first attempt to create order in the hundreds of drawings and plants which Van Reede had brought with him, because Syen soon died.

Syen never published anything on his concepts of classification. This thwarts a conclusive answer to the question as to what method he used in the final arrangement. Moreover, only his commentaries in volume 1 of Hortus Malabaricus are available to us. About a limited number of plants, some fifty, we can at most make a few general observations.

In volume 1, in his address to the reader, Syen stated that he provided the Malabar plants with Bauhinian names. As I have already set forth, this statement is only partly correct. It is more likely that Syen meant that he used Bauhin's *Pinax* as a guide in questions of nomenclature. This raises the question as to whether he did not use the *Pinax* as a classificatory principle in arranging volume 1 as well.

On closer inspection of the sequence in this volume one can clearly distinguish two series of plants derived from Bauhin's classification, to wit, from Book XII and Book XI. The palm-trees and the banana-tree with which volume 1 opens belong all of them to Bauhin's Book XII, section 6. The successive Champacam, Elengi, and Manjapumeram can be derived from Book XII, section 3. Further on in volume 1 we find the Alu group again among the figs in Book XII, section 1, as also Mailanschi among *Oxycantha* in the same section. It would appear that Syen used Book XII, but in a reverse order of the sections, as the axis of his arrangement of volume 1.

The series derived from Book XI is more complicated. Following the sequence of volume 1, we can trace the papaya in section 5, both Schambus in section 6, the successive Conna (in section 2) and the Pulli group (in sections 2 and 6), the successive Mandaru group (in section 2), Canschena-pou (in section 2), Marotti (in section 1), Caniram (in section 6), Nilicamaram (in section 1), and Odallam (in section 6), and finally the separate Ambalam group (in section 6). In the case of Book XI apparently Syen did not, as with Book XII, keep to Bauhin's sequence of the sections, but mixed them up.

Another complication is that Syen did not keep the series derived from Books XII and XI sharply separated. On the contrary, he telescoped the two series together, so that really quite a new sequence of Bauhin's sections was formed. It would be going too far to seek an explanation for Syen's motives for changing Bauhin's sequence, because we do not know his criteria.

At all events it appears from the examples given that Syen was guided in the final arrangement of volume 1 by Bauhin's classification in the *Pinax*, but that on his own authority he made changes in it. This is in agreement with our findings about Syen's nomenclature, in which in addition to identifications with Bauhin's phrase names he also invented new names himself, with reference to Bauhin.

If we survey the scanty information that we have of Syen's editorial work for Hortus Malabaricus, we are bound to establish that he did much more than merely make volume 1 ready for the press. He also laid the basis for the arrangement of volume 2 and moreover made a first draft for the arrangement of volumes 7 to 10.

By making use of all the means available to him, Syen commented and classified the plants of volume 1 to the best of his ability. With the aid of the literature available in his days he placed Van Reede's Malabar plants within the framework of the tropical plants then known from Goa, Batavia, and Brazil. But his method of comparison went beyond the mere consultation of books. He involved the living plants of the Leiden botanical garden as well as the museum objects from its "Ambulacrum" in his research. He introduced an important novelty by comparing Malabar plants with herbaria and other material from the tropics, especially from Ceylon, which he owed to Paul Hermann. He used Bauhin's *Pinax* in an independent way by offering solutions for problems of nomenclature and classification. Van Reede's critical remark that Syen only made comparisons with the European flora is not well-founded. Apparently he failed to see that Syen laid the basis for the methods of commentaries and classification used in Hortus Malabaricus, and thus made a contribution to the foundation of tropical botany as a science. On the other hand it is to be regretted that owing to a misunderstanding between him and Van Reede the original plan of a main division into trees, shrubs, and herbs broke down in advance, in consequence of which the general sequence of the plants in Hortus Malabaricus was lost.

Notes

1. Boerhaave 1720:(28)-(29); Veendorp & Baas Becking 1938: 78-82.
2. Veendorp & Baas Becking 1938:76-78; Lindeboom 1974:3-66.
3. Molhuysen 1918 vol.3:234*,236*: Series Lectionem of February 1671 and of September 1671.
4. Veendorp & Baas Becking 1938:79-80.
5. Hortus Malabaricus vol.3 (1682):(xvi); Heniger 1980:54.
6. Veendorp & Baas Becking 1938:77-78; see also chapter 2, section Cape of Good Hope 1657.
7. Breyne 1678:177-178.
8. Breyne 1678:179-180.
9. See chapter 4, section Cape of Good Hope 1685.
10. Veendorp & Baas Becking 1938:84.
11. Gaymans' herbarium of 1661 of more than 700 exsiccates is in Glasnevin (Kasbeer 1965; Scannel 1979). His herbarium of 1669-1671, of about 1,600 exsiccates, was bestowed on Leiden University in 1984 by California University, and is now in the Rijksherbarium. The research on both herbaria is in progress and is carried out by Ms M.J.P. Scannel (Glasnevin) and Mr. M. Sosef (Leiden) in collaboration with the author.
12. For these catalogues, see Literature sub Leiden.
13. Heniger 1973:39,47.
14. See, for instance, Stevens in Manilal 1980 about the Antillian, Malabar, and Ceylonese elements in the protologue of *Calophyllum calaba* L.
15. VOC 1324:13v, letter of 5 March 1677 of the governor and the council of Ceylon to the Governor-General and the Council of India. This consignment arrived in the Netherlands on 6 September 1677 with the Ceylonese squadron, consisting of the ships "Het Wapen van Goes" and "Roemerswaal" (Bruijn et al. 1979:104-105). The matter was settled by Daniel Desmarets, at the time Walloon clergyman at the court of the Stadtholder William III (VOC 240, Minutes of the Chamber of Amsterdam, n.p., meeting of 13 July 1677).
16. VOC 240, Minutes of the Chamber of Amsterdam, n.p, meeting of 9 September 1675; VOC 320, n.p., letter of 28 September 1675 of Lords XVII to the Governor-General and the Council of India.
17. Breyne 1678:51-54, tab.; Hermann 1687:458; Linnaeus 1737:365; Linnaeus 1753:713.
18. *Acorus Asiaticus radice tenuiore* (Hermann 1687:9); Linnaeus 1737:137; Linnaeus 1753:324.
19. *Blitum monospermum Indicum, aculeatum, capsulâ rotundâ, sive Amaranthus major Zeilanicus spinosus, flore viridi* (Breyne 1680:18-19); Hermann 1687:31-32,33 tab.; Linnaeus 1737:444; Linnaeus 1753:991.
20. *Aeschynomene mitis secunda* (Breyne 1678:47); I could not trace this species.
21. Hermann 1687:129-130: *Cassia Cinamomea*.
22. Breyne 1678:17.
23. Hortus Malabaricus vol.3 (1682):(xv)-(xvi); Heniger 1980: 53-54.

JAN COMMELIN

About Jan Commelin (1629-1692), the second commentator of Hortus Malabaricus, much more is known than about his predecessor Syen. Hunger (1925), Jeurissen & Fournier (1970), and Wijnands (1983) have brought out and discussed in detail a good deal of material about the life and work of Commelin, but a comprehensive biography is still lacking, as does a coherent account about his contributions to horticulture, the Dutch local flora, and exotic botany, in connection with the rise of the Amsterdam botanical garden.

Jan Commelin was born in Leiden as a son of the prominent publisher Isaac Commelin (1598-1676), and he and his family moved to Amsterdam in 1641. In 1652 he married Digna van Wissel (1633-1671), a daughter of the Amsterdam druggist Johannes van Wissel and Petronella Hondius, in consequence of which he became related to the leading Hondius family of publishers, cartographers, and engravers. Despite his twofold relation with publishing families, he chose a different career. The Commelin publishing company was continued later by his younger brother Caspar Commelin (1636-1693), also known as a local historian of Amsterdam, the father of the botanist Caspar Commelin (1667-1731).

Jan Commelin grew to be a wholesale dealer in pharmaceutics. In 1666, upon his nomination as a governor or trustee of the "Spin- en Werkhuis" (house of correction), he entered the lower circles of the Amsterdam municipality. In 1672 he was admitted to the council of Amsterdam, the key to the supreme power in this metropolis. He owed this new position, which he held until his death, to the fall of the republican faction, which had opposed the elevation of William III to the stadtholdership. His membership of the council brought him into contact with Joan Huydecoper van Maarsseveen, the uncle of Joan Bax van Herentals, one of the Amsterdam burgomasters.

Commelin turned out to be an enthusiastic amateur of horticulture. In 1676 he published an essay on the cultivation of citrus fruit, *Nederlantze Hesperides*, in which he paid much attention to the construction of hothouses. Also in 1676 his second wife, Belia Vinck (1630-1697), whom he married in the previous year, bought the estate Zuyder-Hout near Haarlem, where he built up a collection of exotic plants. Several plants cultivated by Commelin were mentioned by Breyne in 1678. Because he continually travelled to and fro between this estate and Amsterdam, and regularly visited Huydecoper's estate Goudenstein, his attention was drawn to the local flora, which culminated in his *Catalogus Plantarum Indigenarum Hollandiae* (1683), the first flora of the province of Holland with many notes on ecology. His friendship with Huydecoper may have been an additional reason why in 1678, after the death of Syen, Van Reede recruited him for the editing of Hortus Malabaricus.

When on 12 November 1682 the council of Amsterdam founded the new Amsterdam Hortus Medicus, the implementation of the decision was entrusted to Huydecoper and Commelin as commissioners. Commelin had the daily direction, which was confirmed by his nomination as commissioner-practicus of the garden on an annuity of 500 guilders in 1690. In a short time the Amsterdam botanical garden grew to be a worthy rival of the Leiden garden; it was generously furnished with municipal funds and vigorously supported by Huydecoper's worldwide colonial relations in the cultivation of native and exotic plants. A first evidence of Commelin's exertions was his garden catalogue, *Catalogus Plantarum Horti Medici Amstelodamensis. Pars Prior* (1689), printed by his brother Caspar Commelin and enumerating about 2,200 kinds of plants.

From 1686 Commelin had the most interesting exotic flowering plants illustrated, thus laying the foundation for the monumental Moninckx Atlas of 420 watercolours, extensively discussed by Wijnands (1983). Commelin conceived the plan to describe these illustrations, but the pressure of his work, among other things the editing of Hortus Malabaricus, prevented him from publishing them. Several years after his death, Frans Kiggelaer (1648-1722), a pharmacist in the Hague, and Frederik Ruysch (1638-1731), professor of botany at the Amsterdam garden, annotated and edited his descriptions: *Horti Medici Amstelodamensis Rariorum Plantarum* (1697). Many of his engravings and descriptions have served Linnaeus in establishing species in *Species Plantarum* (1753).

Above all Commelin was an amateur of botany. He never received any professional training in this science. His development from a keen horticulturist into a respected botanist coincided with his commentaries in Hortus Malabaricus. Van Reede got acquainted with Commelin and learned to appreciate him at the beginning of this development. In 1682, in the preface to volume 3, he wrote:

'the cultivated Johannes Commelinus, a man second to none in botanical knowledge, inflamed by love of botany, illustrated the second volume with very elegant notes'[1].

Van Reede was able to follow Commelin's development from an amateur of botany into a respected botanist largely at a great distance, in Asia. It may even be stated that his commentaries on Malabar plants, which he composed in the years 1678-1692 for the volumes 2 to 12, reflect his growth to mastership.

Commelin's contributions to Hortus Malabaricus form a fascinating episode in the rapid evolution through which exotic botany passed in the last three decades of the 17th century. His fellow-botanists in the Netherlands and elsewhere in Europe followed his publications of the successive volumes with great interest. They were not only fascinated by Van Reede's presentation of many plants not described so far, but Commelin's commentaries, too, provided much food for discussions and reflections, which in turn influenced the latter. The interactions between Commelin and leading botanists such as Paul Hermann, Jacob Breyne, Leonard Plukenet, and John Ray have contributed a great deal to the historical importance of Hortus Malabaricus in botany.

A comparison with Arnold Syen is not altogether fair. In the first place Commelin was able to continue the work on the foundations already laid by Syen in the commentaries of volume 1. And secondly Commelin could use to the full the completely developed and annually increasing resources which the Amsterdam botanical gar-

Resources

The Amsterdam botanical garden

As a wealthy merchant and a member of the powerful council of Amsterdam, seconded by the influential Huydecoper and assisted by a staff of gardeners, Commelin could concentrate completely on the foundation of the Hortus Medicus and could devote himself carefree to the study of botany. He was not taxed by a professorate, as Syen was, for that was discharged by Frederik Ruysch, who lectured on indigenous plants.

On account of Ruysch' duties a part of the garden was set apart for a collection of native and European plants. In a monumental greenhouse more vulnerable plants were accommodated.

The foundation and expansion of the exotic collection in the Amsterdam garden can be inferred from Huydecoper's private letters of 1683-1687 to the various Dutch colonies.

First of all Huydecoper applied, in January 1683, to his cousin Simon van der Stel, commander of the Cape of Good Hope. As has been set forth before, since 1679 Van der Stel privately sent him regularly water-colours, dried plants, bulbs, and seeds of Cape plants[2]. But now Huydecoper wished the consignments to be addressed henceforth to the recently founded garden:

'The enclosed memoranda are sent to Your Honour on behalf of the City, with the request that they [the plants] should be forwarded to her in good condition, which will benefit the embellishment of the Hortus medicus under construction'[3].

In the following years Van der Stel faithfully performed the task entrusted to him and thus laid the foundation of Commelin's Cape collection. In exchange for Van der Stel's contributions to the garden, Huydecoper and Commelin exerted themselves to send the botanist Hendrik Bernard Oldenland to the Cape. To Oldenland we may attribute the many introductions of Cape plants after 1688 into the Amsterdam Hortus Medicus, referred to in Commelin's posthumous *Horti Medici Amstelodamensis*.

It is only in 1685 that there are references to hothouses in the Amsterdam garden, see the engraving made by Bastiaan Stoopendael, one of the engravers of Hortus Malabaricus, in that year[4]. This corresponds to Huydecoper's first summons for tropical plants in 1685, when he requested Isaac Lamotius, chief of Mauritius, for plants from that island[5]. Similarly he requested Andries Cleyer, of Batavia, for a consignment of 'fresh plants'[6]. The latter's generous response of 1688, consisting of Javanese and Japanese plants, has been referred to before. From Batavia, too, contributions to the tropical collection were made by Isaac de Saint-Martin in 1686 and Willem ten Rhijne in 1687, the latter with the precious Sumatran camphor-tree[7].

The presence of Van Reede as commissioner-general in the Western Quarters led to valuable acquisitions from those districts. It was undoubtedly at his recommendation that Elsevier and Wichelman sent bulbs and seeds from Ceylon, for which Huydecoper thanked them in 1687[8]. Well-recorded are several plants presented by Laurens Pijl, governor of Ceylon[9]. Van Reede himself contributed to the Amsterdam garden with plants from Ceylon (1686), Bengal (1686), and Coromandel (1688).

Finally, Huydecoper's influence extended into tropical America. The formal request directed to Surinam since 1684 for plants from this Dutch colony must be ascribed to him. Its governor, Cornelis van Aarssen van Sommelsdijk (1637-1688), pioneered tropical agriculture and horticulture in Surinam in close co-operation with the Amsterdam botanical garden. He founded a botanical garden in Paramaribo, of which little is known. He provided for the consignments of seeds and living plants to Commelin in 1685 and 1686[10]. Van Aarssen's botanical contacts with Jan van Erpecum, director of the Dutch colony of Curaçao, may have led to the consignment of living plants from that island to Caspar Fagel[4] and Simon van Beaumont in 1687, from which Commelin also received a selection[11].

It is amazing to see that in so short a time of several years the exotic collection increased enormously. It is also remarkable that many of the benefactors belonged to Van Reede's own circle of friends and acquaintances.

Nevertheless, Commelin could not draw upon his splendid tropical garden collection at will. In two respects he was curtailed in its use in his work for Hortus Malabaricus. Firstly, it is curious that no living plant from Malabar itself ever reached Commelin. This means that he had to operate carefully in comparative studies of plants from other regions. Secondly, the acquisition of tropical plants in the garden did not keep pace chronologically with the sequence of the volumes to be published. It was not until 1686, when the first tropical plants arrived from Java, Ceylon, and Bengal, that he was able to include them in the commentaries of volume 6 (1686) and the following volumes. And conversely, many tropical plants received from abroad could not be discussed, because the relevant volumes of Hortus Malabaricus had already been published. Several examples of unfortunate coincidences are the following.

In 1686 Commelin received a specimen of *Euphorbium antiquorum* L. from Ceylon, sent by Van Reede with the name Sidracalli. The Malabar plant of the same species, Schadida-calli HM 2:81:42, however, had been published in 1679[12]. The specimen of *Hibiscus cannabinus* L. grown in Amsterdam from seed named Nelta, sent by Van Reede in 1688 from Coromandel, came too late to be discussed at Narinam-poulli HM 6:75-76:44, published in 1686. Commelin restored these omissions in his *Horti Medici Amstelodamensis*[13]. Some of the undated consignments dispatched by Laurens Pijl from Ceylon do not seem to have arrived at a favourable time either. His specimens of *Euphorbia neriifolia* L. and *E. Tirucalli* L. were not discussed at Ela-calli HM 2:83-84:43 and Tirucalli HM 2:85-86:44[14].

On the other hand Commelin did not always involve plants actually present in his commentaries. Thus already in 1687 he had a living specimen of *Pancratium zeylanicum* L., originating from Pijl, from Ceylon, which in spite of the fact that it was illustrated in the Moninckx Atlas was not discussed in 1692 at Catulli pola HM 11:79:40. Commelin seems not to have been aware of the identity of this Ceylon specimen with the Malabar plant, for in *Horti Medici Amstelodamensis* too he did not refer to it[15].

A nice insight into the use of the Amsterdam Hortus Medicus in his work was given by Commelin in his garden catalogue of 1689. Out of the 2,200 kinds of plants then grown in the garden he identified some seventy-five with plants in Hortus Malabaricus. At first sight this may not indeed appear to be an impressive number, only three per cent, but his garden catalogue was not intended as a profound study on nomenclature and synonymy. Evident identifications are, for instance, his Genistas, which according to his commentaries in volume 9 of Hortus Malabaricus, which at that moment was in the press[16], were identical with several representatives of the Tandalecotti group, species of *Crotolaria*. Further Commelin did not include in the catalogue any plants which had indeed been cultivated in the garden before 1689, but which meanwhile had died off. It may therefore be assumed that the number of his garden plants cultivated since 1682 and identical with or related to Malabar plants was in reality much greater than the seventy-five mentioned above.

A more interesting point is that Commelin referred not only to the volumes 2-8 (1679-1688) already published by him, but also to those which were to be published in the near future. Apparently he was several volumes ahead of the commentaries and when walking past his living tropical treasures, tried to make connections with the descriptions and illustrations of Van Reede's plants which he was dealing with at that moment.

In the volumes 2-8, published before 1689, there are already references to plants cultivated in the Amsterdam garden which according to the garden catalogue were still alive in 1689. In 1684 Commelin received a *Cucumis* from Ceylon, which bore fruit after three years. He identified it with Mullen-belleri HM 8:11:6[17]. In 1685 he received seed of a Ceylonese plant which appeared to be Basella HM 7:45:24, *Basella rubra* L.[18]. Also from Ceylon, he cultivated a Nianghala according to his commentary on Mendoni HM 7:107-108:57, *Gloriosa superba* L.[19].

He identified a plant from Surat with Mandsjadi HM 6:25-26:14, *Adenanthera pavonina* L.[20].

A *Cucumis* sent by Isaac Lamotius from Mauritius produced large fruits and seeds in 1686 and was identified by Commelin with Pandi-pavel HM 8:17-18:9, *Momordica Charantia* L. Another species, probably from Ceylon, bearing fruit in 1688, figured in the Moninckx Atlas in that year[21].

From unknown Asiatic regions Commelin grew plants from seed in 1684 and 1685, which he identified with Beloere HM 6:77:45, *Sida asiatica* L.[22]. And in 1686, in the discussion of Thora-paërou HM 6:23-24:13, *Cytisus cajan* L., he referred to an identical plant in his garden[23]. Finally he had a single-flowered form of *Clitorius Ternatea* L., painted by Alida Withoos in 1686 for the Moninckx Atlas, which he presumably identified with Schanga-cuspi HM 8:69-70:38. The double-flowered white form of this species was also pictured in the Atlas[24].

A plant called Anakokke, grown from seed sent by Van Aarssen van Sommelsdijk from Surinam to Amsterdam in 1685, was compared with Konni HM 8:71-72:39, *Glycine Abrus* L.[25]. In the list of seeds accompanying Van Aarssen's letter of 8 March 1685 to the Society of Surinam two kinds of Anakokke were mentioned: 'red beans with black spots, large sort', and 'a small sort growing on a tendril'[26].

From Brazil Commelin had a *Rosa Brasiliensis*, six feet high, in 1684, identified by him with Hinaparetti HM 6:69-72:38-42, *Hibiscus mutabilis* L.[27].

From the West Indies, as he stated, Commelin grew several trees from seed in 1684, which he identified with Tsetti-mandarum HM 6:1-2:1, *Poinciana pulcherrima* L.[28].

A plant of unknown origin is Alpini's *Sambac Lesmin Arabicum*, which was cultivated with difficulty in the Amsterdam garden and identified by Commelin with Nalla-mulla HM 6:87:50, *Nyctanthes multiflora* Burm. f.[29].

In 1689 Commelin cultivated both single and double forms of Hermann's *Nerium Indicum* (etc.). The double form was sent by Laurens Pijl from Ceylon to Amsterdam with the native name Fula Mestica and was afterwards pictured in the Moninckx Atlas. In the garden catalogue of 1689, while volume 9 of Hortus Malabaricus was in the press, Commelin announced the identification of both forms with Belutti-areli HM 9:3:2 and Tsjovanna-aleri HM 9:1-2:1 respectively, *Nerium Oleander* L. or *N. indicum* Mill.[30].

From volume 10 of Hortus Malabaricus, to be published in 1690, Commelin was already cultivating Breyne's *Amarantho affinis Indiae Orientalis* (etc.)[31], which he had pictured in the Moninckx Atlas. Anticipating the publication of this volume, he identified the garden plant with Wadapu HM 10:73-74:37, *Gomphrena globosa* L.[32].

With regard to volume 11, to be published in 1692, Commelin cultivated a number of interesting plants already in 1689. In 1686, from Surinam, and in 1687, from Curaçao, he received some living pine-apple plants. The Curaçao specimen bore fruit in the Amsterdam garden in 1688 and 1689. In the garden catalogue he compared it with Kapa-tsjakka HM 11:1-6:1-2, *Bromelia Ananas* L. or *Ananas comosus* (L.) Merr., and referred to the introduction of the pine-apple from America into India in earlier times[33]. In Bauhin's *Aloe vulgaris*, also pictured in the Moninckx Atlas, he recognized Kadanaku HM 11:7:3, *Aloe perfoliata* L. var. *vera* or *A. vera* (L.) Burm.f.[34]. Of the Kua group of Zingiberaceae, also to be dealt with in volume 11, Commelin grew several representatives in the garden. His specimen of Bauhin's *Zedoaria longa* was identified by him with Kua HM 11:13-14:7. He identified Da Orta's *Zerumbeth*, but renamed by him as *Zinziber latifolium sylvestre*, with Tsjana-kua HM 11:15-16:8, *Costus arabicus* L. Another plant again, Bauhin's *Zedoaria rotunda*, was identified by him with Malan-kua HM 11:17-18:9, *Kaempferia rotunda* L. Commelin connected the *Curcuma Officinarum*, also introduced by Hermann into the Leiden botanical garden from Ceylon, as *Curcuma radice longa*, with Manjella kua HM 11:21:11, *Curcuma longa* L. Finally, for his specimen of Bauhin's *Zinziber* he referred to Inschi HM 11:21-23:12, *Amomum Zingiber* L. Of all these species of the Kua group Commelin had only *Kaempferia rotunda* L. included in the Moninckx Atlas[35].

In the Amsterdam garden of 1689 Commelin had a fine collection of *Araceae*, including several species from Ceylon. His specimen of *Arum humile Arisarum dictum latifolium, Ceylonicum* (etc.), which flowered in August 1686, was identified by him with Nelenschena minor HM 11:33-34:17. His two specimens of *Arum polyphyllum, Dracunculus, & Serpentaria, dictum Ceylonicum* (etc.), both of which he had pictured without flowers in the Moninckx Atlas and which according to Wijnands were the first records of *Amorphophallus paeoniifolius* (Dennst.) Nicolson in cultivation, according to Commelin were identical with Schena HM 11:35-36:18 and Mulenschena HM 11:37:19. Finally, his *Arum Maximum Ceylonicum* was considered by him to be identical with Wel-ila HM 11:43-44:22[36].

From his commentaries in Hortus Malabaricus it appears that several plants grown in the Amsterdam garden were no longer alive in 1689, or at least did not occur in the garden catalogue.

From Ceylon, in 1684, Commelin received seeds of Pungam, which he identified with Pongam seu Minari HM 6:5-6:3. Also from Ceylon, in 1686, he acquired several species of *Dioscorea*, interpreted by him as *Batatta sylvestre* and identified with the Kelengu group HM 7:63-71(2):34-38, and with Mu-kelengu HM 8:97:51.

In 1688 he grew a *Barleria Prionitis* L., Coletta-veetla HM 9:77-78:41, from seed of unknown origin, but presumably from Ceylon, since he identified it with a specimen in Hermann's Ceylon Herbarium.

Furthermore Commelin cultivated several species not mentioned in the catalogue of 1689, which he brought into the discussions of plants in the volumes 10 (1690) and 11 (1691). From Ceylon he had a Jusala, which did not bloom in Amsterdam, but which he could propagate by its tubers, identified with Kurka HM 11:49:25.

In 1698, a misprint for 1689 or 1690, he received seeds of several species from Curaçao. On one of the species, *Capraria biflora* L., grown from them, Commelin perceived some similarity to Tsjeru-parua HM 10:105:53, *Sida acuta* Burm.f.[37], and another species was identified by him with Katu-uren HM 10:107:54.

Other botanical gardens

Commelin did not use exclusively the living tropical collection of the Amsterdam botanical garden as frame of reference for Malabar plants. Many plants which were introduced by the East and West India Companies into the Netherlands rapidly spread through distribution and exchange over the glasshouses of other gardens in Holland, and offered him plentiful additional and even unique material for comparison besides his Amsterdam collection.

The chronology of the first introductions of exotic plants in the second half of the 17th century is largely still an undeveloped field, so that in the survey now following it cannot always be ascertained where and when Commelin may have observed his first specimen of a comparable plant.

The most important additional collection was undoubtedly that of the Leiden botanical garden, directed by Paul Hermann. From some references to Hermann's own Ceylon herbarium (see below) it may be inferred that Commelin regularly visited him in Leiden. His great interest in the Leiden garden also appears from the fact that on behalf of Hermann he presented the latter's garden catalogue of 1687, *Horti Academici Lugduno-Batavi Catalogus*, personally to his fellow-commissioner Huydecoper[38]. This garden catalogue contains a survey of the plants which Hermann cultivated in the years 1681-1686. Unfortunately Hermann was very sparing of information about the years of introduction into his garden and origins of exotic plants, so that with regard to Commelin's observations in the Leiden garden we have to confine ourselves to his commentaries in Hortus Malabaricus from volume 7 (1688) onwards.

In Hermann's *Lilium superbum Ceylanicum* he recognized Van Reede's Mendoni HM 7:107-108:57, *Gloriosa superba* L., specimens of which were also cultivated in the Amsterdam garden and in the private gardens of Beverningk and Fagel[39].

Of *Momordica Charantia* L., of which Commelin cultivated in Amsterdam specimens with large fruits (Pandipavel HM 8:17-18:9), he found in Leiden a specimen with small fruits (Pavel HM 8:19:10)[40].

Hermann's *Bryonia Ceilanica foliis profunde lacinatis,* not known from elsewhere in a Dutch garden, was identified by Commelin with Nehoemeka HM 8:37-38:19, *Bryonia laciniosa* L.[41].

The single and double forms of *Nerium Oleander* L. or *N.indicum* Mill. were known to Commelin not only from the Amsterdam garden (see above), but also from Hermann's garden[42].

In the American plant *Ruellia antipoda* L., grown in Leiden, Commelin thought he recognized Pee-tsjangapulpam HM 9:115:59[43].

Hermann's *Asparagus aculeatus maximus sermentosus Ceylanicus* was identified by him with Schada-velikelangu HM 10:19-20:10, *Asparagus sarmentosus* L.[44].

In his discussion of Inota-inodien HM 10:139:70, *Physalis pubescens* L., and Pee-inota-inodien HM 10:101:71, *Ph. minima* L., Commelin referred to *Solanum vesicarium Indicum minimum* cultivated in Leiden[45].

For the Kua group, several representatives of which were present in Amsterdam as well as in Leiden, he referred to Hermann's *Zerumbeth* in his commentary on Kua HM 11:13-14:7, and he identified the latter's *Curcuma radice longa* with Manjella Kua HM 11:21:11[46].

The Tolabo or *Crinum zeylanicum* (L.) L., which Commelin received as a bulb from Ceylon in 1685 and which was still present in Amsterdam in 1689, was also cultivated in Leiden. He cited it as a conspecies of Belutta pola taly HM 11:75-76:38, *Crinum asiaticum* L.[47].

Commelin disregarded his own *Pancratium zeylanicum* L., of which he had an elegant water-colour in the Moninckx Atlas, pictured after a plant sent by Laurens Pijl from Ceylon to Amsterdam, in his commentary on Catulli pola HM 79:40, and he identified Van Reede's plant with Hermann's *Narcissus Ceylanicus, flore albo hexagono odorato*[48]. He identified Hermann's *Acorus verus Asiaticus radice tenuiore*, already described by Da Orta as *Calamus Aromaticus*, with Va embu HM 11:99:48, *Acorus Calamus* L. var. *verus*[49].

The Amsterdam specimen of *Convolvulus Pes caprae* L. was cultivated from seed only since 1690, but Commelin knew this species from Hermann's *Convolvulus Maritimus Ceylanicus* (etc.), which he identified with Schovanna adamboe HM 11:117:57[50].

Finally he identified *Convolvulus Ceylanicus Villosus* (etc.), grown in Leiden, with Pulli-schovadi HM 11:121:59, *Ipomoea Pes tigridis* L.[51].

Less numerous, but nevertheless interesting, are the plants cultivated in private gardens which Commelin involved in his comparisons with Van Reede's plants. Some species had already become so common in the Netherlands that he did not even refer to a specific cultivator. The Mediterranean *Asclepias gigantea* L., by which he understood Ericu HM 2:53-55:31, was widely grown in Dutch gardens. It flowered here in August, but it did not form fruits or seeds. *Capsicum frutescens* L., which he identified with Capo-molago HM 2:109-110:56, was a common Dutch garden plant, which produced ripe seed before dying off. The *Canna indica* L. or Katu bala HM 11:85-87:43, was known to Commelin from various gardens.

When discussing Poerinsii HM 4:43-44:19, *Sapindus trifoliata* L., Commelin gave particulars about seeds from Barbados, from which he grew a small tree in 1660. In 1674 he received pods with seeds from which originated two small plants which died through the winter cold and which he compared with Caretti HM 2:35-36:22, *Guilandina Bonducella* L.

In the glasshouse of Honselaarsdijk, the palace of William III of Orange, Commelin observed two small trees which he identified as Van Reede's Nandi-ervatam HM 2:105-106:54, and as Todda-panna HM 3:9-14:13-21, *Cycas circinalis* L.

In the private garden Leeuwenhorst of the Grand Pensionary Caspar Fagel, at Noordwijkerhout, a specimen of *Gloriosa superba* L. bloomed in 1686; this was studied by Commelin in connection with Mendoni HM 7:107-108:57[52].

It may be assumed that Commelin also visited the private gardens of Hieronymus Beverningk, the curator of Leiden University, on his estate Oud-Teylingen or Lokhorst near Warmond, in connection with Hortus Malabaricus. Beverningk also grew specimens of *Gloriosa superba* L. and *Momordica Charantia* L., with large fruits, as mentioned before.

Herbarium

Just as in the case of his predecessor Arnold Syen, it is not known whether Jan Commelin possessed a garden herbarium. In the Memorials of the Amsterdam garden there are no references to the making of a collection of dried plants grown in the garden. On the contrary, the Memorials give much information on the production of water-colours and drawings intended for the Moninckx Atlas, so that it is likely that the Atlas was a substitute for a garden herbarium to a certain extent.

However, Huydecoper as well as Commelin owned herbaria privately. Huydecoper's herbarium of Cape plants, presented by his nephew Joan Bax, will have been of little or no use for Commelin's commentaries in Hortus Malabaricus. But just like Syen, Commelin himself had a small Ceylon herbarium of some importance.

In the discussion of Tsjerou-panna HM 4:81:39, *Calophyllum Calaba* L., he referred to a twig in his "Hortus Hyemali seu Herbarium vivum", sent to him by Paul Hermann from Ceylon. And a few pages further on, when annotating Perin-toddali HM 4:85-86:41, *Rhamnus Jujuba* L., 'in the said Herbarium vivum Zeilanicum' he observed a specimen inscribed as *Ziziphus Indica argentea tota cariophylli aromatici flore* with the native name of Waelambillu. Again, in the discussion of Kattu-tagera HM 9:55:30, *Indigofera hirsuta* L., Commelin explicitly cited a plant from his own Ceylon herbarium with the inscription *Astragalus Indicus spicatus, siliquosis copiosis, deorsum spectantibus, non falcatis; seu Polylobos*, which he adopted for this Malabar plant. Apparently Paul Hermann had presented not only Syen, but also Commelin with a collection of Ceylon exsiccates while he still stayed in Ceylon in 1672-1680.

In view of the inscribed phrase names Commelin's Ceylon herbarium can be identified with the herbarium volume in the Institut de France in Paris, described by Lourteig (1966). In later times this volume was in the possession of Johannes Burman, who used it for his *Thesaurus Zeylanicus* (1737). In the preface Burman recalled that it belonged initially to Caspar Commelin, the heir of Jan's scientific legacy. Although Burman knew that the volume had actually been collected by Hermann and, moreover, referred to it as "Herb.Herm.", Herbarium of Hermann, he did not seem to be aware of the fact that Jan Commelin originally used it in the commentaries of Hortus Malabaricus.

There are also other Ceylon exsiccates which Commelin involved in discussions of Malabar plants, but from his statements it is not certain whether he found them in his own herbarium or in Hermann's large Ceylon herbarium in Leiden. A twig of *Cystus Indicus quinque nervis, folio hirsuto & scabro* (etc.), with the Ceylon native name of Mahabotya, was compared with Katou-kadali HM 4:91:43, *Melastoma aspera* L. A 'twig obtained by Hermann' was discussed at Naga-mu-valli HM 8:57:30-31, *Bauhinia scandens* L. In Hermann's herbarium Commelin observed a flowering twig inscribed *Malus limonia pumila sylvestris Zeylanica*, which name he adopted for Mal-naregam HM 4:27-28:12. At Coletta-veetla HM 9:77-78:41, *Barleria Prionitis* L., Commelin adopted both the phrase name of *Eryngium Ceilanicum febrifugum, floribus luteis* and the Ceylon native name of Kathukarohiti from a sheet of a Ceylon exsiccate; he did the same with *Planta bifolia humirepa, Arifolio, flosculis aureis, villosis, articulatis* for Nelam-mari HM 9:161:82, *Hedysarum diphyllum* L. Presumably Commelin also borrowed the phrase *Acacia Tinctoria Hermans* from an inscription on a Hermann sheet for Tsiapangam HM 6:3-4:2, *Caesalpinia Sappan* L.[53].

When surveying Commelin's use of herbarium material from Ceylon, either in his own possession or observed in Hermann's collection in Leiden, one may conclude that he continued Syen's initial method of comparing Malabar plants with Ceylon plants.

The library
In the commentaries of the volumes 2-12 of Hortus Malabaricus Commelin cited some fifty books on botany, natural history, medicine, and travels. In volume 2, published in 1679, starting his commentaries, Commelin consulted some twenty works. From volume 3, published in 1683, after the founding of Hortus Medicus he added another thirty to them.

Besides the library of the Amsterdam Athenaeum Illustre, the institutional library of the Hortus Medicus was also accessible to Commelin. Up to 1706, Wijnands lists one hundred and twenty books on botany belonging at one time to this library, bound in white calf and imprinted *Hortus Medicus Amstelodamensis* in gold, which are nowadays accommodated in the library of the University of Amsterdam[54]. Some ninety books were published before Commelin's death in 1692, so that we may suppose that they were bought during his directorship of the garden. But Wijnands has not clarified the question as to whether the old Hortus Medicus, which was closed before 1682, also had an institutional library which might have been transferred to the new garden in 1682. As long as there is no information about an older botanical library, it is likely that Commelin annotated volume 2 of Hortus Malabaricus with the aid of his private library. Among the books which Wijnands could not trace in the institutional library, Commelin cited in this volume Ferrari's *Flora* (1664) and Recchus' edition of Hernandez' *Rerum Medicarum Novae Hispaniae Thesaurus* (1651). Apparently these two books formed part of Commelin's private library. Other works cited by him in 1679 are, in order of decreasing frequency: Bauhin's *Pinax*, Clusius' *Rariorum Historia* and *Exoticorum Libri Decem*, Piso's and Marcgrav's works on Brazil, Breyne's *Centuria Prima*, Rochefort's natural history of the Antilles, Bontius' *Historia Naturalis*, Veslingi's edition of Alpini's *De Plantis Aegypti*, Dodonaeus' herbal, Cornut's history of Canadian plants, Vallot's catalogue of the Jardin des Plantes in Paris, and accounts of the voyages of Linschoten, Van Neck, and Baldaeus to the East Indies.

Except for Rochefort and the travel books, all of them are also to be found in Wijnands' list of the Hortus library. We cannot decide whether these books were a part of Commelin's private library or of the libraries of the Athenaeum and the older Hortus Medicus.

If we compare the literature cited in volume 2 by Commelin with the literature to which Syen referred in volume 1, we see that Commelin used in broad outline the same frame of reference as his predecessor. Commelin, too, displayed much interest in Bauhin's *Pinax* with respect to the nomenclature and systematics. He largely consulted the same books about Asiatic and American plants and thus continued the comparison of Malabar plants with the neotropical flora started by Syen. But in Commelin's volume 2 a new element also entered. Breyne's *Centuria Prima*, dealing with up-to-date information about newly imported exotic plants, heralded a vivid discussion about tropical plants cultivated in Europe and especially in the Netherlands, to which Commelin himself as commissioner-practicus of the Hortus Medicus was to make important contributions, as witness the Moninckx Atlas and his *Horti Medici Amstelodamensis Rariorum Plantarum* (1697). In his commentaries in the following volumes of Hortus Malabaricus Commelin was to make use increasingly of the results of this discussion.

From volume 3 (1683) up to and including the last volume of Hortus Malabaricus Commelin largely kept to his reference literature used in volume 2, but he gradually implicated another thirty works, published both previously and recently, in his commentaries. Most of them figure in Wijnands' list and may have been acquired by Commelin on behalf of the Hortus library. His most important reference books remained Bauhin's *Pinax*, to which he now added Jean Bauhin's *Historia Plantarum Universalis*, and further the above works by Acosta, Da Orta, Clusius, Piso and Marcgrav, Bontius, and Recchus. Among books printed previously, though scarcely cited, he introduced Cordus and Matthioli (1583) on Dioscorides, Lobelius' herbal (1581), Camerarius' *Hortus Medicus* (1588), Aldini's description of Farnese's garden in Rome (1625), and Parkinson's *Paradisi* about garden plants (1629).

Within the discussion about tropical plants cultivated in the Netherlands Commelin continued to consult Breyne's *Prodromus Fasciculi Rariorum Plantarum* (1680) and *Prodromus ... Secundus* (1689). From Hermann's catalogue of the Leiden botanical garden (1687) he chiefly cited descriptions of Ceylon plants, while he also referred to some phrase names in Hermann's *Prodromus Paradisi Batavi*, edited by William Sherard (1689), listing exotic plants grown in public and private Dutch gardens.

It was only in volume 3 that Commelin first made use of *Istoria Botanica* (1675), in which Zanoni dealt with a number of Matthew's plants. Neither Syen nor Commelin had been able previously to lay hands on this book, for in the first two volumes of Hortus Malabaricus, too, there appear Malabar plants which had already been discussed in Zanoni. Apparently it was not very easy to come by recently published Italian literature, for Vincenzo Maria's *Il Viaggio All'Indie Orientiali* (1672), which contained many references to Indian and Malabar plants and to which Matthew contributed, had never been cited by the commentators of Hortus Malabaricus. This omission was partly made good by Nieuhof's itinerary (1682) describing Malabar plants, which Commelin consulted from volume 4 onwards. Finally, from volume 6, published in 1686, Commelin introduced many references to John Ray's *Methodus Plantarum Nova* (1682) and *Historia Plantarum*, volumes 1 (1686) and 2 (1688). He cited not only the latter's plant descriptions and nomenclature, but also attempted to include the involved Malabar plants in Ray's plant system.

Commelin's commentaries in the volumes 2, 3, and 6 are the best documented ones. Volume 2 counts 22 books and 56 citations on 55 plants, volume 3 counts 15 books and 55 citations on 34 plants, and volume 6 counts 20 books and 47 citations on 54 plants. Only in volume 3 did Commelin attain about the same level as that of Syen's volume 1. Badly documented were volume 5, with 9 books and 12 citations on 63 plants, and volume 10, with 10 books and 10 citations on 94 plants. The worst of all was volume 12, on epiphytes, ferns, and mosses, with 5 books and 10 citations on 77 plants. As was the case for Syen, for Commelin, too, the European literature did furnish points of reference for the interpretation of Malabar trees and shrubs, but on the subject of herbs and cryptogams Commelin had hardly any printed sources at his disposal.

In the above survey of the literature cited by Commelin in his commentaries we can distinguish a development of his working method and thought. At first he adopted Syen's frame of reference, with the emphasis on the nomenclature and systematics of Bauhin and on the comparison of Asiatic with American plants. After the foundation of the Hortus Medicus he built up an institutional library which enabled him not only to enlarge the original frame of reference with books from the earlier Dioscorides tradition, but moreover to renew that frame with recently published works. Although since then he abandoned by no means his interest in Bauhin and the comparison with American plants, henceforth he also took part in the evolving scientific discussion about tropical horticulture and finally was converted to Ray's systematic views. On the other hand, the practical result of his literature references varied a good deal, because many Malabar plants had no affinity with plants already described and accordingly could not be documented at all.

Commelin and Ray

The publications of the English botanist John Ray (1627-1705) in the eighties of the 17th century largely influenced Commelin's systematic views. This was reflected in Hortus Malabaricus, both on the grouping and sequence of Van Reede's plants in the subsequent volumes and on Commelin's nomenclature.

In this context Commelin was highly dependent on the publication scheme of Ray, which did not run parallel to the issues of the volumes of Hortus Malabaricus, in consequence of which his interpretations of the latter's new plant system yielded uncertain results.

Conversely, the volumes of Hortus Malabaricus, in so far as they had been published, served as one of the main sources for Ray to illustrate his own plant system, in which he corrected Commelin's mistakes in taxonomy and nomenclature as much as possible. For a good understanding of the complicated interaction between the two botanists we will discuss here only Ray's influence on Commelin, while the insertion of Van Reede's plants in Ray's botanical writings will be discussed in chapter 12.

In *Methodus Plantarum Nova* (1682) Ray presented a new plant system very useful for that time, based on the general habitus of the plants and on morphological characteristics of fruits, seeds, and petals. He founded the main division into monocotyledons and dicotyledons, and succeeded in distinguishing several natural groups such as *Compositae*, *Umbelliferae*, *Campanulaceae*, *Scrophulariaceae*, *Cruciferae*, *Papilionaceae*, etc. However, for practical reasons he maintained the traditional major division into trees, shrubs, and herbs, thus breaking up his natural groups. In neatly arranged determination tables Ray tried to subdivide the trees, shrubs, and herbs into small units and genera. Ray was not always very fortunate in the choice of his criteria, so that strange aberrations tended to appear in particular in the case of the trees and shrubs and the *Papilionaceae*. Nevertheless, Ray's *Methodus* meant a considerable advance as compared with Bauhin's plant system of forms gradually ascending from simple to complex forms, based upon superficial characteristics.

After the publication of the *Methodus* Ray started his *Historia Plantarum* (1686-1704), a description of all the known plants, including Van Reede's plants, illustrating his method, in which connection he did not shrink from constantly making improvements in and additions to his system. Ultimately with *Methodus Plantarum Emendata*, a complete revision of his method, edited by the Leiden professor of botany Pieter Hotton in 1703, he crowned his systematic achievements surveyed by Linnaeus in *Classes Plantarum* (1738)[55].

In arranging and commenting on Hortus Malabaricus Commelin could make use only of Ray's *Methodus* of

1682, and after that of volume 1 (1686) and volume 2 (1688) of *Historia Plantarum*.

The publication of the *Methodus* came for Commelin exactly at the right moment for him to adjust the volumes 3 and 4 of Hortus Malabaricus, published in 1682 and 1683, to the sequence of Ray's system. In these volumes he referred neither to the *Methodus* nor to Ray's terminology of the divisions, but on the basis of the latter's review of Van Reede's trees and shrubs in the second volume of *Historia Plantarum* a reconstruction of Commelin's sequence can easily be made. In volume 3 Commelin included the *Arbores Pomiferae* dealt with by Ray in the *Methodus* on pp.30-31, Tab.I. In volume 4 Commelin continued Ray's sequence in the *Methodus* with a number of *Arbores pruniferae*, ending with a series of *Arbores bacciferae* of the *Methodus*, pp.34-35, Tab.III.

In discussing Commelin's arrangements in more detail we must first realize that Ray's *Methodus* was mainly based on European genera. It is true that this may serve to distinguish larger divisions in the Malabar plants, but for a more refined subdivision of unknown tropical plants the *Methodus* was insufficient. In the second place Commelin was tied down to a certain degree to tentative arrangements of the materials made by Van Reede and his collaborators in Asia. Especially with regard to Malabar plants which for one reason or another belonged traditionally together Commelin did not break up their coherence, although on the authority of Ray's *Methodus* he might have split up such a group and distributed the subgroups over different volumes of Hortus Malabaricus.

Among Van Reede's drawings only one section illustrates Commelin's problems in rearranging the initial sequence with the aid of Ray's system. This section of drawings bears numbers of another "f" series in pale washed ink, which like some drawings of volume 2 may be attributed to Van Reede or one of his clerks. This series runs, with some interruptions caused by the cutting-off of the paper, from f 1 to f 60.

When we consult the *Methodus* and the second volume of *Historia Plantarum*, we find that this series consists of a chaotic mixture of pomiferous, pruniferous, bacciferous, and nuciferous woods. Logically speaking, Commelin should have devoted volume 3 to Ray's first division of his system, namely *Arbores Pomiferae*, agreeing with Tab.I of the *Methodus*, pp.30-31. In actual fact Commelin selected from Van Reede's initial series some fourteen drawings of pomiferous trees, as far as he could interpret them, and included them in volume 3. Among Van Reede's material he also had ten drawings of the pomiferous *Ficus* (the Alou group, the Teregam group, Tsjela, and Tsjakela), which he added to it. But the number of selected pomiferous trees was not by far enough to fill a complete volume. This may explain why Commelin disturbed his neat arrangement of volume 3 by including obviously diverging groups. According to the *Methodus*, Van Reede's Panja HM 3:59-60:49-51 *Arbor lanigera sive Gossampini Plinii* belonged to Ray's *Arbores lanigerae*, p.39, Tab.VI, which Commelin himself identified as Bontius' *Arbor lanigera*. This tree had no connection at all with the pomiferous trees, but probably Commelin placed it here as a stuffing. A more remarkable fact is the presence of twenty-five engravings at the beginning of volume 3, namely Codda-panna, Todda-panna, and Katou-indel, which must be classified among Ray's pruniferous trees. This disturbance may be ascribed to Commelin's or Van Reede's wish to place these magnificent descriptions and engravings of palms not too far away from the other palms in volume 1.

Commelin placed the remaining drawings of the "f" series in volume 4, beginning with the pruniferous trees and ending with the bacciferous trees in agreement with Ray's second division, *Arbores pruniferae*, pp.32-33, Tab.II, and his third division, *Arbores bacciferae*, pp.34-35, Tab.III respectively of the *Methodus*. In this case problems arose with the Theka group HM 4:57-64:27-30 and the Kalesjam group HM 4:67-72:32-34.

Within the Theka group Van Reede initially numbered the drawings of Theka as f 43, Bentheka as f 44, Tsjerou-theka as f 45, and Katou-theka as f 46. But in Ray's system Theka belonged to the pruniferous trees, Bentheka to the bacciferous trees, Tsjerou-theka to the bacciferous shrubs, and Katou-theka to the pruniferous trees. Instead of splitting up this group, as Ray was to do later in volume 2 of *Historia Plantarum*, Commelin maintained this miscellaneous group. He selected Theka as the type of the group, changed Van Reede's sequence of its representatives by placing the other pruniferous Katou-theka immediately after Theka, followed by both bacciferous species by way of an appendix.

The same occurred with the Kalesjam group, of which Kalesjam (f 47) must be classified among the bacciferous trees and Katou-kalesjam (f 49) among the pomiferous trees, while Ben Kalesjam (f 48) could hardly be interpreted. Logically speaking, Katou-kalesjam ought to be placed in volume 3 of Hortus Malabaricus, as a result of which the group would break up. However, Commelin apparently opted for Kalesjam as the type, and placed the whole group, in spite of its miscellaneous nature, among the bacciferous trees in volume 4.

Again, the pruniferous and bacciferous trees selected from the "f" series did not fill a complete volume. By mixing the two divisions with a number of other woods, Commelin made some curious errors, which finally disturbed the arrangement of volume 4. Thus, among the pruniferous trees he included the Naregam group HM 4:27-32:12-14, which on account of his own nomenclature (kinds of *Malus limonia* or *Limonia*) he ought to have classified among the pomiferous trees in volume 3. The Modagam group, consisting of the pomiferous Modagam HM 4:119:58 and the pruniferous Bella modagam HM 4:121-122:59, were placed among the bacciferous trees at the end of volume 4.

Summarizing, one may say that in the volumes 3 and 4 Commelin tried to use as best as he could Ray's plant system as a guide for his arrangements. He succeeded indeed, in spite of misinterpretations, in broadly following the *Methodus*, but for the finer subdivision he had to take into account the traditional grouping of Malabar plants and the fixed number of pages of the separate volumes of Hortus Malabaricus.

Up to now Commelin had not referred explicitly to Ray's *Methodus* and terminology. But from volume 5 onwards he acknowledged Ray's system more and more openly by quoting his terminology on the title-pages and in the commentaries.

The title-page of volume 5 (1685) reads *De Arboribus et Fruticibus Bacciferis*, corresponding to the *Arbores bacciferae* and the *Frutices bacciferi* in the *Methodus*, pp.34-35, Tab.III, and pp.43-49, Tabs I-III, respectively. Indeed, about three quarters of the trees and shrubs treated in this volume belong to the said divisions. Again Commelin had some difficulties by placing several pruniferous, pomiferous, nuciferous, and siliquosous woods among the bacciferous ones.

The title-page of the next volume 6 (1686) reads *De Arboribus et Fruticibus Siliquosis*, corresponding to the *Arbores siliquosae* in the *Methodus*, pp.40-41, Tab.VII. It appears from the contents of volume 6 that by the term *Frutices Siliquosi* Commelin meant the *Frutices floribus Papilionaceis* in the *Methodus*, pp.52-53, Tab.V. Up to Kada-kandel HM 6:67:37 Commelin correctly placed the plants among the siliquosous trees and shrubs. The title-page, however, did not cover all the plants in this volume, for Commelin continued with several shrubby *Malvaceae* (*Hibiscus* spp.) belonging to Ray's *Frutices qui cum herbis genere proximo conveniunt* in the *Methodus*, pp.54-55, Tab.VI; after this he broke up the order by ending with a large group of bacciferous trees and shrubs which one would rather have expected to find in the volumes 4 and 5. In volume 6 Commelin introduced Ray's terminology into his commentaries for the first time. In the discussion of Niir-pongelion HM 6:53-54:29, which he had difficulty in identifying, he judged that its flowers had only one petal, for which he used Ray's term of *Monopetali*, thus connecting this species with *Herbae flore monopetalo* in the *Methodus*.

In the same year 1686 volume 1 of Hortus Malabaricus was reissued with a new title-page reading *De Varii Generis Arboribus et Fruticibus Siliquosis*, by which Commelin restored this volume to its correct place in Ray's system and moreover connected it with volume 6 containing corresponding plants.

The title-page of volume 7 (1688) reads *De Varii Generis Fruticibus Scandentibus*, corresponding to the *Frutices bacciferi Scandentes* in the *Methodus*, p.46, Tab.II. In this volume Commelin bravely attempted to arrange all kinds of tropical climbers, creepers, and parasites with the aid of Ray's small, defective determination table based upon European and American climbers such as *Hedera, Vitis, Parthenocissus,* and *Lonicera*, so that it is not surprising that even herbaceous, liliaceous climbers, such as the species of *Dioscorea* of HM 7:63-71(2):33-38, and *Gloriosa superba* L., Mendoni HM 7:107-108:57, found a place among the majority of dicotyledonous woody climbers.

With the publication of volume 7 Commelin had meanwhile come to the end of his treatment of all the trees and shrubs in Hortus Malabaricus. So far he had been able to rely only on his own interpretation of the determination tables in the *Methodus*. He broadly followed Ray's sequence of the larger divisions faithfully. His errors, confusions, and make-shift contrivances in the arrangements of the separate volumes are due partly to the incompleteness of Ray's tentative determination tables and partly to the impossibility of fitting in woody plants, which then were still largely unknown, into a fixed plant system.

In the forthcoming volumes of Hortus Malabaricus, on herbs and cryptogams, Commelin's task appeared to be much simpler. In fact, in 1686 Ray had published his first volume of *Historia Plantarum*, dealing with cryptogams and the greater part of the dicotyledonous herbs, which was followed in 1688 by the second volume on the remaining dicotyledonous herbs, the monocotyledonous herbs, anomalous herbs, and the trees and shrubs. Henceforth Commelin no longer depended on the *Methodus* itself. He could now directly consult Ray's elaboration of the latter's own system.

Already in volume 7, in his discussions of Natsjatam HM 7:1-2:1, Valli-kara HM 7:35-36:18, and Cari-villandi HM 7:59-60:31, Commelin borrowed part of his argumentation from Ray's book XIII, dealing with pomiferous and bacciferous herbs. One would now expect that in arranging the last volumes of Hortus Malabaricus Commelin would accurately follow *Historia Plantarum*, in books, sections, and chapters.

In his catalogue of the Amsterdam Hortus Medicus of 1689, listing the cultivated plants in alphabetical order, he accurately quoted the systematical place of each separate genus in *Historia Plantarum* in special paragraphs. In the last volumes of Hortus Malabaricus, however, he referred only seldom to the sections of chapters of *Historia Plantarum* and did not apply Ray's elaborate system to a detailed subdivision of Van Reede's remaining plants.

Just as in the preceding volumes, Commelin confined himself to the formation of larger divisions according to Ray, but he regularly deviated from the sequence, using practical rather than systematical criteria. He broadly divided the herbs into three groups. Contrary to *Historia Plantarum* he treated first the dicotyledons in the volumes 8-11, then the monocotyledons in the volumes 11 and 12, and finally the cryptogams, especially the ferns, in volume 12. This deviation was a logical consequence of Arnold Syen's provisional arrangement of the climbers. After the woody climbers in volume 7 Commelin continued, in volume 8, with the dicotyledonous, herbaceous climbers of the *Cucurbitaceae*, thus fixing the main arrangement of the herbs. Moreover, by interpreting these climbers like Ray's *Herbae Pomiferae* of book XIII, he first finished the Malabar plants belonging to the books XIII-XX and put off the discussion of the relatively few plants belonging to the books IV-XII to volume 10.

The title-page of volume 8 (1688) reads *De Variis Generis Herbis Pomiferis & Leguminosis*. Apart from some minor aberrations, the first part of this volume corresponds to Ray's book XIII, *Herbae Pomiferae*, and the second part of it to book XVIII, *Herbae flore papilionaceo, seu leguminosae*.

The title-pages of the following volumes 9-12 read *De Herbis et diversis illarum speciebus*, without any further indications as to their contents. In volume 9 (1689) Commelin neatly grouped plants belonging to sections of Ray's book XVIII on leguminosae and book XX, *Herbae flore pentapetaloide anomalae*, but in a reverse order. Instead of linking up with the leguminosae of volume 8, he began volume 9 with a number of *Apocynaceae* belonging to book XX. He then proceeded with the remaining leguminosae, which he explicitly classified, in his commentaries, among the papilionaceous plants of book XVIII. In this volume 9 one might also distinguish groups of plants belonging to books XV-XVII, on monopetalous and tetrapetalous plants, but Commelin disturbed its coherence by adding many plants which might have been better classified among Ray's *Herbae anomalae* of book XXIII.

In volume 10 (1690) there is hardly question any more of any recognizable order. In a motley row representatives of nearly all the books of Ray on dicotyledons alternate with each other, so that one receives the impression that this volume constitutes Commelin's appendix of problematic plants. About one half of them was not or only hesitantly annotated by him. He often pointed to the affinity to or similarity with genera known to him. In a number of cases he could only define the shape of the leaf and other characteristics, without being able to connect them with a genus.

In volume 11 (1692) Commelin started with the treatment of the monocotyledons. In the first part of this volume he grouped together plants belonging to Ray's book XXI, *Herbae radice bulbosa*, on bulbous monocotyledons. Among them was a number of aquatic plants, which may explain why he also inserted the dicotyledonous, nymphaeceous groups of Ambel and Tamara, HM 11:51-61:26-31. He ended this volume with a long series of convolvulaceous plants belonging to Ray's book XV, *Herbae flore monopetalo vasculiferae*.

In the last volume of Hortus Malabaricus, published in 1693 after Commelin's death, he hardly made any commentaries, but nevertheless the arrangement of this volume is quite clear. In this Commelin intended to continue his treatment of the monocotyledons with Van Reede's large group of parasitic plants, typified with the native term of maravara, consisting of several orchids belonging to Ray's book XXI, mixed with many ferns belonging to book III. Finally he concluded with a large group of grasses, both *Cyperaceae* and *Gramineae*, in a mixed order belonging to book XXII, including the Tsjurel group of rotangs HM 12:121-125:64-66.

Summarizing, we may state that Jan Commelin's commentaries and arrangements in Hortus Malabaricus are remarkable achievements, certainly if we take into account that he was only an amateur of botany. In the execution of his task Commelin was greatly favoured as regards the available means by the progressing development of exotic botany. For the comparison of Van Reede's plants he could make use much more than Syen of recently published literature on exotic plants.

Strongly supported by the world-wide relations of Joan Huydecoper van Maarsseveen, he brought together plants from all the possessions of the East and West India Companies in the Amsterdam Hortus Medicus founded and directed by him, which he discussed in his treatments of Malabar plants. Moreover he could consult the exotic collections of the rival Leiden botanical garden and of several private gardens.

As regards Syen's methods of commenting and classifying, Commelin adopted the Bauhinian nomenclature, but on the publication of Ray's systematical works he abandoned the Bauhinian systematics and undertook the arrangement of Hortus Malabaricus according to Ray's principles. As a result of different factors he succeeded only partly in this. In the first place Ray's publication scheme did not run parallel to the publication of the successive volumes of Hortus Malabaricus, in consequence of which Commelin made errors in the application of Ray's plant system. Secondly, as a rule he kept to the traditional groupings of Malabar plants, even if this resulted in aberrations with respect to Ray's system. And finally he was faced with great problems in the interpretation of many unknown plants.

Notes

1 Hortus Malabaricus vol.3 (1682):(xvi); Heniger 1980:54.
2 See also chapter 4, section Cape of Good Hope 1685.
3 RAU, Records Huydecoper Family 58, Letter-book 1683-1686, letter of 15 January 1683 of Huydecoper to Simon van der Stel.
4 Wijnands 1983:4. For a reconstruction of the Amsterdam Hortus Medicus in the 17th and 18th centuries, see Van der Pool-Stofkoper 1984.
5 RAU, Records Huydecoper Family 58, Letter-book 1683-1686, letter of 7 December 1685 of Huydecoper to Lamotius.
6 RAU, Records Huydecoper Family 58, ibid., letter of 28 November 1685 of Huydecoper to Andries Cleyer.
7 RAU, Records Huydecoper Family 59, Letter-book 1686-1687, letters of 15 December 1686 and 12 December 1687 of Huydecoper to Isaac de Saint-Martin and Willem ten Rhijne respectively.
8 RAU, Records Huydecoper Family 59, ibid,, letters of 9 December 1687 of Huydecoper to N. Elsevier and Magnus Wichelman.
9 Wijnands 1983:212.
10 For a full discussion of Van Aarssen's botanical activities in Surinam, see Brinkman 1980.
11 RAU, Records Huydecoper Family 59, Letter-book 1686-1687, letter of 20 July 1687 of Huydecoper to F. van Wickevoort. For Simon van Beaumont (1640-1726), secre-

tary of the States of Holland, and his fine collection of Curaçao and other West-Indian plants, see Kuijlen 1977 and Wijnands 1983:208.
12 Wijnands 1983:96-97.
13 Wijnands 1983:143-144.
14 Wijnands 1983:100-101,102.
15 Wijnands 1983:40.
16 Commelin 1689:(v)-(vi),(viii).
17 Commelin 1689:105-106.
18 Commelin 1689:330.
19 Commelin 1689:229; Wijnands 1983:133.
20 Commelin 1689:222.
21 Commelin 1689:47; Wijnands 1983:92.
22 Commelin 1689:18.
23 Commelin 1689:275.
24 Commelin 1689:90; Wijnands 1983:161-162.
25 Commelin 1689:275-276.
26 ARA, Records Society Surinam 213:160, no.38: 'List of seeds', accompanying Van Aarssen's letter to the Society of Surinam at Amsterdam, of 8 March 1685 (ibid.213:90).
27 Commelin 1689:12.
28 Commelin 1689:103.
29 Commelin 1689:171-172.
30 Commelin 1689:247; Wijnands 1983:43-44.
31 Commelin 1689:20.
32 Commelin 1689:20; Wijnands 1983:32.
33 Commelin 1689:23; Wijnands 1983:55. For the sending of the Surinam specimens of ananas, see Van Aarssen's letter to the Society of Surinam at Amsterdam, of 22 April 1686 (ARA, Records Society of Surinam 215:242).
34 Commelin 1689:14; Wijnands 1983:127-128.
35 Commelin 1689:107,371; Wijnands 1983:202.
36 Commelin 1689:36-38; Wijnands 1983:45.
37 Wijnands 1983:187.
38 RAU, Records Huydecoper Family 59, Diary of 1687.
39 Hermann 1687:688-690, tab.; Commelin 1689:229; Wijnands 1983:133.
40 Hermann 1687:664, *Cucumis puniceis Zeylanicus minor, seminibus nigris*; Wijnands 1983:92.
41 Hermann 1687:95-96,97 tab.
42 Hermann 1687:447,449 tab.,450.
43 Hermann 1687:590 tab.,592, *Teucrium Americanum procumbens, Vernonicae aquaticae foliis subrotundis*.
44 Hermann 1687:62,63 tab.,64.
45 Hermann 1687:569-570,571 tab.
46 Hermann 1687:636,637 tab.,638-640, and 208,209 tab.,210-211, respectively.
47 Hermann 1687:682,683 tab.,684; Commelin 1689:201; Wijnands 1983:37.
48 Hermann 1687:691-692,693 tab.; Wijnands 1983:40.
49 Hermann 1687:9.
50 Hermann 1687:174,175 tab.,176-177.
51 Hermann 1687:184,187 tab.
52 Wijnands 1983:133.
53 See also Wijnands 1983:58-59 about the problematic identification of *Erythroxylum japonicum non spinosum coronillae folio*, described and illustrated by Commelin 1697.
54 Wijnands 1983:205-207, Appendix A.
55 For more detailed discussions of Ray's *Methodus* and *Historia Plantarum*, see Raven 1942:192-199,202-241,300-304.

12
Insertion of Hortus Malabaricus in the botanical literature

The identification of the hundreds of plants in Hortus Malabaricus has engaged the attention of many generations of botanists from the appearance of the first volume in 1678 until the present day. Their aim was, and still is, to relate these plants to plants from other regions and to fit them into an up-to-date plant system. Their attempts at complete identification have had varying success, dependent on their means and abilities.

In the 19th century several botanists, such as Dennstedt (1818), Hamilton (1822-1835), Dillwyn (1839), and Hasskarl (1861, 1862, 1867), published keys to Hortus Malabaricus, which were based partly on discussions and citations of the botanical literature since Linnaeus and partly on their own observations in Malabar or elsewhere in Asia. A special place is taken by Dalgado's key (1896), dealing with the medical aspect of Hortus Malabaricus. But these keys largely disregard the exertions of 17th and 18th-century botanists to arrive at reliable identifications, by building on the initial commentaries of Arnold Syen and Jan Commelin.

The present chapter is devoted to botanists who, since the appearance of the first volume of Hortus Malabaricus in 1678 until the end of the Dutch rule of Malabar in 1795, have studied Hortus Malabaricus and made attempts to insert it in the body of botanical knowledge of that time. In this context the question will be brought up in what way, by what means, and to what extent they have discussed Van Reede's plants.

From among the numerous botanists living and working in this period I have selected fifteen according to one or more of the following criteria. In the first place I chose botanists who cover with their publications the whole of Hortus Malabaricus, or at least a considerable part of it, in so far as it was available at a particular moment. For this I preferred botanists, both Dutchmen and foreigners, who had an evident relationship to Dutch colonial botany. Only a few of them were residing or travelling in Asia, especially in India and Ceylon. However, most of them never visited Asia and based comparisons with Malabar plants on herbaria and collections of illustrations, or had at their disposal, or access to, gardens in which Asiatic plants were cultivated.

In the group of botanists thus selected Linnaeus is the central figure, for the publications of most of the selected pre-Linnaean botanists figure as additional references in his editions of *Species Plantarum* and *Systema Naturae*, while the selected Linnaean botanists in turn cited or amended his identifications. A consequence of this is that the description of the process of insertion of Hortus Malabaricus in the botanical literature of the 17th and 18th centuries is based on a closely coherent complex of numberless data. These data are accounted for in the Annotated List of Plants in Part Three of this study. Conversely, the modern taxonomist can use this List as a source of relevant identifications of Malabar plants before 1795.

The first botanist who played a role in the insertion of Hortus Malabaricus in the literature was Jacob Breyne (1637-1697), a merchant and amateur of botany in Danzig[1]. As a young man he received a commercial training in the Netherlands. During that time he entered into relations with Dutch botanists and amateurs of botany, who roused in him a partiality for exotic botany. Through his contacts with the Leiden professors Jacob van Gool, who at first was to edit *Viridarium Orientale* of Matthew of St. Joseph, and Arnold Syen, and through his visits to the private gardens of Hieronymus Beverningk in Warmond and Frans van Sevenhuizen, a chemist in The Hague[2], he collected a large amount of botanical material originating from Asia and the Cape of Good Hope and he got a good insight into the Dutch exotic horticulture of that time. Moreover, via Syen and Beverningk he entered into a correspondence with Andries Cleyer, Willem ten Rhijne, and Herbert de Jager in Batavia and Paul Hermann in Ceylon. Once he had returned to Danzig, he published and illustrated his Dutch observations in *Centuria Prima* (1678), dedicated to Beverningk. During later visits to the Netherlands, in 1679 and 1688, he studied the most recent developments in Dutch horticulture. He published the results of these two journeys in *Prodromus Fasciculi Rariorum Plantarum* (1680) and *Prodromus ... Secundus* (1689).

In *Centuria Prima* Breyne was not yet in a position to involve the first volume of Hortus Malabaricus in his notes on exotic plants, because this volume appeared a few months after his own book[3].

His very first comparisons are to be found in the *Prodromus* of 1680, in which he compared 7 plants from the volumes 1 and 2 with plants growing in Beverningk's garden and with a dried specimen received from Hermann.

In *Prodromus ... Secundus* of 1689 Breyne had at his disposal the volumes 1 to 8 of Hortus Malabaricus, from which he borrowed 47 citations. He compared these plants with newly imported plants in the Amsterdam Municipal Garden directed by Jan Commelin and in the gardens of Beverningk, Caspar Fagel in Noordwijkerhout, Agnes Block in Loenen, William III in Honselaarsdijk, and Simon van Beaumont in The Hague. Moreover, referring to Malabar plants, Breyne discussed various seeds, fruits, and dried plants which he had received in previous years and during his second journey in the Netherlands from Cleyer, Ten Rhijne, Hermann, De Jager, Commelin, Beverningk, and Van Beaumont.

Although Breyne's *Centuria Prima* contains no references to Hortus Malabaricus, in the long run it did begin to play a part in its insertion in the literature, because later botanists identified a number of the descriptions and illustrations in this work with Malabar plants.

Of the pre-Linnaean botanists, John Ray (1627-1705) referred most extensively to Hortus Malabaricus. As a minister in Black Notley he lived in comparative seclusion in the countryside of England, remote from the centres of the study of botany in England and on the continent. Although in 1663, during a continental tour, he stayed a couple of months in the Netherlands, he does not seem, like Breyne, to have sought access to the Dutch circles of exotic botany[4]. Ray's herbarium, mainly consisting of plants collected during this tour, contains only a few Dutch specimens of minor importance[5]. In later days, too, as far as is known, he maintained few, and then superficial, relations with the Netherlands. His correspondence with Pieter Hotton, Hermann's successor in the Leiden chair of botany, is only concerned with booknotices and the preparations for the printing of the second edition of his *Methodus Plantarum*[6]. Ray inserted Hortus Malabaricus in his main work, *Historia Plantarum*. This work appeared in three volumes, which were published in 1686, 1688, and 1704.

The insertion proceeded somewhat chaotically, because, as has already been set forth in the discussion of Commelin's commentaries, Ray's plant system was not analogous to the publication scheme of Hortus Malabaricus. In his first volume indeed he could make use of the volumes 1 to 5, but he only succeeded in fitting in a few plants from the volumes 1 and 2. In his second volume, for which furthermore he had volume 6 at his disposal, he referred to nearly 300 Malabar plants from the volumes 1 to 6. When after a long interruption finally his third volume appeared, the publication of Hortus Malabaricus had meanwhile been completed. In this volume he cited more than 300 plants from the volumes 7 to 12, and moreover he also gave new identifications, additions, and improvements of his references in his earlier volumes. In all, Ray inserted 615 plants (about 85%) from Hortus Malabaricus in his *Historia Plantarum*. In spite of this high score, which even Linnaeus and Nicolaas Laurens Burman were not to achieve, Ray manifestly had difficulties with volume 7, from which he succeeded in including only 45% of the plants.

In the insertion he followed different methods. A part of the Malabar plants could be identified by him with or related to plants which had already been published elsewhere, and could be inserted in his system. For unidentifiable Malabar plants he also attempted to find a logical place within his system. If this was impossible for him, he placed them in an appendix to a section. With respect to questions of nomenclature of these plants he often adopted Syen's and Commelin's botanical names in their commentaries, in so far as he agreed therewith. But in many cases he rejected their nomenclature and proposed new botanical names. But the most important work performed by Ray consisted of the numerous succinct summaries of the descriptions in Hortus Malabaricus, in which he omitted the voluminous details and emphasized the relevant characteristics of the plant. This contributed greatly to the readability and usefulness of Hortus Malabaricus. Thus, Linnaeus used Ray's summaries by the side of the illustrations of Malabar plants and in his references often showed his preference by first citing the summary from *Historia Plantarum*, and only thereafter Hortus Malabaricus itself.

The importance of the Ceylon botanist Paul Hermann (1646-1695) for Hortus Malabaricus has already been referred to repeatedly. Van Reede owed to Hermann, during his visit to Cochin in 1674, the final scheme, while the herbaria of Ceylon plants which Hermann presented to Syen and Commelin have been useful in the early attempts at comparing the floras of Malabar and Ceylon. Conversely, Hermann contributed in different ways to the insertion of Hortus Malabaricus in the botanical literature.

After his stay in Ceylon, 1672-1680, he acted as professor of botany and director of the botanical garden of Leiden University in 1680-1695. In that function he brought the Leiden branch of the study of colonial botany to great prosperity. His large private collection of natural curiosities, mainly gathered in Ceylon, attracted visitors from the Netherlands and from abroad. Thanks to the support of Hieronymus Beverningk, curator of Leiden University, he succeeded in building in Leiden the first hothouses in a Dutch institutional garden, in which he cultivated numerous tropical plants, from America as well as Asia[7]. These included Ceylon plants, which had been grown from seeds which he had partly sent from Ceylon to Leiden, and partly had brought with him upon his repatriation, and partly had later received through the intermediary of Van Reede and others[8].

As I have already set forth, Commelin was able to make extensive use of Hermann's tropical collection for his commentaries in Hortus Malabaricus. In 1687 Hermann published his alphabetical garden catalogue *Horti Academici Lugduno-Batavi Catalogus*, in which he identified 42 of his garden plants with plants from the volumes 1 to 6 of Hortus Malabaricus, which had been published up to that time. Now and then he added remarks in which he pointed to the identity of some Malabar and Ceylon plants.

Hermann devoted a special study to exotic plants

growing in Dutch gardens. For this he did not confine himself to the collections in Leiden and Amsterdam. He also studied the collections of many private gardens, such as those of Beverningk, Fagel, Van Beaumont, Block, and of the prince's garden in Honselaarsdijk. In this study he was assisted by his English pupil William Sherard (1659-1728)[9], who edited in 1689 a preliminary alphabetical list of plant names, *Pauli Hermani Paradisi Batavi Prodromus*. Sherard listed 60 garden plants which he and his teacher considered to be identical with the plants from the volumes 1 to 8 of Hortus Malabaricus. In the nomenclature he usually reverted to the Leiden garden catalogue of 1687, or else he merely quoted the Malayalam name. After the early death of Hermann, Sherard collected the fragmentary notes of this study, which he published in alphabetical order under Hermann's name as *Paradisus Batavus* (1698). In the latter, 31 plants, specifically the Apocynums, were identified with plants from the volumes 1 to 10 of Hortus Malabaricus, which largely meant an increase of the number of identifications in the *Prodromus*. More frequently than in the Leiden garden catalogue, Hermann interwove with these notes remarks about the identity of several Malabar and Ceylon plants. In the nomenclature he followed, though with some modifications, the commentators Syen and Commelin, or he proposed names of his own invention.

In the above-mentioned publications not a single reference to Hermann's own Ceylon herbarium is to be found. Nevertheless, Hermann initially worked on a comparison of his herbarium with Hortus Malabaricus. The catalogue of this herbarium, *Musaeum Zeylanicum*, was published posthumously in 1717. It contains only 9 identifications with plants from the first two volumes of Hortus Malabaricus. This small number can be explained by the fact that Hermann regularly cited in this catalogue his own garden catalogue and Sherard's *Prodromus* without repeating the identifications with Malabar plants mentioned there.

Although Hermann may have contributed comparatively little to the insertion of Hortus Malabaricus, his Ceylon herbarium became an important tool in the hands of Linnaeus in the identification of Van Reede's plants.

The contributions of Jan Commelin (1629-1692) are indissolubly bound up with his activities as a commentator. To avoid unnecessary repetition, reference is made, as to a detailed survey of his methods, means, and results of comparison, to the discussion of his commentaries in the volumes 2 to 12. In his alphabetical *Catalogus Plantarum Horti Medici Amstelodamensis* of 1689 he identified 64 of his garden plants with plants in Hortus Malabaricus. For this he based himself in the first place on the identifications and nomenclature in the volumes 1 to 8 which had then already been published, but at the same time he anticipated volume 9, which was in the press, and the volumes 10 to 12, which were still being edited. In a number of cases he accepted new interpretations published by Breyne (1680, 1689), Ray (1686, 1688), and Hermann (1687) or proposed new ones avoided by them. In his annotations to the genera of his garden plants Commelin attempted to place them in Ray's plant system.

In 1697 there appeared posthumously his *Horti Medici Amstelodamensis Rariorum Plantarum* with full descriptions of a selected number of exotic garden plants and illustrations based on the Moninckx Atlas. In this he compared a modest number of 14 plants, originating from Ceylon, Bengal, and the Cape of Good Hope, with plants from the volumes 2, 3, and 6 to 11, including some new interpretations.

In 1696 Caspar Commelin (1667-1731)[10], Jan's nephew and scientific heir, published *Flora Malabarica*, an alphabetical index to all the volumes of Hortus Malabaricus, in which he enumerated about 550 identifications. This index is chiefly a compilation of all the synonyms which had been published since 1678. In this respect *Flora Malabarica* is a very useful guide to the interpretations of the two commentators and those of Breyne, Hermann, Sherard, Plukenet, and especially Ray. Only in a few cases did Caspar Commelin undertake new interpretations; in a single case he intimated that he had consulted his uncle's Ceylon herbarium.

In the same year 1696 Caspar was nominated botanist of the Amsterdam Hortus Medicus. In this function he continued Jan's work of describing exotic plants in that garden with the publication of *Horti Medici Amstelaedamensis Rariorum Plantarum* in 1701. In this book he discussed only 6 plants in connection with plants from the volumes 2, 5, and 11 of Hortus Malabaricus. Since 1701 Commelin lectured on exotic plants growing in the Amsterdam garden. In his lectures, published in 1703 as *Praeludia Botanica*, he discussed 4 plants from volume 2 of Hortus Malabaricus.

One of the most confusing authors inserting Hortus Malabaricus in their own work was the English botanist Leonard Plukenet (1642-1706). As a royal professor of botany at the court of William III and Mary II and as intendant of the royal gardens of Hampton Court he occupied a prominent position in English botany. At Hampton Court he cultivated exotic plants which he obtained, among others, from Paul Hermann in Leiden[11], while he also studied plants grown in the gardens of Fulham Palace, owned by Henry Compton, bishop of London, who exchanged seeds with the Amsterdam Hortus Medicus. In addition he had access to the Dutch codices of plant drawings in the possession of Compton and Hans Willem Bentinck, earl of Portland[12].

Plukenet published his materials in four volumes. In *Phytographia*, consisting of three parts (1691-1692), he issued his plates, tables 1-328 of which are provided with plant names, with numerous references to Hortus Malabaricus. In *Almagestum Botanicum* (1696), *Almagesti Botanici Mantissa* (1700), and *Amaltheum Botanicum* (1705) he described and annotated his tables and added many supplements, modifications, and augmentations, by which he created great confusion. Without much critical sense he involved Van Reede's plants in his material from Asia, the Cape of Good Hope, and America. In this way, Plukenet discussed and identified, sometimes three or four times, hundreds of plants from all the volumes of Hortus Malabaricus.

A special place taken by Plukenet's attempts at identification was the 8-volume "Malabar" herbarium, consisting of Coromandel plants with "Malabar" or Tamil native names, collected by Samuel Browne, a surgeon of the English East India Company in Madras, who died before 1703. Browne regularly sent seeds and exsiccates from the neighbourhood of Madras to James Petiver (c.1658-1718), the secretary of Hans Sloane, secretary of the Royal Society of London.

In 1699 Petiver published a small collection of Browne's plants in the *Philosophical Transactions*, vol.20, with some references to Hortus Malabaricus. In 1699 Browne's "Malabar" herbarium arrived in London, where the Royal Society asked Petiver to describe it[13]. Even before Petiver could carry out this task, Plukenet had the opportunity to study Browne's plants in a hurry and inserted them in his *Almagesti Botanici Mantissa*, with many references to Hortus Malabaricus. The haste with which Plukenet performed his work appears from his frequent misspellings of the Tamil native names cited by him as "Malabarorum". Petiver published his description of Browne's herbarium in seven instalments in the *Philosophical Transactions*, vols 22 (1702) and 23 (1704), in which he criticized Plukenet's references to Hortus Malabaricus. Altogether Petiver related 138 of the more than 350 exsiccates from Browne's herbarium to van Reede's plants.

Herman Boerhaave (1668-1738), Hermann's second successor as professor of botany and director of the botanical garden of Leiden University, 1709-1728, in his botanical writings made only few references to Hortus Malabaricus. In his catalogue of 1710, *Index Plantarum*, surveying the Leiden garden collections he had taken over from his predecessor Pieter Hotton, he identified only 3 plants in the volumes 1 to 3 of Hortus Malabaricus.

Thanks to an extensive correspondence the Leiden garden since then, under his directorship, grew to be one of the most important botanical gardens of Europe, with a considerable collection of tropical plants. Thus he received many seeds from Ceylon through the intermediary of Antoni van Gesel, surgeon in Colombo, while Caspar Commelin shared with him the annual consignments of seeds by the Ceylon government, for which Van Reede had taken the initiative[14].

Nevertheless, in his second garden catalogue, *Index Alter Plantarum* (1720), Boerhaave identified only 21 garden plants with Van Reede's plants, in which connection he adopted the already existing nomenclature. Adriaan van Royen (1704-1779), Boerhaave's successor as professor of botany and director of the garden[15], too, identified a modest number of 17 plants in his garden catalogue, *Florae Leydensis Prodromus* (1740). On the other hand Boerhaave did contribute indirectly to the insertion by sending his pupil Pieter Hertog to Ceylon. Hertog's herbarium never reached him; it was dealt with by Johannes Burman.

Johannes Burman (1707-1778)[16], the Amsterdam professor of botany and director of the Municipal Garden, was the first author who explicitly compared the Ceylon flora with Hortus Malabaricus. In his *Thesaurus Zeylanicus* (1737) he gave an alphabetical survey of all the Ceylon plants known to him on the basis of published literature, manuscripts, and herbaria.

Burman's principal sources were two Ceylon herbaria. One of these was Jan Commelin's herbarium, originating from Paul Hermann, and the other had been collected by Boerhaave's pupil Pieter Hertog. In addition Burman widely made use of the index to Hermann's own herbarium, *Musaeum Zeylanicum* (1717), and of Hertog's manuscript lists of Ceylon seeds sent to the Netherlands. Burman emphasized the affinity between the Ceylon and Malabar floras by including more than 200 references to Hortus Malabaricus in a special index. Burman also compared his Ceylon plants with *Herbarium Amboinense* of Georg Everhard Rumphius (1628-1702), for the publication of which he had recently received the consent of the Company[17].

Although Rumphius' monumental description of the flora of the Indian Archipelago was completed at the end of the 17th century, it did not play a role in the insertion until more than a generation later. Burman published Herbarium Amboinense in six volumes, 1741-1750, the *Auctuarium* belonging to it in 1755. In spite of his isolated position Rumphius had at his disposal botanical literature, including the volumes 1 to 10 of Hortus Malabaricus. He compared more than 150 plants with Van Reede's plants. In his annotations to Rumphius' descriptions and engravings Burman often discussed these comparisons and added some seventy more references, partly derived from the volumes 11 and 12, which Rumphius had not been able to consult[18].

During his stay in the Netherlands, in 1735-1738, Carolus Linnaeus (1707-1778)[19] could become acquainted extensively in herbaria, libraries, and institutional and private gardens with tropical plants originating from the Dutch colonies. He had the opportunity to assist Burman for a short time with the preparation of *Thesaurus Zeylanicus* and thus to become convinced of the great importance of the missing Ceylon herbarium of Paul Hermann. His frequent visits to the botanical gardens of Amsterdam and Leiden will have given him a vivid picture of tropical plants. Through the intermediary of Boerhaave he became the personal physician of the Amsterdam banker George Clifford (1685-1765), who also entrusted him with the management of his magnificent private botanical garden on his estate Hartekamp in Heemstede. With the aid of Clifford's voluminous library he was able to compile a detailed catalogue of that garden, *Hortus Cliffortianus* (1737). Of the tropical plants cultivated there he identified some 95 with Van Reede's plants, which in this way he largely provided with his own typification. In *Viridarium Cliffortianum* (1737), an excerpt from this garden catalogue, he only repeated a few identifications, which are not included in the Annotated List.

When Linnaeus had returned to Sweden and worked as professor of botany and director of the botanical garden at Uppsala University, unexpectedly the Ceylon herbarium of Hermann that had been imagined to be lost turned up. It was lent to him for some time and he described it in *Flora Zeylanica* (1747). He succeeded in identifying more than 210 of Hermann's exsiccates with Van Reede's plants. In this context he frequently cited Burman's *Thesaurus Zeylanicus*, so that he made a connection between all available Ceylon herbaria and Hortus Malabaricus. In spite of his intensive contacts with Dutch botanists, who regularly provided him with tropical seeds, Linnaeus cultivated only a small collection of tropical plants in the Uppsala garden. In his garden catalogue, *Hortus Upsaliensis* (1748), he referred to ten plants in Hortus Malabaricus, in which context it has to be observed that he omitted a large number of identifications already established by himself.

In *Species Plantarum* (1753), the pivot in the history of taxonomy, Linnaeus identified more than 250 species with Van Reede's plants. Of these he had taken over some 170 identifications from *Flora Zeylanica*. A number of species, however, are based exclusively on Van Reede's descriptions and illustrations. In the second edition of *Species Plantarum* (1762-1763) Linnaeus increased the number of identifications to some 320, for which he based himself, among other things, on Rumphius' *Herbarium Amboinense*. In *Mantissa Plantarum* (1767) and *Mantissa Plantarum Altera* (1771) he added 4 and 25 identifications respectively to this number, so that altogether Linnaeus identified some 350 species with plants in Hortus Malabaricus. It is striking that in the above-mentioned works Linnaeus repeatedly cited Ray's summaries of Van Reede's descriptions. In part this will be due to the fact that it was only at a later moment that a complete set of Hortus Malabaricus was at his disposal.

Nicolaas Laurens Burman (1733-1793)[20], who succeeded his father Johannes as professor of botany and director of the Municipal Garden in Amsterdam, undertook a complete revision of the Asiatic flora, which he published in 1768 as *Flora Indica*. He started from Linnaeus' second edition of *Species Plantarum* and from the newly acquired collections of his father, such as the Java herbaria of Christiaan Kleynhoff and Albert Pryon, the Asiatic herbarium of Laurent Garcin, and the Coromandel drawings of Hendrik Otto van Outgaerden[21]. Of course Burman Jr. was also in a position to study the Ceylon herbaria of Jan Commelin and Pieter Hertog again. With regard to Van Reede's plants he largely agreed with the identifications which Linnaeus had established up to that time. To these he added some 50 species which Linnaeus had not related to Hortus Malabaricus. Moreover, he identified some 30 new species founded by himself with plants in Hortus Malabaricus. Altogether, Burman Jr. succeeded in identifying 385 species with Van Reede's plants[22].

After Burman Jr., the Amsterdam physician and naturalist Martinus Houttuyn (1720-1798)[23] was the last author in the Company's era who was in a position to survey the Asiatic flora. In his *Natuurlyke Historie* (1761-1785), an encyclopaedic work written in Dutch, he described and illustrated numerous animals, plants, and minerals. The botanical section, volume II, in 14 parts, was published during 1773-1783. Houttuyn largely followed Linnaeus' *Systema Naturae*, editions 10 (1758) and 12 (1766-1768), and Murray's edition 13 (1774) of *Systema Vegetabilium*.

With respect to the Asiatic flora he had at his disposal several Java herbaria, which he had received from Richter, Carl Per Thunberg, and Radermacher, a Ceylon herbarium from Thunberg, and a collection of watercolours from Bengal, and he also consulted the Burman collections.

In *Natuurlyke Historie*, Houttuyn identified 338 species with plants in Hortus Malabaricus. Although he was not always consistent in his citations of the literature, it appears from his descriptions that he mainly agreed with the identifications which had been established by Linnaeus and Burman Jr. In a few cases he related new species founded by himself to Van Reede's plants.

It can be stated that since the publication of the first volume of Hortus Malabaricus until the end of the Dutch domination of Malabar the insertion of Hortus Malabaricus in the botanical literature of the 17th and 18th centuries proceeded with varying results. This was dependent on the sometimes widely diverging aims, means, and abilities of the authors.

Ostensibly John Ray and Caspar Commelin were most successful, but their studies were mainly based on the existing botanical literature and to a very modest extent on botanical material from the tropics. Nevertheless, Ray's summaries of the descriptions in Hortus Malabaricus were of use for making the work more accessible, of which later Linnaeus gratefully made use.

The importance of the tropical collections of the botanical gardens for the insertion of Hortus Malabaricus in the literature was at first relatively great, but declined in the 18th century.

On the other hand, after a modest beginning, the importance of herbaria increased greatly. It is striking that in the footsteps of the commentators of Hortus Malabaricus, Syen and Jan Commelin, attempts were made to relate Van Reede's Malabar plants to herbaria originating from neighbouring regions, such as Ceylon by Johannes Burman and Coromandel by James Petiver, as a result of which the affinities between the floras of Malabar, Ceylon, and Coromandel could be firmly founded already in the first half of the 18th century. Thanks to the publication of Rumphius' *Herbarium Amboinense* and the arrival of new Asiatic herbarium material, Linnaeus,

Burman Jr., and Houttuyn could increase the number of identifications in the second half of the 18th century to about half the number of Van Reede's plants.

Notes

1. Kuijlen 1982.
2. In 1697 Sevenhuizen's garden was bought by Simon van Beaumont (Kuijlen 1976:42,159-160,165-166).
3. Syen presented the first volume of Hortus Malabaricus in May 1678 to the Curators of Leiden University. Breyne dated the dedication of *Centuria Prima* to Hieronymus Beverningk on 15 April 1677, while in his request for a patent to the States of Holland, of 12 January 1678, he intimated that his book had already been printed (ARA, Records States of Holland and Westfriesland after 1572, inv.no.1630, Minutes of patents and additional documents 1678).
4. Raven 1942:132.
5. Raven 1942:140-141.
6. Den Tonkelaar 1983:55-64.
7. Veendorp & Baas Becking 1938:84-91.
8. See chapter 4, section South Ceylon 1685.
9. For the relationship between Paul Hermann and William Sherard, see Heniger 1969.
10. Wijnands 1983:10-13.
11. BL, Sloane MSS 4062:218-219,220-220v, letters of 4 November 1686 and 14 February 1686/87 of Patrick Adair to Plukenet.
12. Dandy 1958:84,90; Wijnands 1983:208,209.
13. Dandy 1958:99-102. Browne's "Malabar" herbarium is nowadays distributed over the Sloane herbarium (Petiver's collections) in the British Museum for Natural History.
14. Heniger 1971:36,75,76; for Van Reede's initiative as to the annual consignments from Ceylon, see chapter 4, section South Ceylon 1685.
15. Veendorp & Baas Becking 1938:117-129; Stafleu & Cowan 1983 vol.4:958-960.
16. NNBW vol.4 (1918):353-354; Stafleu & Cowan 1976 vol.1: 413-416.
17. VOC 256, Resolutions of the Chamber of Amsterdam, n.p., meeting of 27 August 1736. On Rumphius, see NNBW vol.3 (1914):1104-1107; Stafleu & Cowan 1983 vol.4:986-990.
18. For a recent interpretation of Rumphius' plants in *Herbarium Amboinense*, see De Wit 1959:339-460.
19. Stafleu 1971; Stafleu & Cowan 1981 vol.3:71-111.
20. NNBW vol.4 (1918):354; Stafleu & Cowan 1976 vol.1:416-417.
21. For Kleynhoff, Pryon, and Van Outgaerden, see Florijn 1985a,b. For Garcin, see Bridel 1857; Stafleu & Cowan 1976 vol.1:912.
22. In 1769 Johannes Burman also published *Flora Malabarica*, an index to all the volumes of Hortus Malabaricus. I do not refer to this index in the Annotated List, because it mainly lists all references to Hortus Malabaricus in his *Thesaurus Zeylanicus*, Linnaeus' second edition of *Species Plantarum*, and Nicolaas Laurens Burman's *Flora Indica*, which have already been included in the List.
23. Stafleu 1971:174-176; Stafleu & Cowan 1979 vol.2:343-345.

Part Three
Annotated list of plants

The annotated list of plants consists of a concordance of the engravings and drawings of Hortus Malabaricus as well as a survey of the references to the botanical literature discussed in chapter 12. The list follows the same sequence as that of the plant descriptions, but sometimes it is interrupted by the insertion of an unpublished drawing.

The annotations of each plant are composed of five elements. The first element consists of two lines. In the first line is mentioned the Malayalam plant name as printed in the heading of the plant description, followed by the numbers of volumes, pages, and engravings of Hortus Malabaricus. In the second line the Malayalam names as engraved on the plate(s) in Roman script are mentioned.

The second element contains the reference to the corresponding drawing(s) in the British Library Additional Manuscripts (BL Add MS).

The third element comprises the prevalent scientific plant name in the commentary by Syen or Commelin in Hortus Malabaricus (Syen in HM, Commelin in HM).

The fourth element contains, in a chronological order, the references to the authors who discussed the relevant plant. Pre-Linnaean and Linnaean botanists had no standardized system of citing author and publication. They used abbreviations of their own invention, which often do not excel in clearness. I have solved a number of more or less obscure abbreviations by adding the correct citations between square brackets in so far as they play a part in discussions elsewhere in this study, notably Caspar Bauhin's *Pinax* and authors who overlooked a literature citation. However, I have not solved obscure abbreviations used by others than the said authors in order to avoid cramming this list with details of minor importance. If a botanical name proved to be identical with that of a previously cited author, this name is not repeated, but replaced by the sign = . Several authors expressed the grade of identification by a scale of terms. I standardized these terms with letters or signs between brackets and will explain them in the small list below.

The fifth element consists of remarks (NB) about errors in the numbering of pages and engravings in Hortus Malabaricus.

SIGNS AND ABBREVIATIONS

c	agrees with
cs	conspecies
?c	possibly agrees with
?cs	possibly a conspecies
cf	confer, compare with
r	refers to
s	simile
v	variety
?	possibly
=	identical
?=	possibly identical
≠	not identical
-	not mentioned

Annotated list of plants

VOLUME 1

Tenga HM 1:1-8:1-4
Tenga
BL Add MS 5028:2-3,4-5,6-7,8-9.
Syen in HM 1:8 Palma Indica coccifera angulosa, Iansiat Indi C.B.P. [Bauhin 1623:508-509].
Hermann 1687:472 = Syen in HM.
Ray 1688 vol.2:1356 = Syen in HM.
Breyne 1689:80 = Syen in HM.
J.Commelin 1689:260 = Syen in HM.
Sherard 1689:361 = Syen in HM.
C.Commelin 1696:50 = Syen in HM.
Plukenet 1696:275 = Syen in HM.
Petiver 1704(1702):1064,no.244 Aumana maraum Malab. Palma Maderaspatana, Oleandri angustissimi folio (s).
Boerhaave 1710:237 = Syen in HM.
Boerhaave 1720 vol.2:170 = Syen in HM (?).
J.Burman 1737:182 = Syen in HM.
Linnaeus 1737:483 Coccus frondibus pinnatis, foliolis ensiformibus, petiolis marginis villosis.
Rumphius 1741 vol.1:1-8,Tab.I Palma indica major (=); J.Burman ibid.:9 (=).
Linnaeus 1747:185,no.391 = Linnaeus 1737.
Linnaeus 1753:1188 Cocos nucifera.
Linnaeus 1763:1658 = Linnaeus 1753.
N.L.Burman 1768:240 = Linnaeus 1763.
Houttuyn 1773 II vol.1:305-344,Pl.III,Fig.5 = Linnaeus 1763.

Caunga HM 1:9-10:5-8
Caunga
BL Add MS 5028:10-11,12-13,14-15,16-17.
Syen in HM 1:10 Palma cujus fructus sessilis Faufel dicitur. Faufel Serapioni: Filfel & Fufel, Avicennae C.B.P. [Bauhin 1623:510].
Ray 1688 vol.2:1363 Areca sive Faufel, sive Avellana Indiana versicolor Park.
Sherard 1689:361 Palma Arecifera nucleo versicolori Nuci Moschatae simili.
C.Commelin 1696:50 = Sherard 1689.
Plukenet 1696:275,tab.309,fig.4 = Sherard 1689.
Petiver 1702(1701):939-940,no.152 Pauck-maraum Malab. Areca seu Faufel Indiae Orientalis.
J.Burman 1737:182-183 Palma Indica minor, fructu Areca dicto, vulgaris. Mus.Zeyl.pag.51 [Hermann 1717:51].
Rumphius 1741 vol.1:26-37,Tab.IV Pinanga (=); J.Burman ibid.:31 (=).
Linnaeus 1747:186,no.392 Arecca frondibus pinnatis, foliolis oppositis lanceolatis plicatis.
Linnaeus 1753:1189 Areca Cathecu.
Linnaeus 1763:1659 = Linnaeus 1753.
N.L.Burman 1768:241 = Linnaeus 1763.

Houttuyn 1773 II vol.1:383-401,Pl.IV,Fig.1 Areca Cathecu Syst.Nat.XII.Tom.II.Gen.1225 [Linnaeus 1767 vol.2:730].

Carim-pana HM 1:11-12:9
Carim-pana
BL Add MS 5028:18-19.
Syen in HM 1:12 Palma Coccifera complicato folio fructu minore.
Ray 1688 vol.2:1366 Palma coccifera, folio plicatili flabelliformi foemina, Carimpana H.M.
C.Commelin 1696:49 = Syen in HM.
Plukenet 1696:277 Palma Barbadensis folio plicatili magno subtus molli & argenteo (?).
J.Burman 1737:187 Palma Indica Tal & Talghala dicta, fructu carnoso, dulci & eduli putamine incluso. Mus.Zeyl.pag.49 [Hermann 1717:49].
Rumphius 1741 vol.1:47,Tab.X Lontarus Domestica (=); J.Burman ibid.:52 (=).
Linnaeus 1747:187,no.395 Borassus frondibus palmatis.
Linnaeus 1753:1187 Borassus flabellifer.
Linnaeus 1763:1657 = Linnaeus 1753.
N.L.Burman 1768:240 = Linnaeus 1763.
Houttuyn 1773 II vol.1:261-274 = N.L.Burman 1768.

Ampana HM 1:10:10
Ampana
BL Add MS 5028:20-21.
Syen in HM 1:10 Palma Malabarica, flosculis stellatis, fructu longo squammato.
Ray 1688 vol.2:1366-1367 Palma coccifera folio flabelliformi mas, Ampana H.M.
C.Commelin 1696:50 = Syen in HM.
Rumphius 1741 vol.1:51 Lontar manneken (=); J.Burman ibid.:52 (=).
Linnaeus 1753:1187 Borassus flabellifer.
Linnaeus 1763:1657 = Linnaeus 1753.
N.L.Burman 1768:240 = Linnaeus 1763.
Houttuyn 1773 II vol.1:261-274 = N.L.Burman 1768.

Schunda-pana HM 1:15-16:11
Schunda-pana
BL Add MS 5028:22-23.
Syen in HM 1:16 Palmae Dactyliferae species.
C.Commelin 1696:50 = Syen in HM.
Plukenet 1696:276-277 Palma Americana pediculis & foliorum carinis, rarioribus, at longissimis spinis aculeata, summis apicibus leviter serratis.
J.Burman 1737:180-181 Palma Indica, vinifera, fructibus urentibus, folio Adianthi saccharum praebens. Mus.Zeyl.pag.44 [Hermann 1717:44].
Rumphius 1741 vol.1:65-66,Tab.XIV Saguaster Major (s); J.Burman ibid.:67 (s).

Linnaeus 1747:187,no.396 Caryota frondibus duplicato-pinnatis, foliolis cuneiformibus obliquis incisis.
Linnaeus 1753:1189 Caryota urens.
Linnaeus 1763:1660 = Linnaeus 1753.
N.L.Burman 1768:241 = Linnaeus 1763.
Houttuyn 1773 II vol.1:428-438 = N.L.Burman 1768.

Bala HM 1:17-20:12-14
Bala
BL Add MS 5028:24-25,26-27,28-29.
Syen in HM 1:20 Palma humilis longis latisque folijs C.B.P. [Bauhin 1623:507].
Hermann 1687:256 Ficoides seu Ficus Indica, longissimo latissimoque folio, fructu longissimo, Musa Serapioni dicta.
Ray 1688 vol.2:1374 Musa arbor J.B.Park.
J.Commelin 1689:127 = Hermann 1687.
C.Commelin 1696:28 = Hermann 1687.
Plukenet 1696:145 Ficus Indica racemosa, foliis & fructu amplissimis, Musa Arabibus dicta.
J.Burman 1737:164-165 Musa Seraphionis. Ficus Indica. Mus.Zeyl.pag.70 [Hermann 1717:70].
Linnaeus 1747:176,no.368 Musa racemo simplicissimo. Hort. cliff.467 [Linnaeus 1737:467].
Rumphius 1747 vol.5:125-129,Tab.LX Musa (=); J.Burman ibid.:129 (=).
Linnaeus 1763:1477 Musa paradisiaca.
N.L.Burman 1768:217 = Linnaeus 1763.
Houttuyn 1776 II vol.6:406-418 = N.L.Burman 1768.

Ambapaja HM 1:21-22:15,1
Amba paja
BL Add MS 5028:30-31.
Syen in HM 1:24 (see Papajamaram following here).
Ray 1688 vol.2:1370 Mamoera mas & foemina Ger.Park.Marggr.
Sherard 1689:362 Pepo arborescens Mas seu sterilis, Mamoeira Mas Marck.
C.Commelin 1696:52 = Sherard 1689.
Plukenet 1696:145 Ficus arbor Utriusque Indiae Platani foliis monosteleches, fructu Mali Cydonii, aut Melonis magnitudine, Phytogr.Tab.278,fig.1.
J.Burman 1737:184 Papaya mas. Boerh.Ind.Hort.Lug.Bat.part. 2.pag.170 [Boerhaave 1720 vol.2:170].
Linnaeus 1737:461 Carica foliorum lobis sinuatis.
Rumphius 1741 vol.1:149,Tab.L Papaja Mas & Femina (=).
Linnaeus 1747:173-174,no.365 = Linnaeus 1737.
Linnaeus 1753:1036 Carica Papaya.
Linnaeus 1763:1466 = Linnaeus 1753.
N.L.Burman 1768:215 = Linnaeus 1763.
Houttuyn 1774 II vol.3:525-528 = N.L.Burman 1768.

Papajamaram HM 1:23-24:15,2
Papaja
BL Add MS 5028:30-31.
Syen in HM 1:24 Arbor Platani folio fructu peponis magnitudine eduli. Mamoera ipsa arbor (...) Clus.cur.post.C.Bauhin [1623:431].
Hermann 1687:483 Pepo arborescens Column. in Hernand. 870.
Ray 1688 vol.2:1370 Mamoeira foemina Marggrav.
J.Commelin 1689:34 = Syen in HM.
Sherard 1689:362 Pepo arborescens foemina sive fertilis Colum. in Hern.
C.Commelin 1696:51-52 Pepo arborescens & papaya orientalis Column. in not. Hernand. 870.
Plukenet 1696:145 Ficus arbor Utriusque Indiae Platani foliis monosteleches, fructu Mali Cydonii, aut Melonis magnitudine, Phytogr.Tab.278.fig.1.
Petiver 1702(1701):936-937,no.145 Poppoi-chedde Malab. Papaia foemina.
Boerhaave 1720 vol.2:170 Papaya; fructu Melopeponis effigie. Plum.659.
J.Burman 1737:184 Papaya, foemina.
Linnaeus 1737:461 Carica foliorum lobis sinuatis.
Linnaeus 1747:173-174,no.365 = Linnaeus 1737.
Linnaeus 1753:1036 Carica Papaya.
Linnaeus 1763:1466 = Linnaeus 1753.
N.L.Burman 1768:215 = Linnaeus 1763.
Houttuyn 1774 II vol.3:525-528 = N.L.Burman 1768.

Ily HM 1:25-26:16
Ily
BL Add MS 5028:32-33.
Syen in HM 1:26 Arundo Indica arborea maxima, cortice spinoso.
C.Commelin 1696:10 = Syen in HM.
Plukenet 1696:53 Arundo arborea Mambu vel Bambu dicta.
Plukenet 1700:28 Arundo arbor Indica procera, fructu Sesami in verticillas densius stipato, Müngell Malabaror.
J.Burman 1737:35-36 Arundo Indica, arborea maxima, cortice spinoso, Tabaxir fundens. Mus.Zeyl.pag.46 [Hermann 1717:46].
Linnaeus 1737:25 Arundo arbor. Bauh.pin.18 [Bauhin 1623:18].
Royen 1740:67 = Linnaeus 1737.
Rumphius 1743 vol.4:10-11 Arundarbor vasaria (?).
Rumphius 1743 vol.4:10-11,Tab.IV Arundarbor fera (?).
Linnaeus 1747:19,no.47 = Linnaeus 1737.
Linnaeus 1753:81 Arundo Bambos.
Linnaeus 1762:120 = Linnaeus 1753.
N.L.Burman 1768:30 = Linnaeus 1762.
Houttuyn 1774 II vol.2:58-62 = N.L.Burman 1768.

Malacca schambu HM 1:27-28:17
Malacca-Schambu
BL Add MS 5028:34-35.
Syen in HM 1:28 Persici ossiculo fructus Malacensis C.Bauhin [1623:441].
Ray 1688 vol.2:1478 Prunus Malabarica fructu umbilicato Pyriformi Jambos dicta minor.
C.Commelin 1696:56 = Ray 1688.
Plukenet 1696:237 = Syen in HM.
J.Burman 1737:124-125 Jambos Malaccensis, fructu aureo, rosam spirante. C.B. Mus.Zeyl.pag.19 [Hermann 1717:19-20].
Rumphius 1741 vol.1:123-124,Tab.XXXVII Jambosa Domestica (=); J.Burman ibid.:124 (=).
Linnaeus 1747:84,no.188 Eugenia foliis integerrimis, pedunculis ramosis terminantibus.
Linnaeus 1753:470 Eugenia Jambos.
Linnaeus 1762:672-673 = Linnaeus 1753.
N.L.Burman 1768:114 = Linnaeus 1762.
Houttuyn 1774 II vol.2:532-533 = N.L.Burman 1768.

Nati-schambu HM 1:29-30:18
Nati-schambu
BL Add MS 5028:36-37.
Syen in HM 1:30 Iambos Sylvatica, fructu cerasi magnitudine, Walgambu Cingalensium D.Hermans.
Ray 1688 vol.2:1478-1479 Jambos prior Acostae (...) Prunus Malabarica fructu umbilicato Pyriformi Jambos dicta, major.
C.Commelin 1696:56 = Ray 1688.
Plukenet 1696:262 = Syen in HM.

J.Burman 1737:125 Jambos sylvestris, fructu rotundo, Cerasi magnitudine. Mus.Zeyl.pag.67 [Hermann 1717:67].
Rumphius 1741 vol.1:123-124,Tab.XXXVII Jambosa Domestica (=); J.Burman ibid.:124 (=).
Rumphius 1741 vol.1:125-126,Tab.XXXVIII Jambosa nigra (-); J.Burman ibid.:126 (=).
Rumphius 1741 vol.1:127-128,Tab.XXXIX Jambosa Silvestris alba (-); J.Burman ibid.:128 (=).
Linnaeus 1747:83-84,no.187 Eugenia foliis integerrimis, pedunculis ramosis infra folia.
Linnaeus 1753:470 Eugenia malaccensis.
Linnaeus 1762:672 = Linnaeus 1753.
N.L.Burman 1768:114 = Linnaeus 1762.
Houttuyn 1774 II vol.2:529-532 = N.L.Burman 1768.

Champacam HM 1:31-32:19
Champacam
BL Add MS 5028:38-39.
Syen in HM 1:32 Champe dicuntur alij flores, quorum etiam magnus apud illos usus est, odore graviori, quam Lilium album, Garz.Fragos. C.Bauhin [1623:470].
Ray 1688 vol.2:1811 Champacca flos Bontii (c).
C.Commelin 1696:70 Uvifera arbor orientalis folio oblongo Par:Bat:P:385 [Sherard 1689:385].
Plukenet 1696:394 = C.Commelin 1696.
J.Burman 1737:31 Arbor procera Hapughala dicta, floribus oblongis, luteis, odoratis. Champacca Malabaris, Pau Fule Lusitanis. Mus.Zeyl.pag.64 [Hermann 1717:64].
Rumphius 1741 vol.2:199-201,Tab.LXVII Sampacca (=).
Linnaeus 1747:60,no.144 Michelia.
Linnaeus 1753:536 Michelia Champaca.
Linnaeus 1762:756 = Linnaeus 1753.
N.L.Burman 1768:124 = Linnaeus 1762.
Houttuyn 1774 II vol.3:75-76 = N.L.Burman 1768.

Elengi HM 1:33-34:20
Elengi
BL Add MS 5028:40-41.
Syen in HM 1:34 Oleae affinis Pyrifolio Malabarica, flore odorifero stellato.
Ray 1688 vol.2:1564-1565 Prunus Malabarica fructu calyculato.
C.Commelin 1696:57 = Ray 1688.
Plukenet 1696:203 Kauki Indorum Breyn.Cent.1.20 [Breyne 1678:20,tab.8].
Plukenet 1700:21 Arbor Indica pallidioribus foliis, fructu conoide ex magno calyce emergente. Cautmogullamaram Malabarorum (?).
Petiver 1702(1701):854-855,no.125 Mogula-maraum. Kauki Zeylanica folio auctiore.
J.Burman 1737:27 Arbor Zeylanica, floribus odoratis, faciem humanam quodammodo referentibus. Mus.Zeyl.pag.39 [Hermann 1717:39].
J.Burman 1737:133 = Plukenet 1696.
Rumphius 1741 vol.2:189-192,Tab.LXIII Flos Cuspidum (=); J.Burman ibid.:191 (=).
Linnaeus 1747:57-58,no.138 Mimusops foliis alternis remotis.
Linnaeus 1753:349 Mimusops Elengi.
Linnaeus 1762:497 = Linnaeus 1753.
N.L.Burman 1768:86 = Linnaeus 1762.
Houttuyn 1774 II vol.2:266-270 = N.L.Burman 1768.

Manjapumeram HM 1:35-36:21
Manjapumeram
BL Add MS 5028:42-43.
Syen in HM 1:36 Arbor tristis folio Myrti C.Bauhin [1623:469].
Ray 1688 vol.2:1698 = Syen in HM.
C.Commelin 1696:9 = Syen in HM.
Plukenet 1696:99 Chrysanthemum Orellanae foliis, impensè scabris, Maderaspatanum, florum petalis quinis, tetragonis, ex calice amplo viridi fistulosis, Phytogr.Tab.83.fig.2; ibid. 1700:46 Poulamulle Malabarorum (s).
Petiver 1702(1701):941,no.157 Poula-mullee Malab. Manja-pu Malabarica flore odoratissimo fugaci.
J.Burman 1737:32 Arbor tristis. Grimm.labor.Ceyl.pag.116 [Grimm 1677:75](c).
Linnaeus 1747:4-5,no.11 Nyctanthes caule tetragono, foliis ovatis, acuminatis, pericarpiis membranaceis compressis.
Linnaeus 1753:6 Nyctanthes arbor tristis.
Linnaeus 1762:8 = Linnaeus 1753.
N.L.Burman 1768:4 = Linnaeus 1762.
Houttuyn 1774 II vol.2:40-41 Nyctanthes arbor tristis Syst.Nat. XII.Tom.II.p.8 [Linnaeus 1767 vol.2:55].

Conna HM 1:37-38:22
Conna
BL Add MS 5028:44-45.
Syen in HM 1:38 Cassia fistula Alexandrina C.Bauhini [1623:403].
Hermann 1687:129 = Syen in HM.
Ray 1688 vol.2:1746-1747 = Syen in HM.
J.Commelin 1689:69-70 = Syen in HM.
C.Commelin 1696:19 = Syen in HM.
Plukenet 1696:89 = Syen in HM.
J.Commelin 1697:215-216,fig.110 = Syen in HM.
Petiver 1704(1702):1064-1065,no.246 Conea maraum Malab. Cassia Fistula Officinarum.
Boerhaave 1720 vol.2:58 = Syen in HM.
J.Burman 1737:56 Cassia Fistula, Zeylanica. Mus.Zeyl.pag. 21,28,& 32 [Hermann 1717:21,28,32].
Rumphius 1741 vol.2:83-87,Tab.XXI Cassia Fistula (=).
Linnaeus 1747:63-64,no.149 Cassia foliolis saepius quinque parium ovatis acuminatis glabris petiolatis.
Linnaeus 1753:377-378 Cassia fistula.
Linnaeus 1762:540 = Linnaeus 1753.
N.L.Burman 1768:96 = Linnaeus 1762.
Houttuyn 1775 II vol.5:19-26 Cassia Fistula [Linnaeus 1767 vol.2:289].

Balam-pulli seu Maderam-pulli HM 1:39-40:23
Balam pulli
BL Add MS 5028:46-47.
Syen in HM 1:40 Siliqua Arabica, quae Tamarindus C.Bauhini [1623:403].
Hermann 1687:588 Tamarindus.
Ray 1688 vol.2:1748-1749 Tamarindus Ger.Park.J.B.
J.Commelin 1689:341 = Syen in HM.
Sherard 1689:380 = Ray 1688.
C.Commelin 1696:65 = Hermann 1687.
Plukenet 1696:361 = Ray 1688.
Petiver 1702(1701):1009,1014,no.197 Pulea chedde Malab. Tamarindus vulgaris Officinarum.
J.Burman 1737:222 Tamarindus. Mus.Zeyl.pag.27 & 60 [Hermann 1717:27,60].
Rumphius 1741 vol.2:90-94,Tab.XXIII Tamarindus (=); J.Burman ibid.:94 (=).
Linnaeus 1737:18 = Ray 1688.
Linnaeus 1747:14,no.33 = Linnaeus 1737.
Linnaeus 1748:15 = Linnaeus 1737.
Linnaeus 1753:34 Tamarindus indica.
Linnaeus 1762:48-49 = Linnaeus 1753.

N.L.Burman 1768:15 = Linnaeus 1762.

Coddam-pulli HM 1:41-42:24
Coddam-pulli
BL Add MS 5028:48-49.
Syen in HM 1:42 Carcapuli, Acostae, Lugd.append.Cast. fructu malo aureo aemulo C.B.P. [Bauhin 1623:437].
Ray 1688 vol.2:1661 = Syen in HM.
C.Commelin 1696:17 Cambogium officinarum, Gutta Gamba, & Gutta Jemon Dale Pharm:443.
Plukenet 1696:41 Arbor Cambodiensis Guttam Gambi fundens, ex dicteriis, Cl.D.Herm. (? ibid.1700:20).
Plukenet 1696:81 = Syen in HM.
J.Burman 1737:27-28 Arbor Indica Gummi Guttam fundens, fructu dulci, rotundo, Cerasi magnitudine, Carcapuli Acostae. Mus.Zeyl.pag.26 [Hermann 1717:26].
Linnaeus 1747:87,no.195 = Cambogia.
Linnaeus 1762:728 Cambogia Gutta.
N.L.Burman 1768:119 = Linnaeus 1762.
Houttuyn 1774 II vol.3:4-6 = N.L.Burman 1768.

Atty-alu HM 1:43-44:25
Atti-alu
BL Add MS 5028:50-51.
Syen in HM 1:44 Ficus Malabarensis folio oblongo, acuminato, fructu vulgari aemulo.
Ray 1688 vol.2:1434 = Syen in HM.
C.Commelin 1696:28 = Syen in HM.
Plukenet 1705:87 Ficus Ind. Or. Lapathi foliis, fructu parvo orbiculari villoso, Atteemaram Malabarorum.
Rumphius 1743 vol.3:127-134,Tab.LXXXIV Varinga latifolia (=).
Linnaeus 1753:1060 Ficus racemosa.
Linnaeus 1763:1515 = Linnaeus 1753.
N.L.Burman 1768:226 = Linnaeus 1763.
Houttuyn 1774 II vol.3:680-682 = N.L.Burman 1768.

Itty-alu HM 1:45-46:26
Itty-alu
BL Add MS 5028:52-53.
Syen in HM 1:46 Ficus Malabarensis, folio densiusculo nitente, fructu parvo, rotundo, coronato.
Ray 1688 vol.2:1436 = Syen in HM.
C.Commelin 1696:29 = Syen in HM.
Plukenet 1696:42 Arbor Sycophora Jamaicensis foliis minoribus (...) Phytogr.Tab.266.fig.2 (?).
Plukenet 1696:145 Ficus arbor densioribus foliis parvis integris Ambonensis, Phytogr.Tab.243.fig.4 (?).
NB In some sets of Hortus Malabaricus the engraving of Ittyalu has been erroneously confounded with that of Nalugu in HM 2:43-44:26!

Arealu HM 1:47-48:27
Arealu
BL Add MS 5028:56-57.
Syen in HM 1:48 Ficus Malebarensis, folio cuspidato, fructu rotundo, parvo, gemino.
Ray 1688 vol.2:1434-1435 = Syen in HM.
C.Commelin 1696:28 = Syen in HM.
Linnaeus 1737:471 Ficus foliis cordatis integerrimis acuminatis.
Rumphius 1743 vol.3:142-145,Tab.XCI,XCII Arbor Conciliorum (=); J.Burman ibid.:145 (=).
Linnaeus 1747:177-178,no.372 = Linnaeus 1737.
Linnaeus 1753:1059 Ficus religiosa.
Linnaeus 1763:1514 = Linnaeus 1753.

N.L.Burman 1768:225 = Linnaeus 1763.
Houttuyn 1774 II vol.3:670-671 Ficus Religiosa [Linnaeus 1767 vol.2:681].

Peralu HM 1:49-50:28
Peralu
BL Add MS 5028:54-55.
Syen in HM 1:50 Ficus Malabarensis folio crassiusculo, majori, fructu gemino, intensè rubente.
Ray 1688 vol.2:1437 = Syen in HM.
C.Commelin 1696:28 = Syen in HM.
Plukenet 1696:144 Ficus Americana, latiori folio venoso ex Curacao (...) Phytogr.Tab.178.fig.1 (?).
Plukenet 1705:87 Ficus Ind. Or. latiori folio Maderaspatan, fructu rotundo, calyculato, sessili. Allamaram Malabarorum.
Linnaeus 1737:471 Ficus foliis ovatis integerrimis obtusis, caule inferne radicato (?).
Rumphius 1743 vol.3:145-148,Tab.XCIII Caprificus Amboinensis (cs).
Linnaeus 1753:1059-1060 Ficus benghalensis.
Linnaeus 1763:1514 = Linnaeus 1753.
Houttuyn 1774 II vol.3:672-674,Pl.XVII,Fig.2 Ficus Benghalensis Burm.Fl.Ind.p.225 [N.L.Burman 1768:225].

Bupariti HM 1:51-52:29
Bupariti
BL Add MS 5028:58-59.
Syen in HM 1:52 Alcea Malabarensis, Abutili folio, flore majore, ex albo flavescente.
Hermann 1687:22 Althaea arborea Indica Populi folio flore ephemero.
Ray 1688 vol.2:1069 = Syen in HM.
Breyne 1689:10 Alcea arborea Indica maxima, Populi folio, flore majore flavescente.
J.Commelin 1689:12 = Syen in HM.
Sherard 1689:307 = Hermann 1687.
C.Commelin 1696:2 = Breyne 1689.
Plukenet 1696:16-17 = Breyne 1689.
J.Burman 1737:136 Ketmia Zeylanica, sempervirens, & florens, Tiliae folio, flore luteo. Hert.Catal.semin.
Rumphius 1741 vol.2:218-221,Tab.LXXIII Novella (=).
Linnaeus 1747:118,no.258 Hibiscus foliis cordatis integerrimis.
Linnaeus 1753:694 Hibiscus populneus.
Linnaeus 1763:976 = Linnaeus 1753.
N.L.Burman 1768:150 = Linnaeus 1763.
Houttuyn 1775 II vol.5:397-398 = N.L.Burman 1768.

Pariti, seu Tali-pariti HM 1:53-54:30
Pariti
BL Add MS 5028:60-61.
Syen in HM 1:54 Alcea Malabarensis, Abutili folio, flore minore ex albo flavescente, exterius subaspero.
Hermann 1687:644 Althaea arbor Indica, Tiliae folio, flore ephemero.
Ray 1688 vol.2:1070 = Syen in HM.
Breyne 1689:10 Alcea arborea Indica, Tiliae folio crassiore, flore minore flavescente.
J.Commelin 1689:12 = Syen in HM.
Sherard 1689:307 = Hermann 1687.
C.Commelin 1696:2 = Breyne 1689.
Plukenet 1696:16 = Syen in HM.
Royen 1740:532 Hibiscus caula arboreo, foliis cordatosubrotundis acuminatis crenatis.
Rumphius 1741 vol.2:223 Novella Rubra (?=).
Linnaeus 1747:118-119,no.259 = Royen 1740.

Linnaeus 1753:694 Hibiscus tiliaceus.
Linnaeus 1763:976 = Linnaeus 1753.
N.L.Burman 1768:150 = Linnaeus 1763.
Houttuyn 1775 II vol.5:398-400 = N.L.Burman 1768.

Cudu-pariti HM 1:55-56:31
Cudu-pariti
BL Add MS 5028:62-63.
Syen in HM 1:56 Alcea Malabarensis Pentaphylla, flores minore ex albo flavescente, semine tomentoso.
Ray 1688 vol.2:1065 Xylon Malabaricum Cudu-Pariti dictum.
Plukenet 1691:Tab.CLXXXVIII,fig.3 Gossipium herbaceum, s. Xylon Maderaspatense, rubicundo flore pentaphyllaeum (?).
C.Commelin 1696:71 = Ray 1688.
Plukenet 1696:172 = Plukenet 1691.
J.Burman 1737:136 Ketmia folio quinquefido, subtus candidante, flore flavescente, semine tomentoso.
Linnaeus 1737:350 Gossypium caule erecto.
Rumphius 1743 vol.4:37,Tab.XIII Gossipium latifolium (cs).
Linnaeus 1747:122,no.267 = Linnaeus 1737.
Linnaeus 1753:693 Gossypium arboreum.
Linnaeus 1763:975 = Linnaeus 1753.
N.L.Burman 1768:150 = Linnaeus 1763.
Houttuyn 1775 II vol.5:390-392 = N.L.Burman 1768.

Chovanna-mandaru HM 1:57-58:32
Chovanna-mandaru
BL Add MS 5028:64-65.
Syen in HM 1:58 Arbor siliquosa Malabarica, foliis bifidis, flore purpurascente striato.
Ray 1688 vol.2:1751 = Syen in HM.
Breyne 1689:19 Arbor Sancti Thomae, Jacobi Zanonii.
C.Commelin 1696:8 = Breyne 1689.
Plukenet 1696:240 = Syen in HM.
Petiver 1702(1701):852,no.117 Aateener chedee Malab. Mandaru Chamberambaca foliis rigidis venosis, subtus pallescentibus (?).
Ray 1704 vol.3:112 = Petiver 1702(1701).
Linnaeus 1753:375 Bauhinia variegata.
Linnaeus 1762:535 = Linnaeus 1753.
N.L.Burman 1768:94 = Linnaeus 1762.
Houttuyn 1774 II vol.2:371-372 = N.L.Burman 1768.

Chovanna-mandaru HM 1:59-60:33
Chovanna-mandaru
BL Add MS 5028:66-67.
Syen in HM 1:60 Arbor siliquosa Malabarica, foliis majoribus bifidis, flore intensius purpurascente striato.
Ray 1688 vol.2:1751 = Syen in HM.
C.Commelin 1696:8 = Syen in HM.
Plukenet 1700:53 Convolvulus maritimus rotundifolius maximus, summo fastigio bisulcus, ex Insula Johanna (r).
Plukenet 1700:240 = Syen in HM.
Linnaeus 1753:375 Bauhinia purpurea.
Linnaeus 1762:536 = Linnaeus 1753.
N.L.Burman 1768:94 = Linnaeus 1762.
Houttuyn 1774 II vol.2:372-373 = N.L.Burman 1768.

Velutta-mandaru HM 1:61-62:34
Velutta-mandaru
BL Add MS 5028:68-69.
Syen in HM 1:62 Arbor siliquosa Malabarica, foliis bifidis, flore candido striato.
Ray 1688 vol.2:1751-1752 = Syen in HM.
Sherard 1689:380 Thomaea arbor siliquosa foliis bifidis minor flore candido.

C.Commelin 1696:8 = Syen in HM.
Plukenet 1696:240 = Syen in HM.
J.Burman 1737:45-46 Bauhinia foliis oblongo-acutis, nervosis, flore albo.
Linnaeus 1737:157 Bauhinia inermis, foliis cordatis fere semibifidis, laciniis acuminato-ovatis erectis dehiscentibus.
Linnaeus 1747:63,no.148 = Linnaeus 1737.
Rumphius 1747 vol.5:1-3,Tab.I Folium Linguae (r).
Linnaeus 1753:375 Bauhinia acuminata.
Linnaeus 1762:536 = Linnaeus 1753.
N.L.Burman 1768:94 = Linnaeus 1762.
Houttuyn 1774 II vol.2:375-376 Bauhinia acuminata Syst.Nat. XII [Linnaeus 1767 vol.2:288].

Canschena-pou HM 1:63-64:35
Canschena pou
BL Add MS 5028:70-71.
Syen in HM 1:64 Arbor siliquosa Malabarica, foliis bifidis minoribus, flore albo flavescente striato.
Ray 1688 vol.2:1752 = Syen in HM.
J.Commelin 1689:34 = Syen in HM.
C.Commelin 1696:8 = Syen in HM.
Plukenet 1696:240 = Syen in HM.
Petiver 1702(1701):856,no.128 Cheru-Mandaree Malab. Mandaru Unaneercoondica,floribus majoribus venosis, Crista Pavonis siliqua (?).
J.Burman 1737:44-45,tab.18* Bauhinia foliis subrotundis, flore flavescente striato.
Linnaeus 1737:157 Bauhinia foliis cordato-subrotundis, laciniis rotundatis.
Linnaeus 1747:62-63,no.147 = Linnaeus 1737.
Linnaeus 1753:375 Bauhinia tomentosa.
Linnaeus 1762:536 = Linnaeus 1753.
N.L.Burman 1768:94 = Linnaeus 1762.
Houttuyn 1774 II vol.2:373-374 = N.L.Burman 1768.

Marotti HM 1:65-66:36
Marotti
BL Add MS 5028:72-73.
Syen in HM 1:66 Laurifolia Malabarica, fructu ossea, nucleos continente.
Ray 1688 vol.2:1670 = Syen in HM.
C.Commelin 1696:40 = Syen in HM.

Caniram HM 1:67-68:37
Caniram
BL Add MS 5028:74-75.
Syen in HM 1:68 Malus Malabarica folio & fructu amaricante, semine plano compresso.
Ray 1688 vol.2:1661-1662 = Syen in HM.
Breyne 1689:92-93 Solanum arboreum Indicum maximum, foliis Oenopliae, sive Napecae majoribus, fructu rotundo duro rubro, semine orbiculari compresso maximis, Nuces vomicae & lignum colubrinum Officinarum ferens.
Sherard 1689:357 Nux Vomica major & officinarum.
C.Commelin 1696:63 = Breyne 1689.
Plukenet 1696:124 Cucurbitifera Malabariensis, Oenopliae subrotundis foliis, fructu orbiculari, rubro, cujus grana sunt Nuces Vomicae Officinarum.
J.Burman 1737:171-172 Nux Vomica Officinarum. Mus.Zeyl. pag.41 [Hermann 1717:41].
Rumphius 1741 vol.1:173-174,Tab.LXVII Vidoricum (-); J.Burman ibid.:174 (cs).
Rumphius 1741 vol.2:121-124,Tab.XXXVIII Lignum Colubrinum (-); J.Burman ibid.:124 (r).
Linnaeus 1747:37-38,no.91 Strychnos foliis ovatis, caule inermis.

Linnaeus 1753:189 Strychnos Nux vomica.
Linnaeus 1762:271 = Linnaeus 1753.
N.L.Burman 1768:58 = Linnaeus 1762.
Houttuyn 1774 II vol.2:116-117 Strychnos Nux Vomica = Linnaeus 1747.

Nilicamaram HM 1:69-70:38
Nilicamaram
BL Add MS 5028:76-77.
Syen in HM 1:70 Arbor Acaciae foliis Malabarica, fructu rotundo, semine triangulo.
Breyne 1680:42 Myrobalano Emblicae affinis, foliis Securidacae.
Ray 1688 vol.2:1556 = Syen in HM.
C.Commelin 1696:45-46 Myrobalanus emblica officinarum Dale Pharm:444.
Plukenet 1696:6,tab.1.fig.5 Acaciae similis non spinosa, ramulis alternis, Maderaspatana (?).
Plukenet 1700:133 Myrobalanus Emblica angustis Acaciae foliis, flosculis Tamarisci racemosis. Nellemaram Malabarorum.
Petiver 1702(1701):847-848,no.106 Nelle maraum Malab. = HM.
J.Burman 1737:5 Acacia Zeylanica floribus luteis, racematim ad foliorum (exortum) extremitates dispositis. Mus.Zeyl.pag.68 [Hermann 1717:68].
Linnaeus 1747:158,no.333 Phyllanthus foliis pinnatis floriferis, caule arboreo, fructu baccato.
Linnaeus 1753:982 Phyllanthus Emblica.
Rumphius 1755:1-2,Tab.I Mirobalanus Embilica (-); J.Burman ibid.:2 (=).
Linnaeus 1763:1393 = Linnaeus 1753.
N.L.Burman 1768:196 = Linnaeus 1763.
Houttuyn 1774 II vol.3:249-252 = N.L.Burman 1768.

Odallam HM 1:71-72:39
Odallam
BL Add MS 5028:78-79.
Syen in HM 1:72 Persicae similis angusti folia, osse cordis humani figura, binos nucleos continente.
Ray 1688 vol.2:1552 Manga fructu venenato, ossiculo cordiformi nucleo gemino.
C.Commelin 1696:43 Manga Sylvestris lactescens venenata, Jasmini flore & odore P.B.Prod.351 [Sherard 1689:351-352].
Plukenet 1696:241 Manghas Orientalis angustifolia ossiculo cordiformi binos nucleos continente.
Hermann 1717:32 Ghonkaduru. Arbor exitiosa lactescens, fructu Aurantii colore bipartito, flore Jasmini albo.
J.Burman 1737:151-152,tab.70,fig.1 Manghas lactescens, foliis Nerii crassis, venosis, Jasmini flore, fructu Persicae simili, venenato (c).
Rumphius 1741 vol.2:243-246,Tab.LXXXI Arbor Lactaria (?=); J.Burman ibid.:246 (c descr.).
Linnaeus 1747:44-45,no.106 Cerbera foliorum nervis transversalibus (?).
Linnaeus 1753:208-209 Cerbera Manghas.
Linnaeus 1762:303-304 = Linnaeus 1753.
N.L.Burman 1768:66-67 = Linnaeus 1762.
Houttuyn 1774 II vol.2:173-176 = N.L.Burman 1768.

Mail-anschi HM 1:73-74:40
Mail-anschi
BL Add MS 5028:80-81.
Syen in HM 1:74 Oxycanthae affinis Malabarica.
Ray 1688 vol.2:1573 Rhamnus Malabaricus fructu racemoso calyculato.
Plukenet 1691:Tab.CCXX,fig.1 Rhamno Malabaricae Mailanschi dictae similis è Maderaspatan.
C.Commelin 1696:58 = Ray 1688.

Plukenet 1696:318 = Plukenet 1691.
Rumphius 1743 vol.4:47,Tab.XVII Cyprus, Alcanna (=); J.Burman ibid.:47 (?=).
Linnaeus 1747:56,no.134 Lawsonia ramis spinosis.
Linnaeus 1753:349 Lawsonia spinosa.
Linnaeus 1762:498 = Linnaeus 1753.
N.L.Burman 1768:88 = Linnaeus 1762.
Houttuyn 1775 II vol.4:478-482 Lawsonia spinosa = Linnaeus 1747.

Cumbulu HM 1:75-75(1):41
Cumbulu
BL Add MS 5028:82-83.
Syen in HM 1:75(1) Nux Malabarica unduosa, flore cucullato.
Ray 1688 vol.2:1664 = Syen in HM.
Breyne 1689:41-42 Dictamno fortè affinis Indica arborescens, Lauri Americanae foliis; sive Ecbolium Zeylanicum foliis Laurinis (r).
J.Commelin 1689:8 Adhadota Zeylanensium, Herm.Cat. [Hermann 1687:642,tab.643] (?).
C.Commelin 1696:48 = Syen in HM.
Linnaeus 1763:868 Bignonia Catalpa (?).
N.L.Burman 1768:131 = Linnaeus 1763 (=).
Houttuyn 1774 II vol.3:104-106 = N.L.Burman 1768.
NB The pagination of the description of Cumbulu has been confused. The reader might be advised to quote HM 1:75-75(1):41.

Canschi HM 1:76-76(1):42
Canschi
BL Add MS 5028:84-85.
Syen in HM 1:76(1) Arbor racemosa Malabarica, fructu triquestro.
Ray 1688 vol.2:1711 = Syen in HM.
C.Commelin 1696:8 = Syen in HM.
Linnaeus 1753:1193 = Trevia nudiflora.
Linnaeus 1763:1661 = Linnaeus 1753.
N.L.Burman 1768:198 = Linnaeus 1763.
Houttuyn 1774 II vol.3:12-13 = N.L.Burman 1768.
NB The pagination of the description of Canschi has been confused here. The reader might be advised to quote HM 1:76-76(1):42.

Palega-pajaneli HM 1:77-78(1):43
Palega-pajaneli
BL Add MS 5028:86-87.
Syen in HM 1:78(1) Arbor siliquosa Malabarica, cordato folio, fructu maximo oblongo, plano, semine membranaceo.
Ray 1688 vol.2:1741 = Syen in HM.
Breyne 1689:34 Clematis arborea Malabarica maxima Juglandis folio, pinnis rotundioribus, flore albicante amplissimo foetido, siliqua compressa latissima.
C.Commelin 1696:20 = Breyne 1689.
Hermann 1698:50 Nerio Similis arbor C.B.P. [Bauhin 1623:464] (cs).
Linnaeus 1747:107,no.236 Bignonia foliis duplicato-pinnatis: foliolis integris acutis utrinque aequalibus.
Linnaeus 1753:625 Bignonia indica.
Linnaeus 1763:871 = Linnaeus 1753.
N.L.Burman 1768:131-132 = Linnaeus 1763.
Houttuyn 1774 II vol.3:108-109 = N.L.Burman 1768.
NB The description of Palega-pajaneli erroneously quotes Fig.44.

Pajaneli HM 1:79-80:44
Pajaneli

BL Add MS 5028:88-89.
Syen in HM 1:80 Arbor siliquosa Malabarica, folio majore mucronato, fructu maximo, oblongo, plano, semine membranaceo.
Ray 1688 vol.2:1741 = Syen in HM.
Breyne 1689:34 Clematis arborea Malabarica maxima foetida, Juglandis folio, pinnis longioribus amplissimis, flore albicante amplo, siliquis compressis latissimis.
C.Commelin 1696:20 = Breyne 1689.
Hermann 1698:50 Nerio Similis arbor C.B.P. [Bauhin 1623:464] (cs).
Linnaeus 1753:625 Bignonia indica β.
Linnaeus 1763:871 = Linnaeus 1753.
Houttuyn 1774 II vol.3:108-109 Bignonia Indica β Burm.Fl.ind. 131 [N.L.Burman 1768:131].
NB The description of Pajaneli erroneously quotes Fig.45.

Pala HM 1:81-82:45
Pala
BL Add MS 5028:90-91.
Syen in HM 1:82 Arbor Malabarica Pentaphyllos, lactescens, siliquis angustis, longissimis.
Ray 1688 vol.2:1749-1750 = Syen in HM.
Breyne 1689:77 Nerium lactescens Malabaricum maximum pentaphyllum polyanthemum, flore minimo racemoso odorato viridi-albicante, siliquis propendentibus longissimis.
C.Commelin 1696:47 = Breyne 1689.
Hermann 1698:44 Apocynum Malabaricum arborescens foliis quinis in orbem congestis.
Ray 1704 vol.3:534 = Hermann 1698.
Linnaeus 1737:76 Tabernaemontana foliis lanceolatis.
Royen 1740:413 = Linnaeus 1737.
Rumphius 1741 vol.2:246-249,Tab.LXXXII Lignum Scholare (=); J.Burman ibid.:249 (=).
Linnaeus 1753:210 Tabernaemontana citrifolia.
Linnaeus 1762:308 = Linnaeus 1753.
N.L.Burman 1768:69 (see Codaga-pala HM 1:85-86:47!).
Houttuyn 1774 II vol.2:188-190 = N.L.Burman 1768 (?).
NB The description of Pala erroneously quotes Fig.46.

Curutu-pala HM 1:83-84:46
Curutu-pala
BL Add MS 5028:92-93.
Syen in HM 1:84 Arbor Malabarica lactescens, fimbriato flore, fructu circa cuspidem reflexo.
Ray 1688 vol.2:1754 = Syen in HM.
C.Commelin 1696:8 = Syen in HM.
Plukenet 1696:43 Arbor Javanica Persicae foliis Herm.Cat.Mss. (?).
Rumphius 1741 vol.2:243-246,Tab.LXXXI Arbor Lactaria (-); J.Burman ibid.:246 (c fig.).
Linnaeus 1753:211 Tabernaemontana alternifolia.
Linnaeus 1762:308 = Linnaeus 1753.
N.L.Burman 1768:69 = Linnaeus 1762.
Houttuyn 1774 II vol.2:191-193 = N.L.Burman 1768.
NB The description of Curutu-pala erroneously quotes Fig.47.

Codaga-pala HM 1:85-86:47
Codaga-pala
BL Add MS 5028:94-95.
Syen in HM 1:86 Arbor Malabarica lactescens, jasmini flore odoro, siliquis oblongis.
Ray 1688 vol.2:1754 = Syen in HM.
Breyne 1689:76 Nerium lactescens Malabaricum, Chamaecerasi Alpinae, sive Mali aurantiae foliis rugosis, flore albo odorato, siliquis propendentibus longis.

C.Commelin 1696:47 = Breyne 1689.
Plukenet 1696:35 Apocynum arboreum, Lauri-folium, siliquis Nerii Maderaspatanum (?).
Hermann 1698:44 Apocynum erectum Malabaricum frutescens Iasmini flore candido.
Ray 1704 vol.3:534 = Hermann 1698.
J.Burman 1737:167,tab.77 Nerium Indicum, siliquis angustis, erectis, longis, geminis.
Linnaeus 1747:45,no.107 Nerium foliis ovatis acuminatis petiolatis.
Linnaeus 1753:209 Nerium antidysentericum.
Linnaeus 1762:306 = Linnaeus 1753.
N.L.Burman 1768:68 = Linnaeus 1762.
N.L.Burman 1768:69 Tabernaemontana citrifolia L. (apparently a mistake for Pala HM 1:81-82:45!).
Houttuyn 1775 II vol.4:391-392 = N.L.Burman 1768:68.
NB The description of Codaga-pala erroneously quotes Fig.48.

Tinda-parua HM 1:87-88:48
Tinda-parua
BL Add MS 5028:96-97.
Syen in HM 1:88 Arbor Malabarica Baccifera, cortice albicante, glomerato flore.
Ray 1688 vol.2:1569-1570 = Syen in HM.
C.Commelin 1696:8 = Syen in HM.
Plukenet 1696:368 = Syen in HM.
Linnaeus 1747:160,no.337 Morus foliis ovato-oblongis utrinque aequalibus serratis.
Linnaeus 1753:986 Morus indica.
Linnaeus 1763:1399 = Linnaeus 1753.
N.L.Burman 1768:62 Ceanothus asiaticus Linn.sp.284 [Linnaeus 1762:284].
N.L.Burman 1768:198 = Linnaeus 1763.
Houttuyn 1774 II vol.3:286-287 = N.L.Burman 1768:198.
NB The description of Tinda-parua erroneously quotes Fig.49.

Ana-parua HM 1:88:no figure

BL Add MS 5028:no drawing.
Plukenet 1700:13 = HM.
NB No description but see Ana-parva HM 7:75-76:40.

Cavalam HM 1:89-90:49
Cavalam
BL Add MS 5028:98-99.
Syen in HM 1:90 Nux Malabarica sulcata, mucilaginosa, fabacea.
Ray 1688 vol.2:1754-1755 Arbor siliquosa Malabarica pluribus ad singulos flores lobis.
C.Commelin 1696:8-9 = Ray 1688.
Plukenet 1696:266 = Syen in HM.
J.Burman 1737:84 Cydonia arbor Balanghas dicta (c).
J.Burman 1737:169-170 Nux Zeylanica, folio multifido, digitato, flore Merdam olente. Par.Bat.Pr.pag.357 [Sherard 1689:357] (cf).
Linnaeus 1747:166,no.350 Sterculia foliis ovalibus integerrimis alternis petiolatis, floribus paniculatis.
Linnaeus 1753:1007-1008 Sterculia Balangas.
Linnaeus 1763:1430 = Linnaeus 1753.
N.L.Burman 1768:307 = Linnaeus 1763.
Houttuyn 1774 II vol.3:435-436 = N.L.Burman 1768.
NB The description of Cavalam erroneously quotes Fig.50.

Amabalam HM 1:91-92:50
Amabalam
BL Add MS 5028:100-101.

Syen in HM 1:92 Mangae affinis, flore parvo stellato, nucleo majori osseo.
Ray 1688 vol.2:1551 = Syen in HM.
Sherard 1689:310 = HM.
C.Commelin 1696:43 = Syen in HM.
Plukenet 1696:307,tab.218,fig.3; ibid.1700:156 Prunus Americanus ossiculo magno ex filamentis lignosis reticulatim conflato Hughesij (?).
Rumphius 1741 vol.1:161-162,Tab.LX Condondum (=); J.Burman ibid.:162 (=).
NB The description of Ambalam erroneously quotes Fig.51.

Cat-ambalam HM 1:93:no figure

BL Add MS 5028:no drawing.
Ray 1688 vol.2:1552 = HM.
Plukenet 1691:Tab.CCIII,fig.7 Mangha sylvestris Indiae Orientalis Hort.Reg.Hampton.(?).
C.Commelin 1696:19 = HM.
Plukenet 1696:241 = Plukenet 1691.
Plukenet 1696:307; ibid.1700:156 Prunifera arbor Americana fructu luteo ovali, ossiculo majore, quorum nuclei ad Porcos saginandos ipsis glandibus referentur (?).
Rumphius 1741 vol.1:162,Tab.LXI Condondum Malaccense (-); J.Burman ibid.:163 (?=).
NB The description of Cat-ambalam erroneously quotes Fig.50, but there is no figure.

Agaty HM 1:95-96:51
Agati
BL Add MS 5028:102-103.
Syen in HM 1:96 Galegae affinis Malabarica arborescens, siliquis majoribus articulatis.
Breyne 1680:47 Sesban affinis Arbor Indiae Orientalis.
Ray 1688 vol.2:1734-1735 = Syen in HM.
C.Commelin 1696:61 = Breyne 1680.
Plukenet 1700:87 = Syen in HM.
Rumphius 1741 vol.1:188-190,Tab.LXVI,LXVII Turia (=).
Linnaeus 1753:722 Robinia grandiflora.
Linnaeus 1763:1060 Aeschynomene grandiflora.
N.L.Burman 1768:169 = Linnaeus 1763.
Houttuyn 1774 II vol.3:183-185 = N.L.Burman 1768.
NB The description of Agaty erroneously quotes Fig.53.

Cada-pilava HM 1:97-98:52
Cada-pilava
BL Add MS 5028:104-105.
Syen in HM 1:98 Conifera, Macandou Javanensium Bont.Hist. Natural.& Med. lib.VI.cap.VII.
Ray 1688 vol.2:1442 Arbor Indica fructu aggregato conoide Cada-Pilava dicta.
C.Commelin 1696:7 = Ray 1688.
Petiver 1702(1701):857,no.131 Noona chedde Malab. Macandou, Arbor Conifera major, Periclymeni flore (?).
Plukenet 1705:27,140 = Syen in HM.
Rumphius 1743 vol.3:158-159,Tab.XCIX Banducus latifolia (-); J.Burman ibid.:159 (=).
Linnaeus 1747:34,no.82 Morinda caule arboreo, pedunculis solitariis.
Linnaeus 1753:176 Morinda citrifolia.
Linnaeus 1762:250 = Linnaeus 1753.
N.L.Burman 1768:52 = Linnaeus 1762.
Houttuyn 1774 II vol.2:102-103 = N.L.Burman 1768.

Appel HM 1:99-100:53
Appel

BL Add MS 5028:106-107.
Syen in HM 1:100 Arbor Malabarica baccifera, flore parvo, umbellato, odoro.
Ray 1688 vol.2:1598 Arbor baccifera Malabarica flore umbellato odoro, fructu rotundo monopyreno. Tetragonia Indica.
C.Commelin 1696:8 = Syen in HM.
Plukenet 1696:38 = Syen in HM.
Petiver 1699(1698):320,no.15 Corain-cheddee Malab. Baccifera racemosa Madraspatana Lauri Ceras foliis floribus parvis numerosissimis (s).

Ameri HM 1:101-102:54
Ameri
BL Add MS 5028:108-109.
Syen in HM 1:102 Isatis Indica folijs Rorismarini, Glasto affinis C.B. [Bauhin 1623:113].
Ray 1686 vol.1:926-927 Nil sive Anil, Glastum Indicum Park.
Hermann 1687:168-169,660 Colutea Indica herbacea, ex qua Indigo.
J.Commelin 1689:93 = Hermann 1687.
C.Commelin 1696:21 = Hermann 1687.
Petiver 1702(1700):703-704,no.57 The true Indigo. Indigo vera Coluteae foliis Utriusq; Indiae.
J.Burman 1737:69 Colutea Indica, humilis, ex qua Indigo folio viridi, Mus.Zeyl.pag.32.& Par.Bat.Pr.pag.325 [Hermann 1717: 32; Sherard 1689:325].
Linnaeus 1737:487 Indigofera foliis nudis.
Linnaeus 1747:125,no.273 Indigofera leguminibus arcuatis incanis, racemis, folio brevioribus.
Rumphius 1747 vol.5:220-225,Tab.LXXX Indicum (?=).
Linnaeus 1748:208-209 = Linnaeus 1747.
Linnaeus 1753:751 Indigofera tinctoria.
Linnaeus 1763:1061-1062 = Linnaeus 1753.
N.L.Burman 1768:170 = Linnaeus 1763.
Houttuyn 1775 II vol.5:540-542 = N.L.Burman 1768.

Colinil HM 1:103-104:55
Colinil
BL Add MS 5028:110-111.
Syen in HM 1:104 Polygala Indica minor, siliquis recurvis.
Ray 1688 vol.2:1734 = Syen in HM.
C.Commelin 1696:55 = Syen in HM.
Plukenet 1696:112 Colutea Indica, fruticescens, foliis supernè glabris, virentibus, subtus sericeo nitore argenteo splendentibus.
Petiver 1702(1700):715-716,no.84 Coolauvalle Malab. = HM.
Plukenet 1705:163 Ornithopodium Indicum, argenties foliis, flore purpureo, siliquâ singulari.
Rumphius 1747 vol.5:220-225,Tab.LXXX Indicum (c).

Schageri-cottam HM 1:105-106:56
Schageri-Cattam
BL Add MS 5028:112-113.
Syen in HM 1:106 Cornus Malabarica, folio cuspidato, ossiculo tomentoso obsito.
Ray 1688 vol.2:1553 = Syen in HM.
C.Commelin 1696:23 = Syen in HM.
Linnaeus 1747:92-93,no.207 Microcos panicula terminatrice.
Linnaeus 1753:514 Microcos paniculata.
Linnaeus 1762:733 = Linnaeus 1753.
N.L.Burman 1768:121 = Linnaeus 1762.
Houttuyn 1774 II vol.3:242-243 Grewia Microcos Syst.Nat.XII. Veg.XIII [Linnaeus 1767 vol.2:602].

Carua HM 1:107-110:57
Carua

BL Add MS 5028:114-115.
Syen in HM 1:110 Canella Malabarica.
Hermann 1687:129-130 Cassia Cinamomea.
Ray 1688 vol.2:1560 Cinnamomum sive Canella Malavarica & Javanensis C.B. [Bauhin 1623:409].
Breyne 1689:18 Arbor canellifera Malabarica, cortice ignobiliore, cujus folium Malabathrum Officinarum.
C.Commelin 1696:19 = Hermann 1687.
Plukenet 1696:88 Cassia Cinamomea, sylvestris, pigrior, Malavarica.
Linnaeus 1737:154-155 Laurus foliis oblongo-ovatis trinerviis nitidis planis.
Linnaeus 1747:61-62,no.146 Laurus foliis lanceolatis trinerviis, nervis basin unitis.
Linnaeus 1753:369 Laurus Cassia.
Linnaeus 1762:528 = Linnaeus 1753.
N.L.Burman 1768:91-92 = Linnaeus 1762.
Houttuyn 1774 II vol.2:327-338 = N.L.Burman 1768.

VOLUME 2

Kaida HM 2:1-2:2-5
Kaida
BL Add MS 5028:118-119,120-121,122-123,124-125.
Ray 1688 vol.2:1442 Frutex Indicus fructu aggregato conoide Kaida dictus.
C.Commelin 1696:5 Ananas Sylvestris, folio aloes, fructu cupressino J:B:T:3.L:XXV:
Plukenet 1696:277-278 Palmae affinis Arbor conifera Mascatensis, longissimo folio tribus ordinibus spinarum munito.
Plukenet 1700:13 = C.Commelin 1696.
Rumphius 1743 vol.4:139-142,Tab.LXXIV Pandanus verus (?c).
Linnaeus 1747:54-55,no.131 Bromelia foliis margine dorsoque aculeatis, caule fulcrato spinoso.

Kaida taddi HM 2:3:1,6
Kaida-taddi, Kaida-Taddi
BL Add MS 5028:117,126-127.
J.Commelin in HM 2:2 Sedum majus arborescens C.Bauhinus [1623:282].
Ray 1688 vol.2:1443 = HM.
C.Commelin 1696:38 = HM.
Plukenet 1696:277-278; ibid.1700:145 Palmae affinis Arbor conifera Mascatensis, longissimo folio tribus ordinibus spinarum munito (?).
J.Burman 1737:20 Ananas sylvestris, arborescens Acostae. Mus.Zeyl.pag.55 [Hermann 1717:55].
Rumphius 1743 vol.4:143-145,Tab.LXXVI Pandanus humilis (?c).

Perin-kaida-taddi HM 2:5:7
Perin-Kaida Taddi
BL Add MS 5028:128-129.
J.Commelin in HM 2:2 Sedum majus arborescens C.Bauhinus [1623:282].
Ray 1688 vol.2:1443 = HM.
C.Commelin 1696:52 = HM.

Kaida tsjerria HM 2:7:8
Kaida-Tsjerria
BL Add MS 5028:130-131.
J.Commelin in HM 2:2 Sedum majus arborescens C.Bauhinus [1623:282].
Ray 1688 vol.2:1443 = HM.
C.Commelin 1696:38 = HM.

Rumphius 1743 vol.4:149,Tab.LXXIX Pandanus Ceramicus (?c).

Panel HM 2:9-10:9
Panel
BL Add MS 5028:132-133.
J.Commelin in HM 2:12 Arbor Indica fraxino similis, oleae fructu Bauhini [1623:416] (s).
Ray 1688 vol.2:1566 Prunifera racemosa Malabarica fructu compresso, nucleo rotundo.
C.Commelin 1696:56 = Ray 1688.

Narum-panel HM 2:11-12:10
Narum-panel
BL Add MS 5028:134-135.
Ray 1688 vol.2:1636-1637 Frutex baccifer fructu ad singulos flores multiplici.
C.Commelin 1696:30 = Ray 1688.
Petiver 1702(1701):851,no.114 Ashoga-maraum Malab. Panel Madraspat. fol. angustissimo mucronato fructu majore Musei Petiver 666 (?).
Hermann 1717:31 Palukaena.
J.Burman 1737:231 Uva Zeylanica, sylvestris, Mali Armeniacae sapore, Uves de Mato Lusitanis. Mus.Zeyl.pag.8 & 31 [Hermann 1717:8,31].
Linnaeus 1747:100,no.224 Uvaria.
Linnaeus 1753:536 Uvaria zeylanica.
Linnaeus 1762:756 = Linnaeus 1753.
N.L.Burman 1768:124 = Linnaeus 1762.
Houttuyn 1774 II vol.3:77-81 = N.L.Burman 1768.

Cara-nosi HM 2:13-14:11
Cara-nosi
BL Add MS 5028:136-137.
J.Commelin in HM 2:14 Piperi similis fructus striatus faemina Casp.Bauhini [1623:412].
Ray 1688 vol.2:1575 Frutex Indicus baccifer fructu calyculato monopyreno, Negundo dicta.
Breyne 1689:104 Vitex trifolia minor Indica.
C.Commelin 1696:69 = Breyne 1689.
Plukenet 1696:390,tab.206,fig.5 = Breyne 1689.
J.Commelin 1697:181-182,fig.93 = C.Commelin 1696.
Boerhaave 1720 vol.2:222 = C.Commelin 1696.
J.Burman 1737:229 Vitex trifolia, Indica, odora, hortensis, floribus coeruleis, racemosis. Mus.Zeyl.pag.48 [Hermann 1717:48].
Rumphius 1743 vol.4:50,Tab.XVIII Lagondium vulgare (=).
Linnaeus 1747:194,no.413 = J.Burman 1737.
Linnaeus 1753:638 Vitex trifoliis.
Linnaeus 1763:890 Vitex trifolia.
N.L.Burman 1768:137 = Linnaeus 1763.
Houttuyn 1775 II vol.5:347-349 = N.L.Burman 1768.

Bem-nosi HM 2:15:12
Bem-nosi
BL Add MSS 5028:138-139.
J.Commelin in HM 2:15 Negunda foemina Acosta.
Ray 1688 vol.2:1575 Negundo mas Garciae & Acostae.
Breyne 1689:104 Vitex trifolia minor Indica serrata.
C.Commelin 1696:69.
Plukenet 1696:390 = C.Commelin 1696.
Plukenet 1696:319; ibid.1700:161 Rhus Afric. trifoliat majus foliis subtùs argenteis, acutis & margine incisis. Phytogr.Tab.219. fig.6.
J.Commelin 1697:179-180,fig.92 = C.Commelin 1696; Kiggelaer & Ruysch ibid.:180 (≠).
Boerhaave 1720 vol.2:222 = C.Commelin 1696.

J.Burman 1737:229 Vitex trifolia, odorata, sylvestris, Indica. Mus.Zeyl.pag.47 [Hermann 1717:47].
Rumphius 1743 vol.4:51,Tab.XIX Lagondium litoreum (=); J.Burman ibid.:51 (r).
Linnaeus 1747:100-101,no.414 = J.Burman 1737.
Linnaeus 1753:638 Vitex Negundo.
Linnaeus 1763:890 = Linnaeus 1753.
N.L.Burman 1768:138 = Linnaeus 1763.
Houttuyn 1775 II vol.5:349-351 = N.L.Burman 1768.

Schetti HM 2:17-18:13
Schetti
BL Add MS 5028:140-141.
Ray 1688 vol.2:1573-1574 Frutex baccifer Malabaricus fructu calyculato, rotundo, rubro, polypyreno.
Breyne 1689:58 Jasminum arborescens Indicum, flore tetrapetalo umbellato phoeniceo, foliis Laurinis latioribus.
Sherard 1689:342 Jasminum Indicum, Laurifolio inodorum umbellatum floribus coccineis.
Plukenet 1691:Tab.LIX,f.2 = Sherard 1689.
C.Commelin 1696:36 = Breyne 1689.
Plukenet 1696:196 = Sherard 1689.
Petiver 1702(1701):1014,no.198 Chegga pu melleha Malab. Schetti Malabar. foliis Laurinis venosis.
J.Burman 1737:125-126,tab.57 Jasminum flore tetrapetalo, Ixora Linnaei, Schetti H.Malab.
Linnaeus 1747:22,no.54 Ixora foliis ovalibus semiamplexicaulibus.
Linnaeus 1753:110 Ixora coccinea.
Linnaeus 1762:159 = Linnaeus 1753.
N.L.Burman 1768:34 = Linnaeus 1762.
Houttuyn 1775 II vol.4:125-127 = N.L.Burman 1768.

Bem-schetti HM 2:19:14
Bem-schetti
BL Add MS 5028:142-143.
Ray 1688 vol.2:1574 = HM.
Plukenet 1691:Tab.CIX,f.2 Schetti album, s. Jasminum Indicum Lauri fol. inodorum, umbellatum, floribus albicantibus P.B.P.['floribus albicantibus' not in Sherard 1689:342].
C.Commelin 1696:37 = Plukenet 1691.
Plukenet 1696:196 = Plukenet 1691.
J.Burman 1737:126-127 Jasminum flore tetrapetalo, flavo, Bem-Schetti H.Malab.
Linnaeus 1747:22-23,no.55 Ixora foliis ovato-lanceolatis.
Linnaeus 1753:110 Ixora alba.
Linnaeus 1762:160 = Linnaeus 1753.
N.L.Burman 1768:34 = Linnaeus 1762.
Houttuyn 1775 II vol.4:127 Ixora alba [Linnaeus 1767 vol.2:120].

Nedum-schetti HM 2:21-22:15
Nedum-schetti
BL Add MS 5028:144-145.
Ray 1688 vol.2:1500 Frutex Indicus baccifer, floribus verticillatis fructu monopyreno.
C.Commelin 1696:31 = Ray 1688.

Scherunam-cottam HM 2:23-24:16
Scherunam-Cottam
BL Add MS 5028:146-147.
J.Commelin in HM 2:24 Corni seu Sorbi species Bont.lib.6.cap.12.
Ray 1688 vol.2:1623 Baccifera Indica fructu tetraspermo in foliorum alis sessili.
C.Commelin 1696:12 = Ray 1688.

Plukenet 1696:43; ibid.1700:21 Arbor Indica Mali aurantiae foliis obtusioribus è Maderaspatan. Phytogr.Tab.142.fig.2.
Linnaeus 1747:175,no.367 Clutia foliis ovalibus petiolatis retusis, floribus racemosis sessilibus.
Linnaeus 1753: Clutia retusa.
Linnaeus 1763:1475-1476 = Linnaeus 1753.
N.L.Burman 1768:217 = Linnaeus 1763.
Houttuyn 1776 II vol.6:402-403 = N.L.Burman 1768.

Schem pariti HM 2:25-26:17
Schem-pariti
BL Add MS 5028:148-149.
J.Commelin in HM 2:26 Rosa-Batavico-Indica Modoru Bont.lib.6.cap.46.
Ray 1688 vol.2:1068 Alcea Javanica arborescens flore pleno rubicundo Breynii [Breyne 1678:121-124,fig.].
C.Commelin 1696:2 = Ray 1688.
Plukenet 1696:14 = Ray 1688.
J.Burman 1737:133-134 Ketmia Sinensis, fructu subrotundo, flore pleno. Tournef.inst.pag.100.
Rumphius 1743 vol.4:24-26,Tab.VIII Flos festalis (=).
Linnaeus 1747:119,no.260 Hibiscus foliis ovatis acuminatis serratis.
Linnaeus 1753:694 Hibiscus Rosa sinensis.
Linnaeus 1763:977 = Linnaeus 1753.
N.L.Burman 1768:151 = Linnaeus 1763.
Houttuyn 1775 II vol.5:400-403 Hibiscus Rosa Sinensis [Linnaeus 1767 vol.2:463].

Belilla HM 2:27-28:18
Belilla
BL Add MS 5028:150-151.
J.Commelin in HM 2:28 Solanum Mexiocanum flore magno Casp.Bauhini [1623:168].
Ray 1688 vol.2:1493-1494 Frutex Indicus baccifer fructu oblongo polyspermo.
C.Commelin 1696:31 = Ray 1688.
Plukenet 1696:106; ibid.1700:49 Cistus sempervirens, Laurifolia, floribus elegantèr, Virginiana, D.Banister. Phytogr.Tab.161.fig.3.
Hermann 1717:36 Mussaenda.
J.Burman 1737:165-166,tab.76 Mussaenda Zeylanica, flore rubro, fructu oblongo, polyspermo, folio ex florum thyrso prodeunte albo. Herb.Hart.
Rumphius 1743 vol.4:112,Tabl.LI Folium principissae (=).
Linnaeus 1747:35,no.84 = Hermann 1717.
Linnaeus 1753:177 Mussaenda frondoso.
Linnaeus 1762:251-252 Mussaenda frondosa.
N.L.Burman 1768:53 = Linnaeus 1762.
Houttuyn 1775 II vol.4:226-227 = N.L.Burman 1768.

Modera-canni HM 2:29-30:19
Modera-Canni
BL Add MS 5028:152-153.
J.Commelin in HM 2:30 Mystax sp.
Ray 1688 vol.2:1570 Frutex baccifer Malabaricus fructu calyculato, rotundo, monopyreno. Mystax dictus.
C.Commelin 1696:30-31 = Ray 1688.
Plukenet 1696:350 Solanum arborescens è Vera-cruce latifolium (?).
Linnaeus 1747:113,no.249 Hugonia spinis oppositis revolutis.
Linnaeus 1753:675 Hugonia Mystrax.
Linnaeus 1763:944-945 = Linnaeus 1753.
N.L.Burman 1768:144 = Linnaeus 1763.
Houttuyn 1774 II vol.3:131-133 = N.L.Burman 1768.

Carim-curini HM 2:31-32:20
Carim-Curini
BL Add MS 5028:154-155.
Ray 1688 vol.2:1709 Frutex Indicus spicatus floribus galeatis, vasculo bivalvi dicocco.
Plukenet 1691:Tab.CLXXI,fig.4 Curini (fortè) prima species (...) à Maderaspatan.
C.Commelin 1696:32 = Ray 1688.
Plukenet 1696:9; ibid.1700:4 Adhatode Zeylonensium, Mus.Zeyl. (...) Phytogr.Tab.173.fig.3 (affinis).
Plukenet 1696:126 Curini forte, prima species (...) à Maderaspatan. Phytogr.Tab.171.fig.4.
Hermann 1717:33 Arbor Zeylanica floribus caeruleis in spica dispositis.
J.Burman 1737:7-8,tab.4,fig.1 Adhatoda spica longissima, flore reflexo.
Linnaeus 1737:9-10 Justicia foliis ovato-lanceolatis, spicis foliolis, florum galea concava.
Linnaeus 1747:6-7,no.17 Justicia arborea, foliis lanceolato-ovatis, bracteis acumine ovatis deciduis, corollarum galea reflexa.
Linnaeus 1753:15 Justicia Ecbolium.
Linnaeus 1762:20-21 = Linnaeus 1753.
N.L.Burman 1768:7 = Linnaeus 1762.
Houttuyn 1775 II vol.4:36-38 = N.L.Burman 1768.

Bem-curini HM 2:33-34:21
Bem-curini
BL Add Ms 5028:156-157.
Ray 1688 vol.2:1709 Frutex Indicus spicatus florum pediculis brevioribus.
Breyne 1689:41-42 Dictamno fortè affinis Indica arborescens, Lauri Americanae foliis; sive Ecbolium Zeylanicum foliis Laurinis.
C.Commelin 1696:32 = Ray 1688.
Boerhaave 1720 vol.1:239 Adhatoda; Indica; folio saligno; florea albo (?).
Linnaeus 1747:7,no.18 Justicia foliis lanceolato-ovatis, bracteis ovatis acuminatis venoso-reticulatis coloratis.
Linnaeus 1753:15 Justicia Betonica.
Linnaeus 1762:21 = Linnaeus 1753.
N.L.Burman 1768:8 = Linnaeus 1762.
Houttuyn 1775 II vol.4:38-39 = N.L.Burman 1768.

Caretti HM 2:35-36:22
Caretti
BL Add MS 5028:158-159.
J.Commelin in HM 2:36 Arbor exotica spinosa folijs lentisci C.B.P. [Bauhin 1623:399].
Breyne 1680:40,fig.5 Inimboy Brasilianorum, Frutex spinosus spicatus, platylobis echnoidibus, Glycyrrhizae foliis.
Ray 1688 vol.2:1743-1744 = J.Commelin in HM.
Breyne 1689:38 Crista pavonis Glycyrrhizae folio, minor, repens, spinosissima, flore luteo spicato minimo, siliquâ latissimâ echinatâ, semine rotundo cinereo, lineis circularibus cincto, majore.
J.Commelin 1689:33 = J.Commelin in HM.
Sherard 1689:348 Lobus echinatus fructu coesio foliis longioribus.
Plukenet 1691:Tab.II,f.2 Acacia gloriosa Leutisci (!) fol: spinos. flore spicato luteo, siliquâ magnâ muricatâ.
C.Commelin 1696:24 = Breyne 1689.
Plukenet 1696:4 = Plukenet 1691.
Petiver 1702(1700):702-703,no.56 Ash-coloured Nicker Tree. Bonduch cinerea fol. longioribus.
J.Burman 1737:4 Acacia, qui lobus echinatus Clusii, oculus Cati Lusitanis. Mus.Zeyl.pag.57 [Hermann 1717:57].
Linnaeus 1737:158 Guilandina caule fructuque aculeatis.
Linnaeus 1747:68,no.156 Guilandina aculeata, foliolis ovalibus cum acumine.
Rumphius 1747 vol.5:89-91 Frutex Globulorum, Tab.XLVIII (-); J.Burman ibid.:91 (v).
Linnaeus 1748:101-102 = Linnaeus 1747.
Linnaeus 1753:381 Guilandina Bonduc.
N.L.Burman 1768:99 Guilandina Bonducella Linn.sp.545 [Linnaeus 1762:545].
Linnaeus 1771:378 Guilandina Bonducella.
Houttuyn 1775 II vol.5:46-48,Pl.XXIV,A = N.L.Burman 1768.

Cupi HM 2:37-38:23
Cupi
BL Add MS 5028:160-161.
Ray 1688 vol.2:1494 Frutex Indicus baccifer floribus umbellatis fructu rotundo polyspermo.
C.Commelin 1696:31 = Ray 1688.
Petiver 1702(1701):1010-1011,no.190 Pautan-chedde Malab. Baccifera racemosa Madraspatana Juglandis folio, nigris maculis eleganter asperis. H.Un.5.Act.Phil.No.244.p.315.5 (?).
Linnaeus 1747:33-34,no.80 Rondeletia foliis petiolatis.
Linnaeus 1753:172 Rondeletia asiatica.
Linnaeus 1762:244 = Linnaeus 1753.
N.L.Burman 1768:51 = Linnaeus 1762.
Houttuyn 1775 II vol.4:194-197,Pl.XX,Fig.1 = N.L.Burman 1768.

Cattu-schiragam HM 2:39-40:24
Cattu-schiragam
BL Add MS 5028:162-163.
Breyne 1680:39-40 Jaceae & Serratulae affinis, capitulis Baccharidis, Trachelii foliis, Zeilanica.
Hermann 1687:334,676,677(fig.) = Breyne 1680.
Ray 1688 vol.2:1443-1444 Scabiosa Indica arborea.
C.Commelin 1696:61 Serratula indica major latifolia mollis Breyn.P.2.90 [Breyne 1689:90].
Plukenet 1696:140 Eupatoria Conyzoides, integro Jacobaeae folio, molli, & incano, Indiae Orient. Phytogr.Tab.177.fig.1 (?).
Hermann 1698:158 = Hermann 1687.
J.Burman 1737:210,tab.95 Scabiosa Conyzoides, foliis latis, dentatis, semine amaro, lumbricos enecante.
Linnaeus 1763:1207 Conyza anthelmintica.
N.L.Burman 1768:178 = Linnaeus 1763.
Houttuyn 1776 II vol.6:89 = N.L.Burman 1768.

Peragu HM 2:41-42:25
Peragu
BL Add MS 5028:164-165.
Ray 1688 vol.2:1571 Frutex baccifer Malabaricus, floribus pentapetalos, binis, una bacca nigra in calyce stelliformiter expanso.
C.Commelin 1696:31 = Ray 1688.
Plukenet 1700:19 Arbor baccifera Abutili foliis lanugine ferruginea villosis, Punnanganarre Malabar. (?).
J.Burman 1737:66-67,tab.29 Clerodendron folio lato & acuminato.
Linnaeus 1747:104-105,no.232 Clerodendrum foliis simplicibus cordatis tomentosis.
Linnaeus 1753:637 Clerodendrum infortunata.
Linnaeus 1763:889 = Linnaeus 1753.
N.L.Burman 1768:137 = Linnaeus 1763.
Houttuyn 1775 II vol.5:339-341 = N.L.Burman 1768.

Nalugu HM 2:43-44:26
Nalugu
BL Add MS 5028:166-167.
Ray 1688 vol.2:1635 Frutex baccifer Malabar. Floribus umbellatis pentapetalis, fructu nigricante polyspermo.
C.Commelin 1696:30 = Ray 1688.
Plukenet 1696:48; ibid.1700:27 Arbor Americana convolvulacea (...) Phytogr.Tab.146.fig.1 (r).
Plukenet 1700:40 Castanae facie diphyllos Arbor Aethiopica Capitis Bonae Spei (?).
Linnaeus 1753:441 Phytolacca asiatica (?).
Houttuyn 1774 II vol.2:223-225 Aralia Chinensis Burm.Fl.Ind.78 [N.L.Burman 1768:78].
NB In some sets of HM the engraving of Nalugu has been erroneously confounded with that of Itty-alu in HM 1:45-46:26!

Niruri HM 2:45-46:27
Niruri
BL Add MS 5028:168-169.
J.Commelin in HM 2:46 Frutex indicus baccifer, Vitis idaeae secundae Clusii foliis Breyne Cent:8 [Breyne 1678:8-9,fig.].
Ray 1688 vol.2:1635-1636 = J.Commelin in HM.
Sherard 1689:384-385 Viti Ideae similis frutex Africanus.
Plukenet 1691:Tab.LXIX,f.3 Vitis Idaeae species Maderaspatana Niruri fortè Malabarensibus dicta.
C.Commelin 1696:31 = J.Commelin in HM.
Plukenet 1696:391 Vitis Idaea Maderaspatensis Niruri (fortè) Malabaraeis dicta Phytogr.Tab.69.fig.3.
Petiver 1704(1703):1458,no.37 Nirouri Madraspat.niger, fructu pyramidali.

Hummatu HM 2:47-48:28
Hummatu
BL Add MS 5028:170-171.
J.Commelin in HM 2:48 Solanum foetidum pomo spinoso oblongo C.B.P. [Bauhin 1623:168].
Ray 1686 vol.1:749 Daturae Malabaricae Hummatu dictae prima species.
Hermann 1687:583 Stramonia, seu Datura, pomo spinoso rotundo longiflore.
J.Commelin 1689:338 Stramonia multis dictum, sive Pomum spinosum I.Bauh.
C.Commelin 1696:64 = Hermann 1687.
Plukenet 1696:358 Stramonia s. Dutroa fructu spinoso rotundo, flore candido.
J.Burman 1737:221 Stramonium Zeylanicum. Mus.Zeyl.pag.69 [Hermann 1717:69].
Linnaeus 1737:55 Datura pericarpiis nutantibus globosis.
Linnaeus 1747:35-36,no.86 = Linnaeus 1737.
Rumphius 1747 vol.5:242-246,Tab.LXXXVII Stramonia Indica (=).
Linnaeus 1748:44 = Linnaeus 1737, 1747.
Linnaeus 1753:179 Datura Metel.
Linnaeus 1762:256 = Linnaeus 1753.
N.L.Burman 1768:53 = Linnaeus 1762.
Houttuyn 1777 II vol.7:634-636 = N.L.Burman 1768.

Nila-hummatu HM 2:49-50:29
Nila hummatu
BL Add MS 5028:172-173.
J.Commelin in HM 2:48 Stramonia sp.
Ray 1686 vol.1:749 Daturae Malabaricae secunda species.
Hermann 1687:584 Stramonia foetida Malabarica, semine pallido, pomo glabro, flore simplici violaceo.
C.Commelin 1696:64 = Hermann 1687.
Plukenet 1700:358 Stramonia Indica, fructu oblongo glabro.
J.Burman 1737:221 Stramonium Zeylanicum. Mus.Zeyl.pag.69 [Hermann 1717:69].

Mudela-nila-hummatu HM 2:51-52:30
Mudela-nila-hummatu
BL Add MS 5028:174-175.
J.Commelin in HM 2:48 Stramonia sp.
Ray 1686 vol.1:750 Daturae Malabaricae tertia species.
Hermann 1687:584 Stramonia foetida Malabarica, semine pallido, pomo glabro, flore duplici triplicive.
C.Commelin 1696:64 = Hermann 1687.
Plukenet 1700:176 Stramonia Indica, fructu oblongo glabro (r).
J.Burman 1737:221 Stramonium Zeylanicum. Mus.Zeyl.pag.69 [Hermann 1717:69].
Rumphius 1747 vol.5:242-246,Tab.LXXXVII Stramonia Indica (=).

Ericu HM 2:53-55:31
Ericu
BL Add MS 5028:176-177.
J.Commelin in HM 2:56 Apocynum AEgyptiacum lactescens siliqua Asclepiadis. Beidel-sar Alpini. C.B.P. [Bauhin 1623:303].
Hermann 1687:52 Apocynum latifolium AEgyptiacum, incanum, erectum, floribus spicatis maximis pallidè violaceis, siliquis folliculatis rugosis.
Breyne 1689:14-15 Apocynum erectum majus latifolium Indicum, flore concavo amplo, carneo-suave-purpurascente (?).
J.Commelin 1689:30 = J.Commelin in HM.
Sherard 1689:313 Apocynum latifolium AEgyptium incanum erectum floribus magnis pallide violaceis.
Plukenet 1691:Tab.CLXXV,fig.3 = Sherard 1689.
C.Commelin 1696:5 = Breyne 1689.
Plukenet 1696:35 = C.Commelin 1696.
Hermann 1698:28-29 Apocynum erectum incanum latifolium Malabaricum floribus ex albo suave purpurascentibus.
Petiver 1702(1701):938-939,no.150 Erca-chedde Malab. Apocynum Malabar. latifol. incanum flore albo.
Hermann 1717:34 Waraghaha. Weraghaha. Apocynum Indicum maximum floribus amplis Janthinis obsoletis. Beidelossar Alpin.
J.Burman 1737:24-25 = Hermann 1717.
Linnaeus 1747:47,no.112 Asclepias foliis oblongo-ovalibus amplexicaulibus.
Linnaeus 1753:214 Asclepias gigantea.
Rumphius 1755:24-26,Tab.XIV,Fig.1 Madorius (=).
Linnaeus 1762:312 = Linnaeus 1753.
N.L.Burman 1768:71 = Linnaeus 1762.
Houttuyn 1777 II vol.7:749-753,Pl.XLIV Asclepias Gigantea [Linnaeus 1767 vol.2:193].

Bel-ericu HM 2:56: no figure

BL Add MS 5028: no drawing.
J.Commelin 1689: Apocynum latifolium Syriacum incanum erectum, floribus umbellatis minoribus obsolete purpurascentibus, siliquis folliculatis rugosis. Herm.Cat. [Hermann 1687:52].
C.Commelin 1696:6 Apocynum majus Syriacum, caule viridi, flore exalbido H:Reg:Par:Cat:135 [Sherard 1689:139].
Plukenet 1696:35 = J.Commelin 1689.
Hermann 1698:29-30 Apocynum erectum incanum latifolium Malabaricum floribus omnino albis.
Petiver 1702(1701):938-939,no.150 Erca-chedde Malab. Apocynum Malabar. fol. incanum flore albo.
J.Burman 1737:25 Apocynum Indicum, sylvestre, inodorum, sili-

quosum, seminibus papposis, floribus albis, amplis. Mus.Zeyl. pag.47 [Hermann 1717:47].

Avanacoe seu Citavanacu HM 2:57-59,64:32
Avanacu, seu Citavanacu
BL Add MS 5028:178-179.
J.Commelin in HM 2:60 Ricinus vulgaris C.B.P. [Bauhin 1623:432].
J.Commelin 1689:302 Ricinus Africanus maximus caule geniculato rutilante, Cat.Hort.Reg.Paris.
C.Commelin 1696:58 = J.Commelin 1689.
Plukenet 1696:319 Ricinus Americanus major, caule virescente HRP.156.
J.Burman 1737:206 Ricinus Americanus, Nhambu-Guacu Pisonis Haeranda Zeylonensibus. Mus.Zeyl.pag.53 [Hermann 1717:53] (c).
Linnaeus 1737:450 Ricinus foliis peltatis palmatis serratis, petiolis glanduliferis.
Rumphius 1743 vol.4:92-96 Ricinus albus (=); J.Burman ibid.: 96 (=).
Linnaeus 1747:161,no.339 = Linnaeus 1737.

Pandi-avanacu HM 2:60: no figure

BL Add MS 5028: no drawing.
J.Commelin in HM 2:60 Ricinus vulgaris C.B.P. [Bauhin 1623:432].
Ray 1688 vol.2:1710 Ricinus vulgaris major.
C.Commelin 1696:58 Ricinus indicus maximus caule geniculato totus ruber Herm:Cat:525 [Hermann 1687:525].
Plukenet 1696:319 Ricinus Africanus maximus, caule geniculato rutilante. HRP.156.
C.Commelin 1703:25 = C.Commelin 1696.
J.Burman 1737:206 Ricinus Americanus, Nhambu-Guacu Pisonis Haeranda Zeylonensibus. Mus.Zeyl.pag.53 [Hermann 1717:53] (c).
Rumphius 1743 vol.4:97,Tab.XLI Ricinus ruber (=).
Linnaeus 1747:272,no.566 = Ray 1688.

Cadel-avanacu HM 2:61-62:33
Cadel-avanacu
BL Add Ms 5028:180-181.
J.Commelin in HM 2:62 Ricinus Indicus minor foliis solani Breyn.Cent.I.cap.54 [Breyne 1678:118-119,fig.] (s).
Ray 1686 vol.1:167 = J.Commelin in HM.
Ray 1688 vol.2:1855-1856 = J.Commelin in HM.
Sherard 1689:370 Ricinus arbor fructu glabro Grana Tiglia Officinis dicto.
C.Commelin 1696:58 = Sherard 1689.
Plukenet 1696:320 Ricinus Orientalis, cujus fructus sunt Pinei nuclei Malucani à nobis putati, & Grana-Tilli Officinarum.
J.Burman 1737:200-201,tab.90 Ricinoides Indica, folio lucido, fructu glabro, Grana Tiglia Officinis dicto. Herb.Hart.
Rumphius 1743 vol.4:98-100,Tab.XLII Granum moluccum (=); J.Burman ibid.:100 (=).
Linnaeus 1747:163,no.343 Croton foliis ovatis glabris acuminatis serratis, caule arboreo.
Linnaeus 1753:1004-1005 Croton Tiglium.
Linnaeus 1763:1426 = Linnaeus 1753.
N.L.Burman 1768:204 = Linnaeus 1763.
Houttuyn 1776 II vol.6:253-256 Croton Tiglium [Linnaeus 1767 vol.2:635].

Codi avanacu HM 2:63:34
Cadi-avanacu

BL Add MS 5028:182-183.
J.Commelin in HM 2:63 Lathyris sp. (s).
Ray 1688 vol.2:1710 An Lathyris fruticescens fructu in foliorum alis echinato.
C.Commelin 1696:21 = Ray 1688 (?).
Plukenet 1696:321 Ricinus Malabaricus fruticescens, Lathyridis facie, fructu in foliorum alis echinato.
Petiver 1704(1703):1456,no.29 Ricinus Malabar. Linariae folio vix serrato.
Hermann 1717:33 Tithymalus tenellus Indicus foliis Linariae raris.
J.Burman 1737:59,tab.25 Chamaelaea foliis linearibus, flosculis spicatis, echinato fructu.
Linnaeus 1747:158-159,no.335 Tragia foliis lanceolatis obtusis integerrimis.
Linnaeus 1753:981 Tragia Chamaelea.
Linnaeus 1763:1391 = Linnaeus 1753.
N.L.Burman 1768:195 = Linnaeus 1763.
Houttuyn 1776 II vol.6:223-224 = N.L.Burman 1768.

Ana-chunda HM 2:65-66:35
Ana-schunda
BL Add MS 5028:184-185.
J.Commelin in HM 2:66 Juripebam foemina Piso Hist.Nat.Bras.lib.4.cap.32 (c).
Breyne 1680:49-50 Solanum spinosum pomiferum Indicum album, sive foliis Hyoscyami albi majoris (s).
Hermann 1687:573 Solanum Pomiferum Indicum candicans maximè tomentosum.
J.Commelin 1689:331 = Hermann 1687.
Sherard 1689:377 Solanum Pomiferum Indicum spinosum tomentosum latissimo folio Hort.Lugd. [?Hermann 1687:573].
C.Commelin 1696:63 = Hermann 1687.
Plukenet 1696:351 Solanum spinosum maximè tomentosum Boccon. de Pl. Sicul. 8. (?).
J.Burman 1737:218-219 Solanum Zeylanicum, spinosum, folio amplo, incano, ad pediculum strictiori. Plukn.Phyt.Tab.226. Fig.6.

Cheru-chunda HM 2:67-68:36
Scheru-schunda
BL Add MS 5028:186-187.
J.Commelin in HM 2:68 Juripeba maris Piso Hist.Nat.Bras.
Breyne 1680:50 Solanum spinosum pomiferum Indicum nigrum (s).
Ray 1688 vol.2:1876 Solanum Indicum spinosum fructu minimo, miniato, glabro.
C.Commelin 1696:63 = Ray 1688.
Plukenet 1696:351 Solanum spinosum profundè laciniatis foliis subtùs lanuginosis Maderaspatanum Phytogr.Tab.316.fig.4 (?).
Boerhaave 1720 vol.2:68 Solanum; fruticosum; Indicum; fructu rubro. T.149.
J.Burman 1737:219 Solanum Indicum, spinosum, flore Borraginis, fructu croceo, rotundo, Persicae magnitudine, Pomum de Hiericho dictum. Mus.Zeyl.pag.43 [Hermann 1717:43] (jungitur).
J.Burman 1737:220,tab.102 Solanum frutescens, villosum, foliis undulatis, mollibus, subtus incanis, spinis flacescentibus, armatum.
Rumphius 1747 vol.5:242,Tab.LXXXVI,Fig.1 Trongum agreste (c).

Chunda HM 2:69-70:37
Schunda

BL Add MS 5028:188-189.
J.Commelin in HM 2:70 Solanum spinosum Malabariensis.
Hermann 1687:573 Solanum Pomiferum Indicum spinosum, Borraginis flore, fructu croceo.
J.Commelin 1689:331 Solanum spinosum, fructo rotundo B.Pin. [Bauhin 1623:167-168].
C.Commelin 1696:63 = Hermann 1687.
Plukenet 1696:350 = J.Commelin 1689.
J.Burman 1737:219 Solanum Indicum, spinosum, flore Borraginis, fructu croceo, rotundo, Persicae magnitudine, Pomum de Hiericho dictum. Mus.Zeyl.pag.43 [Hermann 1717:43].
Rumphius 1747 vol.5:242 Trongum agreste album (?=).

Cattu-gasturi HM 2:71-72:38
Catta-gasturi
BL Add MS 5028:190-191.
J.Commelin in HM 2:72 Bammia Muschata Vesl. in Alpini Obs.Not.
Hermann 1687:25 Althaea Indica magno flore petalis latis & obtusis. Moris.Hist.2.532.
C.Commelin 1696:2 Alcea sive Bannia (!) Muschata Aegyptiaca Alpini & Veslingii Breyn P.1,2 [Breyne 1680:2].
J.Burman 1737:134 Ketmia AEgyptia, semine moschato. Tournef.inst.pag.100.
Linnaeus 1737:349-350 Hibiscus foliis peltato-cordatis septangularibus serratis hispidis.
Rumphius 1743 vol.4:40,Tab.XV Granum Moschatum (=)
Linnaeus 1747:119-120,no.261 = Linnaeus 1737.
Linnaeus 1753:696 Hibiscus Abelmoschus.
Linnaeus 1763:980 = Linnaeus 1753.
N.L.Burman 1768:153 = Linnaeus 1763.
Houttuyn 1775 II vol.5:415-417,Pl.XXVII,Fig.2 = N.L.Burman 1768.

Schorigenam HM 2:73-74:39
Schorigenam
BL Add MS 5028:192-193.
J.Commelin in HM 2:74 Urtica urens pillulas ferens: I.Dioscoridis semine lini C.B.P. [Bauhin 1623:232].
Ray 1686 vol.1:160 Urtica fruticescens Malabarica Schorigenam dicta.
C.Commelin 1696:69 = Ray 1686.
Plukenet 1696:393 Urticae-folia Jamaicensis tricoccos Mus.Cortenian. (?).
Petiver 1699(1698):317-318,no.9 Vella caungerree Malab. Ricinus Altheae folio molli & incano Maderaspatanus & Plukenet Phytogr.Tab.120.Fig.5, Alm.Bot.321 [Plukenet 1691:Tab.CXX, f.5; ibid.1696:321].
Petiver 1702(1701):1009,no.196 Vella caunjerie Malab. Ricinu Malabaricus Urticae folio.
Hermann 1717:31 Kahabilija. Urtica Indica tricoccos Zeylanica.
J.Burman 1737:202-203,tab.92 Ricinokarpos Zeylanica, hirsuta, foliis lanceolatis, serratis.
Royen 1740:20 Croton foliis ovato-lanceolatis serratis hispidis caule fruticoso.
Linnaeus 1747:161,no.340 Acalypha involucris faemineis pentaphyllis pinnatifidis.
Linnaeus 1753:980 Tragia involucrata.
Linnaeus 1763:1391 = Linnaeus 1753.
N.L.Burman 1768:194 = Linnaeus 1763.
Houttuyn 1776 II vol.6:221-222 Tragia involucrata [Linnaeus 1767 vol.2:619].

Batti-schorigenam HM 2:75:40
Batti-schorigenam
BL Add MS 5028:194-195.
J.Commelin in HM 2:75 Pino Pisonis.
Ray 1686 vol.1:159-160 Urtica Brasiliensis, Pino indigenis dicta Marggr. (?).
C.Commelin 1696:69 = Ray 1686.
Plukenet 1696:394 Urticae genus Indianum minimè pungens (?).
J.Burman 1737:231-232,tab.110,fig.1 Urtica pilulifera, foliis majoribus longissimis pediculis, minoribus brevibus pediculis donatis (c).
Linnaeus 1747:165-166,no.336 Urtica foliis alternis ovato-cordatis serratis petiolo subbrevioribus.
Linnaeus 1753:985 Urtica interrupta.
Linnaeus 1763:1395 = Linnaeus 1753.
N.L.Burman 1768:197 = Linnaeus 1763.
Houttuyn 1779 II vol.11:236-237 = N.L.Burman 1768.

Ana-schorigenam HM 2:77-78:41
Ana-schorigenam
BL Add MS 5028:196-197.
J.Commelin in HM 2:78 Urtica major.
Ray 1686 vol.1:160 Urtica Malabarica tertia, Ana-Schorigenam dicta.
C.Commelin 1696:69 = Ray 1686.
Plukenet 1696:393 Urtica urens racemifera major, Ind. Orientalis.
Petiver 1704(1703):1454,no.16 Urticae majoris facie Planta Madraspatana (s).

Valli-schorigenam HM 2:79: no figure

BL Add MS 5028: no drawing.
J.Commelin in HM 2:74 Urtica sp.
C.Commelin 1696:68 = HM.
Ray 1704 vol.3:105 Urtica racemosa scandens angustifolia, fructu tricocco Slon.Cat.Jamaic. (?).

Schadida-calli HM 2:81-82:42
Schadida-calli
BL Add MS 5028:198-199.
J.Commelin in HM 2:82 Euphorbium verum.
Breyne 1689:44-45 Euphorbium Indicum Opuntiae facie, caule geniculato triangulari.
C.Commelin 1696:27-28 = Breyne 1689.
Plukenet 1696:370 Tithymalus aizoides triangularis nodosus & spinosus lacte turgens acri.
J.Commelin 1697:23-24,fig.12 Euphorbium antiquorum verum.
C.Commelin 1701:207-208,fig.104 Tithymalus Aizoides fruticosus Canariensis aphyllus quadrangularis et quinquangularis, spinis geminis aduncis atronitentibus armatus (r).
C.Commelin 1703:21 = Plukenet 1696.
Boerhaave 1710:108 = J.Commelin 1697.
J.Burman 1737:96 Euphorbium trigonum, spinosum, rotundifolium. Act.Reg. Paris.ann.1720.pag.500.
Linnaeus 1737:196 Euphorbia aculeata triangularis subnuda articulata, ramis patentibus.
Rumphius 1743 vol.4:88-91,Tab.XL Ligularia (c).
Linnaeus 1747:89,no.199 = Linnaeus 1737.
Linnaeus 1748:138 = Linnaeus 1737,1747.
Linnaeus 1753:450 Euphorbia antiquorum.
Linnaeus 1762:646 = Linnaeus 1753.
N.L.Burman 1768:110-111 = Linnaeus 1762.
Houttuyn 1777 II vol.8:732-734 Euphorbia Antiquorum Syst. Nat.XII.Gen.609.p.330 [Linnaeus 1767 vol.2:330].

Ela-calli HM 2:83-84:43

Ela-calli
BL Add MS 5028: 200-201.
J.Commelin in HM 2:84 Euphorbium secundum hactenus ignotus.
Ray 1688 vol.2:1888 = HM.
Breyne 1689:45-46 Euphorbio & Tithymalo media affinis aizooïdes Indica arborescens spinosa, Nerii folio.
Sherard 1689:(4) Tithymalus spinosus arborescens Zeylanicus.
Plukenet 1691:Tab.CCXXX,fig.4 = Sherard 1689 (?).
C.Commelin 1696:28 = J.Commelin in HM.
Plukenet 1696:369-370 = Plukenet 1691.
J.Commelin 1697:25-26,fig.13 Titymalus Indicus arborescens spinosus Nerii folio.
C.Commelin 1703:22,56,fig.6 Tithymalus aizoides arborescens spinosus, caudice rotundo, nerii folio (r).
Petiver 1704(1703):1455,no.24 Ricinus Malabaricus spinosus Phyllitidis folio.
J.Burman 1737:95-96 Euphorbio-Tithymalus spinosus, caule rotundo, & anguloso, foliis Nerii latioribus, & angustioribus.
Linnaeus 1737:196 Euphorbia aculeata seminuda, angulis oblique tuberculatis.
Rumphius 1743 vol.4:88-91,Tab.XL Ligularia (=).
Linnaeus 1747:89-90,no.200 = Linnaeus 1737.
Linnaeus 1753:451-452 Euphorbia neriifolia.
Linnaeus 1762:648 = Linnaeus 1753.
N.L.Burman 1768:111 = Linnaeus 1762.
Houttuyn 1777 II vol.8:738-739 Euphorbia Neriifolia [Linnaeus 1767 vol.2:330].

Tiru-calli HM 2:85-86:44
Tiru-calli
BL Add MS 5028:202-203.
J.Commelin in HM 2:86 Tithymalus myrtifolius arboreus C.B.P. [Bauhin 1623:290].
Ray 1688 vol.2:1710 Tithymalus Indicus fruticescens.
C.Commelin 1696:66 Tithymalus ramosissimus frutescens pene aphyllos Par:Bat:P:381 [Sherard 1689:381].
Plukenet 1696:368-369 Tithymalus arborescens, caule aphyllo Phytogr.Tab.319.fig.6.
J.Commelin 1697:27-28,fig.14 = Ray 1688.
C.Commelin 1703:22-23 = C.Commelin 1696.
Boerhaave 1720 vol.1:257 = Ray 1688.
J.Burman 1737:223 = C.Commelin 1696.
Linnaeus 1748:139-140 Euphorbia inermis fruticosa subnuda filiformis erecta, ramis patulis determinate confertis. Hort.cliff.197. Fl.zeyl.197 [Linnaeus 1747:88,no.196].
Linnaeus 1753:452 Euphorbia Tirucalli.
Rumphius 1755:62-63,Tab.XXIX Ossifraga lactea (=).
Linnaeus 1762:649 = Linnaeus 1753.
N.L.Burman 1768:111 = Linnaeus 1762.
Houttuyn 1777 II vol.8:741-742 Euphorbia Tirucalli [Linnaeus 1767 vol.2:331].

Bahel-schulli HM 2:87-88:45
Bahel-schulli
BL Add MS 5028:204-205.
J.Commelin in HM 2:88 Genista spinosa major longioribus aculeis C.B.P. [Bauhin 1623:394-395].
Ray 1688 vol.2:1731 Genista spinosa Indica verticillata, flore purpuro-coeruleo, seu Spartium spinosum siliqua geminata.
Plukenet 1691:Tab.CXIX,f.5 Melampyro cognata Maderaspatana spinis horrida (?).
C.Commelin 1696:33 = Ray 1688.
Plukenet 1696:245 = Plukenet 1691.
Petiver 1699(1698):316,no.6 Neer Mulle Malab. Adhatoda Malabarica Spinosa Echii folio.

Nir-schulli HM 2:89-90:46
Nir-schulli
BL Add MS 5028:206-207.
J.Commelin in HM 2:90 Teucrium sp. or Genista sp.
Ray 1688 vol.2:1767 Frutex Indicus flore dipetalo labiato, siliquâ geminatâ aculeata.
Plukenet 1691:Tab.XLIX,f.3 = HM (s).
C.Commelin 1696:31 = Ray 1688.
Plukenet 1696:180,ibid.1700:99 Gratiolae affinis Maderaspatana, Digitalis aemula, folio Clinopodij, capsulis in verticillas positis, Phytogr.Tab.193.fig.3 (s).
Plukenet 1696:264 = Plukenet 1691.

Cara-schulli HM 2:91-92:47
Cara-schulli
BL Add MS 5028:208-209.
Ray 1688 vol.2:1755 Frutex Indicus spinosus Capparis forma, Siliqua bivalvi brevi.
C.Commelin 1696:32 = Ray 1688.
Plukenet 1696:80 Capparis forma, Frutex spinosus Malabaricus.
J.Burman 1737:53 Capparis spinosa, foliis oblongis. Mus.Zeyl. pag.7 [Hermann 1717:7] (spectat).
Linnaeus 1753:636 Barleria buxifolia.
Linnaeus 1763:887 = Linnaeus 1753.
N.L.Burman 1768:136 = Linnaeus 1763.
Houttuyn 1779 II vol.9:583 Barleria Buxifolia [Linnaeus 1767 vol.2:425].

Paina-schulli HM 2:93-94:48
Paina-schulli
BL Add MS 5028:210-211.
J.Commelin in HM 2:94 Ruscus sylvestris (r).
Ray 1688 vol.2:1766-1767 Frutex Indicus spinosus foliis Agrifolii siliqua geminata brevi.
C.Commelin 1696:32 = Ray 1688.
Plukenet 1696:38 Aquifoliae facie arbor Malabarica, Acanthii flore albo cucullato. Phytogr.Tab.261.fig.2.
Linnaeus 1753:639 Acanthus ilicifolius.
Linnaeus 1763:892 = Linnaeus 1753.
N.L.Burman 1768:138 = Linnaeus 1763.
Houttuyn 1778 II vol.9:589-590 Acanthus Ilicifolius [Linnaeus 1767 vol.2:427].

Carambu HM 2:95-96:49
Carambu
BL Add MS 5028:213(!)-213.
J.Commelin in HM 2:96 Caryophyllus spurius Malabariensis flore luteo minore.
Hermann 1687:396 Lysimachia Indica non papposa, flore luteo minimo siliquis Caryophyllum aromaticum aemulante.
Ray 1688 vol.2:1510 = J.Commelin in HM.
J.Commelin 1689:214 = Hermann 1687.
C.Commelin 1696:41 = Hermann 1687.
Plukenet 1696:235 = Hermann 1687.
Petiver 1699(1698):326,no.26 Neer Ureevee Mal. Lysimachia non papposa humilis Maderaspatana Clinopodii Virginiani lutei foliis fructu Caryophylloide parvo, Pluk.Tab.203.Fig.5. & Alm.Bot.236 [Plukenet 1691:Tab.CCIII,fig.5; ibid.1696:236] (?).
Royen 1740:252 Ludwigia capsulis oblongis uncialibus.
Linnaeus 1747:27,no.66 Ludwigia caule diffuso, foliis lanceolatis, capsulis pedunculatis folio dimidio brevioribus.
Rumphius 1750 vol.6:49-50,Tab.XXI,Fig.1 Herba vitiliginum (?=).
Linnaeus 1753:119 Ludvigia perennis; ibid.:388 Jussiaea suffruticosa.

Linnaeus 1762:113-114 = Linnaeus 1753:119; ibid.:555-556 = Linnaeus 1753:388.
N.L.Burman 1768:37 = Linnaeus 1762:113-114.
N.L.Burman 1768:103 = Linnaeus 1762:555-556.
Houttuyn 1777 II vol.8:532-533 = N.L.Burman 1768:103.

Cattu-carambu HM 2:97:50
Cattu-Carambu
BL Add MS 5028:214-215.
J.Commelin in HM 2:97 Caryophyllus spurius Malabariensis flore luteo.
Ray 1688 vol.2:1510 Lysimachia Indica non papposa lutea flore fructuque majore Caryophylloide.
J.Commelin 1689:214 Lysimachia Indica lutea corniculata, non papposa, caule altissimo flore odorato.
C.Commelin 1696:41 = J.Commelin 1689.
Linnaeus 1747:75,no.170 Jussiaea erecta, floribus tetrapetalis octandris sessilibus.
Rumphius 1750 vol.6:49-50,Tab.XXI,Fig.1 Herba vitiliginum (-); J.Burman ibid.:50 (=).
N.L.Burman 1768:103 Jussiaea suffruticosa Linn.sp.555 [Linnaeus 1762:555-556] β.

Nir-carambu HM 2:99-100:51
Nir-Carambu
BL Add MS 5028:216-217.
Ray 1688 vol.2:1510-1511 Lysimachia Indica non papposa repens flore pentapetalo, fructu Caryophylloide.
C.Commelin 1696:41 = Ray 1688.
J.Burman 1737:146-147 Lysimachiae species, fructu Caryophylloideo, Kikirindia Zeylonensibus. Mus.Zeyl.pag.17 [Hermann 1717:17].
Linnaeus 1747:75,no.169 Jussiae repens, floribus pentapetalis decandris, pedunculis folio longioribus.
Linnaeus 1753:388 Jussiaea repens.
Linnaeus 1762:555 = Linnaeus 1753.
N.L.Burman 1768:103 = Linnaeus 1762.
Houttuyn 1777 II vol.8:531 = N.L.Burman 1768.

Ponnam-tagera seu Ponna-virem HM 2:101-102:52
Ponnam-tagera, seu Ponna-virem
BL Add MS 5028:218-219.
Breyne 1680:51 Sophera Indiae Orientalis, Ponnam-tagera Malabaris dicta.
Hermann 1687:557 Senna Orientalis fruticosa, Sophera dicta.
Breyne 1689:29 Chamaecassia angusti-folia Indica.
J.Commelin 1689:136 Galega affinis Sophera dicta B.Pin. [Bauhin 1623:352].
C.Commelin 1696:61 = Hermann 1687.
Plukenet 1696:342 = Hermann 1687.
Hermann 1717:32 Aehala. Cassia fistula Zeylanica.
Boerhaave 1720 vol.2:57 = Hermann 1687.
J.Burman 1737:213-214,tab.98 Senna vigintifolia, siliquis teretibus.
Rumphius 1743 vol.4:63,Tab.XXIII Flos flavus (?=); J.Burman ibid.:63 (≠).
Linnaeus 1747:64,no.150 Cassia foliolis decem parium lanceolatis, glandula baseos oblonga.
Rumphius 1747 vol.5:283-284,Tab.XCVII,Fig.1 Gallinaria acutifolia (-); J.Burman ibid.:284 (c).
Linnaeus 1753:379 Cassia Sophera.
Linnaeus 1762:542 = Linnaeus 1753.
N.L.Burman 1768:97 = Linnaeus 1762.
Houttuyn 1775 II vol.5:34 = N.L.Burman 1768.

Tagera HM 2:103:53
Tagera
BL Add MS 5028:220-221.
Ray 1688 vol.2:1743 Sena spuria Malabarica.
Breyne 1689:29 Chamaecassiae affinis tetraphylla, siliquis tenuissimis, semine tereti, apicibus obtusis, quasi abscissis, suâ longitudine, secundum longitudinem siliquae posito.
C.Commelin 1696:61 = Ray 1688.
Plukenet 1696:342 Sena spuria Orientalis tenuissimis siliquis tetraphylla.
J.Burman 1737:213,tab.98 Senna vigintifolia, siliquis teretibus (jungitur).
Rumphius 1743 vol.4:63,Tab.XXIII Flos flavus (s); J.Burman ibid.:63 (≠).
Rumphius 1747 vol.5:283-284,Tab.XCVII,Fig.1 Gallinaria acutifolia (=); J.Burman ibid.:284 (≠).
Linnaeus 1753:376 Cassia Tagera (?).
Linnaeus 1762:538 = Linnaeus 1753 (?).
N.L.Burman 1768:95 = Linnaeus 1762 (=).
Houttuyn 1775 II vol.5:16 = N.L.Burman 1768 (?).

Nandi-ervatam HM 2:105-106:54
Nandi-ervatam major
BL Add MS 5028:222-223.
J.Commelin in HM 2:106 Syringa Malabar. lactescens flore niveo pleno odoratissimo.
Ray 1688 vol.2:1785-1786 = J.Commelin in HM.
Sherard 1689:351-352 Manghas sylvestris lactescens venenata Jasmini flore & odore.
C.Commelin 1696:65 = J.Commelin in HM.
Plukenet 1696:197 Jasminum Indicum odoratum, Aurantiae foliis album, flore multiplici roseo, è Maderaspatan. (?).
J.Burman 1737:129-130,tab.59 Jasminum Zeylanicum, folio oblongo, flore albo, pleno, odoratissimo. Herb.Hart.
Rumphius 1743 vol.4:87,Tab.XXXIX Flos manilhanus (?=); J.Burman ibid.:87 (=).
N.L.Burman 1768:5 Nyctanthes acuminata.
Houttuyn 1775 II vol.4:16-17 = N.L.Burman 1768.

Nandi-ervatam minor HM 2:107:55
Nandi-ervatam minor
BL Add MS 5028:224-225.
J.Commelin in HM 2:107 Jasminum Malabaricum foliis mali Aurantii, flore niveo odoratissimo.
Ray 1688 vol.2:1786 = J.Commelin in HM.
C.Commelin 1696:37 = J.Commelin in HM.

Capo-molago HM 2:109-110:56
Capo-molago
BL Add MS 5028:226-227.
J.Commelin in HM 2:110 Piper Indicum siliqua flava, vel aurea C.B.P. [Bauhin 1623:102].
C.Commelin 1696:54 = J.Commelin in HM.
Plukenet 1696:353 Solanum mordens fructu oblongo pendulo minore.
Linnaeus 1737:60 Capsicum frutescens.
Royen 1740:426 Capsicum caule fruticoso.
Rumphius 1747 vol.5:247-252,Tab.LXXXVIII Capsicum Indicum (=).
Linnaeus 1753:189 Capsicum frutescens.
N.L.Burman 1768:57-58 Capsicum annuum Linn.sp.270 [Linnaeus 1762:270-271].

VOLUME 3

Codda-panna, sive Palma montana Malabarica HM 3:1-6:1-12

Codda-pana
BL Add MS 5029:1-24.
J.Commelin in HM 3:6 Palma montana Malabarica; folio magno, complicato, acuto, flore albo racemoso, fructu rotundo.
Ray 1688 vol.2:1367-1368 Palma montana folio plicatili flabelliformi maximo, semel tantum frugifera.
C.Commelin 1696:50 = J.Commelin in HM.
Plukenet 1696:277 = Ray 1688.
J.Burman 1737:181-182 Palma Zeylanica, folio longissimo & latissimo, Tala & Talaghas dicta. Mus.Zeyl.pag.54 [Hermann 1717:54].
Linnaeus 1737:482 Corypha frondibus pinnato-palmatis plicatis interjecto digitis filo.
Rumphius 1741 vol.I:44 (App),Tab.VIII Saribus (c); J.Burman ibid.:44 (c).
Van Royen 1740:4 = Linnaeus 1737.
Linnaeus 1747:186-187,no.394 = Linnaeus 1737.
Linnaeus 1753:1187 Corypha umbraculifera.
Linnaeus 1763:1657 = Linnaeus 1753.
N.L.Burman 1768:240 = Linnaeus 1763.
Houttuyn 1773 II vol.1:275-282,Pl.III,Fig.1-2 = N.L.Burman 1768.

Niti-panna HM 3:7: no figure

BL Add MS 5029: no drawing.
C.Commelin 1696:48 = HM.

Todda-panna HM 3:9-14:13-21
Todda-panna
BL Add MS 5029:25-42.
J.Commelin in HM 3:14 Palma referens arbor farinifera C.B.P. [Bauhin 1623:508].
Hermann 1687:472 Palma prunifera Japponica.
Ray 1688 vol.2:1360 Palma Indica caudice in annulos protuberantes distincto, fructu Pruniformi.
Breyne 1689:81 Palma farinifera Japonica, Sotitsou Japonensibus.
J.Commelin 1689:260 = Hermann 1687.
Sherard 1689:361 Palma Japonica spinosis pediculis Polypodii folio.
C.Commelin 1696:50 = Ray 1688.
Plukenet 1696:276 = J.Breyne 1689.
C.Commelin 1701:223 = Hermann 1687.
Boerhaave 1720 vol.2:170 = Sherard 1689.
Linnaeus 1737:482 Cycas frondibus pennatis, foliolis linearilanceolatis, petiolis spinosis.
Rumphius 1741 vol.I:90 (App),Tab.XX-XXIII Olus Calappoides (c).
Rumphius 1741 vol.I:92,Tab.XXIV (Auct) Arbor Calappoides Sinensis (r).
Linnaeus 1747:186,no.393 = Linnaeus 1737.
Linnaeus 1753:1188 Cycas circinalis.
Linnaeus 1763:1658 = Linnaeus 1753.
N.L.Burman 1768:240 = Linnaeus 1763.
Houttuyn 1773 II vol.1:299-304,Pl.III,Fig.3-4 = N.L.Burman 1768.

Katou-indel HM 3:15-16:22-25
Katou-indel
BL Add MS 5029:43-44,45-46,47-48,49-50.
J.Commelin in HM 3:16 Palma sylvestris Malabarica, folio acuto, fructu pruni facie.
Ray 1688 vol.2:1364-1365 = J.Commelin in HM.
C.Commelin 1696:50 = J.Commelin in HM.
Plukenet 1696:276 = J.Commelin in HM.
J.Burman 1737:183-184 Palma Dactylifera, minor, humilis, sylvestris, fructu minore. Par.Bat.Pr.pag.361 [Sherard 1689:361].
Linnaeus 1737:483 Phoenix frondibus pinnatis, foliolis alternis ensiformibus basi complicatis, petiolis compressis dorso rotundatis (?).
Linnaeus 1747:187,no.397 = J.Burman 1737.
Linnaeus 1753:1189 Elate sylvestris.
Linnaeus 1763:1659 = Linnaeus 1753.
N.L.Burman 1768:241 = Linnaeus 1763.
Houttuyn 1773 II vol.1:406-424,Pl.IV,Fig.2 = N.L.Burman 1768.

Tsjaka-maram, sive Jaca, vel Jaaca HM 3:17-20:26-28
Tsjakamaram, aut. Pilau
BL Add MS 5029:51-52,53-54,55-56.
J.Commelin in HM 3:20 Jaaca Garcia ab Horto libr.2.cap.11 (etc.).
Ray 1688 vol.2:1440-1441 Jaca Indica J.B. (etc).
C.Commelin 1696:35 Jaca arbor J:B:T:I.L.I.115 (etc.).
Rumphius 1741 vol.1:106-107 (App),Tab.XXX Saccus arboreus major (=); J.Burman ibid.:107 (=).

Atamaram HM 3:21-22:29
Atamaram
BL Add MS 5029:57-58.
J.Commelin in HM 3:22 Ahate de Panucho Recchus de plantis Mexicanis pag.348 (s).
Ray 1688 vol.2:1650 Pomifera Indica fructu conoide squamoso, viridi.
Sherard 1689:312 Anona Indica pomo viridi.
Plukenet 1691:tab.135,fig.2 Anona Indica fructu conoide viridi squamis veluti aculeato.
C.Commelin 1696:5 = Sherard 1689.
Plukenet 1696:32 = Plukenet 1691.
Petiver 1702(1701):1008,no.193 Cheta paulum Malab. Anona folio obtusiore, fructu glanduloso conoide.
Linnaeus 1737:222 Magnolia foliis ovato-lanceolatis (cs).
Rumphius 1741 vol.1:138 (App),Tab.XLV Anona (r).
Rumphius 1741 vol.1:138-139,Tab.XLVI Anona Tuberosa (-); J.Burman ibid.:139 (=).
Linnaeus 1747:217,no.508 = Sherard 1689.
Linnaeus 1762:757 Annona squamosa.
N.L.Burman 1768:125 = Linnaeus 1762.
Houttuyn 1774 II vol.3:89-93,Pl.XII,Fig.2 = N.L.Burman 1768.

Anona-maram, an Anon Oviedi HM 3:23:30-31
Anona-maram
BL Add MS 5029:59-60,61-62.
J.Commelin in HM 3:23 Quautzapotl seu Anona Recchus libr.3.cap.60.
Hermann 1687:645 Anona Indica, fructu parvo violaceo.
Ray 1688 vol.2:1650-1651 = Hermann 1687 (?).
Sherard 1689:312 Anona Indica pomo coeruleo.
Plukenet 1691:tab.134,fig.4 Anona indic. angustif. fructu coerul. cortice squamato glabro.
C.Commelin 1696:5 = Hermann 1687.
Plukenet 1696:32 = Plukenet 1691.
Petiver 1702(1701):1008,no.193,1 Ata paulum Malab. Anona folio acutiore, fructu majore.
Rumphius 1741 vol.1:138 (App),Tab.XLV Anona (r).
Linnaeus 1767 vol.2:374 Anona reticulata.
Houttuyn 1774 II vol.3:93-96 = Linnaeus 1767.

Ansjeli HM 3:25-27:32

Ansjeli
BL Add MS 5029:63-64.
J.Commelin in HM 3:27 Angelina arbor Zanoni.
Ray 1688 vol.2:1384-1385 Castanea Malabarica Angelina dicta.
C.Commelin 1696:19 = Ray 1688.
Rumphius 1741 vol.1:109 (App),Tab.XXXI Saccus arboreus minor (=); J.Burman ibid.:109 (≠).
Rumphius 1741 vol.1:110-112,Tab.XXXII Soccus Lanosus (-); J.Burman ibid.:112 (=).

Katou Tsjaca HM 3:29-30:33
Katou Tsjaka
BL Add MS 5029:65-66.
Ray 1688 vol.2:1441 Arbor Indica fructu aggregato globoso, Katu Tsjaka dicta.
C.Commelin 1696:7 = Ray 1688.
Plukenet 1696:203 Arbor Indica, floribus & fructu in globum aggregatis.
Plukenet 1696:336,tab.77,fig.4 Scabiosa dendroides Americana, ternis foliis circa caulem ambientibus, floribus ochroleucis; ibid.1700:168 (?).
Royen 1740:187 Cephalanthus foliis oppositis.
Rumphius 1743 vol.3:37 (App),Tab.XIX Samama (=); J.Burman ibid.:37 (r).
Linnaeus 1747:22,no.53 Cephalanthus foliis oppositis.
Linnaeus 1753:95 Cephalanthus orientalis.
Linnaeus 1762:243 Nauclea orientalis.
N.L.Burman 1768:51 = Linnaeus 1762.
Houttuyn 1774 II vol.2:79-80 Nauclea Orientalis. Syst.Nat.XII. Tom.II.Gen.222 [Linnaeus 1767 vol.2:163].

Pela seu Guajabor pomifera Indica BPin. HM 3:31-32:34
Pela
BL Add MS 5029:67-68.
J.Commelin in HM 3:32 Guyabo pomifera Indica C.B.P. [Bauhin 1623:437].
Hermann 1687:305 Guajana alba dulcis.
Ray 1688 vol.2:1455 Guayava Park.Ger.(etc).
J.Commelin 1689:150 = J.Commelin in HM.
Sherard 1689:339 = Hermann 1687.
C.Commelin 1696:34 Guajava alba dulcis fructu longiori Herm:Cat.305 [Hermann 1687:305].
Plukenet 1696:181 Guaiava alba acida, fructu rotundiori, Ind. Orient. Phytogr.Tab.193,fig.4.
J.Commelin 1697:121-122, fig.63 = Sherard 1689.
Boerhaave 1710:269 Guajava Clus.Hist.
J.Burman 1737:112 Guajavos fructu pallido, dulci Mus.Zeyl. pag.3 [Hermann 1717:3].
Linnaeus 1737:184 Psidium caule quadrangulo.
Rumphius 1741 vol.1:141 (App),Tab.XLVII Cujavus Domestica (=).
Linnaeus 1747:85,no.192 = Linnaeus 1737.
Linnaeus 1762:672 Psidium pyriferum.
N.L.Burman 1768:113 = Linnaeus 1762.
Houttuyn 1774 II vol.2:525-527 = N.L.Burman 1768.

Malacka-pela HM 3:33:35
Malacka pela
BL Add MS 5029:69-70.
J.Commelin in HM 3:33 Xalxocoptl prima Recchi.
Hermann 1687:305,671 Guajana rubra acida fructu rotundiori.
Ray 1688 vol.2:1455 Guayava Pisoni.
J.Commelin 1689:150 Guajava Indica fructu Mali facie I.Bauh.
Plukenet 1691:tab.193,fig.4 Guaiava alba acida, fructu rotundiori, Ind.Or.
C.Commelin 1696:33-34 = Hermann 1687.
Plukenet 1696:181 = Plukenet 1691.
Linnaeus 1737:184 Psidium caule quadrangulo.
Rumphius 1741 vol.1:141 (App),Tab.XLVIII Cujavus agrestis (=); J.Burman ibid.:144 (=).
Linnaeus 1762:672 Psidium pomiferum.
N.L.Burman 1768:115 = Linnaeus 1762.
Houttuyn 1774 II vol.2:527-528 = N.L.Burman 1768.

Pelou seu Guayabo sylvestre HM 3:35-36:36
Pelou
BL Add MS 5029:71-72.
J.Commelin in HM 3:36 Guayavo Sylvestre.
Ray 1688 vol.2:1455-1456 = J.Commelin in HM.
C.Commelin 1696:35 = J.Commelin in HM.

Covalam seu Cydonia exotica. Casp. Bauh. in Pin. HM 3:37-38:37
Covalam
BL Add MS 5029:73-74.
J.Commelin in HM 3:38 Cydonia exotica C.B.P. [Bauhin 1623:435].
Ray 1688 vol.2:1665-1666 Cucurbitifera trifolia Indica fructus pulpa Cydonii aemula.
Sherard 1689:330 = J.Commelin in HM.
Plukenet 1691:tab.170,fig.5 = Ray 1688.
C.Commelin 1696:25 = Ray 1688.
Plukenet 1696:125 = Plukenet 1691.
Petiver 1702(1701):848-849,no.108 Ville-Vittree Malab. Covaalam trifoliatum minus e Madraspatan.
J.Burman 1737:84-85 Cydonia exotica, quae Marmelos arbor.Mus.Zeyl.pag.60 [Hermann 1717:60].
Rumphius 1741 vol.1:197-199,Tab.LXXXI Bilacus; J.Burman ibid.:199 (=).
Rumphius 1741 vol.1:200 (Auct), Tab.LXXXII Bilacus Taurinus (r).
Linnaeus 1747:95,no.212 Crateva spinosa.
Linnaeus 1753:444 Crateva Marmelos.
Linnaeus 1762:637 = Linnaeus 1753.
N.L.Burman 1768:109 = Linnaeus 1762.
Houttuyn 1774 II vol.2:518-519 = N.L.Burman 1768.

Syalita HM 3:39-42:38-39
Sijalita
BL Add MS 5029:75-76,77-78.
J.Commelin in HM 3:42 Malus rosea Malabarica.
Ray 1688 vol.2:1707-1708 Arbor Indica flore maximo, cui multae innascuntur siliquae.
C.Commelin 1696:7-8 = Ray 1688.
Plukenet 1700:124-125 Malus rosea Malabarica Syalita dicta.
Linnaeus 1737:221-222 Dillenia.
Rumphius 1741 vol.2:140-141,Tab.XLV Songium (-); J.Burman ibid.:141 (cs).
Rumphius 1741 vol.2:142-143 (Auct),Tab.XLVI Sangius (=).
Linnaeus 1753:535 Dillenia indica.
Linnaeus 1762:754-755 = Linnaeus 1753.
N.L.Burman 1768:124 = Linnaeus 1762.
Houttuyn 1774 II vol.3:60-63 = N.L.Burman 1768.

Blatti seu Jambos sylvestris HM 3:43-44:40
Blatti
BL Add MS 5029:79-80.
J.Commelin in HM 3:44 Jambos Sylvestris.
Ray 1688 vol.2:1479 = J.Commelin in HM.
C.Commelin 1696:36 = J.Commelin in HM.

Rumphius 1741 vol.1:129-130 (Auct),Tab.XL Jambosa Silvestris Parvifolia (-); J.Burman ibid.:130 (?).
Rumphius 1743 vol.3:105 (App),111-115,Tab.LXXIII-LXXV Mangium Caseolare (=); J.Burman ibid.:115 (=).

Panitsjika-maram HM 3:45-47:41
Panitsjika Maram
BL Add MS 5029:81-82.
J.Commelin in HM 3:47 Janipaba Piso Hist.Nat.& Med.Brasiliens.lib.4.cap.16.
Ray 1688 vol.2:1666-1667 Pomifera Indica tinctoria Janipaba dicta (?).
C.Commelin 1696:55 = Ray 1688 (=).
Plukenet 1696:123-124 Cucurbitifera arbor, Americana, folio longo, mucrunato, fructu orbiculari granis cordiformibus, pulpa nigra involutis, Phytogr.Tab.171.fig.1; ibid.1700:59 (?=).
Plukenet 1696:180 Guaiacana Loto arbori, s. Guaiaco Patavino affinis Virginiana, Pishamin dicta Parkinsono, Phytogr.Tab. 244.fig.5; ibid.1700:99 (?).
Rumphius 1741 vol.1:132-134,Tab.XLIII Mangostana; J.Burman ibid.:134 (cs).
Rumphius 1741 vol.1:134-135,Tab.XLIV Mangostana Celebica (=); J.Burman ibid.:135 (=).

Niirvala HM 3:49-50:42
Niirvala
BL Add MS 5029:83-84.
J.Commelin in HM 3:50 Tapia Piso Hist.Nat.& Med.Brasil.lib. 3.cap.16. (s).
Ray 1688 vol.2:1644 Pomifera Indica trifolia fructu Pruniformi caudato.
Plukenet 1691:tab.229,fig.1: Therebinth; Americ.trifolia lucida Palamalatta vulgo PBP [Sherard 1689:380] (?).
C.Commelin 1696:55 = Ray 1688.
Plukenet 1696:34 Apioscorodon, s.Arbor Americana triphyllos, Allii odore, poma ferens. Phytogr.Tab.137.fig.7 (?); ibid.1696: 181 Hederae Virginianae triphyllae quodammodò accedens Arbor Jamaicensis (?).
Plukenet 1696:363 = Plukenet 1691.
Plukenet 1705: Arbuscula Tiliae folio trifoliata, Aurantiae fructu, Maderaspatana. Movenam-comboo Malabarorum (?).
Linnaeus 1737:484 Crateva.
Linnaeus 1747:94,no.211 Crateva inermis.
Linnaeus 1753:444 Crateva tapia.
Linnaeus 1762:637 = Linnaeus 1753.
N.L.Burman 1768:109 = Linnaeus 1762.
Houttuyn 1774 II vol.2:516-518 = N.L.Burman 1768.

Tamara-tonga, seu Carambolas HM 3:51-53:43-44
Tamara-tonga
BL Add MS 5029:85-86,87-88.
J.Commelin in HM 3:53 Carambolas Garzia ab Horto lib.2.Cap. 15 (etc.).
Ray 1688 vol.2:1449 Malus Indica pomo anguloso Carambolas dicta.
C.Commelin 1696:42 = Ray 1688.
Plukenet 1696:238 = Ray 1688.
Plukenet 1700:36 Carambola Malabarorum Tamaratonga, Hort.Malab.Part.3.Tab.43,44.
Petiver 1702(1701):1016-1017,no.207 Tammerten cheddee Malab. Carambola Malabarica fructu pentagono.
J.Burman 1737:148 Malus Indica, foliis Sennae Occidentalis, fructu acido, flavo, pentagono, sulcato, floribus rubris. Herb.Hert.
Rumphius 1741 vol.1:118 (App),Tab.XXXV Prunum Stellatum (=); J.Burman ibid.:118 (=).

Linnaeus 1747:79-80,no.178 Averrhoa foliorum axillis fructificantibus, pomis oblongis acute angulatis.
Linnaeus 1753:428 Averrhoea Carambola.
Linnaeus 1762:613 = Linnaeus 1753.
N.L.Burman 1768:106 = Linnaeus 1762.
Houttuyn 1774 II vol.2:477-479 = N.L.Burman 1768.

Bilimbi HM 3:55-56:45-46
Bilimbi
BL Add MS 5029:89-90,91-92.
J.Commelin HM 3:56 Billingbing Bontius libr.4.cap.42.
Ray 1688 vol.2:1449-1450 Malus Indica fructu pentagono, Bilimbi dicta.
C.Commelin 1696:42 = Ray 1688.
Plukenet 1696:238 = Ray 1688.
Plukenet 1700:36 Carambola parva Ind.Orientalis. Billingbing Bontij lib.4.cap.42.
Petiver 1704(1702):1058,no.232 Coche Tammartia Malab.Bilimbi Malabarica Pajomiriobae folio.
J.Burman 1737:147 = Ray 1688.
Rumphius 1741 vol.1:118,Tab.XXXV (App) Prunum Stellatum (=).
Rumphius 1741 vol.1:118-120,Tab.XXXVI Blimbingum Teres (-); J.Burman ibid.:120 (=).
Linnaeus 1747:79,no.177 Averrhoa caudice nudo fructificante, pomis oblongis obtuse angulatis.
Linnaeus 1753:428 Averrhoea Bilimbi.
Linnaeus 1762:613 = Linnaeus 1753.
N.L.Burman 1768:106 = Linnaeus 1762.
Houttuyn 1774 II vol.2:475-477 = N.L.Burman 1768.

Neli-pouli seu Bilimbi altera minor HM 3:57-58:47-48
Neli-pouli
BL Add MS 5029:93-94; the drawing of tab.48 is missing!
J.Commelin in HM 3:58 Bilimbi altera minor.
Ray 1688 vol.2:1450 = J.Commelin in HM.
C.Commelin 1696:16 = J.Commelin in HM.
Plukenet 1696:45 Arbor Malabarica Fraxini ferè folio, Ossiculo fructus octangulari. Phytogr.Tab.269.fig.2 (?).
Petiver 1704(1702):1058-1059,no.233 Paringe Nellekai Malab. Charamei Malabar. Cassiae fistulae folio.
J.Burman 1737:148 Malus Indica, fructu parvo, rotundo, acido, striato.
Rumphius 1741 vol.1:151-153,Tab.LIV Lansium (-); J.Burman ibid.:153 (cs).
Linnaeus 1747:80,no.179 Averrhoa ramis nudis fructificantibus, pomis subrotundis.
Linnaeus 1753:428 Averrhoea acida.
Rumphius 1755:34-35,Tab.XVII,Fig.2 Cheramela (=).
Linnaeus 1762:613 = Linnaeus 1753.
N.L.Burman 1768:106 = Linnaeus 1762.
Houttuyn 1774 II vol.2:479-480 = N.L.Burman 1768.

Panja, Panjala. Sive Arbor Lanigera. Bonti. HM 3:59-60:49-51
Panja
BL Add MS 5029:95-96,97-98,99-100.
J.Commelin in HM 3:60 Arbor Lanigera sive Gossampini Plinii Bontius lib.6.cap.14.
Hermann 1687:294 Gossipium arboreum Orientale, foliis Salicis digitatis latioribus.
Ray 1688 vol.2:1899-1900 Gossipium Indicum Salicis folio, fructu quinquecapsulari.
J.Commelin 1689:147 = Ray 1688.
Sherard 1689:337 = Ray 1688.
C.Commelin 1696:34 = Hermann 1687.
Linnaeus 1737:75 Xylon caule inermi.

Rumphius 1741 vol.1:196 (App),Tab.LXXX Eriophoros Javana
    (= fig.51); J.Burman ibid.:197 (r figs 49-52).
Linnaeus 1747:98,no.220 Xylon foliis digitatis, caule inermi.
Linnaeus 1748:148 = Linnaeus 1747.
Linnaeus 1753:511 Bombax pentandrum.
Linnaeus 1763:959 = Linnaeus 1753.
N.L.Burman 1768:145 = Linnaeus 1763.
Houttuyn 1774 II vol.3:147-151 = N.L.Burman 1768.

Moul-elavou Sive Arbor lanigera spinosa HM 3:61-62:52
Moul-elavou
BL Add MS 5029:101-102; this drawing is in bad condition and has been pasted on a new sheet afterwards.
J.Commelin in HM 3:62 Arbor lanigera spinosa Malabarica.
Ray 1688 vol.2:1899 Gossipium arboreum caule spinoso C.B. [Bauhin 1623:430].
C.Commelin 1696:33-34 = Ray 1688.
Plukenet 1696:172 Gossipium s. Xylon arbor Occidentale digitatis foliis per marginem crenatis, fructu conoide quinque capsulari, lanugine leucophaeâ referto, Phytogr.Tab.189.fig.1 (?).
Linnaeus 1737:75 Xylon caule aculeato.
Rumphius 1741 vol.1:196 (App),Tab.LXXX Eriophoros Javana (=); J.Burman ibid.:197 (r figs 49-52).
Linnaeus 1747:98,no.221 Xylon foliis digitatis, caule aculeato.
Linnaeus 1748:148 = Linnaeus 1747.
Linnaeus 1753:511 Bombax Ceiba.
Linnaeus 1763:959-960 = Linnaeus 1753.
Linnaeus 1767 vol.2:457 Bombax heptaphyllum.
N.L.Burman 1768:145 = Linnaeus 1763.
Houttuyn 1774 II vol.3:153-154 = Linnaeus 1767.

Bellutta tsjampakam Sive Castanea rosea Indica HM 3:63-64:53
Belutta-tsiampakam
BL Add MS 5029:103-104.
J.Commelin in HM 3:64 Castanea rosea.
Ray 1688 vol.2:1680 Castanea Indica florida.
C.Commelin 1696:19 = Ray 1688.
Plukenet 1696:90 Castanea rosea, Indica.
Linnaeus 1747:91,no.203 Mesua foliis lanceolatis β.
Linnaeus 1753:515 Mesua ferrea.
Linnaeus 1762:734 = Linnaeus 1753.
N.L.Burman 1768:121 = Linnaeus 1762.
Houttuyn 1774 II vol.3:158-160 = N.L.Burman 1768.

Kapa-mava sive Acajous vel Anacardi alia species B.Pin. HM 3:65-67:54
Kapa-mava
BL Add MS 5029:105-106.
J.Commelin in HM 3:67 Anacardij alia species C.B.P. [Bauhin 1623:512].
Hermann 1687:36 Anacardium occidentale Cajous dictum ossiculo reni leporis figurâ.
Ray 1688 vol.2:1649 Pomifera seu potius prunifera Indica nuce reniformi summo pomo innascente, Cajous dicta.
J.Commelin 1689:21 = J.Commelin in HM.
Sherard 1689:310-311 = Hermann 1687.
C.Commelin 1696:4 = Hermann 1687.
Plukenet 1696:28-29 = J.Commelin in HM.
Hermann 1711:10,no.129 Acajous vel Anacardi alterius speciei Cas.Bauh. [Bauhin 1623:512].
Boerhaave 1720 vol.2:262 Acajou. Thev.Franc.Autaret.101.
J.Burman 1737:19 = Hermann 1687.
Linnaeus 1737:161 Anacardium.
Rumphius 1741 vol.1:178,Tab.LXIX (App) Cassuvium (=).
Linnaeus 1747:73-74,no.165 = Linnaeus 1737.

Linnaeus 1753:383 Anacardium occidentale.
Linnaeus 1762:548 = Linnaeus 1753.
N.L.Burman 1768:100 = Linnaeus 1762.
Houttuyn 1774 II vol.2:404-411, Pl.IX,Fig.1 = N.L.Burman 1768.

Itti-arealou HM 3:69-70:55
Itta-arealou
BL Add MS 5029:107-108.
J.Commelin in HM 3:70 Ficus Malabarica, Folio mali cotonei, Fructu exiguo, plano-rotundo, sanguineo.
Ray 1688 vol.2:1436-1437 = J.Commelin in HM.
C.Commelin 1696:29 = J.Commelin in HM.
Rumphius 1743 vol.3:142 (App), Tab.XC Varinga parvifolia (=).

Tsjerou-meer-alou HM 3:71:56
Tsjerou-meerallou
BL Add MS 5029:109-110.
J.Commelin in HM 3:56 Ficus Malabarica, folio & fructu minore.
Ray 1688 vol.2:1437 = J.Commelin in HM.
C.Commelin 1696:29 = J.Commelin in HM.

Katou-alou HM 3:73-74:57
Catu-alu
BL Add MS 5029:111-112.
J.Commelin in HM 3:74 Ficus Indica Clusii Exotic.libr.1.cap.1 & Ficus Indica foliis mali Cotonei similibus, fructu ficubus simili C.B.P.lib.12.sect.1 [Bauhin 1623:457-458].
Ray 1688 vol.2:1437 = Ficus Indica J.B.
Plukenet 1691:tab.178,fig.4 Ficus arbor Americana, Arbuti folijs, non serrata, fructu Pisi magnitudine, funiculis è ramis ad terram demissis prolifera (?) = ibid.1696:144.
C.Commelin 1696:28 = J.Commelin in HM.
Linnaeus 1753:1060 Ficus indica.
Linnaeus 1763:1514 = Linnaeus 1753.
N.L.Burman 1768:225-226 = Linnaeus 1763.
Houttuyn 1774 II vol.3:674-680 = N.L.Burman 1768.

Atti-meer-alou HM 3:75-76:58
Attii-meer-alou
BL Add MS 5029:113-114.
Ray 1688 vol.2:1438 Ficus Indica fibris ex ipso trunco exeuntibus eique accrescentibus augescente.
C.Commelin 1696:28 = Ray 1688.
Houttuyn 1774 II vol.3:682-683 Ficus benghalensis (r).

Handir-alou HM 3:77-78:59
Handur-alou
BL Add MS 5029:115-116.
J.Commelin in HM 3:78 Arbor Peregrina, Fructu ficui similis gerens Clusius.
Ray 1688 vol.2:1438 Ficus Indica fibris ex ipso trunco exeuntibus eique accrescentibus augescente 2.
C.Commelin 1696:28 = Ray 1688.
Rumphius 1743 vol.3:153-155,Tab.XCVI Ficus Septica (-); J.Burman ibid.:155 (=).
N.L.Burman 1768:226 Ficus septica.
Houttuyn 1774 II vol.3:684 = N.L.Burman 1768.

Teregam HM 3:79:60
Teregam
BL Add MS 5029:117-118.
J.Commelin in HM 3:79 Ficus Malabarica, foliis rigidis, Fructu rotundo, lanuginoso, flavescente, cerasi magnitidine.
Ray 1688 vol.2:1435 = J.Commelin in HM.

C.Commelin 1696:29 = J.Commelin in HM.
Rumphius 1743 vol.4:129 (App),Tab.LX Folium politorium (cs); J.Burman ibid.129 (r).
Rumphius 1743 vol.3:150-152,Tab.XCIV Caprificus aspera (-); J.Burmaṇ ibid.:152 (? =).
N.L.Burman 1768:226 Ficus ampelos.
Houttuyn 1774 II vol.3:683 = N.L.Burman 1768.

Perin teregam HM 3:81:61
Perim-teregam
BL Add MS 5029:119-120.
Ray 1688 vol.2:1435-1436 Ficus Malabarica foliis asperis major, fructu itidem roṭundum, lanuginoso majore.
C.Commelin 1696:29 = Ray 1688.
Rumphius 1743 vol.3:152-153, Tab.XCV Caprificus viridis (-); J.Burman ibid.:153 (=).

Valli-teregam HM 3:83:62
Valli-teregam
BL Add MS 5029:121-122.
Ray 1688 vol.2:1530 Convolvulus Indicus arborescens fructu Pruni aemulo.
Plukenet 1691:tab.237,fig.4 Uvifera arbor Americana convolvulacea fructu punctato Barbadensibus Checquer Grape (?) = ibid.1696:394.
C.Commelin 1696:21-22 = Ray 1688.
N.L.Burman 1768:227 Ficus grossularioides β.

Tsjela HM 3:85:63
Tsiela
BL Add MS 5029:123-124.
Ray 1688 vol.2:1435 Ficus Malabarica fructu Ribesii forma & magnitudine, Tsjela dicta.
Plukenet 1691:tab.143,fig.4 Arbor Maderaspatana Galactoxyli Americani foliorum aemula (?) = ibid. 1696:41.
C.Commelin 1696:29 = Ray 1688.
Plukenet 1696:145 = Ray 1688.
Plukenet 1700:75 Ficus Indica Mali Limoniae folio, subtùs canescente, fructu exiguo, cortici adnato; Sunutperai Malabarorum.
Linnaeus 1753:1060 Ficus indica β.
Linnaeus 1763:1514-1515β = Linnaeus 1753.
N.L.Burman 1768:225-226β = Linnaeus 1763.
Houttuyn 1774 II vol.3:674-680 = N.L.Burman 1768.

Tsjakela HM 3:87:64
Tsiela (!)
BL Add MS 5029:125-126.
Ray 1688 vol.2:1435 Ficus Malabarica semel in anno fructifera, fructu minimo, Tsjakela dicta.
C.Commelin 1696:29 = Ray 1688.
N.L.Burman 1768:227 Ficus Tsjakela.
Houttuyn 1774 II vol.3:684 = N.L.Burman 1768.

VOLUME 4

Mao, seu Mau, vel Mangas HM 4:1-4:1-2
Mau
BL Add MS 5029:127-128,129-130.
J.Commelin in HM 4:4 Persicae similis putamine villoso Casparo Bauhino pin.lib.11.sect.6 [Bauhin 1623:440].
Ray 1688 vol.2:1550-1551 Manga Indica fructu magno reniformi.
Sherard 1689:351 Manghas domestica.
C.Commelin 1696:43 = Ray 1688.

Plukenet 1696:43 Arbor lonchifolia Ind.Or.comantibus floribus, & ferè corymbosis. Phytogr.Tab.142.fig.1 (r).
Plukenet 1696:241 Manghas domestica.
J.Burman 1737:152-153 Mangifera arbor. Mus.Zeyl.pag.59 & 66 [Hermann 1717:59,66].
Rumphius 1741 vol.1:97 (App),Tab.XXV Manga Domestica (=); J.Burman ibid.:97 (=).
Linnaeus 1747:211,no.471 = J.Burman 1737.
Linnaeus 1753:200 Mangifera indica.
Linnaeus 1762:290 = Linnaeus 1753.
N.L.Burman 1768:62 = Linnaeus 1762.
Houttuyn 1774 II vol.2:160-164 = Mangifera Indica Syst.Nat. XII.Tom.II.Gen.276.p.183 [Linnaeus 1767 vol.2:183].

Adamaram HM 4:5-7:3-4
Adamaram
BL Add MS 5029:131-132,133-134.
J.Commelin in HM 4:7 Amygdala Indica Joh.Nieuhof in Itinerario.
Ray 1688 vol.2:1521 = J.Commelin in HM.
C.Commelin 1696:4 = J.Commelin in HM.
Plukenet 1696:28 = J.Commelin in HM.
Plukenet 1700:12 Hommottete Hottenttorum; ibidem 1696:28 Amygdala Benghalensis, cristato fructu magno, figurâ fermè primoidali.
Plukenet 1700:156; ibid.1696:306 Prunifera Fago similis arbor Gummi Elemi fundens, fructu figurâ & magnitudine Olivae ex Insulâ Barbadensi Phytogr.Tab.217.fig.4 (cf).
Petiver 1702(1701):1007-1008,no.181 Pan-neer Maraum Malab. Arbor Salawaccensis Catappae Malabaricae folio, flore extus sericeo.
Rumphius 1741 vol.1:176 (App),Tab.LXVIII Catappa (=).
Linnaeus 1767:128 Terminalia Catappa.
Houttuyn 1774 II vol.3:574-576 Terminalia Catappa Syst.Nat. XII.Gen.1283.p.638 [Linnaeus 1767 vol.2:674].

Panem-palka HM 4:9-10:5
Panam-pálca
BL Add MS 5029:135-136.
J.Commelin in HM 4:10 Nux moschata fructu oblongo. C.B. [Bauhin 1623:407].
Ray 1688 vol.2:1524-1525 Nux myristica major spuria Malabarica.
C.Commelin 1696:48 = Ray 1688.
Plukenet 1696:265 Nux Myristica spuria.
J.Burman 1737:172 Nux Myristica, oblonga, Malabarica, Mus. Zeyl.pag.59 [Hermann 1717:59].
Rumphius 1741 vol.2:24,Tab.V Nux Myristica mas (c); J.Burman ibid.:25 (=).
Linnaeus 1747:229,no.588 Myristica fructu inodoro.

Samstravadi, seu Caipa-tsjambu HM 4:11-13:6
Samstravari
BL Add MS 5029:137-138.
J.Commelin in HM 4:13 Tsjambou part.1.Tab.17.&18 (c) [Malacca schambu HM 1:27-28:17; Nati-schambu HM 1:29-30:18].
Ray 1688 vol.2:1479-1480 Jambos sylvestris Malabarica Samstravadi dictus.
C.Commelin 1696:36 = Ray 1688.
Plukenet 1700:137 Nucipomifera Arbor, foliis densioribus subtùs argenteis, floribus in praelongam spicam dispositis, fructu tetragono (...) Neereaddumba Malabarorum (?).
Rumphius 1741 vol.1:126-127,Tab.XXXVIII,Fig.2 Jambosa aquea (-); J.Burman ibid.:127 (c).
Lınnaeus 1747:85,no.191 Eugenia foliis crenatis, pomis ovatis, racemo longissimo.

Linnaeus 1753:471 Eugenia racemosa.
Linnaeus 1762:673 = Linnaeus 1753.
N.L.Burman 1768:115 = Linnaeus 1762.
Houttuyn 1774 II vol.2:535-536 = N.L.Burman 1768.

Tsjeria samstravadi HM 4:15:7
Sjería-samstravádi
BL Add MS 5029:139-140.
J.Commelin in HM 4:15 Tsjambou (s).
Ray 1688 vol.2:1480 Jambos sylvestris Samstravadi dictus alter.
C.Commelin 1696:36 = Ray 1688.
Plukenet 1696:266 Nuci pomifera Arbor Orientalis Castaneae equinae foliis, fructu longo corticoso crasso, tetragono, summo apice (Pomi in modum) umbilicato, nucleum nudum angulosum includente.
Petiver 1702(1701):937-938,no.147 Neer caddumba Malab. Samstravadi Malab. Hippocastanei foliis vix serratis.
Linnaeus 1747:85,no.190 Eugenia foliis crenatis, pedunculis terminantibus, pomis oblongis acutangulis.
Linnaeus 1753:471 Eugenia acutangula.
Linnaeus 1762:673 = Linnaeus 1753.
N.L.Burman 1768:114 = Linnaeus 1762.
Houttuyn 1774 II vol.2:534-535 = N.L.Burman 1768.

Malla-katou-tsjambou HM 4:17-18:8
Catu-tsjámbu
BL Add MS 5029:141-142.
J.Commelin in HM 4:18 Malacca-scambu & Nati-schambu part.1.Tab.17.&18 (s).
Ray 1688 vol.2:1480 Jambos sylvestris montana.
C.Commelin 1696:36 = Ray 1688.
Plukenet 1700:23 Arbor Indica Pyri densioribus & subrotundis foliis, fructu Nucis Moschatae magnitudine summo vertice coronato. Cammaulmaram Malabarorum (c).
J.Burman 1737:125 Jambos sylvestris & montana, fructu Cerasi magnitudine.
Rumphius 1741 vol.1:127-128,Tab.XXXIX Jambosa Silvestris alba (-); J.Burman ibid.:128 (=).
Linnaeus 1747:215,no.501 Maharatombola.Herm.zeyl.68 [Hermann 1717:68].

Katou-tsjeroe, seu Cheru HM 4:19-21:9
Cáttu-tsjéru
BL Add MS 5029:143-144.
Ray 1688 vol.2:1547 Prunifera Malabarica fructu racemoso parvo, acri, succo tinctorio.
Plukenet 1691:Tab.CCXVIII,fig.1 Prunifera arbor s. Nuci-prunifera folio dodrantali longitudine, laevi mollitie praedito Barbadensibus Bully-Bay nuncupatur (?).
C.Commelin 1696:56 = Ray 1688.
Plukenet 1696:306 = Plukenet 1691.

Tani HM 4:23-24:10
Táni
BL Add MS 5029:145-146.
J.Commelin in HM 4:24 Fructus in insula S.Mariae, pyra majora referens, intus mucculentum Casparo Bauhino pinac.libr.11. sect.6 [Bauhin 1623:439] (s).
Ray 1688 vol.2:1547 Prunus Indica racemosa, fructu pyriformi.
C.Commelin 1696:57 = Ray 1688.
Plukenet 1700:133 Myrobalanus Bellerica Officinar. Taunecai Malabarorum (?).
Petiver 1702(1701):844,no.96 Tauneekia Malab.Myrobalanus Bellerica Officinarum Dale Pharmacolog.p.444 (s).
Petiver 1704(1702):1056-1057,no.228 = Petiver 1702(1701) (?).
J.Burman 1737:197 Prunus Indica, sylvestris, fructu flavo, pyriformi, Dematha Zeylonensibus. Mus.Zeyl.pag.3 [Hermann 1717:3].
Rumphius 1741 vol.1:170-171,Tab.LXV Gajanus (-); J.Burman ibid.:171 (?).

Tsjem-tani HM 4:25-26:11
Tsjem-táni
BL Add MS 5029:147-148.
Ray 1688 vol.2:1556 Myxa pyriformis ossiculo trispermo.
C.Commelin 1696:46 = Ray 1688.
Plukenet 1696:306 Prunus Sebestenae similis Americana PBP. 374 [Sherard 1689:374] (?).
Plukenet 1705:178 Pruno affinis Maderaspatana, Tiliae ferè foliis, fructu racemoso pyriformi, ossiculo trinucleo.
Linnaeus 1753:1193 Rumphia amboinensis.
Linnaeus 1762:49 = Linnaeus 1753.
N.L.Burman 1768:16 = Linnaeus 1762.
Houttuyn 1774 II vol.2:53-54 = N.L.Burman 1768.

Mal-naregam, seu Malus limonia pumila sylvestris Zeylanica Dn. Hermani HM 4:27-28:12
Cátu Tsjéru Naregam
BL Add MS 5029:149-150.
J.Commelin in HM 4:28 Malus limonia pumila sylvestris Zeylanica.
Ray 1688 vol.2:1657 = J.Commelin in HM.
C.Commelin 1696:43 = J.Commelin in HM.
Petiver 1699(1698):333,no.43 Curuta chedde Mal. Limo Madraspatanus apicibus foliorum ferè sinuatis, fructu cuspidato (?).
Plukenet 1700:57 Coru Indorum Mali aureae foliis, floribus albis; Parencoruttee Malabarorum (?).
Petiver 1702(1701):848,no.107 Corutree Malab. = Petiver 1699 (1698).
Ray 1704 vol.3:80(2) = Petiver 1702(1701).
J.Burman 1737:143 Limonia Malus, sylvestris, Zeylanica, fructu pumilo (...). Mus.Zeyl.pag.63 [Hermann 1717:63].

Katou-naregam HM 4:29-30:13
Cátu-Náregam
BL Add MS 5029:151-152.
J.Commelin in HM 4:30 Limonia sp.
Ray 1688 vol.2:1463 Malus Limonia Malabarica fructu umbilicato.
Sherard 1689:338 Granata malus Zeylanica spinosa.
Plukenet 1691:tab.90,fig.6 Malus Granata Zeylanensis aculeata (?).
C.Commelin 1696:42-43 = Ray 1688.
Plukenet 1700:115; ibid.1696:210-211 Lauri folia minor ex Iava àn Laurifolia Iavanensis. CBP.461 [Bauhin 1623:461].
Ray 1704 vol.3:19(2) = Plukenet 1700.
J.Burman 1737:111 = Sherard 1689.
Rumphius 1741 vol.1:144,Tab.XLVIII Cujavus agrestis (-); J.Burman ibid.:144 (c).

Tsjerou-katou-naregam HM 4:31-32:14
Tsjéru-Catu-Náregam
BL Add MS 5029:153-154.
J.Commelin in HM 4:32 Limonia sp.
Ray 1688 vol.2:1658 Malus Limonia Indica fructu pusillo.
C.Commelin 1696:42 = Ray 1688.
Plukenet 1696:239 Malus Limonia Lentisci foliis Zeylanica, fructu minimo, Uvarum magnitudine aemulo.
Petiver 1702(1701):938,no.148 Valanga Malab. Limo Eremitana fere hexaphylla, caule alato (?).
Linnaeus 1737:489 Schinoides petiolis subtus aculeatis (?).

Linnaeus 1747:77-78,no.175 Schinus foliis pinnatis, rachi membranaceo-articulata, spicis axillaribus solitariis.
Linnaeus 1753:389 Schinus Limonia.
Linnaeus 1762:554 Limonia acidissima.
N.L.Burman 1768:102 = Linnaeus 1762.
Rumphius 1741 vol.2:133,Tab.XLIII Boa Balangan (c); J.Burman ibid.:134 (c).
Houttuyn 1774 II vol.2:441-443 Limonia pinnatifolia.

Paenoe HM 4:33-35:15
Paenù
BL Add MS 5029:155-156.
J.Commelin in HM 4:35 Jetaiba Piso rerum natural.Brasiliens. lib.4.cap.9 (r).
Ray 1688 vol.2:1482 Amygdalae affinis Indica, fructu umbilicato, nucleo nudo, cortice pulvinato trifido tecto.
C.Commelin 1696:4 = Ray 1688.
Plukenet 1696:28 = Ray 1688.
J.Burman 1737:28 Arbor Kaekuriaghaha odorata, ex qua fluit Gumm.Elemi. Mus.Zeyl.pag.52 [Hermann 1717:52].
Rumphius 1741 vol.2:151-153,Tab.XLVIII Canarium Zephyrinum, sive Silvestre primum (-); J.Burman ibid.:153 (cs).
Linnaeus 1747:91-92,no.204 Vateria.
Linnaeus 1753:515 Vateria indica.
Linnaeus 1762:734 = Linnaeus 1753.
N.L.Burman 1768:122 = Linnaeus 1762.
Houttuyn 1774 II vol.3:42-44 Vateria Indica Syst.Nat.XII.Tom. II.Gen.666.p.364 [Linnaeus 1767 vol.2:364].

Nyalel HM 4:37-38:16
Niálel
BL Add MS 5029:157-158.
J.Commelin in HM 4:38 Sambuco Indica Bontii (s).
Ray 1688 vol.2:1606-1607 Arbor baccifera racemosa, fructu corticoso, dipyreno.
Plukenet 1691:Tab.CCXXXVII,fig.5 Uvifera arbor Americana per funiculos è summis ramis ad terram usque demissis prolifera (?).
Plukenet 1696:394 = Plukenet 1691.
C.Commelin 1696:6 = Ray 1688.

Angolam HM 4:39-40:17
Angólam
BL Add MS 5029:159-160.
Ray 1688 vol.2:1497 Arbor Indica baccifera fructu umbilicato rotundo Cerasi magnitudine dicocco.
C.Commelin 1696:12 = Ray 1688.
Plukenet 1696:31 = Ray 1688 (?).
Petiver 1699(1698):313,no.1 Vellaiengeel maraum Malabar. Arbor Madraspatana floribus hexapetalis heptapetalisve, fructu coronato (cs).

Idou-moulli HM 4:41-42:18
Idú-múlli
BL Add MS 5029:161-162.
Ray 1688 vol.2:1480 Prunus fructu umbilicato, pyriformi spinosa, racemosa.
Plukenet 1691:Tab.LXIX,fig.7 Wadoukae Malabaricae haùd multum dispar, Frutex aculeatus è Maderaspatan (?).
C.Commelin 1696:56-57 = Ray 1688.
Plukenet 1696:395 = Plukenet 1691.
Plukenet 1700:133 Myrobalanus Bellerica Officinar. Taunecai Malabarorum (cs).

Poerinsii HM 4:43-44:19

Púrinsji
BL Add MS 5029:163-164.
Ray 1688 vol.2:1548 Prunifera fructu racemoso parvo, nucleo saponario.
C.Commelin 1696:60 Saponaria arbor trifolia indica Herm.Cat. 536 [Hermann 1687:536].
Plukenet 1696:265 Nux Portoricensis amplissimis foliis venosis & laete virentibus. Phytogr.Tab.208.fig.2 (?).
J.Burman 1737:209 Saponaria arbor, Zeylanica, trifolia, semine Lupini. Par.Bat.Pr.pag.373 [Sherard 1689:373] (cf).
Rumphius 1741 vol.2:134-135 Saponaria (cs); J.Burman ibid.:135 (c).
Linnaeus 1753:367 Sapindus trifoliata.
Linnaeus 1762:526 = Linnaeus 1753.
N.L.Burman 1768:91 = Linnaeus 1762.
Houttuyn 1774 II vol.2:316-317 = N.L.Burman 1768.

Adamboe HM 4:45-46:20-21
Adamboe
BL Add MS 5029:165-166,167-168.
J.Commelin in HM 4:47 Malvae seu Altheae arboreae species.
Ray 1688 vol.2:1902 Alcea Indica arborea pericarpio carnoso in plura loculamenta partito.
Breyne 1689:11 Alcea Indica arborescens, Lauri folio prima (etc.).
Sherard 1689:307 Althaea arborea Indica floribus violaceo purpureis spicatis.
C.Commelin 1696:2-3 = Ray 1688.
Plukenet 1696:16 = Ray 1688.
Hermann 1717:61 Murtughas = Sherard 1689.
J.Burman 1737:137 Ketmia Indica, foliis Laurinis, flore violaceo, spicato.
Linnaeus 1747:221-222,no.533 = Hermann 1717.

Katou-adamboe HM 4:47:22
Adamboe
BL Add MS 5029:169-170.
J.Commelin in HM 4:47 Malvae seu Altheae arboreae species.
Ray 1688 vol.2:1902 Alcea Indica arborea elatior pericarpio carnoso subaspero.
Breyne 1689:11 Alcea Indica arborescens, Lauri folio 2 (etc.).
C.Commelin 1696:3 = Ray 1688.
Plukenet 1696:16 Alcaea Indicae arboreae, genus peculiare, foliis Beidel Ossaris, Alpini, fructu intùs carnoso.

Karin-kara HM 4:49-50:23
Kariṅ-kára
BL Add MS 5029:171-172.
Ray 1688 vol.2:1663 Malus Indica, pomo corticoso Juglandi pari, monopyreno.
C.Commelin 1696:42 = Ray 1688.
Rumphius 1741 vol.2:94-95,Tab.XXIV,Fig.1 Malum Granatum (-); J.Burman ibid.:95 (=).

Perin-kara HM 4:51-52:24
Periṁkára
BL Add MS 5029:173-174.
Ray 1688 vol.2:1546 Olea sylvestris Malabarica, fructu dulci.
C.Commelin 1696:48 = Ray 1688.
Plukenet 1700:175; ibid.1696:355 Sorbi Alpinae (fortè) species Arbor Americana durioribus serratis foliis ex-Insulâ Jamaicae Phytogr.Tab.318.fig.1 (?).
J.Burman 1737:93,tab.40 Elaiocarpos folio Lauri serrato, floribus spicatis.
Linnaeus 1747:92,no.206 Elaeocarpus.

Linnaeus 1753:515 Elaeocarpus serrata.
Linnaeus 1762:734 = Linnaeus 1753.
N.L.Burman 1768:121 = Linnaeus 1762.
Houttuyn 1774 II vol.3:36-39 = N.L.Burman 1768.

Manyl-kara HM 4:53-54:25
Manil-kára
BL Add MS 5029:175-176.
J.Commelin in HM 4:54 Pruno similis fructus Chinensis Casparus Bauhinus pinacis lib.11.sect.6 [Bauhin 1623:444].
Ray 1688 vol.2:1565 Prunus Chinensis duplici in fructu ossiculo.
C.Commelin 1696:57 = Ray 1688.
Rumphius 1743 vol.3:19-21,Tab.VIII Metrosideros Macassarensis (=); J.Burman ibid.:21-22 (=).

Kara-angolam HM 4:55-56:26
Kara angolam
BL Add MS 5029:177-178.
J.Commelin in HM 4:56 Angolam Tab.17 species [HM 4:39-40: 17].
Ray 1688 vol.2:1483 Arbor Indica Prunifera fructu umbilicato corticoso Persici simili.
C.Commelin 1696:56 = Ray 1688.
Petiver 1699(1698):313,no.1 Vellaiengeel maraum Malabar. Arbor Madraspatana floribus hexapetalis heptapetalisve, fructu coronato (cs).
Plukenet 1700:307 Prunifera Arbor, foliis per siccitatem rugosis, floribus purpureis stamineis, fructu calyce eleganti coronato (...) Odingee s. Odinjee Malabarorum (?).

Theka HM 4:57-58:27
Tekka
BL Add MS 5029:179-180.
J.Commelin in HM 4:58 Kyati seu Quercus Indicae Jacobus Bontius lib.6.cap.16.
Ray 1688 vol.2:1565-1566 Prunifera vesicaria fructu multiplici, ossiculo in singulis quadrato.
C.Commelin 1696:56 = Ray 1688.
Plukenet 1696:361 = HM 4.
Plukenet 1700:178 = J.Commelin in HM.
Rumphius 1743 vol.3:35-36 (App),Tab.XVIII Jatus (=).

Katou-theka HM 4:59-60:28
Catútekka
BL Add MS 5029:181-182.
Ray 1688 vol.2:1482 Prunifera Indica fructu umbilicato racemoso Avellanae magnitudine.
C.Commelin 1696:56 = Ray 1688.
Petiver 1702(1700):708,no.68 Terrane Malab. Tekka Laurocerasi folio baccis coronatis (?).
Ray 1704 vol.3:25(2) = Petiver 1702(1700).
J.Burman 1737:159,tab.74 Microcos foliis alternis, oblongis, acuminatis (?).

Tsjerou-theka HM 4:61-62:29
Tsjéru-Téka
BL Add MS 5029:183-184.
Ray 1688 vol.2:1501 Baccifera Indica trifolia, racemosa, acinis umbilicatis tricoccis.
C.Commelin 1696:11 = Ray 1688.
Plukenet 1700:26; ibid.1696:48 Arbuscula Barbadensis amplexicaulis triphyllos, Mail-elou Malabaricae foliis tenerioribus, & petiolis alatis. Phytogr.T.145.fig.4.

Bentheka HM 4:63-64:30
Ben-téka
BL Add MS 5029:185-186.
Ray 1688 vol.2:1633 Baccifera racemosa acinis oblongis polypyrenis Arecae.
C.Commelin 1696:13 = Ray 1688.

Iripa HM 4:65-66:31
Irípa
BL Add MS 5029:187-188.
J.Commelin in HM 4:66 Clusius Exoticor.lib.2.cap.25.
Ray 1688 vol.2:1675-1676 Malus Indica pomo cucurbitaeformi monopyreno.
C.Commelin 1696:42 = Ray 1688.
Rumphius 1741 vol.1:167,Tab.LXIII Cynomorium Silvestre (-); J.Burman ibid.:167 (=).
Linnaeus 1747:74,no.167 Cynometra ramis floriferis. Act.ups 1741.p.79.
Linnaeus 1753:382-383 Cynometra ramiflora.
Linnaeus 1762:547 = Linnaeus 1753.
N.L.Burman 1768:100 = Linnaeus 1762.
Houttuyn 1774 II vol.2:403-404 = N.L.Burman 1768.

Kalesjam HM 4:67-68:32
Calesani
BL Add MS 5029:189-190.
Ray 1688 vol.2:1597 Arbor Baccifera racemosa, Vitis floribus acinis oblongis, compressis monopyrenis.
C.Commelin 1696:6 = Ray 1688.

Katou-Kalesjam HM 4:69-70:33
Catú-Calesjam
BL Add MS 5029:191-192.
J.Commelin in HM 4:70 Sorbus Malabarica.
Ray 1688 vol.2:1643 Sorbus Spuria Malabarica Katou-Kalesiam dicta.
C.Commelin 1696:64 = Ray 1688.
Plukenet 1696:355 = Ray 1688.

Ben Kalesjam HM 4:71-72:34
Ben-Calesam
BL Add MS 5029:193-194.
Ray 1688 vol.2:1786 Arbor Indica foliis alatis, flore & fructu vidua.
C.Commelin 1696:7 = Ray 1688.
Plukenet 1696:66 Arbor Indica foliis alatis, floribus & fructu nondùm compertis.

Ponga HM 4:73-74:35
Pongù
BL Add MS 5029:195-196.
J.Commelin in HM 4:74 Jaca minor sylvestris Malabarica.
Ray 1688 vol.2:1441 = J.Commelin in HM.
C.Commelin 1696:35 = J.Commelin in HM.
Plukenet 1696:92 Cenchramidea arbor pilulifera, fructu tuberculis inaequali, ex granulis coniformibus in orbem glomerato, non capsularis, Phytogr.Tab.156.fig.3 (cs).
Plukenet 1696:265 Nucifraga s. Nux Ind. Orient. Arbutus fructus similitudine (cf).
Petiver 1699(1698):329,no.34 Rutrashacaudumba Mal. Jaca Madraspatana fructu Sparganii (?).
Rumphius 1741 vol.1:154-157,Tab.LVII Cussambium (-); J.Burman ibid.:157 (cs).

Kariil HM 4:75-76:36
Káril

BL Add MS 5029:197-198.
J.Commelin in HM 4:76 Arbor prunifera Malabarica pentaphylloides, fructu calyci insidente.
Ray 1688 vol.2:1564 Prunus pentaphyllos Malabarica fructu calyci insidente.
Plukenet 1691:Tab.CCXVIII,fig.4 = Ray 1688.
C.Commelin 1696:57 = Ray 1688.
Plukenet 1696:306 = Plukenet 1691.
J.Burman 1737:169-170 Nux Zeylanica, folio multifido, digitato, flore Merdam olente. Par.Bat.Pr.pag.357 (cf).
Rumphius 1743 vol.3:169-171,Tab.CVII Clompanus minor (-); J.Burman ibid.:171 (r).
Linnaeus 1747:166,no.349 Sterculia foliis digitatis.
Linnaeus 1753:1008 Sterculia foetida.
Linnaeus 1763:1431 = Linnaeus 1753.
N.L.Burman 1768:207 = Linnaeus 1763.
Houttuyn 1774 II vol.3:437-438 = N.L.Burman 1768.

Vidimaram HM 4:77-78:37
Vidímaram
BL Add MS 5029:199-200.
J.Commelin in HM 4:78 Arbor prunifera (r).
Ray 1688 vol.2:1563-1564 Prunus Malabarica fructu racemoso, calyce excepto.
Breyne 1689:88 Sebestena domestica, Caspar. Bauhini pin. [Bauhin 1623:446].
Sherard 1689:374 = Breyne 1689 (?).
Plukenet 1691:Tab.CCXVII,fig.3 Prun; Sebestena longiori folio Maderaspatensis.
C.Commelin 1696:60 Sebesten officinarum Dale Pharm. 424.
Plukenet 1696:306 = Plukenet 1691.
Petiver 1702(1700):720,no.90 Prunifera Madraspatana, fructu mucilaginoso calyce magno (s).
Linnaeus 1737:63 Cordia foliis subovatis serratodentatis.
Rumphius 1743 vol.3:156 (App),Tab.XCVII Arbor Glutinosa (=).
Linnaeus 1753:190 Cordia Myxa.
Linnaeus 1762:273 = Linnaeus 1753.
N.L.Burman 1768:58-59 = Linnaeus 1762.
Houttuyn 1774 II vol.2:122-123 = N.L.Burman 1768.

Ponna HM 4:79-80:38
Púnna
BL Add MS 5029:201-202.
Ray 1688 vol.2:1525 Prunifera seu Nucifera Malabarica foliis Nymphaeae fructu rotundo, cortice pulvinato.
Sherard 1689:368 = HM.
Plukenet 1691:Tab.CXLVII,fig.3 Arbor Indicae Mali Medicae amplioribus folijs Maderaspatana (?).
C.Commelin 1696:56 = Ray 1688.
Plukenet 1696:41 = Plukenet 1691.
Plukenet 1700:136 Nucifera arbor sempervirens Indiarum, praelongis foliis venustè venosis, cujus lignum Redwood, (i.e.) Erythroxylon Barbadensibus, Nux verò Dhumba Ceylonensibus dicta (cf).
Petiver 1702(1700):710,no.73 Punne Maraum Malab. Ponna Malabarica major, folio pulchre venoso fructu globoso.
J.Burman 1737:131 Inophyllum flore octifido.
Rumphius 1741 vol.2:215 (App),Tab.LXXI Bintangor Maritima (c); J.Burman ibid.:216 (=).
Linnaeus 1747:90,no.201 Calophyllum foliis ovalibus.
Linnaeus 1753:513-514 Calophyllum Inophyllum.
Linnaeus 1762:732 = Linnaeus 1753.
N.L.Burman 1768:120 = Linnaeus 1762.
Houttuyn 1774 II vol.3:22-24 = N.L.Burman 1768.

Tsjerou-ponna HM 4:81:39
Tsjéru-púnna
BL Add MS 5029:203-204.
J.Commelin in HM 4:81 Kina P.Hermann.
Ray 1688 vol.2:1537 Cornus Malabarica foliis Nymphaeae.
C.Commelin 1696:23 = Ray 1688.
Plukenet 1700:57 = Ray 1688.
J.Burman 1737:130-131,tab.60 Inophyllum flore quadrifido.
Linnaeus 1737:206 Calophyllum foliis ovatis obtusis striis parallelis transversalibus.
Rumphius 1743 vol.3:98 (App),Tab.LXIV Lignum Clavorum (r).
Linnaeus 1747:90-91,no.202 Calophyllum foliis ovatis obtusis.
Linnaeus 1753:514 Calophyllum Calaba.
Linnaeus 1762:732 = Linnaeus 1753.
N.L.Burman 1768:120 = Linnaeus 1762.
Houttuyn 1774 II vol.3:24-27 = N.L.Burman 1768.

Mallam-toddali HM 4:83-84:40
Mallani-taddali
BL Add MS 5029:205-206.
Ray 1688 vol.2:1597 Baccifera Indica racemosa, florum staminulis binis, acinis monopyrenis.
C.Commelin 1696:11 = Ray 1688.
Plukenet 1696:237 = Ray 1688.
Plukenet 1696:329-330 Salvifolia arbor Orientalis foliis tenuisimè crenatis Phytogr.Tab.221.fig.4 (?).
Petiver 1702(1701):933-934,no.138 Maula poo Malab. Lotodendron Madraspat. folio longiore pubescente Mus.Petiver 656 (?).
Linnaeus 1737:83 Ulmus fructu baccato.
Linnaeus 1747:176,no.369 Celtis foliis oblique cordatis serratis: subtus villosis.
Linnaeus 1753:1044 Celtis orientalis.
Linnaeus 1763:1478 = Linnaeus 1753.
N.L.Burman 1768:60 Rhamnus Napaea Linn.sp.282 (sic); ibid.: 218 = Linnaeus 1763.
Houttuyn 1774 II vol.3:571-573 = N.L.Burman 1768:218.

Perin-toddali, seu Jujube Indica, B.Pin. HM 4:85-86:41
Perim-toddali
BL Add MS 5029:207-208.
J.Commelin in HM 4:86 Jujube Indica Casparus Bauhinus in pinace [Bauhin 1623:444].
Ray 1688 vol.2:1535 = J.Commelin in HM.
Breyne 1689:60 Jujuba Indica rotundifolia spinosa, foliis majoribus subtus lanuginosis & incanis.
J.Commelin 1689:178-179 = J.Commelin in HM.
C.Commelin 1696:38 = Breyne 1689.
Plukenet 1696:199 = J.Commelin in HM.
Petiver 1702(1700):706,no.64 Caut-Yellendae Malab. Toddali Madraspat. spinosus, folio rigido lobato subtus incano (cs).
Petiver 1702(1701):941,no.158 Yellenda-maraum Malab. Jujuba Madraspat. spinosa, foliis subtus ferrugineis (?).
J.Burman 1737:92-93,tab.39,fig.2 Elaeachnus foliis rotundis, maculatis (cf).
J.Burman 1737:131-132,tab.61 Jujuba aculeata, nervosis foliis, infra sericeis, flavis.
Rumphius 1741 vol.2:117-119,Tab.XXXVI Malum Indicum (-); J.Burman ibid.:119 (c).
Linnaeus 1762:282 Rhamnus Jujuba.
N.L.Burman 1768:60 = Linnaeus 1762.
Houttuyn 1775 II vol.4:293-294 = N.L.Burman 1768.

Kadali HM 4:87-88:42
Kadali

BL Add MS 5029:209-210.
J.Commelin in HM 4:88 Cysti species.
Ray 1688 vol.2:1493 Baccifera Indica fructu umbilicato quinquecapsulari polyspermo.
Breyne 1689:97 Stramonia, sive Datura cistioides frutescens, hirsuta major Indica.
C.Commelin 1696:12 = Ray 1688.
J.Burman 1737:155-156,tab.73 Melastoma quinquenervia, hirta, major, capitulis sericeis, villosis (cs).
Rumphius 1743 vol.4:137-138 (App),Tab.LXXII Fragarius niger (s).
Linnaeus 1747:76,no.171 Melastoma foliis lanceolato-ovatis scabris quinquenervis.
Linnaeus 1753:390 Melastoma malabathrica.
Linnaeus 1762:559 = Linnaeus 1753.
N.L.Burman 1768:104 = Linnaeus 1762; ibid.:105 Melastoma aspera Linn.sp.560.
Houttuyn 1774 II vol.2:450-451 = N.L.Burman 1768:104.

Ben kadali HM 4:89:no figure

BL Add MS 5029:no drawing.
Ray 1688 vol.2:1493 Bacccifera indica flore albicante, fructu viridi, pulpa albicante.
C.Commelin 1696:12 = Ray 1688.

Katou-kadali HM 4:91:43
Kalóù-Kadáli
BL Add MS 5029:211-212.
J.Commelin in HM 4:91 Cysti Malabarici genus.
Ray 1688 vol.2:1493 Baccifera indica floribus minoribus, fructus cortice aspero.
Plukenet 1691:Tab.CLXI,fig.2 Cistus Chamaerhododendros, s. Ledum Orientale pentaneuros, foliis brevioribus, ferrugineâ, & molli lanugine villosis (?).
C.Commelin 1696:12 = Ray 1688.
Plukenet 1696:106 = Plukenet 1691.
J.Burman 1737:154-155 Melastoma quinquenervia, minor, capitulis villosis.
J.Burman 1737:155-156, tab.73 Melastoma quinquenervia, hirta, major, capitulis sericeis, villosis.
Rumphius 1743 vol.4:135-136,Tab.LXXI Fragarius ruber (-); J.Burman ibid.:136 (=).
Linnaeus 1747:76,no.172 Melastoma foliis lanceolatis trinerviis scabris.
Linnaeus 1753:391 Melastoma aspera.
Linnaeus 1762:560 = Linnaeus 1753.
N.L.Burman 1768:104-105 Melastoma hirta Linn.sp.559.
Houttuyn 1774 II vol.2:453-454 Melastoma aspera [Linnaeus 1767 vol.2:298].

Tsjerou Kadali HM 4:93:44
Tsjérou-Kadali
BL Add MS 5029:213-214.
J.Commelin in HM 4:93 Cysti Indici species.
Ray 1688 vol.2:1493 Baccifera indica foliis, floribus & fructibus minoribus.
C.Commelin 1696:12 = Ray 1688.
Plukenet 1700:49 Cistus Orientalis pulpifer Jujubinis foliis trinervis, capsulâ parvâ.

Oepata HM 4:95-96:45
Oépata
BL Add MS 5029:215-216.
J.Commelin in HM 4:96 Anacardo (s).
Ray 1688 vol.2:1566-1567 Arbor Indica fructu conoide, cortice pulvinato nucleum unicum, nullo ossiculo tectum claudente.
C.Commelin 1696:4 Anacardium B:Pin:511. & variorum [Bauhin 1623:511-512].
Plukenet 1696:28 Anacardium Orientale.
Linnaeus 1747:23,no.57 Avicennia.
Linnaeus 1753:110-111 Avicennia officinalis.
Linnaeus 1763:891 Bontia germinans.
N.L.Burman 1768:138 = Linnaeus 1763.

Wadouka HM 4:97:46
Wadoúka
BL Add MS 5029:217-218.
Ray 1688 vol.2:1607 Baccifera spinosa Indica racemosa acinis globosis, gemino in singulis nucleo.
C.Commelin 1696:13 = Ray 1688.

Rava-pou HM 4:99-100:47-48
Ráva-pù
BL Add MS 5029:219-220,221-222.
J.Commelin in HM 4:100 Jasminum Indicum Bacciferum flore albo majore, noctu olente. Seu Arbor tristis de die altera.
Ray 1688 vol.2:1602 = J.Commelin in HM.
C.Commelin 1696:37 = Ray 1688.
Linnaeus 1737:5 Nyctanthes caule volubili, foliis subovatis acutis (cs).
Linnaeus 1747:85,no.190 Eugenia foliis crenatis, pedunculis terminantibus, pomis oblongis acutangulis.
Linnaeus 1753:6 Nyctanthes hirsuta.
Linnaeus 1762:8 = Linnaeus 1753.
N.L.Burman 1768:4 = Linnaeus 1762.
Houttuyn 1774 II vol.2:41-42 = N.L.Burman 1768.

Anavinga HM 4:101-102:49
Ana-vínga
BL Add MS 5029:223-224.
Ray 1688 vol.2:1632 Baccifera Indica fructu rotundo, cuspidato, Cerasi magnitudine, polypyreno.
C.Commelin 1696:13 = Ray 1688.
J.Burman 1737:111-112,tab.48 Grossularia spinis vidua, baccis in racemo congestis, spadiceis, foliis crenatis, ovato-acuminatis (?).

Courondi HM 4:103-104:50
Corondi
BL Add MS 5029:225-226.
J.Commelin in HM 4:104 Corundi Zanonius hist. Botanic. cap. 13.
Ray 1688 vol.2:1664 Arbor Indica fructu rotundo, cortice molli nucleum unicum nudum glandi similem continente.
C.Commelin 1696:7 = Ray 1688.
Plukenet 1700:156; ibid.1696:307 Prunifera arbor Americana fructu luteo ovali, ossiculi majore, quorum nuclei ad Porcos saginandos ipsis glandibus praeferuntur (s).

Bengieri HM 4:105-106:51
Bengíri
BL Add MS 5029:227-228.
J.Commelin in HM 4:106 Avenacou seu Ricini species part.2.Tab.32.33.34 [HM 2:57-63:32-34].
Ray 1688 vol.2:1856 Ricini Indici species.
C.Commelin 1696:58 = Ray 1688.
Plukenet 1696:320 Ricinus Indicus Patsjoti Malabaricae foliis, fructu majore rotundo hexagono, Nilicamaram aemulo.

Aria-bepou. s: nimbo Acostae HM 4:107-108:52
Aria-Bepoú

BL Add MS 5029:229-230.
J.Commelin in HM 4:108 Nimbo Acosta cap.39.& Garcia ab Horto lib.2.cap.2.
Hermann 1687:652 Azedarach floribus albis semper virens.
Ray 1688 vol.2:1545 Olea Malabarica Nimbo dicta, fructu racemoso oblongo.
Breyne 1689:21 Azadirachta Indica, foliis Fraxini, sive non ramosis majoribus, flore minore albo.
J.Commelin 1689:33-34 Arbor Indica Fraxino similis, Oleae fructu B.Pin. [Bauhin 1623:416].
Plukenet 1691:Tab.CCXLVII,fig.1 Olea Malabarica Fraxineo folio è Maderaspatan.
C.Commelin 1696:11 = Breyne 1689.
Plukenet 1696:269 = Plukenet 1691.
Petiver 1702(1701):945-946,no.179 Waapa maraum Malab. Azedarach Malabarica Fraxini tenuiore folio.
Boerhaave 1720 vol.2:236 Azeradach; sempervirens, & florens.T.616.
J.Burman 1737:40 Azedarach fructu polypyreno, Mangoseros sylvestris. Mus.Zeyl.pag.3 [Hermann 1717:3] (cs).
J.Burman 1737:40-41,tab.15 Azedarach foliis falcato serratis (cs).
Linnaeus 1737:161 Melia foliis pinnatis.
Linnaeus 1747:71-72,no.161 = Linnaeus 1737.
Linnaeus 1753:385 Melia Azadirachta.
Linnaeus 1762:550 = Linnaeus 1753.
N.L.Burman 1768:101 = Linnaeus 1762.
Houttuyn 1774 II vol.2:433-434 = N.L.Burman 1768.

Karibepou. s. Nimbo altera HM 4:109-110:53
Karì-bépu
BL Add MS 5029:231-232.
J.Commelin in HM 4:110 Arbor Indica Fraxino similis altera, flore albo, fructu rotundo nigro, seu Nimbo altera.
Ray 1688 vol.2:1545 Olea Malabarica Nimbo dicta, fructu racemoso rotundo.
Breyne 1689:21 Azadirachta Indica, foliis Fraxini, sive non ramosis minoribus, flore parvo albo.
C.Commelin 1696:11 = Ray 1688.
Plukenet 1696:269 Olea Malabarica Nimbo dicta fructu rotundiore.

Kari-vetti HM 4:111-112:54
Karì-vettì
BL Add MS 5029:233-234.
J.Commelin in HM 4:112 Pevetti (s) [HM 4:113:55].
Ray 1688 vol.2:1596 Arbor baccifera Indica racemosa, acinis oblongis monopyrenis, flore tetrapetaloide.
Plukenet 1691:Tab.CCVI,fig.6 Olea Laurino folio Portoricensis, summo margine crenato (s).
C.Commelin 1696:6 = Ray 1688.
Plukenet 1696:269 = Plukenet 1691.

Pevetti HM 4:113:55
Pêe-vettì
BL Add MS 5029:235-236.
J.Commelin in HM 4:55 Solanum somniferum antiquorum Prosper Alpinus de plantis exoticis libr.1.cap.33 (s).
Ray 1688 vol.2:1632 Baccifera Indica floribus ad foliorum exortus, fructu sulcato decapyreno.
J.Commelin 1689:333 = J.Commelin in HM.
C.Commelin 1696:64 Solanum somniferum verticillatum B.Pin. 166 [Bauhin 1623:166].
Plukenet 1696:352 Solanum Verticillatum JB Tom.3.610.
J.Burman 1737:10 Alkekengi somniferum Cydoniae folio, flore & fructu rubris.

Linnaeus 1737:62 Physalis caule fruticoso tereti, foliis ovatis integerrimis, floribus confertis.
Linnaeus 1747:40,no.96 = Linnaeus 1737.
Linnaeus 1753:182 Physalis flexuosa.
Linnaeus 1762:261 = Linnaeus 1753.
N.L.Burman 1768:54 = Linnaeus 1762.
Houttuyn 1775 II vol.4:236 = N.L.Burman 1768.

Noeli-tali. seu Berberis Indica aurantiae folio HM 4:115-116:56
Núli-táli
BL Add MS 5029:237-238.
J.Commelin in HM 4:116 Berberis Indica aurantiae folio.
Ray 1688 vol.2:1606 = J.Commelin in HM.
C.Commelin 1696:15 = J.Commelin in HM.
Plukenet 1696:67 Berberis Indica Aurantiae folio.
J.Burman 1737:22,tab.10 Antidesma spicis geminis.
Linnaeus 1747:169,no.357 Antidesma.
Linnaeus 1753:1027 Antidesma alexiteria.
Linnaeus 1763:1455-1456 = Linnaeus 1753.
N.L.Burman 1768:212 = Linnaeus 1763.
Houttuyn 1774 II vol.3:510-512 = N.L.Burman 1768.

Poutaletsje HM 4:117-118:57
Poútaletsje
BL Add MS 5029:239-240.
J.Commelin in HM 4:118 Ligustri species.
Ray 1688 vol.2:1634 Baccifera Indica baccis oblongis in umbellae formam dispositis.
C.Commelin 1696:12 = Ray 1688.
Plukenet 1696:305 Poutaletsjae Malabarorum, similis, Arbuscula Maderaspatensis. Phytogr.Tab.54.fig.1.
J.Burman 1737:142 Ligustrum Indicum, seu Alcanna Mus.Zeyl. pag.65 [Hermann 1717:65] (?).
Rumphius 1741 vol.1:153-154,Tab.LV Lansium Silvestre (-); J. Burman ibid.:154 (cs).
Linnaeus 1747:56-57,no.135 Lawsonia ramis inermibus.
Linnaeus 1753:349 Lawsonia inermis.
Linnaeus 1762:498 = Linnaeus 1753.
N.L.Burman 1768:88 = Linnaeus 1762.
Houttuyn 1775 II vol.4:474-478 Lawsonia inermis Syst.Nat. XII.Gen.477.p.267 [Linnaeus 1767 vol.2:267].

Modagam HM 4:119:58
Módagam
BL Add MS 5029:241-242.
Ray 1688 vol.2:1644 Pomifera Indica flore Rhododendri, fructu Pyriformi.
C.Commelin 1696:55 = Ray 1688.
Petiver 1699(1698):330-331,no.37 Neer Caudumba. Pentaphlora Madraspatana arborescens Benzoini foliis (s).

Bella modagam HM 4:121-122:59
Béla-Módagam
BL Add MS 5029:243-244.
Hermann 1687:652 = HM.
Ray 1688 vol.2:1481 Prunifera Indica fructu umbilicato sulcato, Grossulariae simili.
J.Commelin 1689:49 = HM.
Sherard 1689:317 = HM.
C.Commelin 1696:56 = Ray 1688.
Plukenet 1696:361 Takkada Frutex Zeylanensium (?).
Hermann 1717:45 Takkada. Arbor exitiosa marina lactescens Indica Takkada vocata fructu Cerasi magnitudine incarnato striato.
J.Burman 1737:29-30 = Hermann 1717.

Rumphius 1743 vol.4:118 (App),Tab.LIV Buglossum litoreum (r); J.Burman ibid.:118 (r).
Linnaeus 1747:213,no.489 = Hermann 1717.

Tondi teregam HM 4:123-124:60
Tóndi-Teregam
BL Add MS 5029:245-246.
Ray 1688 vol.2:1787-1788 Arbor flore tetrapetalo odorato, fructu nullo.
C.Commelin 1696:7 = Ray 1688.

Not in HM 4

BL Add MS 5029:247-248; an unfinished drawing.

Ramena-pou-maram HM 4:125:61
Ramena-pú-maram
BL Add MS 5029:249-250.
Ray 1688 vol.2:1635 Baccifera Indica umbellata flore pallido pentapetalo, raro fructus ferens.
C.Commelin 1696:12 = Ray 1688.

VOLUME 5

Mail-elou HM 5:1-2:1
Mail-Eloú
Bl Add MS 5030:2-3.
J.Commelin in HM 5:4 Tini species.
Ray 1688 vol.2:1557-1558 Arbor baccifera trifolia Malabarica simplici ossiculo cum pluribus nucleis.
Breyne 1689:104 Vitex trifolia major Indica, fructu carnoso, floribus minoribus & rarioribus.
C.Commelin 1696:69 = Breyne 1689.
Plukenet 1696:391 = Breyne 1689.

Katou-mail-elou HM 5:3-4:2
Katoú-Mail-Eloú
BL Add MS 5030:4-5.
J.Commelin in HM 5:4 Tini species.
Ray 1688 vol.2:1558 Arbor baccifera Malabarica, folio pinnato, flor. umbellatis simplici ossiculo cum pluribus nucleis.
Breyne 1689:105 Vitex maxima Indica, fructu carnoso, floribus majoribus & densioribus.
C.Commelin 1696:69 = Breyne 1689.
Plukenet 1696:48 Arbor Venenatae quorundum similis trifolia, è Maderaspatan (?).
Petiver 1702(1701):847,no.105 Punnunga Narree Malab. Coccifera Madraspat. racemosa Urucu folio molli flavescente Mus.Petiver.377 (?).

Parili HM 5:5:3
Parili
BL Add MS 5030:6-7.
Ray 1688 vol.2:1557 Arbor baccifera Malabarica ossiculo fructus trispermo.
C.Commelin 1696:6 = Ray 1688.

Beênel HM 5:7-8:4
Beênel
BL Add MS 5030:8-9.
Ray 1688 vol.2:1557 Frutex baccifer Malabaricus floribus umbellatis simplici ossiculo tetraspermo.
C.Commelin 1696:30 = Ray 1688.
Plukenet 1696:64 = Ray 1688.

Plukenet 1700:153; ibid. 1696:300 Pneumatoxylum Arbor baccifera calyculata, tetrapyrene, alatis foliis Americana; Barbadensibus Spirit-wood vocata. Phytogr.Tab.215.fig.5 (s).
N.L.Burman 1768:206,Tab.62,f.2 Croton racemosum β.

Patsjotti HM 5:9-10:5
Patsjotti
BL Add MS 5030:10-11.
J.Commelin in HM 5:10 Myrtus americana sylvestris Pisonis (r).
Ray 1688 vol.2:1498 Baccifera Indica fructu umbilicato racemoso, monopyreno.
C.Commelin 1696:12 = Ray 1688.
N.L.Burman 1768:203,Tab.61,f.2 Acalypha spiciflora γ.

Perin-Patsjotti HM 5:11:6
Perin-Patsjóti
BL Add MS 5030:12-13.
Ray 1688 vol.2:1624 Frutex baccifer Malabaricus flosculis pentapetalis ramis inhaerentibus, fructu tetrapyreno.
C.Commelin 1696:30 = Ray 1688.

Katou-Patsjotti HM 5:13:7
Katoú-Patsjótti
BL Add MS 5030:14-15.
Ray 1688 vol.2:1572 Frutex baccifer Malabaricus fructu calyce excepto, sulcato tripyreno.
C.Commelin 1696:30 = Ray 1688.
Petiver 1702(1701):845,no.98 Tirnama pollee Malab. Patsjotti Zeylanica Lauro Cerasi folio Leviter serrato (s).
J.Burman 1737:185 = Petiver 1702(1701).
N.L.Burman 1768:205,Tab.64,f.1 Croton castaneifolium β.

Acara-Patsjotti HM 5:15:8
Akára-Patsjóti
BL Add MS 5030:16-17.
Ray 1688 vol.2:1572-1573 Baccifera Indica fructu calyculato, subrotundo, cuspidato, quadripartito, tetrapyreno.
Plukenet 1691:Tab.CXLII,fig.2 Arbor Indica Mali aurantiae folijs obtusioribus è Maderaspatan. (?).
C.Commelin 1696:11 = Ray 1688.
Plukenet 1696:43 = Plukenet 1691.
N.L.Burman 1768:121 = Calophyllum Akara.

Mala poenna HM 5:17-18:9
Mála-Poenna
BL Add MS 5030:18-19.
Ray 1688 vol.2:1596 Arbor Baccifera Indica fructu cuspidato monopyreno.
C.Commelin 1696:6 = Ray 1688.
Plukenet 1696:237 = Ray 1688.

Pavetta, seu Malleamothe HM 5:19-20:10
Pavetta
BL Add MS 5030:20-21.
J.Commelin in HM 5:20 Arbor Malabarensium fructu Lentisci C.Bauhin [1623:399].
Ray 1688 vol.2:1581-1582 Pavate Park. Acostae.
C.Commelin 1696:51 Pavate arbor foliis mali aureae J:B:T:2.L: XV.102.
Plukenet 1696:283 = Ray 1688.
Linnaeus 1747:23,no.56 Pavetta.
Linnaeus 1753:110 Pavetta indica.
Linnaeus 1762:160 = Linnaeus 1753.
N.L.Burman 1768:35 = Linnaeus 1762.
Houttuyn 1775 II vol.4:129-130 = N.L.Burman 1768.

Tsjeriam-cottam HM 5:21-22:11
Tsjeriaṁ-Cottaṁ
BL Add MS 5030:22-23.
Ray 1688 vol.2:1596 Frutex Indicus baccifer fructu racemoso, cuspidato, Ribium simili, monopyreno.
C.Commelin 1696:31 = Ray 1688.
Plukenet 1696:378 = Ray 1688.
Plukenet 1700:22 Arbor Indica Ovali folio flosculis plurimis in spicis summo ramulo dispositis acinifera, Pulicheemaram Malabarorum (?).
Petiver 1702(1701):853-854,no.122 Pulichee-maraum Malab. Baccifera Madraspat. Ribis more, floribus muscosis Juli instar Mus.Petiver.621 (?).

Basaal HM 5:23-24:12
Besââl
BL Add MS 5030:24-25.
Ray 1688 vol.2:1570 Baccifera Indica fructu cuspidato monopyreno, calyce residuo excepto.
C.Commelin 1696:11 = Ray 1688.

Kare-kandel HM 5:25-26:13
Kare-Kandel
BL Add MS 5030:26-27.
J.Commelin in HM 5:26 Kandol Zanonius cap.13.
Ray 1688 vol.2:1498 Baccifera Indica umbellata, fructu umbilicato, striato, monopyreno.
C.Commelin 1696:11 = Ray 1688.

Corinti Panel HM 5:27:14
Corinti-Panel
BL Add MS 5030:28-29.
J.Commelin in HM 5:27 Nimbo (?).
Ray 1688 vol.2:1594 Frutex Indicus baccifer, flore hexapetalo, fructu rotundo, racemoso, monopyreno.
C.Commelin 1696:31 = Ray 1688.
Plukenet 1696:120 Frutex Malabaricus baccifer, floribus hexapetalis, racematim dispositis, fructu rubro, monopyreno.

Perin-Panel HM 5:29-30:15
Perin-Panel
BL Add MS 5030:30-31.
Ray 1688 vol.2:1625 Frutex Indicus baccifer, floribus racemosis fructu oblongo, tetrapyreno.
Plukenet 1691:Tab.CXL,fig.2 Arbor Orientalis baccifera, Laurifolijs crassis, & venosis, per siccitatem atronitentibus, quasi Vernice tinctis, polypyrene (?).
C.Commelin 1696:31 = Ray 1688.
Plukenet 1696:42 = Plukenet 1691.
Plukenet 1696:288 = Ray 1688.
Plukenet 1700:19 Arbor baccifera Indica Lauri subrotundo folio flosculis numerosis ad Lichenis petraei stellata capitula propè accedentibus Cnraumchidde Malabaror. (?).

Tsjérou-pánel HM 5:31:16
Tsjérou-Pánel
BL Add MS 5030:32-33.
Ray 1688 vol.2:1594 Frutex Indicus baccifer hexapetalos, fructu rotundo monopyreno nigro.
C.Commelin 1696:31 = Ray 1688.
Petiver 1702(1701):940,no.156 Malle-cungee Malab. Panel Eremitana Lauri folio venoso (?).

Kaltsjerou-panel, seu Panel montana minor HM 5:33:17
Kal Tsjeroú Panel
BL Add MS 5030:34-35.
Ray 1688 vol.2:1594 Panella minor foliis minoribus aromaticis.
C.Commelin 1696:50 = Ray 1688.

Katsjau-Panel, seu Panel sylvestris HM 5:35:18
Kasjoú Panel
BL Add MS 5030:36-37.
Ray 1688 vol.2:1594 Praecedentis fruticis [Tsjerou-panel HM 5:31:16] minor species.
C.Commelin 1696:50 = HM.
Plukenet 1696:46 Arbuscula Maderaspatana Ligustri facie, folio Visci arborei colore & consistentiâ pari. Phytogr.Tab.143.fig.1 (?).
Plukenet 1696:90 = Ray 1688.

Kasjavo maram HM 5:37:19
Kasjavo-Maram
BL Add MS 5030:38-39.
Ray 1688 vol.2:1595 Arbor baccifera Indica racemosa tetrapetalo flore fructu rotundo monopyreno.
C.Commelin 1696:6 = Ray 1688.

Belluta-kanneli HM 5:39-40:20
Bellútta-Kanneli
BL Add MS 5030:40-41.
Ray 1688 vol.2:1498-1499 Baccifera Indica fructu umbilicato, racemoso, candido monopyreno, rotundo.
C.Commelin 1696:11 = Ray 1688.
Plukenet 1700:39 Caryophyllus aromaticus Ind. Orient. acuminato folio minore, fructu rotundo monopyrene. Nauvelmaram Malabarorum.

Ponnagam HM 5:41-42:21
Ponnagàm
BL Add MS 5030:42-43.
Ray 1688 vol.2:1496 Baccifera Indica fructu glabro tribus loculis terna semina continente.
C.Commelin 1696:12 = Ray 1688.

Tsjerou-ponnagam HM 5:43:22
Tsjerú-Ponnagam
BL Add MS 5030:44-45.
Ray 1688 vol.2:1496 Baccifera Indica fructu lanuginoso, cum ossiculo tribus loculis terna semina continente.
C.Commelin 1696:12 = Ray 1688.

Pee-tsjerou-ponnagam HM 5:45:23
Pee-Tsjeroú-Ponnagam
BL Add MS 5030:46-47.
Ray 1688 vol.2:1611 = HM.
Plukenet 1700:22 Arbuscula baccifera Asclepiadis folio fructu Tiliae tetragono. Chermurre Malabarorum (r).
Petiver 1704(1703):1455,no.25 Ricinus Madraspat. pediculus apiculatis folijs Aurantij (?).
N.L.Burman 1768:203,Tab.61,f.2 Acalypha spiciflora β.

Pee-ponnagam HM 5:47:24
Peê-Ponnagaṁ
BL Add MS 5030:48-49.
Ray 1688 vol.2:1495 Baccifera Indica fructu cum quadripartito in vertice umbilico, quadricapsulari.
C.Commelin 1696:12 = Ray 1688.
Plukenet 1696:284 = Ray 1688.
NB The pagination of the description has been misprinted as 37.

Molago-maram HM 5:49:25
Molagò-Maraṁ

BL Add MS 5030:50-51.
Ray 1688 vol.2:1593-1594 Baccifera Indica trifolia, fructu rotundo monopyreno, pediculo longo.
C.Commelin 1696:11 = Ray 1688.
Plukenet 1696:48 Arbuscula Jamaicensis baccifera, Hederae Virginianae triphyllae foliis crassioribus subtùs lanuginosis. Phytogr.Tab.267.fig.1 (c).
Plukenet 1696:94 = Ray 1688.
Plukenet 1700:171 Serpentaria trifoliata, s. Saururus Indiae Orientalis triphyllos, foliis subtùs lanatis, spicis ex alis summo apice compressiusculis, & tridentatis. Shalamaram Malabarorum (s).
Petiver 1702(1701):1013,no.186 Shala-maraum Malab. Rus trifoliatae facie Frutex Salawaccensis floribus Juliformibus Mus.Petiver.678 (?).
Linnaeus 1762:381-382 Rhus Cominia.
N.L.Burman 1768:75 = Linnaeus 1762.
Houttuyn 1774 II vol.2:219 = N.L.Burman 1768.

Mail-ombi HM 5:51:26
Mâil-Kómbi
BL Add MS 5030:52-53.
J.Commelin in HM 5:51 Berberis species.
Ray 1688 vol.2:1500 Baccifera Indica racemosa fructu umbilicato rotundo monopyreno.
C.Commelin 1696:11 = Ray 1688.
Plukenet 1700:31 Berberidis affinis.
Rumphius 1741 vol.1:183-184,Tab.LXXIII Gnemon Silvestris (-); J.Burman ibid.:184 (cs).

Njara HM 5:53:27
Njara
BL Add MS 5030:54-55.
Ray 1688 vol.2:1499 Baccifera Indica umbellata, fructu umbilicato nigro, ossiculo intus unico.
Plukenet 1691:Tab.XLIX,f.2 = HM.
C.Commelin 1696:11 = Ray 1688.
Plukenet 1696:263 = Plukenet 1691.

Kaka-njara HM 5:55:28
Káka Njara
BL Add MS 5030:56-57.
Ray 1688 vol.2:1571 Baccifera Indica fructu oblongo, calyci insidente monopyreno, ossiculo compresso.
C.Commelin 1696:11 = Ray 1688.

Perin-Njara HM 5:57-58:29
Perin-Njara
BL Add MS 5030:58-59.
Ray 1688 vol.2:1499 Baccifera Malabarica fructu umbilicato Pruniformi, unico intus nucleo.
C.Commelin 1696:13 = Ray 1688.
Plukenet 1700:39; ibid. 1696:88 Caryophyllus languescente vi aromaticus, Malabariensis, folio, & fructu maximo, Phytogr. Tab.274.fig.2.
Petiver 1702(1701):1012-1013,no.184 Nauvel-maraum Malab. Pimenta Malabarica Caryophylli aromatici folio.

Vetadagou HM 5:59:30
Vetadagú
BL Add MS 5030:60-61.
Ray 1688 vol.2:1633 Baccifera Indica fructu rotundo atropurpureo pentacocco.
C.Commelin 1696:13 = Ray 1688.
Plukenet 1696:391 Vitis Idaeae species Malabarica.

Kal-vetadagou HM 5:61:31
Kal-Vetadagoú
BL Add MS 5030:62-63.
J.Commelin in HM 5:61 Vitis Idaeae species.
Ray 1688 vol.2:1633 = HM.
C.Commelin 1696:69 = J.Commelin in HM.
Plukenet 1696:391 Vitis Idaea Malabarica baccis aurantiacis, Kal-Vetadagou dicta.

Watta-tali HM 5:63-64:32
Wátta Tály
BL Add MS 5030:64-65.
Ray 1688 vol.2:1595 Baccifera Malab. floribus spicatis dipetalis, fructu monopyreno.
C.Commelin 1696:13 = Ray 1688.
Plukenet 1696:395 = Ray 1688.
N.L.Burman 1768:203,Tab.61,fig.1 Acalypha hispida.
Houttuyn 1776 II vol.6:321-324 Caturus Spiciflorus Syst.Nat. XII.Gen.1280,p.650 [Linnaeus 1767 vol.2:650].

Karetta amelpodi HM 5:65-66:33 (right)
Káreta-Amél-Podì (tab.33 right)
BL Add MS 5030:66.
Ray 1688 vol.2:1611 Baccifera Indica floribus umbellatis, fructu rotundo tricocco.
C.Commelin 1696:13 = Ray 1688.
C.Commelin 1701:1 Alaternoides species (r).

Katou bellutta amelpodi HM 5:66:33 (left)
Katoú Velútta Amél-Podì (tab.33 left)
BL Add MS 5030:67.
Ray 1688 vol.2:1612 = HM.
C.Commelin 1696:39 = HM.
C.Commelin 1701:1 Alaternoides species (r).

Mouli-ila, seu Moul elavou HM 5:67-68:34
Mouli-ila
BL Add MS 5030:68-69.
Ray 1688 vol.2:1658 Malus Limonia Indica, floribus umbellatis, fructu parvo.
Plukenet 1691:Tab.CCXXXIX,fig.6 Zanthoxylum Americanum s. Hercules Arbor aculeata major, Juglandis foliis alternis (?).
C.Commelin 1696:42 = Ray 1688.
Plukenet 1696:396 = Plukenet 1691.
Plukenet 1696:396 Zanthoxylum aculeatum Fraxini smuosis (sic), & punctatis foliis Americanum Prickly Pellow-wood Barbadensibus dictum, & interdùm à foliorum vi causticâ Pellitory of Spain Tree Phytogr.Tab.239.fig.4.
Plukenet 1705:77-78 Euonymo adfinis aromatica, s. Zanthoxylum latiore Fraxini folio conjugato, minus spinosum; ex Insulâ Cheusán.

Ben-kara HM 5:69-70:35
Benkara
BL Add MS 5030:70-71.
J.Commelin in HM 5:74 Rhamnus species.
Ray 1688 vol.2:1494-1495 Baccifera Indica, floribus spicatis fructu rotundo, nigricante, polyspermo.
C.Commelin 1696:12 = Ray 1688.
Plukenet 1700:57 Coru Indorum Limoniae folio grandioribus aculeis horrida. Perencalla Malabareis.
Petiver 1702(1701):1013,no.185 Peren-calla Malab. Kara Salawaccensis major, fol.longiore venoso (?).

Kanden-kara HM 5:71-72:36

Kanden-kara
BL Add MS 5030:72-73.
Ray 1688 vol.2:1606 Baccifera Indica floribus racemosis, fructu plano-rotundo, dipyreno.
C.Commelin 1696:11 = Ray 1688.
Petiver 1702(1701):843,no.93 Tetum-cootan Malabar. Lycium Chamberambacum Laurifoliis floribus comosis (?).
Plukenet 1705:137 Lycium Prunifolium Africanum, spinis carens, fructu Guaiaci compresso, binis ossiculis, Sebestenae sapore (s).

Tsjerou kara HM 5:73-74:37
Tsjéru-Kára
BL Add MS 5030:74-75.
J.Commelin in HM 5:74 Rhamnus species.
Ray 1688 vol.2:1497 Baccifera Indica flosculis ad foliorum exortum confertis, fructu dicocco.
Plukenet 1691:Tab.XCVII,f.3 Lycium Bisnagaricum acuminatis minus durioribus foliis, & aculeis ex opposito binis.
C.Commelin 1696:12 = Ray 1688.
Plukenet 1696:234 = Plukenet 1691.
Petiver 1702(1701):1019,no.216 Caurai chedde Malab. Lycium Salawaccense spinis binis distentis, nucleo gemino vel triquetro (?).
Ray 1704 vol.3:33(2) = Petiver 1702(1701) (?).

Taliir-kara HM 5:75:38
Talir-Cára
BL Add MS 5030:76-77.
Ray 1688 vol.2:1787 Arbor Indica spinosa flore & fructu vidua.
C.Commelin 1696:7 = Ray 1688.

Courou-moelli HM 5:77:39
Coúroú-Moelli
BL Add MS 5030:78-79.
J.Commelin in HM 5:77 Rhamnus species.
Ray 1688 vol.2:1634 Frutex Indicus baccifer fructu rotundo Polypyreno.
C.Commelin 1696:31 = Ray 1688.
Plukenet 1700:122 Lycium Africanum, non serratum, Prunifolium, minus, ex Insulâ Rhodiensi (?).
Linnaeus 1737:69 Sideroxylum spinosum.
Linnaeus 1753:193 Sideroxylon spinosum.
Linnaeus 1762:279 = Linnaeus 1753.
N.L.Burman 1768:59-60 = Linnaeus 1762.
Houttuyn 1774 II vol.2:157-158 = N.L.Burman 1768.

Sondari HM 5:79-80:40
Sundari
BL Add MS 5030:80-81.
J.Commelin in HM 5:80 Undari Zanonius cap.13.
Ray 1688 vol.2:1624 Frutex Indicus baccifer floribus umbellatis, fructu tetracocco.
C.Commelin 1696:31 = Ray 1688.
Plukenet 1700:153; ibid. 1696:300 Pneumatoxylum Arbor baccifera calyculata, tetrapyrene, alatis foliis Americana; Barbadensibus Spirit-wood vocata. Phytogr.Tab.215.fig.5 (?).

Kaka-toddali HM 5:81-82:41
Kaka-Toddalij
BL Add MS 5030:82-83.
Ray 1688 vol.2:1612 Frutex baccifer Indicus spinosus, trifolius floribus spicatis, fructu plano-rotundo tricocco.
Plukenet 1691:Tab.XCV,f.5 = Ray 1688.
C.Commelin 1696:30 = Ray 1688.

Plukenet 1696:202 = Plukenet 1691.
C.Commelin 1701:1 Alaternoides species (r).
Petiver 1702(1701):1010,no.189 Mella Kurni Malab. Chamaelea Malabarica trifoliata spinosa Mus.Petiver.41.
J.Burman 1737:32 Arbuscula Zeylanica, tricapsularis, & tricoccos, Keembya dicta. Mus.Zeyl.pag.69 [Hermann 1717:69] (?).
J.Burman 1737:58-59,Tab.24 Chamaelaea trifolia, aculeata, floribus spicatis.
Linnaeus 1747:60,no.143 Paullinia foliis ternatis, caule aculeatis cirrhis nullis.
Linnaeus 1753:365 Paullinia asiatica.
Linnaeus 1762:524 = Linnaeus 1753.
N.L.Burman 1768:90 = Linnaeus 1762.
Houttuyn 1775 II vol.4:557-559 = N.L.Burman 1768.

Bruxaneli HM 5:83-84:42
Brúxáneli
BL Add MS 5030:84-85.
Ray 1688 vol.2:1497-1498 Baccifera Indica flosculis umbellatis, baccis umbilicatis dicoccis.
C.Commelin 1696:12 = Ray 1688.

Perin nirouri, seu Ma nirouri HM 5:85-86:43
Perin Niroúri, Ma Niroúri
BL Add MS 5030:86-87.
Ray 1688 vol.2:1558-1559 Frutex baccifer Malabaricus ossiculo fragili cum sex intus nucleis.
C.Commelin 1696:30 = Ray 1688.
Plukenet 1700:188 Vitis Idaea Malabarica, fructu majori.
Petiver 1704(1703):1457,no.35 Nirouri Malabarica fructu & calyce maximo.

Tsjeria nirouri et Katou nirouri HM 5:87:44
Katoú-Niroúri
BL Add MS 5030:88-89.
Ray 1688 vol.2:1559,1636 = HM.
C.Commelin 1696:67 = HM.
Plukenet 1700:84 Keelanelle Malabareis; ibid.1696:159 Fruticulus foliis brevioribus; subrotundis, & densius stipatis, Phytogr. Tab.183.fig.5 (cs).
Plukenet 1700:188 Vitis Idaea Malabarica, fructu minore.
Petiver 1704(1703):1457-1458,no.36 Nirouri Malabar floribus binis ternisve.

Cammetta HM 5:89-90:45
Camettì
BL Add MS 5030:90-91.
J.Commelin in HM 5:90 Tithymalus arborescens.
Ray 1688 vol.2:1496 Baccifera Indica floribus spicatis, fructu umbilicato tricocco, lacte acerrîmo manante.
Plukenet 1691:Tab.CXIII,f.1 Tithymal; exotic; Carobefolijs aliquatenùs accedens. (?).
C.Commelin 1696:66 = J.Commelin in HM.
Plukenet 1696:369 = Plukenet 1691.
Plukenet 1700:162 Ricinus Tithymaloides lactifluus, Arbor Malabarica.

Pai-paroea, seu Couradi HM 5:91-92:46
Pái-Pároea
BL Add MS 5030:92-93.
Ray 1688 vol.2:1624 Frutex baccifer Malab. fructu planorotundo piloso tetrapyreno.
Plukenet 1691:Tab.L,f.4 = Ray 1688.
C.Commelin 1696:30 = Ray 1688.
Plukenet 1696:275 = Plukenet 1691.

Petiver 1702(1700):715,no.83 Poon-nasai Malab. Angola Malabarica Ulmi folio (?).
Linnaeus 1737:433-434 Grewia corollis acutis (r).
Royen 1740:477 Grewia corollis obtusis.
Linnaeus 1747:154,no.324 Grewia foliis sublanceolatis.
Linnaeus 1753:964 Grewia orientalis.
Linnaeus 1763:1367 = Linnaeus 1753.
N.L.Burman 1768:192 = Linnaeus 1763.
Houttuyn 1774 II vol.3:240-241 = N.L.Burman 1768.

Katapa HM 5:93-94:47
Katapa
BL Add MS 5030:94-95.
Ray 1688 vol.2:1495 Baccifera Indica ad foliorum alas florida, fructu umbilicato & calyculato tricocco.
C.Commelin 1696:12-13 = Ray 1688.
Plukenet 1700:3 Acetosae arboreae foliis minoribus & magis acuminatis. Chittramullum Malabarorum, & Coadevalle Gentilium Indorum (?).

Tsjocatti HM 5:95-96:48
Tsjócati
BL Add MS 5030:96-97.
Ray 1688 vol.2:1572 Frutex baccifer Malab. fructu calyculato, tetracocco, umbellato.
C.Commelin 1696:31 = Ray 1688.
Plukenet 1696:378 = Ray 1688.

Niir-notsjiil HM 5:97-98:49
Nir-Notsjil
BL Add MS 5030:98-99.
Ray 1688 vol.2:1573 Baccifera Malab. fructu oblongo, tetracocco, calyculato.
C.Commelin 1696:13 = Ray 1688.
Plukenet 1700:26; ibid. 1696:48 Arbuscula Barbadensis amplexicaulis triphyllos, Mail-elou Malabarica foliis tenerioribus, & petiolis alatis. Phytogr.T.415.fig.4 (r).
Plukenet 1705:167 Periclymeni similis Myrtifolia arbor Maderaspatensis (...) Changum-goupee Malabarorum.
N.L.Burman 1768:136-137 Volkameria inermis Linn.sp.889 [Linnaeus 1763:889].
Houttuyn 1775 II vol.5:336-337 = N.L.Burman 1768.

Tsjerou kanneli HM 5:99:50
Tsjeroú-Kanneli
BL Add MS 5030:100-101.
Ray 1688 vol.2:1499 = HM.
C.Commelin 1696:67 = HM.
Plukenet 1696:378 Arbuscula Indica baccifera, floribus hexapetalis plurimis simùl in surculis, fructu tricocco.
Plukenet 1700:25 Arbor Indica Euonymi brevioribus foliis, floribus ad stipitem racemosis. Cungee Malabarorum (?).
Petiver 1702(1700):709,no.71 Davadarree Malab. Berberidis facie, arbor Madraspat. foliis non serratis. Mus.Petiv.623 (?).

Amelpodi HM 5:101:51
Amel-Podì
BL Add MS 5030:102-103.
Ray 1688 vol.2:1787 Arbor Indica akarpon floribus umbellatis, tetrapetalis.
C.Commelin 1696:7 = Ray 1688.
Plukenet 1700:25 Arbor Sirinamensis Lauri densioribus foliis, media costâ per longitudinem tuberculis veluti perforatis conspicuâ (?).

Poeatsjetti HM 5:103-104:52

Púà-Tsjettì
BL Add MS 5030:104-105.
J.Commelin in HM 5:104 Euonymus prima Clusii.
Ray 1688 vol.2:1711 Frutex Malabar. flore pentapetalo stellato, fructu pentagono & pentacocco.
C.Commelin 1696:31 = Ray 1688.
Plukenet 1705:79 Euonymo adfinis Malabariensis, summis virgulis plumosis.

Katou-karua HM 5:105-106:53
Katoú-Kárua
BL Add MS 5030:106-107.
J.Commelin in HM 5:106 Canella sylvestris.
Ray 1688 vol.2:1562 Canella sylvestris Malabarica.
C.Commelin 1696:17 = Ray 1688.
J.Burman 1737:63-64,Tab.28 Cinnamomum perpetuo florens, folio tenuiore, acuto.
N.L.Burman 1768:92 Laurus malabatrum.

Amvetti, seu Vetti-tali HM 5:107-108:54
Amveti
BL Add MS 5030:108-109.
Ray 1688 vol.2:1712 Arbor Indica floribus spicatis, seminibus parvis in vasculis siccis.
C.Commelin 1696:7 = Ray 1688.

Mala-elengi HM 5:109-110:55
Mála Elengi
BL Add MS 5030:110-111.
Ray 1688 vol.2:1637 Baccifera Indica flore composito.
Plukenet 1691:Tab.CLIII,fig.7 Carappa arbor ex Curassaviâ PBP [Sherard 1689:320] (?).
C.Commelin 1696:12 = Ray 1688.
Plukenet 1700:156; ibid. 1696: Prunifera Americana Laurifolia diphyllos, Glycoxylum (i.e.) Sweetwood Barbadensibus dictum (r).
Rumphius 1741 vol.1:181-183,Tab.LXXI-LXXII Gnemon Domestica (-); J.Burman ibid.:183 (=).

Tsjerou poeam HM 5:111:56
Tsjeroú-Pôeam
BL Add MS 5030:112-113.
Ray 1688 vol.2:1571-1572 Baccifera Malab. racemosa, tripetala fructu oblongo tricocco, calyce excepto.
C.Commelin 1696:13 = Ray 1688.

Ben-moenja HM 5:113:57
Bem-Móenja
BL Add MS 5030:114-115.
Ray 1688 vol.2:1787 Arbor Indica foliis alatis, flore & fructu vidua.
C.Commelin 1696:7 = Ray 1688.
Plukenet 1696:66 = Ray 1688.
Rumphius 1741 vol.1:191-192,Tab.LXXVIII Olus album (-); J.Burman ibid.:192 (?=).

Biti HM 5:115:58
Biti
BL Add MS 5030:116-117.
Ray 1688 vol.2:1735 Biti Malabarensibus.
C.Commelin 1696:16 = HM.
Plukenet 1700:168-169 Scorpioides arbor, Biti folio singulari, siliquâ nodosâ intortâ. Muddemaermoodde Malabarorum.

Asjogam HM 5:117:59
Asjógam

BL Add MS 5030:118-119.
Ray 1688 vol.2:1786 Arbor Indica foliis adversis, flore flavescente tetrapetaloide, odorato, fructu nondum comperto.
C.Commelin 1696:7 = Ray 1688.
Plukenet 1700:21 Arbor Indica longis mucronatis integris foliis, fructu albicante Nucis Palmae Indel dictae aemulo; Ashogamaram Malabarorum (?).

Beesha HM 5:119-120:60
Beêsha
BL Add MS 5030:120-121.
J.Commelin in HM 5:120 Arundo species.
Ray 1688 vol.2:1316 Arundo Indica Bambu species altera.
C.Commelin 1696:10 = Ray 1688.
Plukenet 1696:53 = Ray 1688.

Nola-ily HM 5:119-120: no figure

BL Add MS 5030: no drawing.
J.Commelin in HM 5:120 Arundo species.
Ray 1688 vol.2:1316 Arundo Indica Bambu dicta tertia.
C.Commelin 1696:10 = Ray 1688.
Plukenet 1700:28 Arundo arbor Indica procera, fructu Sesami in verticillas densius stipato, Mungell Malabaror. (?).

VOLUME 6

Tsetti-mandarum HM 6:1-2:1
Tsjétti-Mandáru
BL Add MS 5030:126-127.
J.Commelin in HM 6:2 Frutex pavoninus, sive Crista pavonis J.Breyne Cent.I.Exotic.cap.XXII [Breyne 1678:61-64,tab.].
Hermann 1687:192,663 = J.Commelin in HM.
Breyne 1689:37 Crista pavonis Coronillae folio I, sive floribus spicatis amplissimis ex aureo & coccineo variegatis, siliqua pisi (?).
J.Commelin 1689:103 = J.Commelin in HM.
C.Commelin 1696:23 = Breyne 1689.
Petiver 1704(1702):1060-1061,no.237 Mail Conei Malab. Pavoninus utriusque Indiae, flore variegato.
J.Burman 1737:79 Crista pavonis flore elegantissimo, variegato.
Linnaeus 1737:158 Poinciana foliis duplicato-pinnatis, foliolis oppositis oblongo-ovalibus, caule inermi.
Royen 1740:466 = Linnaeus 1737.
Rumphius 1743 vol.4:55 (App),Tab.XX Crista pavonis (=).
Linnaeus 1747:70,no.159 = Linnaeus 1737.
Linnaeus 1748:101 Poinciana aculeis geminis.
Linnaeus 1753:380 Poinciana pulcherrima.
Linnaeus 1762:544 = Linnaeus 1753.
N.L.Burman 1768:98-99 = Linnaeus 1762.
Houttuyn 1775 II vol.5:42-44 Poinciana Pulcherrima [Linnaeus 1767 vol.2:290].

Tsiapangam HM 6:3-4:2
Tsjam Pángam
BL Add MS 5030:128-129.
J.Commelin in HM 6:4 Acacia Tinctoria Hermans.
Ray 1688 vol.2:1737 Ligno Brasiliano simile, seu Lign. Sapou, lanis tingendis percommodum C.B. [Bauhin 1623:393].
Breyne 1689:37 Crista pavonis Coronillae foliis 2, sive tinctoria Indica, flore luteo racemoso minore, siliqua latissima glabra, lignum rubrum, Sappan dictum, ferens.
Sherard 1689:332-333 Erythroxylum seu lignum rubrum Indicum spinosissimum Coluteae foliis, floribus luteis, siliquis maximis.
C.Commelin 1696:23 = Breyne 1689.
Plukenet 1696:5 Acacia gloriosa Zeylanica, tinctoria, amplioribus foliis, spinosa.
J.Burman 1737:3-4 Acacia major, tinctoria, Zeylanica, Pansapan dicta. Mus.Zeyl.pag.42 [Hermann 1717:42].
Rumphius 1743 vol.4:59 (App),Tab.XXI Lignum Sappan (=).
Linnaeus 1747:69,no.158 Caesalpinia aculeis recurvis, foliolis emarginatis, filamentis lanatis.
Linnaeus 1753:381 Caesalpinia Sappan.
Linnaeus 1762:545 = Linnaeus 1753.
N.L.Burman 1768:99 = Linnaeus 1762.
Houttuyn 1774 II vol.2:386-388 = N.L.Burman 1768.

Pongam seu Minari HM 6:5-6:3
Púngam
BL Add MS 5030:130-131.
J.Commelin in HM 6:6 Faba species Clusius Exotic.L.III.c.XIII.
Ray 1688 vol.2:1733 Arbor Siliquosa flore papilionaceo: fabis longis & latis plana parte sibi invicem incumbentibus.
Breyne 1689:39 Crista pavonis monospermos 3, sive Arbor vespertilionis maxima Indica, Juglandis folio majore, floribus spicatis albicantibus odoratis, siliqua nonnihil falcata, semine renali latissimo.
C.Commelin 1696:24 = Breyne 1689.
Plukenet 1696:294 Phaseolis accedens Malabarica alatis foliis glabris monospermos, siliquâ latiore brevi. Phytogr.Tab.310. fig.3 (?).
Petiver 1699(1698):324-325,no.24 Punga maraum Mal. = Breyne 1689.
Petiver 1702(1700):710-711,no.74 Punga-maraum Malab. Minari Malabarica, fl. Roseo-albicante, siliqua ovale compressa.

Intsia HM 6:7-8:4
Insja
BL Add MS 5030:132-133.
J.Commelin in HM 6:8 Acacia Malabarica flore globoso siliquis latis.
Ray 1688 vol.2:1739 Acacia Malabarica globosa Intsia dicta.
Breyne 1689:73 Mimosa arborea Javanensis spinosa, tenuissimis pinnis, siliqua lata pergrandi spadicea (?).
Sherard 1689:304 Acacia Malabarica sulcata floribus globosis albis.
Plukenet 1691:Tab.CXXII,f.1 Acacia Maderaspatana spinosa pinnis veluti lunulatis acutioribus Myrti aemulis nervo pinnularum ad unum latum vergente, siliqua lata (?).
C.Commelin 1696:1 = J.Commelin in HM.
Plukenet 1696:4 = Plukenet 1691.
J.Commelin 1697:205-206,Fig.105 Acacia Javanica sulcata, caule et foliorum costis spinosis. Cat.Hort.Beaum. [Kiggelaer 1690:2] (?).
J.Burman 1737:3 Acacia Zeylanica sarmentosa, flore luteo globoso. Mus.Zeyl.pag.34 [Hermann 1717:34] (s).
Linnaeus 1737:209 Mimosa aculeis undique sparsis solitariis, foliis duplicato-pinnatis, caule angulato.
Royen 1740:471 = Linnaeus 1737.
Linnaeus 1753:522 Mimosa Intsia.
Linnaeus 1763:1508 = Linnaeus 1753.
N.L.Burman 1768:224 = Linnaeus 1763.
Houttuyn 1776 II vol.6:460-461 Mimosa Intsia [Linnaeus 1767 vol.2:678].

Waga HM 6:9:5
Wága

BL Add MS 5030:134-135.
J.Commelin in HM 6:9 Acacia Malabarica altera spinis carens.
Ray 1688 vol.2:1766 Arbor Indica siliquosa flore tetrapetalo stellato, siliquis bipalmaribus planis.
C.Commelin 1696:1 = J.Commelin in HM.
Petiver 1702(1701):936,no.*144 Caut Wallee Malab. Waga Madraspatana Senae foliis siliqua lata compressa, ubi seminibus inflata Mus.Petiver.697 (?).
J.Burman 1737:5 Acacia Malabarica, flosculis spicatis, siliquis compressis, latissimis. Mus.Zeyl.pag.59 [Hermann 1717:59] (?).
Linnaeus 1747:217,no.506 = J.Burman 1737:5 (=).

Moullava HM 6:11:6
Mulláva
BL Add MS 5030:136-137.
J.Commelin in HM 6:11 Acacia cognata planta J.Bauh.P.I.Libr. p.432 (?).
Ray 1688 vol.2:1752 Siliquosa Indica flore pentapetalo luteo, siliquis lenibus tetraspermis fere.
C.Commelin 1696:62 = Ray 1688.
Plukenet 1696:254 = Ray 1688.
Plukenet 1700:2; ibid.1696:4 Acaciae fortè cognatus è Maderaspatan (?).

Mouricou HM 6:13-14:7
Murícu
BL Add MS 5030:138-139.
J.Commelin in HM 6:14 Siliqua sylvestris spinosa arbor Indica Bauhini (Pinac.L.XI.Sect.II) [Bauhin 1623:402].
Hermann 1687:189,661 Coral, arbor siliquosa I.B.I.L.12.426.
Ray 1688 vol.2:1736 = J.Commelin in HM.
J.Commelin 1689:100 = Hermann 1687.
Sherard 1689:327 Coral arbor spinosa orientalis fructu obscure rubente.
C.Commelin 1696:22 = Sherard 1689.
Plukenet 1696:293 Phaseolus accedens Coral arbor spinosa Orientalis fructu obscurè rubente.
J.Burman 1737:74-75 Corallodendron triphyllum, Americanum, spinosum, flore ruberrimo. Tournef.inst.pag.661.
Linnaeus 1737:354 Erythrina foliis ternatis, caule spinoso.
Rumphius 1741 vol.2:233 (App),Tab.LXXVI Gelala Litorea (=).
Linnaeus 1747:126,no.275 = Linnaeus 1737.
Linnaeus 1753:706 Erythrina corallodendrum β orientalis.
Linnaeus 1763:992-993 = Linnaeus 1753.
N.L.Burman 1768:154 = Linnaeus 1763.
Houttuyn 1774 II vol.3:170-172 = N.L.Burman 1768.

Kal-toddavaddi HM 6:15-16:8
Kall-Todda-Váddi
BL Add MS 5030:140-141.
Ray 1688 vol.2:1740 Mimosa Malabarica flore pentapetalo, siliquis lanuginosis.
Breyne 1689:37-38 Crista pavonis Coronillae folio quarta, sive sensitiva, sive Crista pavonis minor Indica spinosissima, foliis Coronillae sensibilibus, flore luteo amplo ad foliorum alas, siliqua lata parva lanuginosa.
C.Commelin 1696:23-24 = Breyne 1689.

Wellia tagera HM 6:17-18:9-10
Wellia-Tagerà
BL Add MS 5030:142-143,144-145.
Ray 1688 vol.2:1746 Siliquosa Malabarica flore pentapetalo, siliquis longis, planis, diaphragmatis semina secludentibus interceptis.

Breyne 1689:30 Chamaecristae pavonis affinis Indica, Cassiae folio, siliquis compressis latissimis propendentibus, floribus amplissimis in summo ramorum spicatis.
J.Commelin 1689:370 = Ray 1688.
C.Commelin 1696:19 = Breyne 1689.
Plukenet 1696:342 Sena spuria Orientalis faetidi glabra, foliis monetam Japonicam Kupang dictam referentibus, major PBP.375 [Sherard 1689:375].
Petiver 1702(1700):702,no.55 = HM.

Mouringou HM 6:19-20:11
Múríngú
BL Add MS 5030:146-147.
J.Commelin in HM 6:20 Arbor exotica Lentisci foliis Bauh. Pinac.L.II.Sect.II [Bauhin 1623:399].
Ray 1688 vol.2:1745 Moringa Lentisci folio, fructu magno anguloso, in quo semina Ervi J.B.
Breyne 1689:22 Balanus myrepsica, siliquá triangulari, semine minore alato.
J.Commelin 1689:237 Moringa, Acosta.
Sherard 1689:357 Nux Been Zeylanica, siliquâ triangulâ seminibus alatis Hort.Lugd. [Hermann 1687:692].
C.Commelin 1696:13 = Breyne 1689.
Plukenet 1696:253 = Ray 1688.
J.Burman 1737:162-164,Tab.75 Moringa Zeylanica, foliorum pinnis pinnatis, flore majore, fructu anguloso (c).
Rumphius 1741 vol.1:184-188,Tab.LXXIV-LXXV Morunga (-); J.Burman ibid.:188 (=).
Linnaeus 1747:67,no.155 Guilandina inermis, foliis triplicatopinnatis, foliolis infimis ternatis.
Linnaeus 1753:381 Guilandina Moringa.
Linnaeus 1762:546 = Linnaeus 1753.
N.L.Burman 1768:100 = Linnaeus 1762.
Houttuyn 1774 II vol.2:389-393 Guilandina Moringa [Linnaeus 1767 vol.2:291].

Katou-conna HM 6:21:12
Kátu-Cónna
BL Add MS 5030:148-149.
J.Commelin in HM 6:21 Cassia fistula species?
Ray 1688 vol.2:1746 Arbor Indica siliquosa flore pentapetalo, siliquis in spiram contortis lanuginosis.
C.Commelin 1696:7 = Ray 1688.
Plukenet 1696:95 Ceratiae quodammodò affinis majoribus, & longioribus foliis, ex Indiâ Orientali, fabis atro-nitentibus (?).
Linnaeus 1747:97-98,no.218 Mimosa foliis bigeminis (acuminatis).
Linnaeus 1753:517 Mimosa bigemina.
Linnaeus 1763:1499 = Linnaeus 1753.
N.L.Burman 1768:222 = Linnaeus 1763.
Houttuyn 1774 II vol.3:608-609 = N.L.Burman 1768.

Thora-paërou HM 6:23-24:13
Thóra-Paeraú
BL Add MS 5030:150-151.
Hermann 1687:694 Phaseolus arbor Indica, incana, siliquis torosis. Kayan dicta.
J.Commelin 1689:275 = Hermann 1687.
Sherard 1689:311 Anagyris Indica leguminosa siliquis torosis.
Plukenet 1691:Tab.CCXIII,fig.3 = Hermann 1687.
C.Commelin 1696:53 = Hermann 1687.
Plukenet 1696:293 = Plukenet 1691.
Petiver 1699(1698):327-328,no.30 Toura Mal.Laburnum humilius siliqua inter grana & grana juncta semine esculento Cat.Plant.Jam.139.

J.Burman 1737:86,Tab.37 Cytisus folio molli, incano, siliquis Orobi contortis, & acutis.
Linnaeus 1737:354-355 Cytisus foliolis ovato-lanceolatis, intermedio petiolato, pedunculo ex alis multifloro.
Linnaeus 1747:128-129,no.279 Cytisus foliis sublanceolatis tomentosis: intermedio longius petiolato, pedunculis axillaribus erectis.
N.L.Burman 1768:163 Cytisus cajan Linn.sp.1041 [Linnaeus 1763:1041].

Mandsjadi HM 6:25-26:14
Mantsjadi
BL Add MS 5030:152-153.
J.Commelin in HM 6:26 Pisum virulentum Chinense Bauhinus in Pinace [Bauhin 1623:343].
Hermann 1687:495,694 Phaseolus alatus arboreus, fructu orbiculato compresso coccineo.
Ray 1688 vol.2:1752-1753 Arbor siliquosa Indica flore spicato pentapetalo, siliquis longis, nodosis fabis coccineis; ibid.1704 vol.3:112(2).
Breyne 1689:38-39 Crista pavonis Glycyrrhizae folio, maxima Indica, flore subluteo minimo spicato, siliquis angustis longissimis, ubi semina occultantur, protuberantibus, semine orbiculato compresso sanguineo.
J.Commelin 1689:222 = Ray 1688.
Sherard 1689:352 = HM.
C.Commelin 1696:24 = Breyne 1689.
Plukenet 1696:294-295 Phaseolus alatus arbor Indica, fructu coccineo ferè orbiculari medio utrinque tumido.
Petiver 1702(1701):937,no.146 Tande maraum Malab.Mandsjadi Malab. Glycyrrhizae folio sem.coccineis.
J.Burman 1737:79-80 Crista Pavonis arbor, foliis subrotundis, alternis flore spicato, pentapetalo, flavo, lobis longis, fructu orbiculato, coccineo.
Rumphius 1743 vol.3:174 (App),Tab.CIX Corallaria parvifolia (=); J.Burman ibid.:174 (=).
Linnaeus 1747:70-71,no.160 Adenanthera foliis decompositis. Linn.virid.36.
Linnaeus 1753:384 Adenanthera pavonina.
Linnaeus 1762:550 = Linnaeus 1753.
N.L.Burman 1768:100-101 = Linnaeus 1762.
Houttuyn 1774 II vol.2:419-423 = N.L.Burman 1768.

Pongelion sive Perimaram HM 6:27-28:15
Pongelioṅ
BL Add MS 5030:154-155.
Ray 1688 vol.2:1753 Arbor Indica siliquosa, floribus racemosis, pentapetalis, siliquis foliaceis, ad singulos flores ternis.
C.Commelin 1696:7 = Ray 1688.

Plaso HM 6:29-30:16-17
Pláso
BL Add MS 5030:156-157,158-159.
J.Commelin in HM 6:30 Lobus species.
Ray 1688 vol.2:1721 Arbor siliquosa trifolia Indica, flore papilionaceo siliqua grandi pilosa unicam intus fabam continente.
C.Commelin 1696:8 = Ray 1688.

Karin-njoti HM 6:31-32:18
Kariṅ-Njóta
BL Add MS 5030:160-161.
J.Commelin in HM 6:32 Lobus Clusius Exotic.l.3.Cap.16 (s).
Ray 1688 vol.2:1766 Arbor Malabarica flore vario, siliqua nuciformi.
C.Commelin 1696:8 = Ray 1688.

Kaka mullu, vel Kaka moullou HM 6:33:19
Cácu-Mullù
BL Add MS 5030:162-163.
J.Commelin in HM 6:33 Loborum membranaceorum genera Clusius Exotic.L.III.C.XVI (cs).
Ray 1688 vol.2:1745-1746 Siliquosa Indica flore Papilionaceo decapetalo siliquis latis monospermis.
Breyne 1689:39 Crista pavonis monospermos secunda; sive Arbor vespertilionis Indica, lentisci folio, spinosissima, flore racemoso, luteo parvo odorato, siliqua & semine glabro.
C.Commelin 1696:24 = Breyne 1689.
Plukenet 1696:202 = Ray 1688.
J.Burman 1737:4-5 Acacia gloriosa Coluteae folio Chinensis, rachi medio tam ad genicula, quam ad internodia spinis curtis duplicatis, deorsum inflexis munito, an Wawoulaethya Cinghalensibus dicta. Plukn.Almag.pag.5 [Plukenet 1696:5].
Linnaeus 1747:214,no.495 Wawulaethya Herm.zeyl.40 [Hermann 1717:40].
Rumphius 1747 vol.5:94-95,Tab.L Nugae silvarum (-); J.Burman ibid.:95 (=).

Bankaretti HM 6:35:20
Ban Carettì
BL Add MS 5030:164-165.
Ray 1688 vol.2:1744 Arbor spinosa Indica siliquis villosis monospermis.
Breyne 1689:39 Crista pavonis monospermos I, sive Arbor vespertilionis Juglandis folio minore spinosissima, flore viridi-luteo ad foliorum alas, semine & siliquâ villosâ.
C.Commelin 1696:24 = Breyne 1689.
Plukenet 1696:294 Phaseolus accedens alatis foliis monospermos, Arbor Indica spinosa.

Nagam HM 6:37:21
Nágam
BL Add MS 5030:166-167.
Ray 1688 vol.2:1753 Siliquosa flore umbellato pentapetalo, siliquis digitalibus, spadiceis, monospermis, ad unum florem pluribus.
C.Commelin 1696:62 = Ray 1688.
Plukenet 1696:250 = Ray 1688.
Plukenet 1700:12 Amygdalus amara Indorum putamine fungoso amicta, flore umbellato, nucibus ad unum florem plurimis, Samandra Bengalensibus, & Coemburn Zeylanensibus dicta.
J.Burman 1737:19 = Plukenet 1700.
Linnaeus 1747:202-203,no.433 Samandura Herm.zeyl.5.11 [Hermann 1717:5,11].

Noëlvalli & Pannivalli HM 6:39:22
Nul-Válli
BL Add MS 5030:168-169.
J.Commelin in HM 6:39 Orchis species (c).
Ray 1688 vol.2:1734 Siliquosa Indica flore palilionaceo. Siliqui planis brevibus duo aut tria semina isthmis distincta continentibus.
Breyne 1689:53 Glycyrrhiza, vel (si mavis) Glycyrrhizae affinis arborescens Americana, floribus ex luteo & rubro variegatis, folio acuminato, siliquâ latissimâ (?).
C.Commelin 1696:62 = Ray 1688.
Plukenet 1696:264 Frutex alatus siliquosus Orchidis floribus aliquatenùs accedens, siliquâ isthmis distinctâ è Maderaspatan.

Katou-pulcolli HM 6:41:23
Katú-Púl-Colli
BL Add MS 5030:170-171.

J.Commelin in HM 6:41 Fabago Belgarum Dodonei (r).
Ray 1688 vol.2:1765 Frutex Indicus flore dipetalo capsula oblonga, binis cellulis bina semina continente.
C.Commelin 1696:31 = Ray 1688.
Plukenet 1696:202-203 Frutex Indicus siliquosus Alsines facie diphyllos.

Perincourigil HM 6:43-44:24
Periṁ-Curigil
BL Add MS 5030:172-173.
J.Commelin in HM 6:44 XII. exoticorum fructuum speciebus Clusius Exotic.L.2.C.16 (c).
Ray 1688 vol.2:1750 Siliquosa Malabarica pentapetala, siliquis tomentosis.
C.Commelin 1696:62 = Ray 1688.
Plukenet 1696:288 = Ray 1688.

Karin-tagera HM 6:45:25
Kariṅ-Tagerà
BL Add MS 5030:174-175.
Ray 1688 vol.2:1735 = HM.
Plukenet 1691:Tab.CXXXIX,fig.5 Arbuscula Indica siliquosa, Corylii foliis parvis serratis, summâ parte sinuatis.
C.Commelin 1696:38 = HM.
Plukenet 1696:44 = Plukenet 1691.

Padri HM 6:47-48:26
Pádri
BL Add MS 5030:176-177.
J.Commelin in HM 6:48 quadrangularis Lobi Clusius Exotic.L.III.C.XVII (?).
Ray 1688 vol.2:1750 Siliquosa flore pentapetalo, siliquis longis angustis, quadratis intortis.
Breyne 1689:34 Clematis arborea Indica Juglandis folio, flore luteo spicato minimo, siliquis longis, angustis, quadratis atque contortis, seminibus substantia lignosa obductis.
C.Commelin 1696:20 = Breyne 1689.

Kedangu HM 6:49-50:27
Kedángu
BL Add MS 5030:178-179.
J.Commelin in HM 6:50 Colutea Siliquosa Malabariensis.
Ray 1688 vol.2:1735 Siliquosa Malabarica siliquis spithamaeis angustissimis contortis.
C.Commelin 1696:21 = J.Commelin in HM.
Plukenet 1696:112 = J.Commelin in HM.
J.Burman 1737:93-94,Tab.41 Emerus siliquis geminatis, longissimis.
Rumphius 1741 vol.1:188-190,Tab.LXXVI-LXXVII Turia (-); J.Burman ibid.:190-191 (=).

Alpam HM 6:51-52:28
Alpaṁ
BL Add MS 5030:180-181.
Ray 1688 vol.2:1765-1766 Siliquosa Indica flore tripetalo siliquis teretibus pulpa absque seminibus repletis.
C.Commelin 1696:62 = Ray 1688.
Plukenet 1700:115; ibid.1696:211 Laurifolia fructu parvo typhoide non papposo gemello, siliquas aemulante, Arbuscula è Maderaspatan, Phytogr.Tab.96.fig.7 (cs).

Niir-pongelion HM 6:53-54:29
Nir-Pongelioṅ
BL Add MS 5030:182-183.
Ray 1688 vol.2:1764 Arbor siliquosa Indica siliquis longis contortis, in quatuor cellulas per longum divisis.
Breyne 1689:34 Clematis arborea Indica, Juglandis folio, longissimis floribus, siliquis teretibus longissimis contortis, semine substantia ex lignea fungosa incluso.
C.Commelin 1696:20 = Breyne 1689.
Rumphius 1743 vol.3:74 (App),Tab.XLVI Lignum Equinum (=).

Isora-murri HM 6:55-56:30
Isoca-Múri
BL Add MS 5030:184-185.
J.Commelin in HM 6:56 Iger-Murri Zanonius 13.Cap.Historiae Botanicae.
Ray 1688 vol.2:1765 Frutex Indicum, fructu è styli apice egresso, sextuplici funiculo, in spiram convoluto constante.
Plukenet 1691:Tab.CCXLV,fig.2 Helicteres arbor Indiae Orientalis siliquâ varicosâ & Funiculi in modum contortuplicatâ.
C.Commelin 1696:31 = Ray 1688.
Plukenet 1696:181 = Plukenet 1691.
Plukenet 1700:100; ibid.1696:182 Helicteres arbor Indiae Occident.fructu majore; Jamaicensibus nostratibus Button wood dicta, Phytogr.Tab.245.fig.3.
Petiver 1702(1701):938,no.149 Coodee-wengee Malab.Helicteres Indiae utriusque Coryli folio.
Ray 1704 vol.3:521 Abutilo affinis arbor Althaeae folio, cujus fructus est styli apex auctus, quatuor vel quinque siliquis hirsutis, funis ad instar in spiram convolutis constans Slon.Cat.Jamaic. (?).
Linnaeus 1737:433 Helicteres β.
Linnaeus 1753:963 Helicteres Isora.
Rumphius 1755:32-33,Tab.XVII, Fig.1 Fructus Regis (c).
Linnaeus 1763:1366 = Linnaeus 1753.
N.L.Burman 1768:192 = Linnaeus 1763.
Houttuyn 1774 II vol.3:230-232 Helicteres Isora Syst.Nat.XII. Tom.II.Gen.1025.p.601 [Linnaeus 1767 vol.2:601].

Candel HM 6:57-58:31-32
Kandel, Kandèl
BL Add MS 5030:186-187,188.
J.Commelin in HM 6:67 Kandel species Zanonius Cap.13.
Ray 1688 vol.2:1769-1770 Frutex Indicus ramis demissis radices agentibus se multiplicans, fructu oblongo, terete, corticoso.
Plukenet 1691:Tab.CCIV,fig.3 Mangle arbor Pyrifolia salsis & uliginosis locis in Americâ proveniens, fructu oblongo tereti, summis ramis radicosa (?).
C.Commelin 1696:32 = Ray 1688.
Plukenet 1696:241 = Plukenet 1691.
Rumphius 1743 vol.3:105 (App),Tab.LXVIII Mangium celsum (=); J.Burman ibid.:106 (=).
Rumphius 1743 vol.3:107-108,Tab.LXX Mangium digitatum (-); J.Burman ibid.:108 (c).
Linnaeus 1753:443 Rhizophora gymnorhiza.
Linnaeus 1762:634 = Linnaeus 1753.
N.L.Burman 1768:108 = Linnaeus 1762.
Houttuyn 1774 II vol.2:491-494 = N.L.Burman 1768.

Karii-kandel seu Kanil-kandel HM 6:59:33
Carì-Candèl
BL Add MS 5030:189-190.
J.Commelin in HM 6:67 Kandel species Zanonius Cap.13.
Ray 1688 vol.2:1770 Candela arbor floribus in eodem pediculo ternis, fructu angustiore.
Plukenet 1691:Tab.CCIV,fig.3 Mangle arbor Pyrifolia salsis & uliginosis locis in Americâ proveniens, fructu oblongo tereti, summis ramis radicosa (?).
C.Commelin 1696:17 = Ray 1688.
Plukenet 1696:241 = Plukenet 1691.

Plukenet 1700:62; ibid.1696:127 Cynoxylum Americanum, folio crassiusculo, molli, & tenaci, Phytogr.Tab.172.fig.6 (?).
Plukenet 1700:68 Epidendron Ind.Orient.floribus ex alis in spicam, prodeuntibus. Cheddemeelchedde Malabarorum (?).
Ray 1704 vol.3:115(2) Mangle Laurocerasi foliis, flore albo tetrapetalo Slon.Cat.Jamaic.(?).
Rumphius 1743 vol.3:106-107,Tab.LXIX Mangium Minus (-); J.Burman ibid.:107 (=).
Linnaeus 1753:443 Rhizophora cylindrica.
Linnaeus 1762:635 = Linnaeus 1753.
N.L.Burman 1768:108 = Linnaeus 1762.
Houttuyn 1774 II vol.2:503-504 Mangle cylindrica [misprint of Rhizophora cylindrica = N.L.Burman 1768].

Pee-kandel HM 6:61:34
Peê-Candèl
BL Add MS 5030:191-192.
J.Commelin in HM 6:67 Kandel species Zanonius Cap.13.
Ray 1688 vol.2:1770 Candela Indica fructu longiore & crassiore, flore tetrapetalo.
Plukenet 1691:Tab.CCIV,fig.3 Mangle arbor Pyrifolia salsis & uliginosis locis in Americâ proveniens, fructu oblongo tereti, summis ramis radicosa (?).
C.Commelin 1696:17 = Ray 1688.
Plukenet 1696:241 = Plukenet 1691.
Rumphius 1743 vol.3:105 (App),108-111,Tab.LXXI-LXXII Mangium Candelarium (?=); J.Burman ibid.:111 (=).
Linnaeus 1753:443 Rhizophora Mangle.
Linnaeus 1762:634 = Linnaeus 1753.
N.L.Burman 1768:108 = Linnaeus 1762.
Houttuyn 1774 II vol.2:495-503 = N.L.Burman 1768.

Tsjerou-kandel HM 6:63:35
Tsjéru-kandel
BL Add MS 5030:193-194.
J.Commelin in HM 6:67 Kandel species Zanonius Cap.13.
Ray 1688 vol.2:1770 Candela Indica humilior flore exalbido pentapetalo, fructu majore.
C.Commelin 1696:17 = Ray 1688.
Rumphius 1743 vol.3:105 (App),108-111,Tab.LXXII Mangium Candelarium (?=).
Linnaeus 1753:443 Rhizophora Candel.
Linnaeus 1762:634 = Linnaeus 1753.
N.L.Burman 1768:108 = Linnaeus 1762.
Houttuyn 1774 II vol.2:494 = N.L.Burman 1768.

Pou-kandel HM 6:65:36
Pù-Candèl
BL Add MS 5030:195-196.
J.Commelin in HM 6:67 Kandel species Zanonius Cap.13.
Ray 1688 vol.2:1770 Candela Indica floribus pentapetalis odoratiss.fructu minore incurvo.
Plukenet 1691:Tab.CCIX,fig.4 Mangle alba Coriaria, folio densiusculo subrotundo glabro, fructu formâ Caryophylli aromatici majore (?).
C.Commelin 1696:17 = Ray 1688.
Plukenet 1696:241 = Plukenet 1691.

Kada-kandel HM 6:67:37
Kádá-Kandèl
BL Add MS 5030:197-198.
J.Commelin in HM 6:67 Kandel species Zanonius Cap.13.
Ray 1688 vol.2:1771 = HM.
C.Commelin 1696:38 = HM.

Ray 1704 vol.3:115(2) Mangle julifera, foliis ellipticis, ex adverso nascentibus Slon.Cat.Jamaic. (?).

Hina paretti HM 6:69-72:38-42
Hina Paretti
BL Add MS 5030:199-200,201-202,203-204,205-206,207-208.
J.Commelin in HM 6:72 Althaea Rosea Sinensis Morisoni Histor.Oxoniens.p.530.
Hermann 1687:22,644 Althaea arborea Rosea Sinensis flore multiplici.
Breyne 1689:10 Alcea arborescens Japonica, pampineis foliis subasperis, flore mutabili, sive colorem mutante.
J.Commelin 1689:12 Alcea arborea rosea Sinensis.
Sherard 1689:307 Althaea arborea Rosea Sinensis flore pleno Ferr.Fl.Cult.
C.Commelin 1696:2 = Breyne 1689.
Plukenet 1696:14 = Breyne 1689.
Linnaeus 1737:349 Hibiscus foliis cordato-quinquangularibus obsolete serratis.
Rumphius 1743 vol.4:28 (App),Tab.IX Flos horarius (=).
Linnaeus 1753:694 Hibiscus mutabilis.
Rumphius 1755:26-27,Tab.XIV,Fig.2 Catsjopiri (-); J.Burman ibid.:27 (=).
Linnaeus 1763:977-978 = Linnaeus 1753.
N.L.Burman 1768:151 = Linnaeus 1763.
Houttuyn 1775 II vol.5:404-405 = N.L.Burman 1768.

Ain-pariti HM 6:73:43
Aiǹ-Paritì
BL Add MS 5030:209-210.
J.Commelin in HM 6:73 Rosa Batavico-Indica inodora, seu malva frutescens Bontius Histor.Natur & Med.Ind.Orient.Lib.VI. c.XLVI.
Ray 1688 vol.2:1900 Alcea Indica fruticosa flore coccineo petalis crispis.
C.Commelin 1696:3 = Ray 1688.
Plukenet 1696:14 Alcea Javanica arboresc.flore rubicundo, simplici. Breyn.Cent.I.122.
J.Burman 1737:133-134 Ketmia Sinensis, fructu subrotundo, flore pleno. Tournef.inst.pag.100.
Rumphius 1743 vol.4:26 (App),Tab.VIII Flos festalis (=).

Narinam-poulli HM 6:75-76:44
Narinámpuli
BL Add MS 5030:211-212.
J.Commelin in HM 6:76 Alcea Indica spinosa magno flore ex albido flavescente.
Ray 1688 vol.2:1901 = J.Commelin in HM.
C.Commelin 1696:3 = J.Commelin in HM.
Plukenet 1696:15 = J.Commelin in HM.
J.Commelin 1697:35-36,Fig.18 Alcea Bengalensis spinosissima Acetosae sapore, flore luteo-pallido, umbone purpurascente (r).
J.Burman 1737:135 Ketmia Indica, Gossypii folio, Acetosae sapore. Tournef.inst.pag.100.
J.Burman 1737:135 Ketmia Indica, spinulosa, profunde laciniata, Acetosae sapore (c).
Linnaeus 1747:121,no.264 Hibiscus foliis palmato-digitatis quinquepartitis: laciniis lanceolatis, caule aculeato.
Linnaeus 1753:695-696 Hibiscus Sabdariffa γ.
Linnaeus 1763:979 Hibiscus surattensis β.
N.L.Burman 1768:152 = Linnaeus 1763.
Houttuyn 1774 II vol.5:413-414 = N.L.Burman 1768.

Beloere HM 6:77:45

Belluren
BL Add MS 5030:213-214.
J.Commelin in HM 6:77 Abutilon species.
Hermann 1687:24,644 Althaea Theophrasti similis C.B.P.316 [Bauhin 1623:316].
Ray 1688 vol.2:1880 Abutilo Indico Camerarii simile si non idem.
J.Commelin 1689:18 = Ray 1688.
C.Commelin 1696:4 = J.Commelin in HM.
Plukenet 1696:17 Alcea Indica. Abutilon dicta, minor, petiolo ad florem, geniculato.
Rumphius 1743 vol.4:30 (App),Tab.X Abutilon hirsutum (=); J.Burman ibid.:30 (v).
Rumphius 1743 vol.4:31-33,Tab.XI Abutilon laeve sive agreste (-); J.Burman ibid.:33 (=).
Linnaeus 1767 vol.2:458 Sida asiatica.
Houttuyn 1779 II vol.10:43-44 = Linnaeus 1767.

Katu-beloeren HM 6:79:46
Katú-Belluren
BL Add MS 5030:215-216.
J.Commelin in HM 6:79 Alcea species.
Ray 1688 vol.2:1880-1881 Abutilon Indicum quinque ad singulos flores thecis.
Sherard 1689:308 Althaea Indica Vitis folio flore amplo flavo pendente Hort.Lugd. [Hermann 1687:26-28,tab.].
C.Commelin 1696:3 = Sherard 1689.
Plukenet 1696:15 Alcea Indica, vitis folio flore amplo flavescente, quinis ad singulos flores thecis.
J.Burman 1737:137 Ketmia Indica, Vitis folio, flore amplo, flavo.
Linnaeus 1747:121-122,no.265 Hibiscus foliis quinquangularibus acutis serratis, caule inermi, floribus pendulis.
Linnaeus 1753:696-697 Hibiscus vitifolius.
Linnaeus 1763:980-981 = Linnaeus 1753.
N.L.Burman 1768:153 = Linnaeus 1763.
Houttuyn 1779 II vol.10:69 = N.L.Burman 1768.

Sjouanna-amelpodi HM 6:81-82:47
Tsjovánna-Amel-Podì
BL Add MS 5030:217-218.
J.Commelin in HM 6:82 Jasminum inodorum Bacciferum.
Ray 1688 vol.2:1607 Frutex Indicus pentapetalos gemina bacca, calyce exceptâ.
C.Commelin 1696:32 = Ray 1688.
Plukenet 1696:196 Jasminum Indicum inodorum flore pentapetalo fructu gemello quasi in unam baccam coalito.
Plukenet 1700:20 Arbor Indica Lauri amplioribus foliis obtusis è regione binis, floribus Jasmini, summo ramulo umbellatim positis, ex Insula Johanna.
J.Burman 1737:168,Tab.78,Fig.1 Nerium Indicum, folio subrotundo, undulato, crasso, flore dilute rubente (r).

Belutta-amelpodi HM 6:83:48
Velútta-Amèl-Podì
BL Add MS 5030:219-220.
J.Commelin in HM 6:83 Jasminum Sylvestris inodorum species.
Ray 1688 vol.2:1787 Frutex Indicus akarpon foliis binis adversis, floribus pentapetalis candidis, unguibus luteis.
C.Commelin 1696:31 = Ray 1688.
Plukenet 1696:196 Jasminum Indicum sylvestre inodorum floribus albis ad unguiculos croceo tinctis.

Tsjeni-mulla HM 6:85:49
Tsjéru-Mullà
BL Add MS 5030:221-222.
J.Commelin in HM 6:85 Jasminum Arabicum species.
Ray 1688 vol.2:1601 Jasminum Indicum flore albo odoratissimo.
Breyne 1689:59 Jasminum Indicum Myrti Laureae foliis, flore albo.
C.Commelin 1696:36 = Breyne 1689.
Plukenet 1696:196 Jasminum Indic: suavissimum flore albo pentapeloide quinis antheris grandiusculis erectis ornato.

Nalla-mulla HM 6:87:50
Nálla-Mullà
BL Add MS 5030:223-224.
J.Commelin in HM 6:87 Sambac Lesmin Arabicum Prosp.Alpini Plant.Aegypt. Cap.XIX.& Syringa Arabica foliis mali aurantii.
Hermann 1687:586,697 Syringha Arabica, foliis Mali Aurantii C.B.P.398 [Bauhin 1623:398].
Ray 1688 vol.2:1600-1601 Sambac Arabicum sive Gelseminum Arabicum Alpin. plant.Aegypt.
J.Commelin 1689:171-172 Jasminum Arabicum, sive Syringa Arabica, Clusii cur.post.
Sherard 1689:379 = Hermann 1687.
C.Commelin 1696:36 Jasminum Indicum Mali aurantiae foliis, flore albo pleno minore Breyn:P.59 [Breyne 1689:58-59].
Plukenet 1696:196 Jasminum s.Sambac Arabum Alpino, J.B.T.2.L.15.102.
Boerhaave 1720 vol.2:217 = Plukenet 1696.
J.Burman 1737:128 Jasminum Zeylanicum inodorum, maximum. Maharatambala Zeylonensibus.P.Hermans.apud Ray. Dendr.part.3.pag.63 [Ray 1704 vol.3:63(2)].
Linnaeus 1737:5 Nyctanthes caule volubili, foliis subovatis acutis.
Linnaeus 1747:5,no.12 = Linnaeus 1737.
Rumphius 1747 vol.5:52-55,Tab.XXX Flos Manorae (=).
N.L.Burman 1768:5,Tab.3,f.1 Nyctanthes multiflora.
Houttuyn 1775 II vol.4:14-15 = N.L.Burman 1768.

Kudda-mulla HM 6:89:51
Cudda-Múllà
BL Add MS 5030:225-226.
Ray 1688 vol.2:1601 Gelseminum vel Jasminum Catalonicum multiplex Park.
Breyne 1689:59 Jasminum Indicum Mali aurantiae foliis, flore albo pleno amplissimo (?).
Sherard 1689:379 Syringha Arabica foliis Mali Aurantii flore pleno foliis ternis.
C.Commelin 1696:36 = Ray 1688.
Plukenet 1696:196 Jasminum Arabicum flore amplo pleno odoratissimo, foliis ad genicula ternis.
Hermann 1711:8,no.95 Ramulus cum floribus & foliis Mogori seu Kuddae mullae Hort.Mal.
J.Burman 1737:128-129,Tab.58,Fig.2 Jasminum Limonii folio conjugato, flore odorato, pleno, vario (r).

Pitsjegam-mulla HM 6:91:52
Pilsjegam-Mullà
BL Add MS 5030:227.
J.Commelin in HM 6:91 Jasminum humilius flore magno Bauh.Pin.L.X.Sect.II [Bauhin 1623:398].
Hermann 1687:337,678 = J.Commelin in HM.
Ray 1688 vol.2:1600 = J.Commelin in HM.
J.Commelin 1689:171 = J.Commelin in HM.
C.Commelin 1696:36 = J.Commelin in HM.
Plukenet 1696:196 = J.Commelin in HM.
Linnaeus 1737:5 Jasminum foliis oppositis pinnatis α.

Linnaeus 1762:9 Jasminum grandiflorum.
N.L.Burman 1768:5-6 = Linnaeus 1762.
Houttuyn 1775 II vol.4:19-20 = N.L.Burman 1768.

Catu pitsjegam-mulla HM 6:93:53
Katu-Pilsjegam-Mulla
BL Add MS 5030:228-229.
J.Commelin in HM 6:93 Jasminum vulgatius flore albo Bauh.Pin. [Bauhin 1623:397] (c).
Ray 1688 vol.2:1602 Jasminum Indicum flore polypetalo, candido oris rufescentibus.
C.Commelin 1696:36-37 = Ray 1688.
Plukenet 1696:196 = Ray 1688.
Petiver 1702(1701):942,no.161,1 Pecalah Malab.Jasminum Eremitana polypetalon, Pervincae folio (?).
Linnaeus 1737:5 Nyctanthes caule volubili, foliis subovatis acutis.
Linnaeus 1753:6 Nyctanthes angustifolia.
Linnaeus 1762:8-9 = Linnaeus 1753.
N.L.Burman 1768:5 = Linnaeus 1762.
Houttuyn 1775 II vol.4:15-16 = N.L.Burman 1768.

Katu tsjiregam-mulla HM 6:95:54
Katu-Tsjiregam-Mullà
BL Add MS 5030:230-231.
J.Commelin in HM 6:95 Jasminum sylvestre species.
Ray 1688 vol.2:1602 Jasminum Indicum flore polypetalo candidissimo, fructu majore.
C.Commelin 1696:37 = Ray 1688.
Plukenet 1696:196 Jasminum Indicum sylvestre inodorum Lauri folio, flore polypetaloide summis ramulis innixo monococcon.
Petiver 1702(1701):942,no.161,2 Coode-woola-checa Malab.Jasminum Eremitanum polypetalon Myrti fol.acuto (?).
Linnaeus 1737:5 Nyctanthes caule volubili, foliis subovatis acutis (cs).
N.L.Burman 1768:4 Nyctanthes Sambac Linn.sp.8 [Linnaeus 1762:8].
Houttuyn 1775 II vol.4:10-13 = N.L.Burman 1768.

Tsjiregam-mulla HM 6:97:55
Tsjíregaṁ-Mullà
BL Add MS 5030:232-233.
J.Commelin in HM 6:97 Jasminum species.
Ray 1688 vol.2:1601-1602 Jasminum Indicum flore polypetalo exalbido, fructu minori.
C.Commelin 1696:36 = Ray 1688.
Plukenet 1696:195-196 Jasminum Indicum odoratum, flore polypetaloide exalbido, ad geniculos prodeunte, fructu Cerasi minoris gemello.
Linnaeus 1737:5 Nyctanthes caule volubili, foliis subovatis acutis (cs).
Rumphius 1747 vol.5:86-87,Tab.XLVI Jasminum Litoreum (-); J.Burman ibid.:87 (c).
Linnaeus 1753:6 Nyctanthes undulata.
Linnaeus 1762:8 = Linnaeus 1753.
N.L.Burman 1768:4 = Linnaeus 1762.
Houttuyn 1775 II vol.4:13-14 = N.L.Burman 1768.

Katu-mulla HM 6:99:56
Katu-Mullà
BL Add MS 5030:234-235.
J.Commelin in HM 6:99 Jasminum species.
Ray 1688 vol.2:1602 Jasminum Indicum flore pentapetalo candidissimo fructu.
C.Commelin 1696:37 = Ray 1688.

Plukenet 1696:196 Jasminum Indicum inodorum, flore pentapetaloide candidissimo dicoccon.
J.Burman 1737:128 Jasminum Indicum, sylvestre, bacciferum, inodorum. Mus.Zeyl.pag.60 [Hermann 1717:60].

Badukka HM 6:105:57
Badúkka
BL Add MS 5030:236-237.
J.Commelin in HM 6:103 Capparis species.
Ray 1688 vol.2:1630 Capparis arborescens Indica Badukka dicta, flore tetrapetalo.
C.Commelin 1696:18 = Ray 1688.
Plukenet 1700:36 Muladundee Malabarorum (?); ibid.1696:80 Capparis (fortè) Maderaspatensis, Euonymi pallentibus foliis, spinis brevibus duplicatis, sub titulo Rhamni talis Icon exhibetur, Phytogr.Tab.107.fig.3.
Linnaeus 1737:204 Capparis inermis, foliis ovato-oblongis per spatia confertis perennantibus β.
Linnaeus 1753:504 Capparis Baducca.
Linnaeus 1762:720-721 = Linnaeus 1753.
N.L.Burman 1768:118-119 = Linnaeus 1762.
Houttuyn 1775 II vol.5:236-237 = N.L.Burman 1768.

Solda HM 6:103:58
Solda
BL Add MS 5030:238-239.
J.Commelin in HM 6:103 Capparis species.
Ray 1688 vol.2:1630 Capparis arborescens Indica flore pentapetalo Solda dicta.
C.Commelin 1696:18 = Ray 1688.
Plukenet 1696:80 = Ray 1688.

Sida-pou HM 6:109:59
Sida Pou
BL Add MS 5030:240-241.
Ray 1688 vol.2:1788 Arbor Indica akarpon, floribus odoratis pentapetalis, petalo uno luteo, reliquis candidiss.
C.Commelin 1696:7 = Ray 1688.
Plukenet 1700:22 Arbusculâ corymbosa Ind.Or.folio fermè Ovali figura. Sirrunarvelle Malabaror. (r).

Tsjude-maram HM 6:111:60
Tsjude-Maram
BL Add MS 5030:242-243.
Ray 1688 vol.2:1733 Frutex Indicus, flore papilionaceo sanguineo, foliis pinnatis, fructu viduus; ibid.1704 vol.3:108(2).
Breyne 1689:82 Periclymenum Sinense variegatum, flore sangvineo amplo.
C.Commelin 1696:52 = Breyne 1689.
Plukenet 1700:287 Periclymenum Zeylanicum herbaceum foliis variegatis, diversi-coloribus maculis ornatis PBP.363 [Sherard 1689:363].
J.Burman 1737:186-187 Periclymenum Indicum, foliis maculatis, latioribus, Laurinis.
Rumphius 1743 vol.4:74 (App),Tab.XXX Folium bracteatum (c).
Linnaeus 1762:21 Justicia picta.
N.L.Burman 1768:7 = Linnaeus 1762.
Houttuyn 1775 II vol.4:40-41 = N.L.Burman 1768.

Tsere-maram HM 6:109:61
Tsjéra-Maraṁ
BL Add MS 5030:244-245.
Ray 1688 vol.2:1733 = HM.
Breyne 1689:82 Periclymeno Sinensi variegato similis frutex, longo angusto folio variegato Sinicus.
C.Commelin 1696:52 = Breyne 1689.

J.Burman 1737:187 Periclymenum Indicum, foliis maculatis, angustioribus, Salicinis (?).
Linnaeus 1763:1424 Croton variegatum.
N.L.Burman 1768:203 = Linnaeus 1763.
Houttuyn 1776 II vol.6:241-241 = N.L.Burman 1768.

VOLUME 7

Natsjatam; Natsjatám-crua; aut Batta valli HM 7:1-2:1
Natsjátam
BL Add MS 5031:1-2.
J.Commelin in HM 7:2 Raji, Hist.Plant.Tom.I.lib.13.part 2.cap.21.species Solani; Flores Monopetali, quinque partiti aut stellati sunt, apicibus in umbilico oblongis, &c. [Ray 1686 vol.1:671 seq.].
Breyne 1689:19 Arbor Indica Cocculos Officinarum ferens.
C.Commelin 1696:7 = Breyne 1689.
Plukenet 1696:349 Solanum lignosum s.scandens rotundifolium Indicum, fructu duriore racemosa.
Plukenet 1700:52 Cocculus Officinarum. Cautacudde Malabarorum (?).
Ray 1704 vol.3:120(2) Arbuscula Coculos hosce proferens.
Plukenet 1705:61 Cocculi Indi species, tenerioribus subrotundis foliis, incanis; Caulcadde Malabarorum (c).
Rumphius 1747 vol.5:35-37,Tab.XXII Tuba Baccifera (=); J.Burman ibid.:37 (=).
Linnaeus 1753:340-341 Menispermum Cocculus (?).
Linnaeus 1763:1468 = Linnaeus 1753.

Tsjereu-cániram vel Tsjerou-cansjeram HM 7:3:2
Tsjèru-Càniram
BL Add MS 5031:3-4.
C.Commelin 1696:67 = HM.
Plukenet 1696:395 Xylomastix arbor Americana Whip-wood (i.e.) Arbor flagellifera Barbadensibus dicta Phytogr.Tab.238.fig.2 (?).

Válli-cániram HM 7:5-6:3
Válli-cániram
BL Add MS 5031:5-6.
C.Commelin 1696:68 = HM.

Schéru-válli-caniram, vel Sjérou-válli-kansjeram HM 7:7-8:4
Schèru-vàlli-càniram
BL Add MS 5031:7-8.
C.Commelin 1696:60 = HM.
Plukenet 1696:395 Xylomastix arbor Americana Whip-wood (i.e.) Arbor flagellifera Barbadensibus dicta Phytogr.Tab.238.fig.2 (c).
Plukenet 1700:68 Epidendron Ind.Orient.floribus ex spicam, prodeuntibus. Cheddemeelchedde Malabarorum (r).

Schéru-katu-válli-cániram, vel Tsjeru-Katu-valli-Kansje-ram HM 7:9:5
Tsjerù-katu-vàlli-Cániram
BL Add MS 5031:9-10.
Breyne 1689:93 Solanum arborescens Indicum, foliis Napecae minoribus, fructu rotundo duro & semine orbiculari compresso minoribus (?).
C.Commelin 1696:63 = Breyne 1689.
Plukenet 1700:60 Cucurbitifera Ind.Or.Canellae foliis laetè virentibus fructu minore. Nellawattachedde Malabarorum.
Petiver 1702(1701):935,no.142 Nella watta chedde Malab.Vomica Nux Emeritana, folio angustiore, fructu minore globosa (?).

Rumphius 1741 vol.2:121-124,Tab.XXXVIII Lignum Colubrinum (-); J.Burman ibid.:124 (r).
Linnaeus 1762:271 Strychnos colubrina.

Schémbra-válli HM 7:11-12:6
Schèmbra-vàlli
BL Add MS 5031:11-12.
J.Commelin in HM 7:13 Vitis sylvestris species.
C.Commelin 1696:69 = J.Commelin in HM.
J.Burman 1737:230 Vitis folio Cucurbitae, seu non sinuato. Mus.Zeyl.pag.7 [Hermann 1717:7] (?).
Rumphius 1747 vol.5:452,Tab.CLXVII Labrusca Molucca (c).
Linnaeus 1762:293 Vitis indica.
N.L.Burman 1768:62 = Linnaeus 1762.
Houttuyn 1775 II vol.4:375-376 = N.L.Burman 1768.

Vállia-píra-pítica HM 7:13:7
Wàllia-pìra-pitica
BL Add MS 5031:13-14.
J.Commelin in HM 7:13 Vitis sylvestris species.
C.Commelin 1696:69 = J.Commelin in HM.

Válliá-tsjórí-valli HM 7:15:8
Wàllia-Tsjori-vàlli
BL Add MS 5031:15-16.
C.Commelin 1696:68 = HM.

Tsjóri-válli HM 7:17:9
Tsjòri-vàlli
BL Add MS 5031:17-18.
Plukenet 1691:Tab.CLXIII,fig.1 Clematitis triphylla arborescens, è Maderaspatan (?).
C.Commelin 1696:20 Clematis triphylla arborescens à Maderaspatan Pluken. P.3.T.CLXIII.
Ray 1704 vol.3:36(2) Hedera Indica trifolia, claviculata, racemosa, fructu plano-rotundo, nigro.
Rumphius 1747 vol.5:450-451,Tab.CLXVI,Fig.2 Folium Causonis I-III (=).

Belútta-tsjóri-válli HM 7:19:10
Belùtta-tsjòri-vàlli
BL Add MS 5031:19-20.
C.Commelin 1696:15 = HM.
Ray 1704 vol.3:37(2) Hedera Indica folio composito, claviculata, racemosa, uvis plano-rotundis, eburneis.
Plukenet 1705:114 Hedera Quinquefolia Canadensia similes scandens, ex Insula Cheusán (?).
Rumphius 1747 vol.5:446-448,Tab.CLXV Funis Crepitans III (-); J.Burman ibid.:448 (c).
N.L.Burman 1768:75 Sambucus canadensis Linn.sp.385 [Linnaeus 1762:385].

Schunámbu-válli HM 7:21-22:11
Schunámbu-válli
BL Add MS 5031:21-22.
C.Commelin 1696:60 = HM.
Petiver 1702(1701):846,no.101 Perreaurulla Malab.Vitis Madraspat.fructu azureo folio subrotundo & anguloso Mus.Petiver 696 (?).
Ray 1704 vol.3:37(2) Hedera Indica folio simplici, integro, claviculata, racemosa, racemis laxis, uvis rotundis nigris.
Linnaeus 1747:24-25,no.60 Cissus.
Rumphius 1747 vol.5:446-448,Tab.CLXIV,Fig.1 Funis Crepitans I (-); J.Burman ibid.:448 (c).
Rumphius 1747 vol.5:479,Tab.CLXXVIII,Fig.1 Oculus Astaci (?c).

Linnaeus 1753:117 Cissus vitiginea.
Linnaeus 1762:170 = Linnaeus 1753.
Linnaeus 1767 vol.2:124 Cissus Sicyoides.
N.L.Burman 1768:35-36 = Linnaeus 1762.
Houttuyn 1775 II vol.4:144-147 = Linnaeus 1767.

Molagò-codí HM 7:23-24:12
Malagó-codí
BL Add MS 5031:23-24.
J.Commelin in HM 7:24 Piper rotundum nigrum apud Bauh.in Pinac: [Bauhin 1623:411-412].
C.Commelin 1696:54 = J.Commelin in HM.
Plukenet 1700:151; ibid.1696:297 Piper rotundum nigrum similiter ac album CBP.411 [Bauhin 1623:411-412].
J.Burman 1737:193-194 Piper rotundum ex Malabara, foliis latis, quinquenerviis, albicantibus. Mus.Zeyl.pag.32 [Hermann 1717:32].
Linnaeus 1737:6-7,Tab.IV Piper foliis cordatis, caule procumbente (cs).
Linnaeus 1747:10-11,no.26 Piper foliis ovatis subseptinerviis glabris, petiolis simplicissimis.
Linnaeus 1753:28 Piper nigrum.
Linnaeus 1762:40 = Linnaeus 1753.
N.L.Burman 1768:13 = Linnaeus 1762.
Houttuyn 1775 II vol.4:68-72 Piper nigrum Syst.Nat.XII.Gen. 43.p.67 [Linnaeus 1767 vol.2:67].

Cáttu-mólago HM 7:25:13
Cáttu-mólago
BL Add MS 5031:25-26.
J.Commelin in HM 7:25 Pipere saemineo Pisonis (c).
C.Commelin 1696:19 = HM.
Plukenet 1700:151; ibid.1696:297 Piper ignobile Canari Malabaraeis dictum Pison.Mantiss.182 (?).

Cattu-tírpali HM 7:27-28:14
Cáttu-tirpali
BL Add MS 5031:27-28.
J.Commelin in HM 7:28 Piper longum Orientale Bauh.in Pinac. [Bauhin 1623:412].
C.Commelin 1696:54 = J.Commelin in HM.
Plukenet 1700:151 Piper longum Orientale, foliis amplioribus quinquenerviis, petiolis oblongis insidentibus.
Petiver 1704(1702):1063-1064,no.243 Tipplelee Malab.Piper longum Officinarum.
Rumphius 1747 vol.5:333-335,Tab.CXVI,Fig.1 Piper longum (=); J.Burman ibid.:335 (=).
Linnaeus 1753:29 Piper longum.
Linnaeus 1762:41 = Linnaeus 1753.
N.L.Burman 1768:14 = Linnaeus 1762.
Houttuyn 1775 II vol.4:76-78 = N.L.Burman 1768.

Beëtla-codí HM 7:29:15
Beetla-codí
BL Add MS 5031:29-30.
J.Commelin in HM 7:29 Piper longum (cs).
C.Commelin 1696:15 Betre sive Tembul B:Pin:410 [Bauhin 1623:410].
J.Burman 1737:46-47 = C.Commelin 1696.
Linnaeus 1747:11-12,no.27 Piper foliis ovatis oblongiusculis acuminatis septinerviis, petiolis bidentatis.
Rumphius 1747 vol.5:336-340,Tab.CXVI,Fig.2 Sirii folium (=).
Linnaeus 1753:28-29 Piper Betle.
Linnaeus 1762:40 = Linnaeus 1753.
N.L.Burman 1768:14 = Linnaeus 1762.
Houttuyn 1775 II vol.4:72-73 = N.L.Burman 1768.

Amólago HM 7:31:16
Amálago
BL Add MS 5031:31-32.
J.Commelin in HM 7:31 Piper longum sp.
Plukenet 1691:Tab.CCXV,fig.2 Piper Frutex Americ.spicâ longâ gracili.
C.Commelin 1696:54 Piper longum angustissimum ex Florida B:Pin:412 [Bauhin 1623:412].
Plukenet 1696:297 = Plukenet 1691.
J.Burman 1737:193,Tab.83,Fig.2 Piper, qui Saururus foliis septinerviis, oblongo-acuminatis.
Linnaeus 1747:12,no.28 Piper foliis ovatis acutiusculis subtus scabris: nervis quinque subtus elevatis.
Rumphius 1747 vol.5:46-48,Tab.XXVIII,Fig.1 Sirium arborescens tertium (=); J.Burman ibid.:48 (r).
Linnaeus 1753:29 Piper Malamiris.
Linnaeus 1762:41 = Linnaeus 1753.
N.L.Burman 1768:14 = Linnaeus 1762.
Houttuyn 1775 II vol.4:74-75 = N.L.Burman 1768.

Katu-kára-walli HM 7:33:17
Katu-Kára-Walli
BL Add MS 5031:33-34.
J.Commelin in HM 7:33 Rhamnus species.
Plukenet 1691:Tab.CVIII,f.2 Rhamnus àn potiùs Lycium Fingrego Jamaicensibus dictum (?).
C.Commelin 1696:58 = J.Commelin in HM.
Linnaeus 1753:1026-1027 Pisonia mitis.
Linnaeus 1763:1511 = Linnaeus 1753.
N.L.Burman 1768:224 = Linnaeus 1763.
Houttuyn 1774 II vol.3:645-646 = N.L.Burman 1768.

Válli-kára HM 7:35-36:18
Valli-kara
BL Add MS 5031:35-36.
J.Commelin in HM 7:36 Dulc-Amara species.
C.Commelin 1696:27 = J.Commelin in HM.

Pee-ámerdu HM 7:37-38:19-20
Peê-ámerdu, Pee-amerdu
BL Add MS 5031:37-38,39-40.
C.Commelin 1696:51 = HM.

Cit-amerdu HM 7:39:21
Citámerdu
BL Add MS 5031:41-42.
C.Commelin 1696:20 = HM.
Rumphius 1747 vol.5:482,Tab.CLXXX Olus Sanguinis (-); J.Burman ibid.:482 (?c).
N.L.Burman 1768:216 Menispermum glabrum.

Ula HM 7:41:22
Ula
BL Add MS 5031:43-44.
C.Commelin 1696:69 = HM.
Rumphius 1747 vol.5:12-13,Tab.VIII Gnemon funicularis (=).

Pee-Ula HM 7:43:23
Pee-Ula
BL Add MS 5031:45-46.
C.Commelin 1696:51 = HM.
Plukenet 1696:349 Solani lignosi s.Dulc-amarae alatis foliis non auritis Frutex Indicus per summos ramos floridus, Phytogr. Tab.317.fig.1 (?).
Ray 1704 vol.3:654-655 Frutex Indicus, foliis adversis, floribus racemosis octapetalis, fructu nullo.

Basella HM 7:45:24
Basélla
BL Add MS 5031:47-48.
J.Commelin in HM 7:45 Bryonia species.
Breyne 1689:23 Beta baccifera aizooïdes rotundifolia Zeylanica.
J.Commelin 1689:330 Solanum scandens Malabaricum, Betae folio.
Sherard 1689:317 = HM.
Plukenet 1691:Tab.LXIII,f.1 Mirabili Peruvianae affinis tinctoria Betae folio scandens, Cochin-Chinensis flosculis puniceis, muscosis spicatis, fructu intùs cochleato.
C.Commelin 1696:63 = J.Commelin 1689.
Plukenet 1696:252-253 = Plukenet 1691.
Ray 1704 vol.3:358 Solanum peregrinum: Betae folio hist.nost. [Ray 1688 vol.2:1876].
Boerhaave 1720 vol.2:266 = Breyne 1689.
J.Burman 1737:44 = J.Commelin 1689.
Linnaeus 1737:39 Cuscuta foliis subcordatis.
Royen 1740:400-401 = Linnaeus 1737.
Rumphius 1747 vol.5:417-418,Tab.CLIV,Fig.2 Gandola (-); J.Burman ibid.:418 (=).
Linnaeus 1747:50,no.119 = HM.
Linnaeus 1753:272 Basella rubra.
Linnaeus 1762:390 = Linnaeus 1753.
N.L.Burman 1768:272 = Linnaeus 1762.
Houttuyn 1777 II vol.8:249-251 Basella rubra Syst.Nat.XII.Gen. 379.p.221 [Linnaeus 1767 vol.2:221].

Curinil vel Curiginil HM 7:47:25
Curiníl
BL Add MS 5031:49-50.
C.Commelin 1696:25 = HM.
Ray 1704 vol.3:357 Baccifera Indica, foliis ex adverso binis, ad Solanum accedens, fructu monopyreno, Foribus (sic) pentapetalis, in communi pediculo pluribus.

Curigi-taly HM 7:49:26
Curigi-táli
BL Add MS 5031:51-52.
C.Commelin 1696:25 = HM.
Plukenet 1705:31 Arbuscula trifolia, & quinque folia Ind.Or.siliquâ Epiglottidi simili, unicum semen continente (?).

Páda-vára HM 7:51-52:27
Páda-vára
BL Add MS 5031:53-54.
C.Commelin 1696:49 = HM.
Plukenet 1696:287 Periclymenum Americanum è cujus radice fit Atramentum Phytogr.Tab.212.fig.4 (?).
Plukenet 1700:148 Periclymenum Indicum elatius, cuspidato folio majori glabro.
Ray 1704 vol.3:74(2) Frutex baccifer Indicus foliis adversis, 5 in summis ramulis petiolis floriferis, sex septémve flores & dein baccas polypyrenas confertas sustinentibus.

Unjala HM 7:53:28
Unjála
BL Add MS 5031:55-56.
C.Commelin 1696:69 = HM.
Plukenet 1700:189 Frutex Indicus, foliis rigidioribus ternis, quaternis, immò quinis, fructu orbiculari, compresso, fungoso, unicum semen complectente.

Itti-canni HM 7:55:29
Itti-Cánni
BL Add MS 5031:57-58.
J.Commelin in HM 7:55 Species Caprifolii vel Periclymenii.
Plukenet 1691:Tab.CCXII,fig.5 Periclymenum surrectum Persicae foliis Maderaspatanum.
C.Commelin 1696:52 Periclymenum indicum flore flavescente Turnef:El:Bot:481.
Plukenet 1696:287 = Plukenet 1691.
Linnaeus 1747:34-35,no.83 Lonicera pedunculis multifloris, involucris pentaphyllis, foliis ovato-lanceolatis petiolatis.
Linnaeus 1753:175 Lonicera parasitica.
Linnaeus 1762:473 Loranthus Lonicerioides.
N.L.Burman 1768:84 = Linnaeus 1762.
Houttuyn 1775 II vol.4:459 = N.L.Burman 1768.

Tiri-itti-canni HM 7:57:30
Tirí-Itti-Canni
BL Add MS 5031:59-60.
C.Commelin 1696:66 = HM.
Plukenet 1696:343 Serpentaria repens floribus stamineis spicatis, Bryoniae nigrae folio ampliore pingui, Phytogr.Tab.117. fig.3,&4 (cs).
Ray 1704 vol.3:655 Frutex parasiticus spinosus, florum pentapetaloon spicis oblongis è ramorum nodulis erumpentibus, fructu nullo.

Cari-villandi HM 7:59-60:31
Kari-Vilandi
BL Add MS 5031:61-62.
J.Commelin in HM 7:60 Smilax aspera species.
C.Commelin 1696:62-63 = J.Commelin in HM.
Plukenet 1705:194 Smilax viticulis asperis, leni Canellae folio, ex Indiâ Orientali.
Boerhaave 1720 vol.2:268 Smilax; Americana; Menthonicae folio (?).
Rumphius 1747 vol.5:437-440,Tab.CLXI Pseudochina Amboinensis (cs); J.Burman ibid.:440 (cs).
N.L.Burman 1768:213 Smilax indica.

Wattou-valli HM 7:61:32
Wattou-valli
BL Add MS 5031:63-64.
C.Commelin 1696:70 = HM.
Plukenet 1696:37 Apocynum scandens Scammoniae Monspeliacae foliis, Curassavicum (?).

Tjageri-nuren HM 7:63:33
Tsjagerí-Núren
BL Add MS 5031:65-66.
J.Commelin in HM 7:71(2) species Batatti sylvestris.
C.Commelin 1696:66 = HM.
Ray 1704 vol.3:133 Battata sylvestris spinosa trifolia, flore vidua, verrucosa, fructibus in spicis longis, rarioribus, triangularibus, ad pediculum reflexis.
Plukenet 1705:184 Ricophora Maderaspatana triphyllos, oblongo fructu triangulari (?).
Linnaeus 1737:459 Dioscorea foliis ternatis.
Rumphius 1747 vol.5:361-364,Tab.CXXVIII Ubium silvestre (-); J.Burman ibid.:364 (c).
Linnaeus 1753:1032 Dioscorea triphylla.
Linnaeus 1763:1462 = Linnaeus 1753.
N.L.Burman 1768:214 = Linnaeus 1763.
Houttuyn 1779 II vol.11:359-360 Dioscorea Triphylla [Linnaeus 1767 vol.2:656].

Katu-nuren-kelengu HM 7:63:34

Kátu-Núren-Kelèngú
BL Add MS 5031:67-68.
J.Commelin in HM 7:71(2) species Batatti sylvestris.
C.Commelin 1696:39 = HM.
Ray 1704 vol.3:133 Battata sylvestris verrucosa, spinoso caule, foliis digitatis, fructu carens.
Linnaeus 1737:459 Dioscorea foliis digitatis, caule spinoso.
Rumphius 1747 vol.5:359-361,Tab.CXXVII Ubium quinquefolium (?c).

Nurem-kelengu HM 7:67:35
Núren-Kelèngú
BL Add MS 5031:69-70.
J.Commelin in HM 7:71(2) species Batatti sylvestris.
C.Commelin 1696:48 = HM.
Plukenet 1696:321 Ricophora pentaphyllos caule spinoso, fructu oblongo triquetro.
Ray 1704 vol.3:133 Battata Indica, caule spinoso, foliis digitatis, fructu triangulo, cum tribus intus seminibus compressis.
Linnaeus 1737:459 Dioscorea foliis digitatis, caule spinoso.
Rumphius 1747 vol.5:361-364,Tab.CXXVIII Ubium silvestre (?c); J.Burman ibid.:364 (≠).
Linnaeus 1753:1032 Dioscorea pentaphylla.
Linnaeus 1763:1462 = Linnaeus 1753.
N.L.Burman 1768:213 = Linnaeus 1763.
Houttuyn 1779 II vol.11:359-360 = N.L.Burman 1768.

Katu-katsjil HM 7:69:36
Kátu-Kátsjil
BL Add MS 5031:71-72.
J.Commelin in HM 7:71(2) species Batatti sylvestris.
Plukenet 1691:Tab.CCXX,fig.6 Ricophora Indica Bryoniae nigrae similis, ad foliorum ortum verrucosa HRH (?).
C.Commelin 1696:39 = HM.
Plukenet 1696:321 = Plukenet 1691.
Plukenet 1700:185 Triopteris Americana scandens, fructu fulgente, majore aureo (cs).
Hermann 1698:217-218 Rizophora Zeylanica Scammonii folio singulari radice rotunda.
Petiver 1704(1703):1460,no.51 Triopteris Malabaric.scan-dens Inhame folio (?).
Ray 1704 vol.3:133 Battata sylvestris foliis Smilacis nervosis, flore vidua, fructu triangulari compresso dispermo.
J.Burman 1737:207 = Hermann 1698.
Linnaeus 1737:459 Dioscorea foliis cordatis, caule bulbifero.
Linnaeus 1747:170,no.359 = Linnaeus 1737.
Rumphius 1747 vol.5:354-355,Tab.CXXIV Ubium pomiferum (-); J.Burman ibid.:355 (=).
Linnaeus 1753:1033 Dioscorea bulbifera.
Linnaeus 1763:1463 = Linnaeus 1753.
N.L.Burman 1768:214 = Linnaeus 1763.
Houttuyn 1779 II vol.11:360-361 Dioscorea Bulbifera [Linnaeus 1767 vol.2:656].

Káttu-kelángu HM 7:71:37
Káttu-Kelèngú
BL Add MS 5031:73-74.
J.Commelin in HM 7:71(2) species Batatti sylvestris.
C.Commelin 1696:39 = HM.
Ray 1704 vol.3:133-134 Battata sylvestris, spinosa, Smilacis folio, floribus stamineis racemosis, pro fructu verrucosam excrescentiam protrudens.
Plukenet 1705:184 Ricophorae (fortè) genus Bryoniae nigrae foliis heptaneuros; nervis per longitudinem profundius punctatis, plurimis conspicuis venis in parallelos venustè transcurrentibus; Payn-cheddy Malabarorum (?).

J.Burman 1737:218,Tab.101 Smilax foliis peltatis, cordato-oblongis, floribus minutissimis & copiosissimis.
Linnaeus 1737:459 Dioscorea foliis cordatis, caule aculeato bulbifero.
Rumphius 1747 vol.5:356-357,Tab.CXXV Ubium Ovale (?c).
Linnaeus 1753:1033 Dioscorea aculeata.
Linnaeus 1763:1462 = Linnaeus 1753.
N.L.Burman 1768:214 = Linnaeus 1763.
Houttuyn 1779 II vol.11:360-361 Dioscorea aculeata [Linnaeus 1767 vol.2:656].

Katsji kelengu HM 7:71(2):38
Katsjil-Kelèngú
BL Add MS 5031:75-76.
J.Commelin in HM 7:71(2) species Batatti sylvestris.
Sherard 1689:370-371 Ricophora Indica sive Inhame rubra caule alato Scammonii foliis nervosis conjugatis.
C.Commelin 1696:58 = Sherard 1689.
Plukenet 1696:321 = Sherard 1689.
Ray 1704 vol.3:134 Battata sylvestris Indica, foliis Smilacis, laevibus, caulibus alatis, flore vidua, pro fructu glandes tuberosas ad exortum foliorum emittens.
J.Burman 1737:206-207 = Sherard 1689.
Linnaeus 1737:171,no.360 Dioscorea foliis cordatis, caule alato bulbifero.
Rumphius 1747 vol.5:346-349,Tab.CXX Ubium vulgare (r).
Linnaeus 1753:1033 Dioscorea alata.
Linnaeus 1763:1462 = Linnaeus 1753.
N.L.Burman 1768:214 = Linnaeus 1763.
Houttuyn 1779 II vol.11:360-361 Dioscorea Alata [Linnaeus 1767 vol.2:656].

Erima-táli HM 7:73:39
Erima-tály
BL Add MS 5031:77-78.
Plukenet 1691:Tab.CCXXXVIII,fig.2 Xylomastix arbor Americana Whip-wood (i.e.) Arbor flagellifera Barbadensibus dicta (r).
C.Commelin 1696:27 = HM.
Plukenet 1696:395 = Plukenet 1691.

Ána-parva HM 7:75-76:40
Aña-párua
BL Add MS 5031:79-80.
C.Commelin 1696:5 = HM.
Ray 1704 vol.3:654 Frutex Parasiticus Aurantiae ferè foliis, flore carens, fructu ruberculis aspero, numerosa intus semina continente, ê foliorum alis exeunte.
Rumphius 1747 vol.5:490 (Auct),Tab.CLXXXIV,Fig.1-3 Adpendix duplo folio, seu Tapanawa Kitsjil (-); J.Burman ibid.:490 (c).
Linnaeus 1763:1374 Pothos scandens.
N.L.Burman 1768:193 = Linnaeus 1763.
NB See also Ana-parua HM 1:88.

Tsjangelam-parenda HM 7:77:41
Tsjangelam Parenda
BL Add MS 5031:81-82.
J.Commelin in HM 7:77 Sedum repens species.
C.Commelin 1696:61 = J.Commelin in HM.
Plukenet 1696:298 Planta baccifera scandens, Epidendros Maderaspatana, geniculato & quadripinnato caule, flosculis exiguis ad genicula capreolis donata. Phytogr.Tab.310.fig.6 (?).
Ray 1704 vol.3:661 = J.Commelin in HM

[no name] HM 7:79:42

Pu-Wálli
BL Add MS 5031:83-84.
C.Commelin 1696:57 = HM.
Plukenet 1705:179 Arbor Maderaspatana, Ella-cooty-maram Malabarorum (s).

Pu-pal-valli HM 7:81:43
Pu-pal-válli
BL Add MS 5031:85-86.
J.Commelin in HM 7:81 Alcaea Carpini Folio, semine rostrato (s).
C.Commelin 1696:57 = HM.
Plukenet 1696:113 Lappula Malabarica spicata, Betae foliis Antagonistis, floribus interiùs lanuginosis (?).
Plukenet 1696:206 Lappula Eupatoria maximè odorata (?).
Petiver 1704(1703):1454,no.21 Lappula Madraspat.minor (?).
Plukenet 1705:130 Lappula Malabarica pentapetalos spicata, foliis oblongo rotundis, floribus villosis monopyrene.

Acátsja-valli ceu Mudila tali HM 7:83-84:44
Acátsja-válli
BL Add MS 5031:87-88.
J.Commelin in HM 7:84 Cuscuta species.
Plukenet 1691:Tab.CLXXII,f.2 Cuscuta baccifera Barbadensium à maritimis (?).
C.Commelin 1696:25 = J.Commelin in HM.
Plukenet 1696:126 = Plukenet 1691.
Petiver 1702(1701):1022,no.226 Cootan Malab.Cuscuta baccifera Salawaccensis fructu coronato.
Ray 1704 vol.3:551 Cuscuta Indica rarius foliosa, flosculis spicatis hexapetalis, fructu coronato rotundo monospermo.
J.Burman 1737:84 Cuscuta Indica, floribus albis, stellatis. Mus.Zeyl.pag.67 [Hermann 1717:67] (?).
Rumphius 1747 vol.5:491-492,Tab.CLXXXIV,Fig.4 Cussuta (=).
Linnaeus 1753:35-36 Cassytha filiformis.
Linnaeus 1762:530-531 = Linnaeus 1753
N.L.Burman 1768:92 = Linnaeus 1762.
Houttuyn 1777 II vol.8:496-497 Cassytha Filiformis Syst.Nat.XII.Gen.500.p.281 [Linnaeus 1767 vol.2:281].

Káreta-tsjóri-válli HM 7:85:45
Kareta-Tsjori-válli
BL Add MS 5031:89-90.
J.Commelin in HM 7:85 Hedera Trifolia Baccifera Malabarensis.
C.Commelin 1696:35 = J.Commelin in HM.
Ray 1704 vol.3:37(2) = J.Commelin in HM.
Plukenet 1705:114 = J.Commelin in HM.
Rumphius 1747 vol.5:450-451,Tab.CLXVI,Fig.2 Folium Causonis I-III (cs?).
N.L.Burman 1768:63 Vitis trifolia Linn.sp.293 [Linnaeus 1762:293].

Modira-valli HM 7:87:46
Módira-válli
BL Add MS 5031:91-92.
C.Commelin 1696:45 = HM.
Ray 1704 vol.3:653 Frutex Parasiticus Indicus, spinosus, flore fructuque carens, foliis longis, cuspidatis, spissis, glabris, per margines aequalibus.

Válli-módagam HM 7:89:47
Válli-Modagam
BL Add MS 5031:93-94.
C.Commelin 1696:68 = HM.
Ray 1704 vol.3:653 Frutex parasiticus Indicus, flore fructuque carens, foliis latioribus, costa media grandi.

Meriám-pulli HM 7:91:48
Neriam-pullí
BL Add MS 5031:95-96.
C.Commelin 1696:47 = HM.
Rumphius 1747 vol.5:446-448 Funis Crepitans IV (?c).

Pada-valli; ceu Pada-kelengu HM 7:93:49
Páda-kelángú. seu Muschega-páda. seu. Pádaválli
BL Add MS 5031:97-98.
J.Commelin in HM 7:93 Species Bryoniae levis Malabariensis.
Breyne 1689:20 Arbuscula exotica foliis umbilicatis Pada Kelangu, Nuculam Nuculae Indicae sonorae Johannis Bauhini similem ferens.
C.Commelin 1696:16 = J.Commelin in HM.
Plukenet 1696:71 Bryonia Coromandeliensis umbilicatis foliis, crassis, acuminatis, villosa.
J.Burman 1737:171 Nux Zeylanica, umbilicatis foliis. H. Beaum.pag.31 [Kiggelaer 1690:31].

Kappa-keléngú HM 7:95:50
Kappá-keléngú
BL Add MS 5031:99-100.
J.Commelin in HM 7:95 Camotes Hispanorum Clusii, Bauh.Pin. [Bauhin 1623:91].
Sherard 1689:325-326 Convolvulus Indicus radice tuberosâ eduli cortice rubro Batattas dictus.
C.Commelin 1696:22 = Sherard 1689.
Plukenet 1696:114 Convolvulus angulosis foliis, Malabaricus, radice tuberosâ, eduli.
Linnaeus 1737:67 Convolvulus foliis cordatis angulatis, radice tuberosa.
Rumphius 1747 vol.5:367-370,Tab.CXXX Batatta (=); J.Burman ibid.:370 (=).
Linnaeus 1753:154 Convolvulus Batatas.
Linnaeus 1762:220 = Linnaeus 1753.
N.L.Burman 1768:44 = Linnaeus 1762.
N.L.Burman 1768:215 Dioscorea cylindrica.
Houttuyn 1777 II vol.7:541-544 Convolvulus Batatas [Linnaeus 1767 vol.2:156].
Houttuyn 1779 II vol.11:364 = N.L.Burman 1768:215.

Pódava-keléngu HM 7:97:51-52
Pódava keléngú
BL Add MS 5031:101-102,103-104.
J.Commelin in HM 7:97 Rapum Brasilicanum (sic) seu Americanum alterum Bauh.Pin.lib.3.sect.1 [Bauhin 1623:89].
Sherard 1689:371 Ricophora sive Inhame Malabarica spinosa.
C.Commelin 1696:58-59 = Sherard 1689.
Plukenet 1696:321 = Sherard 1689.
Ray 1704 vol.3:134 = Sherard 1689.
Rumphius 1747 vol.5:357-359,Tab.CXXVI Combilium (-); J.Burman ibid.:359 (cs,v).

Panámbu-valli HM 7:99:53
Panamboe Pamboe-válli
BL Add MS 5031:105-106.
J.Commelin in HM 7:99 Arundo Indica Volubilis sp.
C.Commelin 1696:10 = J.Commelin in HM.
Ray 1704 vol.3:573 Canna Indica florida, baccifera, fructu umbilicato, monopyreno, floribus racemosis.
Plukenet 1705:101 Frutex Indicus scandens, folio Mendoni instàr in capreolam definente, fructu parvo coronato ... Shevenar Malabarorum (?).
Plukenet 1705:146 Methonicae foliis in claviculam definentibus, Planta Maderaspatana, Shevenar Corungo Malabarorum (?).
J.Burman 1737:35 Arundo sarmentosa, Indica, baccifera, foliis

in extremo capreolatis. Mus.Zeyl.pag.6 [Hermann 1717:6].
Linnaeus 1747:55 Flagellaria.
Rumphius 1747 vol.5:120-121,Tab.LIX,Fig.1 Palmijuncus laevis (=); J.Burman ibid.:121 (=).
Linnaeus 1753:333 Flagellaria indica.
Linnaeus 1762:475-476 = Linnaeus 1753.
N.L.Burman 1768:85 = Linnaeus 1762.
Houttuyn 1775 II vol.4:465-466 = N.L.Burman 1768.

Píripu HM 7:101:54
Piripú
BL Add MS 5031:107-108.
C.Commelin 1696:54 = HM.
N.L.Burman 1768:122,Tab.37,f.1 Delima sarmentosa Linn.sp.736 [Linnaeus 1762:736].
Linnaeus 1771:402 = N.L.Burman 1768.
Houttuyn 1775 II vol.5:244 = N.L.Burman 1768.

Tsjeria-pu-pupal-válli HM 7:103:55
Tsjeria-pupal-válli
BL Add MS 5031:109-110.
C.Commelin 1696:67 = HM.
Ray 1704 vol.3:654 Frutex Indicus, foliis adversis, floribus in summis caulibus racemosis, ê tubo oblongo in quinque segmenta cuspidata expansis, fructu nullo.

Tálu-dáma HM 7:105:56
Tálu-dáma
BL Add MS 5031:111-112.
Breyne 1689:99 Thalictro affinis Indica, Alni folio, semine striato aspero.
C.Commelin 1696:68 = Hermann 1698.
Hermann 1698:237-238 Valerianella Corassavica semine aspero viscoso. Par.Bat.prodr. [Sherard 1689:382].
Plukenet 1700:185; ibid.1696:379 Tulipifera Caroliniana foliis productioribus magis angulosis. Mus.Corten.Phytogr.Tab.68. fig.3 (?).
Ray 1704 vol.3:390 Herba Indica foliis subrotundis, adversis, flosculis pluribus simul, cyathiformibus cirum oras crenatis, vasculo oblongo-rotundo, monospermo.
Plukenet 1705:207 Valerianella Curassavica, semine aspero viscoso, Almag.Bot.381 [Plukenet 1696:381]. Mocaretty, & aliquando Mokretta Malabarorum.
Linnaeus 1737:17 Boerhaavia foliis ovatis.
Royen 1740:234 = Linnaeus 1737.
Linnaeus 1747:4,no.10 Boerhavia diffusa.
Linnaeus 1753:3 Boerhaavia diffusa.
Linnaeus 1762:4 = Linnaeus 1753.
N.L.Burman 1768:3 = Linnaeus 1762.
Houttuyn 1777 II vol.7:43-44 Boerhaavia diffusa [Linnaeus 1767 vol.2:51].

Mendoni HM 7:107-108:57
Mendóni
BL Add MS 5031:113-114.
J.Commelin in HM 7:108 Lilium superbum Ceylanicum Paulus Hermans [Hermann 1687:688-690,tab.].
J.Commelin 1689:229 = J.Commelin in HM.
Sherard 1689:353 Methonica Malabarorum Hort.Lugd. [Hermann 1687:688-690,tab.].
Plukenet 1691:Tab.CXVI,f.3 Methonica Malabarorum; Niengha la Zeynanensium Mus.Zeyl.
C.Commelin 1696:44 = J.Commelin in HM.
Plukenet 1696:249 = Plukenet 1691.
J.Commelin 1697:69-70,Fig.35 = J.Commelin in HM.
Petiver 1702(1701):940,no.155 Calapeecalunga Malab.Methonica Malabarorum. Nienghala Zeylonensium. Lilium Zeylanicum superbum vulgo H.Lugd.Bat.689 [Hermann 1687:688-690, tab.].
Petiver 1704(1702):1061,no.238 Shevanar calunga Malab. Lilium Zeylanicum superbum Hort.Amst.V.1.p.69.Fig.35 [J.Commelin 1697].
Boerhaave 1720 vol.2:134 Methonica; Malabarorum H.L.688 [Hermann 1687:688-690,tab.].
J.Burman 1737:158-159 Methonica gloriosa, foliis capreolatis, floribus fimbriatis, reflexis.
Linnaeus 1737:121 Gloriosa.
Linnaeus 1747:51-52,no.122 = Linnaeus 1737.
Linnaeus 1753:305 Gloriosa superba.
Linnaeus 1762:437 = Linnaeus 1753.
N.L.Burman 1768:82 = Linnaeus 1762.
Houttuyn 1780 II vol.12:260-262 Gloriosa superba Syst.Nat.XII. Gen.409 [Linnaeus 1767 vol.2:241].

Veetla-caitu HM 7:109:58
Vêetla-caitú
BL Add MS 5031:115-116.
C.Commelin 1696:68 = HM.
Plukenet 1696:135 Ephemerum phalangoides, Malabariense, repens, hexapetalon, capsulâ seminali triquetrâ, & in longum striatâ.
Hermann 1698:148 Ephemerum Malabaricum procumbens subrotundia foliis.
Ray 1704 vol.3:566 = Hermann 1698.
Ray 1704 vol.3:567 Nervifolia Indica repens, Ephemeri Phalangoidis folio, flore hexapetalo coeruleo, vasculo tricapsulari.
N.L.Burman 1768:18 Tab.7,f.4 Commelina cristata Linn.sp.62 [Linnaeus 1762:62].

Pongolam HM 7:111:59
Pongólam
BL Add MS 5031:117-118.
C.Commelin 1696:56 = HM.
Plukenet 1700:24 Arbor Africana foliis Arbuti Spiraeae floribus in spicam densiorem adactis (c).
Plukenet 1700:159; ibid.1696:Raphanus aquaticus alter CBP.97. & Prodr.38 [Bauhin 1623:97; ibid.1620:38]. (?).
Ray 1704 vol.3:654 Rhaphanus sylvestris Malabaricus, fructu carens.
Rumphius 1747 vol.5:73-74,Tab.XXXIX,Fig.1 Sinapister (?c).

VOLUME 8

Bela-schora HM 8:1-2:1
Belá-schorá
BL Add MS 5031:119-120.
J.Commelin in HM 8:2 Pepo vulgaris Raji Hist.Plant.lib.13.cap.2 [Ray 1686 vol.1:639-640].
C.Commelin 1696:52 = J.Commelin in HM.
Plukenet 1696:286 = J.Commelin in HM.
Rumphius 1747 vol.5:398-399,Tab.CXLV Cucurbita Indica vulgaris, & Pepones.

Schakeri-schora HM 8:3:2
Schakerí-schorá
BL Add MS 5031:121-122.
J.Commelin in HM 8:3 Pepo Indicus compressus maximus, Morison.Hist.Oxon.Sect.1.Cap.3.
C.Commelin 1696:52 Pepo compressus major B:Pin:311 [Bauhin 1623:311].
Plukenet 1696:286 = J.Commelin in HM.

Cumbulam HM 8:5-6:3
Cumbulám
BL Add MS 5031:123-124.
J.Commelin in HM 8:6 Pepo oblongus Moriss.Hist.Oxon.Sect.1. cap.3.
C.Commelin 1696:52 Pepo oblongus B:Pin:311 [Bauhin 1623:311].
Plukenet 1696:286 = J.Commelin in HM.
Rumphius 1747 vol.5:395-396,Tab.CXLIII Camolenga (?c).

Caca-palam HM 8:7-8:4
Cacá-palám
BL Add MS 5031:125-126.
J.Commelin in HM 8:8 Colocynthus oblonga B.Pin. [Bauhin 1623:313].
C.Commelin 1696:21 = J.Commelin in HM.
Ray 1704 vol.3:332-333 = J.Commelin in HM.
Rumphius 1747 vol.5:397-398 Calabassa Utan (?=).

Caipa-schora HM 8:9:5
Caipá-schorá
BL Add MS 5031:127-128.
J.Commelin in HM 8:9 Colochyntis Piriformis, seu Pepo Amarus Bauh.Pin.Sect.4.lib.8. [Bauhin 1623:313-314].
C.Commelin 1696:21 = J.Commelin in HM.
Plukenet 1696:286 = J.Commelin in HM.
Ray 1704 vol.3:333 = J.Commelin in HM.
Rumphius 1747 vol.5:397-398,Tab.CXLIV Cucurbita lagenaria (?).

Bilenschora HM 8:9: no figure

BL Add MS 5031: no drawing.
C.Commelin 1696:16 = HM.

Mullen-belleri HM 8:11:6
Mullén-bellerí
BL Add MS 5031:129-130.
J.Commelin in HM 8:11 Cucumeres vulgares.
J.Commelin 1689:105-106 Cucumis sativus Malabariensis, fructu mucronato.
C.Commelin 1696:25 = J.Commelin 1689.
Plukenet 1696:123 Cucumis Malabariensis, fructu longo, gracili.
Ray 1704 vol.3:334 Cucumis Indicus, caulibus asperis, vulgari longior & angustior, carne densiore.

Picinna HM 8:13-14:7
Pičinna
BL Add MS 5031:131-132.
J.Commelin in HM 7:14 Cucumis sylvestris Malabaricus, fructu striato amaro.
Plukenet 1691:Tab.CLXII,fig.1 Cucumis Indic; striatus, operculo donatus, corticoso putamine tectus (?).
C.Commelin 1696:25 = J.Commelin in HM.
Plukenet 1696:123 = Plukenet 1691.
Petiver 1704(1702):1061-1062,no.239 Peape pingkai Malab.Luffa Malabarica fructu reticulato sem.nigro.
Ray 1704 vol.3:335 = J.Commelin in HM.
Rumphius 1747 vol.5:407,408,Tab.CXLVIII Petola Anguina.
Rumphius 1747 vol.5:408,Tab.CXLIX Petola Bengalensis (c); J.Burman ibid.:408 (=).

Cattu-picinna HM 8:15:8
Cáttu-picinna
BL Add MS 5031:133-134.
J.Commelin in HM 8:15 Cucumis sylvestris sp.
C.Commelin 1696:25 = J.Commelin in HM.
Plukenet 1696:123 Cucumis Malabaricus, amarus, fructu sine costulis (?).
Ray 1704 vol.3:335 Cucumis sylvestris fructu minore, sine costis, amaro.
Rumphius 1747 vol.5:409-410,Tab.CL Petola silvestris (?c).

Pandi-pavel HM 8:17-18:9
Pándi-páuel
BL Add MS 5031:135-136.
J.Commelin in HM 8:18 Cucumis Puniceus Indicus fructu majore ex flavo rubente.
Breyne 1689:73 Momordica sive Charantia Indiae orientalis fructu aurantio majore, oblongo, seminibus albis.
J.Commelin 1689:47 Balsamina Cucumerina Ceylanica, fructu flavescente.
Sherard 1689:329 Cucumis puniceis Indicus major.
C.Commelin 1696:13 = J.Commelin 1689.
Plukenet 1696:123 = Sherard 1689.
Ray 1704 vol.3:336 = Breyne 1689.
J.Burman 1737:161-162 Momordica Zeylanica, pampinea fronde, fructu longiore. Tournef.inst.pag.103.
Linnaeus 1737:451 Momordica pomis angulatis tuberculatis, foliis villosis longitudinaliter palmatis.
Linnaeus 1747:166-167,no.351 = Linnaeus 1737.
Rumphius 1747 vol.5:410-412,Tab.CLI Amara Indica (=).
Linnaeus 1753:1009 Momordica Charantia.
Linnaeus 1763:1433 = Linnaeus 1753.
N.L.Burman 1768:208 = Linnaeus 1763.
Houttuyn 1779 II vol.11:295-297 = N.L.Burman 1768.

Pavel HM 8:19:10
Pável
BL Add MS 5031:137-138.
J.Commelin in HM 8:19 Cucumis Malabaricus fructu ex flavo rubente minor.
J.Commelin 1689:47 Balsamina Cucumerina puniceo Ceylonica major.
Sherard 1689:329 Cucumis puniceis Indicus minor.
C.Commelin 1696:14 = J.Commelin 1689.
Plukenet 1696:123 = Sherard 1689.
Ray 1704 vol.3:336 Cucumis puniceis Zeylanicus minor, seminibus nigris Herman.Cat.Hort.Acad.Lugd.Bat. [Hermann 1687:664].
Boerhaave 1720 vol.2:77 Momordica; Zeylanica; pampineâ fronde; fructu breviori. T.103.
J.Burman 1737:162 = Boerhaave 1720.
Linnaeus 1737:451 Momordica pomis angulatis tuberculatis, foliis villosis longitudinaliter palmatis α.
Linnaeus 1747:166-167,no.351 β = Linnaeus 1737.
Rumphius 1747 vol.5:410-412,Tab.CLI Amara Indica (=).
Linnaeus 1753:1009 Momordica Charantia β.
Linnaeus 1763:1433 = Linnaeus 1753.
N.L.Burman 1768:208 = Linnaeus 1763.
Houttuyn 1779 II vol.11:295-297 = N.L.Burman 1768.

Balia-mucca-piri HM 8:21-22:11
Bália-mucca-piri
BL Add MS 5031:139-140.
J.Commelin in HM 8:22 Balsamina Cucumerina Indica folio integro fructu variegato.
J.Commelin 1689:47 Balsamina rotundifolia repens seu mas B.Pin. [Bauhin 1623:306].
C.Commelin 1696:14 = J.Commelin in HM.

Ray 1704 vol.3:334-335 = J.Commelin in HM.

Erima-pavel HM 8:23:12
Erima-páuel
BL Add MS 5031:141-142.
J.Commelin in HM 8:23 Balsamina Cucumerina foliis integris fructu aculeato.
C.Commelin 1696:13 = J.Commelin in HM.
Ray 1704 vol.3:336 = J.Commelin in HM.

Mucca-piri HM 8:25-26:13
Muccá-pirí
BL Add MS 5031:143-144.
J.Commelin in HM 8:26 Balsamina Cucumerina punicea glabra, maculis albis notata.
Plukenet 1691:Tab.XXVI,f.4 Cucumis Brijonoides Bisnagarica fructu parvo, florum calijce muricato (?).
C.Commelin 1696:14 = J.Commelin in HM.
Plukenet 1696:123 = Plukenet 1691.
Ray 1704 vol.3:335 Cucumis Bryonoides Bisnagarica, fructu parvo, florum calice muricato Pluk.Phyt.T.26.F.4 (?).
Ray 1704 vol.3:336 Balsamina Cucumerina pilis albicantibus obsita, Commelin.in Not. (sic).

Covel HM 8:27-28:14
Covel
BL Add MS 5031:145-146.
J.Commelin in HM 8:28 Balsamina Cucumerina pinacea glabra, maculis albis notata.
C.Commelin 1696:14 = J.Commelin in HM.
Ray 1704 vol.3:336-337 = J.Commelin in HM.

Padavalam HM 8:29-30:15
Padávalám
BL Add MS 5031:147-148.
J.Commelin in HM 8:30 Balsamina Cucumerina folio amplo, flore candido, fructu glabro ex flavo rubescente.
C.Commelin 1696:13-14 = J.Commelin in HM.
Plukenet 1696:71 Bryonia Curassavica, floribus albis Jasmini. PBP.318 [Sherard 1689:318] (?).
Plukenet 1700:59 Cucumis puniceis Indicus, subrotundis foliis, fructu pyramidali laevi, seminibus ad oras angulatis.
Ray 1704 vol.3:337 = J.Commelin in HM.
Linnaeus 1753:1008 Trichosanthes cucumerina.
Linnaeus 1763:1432 = Linnaeus 1753.
N.L.Burman 1768:208 = Linnaeus 1763.
Houttuyn 1779 II vol.11:293 Trichosanthes Cucumerina [Linnaeus 1767 vol.2:638].

Scheru-padavalam HM 8:31:16
Scherú-pádávalám
BL Add MS 5031:149-150.
J.Commelin in HM 8:31 Balsamina Cucumerina, flore candido muscoso, fructu glabro.
C.Commelin 1696:13 = J.Commelin in HM.
Ray 1704 vol.3:337 = J.Commelin in HM.

Tota-piri HM 8:33:17
Tóti-píra
BL Add MS 5031:151-152.
J.Commelin in HM 8:33 Balsamina cucumerina fructu viridi striis albis.
Breyne 1689:74 Momordica flore fistuloso radiato albo, Hederae foliis, Malabarica (?).
C.Commelin 1696:14 = J.Commelin in HM.

Plukenet 1696:71 Bryonia Curassavica, floribus albis Jasmini. PBP.318 [Sherard 1689:318] (?).
Ray 1704 vol.3:337-338 Balsamina Cucumerina, foliis integris, cordiformibus, cuspidatis, floribus candidis filamentosis, fructu Cucumerino, oblongo, viridi.
J.Burman 1737:162 = Breyne 1689.
Linnaeus 1753:1008 Trichosanthes nervifolia.
Linnaeus 1763:1432 = Linnaeus 1753.
N.L.Burman 1768:208 = Linnaeus 1763.
Houttuyn 1779 II vol.11:293 Trichosanthes Nervifolia [Linnaeus 1767 vol.2:638].

Bem-pavel HM 8:35-36:18
Bén-pável
BL Add MS 5031:153-154.
J.Commelin in HM 8:36 Balsamina cucumerina sp.
C.Commelin 1696:14 = J.Commelin in HM.
Ray 1704 vol.3:338 Balsamina Cucumerina, radice tuberosa, flore hexapetalo, fructu nullo.

Nehoemeka HM 8:37-38:19
Nehoemeka
BL Add MS 5031:155-156.
J.Commelin in HM 8:38 Bryonica Ceylanica foliis profunde lacinatis Catal.Hort.Acad.Lugd.Batav.D°. Hermanni [Hermann 1687:95-96,97 tab.].
Breyne 1689:74 Momordica minor Indiae orientalis, Bryoniae facie, fructu glabro molli Uvae crispae magnitudine & effigie, suavepurpurascente, lineis niveis striato.
J.Commelin 1689:47 Balsamina Cucumerina Malabarica; fructu minore oblongo rotundo glabro.
Sherard 1689:356 Bryonia sp.
Plukenet 1691:Tab.XXV,f.3 Convolvulo similis Bryoniae albae folijs villosis, ex India Orientali (?).
C.Commelin 1696:14 = J.Commelin 1689.
Plukenet 1696:71 = J.Commelin in HM (≠).
Plukenet 1696:115 = Plukenet 1691.
Ray 1704 vol.3:343 = J.Commelin in HM.
J.Burman 1737:50 = J.Commelin in HM.
Linnaeus 1737:451 Bryonia foliis palmatis scabris, laciniis lanceolatis serratis: lateralibus minimis (r).
Linnaeus 1763:1438 Bryonia laciniosa.
N.L.Burman 1768:210 = Linnaeus 1763.
Houttuyn 1779 II vol.11:332-333 = N.L.Burman 1768.

Modecca HM 8:39-46(40):20
Modékka
BL Add MS 5031:157-158.
Breyne 1689:48 Flos passionis spurius Malabaricus, quinquifido folio primus, sive spinosus, fructu rotundo.
Sherard 1689:354 Modecca Prima Hort.Mal.fructu majori flore ex albo flavescente.
C.Commelin 1696:29 = Breyne 1689.
Plukenet 1696:283 Passiflora spuria Bryonoides quinquifido folio Malabarensis.
Ray 1704 vol.3:343-344 Pomifera Indica, claviculata, foliis in 3 aut 5 lacinias sectis, floribus pentapetalis, fructu magno, tripartito, seminibus singulis sacculis inclusis.
J.Burman 1737:72-73 Convolvulus Indicus, maximus, capreolatus, tuberosus, foliis profunde incisis seu tripartitis. Mus.Zeyl. pag.63 [Hermann 1717:63].
Linnaeus 1747:230,no.591 = Sherard 1689.
N.L.Burman 1768:45 Convolvulus paniculatus Linn.sp.223 [Linnaeus 1762:223].
Linnaeus 1771:336 = N.L.Burman 1768.

Palmodecca HM 8:41:21
Modékka-altera
BL Add MS 5031:159-160.
Breyne 1689:48-49 Flos passionis spurius Malabaricus, quinquifido folio, fructu rotundo secundus.
Sherard 1689:354-355 Modecca secunda seu Malmodecca Hort.Mal.Fructu majori flore ex albo viridante.
C.Commelin 1696:29 = Breyne 1689.
Plukenet 1696:283 Passiflora spuria Brynoides (sic) Malabarensis folio trifido & quinquifido.
Ray 1704 vol.3:344 Modecca altera, floribus è collo oblongo cyathiformi in 5 cuspides expanso.

Motta-modecca HM 8:43:22
Mótta-modékka
BL Add MS 5031:161-162.
Breyne 1689:49 Flos passionis spurius Malabaricus, quinquifido folio, tertius, fructu vario.
Sherard 1689:355 Modecca tertia seu Motamodecca Hort.Mal.fructu minori flore pallide virente.
C.Commelin 1696:29 = Breyne 1689.
Plukenet 1696:283 Passiflora spuria Bryonoides Malabarensis foliis variè scissis, fructu diverso.
Ray 1704 vol.3:344 Modecca fructu vario, alio oblongo bivalvi, alio rotundo tripartito.

Orela-modecca HM 8:45:23
Oŕela-modékka
BL Add MS 5031:163-164.
Breyne 1689:49 Flos passionis spurius Malabaricus, folio integro, fructu rotundo.
C.Commelin 1696:29 = Breyne 1689.
Ray 1704 vol.3:344 Modecca foliis plerunque integris, rarius incisis.

Modira-caniram HM 8:47:24
Módira-Cániram
BL Add MS 5031:165-166.
J.Commelin in HM 8:47 Nux vomica officinarum Bauhin.in Pinac.lib.12.sect.6 [Bauhin 1623:511].
Breyne 1689:93 Solanum arborescens Indicum, foliis Napecae majoribus magis mucronatis, fructu rotundo duro spadiceonigrescente, semine orbiculari compressis maximis.
C.Commelin 1696:63 = Breyne 1689.
Plukenet 1696:124 Cucurbitifera Malabariensis, Loti arboris folio, mucronato, spadiceo-nigrescente, fructu orbiculari, granis Nucis Vomicae similibus, ex quâ Lignum Colubrinum Officinarum.
J.Burman 1737:171-172 Nux Vomica Officinarum. Mus.Zeyl. pag.41 [Hermann 1717:41].
Rumphius 1741 vol.2:121-124,Tab.XXXVIII Lignum Colubrinum (-); J.Burman ibid.:124 (r).
Linnaeus 1753:189 Strychnos colubrina.
Linnaeus 1762:271 = Linnaeus 1753.
N.L.Burman 1768:58 = Linnaeus 1762.

Careloe-vegon HM 8:49:25
Carelú-vágon
BL Add MS 5031:167-168.
J.Commelin in HM 8:49 Aristolochia Clematitis Indica, flore albicante, fructu majore.
Sherard 1689:(387) Aristolochia Clematitis Indica odorata.
C.Commelin 1696:9 = J.Commelin in HM.
Plukenet 1696:49 Aristolochia Clematitis Gangetica, foliis in aculeum abeuntibus, radice odoratissimâ.
Plukenet 1696:114 Convolvulus Malabariensis densiore folio, polyanthos, Phytogr.Tab.276.fig.3 (r).
Plukenet 1705:37 = J.Commelin in HM.
J.Burman 1737:32-33 Aristolochia longa, Indica, aromatica, odorata, Mus.Zeyl.pag.9 [Hermann 1717:9].
Linnaeus 1737:433 Aristolochia caule volubili, foliis cordato-oblongis planis, fructu pendulo, pedunculis ramosis.
Linnaeus 1747:153-154,no.323 Aristolochia foliis cordato-oblongis, caule volubili, pedunculis multifloris.
Rumphius 1747 vol.5:474-475 Peponaster (r).
Rumphius 1747 vol.5:476-478,Tab.CLXXVII Radix Puloronica (r).
Linnaeus 1753:960-961 Aristolochia indica.
Linnaeus 1763:1362 = Linnaeus 1753.
N.L.Burman 1768:191 = Linnaeus 1763.
Houttuyn 1776 II vol.6:206 = N.L.Burman 1768.

Karivi-valli HM 8:51:26
Karivi-válli
BL Add MS 5031:169-170.
J.Commelin in HM 8:51 Aristolochia sp.
C.Commelin 1696:9 = J.Commelin in HM.
Ray 1704 vol.3:338-339 Balsaminae Cucumerinae affinis, flore parvo quinquepartito, fructu oblongo cuspidato.
Plukenet 1705:36 Aristolochia Maderaspatana Sonchi foliis integris, Ari in modum auriculatis, Neer-Covy Malabareis.

Kudici-valli HM 8:51(2):27
Kudicí-válli
BL Add MS 5031:171-172.
C.Commelin 1696:40 = HM.
Plukenet 1700:55 Convolvulus Indicus capreolis donatus, foliis Bryoniae albae divisuris, Sandaracae colorem referentibus, floribus striatis, per limbum flavis, & in collo ex viridi albicantibus.
Ray 1704 vol.3:388 Campanula Indica capreolata, foliis ad Bryoniam album accedentibus, fructu quadri-capsulari, duobus in singulis capsulis seminibus.
Plukenet 1705:8 Alceae Indicae affinis Herba Malabarica, capreolis ascendens, Bryoniae albae folio minore.
J.Burman 1737:73 = Plukenet 1700.

Ulinja HM 8:53-54:28
Ulínja
BL Add MS 5031:173-174.
J.Commelin in HM 8:54 Pisum vesicarium, fructu nigro alba macula notato, B.Pinac. [Bauhin 1623:343].
C.Commelin 1696:23 Cor-indum ampliore folio fructu majore Turn:El:Bot:342.
Plukenet 1696:120 Cor Indum fructu majore.
Ray 1704 vol.3:388 Pisum cordatum vulgo dictum.
J.Burman 1737:76 = C.Commelin 1696.
Linnaeus 1737:151 Cardiospermum.
Linnaeus 1747:59,no.142 = Linnaeus 1737.
Rumphius 1750 vol.6:62,Tab.XXVI,Fig.2 Halicacabus baccifer (cs).

Naga-valli Vel Mandaru-valli HM 8:55:29
Nága-válli vel Mandaru-válli
BL Add MS 5031:175-176.
J.Commelin in HM 8:57 Clematitis sp.
C.Commelin 1696:46 = HM.
Ray 1704 vol.3:328,655 Clematitis Indica folio bifido, flore fructuque carens.
Rumphius 1747 vol.5:1-3,Tab.I Folium Linguae (=).

Linnaeus 1753:374 Bauhinia scandens.
Linnaeus 1762:535 = Linnaeus 1753.
N.L.Burman 1768:94 = Linnaeus 1762.

Naga-mu-valli HM 8:57:30-31
Nága-mu-válli
BL Add MS 5031:177-178,179-180.
J.Commelin in HM 8:57 Clematitis sp.
C.Commelin 1696:46 = HM.
Ray 1704 vol.3:328 Clematitis Indica, folio bifido, flore fructuque carens, arbores transcendens, spipite lato, crasso, lignoso, spinoso.
Plukenet 1705:60 Clematis arborea, sumino folio bicorni, ex Insulâ Cheusan.
Houttuyn 1775 II vol.5:12-13 Bauhinia scandens Burm.Fl.Ind. p.94 [N.L.Burman 1768:94].

Perim-kaku-valli HM 8:59-60:32-34
Perím-káku-válli
BL Add MS 5031:181-182,183-184,185-186.
J.Commelin in HM 8:60 Phaseoli novi orbis sive Fabae purgatricis latiβima variatio longioris Joh.Bauhinus, tom.2.lib.17.
Breyne 1689:63 Lens phaseoloides maxima Indica, Cassiae foliis, semine maximo cordiformi.
Sherard 1689:365 Phaseolus alatus Indicus fructu fusco orbiculato maximo lobis latissimis & longissimis Hort.Lugd. [Hermann 1687:494-495].
C.Commelin 1696:53-54 = Sherard 1689.
Plukenet 1696:295 Phaseolus Utriusque Indiae arboreus, alatis foliis, fructu magno cordiformi, lobis longissimis nodosis, plerumque intortis.
Ray 1704 vol.3:116(2) Faba arborea Indica, siliquis 7, 7ve pedes longis, palmam latis, Fabis intus plurimus ad 30 usque, oblongis, planis, cordiformibus.
J.Burman 1737:139 Lens Phaseoloides foliis subrotundis, oppositis, flore spicato, pentapetalo, lobis latissimis, fructu orbiculato, fusco.
Linnaeus 1747:236-237,no.644 Pusaetha.Puswael.Herm.zeyl.44 [Hermann 1717:44].
Rumphius 1747 vol.5:5-8,Tab.IV Faba marina (=).
Linnaeus 1763:1501-1502 Mimosa scandens.
N.L.Burman 1768:222-223 = Linnaeus 1763.
Houttuyn 1776 II vol.6:441-445 = N.L.Burman 1768.

Nai-corana HM 8:61-62:35
Nái-coraná
BL Add MS 5031:187-188.
J.Commelin in HM 8:62 Phaseolus Zurattensis siliqua hirsuta, pungente, Cohuge dicta, Raji Hist.plant.lib.18.cap.1 [Ray 1686 vol.1:887].
J.Commelin 1689:275 = J.Commelin in HM.
Sherard 1689:364 Phaseolus utriusque Indiae lobis villosis pungentibus minor.
C.Commelin 1696:53 = J.Commelin in HM.
Plukenet 1696:292 Phaseolus Orientalis pruritum excitans hirsutie siliquarum, fructu nigro splendente.
Ray 1704 vol.3:444 = J.Commelin in HM.
Hermann 1711:8,no.110 Phaseolus utriusque Indiae lobis villosis pungentibus minor.
Linnaeus 1747:223-224,no.544 Phaseolus indicus hortensis, aureo fructu: spadiceo hilo oblongo albo prominente.Burm. zeyl.191 [J.Burman 1737:191].
Rumphius 1747 vol.5:393-394,Tab.CXLII Cacara pruritus (=).
Linnaeus 1763:1019 Dilochos pruriens.
N.L.Burman 1768:159 = Linnaeus 1763.

Houttuyn 1779 II vol.10:149 = N.L.Burman 1768.

Kaku-valli HM 8:63-65:36
Káku-válli
BL Add MS 5031:189-190.
J.Commelin in HM 8:65 Phaseolus Indicus siliqua majore pungente.
C.Commelin 1696:53 = J.Commelin in HM.
Plukenet 1696:292-293 Phaseolus Indiae Orientalis, siliquis plurimis hirsutie pungentibus, fabâ magna Ostream referente (?).
Ray 1704 vol.3:444 Mucunae Marcgravii valde similis, vel planè eadem.
J.Burman 1737:191 Phaseolus Indicus, lobis villosis, pruritum excitantibus. Mus.Zeyl.pag.60 [Hermann 1717:60].
Rumphius 1747 vol.5:10-11,Tab.VI Lobus Litoralis (=); J.Burman ibid.:11 (r).
Linnaeus 1763:1020 Dolichos urens.
N.L.Burman 1768:159 = Linnaeus 1763.
Houttuyn 1779 II vol.10:148-149 Dolichos Altissimus [Linnaeus 1767 vol.2:482].
Houttuyn 1779 II vol.10:149-151 = N.L.Burman 1768.

Putsja-paeru HM 8:67-68:37
Pútsja-paerú
BL Add MS 5031:191-192.
Breyne 1689:82 Phaseolus Indicus hirsutus, flore luteo, siliquâ angustâ parvâ hispidâ.
C.Commelin 1696:53 = Breyne 1689.
Plukenet 1696:290 Phaseolus minimus fructu viridi exiguo ovato. Moris.Hist.2.70 (?).
Ray 1704 vol.3:444 Phaseolus Indicus, siliquis angustis, hirsutis, fabis spadiceo-fuscis, membranis intercedentibus sejunctis.
J.Burman 1737:191-192 Phaseolus Indicus, herbaceus, lobis villosis pungentibus, minor. Mus.Zeyl.pag.69 [Hermann 1717: 69].
Rumphius 1747 vol.5:381-382,Tab.CXXXVIII Cacara nigra (-); J.Burman ibid.:382 (=).
Rumphius 1747 vol.5:392 Cacara pilosa (-); J.Burman ibid:392 (?=).

Schanga-cuspi HM 8:69-70:38
Schánga-cuspi
BL Add MS 5031:193-194.
J.Commelin in HM 8:70 Flos Clitorius Tarnatensibus Breyni.Centur.I. [Breyne 1678:76-77,tab.].
J.Commelin 1689:90 Clitorius flos.caeruleus Breyn.Cent. [Breyne 1678:76-77,tab.].
J.Commelin 1689:276 Phaseolus Indicus Glycyrrhizae foliis, flore caeruleo amplo, sive Flos Clitoridis Breyni Prodrom. [Breyne 1680:43-44].
Sherard 1689:365 Phaseolo adfinis Glyzyrrhyzae Germanicae foliis Orientalis flore albo.
C.Commelin 1696:54 = Sherard 1689.
Plukenet 1696:294 Phaseolus alatus Ternatensium floribus albis.
J.Commelin 1697:47-48,Fig.24 Flos Clitorius Breynii [Breyne 1678:76-77,tab.].
Petiver 1699(1698):323,no.20 Vela cacaunha Mal. = Plukenet 1696.
Petiver 1702(1701):1011,no.191 Vela Cacuanha Malab.Clitorius Malabaricus, fl.albo.
Ray 1704 vol.3:443 = J.Commelin in HM.
J.Burman 1737:101 Flos Clitorius, flore albo.
Linnaeus 1737:360 Clitoria foliis pinnatis.
Linnaeus 1747:130,no.283 = Linnaeus 1737.
Rumphius 1747 vol.5:56-57,Tab.XXXI Flos Coeruleus (=).

Linnaeus 1753:753 Clitoria Ternatea.
Linnaeus 1763:1025-1026 = Linnaeus 1753.
Houttuyn 1779 II vol.10:170-171 Clitoria Ternatea Syst.Nat.XII. Gen.869.p.484 [Linnaeus 1767 vol.2:484].

Schanga-cuspi alia species HM 8:70: no figure

BL Add MS 5031: no drawing.
Petiver 1699(1698):322-323,no.19 Carpa Cacuanna Mal.Flos Clitoridis Ternatensibus Breyn.Cent.p.76.Cap.31.Fig. [Breyne 1678:76-77,tab.].
Petiver 1702(1701):1011,1008,no.192 Carpa Cacuanna Malab.Clitorius Malabaricus, fl.caeruleo.

Konni HM 8:71-72:39
Kónni
BL Add MS 5031:195-196.
J.Commelin in HM 8:72 Phaseolus alatus major, fructu coccineo, macula nigra notata.
J.Commelin 1689:275-276 Phaseolus alatus, sive Abrus minor, fructu coccineo, macula nigra notato Moris.Hist.
Sherard 1689:337 Glyzyrrhiza Indica siliquis & seminibus Pisi coccineis hilo nigro notatis.
Plukenet 1691:Tab.CCXIV,fig.5 Phaseolus arboreus alat. & volubilis maj. Ind.Or.fructu coccineo, hilo nigro, notato.
C.Commelin 1696:53 = J.Commelin in HM.
Plukenet 1696 = Plukenet 1691.
Petiver 1699(1698):332-333,no.41 Shega pu coondamonee Mal.Abrus minor Indiae Orientalis siliquis majoribus.
Ray 1704 vol.3:447 Phaseolus bicolor Anacock dictus J.B.
J.Burman 1737:177-178 Orobus Indicus, Abrus Alpini dictus, fructu coccineo, macula nigra notato.
Linnaeus 1737:488 Abrus.Dalech.suppl.193.
Linnaeus 1747:130-132,no.284 Glycine foliis pinnatis conjugatis, foliolis paribus ovali-oblongis obtusis.
Rumphius 1747 vol.5:57-60,Tab.XXXII Abrus frutex (=).
Linnaeus 1748:228 = Linnaeus 1747.
Linnaeus 1753:753-754 Glycine Abrus.
Linnaeus 1763:1025 = Linnaeus 1753.
N.L.Burman 1768:161 = Linnaeus 1763.
Houttuyn 1775 II vol.5:436-439,Pl.XXVII,Fig.3 Abrus precatoria Syst.Nat.XII.Gen.1286.p.472 [Linnaeus 1767 vol.2:472] = N.L.Burman 1768.

Ana-mullu HM 8:73-74:40
Ána-mullú
BL Add MS 5031:197-198.
C.Commelin 1696:5 = HM.
Plukenet 1696:29 Phaseolus alatus Arbor Indica, validissimis aculeis horrida dispermos; siliquis latis valdè compressis, semine parvo lunulato.
Ray 1704 vol.3:443-444 Frutex leguminosus stipite crasso, longis & robustis spinis armato, foliis pinnatis, floribus papilionaceis, parvis, siliquis latis compressis, fabis planis lunulatis.

Paeru HM 8:75-77:41
Paerú
BL Add MS 5031:199-200.
J.Commelin in HM 8:81 Phaseolus Indicus trifoliatus minor, flore caeruleo, fructu minore.
C.Commelin 1696:54 = J.Commelin in HM.
Plukenet 1696:290 Phaseolus Indicus flore amplo caeruleo siliquis erectis gemellis, fructu parvo ruffescente.
Ray 1704 vol.3:444-445 Phaseolus Indicus trifoliatus minor, flore coeruleo, siliqua longa angustissima, fabis membranis transversis dissepris.

J.Burman 1737: Phaseolus sylvaticus, siliquis angustis. Mus. Zeyl.pag.43 [Hermann 1717:43] (?).
Rumphius 1747 vol.5:383-385,Tab.CXXXIX Phaseolus ruber (c); J.Burman ibid.:385 (c).
N.L..Burman 1768:161 Dolichos catjang.
Houttuyn 1779 II vol.10:161 = N.L.Burman 1768.

Catu-paeru HM 8:79-81:42
Kátu-paerú
BL Add MS 5031:201-202.
J.Commelin in HM 8:81 Phaseolus Indicus trifoliatus sylvestris flore purpurascente, fructu majore.
C.Commelin 1696:54 = J.Commelin in HM.
Plukenet 1696:291 = J.Commelin in HM.
Ray 1704 vol.3:445 = J.Commelin in HM.
Rumphius 1747 vol.5:378-379,Tab.CXXXVI Cacara, sive Phaseolus perennis (-); J.Burman ibid.:379 (?c).
Rumphius 1747 vol.5:390-391,Tab.CXLI,Fig.1 Cacara litorea (c).

Catu-tsjandi HM 8:83-84:43
Kátu-tsjándi
BL Add MS 5031:203-204.
J.Commelin in HM 8:88 Faba peregrina sp.
Plukenet 1691:Tab.LI,f.2 Phaseolus maritimus purgans, radice vivaci, foliis crassis, subrotundis, Bisnagaric.
C.Commelin 1696:53 = Plukenet 1691.
Plukenet 1696:292 = Plukenet 1691.
Plukenet 1700:149; ibid.1696:292 Phaseolus Indicus maritimis perennis, foliis crassis subrotundis minor PBP.364 [Sherard 1689:364].
Petiver 1702(1701):1015,no.201 Mooellee Malab.Phaseolus utriusque Indiae foliis rotundis.
Ray 1704 vol.3:445 Phaseolus Indicus trifolius, siliqua longa grandi, lata, ventre convexo, dorso concavo.
Rumphius 1747 vol.5:390-391,Tab.CXLI,Fig.1 Cacara litorea (-); J.Burman ibid.:391 (=).

Bara-mareca HM 8:85:44
Baramáreca
BL Add MS 5031:205-206.
J.Commelin in HM 8:88 Faba peregrina sp.
C.Commelin 1696:14 = HM.
Plukenet 1696:292 Phaseolus Indicus siliquâ magnâ falcatâ quaternis in dorso nervis, cum eminentiis plurimis verrucosis secundùm longitudinem insignita, fructu amplo niveo hilo croceo (?).
Ray 1704 vol.3:445 Phaseolus Indicus trifolius, siliqua longa grandi fabis obscurentis.
J.Burman 1737:189 Phaseolus Zeylanicus, sylvestris, maximus, Mus.Zeyl.pag.44 [Hermann 1717:44].
Rumphius 1747 vol.5:376,Tab.CXXXV,Fig.1 Lobus Machaeroides (=).
Linnaeus 1763:1022 Dolichos ensiformis.
N.L..Burman 1768:160 = Linnaeus 1763.
Houttuyn 1779 II vol.10:146-147 Dolichos Ensiformis [Linnaeus 1767 vol.2:482].

Catu-baramareca HM 8:87-88:45
Kátu-baramáreca
BL Add MS 5031:207-208.
J.Commelin in HM 8:88 Faba peregrina sp.
Sherard 1689:364 Phaseolus Indicus maritimus perennis, foliis crassis subrotundis minor.
C.Commelin 1696:53 = Sherard 1689.
Plukenet 1696:292 = Sherard 1689 (?).
Plukenet 1700:149; ibid.1696:292 Phaseolus maritimus purgans,

radice vivaci, foliis crassis subrotundis Bisnagaricus.Phytogr.Tab.51.fig.2.
Ray 1704 vol.3:446 Phaseolus trifolius Indicus, siliqua grandi, recurva, fabis crassis ovalibus, noxiis.
J.Burman 1737:189-190 Phaseolus Zeylanicus, marinus, folio pingui, & crasso. Mus.Zeyl.pag.43 [Hermann 1717:43].
Linnaeus 1747:223,no.541 = J.Burman 1737.

Tsjeria-cametti-valli HM 8:89-90:46
Tsjería-Camétti-válli
BL Add MS 5031:209-210.
C.Commelin 1696:67 = HM.
Plukenet 1696:290 Phaseolus Malabariensis Utriculatus monospermos.
Ray 1704 vol.3:446 Phaseolus Indicus, flore dilutè rubente, siliqua brevi, subrotunda, Monosperma, transparente.
Plukenet 1705:169 Phaseolo affinis Malabariensis trifoliata, & quinquefoliata, siliquâ latâ membranaceâ brevi, unicum semen continente.

Penar-valli. Foemina HM 8:91-92:47-48
Penár-valli, Penar-válli
BL Add MS 5031:211-212,213-214.
J.Commelin in HM 8:92 Ahoray Theveti (c).
C.Commelin 1696:51 = HM.
Linnaeus 1753:1028 Zanonia.
Linnaeus 1763:1457 Zanonia indica.
N.L.Burman 1768:212 = Linnaeus 1763.
Houttuyn 1779 II vol.11:353-354 = N.L.Burman 1768.

Penar-valli. Masc. HM 8:93:49
Penar-válli Mas
BL Add MS 5031:215-216.
C.Commelin 1696:51 = HM.
Linnaeus 1753:1028 Zanonia.
Linnaeus 1763:1457 Zanonia indica.
N.L.Burman 1768:212 = Linnaeus 1763.
Houttuyn 1779 II vol.11:353-354 = N.L.Burman 1768.
NB The pagination of the description has been misprinted as 39.

Katu-ulinu HM 8:95:50
Kátu-ulunu
BL Add MS 5031:217-218.
C.Commelin 1696:39 = HM.
Plukenet 1700:149 Phaseolus Indicus, flore flavo, caule, & siliquâ tereti, pilis rarioribus obsitis.
Petiver 1702(1701):857-858,no.132 Surru pierru Malab.Phaseolus Malab. pilosus auriculatus, fl.flavo.
Ray 1704 vol.3:446 Phaseolus Indicus foliis ad pediculum reflexis, siliquis teretiusculis, angustis, pilis rarioribus oblongis asperis, fabis etiam teretibus.

Mu-kelengu HM 8:97:51
Múkeléngu
BL Add MS 5031:219-220.
J.Commelin in HM 8:97 Batatas sylvestris sp.
Sherard 1689:371 Ricophora sive Inhame Malabarica folio rotundo in acutissimum apicem abeunte.
C.Commelin 1696:59 = Sherard 1689.
Plukenet 1696:321 = Sherard 1689.
Ray 1704 vol.3:134 = J.Commelin in HM.
Linnaeus 1737:459 Dioscorea foliis cordatis, caule laevi.
Linnaeus 1747:170,no.358 = Linnaeus 1737.
Linnaeus 1753:1033 Dioscorea sativa.
Linnaeus 1763:1463 = Linnaeus 1753.
N.L.Burman 1768:215 = Linnaeus 1763.

Houttuyn 1779 II vol.11:361-362 Dioscorea sativa [Linnaeus 1767 vol.2:656].

VOLUME 9

Tsjovanna-aleri vel Fula mestica incarnata HM 9:1-2:1
Tsjovánna-arelí
BL Add MS 5032:1-2.
J.Commelin in HM 9:2 Nerium Indicum latifolium, floribus plenis odoratis. Hermanni Cat.Hort.Lugd.Batav. [Hermann 1687:447,449 tab.,450].
J.Commelin 1689:247 = J.Commelin in HM.
C.Commelin 1696:47 = J.Commelin in HM.
Plukenet 1696:262-263 Nerium Indicum flore rubescente pleno Breyn.Prodr.2.76 [Breyne 1689:76].
J.Commelin 1697:45-46,Fig.23 Nerium lati-folium Indicum flore variegato, odorato, pleno.
Hermann 1698:48-49 = J.Commelin in HM.
J.Burman 1737:167 Nerium Zeylanicum, floribus roseis, amplis, plenis. Mus.Zeyl.pag.46 [Hermann 1717:46].
Linnaeus 1737:76-77 Nerium foliis lineari-lanceolatis γ.
Linnaeus 1747:45-46,no.108 γ = Linnaeus 1737.
Linnaeus 1762:305-306 Nerium Oleander.
N.L.Burman 1768:67 = Linnaeus 1762.

Belutta-areli aut Fula-mestica alba HM 9:3:2
Belútta-arelí
BL Add MS 5032:3-4.
J.Commelin in HM 9:3 Nerium Indicum angusti folium, floribus odoratis simplicibus. Hermann:Catal.Hort.Lugd.Batav. [Hermann 1687:447].
J.Commelin 1689:247 = J.Commelin in HM.
C.Commelin 1696:47 = J.Commelin in HM.
Plukenet 1696:262 = J.Commelin in HM.
J.Burman 1737:167,Tab.77 Nerium Indicum, siliquis angustis, erectis, longis, geminis.
Linnaeus 1737:76-77 Nerium foliis lineari-lanceolatis β.
Linnaeus 1747:45-46,no.108 β = Linnaeus 1737.
Linnaeus 1762:305-306 Nerium Oleander.
N.L.Burman 1768:67 = Linnaeus 1762.

Nelem pala HM 9:5:3-4
Nelam-pala, Nélem-pála
BL Add MS 5032:5-6,7-8.
J.Commelin in HM 9:5 Apocynum Malabaricum Arborescentis, Limonii folio, flore ex flavo rubescente, siliquis maximis.
C.Commelin 1696:6 = J.Commelin in HM.
Plukenet 1696:35 Apocynum arboreum ad Elaeagni faciem accedens Canariense, siliquis binis Nerii aemulis, (Cornicar Insulanis vulgò) apicibus recurvis. Phytogr.Tab.260.fig.3 (?).
Hermann 1698:44 Apocynum Malabaricum arborescens Nerii siliquis & semine floribus ex flavo rubescentibus.
Plukenet 1700:16 Apocynum arboreum Ind.Or.Laurinis foliis, rostrato Borraginis flore, praegrandi siliquâ, seminibus longis gtacilibus (sic) sericeis pappis involutis. Addeweepaula Gentilibus Indis, Bupaulemaram Malabareis vocitatur (?).
Petiver 1702(1700):700,no.50 Apocynum arboreum, Lauri folio, Dulcamarae flore. Addeweepaula Gent.Bupaulemaraum Malab. (?).
Ray 1704 vol.3:536 = J.Commelin in HM.
J.Burman 1737:168,Tab.78,Fig.1 Nerium Indicum, folio subrotundo, undulato, crasso, flore dilute rubente.

Belutta-kaka-kodi HM 9:7-8:5-6
Belútta-káka-kodí

BL Add MS 5032:9-10,11-12.
J.Commelin in HM 9:8 Apocynum Indicum maximum, folio amplo, rotundo, flore candido, siliquis longis.
C.Commelin 1696:6 = J.Commelin in HM.
Hermann 1698:62 Apocynum Scandens Malabaricum fruticosum floribus Nerii caryophyllos redolentibus.
Ray 1704 vol.3:542 = J.Commelin in HM.

Ada-kodien HM 9:9-10:7
Áda-kodień
BL Add MS 5032:13-14.
J.Commelin in HM 9:10 Apocynum scandens, flore variegato, siliquis Ericu similibus.
C.Commelin 1696:6 = J.Commelin in HM.
Hermann 1698:62-63 Apocynum Scandens Malabaricum, flore vario folliculis triplici costa insignitis.
Ray 1704 vol.3:542 = J.Commelin in HM.

Kaka kodi HM 9:11-12:8
Káka-kodí
BL Add MS 5032:15-16.
J.Commelin in HM 9:12 Apocynum Indicum floribus parvis, flavescentibus, umbellatim dispositis, siliquis nigricantibus, oblongè striatis.
C.Commelin 1696:5 = J.Commelin in HM.
Hermann 1698:63 Apocynum Scandens Malabaricum fruticosum siliquis bifidis expansis flosculis flavis omnium minimis.
Ray 1704 vol.3:542-543 = J.Commelin in HM.

Kudici kodi HM 9:13:9
Kudicí-kodí
BL Add MS 5032:17-18.
J.Commelin in HM 9:13 Apocynum Indicum, folio oblongo, glabro, flore ex albo flavescente.
C.Commelin 1696:5-6 = J.Commelin in HM.
Hermann 1698:63 Apocynum Scandens Malabaricum siliquis singularibus floribus flavis stellatis.
Ray 1704 vol.3:543 = J.Commelin in HM.

Wallia-pal-valli HM 9:15:10
Wállia-pal-válli
BL Add MS 5032:19-20.
J.Commelin in HM 9:15 inter Periplocas vel species Apocyni.
C.Commelin 1696:6 = J.Commelin in HM.
Hermann 1698:63 Apocynum Scandens Malabaricus siliquis singularibus floribus flavis stellatis absque staminibus.
Ray 1704 vol.3:543 Apocynum scandens siliquis angustis longissimis, floribus staminibus carentibus, flavis, stellatis.

Katu-pal-valli HM 9:17-18:11
Kátu-pal-válli
BL Add MS 5032:21-22.
J.Commelin in HM 9:18 Apocynum Indicum scandens, latifolium, flore viridi diluto, siliquis oblongis, cuspidatis.
C.Commelin 1696:6 = J.Commelin in HM.
Hermann 1698:63-64 Apocynum Scandens Malabaricum siliquis bifidis expansis floribus minoribus viridantibus.
Plukenet 1700:17; ibid.1696:37 Apocynum, s.Periploca, scandens folio longo, flore purpurante. JB.T.2.l.15.134.
Ray 1704 vol.3:543-544 Apocynum Indicum scandens, latifolium, flore viridi-diluto, siliquis oblongis binis ad basin junctis, cuspidatis.
N.L.Burman 1768:70 Periploca dubia.

Pal-valli HM 9:19:12
Pal-válli

BL Add MS 5032:23-24.
J.Commelin in HM 9:19 Apocynum Malabaricum scandens, latifolium, flore rubescente, siliquis, longiβimis & angustiβimis.
C.Commelin 1696:6 = J.Commelin in HM.
Plukenet 1696:37 = J.Commelin in HM.
Hermann 1698:64 Apocynum Scandens Malabaricum siliquis bifidis longissimis & angustissimis floribus nonnihil villosis.
Ray 1704 vol.3:544 = J.Commelin in HM.
J.Burman 1737:168-169,Tab.78,Fig.2 Nerium Indicum scandens, flore albo, siliquis geminis.

Nansjera-patsja HM 9:21-22:13
Nansjerá-patsjá
BL Add MS 5032:25-26.
J.Commelin in HM 9:22 Apocyni similis, planta scandens, succo limpido turgens.
C.Commelin 1696:6 = J.Commelin in HM.
Hermann 1698:64 Apocynum Scandens Malabaricum siliquis bifidis expansis floridantibus majoribus viribus umbone rubescente.
Ray 1704 vol.3:544 Apocynum Indicum scandens, siliquis ad singulos flores binis, succo limpido turgens.
Rumphius 1747 vol.5:467-468,Tab.CLXXIII,Fig.1 Sussuela esculenta I (-); J.Burman ibid.:468 (c).
Rumphius 1747 vol.5:469,Tab.CLXXIV Olus crepitans (?c).

Kametti-valli HM 9:23-24:14
Kámetti-válli
BL Add MS 5032:27-28.
J.Commelin in HM 9:24 Apocynum Indicum latifolium, flore rubescente, interius candido, siliquis oblongis, angustis, venis in longitudine striatis.
C.Commelin 1696:6 = J.Commelin in HM.
Hermann 1698:64 Apocynum Scandens Malabaricum siliquis bifidis expansis floribus jasmini intus candidus extus rubris.
Ray 1704 vol.3:544 Apocynum Indicum latifolium scandens, flore rubescente, interiùs candido, siliquis longis angustis, binis ad basin junctis, venis per longum striatis.

Watta-kaka-codi HM 9:25-26:15
Wátta-kakacodi
BL Add MS 5032:29-30.
J.Commelin in HM 9:26 Apocynum Malabaricum, folio cordis humani forma, flore viridis diluto, umbellatim disposito, siliquis oblongis, latis obtusis.
Hermann 1698:64 Apocynum Scandens Malabaricum siliquis bifidis expansis profunde liratis floribus viridantibus.
Petiver 1699 (1698):320-321,no.16 Serrufaulee Malab.Periploca Madraspatana Smilacis folio (s).
Ray 1704 vol.3:544-545 = J.Commelin in HM.
Rumphius 1747 vol.5:470-471,Tab.CLXXV,Fig.2 Nummularia lactea minor (-); J.Burman ibid.:471 (c).

Njota-njodem-valli HM 9:27-28:16
Njota-njodien-válli
BL Add MS 5032:31-32.
J.Commelin in HM 9:28 Apocyni repentes species.
C.Commelin 1696:6 = J.Commelin in HM.
Hermann 1698:65 Apocynum Scandens Malabaricum siliquis singularibus floribus Aristolochiae ad oras quinque partitis.
Ray 1704 vol.3:545 Apocynum Indicum scandens, seu Periploca, succo limpido, floribus multis in communis pediculi summitate coronae forma dispositis, reflexis, matracii figura, siliquis longis angustis.
Linnaeus 1747:46,no.110 Ceropegia pedunculis bifloris (r).

Linnaeus 1753:211 Ceropegia Candelabrum.
Linnaeus 1762:309 = Linnaeus 1753.
N.L.Burman 1768:69 = Linnaeus 1762.
Houttuyn 1777 II vol.7:731 = N.L.Burman 1768.

Parparam HM 9:29:17
Parparaṁ
BL Add MS 5032:33-34.
J.Commelin in HM 9:29 Apocynum (c).
C.Commelin 1696:6 = HM.
Hermann 1698:42-43 Apocynum Malabaricum erectum flore tetrapetaloide.
Ray 1704 vol.3:534 = Hermann 1698.

Neli-tali HM 9:31-32:18
Néli-táli
BL Add MS 5032:34-35.
J.Commelin in HM 9:32 Ferrum Equinum majus Malabariense, siliquis in summitate.
C.Commelin 1696:28 = J.Commelin in HM.
Plukenet 1696:270 Onobrychis Orientalis Mimosae brevioribus foliis, falcatis & dentatis siliquis, ad fastigia ramulorum orris. Phytogr.Tab.309 (?).
Ray 1704 vol.3:481 Aeschynomene flore papilionaceo, Ferri equini siliquis.
Royen 1740:385 Aeschynomene caule laevi herbaceo, foliolis obtusis, leguminibus laevibus hinc obsolete torosis.
Linnaeus 1753:713-714 Aeschynomene indica.
Linnaeus 1763:1061 = Linnaeus 1753.
N.L.Burman 1768:169 = Linnaeus 1763.
Houttuyn 1779 II vol.10:229-230 = N.L.Burman 1768.

Todda-vaddi HM 9:33-34:19
Tódda-váddi
BL Add MS 5032:36.
J.Commelin in HM 9:34 Herba Mimosa, foliis faenugraeci silvestris Bauhino in Pinace [Bauhin 1623:360].
C.Commelin 1696:35 Herba vivi foliis polypodii B:Pin:360 [Bauhin 1623:359].
Plukenet 1696:252 Mimosa humilis, Ind.Orient.simpliciter pinnatis Tamarindi foliis, floribus coronariis flavis, lituris rubris elegantèr striatis.
Ray 1704 vol.3:481 Herba viva Acostae cap.5. Totta-Vari Zanon.
J.Burman 1737:178-179 Oxys Indica, Tamarindi foliis, floribus umbellatis.
Linnaeus 1747:80,no.180 Oxalis pedunculis multifloris, foliis pinnatis.
Rumphius 1747 vol.5:301-305,Tab.CIV,Fig.2 Herba Sentiens (c).
Linnaeus 1753:434 Oxalis sensitiva.
Linnaeus 1762:622 = Linnaeus 1753.
N.L.Burman 1768:107 = Linnaeus 1762.
Houttuyn 1777 II vol.8:659-661 = N.L.Burman 1768.

Niti-todda-vaddi HM 9:35-36:20
Niti-tódda-váddi
BL Add MS 5032:37-38.
J.Commelin in HM 9:36 Herba casta Zeylanica, siliquis latis, compressis, minoribus. Breyni Centuri I.fol.47 [Breyne 1678:47-48].
C.Commelin 1696:2 = J.Commelin in HM.
Plukenet 1696:252 Mimosa Orientalis non spinosa rarioribus ramis floribus spicatis, Phytogr.Tab.307.fig.4 (?).
Petiver 1704(1702):1059-1060,no.235 Poon chedde Malab.Waga sensitiva aquatica Malabarica non ramosa.
Ray 1704 vol.3:481-482 = J.Commelin in HM.

J.Burman 1737:120 Hedysarum annuum, minus, Zeylanicum, Mimosae foliis. Tournef.inst.pag.56.
J.Burman 1737:160 Mimosa herba, Zeylanica, siliquis compressis, hispidis, articulatis, minor. Mus.Zeyl.pag.40 [Hermann 1717:40].
Linnaeus 1747:116-117,no.505 Mimosa inermis, foliis duplicato pinnatis, spicis cernuis, floribus decandris: inferioribus castratis apetalis β.
Linnaeus 1763:1061 Aeschynomene pumila.
N.L.Burman 1768:223 = Linnaeus 1763.
Linnaeus 1771:503 Mimosa virgata [Linnaeus 1763:1502].
Houttuyn 1776 II vol.6:449 = N.L.Burman 1768.

Malam-todda-vaddi HM 9:37:21
Malam-Tódda-váddi
BL Add MS 5032:39.
J.Commelin in HM 9:37 Mimosa di Jamaica Zanoni Cap.77.
C.Commelin 1696:42 = HM.
Plukenet 1696:251 = J.Commelin in HM (?).
Plukenet 1700:131; ibid. 1696:252 Mimosa, s.Aeschynomene mitis, pinnulis amplioribus, Chamaefilicis marinae aemulis de Tunquin (?).
Petiver 1704(1702):1060,no.236 Chaddai lackaree Malab.Waga sensitiva, Maderaspatana, ramosa, virgulis lignosis (?).

Man-todda-vaddi HM 9:39:22
Man-Tódda-váddi
BL Add MS 5032:40.
J.Commelin in HM 9:39 Mimosa Malabarica, foliis pilosis, non spinosa, minor, flore papilionaceo luteo, siliquis angustis, articulatis.
C.Commelin 1696:44 = J.Commelin in HM.
Plukenet 1700:53 Colutea parva Malabarica foliis hirsutis, siliquâ scorpioide brevi.
Ray 1704 vol.3:482 = J.Commelin in HM.
Plukenet 1705:62 Colutea parva Malabarica, foliis pubescentibus, siliquâ Scorpioide brevi.

Aria-veela HM 9:41:23
Ária-veêla
BL Add MS 5032:41-42.
C.Commelin 1696:9 = HM.
Plukenet 1696:280 Papaver corniculatum acre triphyllon Indicum floribus luteis viscosum, Ranmanissa Cochinensibus dictum.
Ray 1704 vol.3:427 Tetrapetala anomala trifolia aut quadrifolia, siliquis longis angustis, seminibus innumeris repletis.
J.Burman 1737:215 Sinapistrum Zeylanicum, triphyllum, & pentaphyllum, viscosum, flore flavo. Boerhav. apud Martin.Cent. I.Dec.3.pag.25 [Martyn 1728:25,tab.].
Linnaeus 1737:341 Cleome floribus dodecandris.
Linnaeus 1747:109,no.241 Cleome floribus dodecandris, foliis ternatis.
Rumphius 1747 vol.5:280-281,Tab.XCVI,Fig.2 Lagansa alba (c).
Linnaeus 1753:672 Cleome viscosa.
Linnaeus 1763:938-939 = Linnaeus 1753.
N.L.Burman 1768:141 = Linnaeus 1763.
Houttuyn 1778 II vol.9:753 = N.L.Burman 1768.

Cara-veela HM 9:43:24
Cára-veêla
BL Add MS 5032:43-44.
J.Commelin in HM 9:43 Pentaphyllum siliquosum Malabariense.
C.Commelin 1696:51 = J.Commelin in HM.

Plukenet 1696:280 Papaver corniculatum acre quinquefolium Aegyptiacum minus flore carneo non spinosum.
Petiver 1699(1698):316-317,no.7 Nella Walle Malab. Sinapistrum Indicum pentaphyllum flore carneo minus non spinosum Herm.Hort.Lugd.Bat. [Hermann 1687:564].
J.Burman 1737:216 = Petiver 1699(1698).
Linnaeus 1737:341 Cleome floribus gynandris.
Linnaeus 1747:108-109,no.239 = Linnaeus 1737.
Rumphius 1747 vol.5:280-281,Tab.XCVI,Fig.3 Lagansa rubra (c).
Linnaeus 1753:671 Cleome gynandra.
Linnaeus 1763:938 Cleome pentaphylla.
N.L.Burman 1768:141 = Linnaeus 1763.
Houttuyn 1778 II vol.9:751-752 = N.L.Burman 1768.

Tandale-cotti HM 9:45-46:25
Tandalé-cottí
BL Add MS 5032:45-46.
J.Commelin in HM 9:46 Genista Malabarica, folio singulari, floribus luteis, siliquis bullatis.
C.Commelin 1696:33 = J.Commelin in HM.
Plukenet 1696:122 Crotalaria Asiatica, folio singulari cordiformi, floribus luteis, Hort.Leyd.200 [Hermann 1687:200,201 tab.,202].
Petiver 1702(1700):591,no.38 Vatakelugelepe Mal. Geleka-chittu Gent. = Plukenet 1696.
Ray 1704 vol.3:464 = Plukenet 1696.
J.Burman 1737:80-81 = J.Commelin in HM.
Linnaeus 1747:127,no.276 Crotalaria foliis simplicibus oblongis cuneiformibus retusis.
Rumphius 1747 vol.5:278-279,Tab.XCVI,Fig.1 Crotalaria major (c); J.Burman ibid.:279 (c).
Linnaeus 1753:715 Crotolaria retusa.
Linnaeus 1763:1004 = Linnaeus 1753.
N.L.Burman 1768:155 = Linnaeus 1763.
Houttuyn 1779 II vol.10:100 = N.L.Burman 1768.

Tandale-cotti, No.2 HM 9:47:26
Katoú-Tandalé-cottí
BL Add MS 5032:47-48.
J.Commelin in HM 9:47 Genista Malabarica, folio singulari oblongo, flore flavodilutiore, siliquis bullatis.
Plukenet 1691:Tab.CLXIX,fig.5 Crotalaria Benghalensis, folijs Genistae subhirsutis PBP [Sherard 1689:329].
C.Commelin 1696:33 = J.Commelin in HM.
Plukenet 1696:122 = Plukenet 1691.
Petiver 1702(1700):592-593,no.43 Hoary Willow-leaved Malabar yellow Rattle-Broom. Janapachidde Malab.
Ray 1704 vol.3:464 Crotalaria Malabarica sylvestris, foliis singularibus, majoribus luteis.
J.Burman 1737:82 Crotalaria Asiatica, folio singulari, floribus luteis. Hert.Catal.semin.
Linnaeus 1737:357 Crotalaria foliis lanceolatis, petiolis sessilibus, caule striato.
Linnaeus 1763:1004 Crotolaria juncea.
N.L.Burman 1768:155 = Linnaeus 1763.
Houttuyn 1779 II vol.10:99 = N.L.Burman 1768.

Nellia-tandale-cotti HM 9:49:27
Nella-Tandalé-cottí
BL Add MS 5032:49-50.
J.Commelin in HM 9:49 Genista trifolia Malabarica, floribus luteis amplis, siliquis bullatis.
C.Commelin 1696:33 = J.Commelin in HM.
Plukenet 1696:122 Crotalaria Asiatica, trifolia subhirsuta, H.Leyd.app.663 [Hermann 1687:663].
Ray 1704 vol.3:465 Crotalaria Asiatica frutescens trifolia floribus luteis amplis Herman.Hort.Ac.Lugd.Bat. [Hermann 1687:196,197 tab.,198].
J.Burman 1737:21-22 Anonis Asiatica, frutescens, floribus luteis, amplis. Tournef.inst.pag.409. & Herb.Hert.
Linnaeus 1747:127-128,no.278 Crotalaria foliis ternatis, foliolis ovatis acuminatis, stipulis nullis.
Linnaeus 1753:715-716 Crotolaria laburnifolia.
Linnaeus 1763:1005 = Linnaeus 1753.
N.L.Burman 1768:156 = Linnaeus 1763.
Houttuyn 1779 II vol.10:102-103 Crotalaria Laburnifolia [Linnaeus 1767 vol.2:477].

Wellia-tandale-cotti HM 9:51:28
Wellia-tandalé-cottí
BL Add MS 5032:51-52.
J.Commelin in HM 9:51 Genista Malabarica pentaphylloïdes, flore amplo aureo flavescente, siliquis bullatis.
C.Commelin 1696:33 = J.Commelin in HM.
Plukenet 1696:122 Crotalaria pentaphylloides, Maderaspatana, floribus luteis.
Petiver 1699(1698):324,no.23 Neer kille gelippe Mal. = J.Commelin in HM.
Ray 1704 vol.3:465 Crotalaria pentaphylla, siliquis latis, valde tumidis.
Rumphius 1747 vol.5:278-279 Crotalaria minor (c); J.Burman ibid.:279 (c).
Linnaeus 1753:716 Crotolaria quinquefolia.
Linnaeus 1763:1005 = Linnaeus 1753.
N.L.Burman 1768:157 = Linnaeus 1763.
Houttuyn 1779 II vol.10:104 = Crotalaria quinquefolia [Linnaeus 1767 vol.2:478].

Pee-tandale-cotti HM 9:53:29
Pee-tandalé-cottí
BL Add MS 5032:53-54.
J.Commelin in HM 9:53 Genista Indica, Alni folio, floribus caeruleis, siliquis bullatis.
C.Commelin 1696:32-33 = J.Commelin in HM.
Plukenet 1696:122 Crotalaria Asiatica, folio singulari, verrucoso, floribus caeruleis. Hort.Leyd.200 [Hermann 1687:198,199 tab.,200].
Petiver 1702(1700):592,no.41 Dr Hermans blue Malabar Rattle-Broom.
Ray 1704 vol.3:465 = Plukenet 1696.
J.Burman 1737:81 = Plukenet 1696.
Linnaeus 1737:357 Crotalaria foliis ovatis, petiolis stipula duplici auctis, ramis tetragonis.
Linnaeus 1747:127,no.277 Crotalaria foliis simplicibus ovatis, stipulis semicordatis, ramis tetragonis.
Linnaeus 1753:715 Crotolaria verrucosa.
Linnaeus 1763:1005 = Linnaeus 1753.
N.L.Burman 1768:156 = Linnaeus 1763.
Houttuyn 1779 II vol.10:101 Crotalaria Verrucosa [Linnaeus 1767 vol.2:477].

Kattu-tagera HM 9:55:30
Kátu-tagerá
BL Add MS 5032:55-56.
J.Commelin in HM 9:55 Astragalus Indicus spicatus, siliquis copiosis deorsum spectantibus, non falcatis; seu Polybos.
C.Commelin 1696:10 = J.Commelin in HM.
Plukenet 1696:113 Colutea Orientalis plerumque heptaphyllos, hirsuta, floribus spicatis, saturatè purpureis, siliquis plurimis quadratis, valdè pilosis, summo surculorum dependentibus.

J.Burman 1737:37-38,Tab.14 Astragalus spicatus, siliquis pendulis hirsutis, foliis sericeis (r).
Linnaeus 1747:124-125,no.272 Indigofera leguminibus pendulis lanatis tetragonis.
Rumphius 1747 vol.5:283-284,Tab.XCVII,Fig.2 Gallinaria rotundifolia (?c); J.Burman ibid.:284 (≠).
Linnaeus 1753:751 Indigofera hirsuta.
Linnaeus 1763:1062 = Linnaeus 1753.
N.L.Burman 1768:170 = Linnaeus 1763.
Houttuyn 1775 II vol.5:538-539 Indigofera Hirsuta.

Kondam-pallu HM 9:57:31
Kondam-pullú
BL Add MS 5032:57.
C.Commelin 1696:40 = HM.
Petiver 1702(1700):586-587,no.20 Neer-pundo Gent. Neruchadday Malab. = J.Commelin in HM.
Ray 1704 vol.3:404 Dipetalos Indica purpuro-coerulea, siliquosa, foliis longis angustis ex adverso binis.
Plukenet 1705:117 Hesperidi affinis dipetalos, siliqua brevi Malabariensis.
Rumphius 1750 vol.6:39-40,Tab.XVI,Fig.1 Herba admirationis (c).
Linnaeus 1753:937 Impatiens oppositifolia (?).
Linnaeus 1763:1328 = Linnaeus 1763.
N.L.Burman 1768:187 = Linnaeus 1763.
Houttuyn 1779 II vol.11:149 Impatiens oppositifolia [Linnaeus 1767 vol.2:586].

Nir-murri HM 9:59:32
Nir-múrri
BL Add MS 5032:58.
J.Commelin in HM 9:59 Lateri species.
C.Commelin 1696:47 = HM.
Plukenet 1700:114 Lathyris congener spicata, foliis integris gramineis, s. Catanance leguminosa Indorum microlobos, semine quadrato.
Ray 1704 vol.3:474 Legumen erectum foliis simplicibus, floribus in summis caulibus & ramulis, velut in spica; seminibus quadratis.
Plukenet 1705:189 Scorpioidi affinis, angustioribus foliis, flore luteo Maderaspatana (?).

Cupa-vela HM 9:61:33
Cúpa-véela
BL Add MS 5032:59-60.
Plukenet 1691:Tab.CXIX,f.7 Sinapistrum Indicum diphyllon, siliquis ad alas ex uno pediculo binis.
C.Commelin 1696:62 = Plukenet 1691.
Plukenet 1696:280 = Plukenet 1691.
Petiver 1702(1700):591,no.40 Sinapistrum Malabaricum diphyllon. Kaukaupoondoo Malab.
Ray 1704 vol.3:527 Herba siliquosa Indica, pentapetala, foliis adversis, siliquis longis angustis, in foliorum alis, seminibus oblongis.

Tsjeru-vela HM 9:63:34
Tsjéru-véla
BL Add MS 5032:61.
C.Commelin 1696:67 = HM.
Plukenet 1696:280 Papaver corniculatum acre singulari folio Malabaricum.
Ray 1704 vol.3:427 Herba siliquosa Indica tetrapetala, siliquis longis angustis, è singulis foliorum alis singulis.
J.Burman 1737:217,Tab.100,Fig.2 Sinapistrum Zeylanicum, viscosum, folio solitario, flore flavo, siliqua tenui.

Linnaeus 1747:110,no.243 Cleome floribus hexandris, foliis simplicibus, ovato lanceolatis.
Linnaeus 1753:672 Cleome monophylla.
Linnaeus 1763:940 = Linnaeus 1753.
N.L.Burman 1768:142 = Linnaeus 1763.
Houttuyn 1778 II vol.9:759 = N.L.Burman 1768.

Vallia-capo-molago HM 9:65:35
Vállia-capó-mólago
BL Add MS 5032:62-63.
J.Commelin in HM 9:65 Capsicum sp.
C.Commelin 1696:54 Piper indicum siliqua flava vel aurea, figura latiore B:Pin:102 [Bauhin 1623:102].
Plukenet 1696:353 Solanum mordens, fructu propendente oblongo crasso & obtuso.
J.Burman 1737:53 Capsicum Indicum. Mus.Zeyl.pag.61 [Hermann 1717:61].
Linnaeus 1747:38,no.92 Capsicum annuum. Hort.cliff.59 [Linnaeus 1737:59-60].
Rumphius 1747 vol.5:252 Capsicum obtusum (?c).
Linnaeus 1753:188-189 Capsicum annuum.
Linnaeus 1762:270-271 = Linnaeus 1753.
N.L.Burman 1768:58 Capsicum frutescens Linn.sp.271 [Linnaeus 1762:271].
Houttuyn 1777 II vol.7:682-684 Capsicum annuum [Linnaeus 1767 vol.2:174].

Nir-pullari HM 9:67:36
Nir-pularí
BL Add MS 5032:64.
C.Commelin 1696:47 = HM.
Plukenet 1696:290 Phaseolus minor Bisnagaricus foliis argenteo villosis, siliquis torosis brevibus spadiceâ hirsutie pubescentibus, fructu parvo Scaraboide nigro. Phytogr.Tab.52.f.3 (?).
Ray 1704 vol.3:470-471 = HM.
Plukenet 1705:143 Meliloti similis Maderaspatana, Naur-velle Malabarorum (?).
Plukenet 1705:162 Onobrychis parva trifoliata ex Oris Malabaricis (?).
J.Burman 1737:188 Phaseolus Zeylanicus, tomentosus, Salviae foliis, lobis parvis, oblique articulatis, alter. Mus.Zeyl.pag.7 [Hermann 1717:6-7] (?).
Linnaeus 1747:125-126,no.274 Indigofera leguminibus pendulis lanatis tetragonis.
Linnaeus 1753:751 Indigofera glabra.
Linnaeus 1763:1062 = Linnaeus 1753.
N.L.Burman 1768:171 = Linnaeus 1753.
Houttuyn 1779 II vol.10:249-250 = N.L.Burman 1768.

Manneli HM 9:69:37
Mánneli
BL Add MS 5032:65-66.
J.Commelin in HM 9:73 Genistellae tinctoriae species.
Plukenet 1691:Tab.CCI,fig.2 Lotus tenuifolius Maderaspatanus, siliquâ singulari glabrâ.
C.Commelin 1696:41 = Plukenet 1691.
Plukenet 1696:225-226 = Plukenet 1691.
Petiver 1699(1698):318,no.10 Shevanar weamboo Malab. Anil Maderaspatana foliis minimis confertis (c).
Plukenet 1700:119 Lotus Aethiopicus fruticosus elegans, cauliculis rubicundis (?).
Ray 1704 vol.3:471 Dorycnium Indicum, floribus singularibus rubris in pedicellis oblongis, siliquis perexiguis.
Plukenet 1705:133 Lotus fruticosus Thymi montani foliis, cauliculis cinereis, è Maderaspatan. Shevanar-Weembooe Malabarorum (s).

Plukenet 1705:134 Loto affinis altera species sive varietas, foliolis aliquantulum latioribus. Bongela-chedde Malabarorum.
J.Burman 1737:89 Dorycnium Zeylanicum, folio minutissimo. Hertog in Catal.semin.
Linnaeus 1747:124,no.271 Genista foliis quinatis sessilibus, pedunculis axillaribus capillaribus unifloris, floribus minimis.
Linnaeus 1753:712 Aspalathus indica.
Linnaeus 1763:1001-1002 = Linnaeus 1753.
N.L.Burman 1768:155 = Linnaeus 1763.
Houttuyn 1775 II vol.5:477 = N.L.Burman 1768.

Tsjovanni-manneli HM 9:71:38
Tsjovánna-mannelí
BL Add MS 5032:67-68.
J.Commelin in HM 9:73 Mimosae genus.
C.Commelin 1696:44 = J.Commelin in HM.
Plukenet 1700:120 Loti corniculati & Coluteae media, pilosis foliis, siliquâ quadratâ; Codeserupaulada Malabarorum (?cs).
Ray 1704 vol.3:480 Mimosa pumila non spinosa, flore papilionaceo flavo, foliolis pinnatim ad mediam costam annexis paucis, siliquis parvis brevibus.
Plukenet 1705:133 Loto & Coluteae compos, in costa foliorum, & margine pilosa.
N.L.Burman 1768:155 Aspalathus persica β.

Tsjeru-manneli HM 9:73:39
Tsjéru-mannelí
BL Add MS 5032:69.
J.Commelin in HM 9:73 Mimosa sp.
C.Commelin 1696:67 = HM.
Ray 1704 vol.3:635 Herba aquatica foliolis minimis, flosculis candidis è tubo longiusculo in 5 segmenta cuspidata expansis.

Suendadi-pullu HM 9:75:40
Suendádi-pullú
BL Add MS 5032:70-71.
J.Commelin in HM 9:75 Melilotus flore albo (c).
Plukenet 1691:Tab.XLV,f.5 Melilot. Indic. humil. erecta florib. exiguis odorat. albis, pericarpijs majorib, spicatim densius stipatis.
C.Commelin 1696:44 = Plukenet 1691.
Plukenet 1696:247 = Plukenet 1691.
Petiver 1702(1700):582,no.5 Yelanaiureve Malab. (s).
J.Burman 1737:157 Melilotus Indica, hortensis, sativa, floribus odoratis, albis. Mus.Zeyl.pag.64 [Hermann 1717:64].
Linnaeus 1747:224-225,no.552 Trifolium erectum, caule teretiusculo, pedunculus compressoangulatis, pericarpiis racemosis rotundis rugosis nudiusculis monospermis.
N.L.Burman 1768:172-173 Trifolium M. indica Linn.sp.1077 [Linnaeus 1763:1077].

Coletta-veetla HM 9:77-78:41
Colétta-vêetla
BL Add MS 5032:72-73.
J.Commelin in HM 9:78 Eryngium Ceylanicum febrifugum, floribus luteis; Cingalenses dicunt Kathukarohiti.
Plukenet 1691:Tab.CXIX,f.5 Melampyro cognata Maderaspatana spinis horrida (?).
C.Commelin 1696:44 = Plukenet 1691.
Plukenet 1696:245 = Plukenet 1691.
Petiver 1699(1698):319,no.13 Varamullee Malab. Adhatoda Malabarica tetracantha.
Petiver 1702(1701):1021,no.223 Varanna Mullee Malab. = Petiver 1699(1698).
Ray 1704 vol.3:241 = J.Commelin in HM.

Ray 1704 vol.3:402 = Plukenet 1691.
Boerhaave 1720 vol.2:263 = J.Commelin in HM (?).
J.Burman 1737:8 Adhatoda ad alas spinosa & florifera.
Linnaeus 1737:487 Prionitis.
Royen 1740:291 Barleria foliis integerrimis, spinis lateralibus.
Linnaeus 1747:105-106,no.233 = Royen 1740.
Rumphius 1750 vol.6:163-164,Tab.LXXI Aquifolium Indicum (mas, femina) (=); J.Burman ibid.:164 (femina ≠).
Linnaeus 1753:636 Barleria Prionitis.
Linnaeus 1763:887 = Linnaeus 1753.
N.L.Burman 1768:135-136 = Linnaeus 1763.
Houttuyn 1778 II vol.9:583 = N.L.Burman 1768.

Vada-kodi HM 9:79-80:42
Váda-kodí
BL Add MS 5032:74-75.
J.Commelin in HM 9:80 Persicaria Malabarica, flore galeato & labiato monospermon.
C.Commelin 1696:52 = J.Commelin in HM.
Petiver 1699(1698):319,no.12 Carennucheel Malab. Adhatoda Madraspatana Hydropiperis folio (s).
Petiver 1702(1701):853,no.121 Nuchulee Malab. Vairelchetto Gent. Vitex Madraspat. foliis latioribus digitalis, floribus racemosis Act.Phil.No.244.p.315.3 [Petiver 1699(1698):315, no.3].
Ray 1704 vol.3:250 = J.Commelin in HM.
Plukenet 1705:206 Vadae-kodi, H.Malab.part.9.tab.42. similis, Maderaspatana, & fortè Idem. Cara-Nuchel Malabarorum.
Rumphius 1750 vol.6:163-164,Tab.LXXI,Fig.2 Aquifolium Indicum femina (-); J.Burman ibid.:164 (cs).
N.L.Burman 1768:10 Justicia Gendarussa.
Houttuyn 1775 II vol.4:48-50 = N.L.Burman 1768.

Adel-odagam HM 9:81:43
Ádel-ódagam
BL Add MS 5032:76-77.
Plukenet 1691:Tab.XCIX,f.3 Melampyro affinis tetraphylla Gangetica, floribus inter folia sparsis (?).
C.Commelin 1696:2 = HM.
Plukenet 1696:33 Anonymus Maderaspatan. Senecionis Hieracii folio similis, spicâ foliaceâ & Hormini adinstàr squamatâ. Phytogr.Tab.CCXLI,fig.1 (?).
Plukenet 1696:245 = Plukenet 1691.
Plukenet 1700:62 Cynorynchium Nov' anglicum Digitali accedens (?).
Ray 1704 vol.3:249-250 Spicata Indica fruticosa, spicis è foliorum adversorum alis egressis, flore galeato & labiato, monospermos, foliolis floribus intermixtis.
Ray 1704 vol.3:402 = Plukenet 1691.
Linnaeus 1762:23-24 Justicia bivalvis.
N.L.Burman 1768:10 = Linnaeus 1762.
Houttuyn 1777 II vol.7:116-117 Justicia bivalvis [Linnaeus 1767 vol.2:60].

Katu-karivi HM 9:83:44
Kátu-karivi
BL Add MS 5032:78-79.
C.Commelin 1696:39 = HM.
Ray 1704 vol.3:402 Spicata Indica, flore galeato & labiato, spitis tum in summis caulibus, tum è foliorum adversorum sinubus exeuntibus, nullis floribus intermixtis foliolis.
Plukenet 1705:181 Rapuntii genus Malabaricum, Lychnidis latiore folio molli, & incano, flore versicolore (?).

Valli-upu-dali HM 9:85:45

Válli-upú-dalí
BL Add MS 5032:80-81.
J.Commelin in HM 9:85 Digitalis affinis Malabarica, folio piloso.
C.Commelin 1696:26 = J.Commelin in HM.
Plukenet 1700:99 Gratiolae affinis Coromandeliana, Digitalis aemula, folio Clinopodij, capsulis bivalvibus, in laxiorem spicam congestis (?).
J.Burman 1737:87 Digitalis Zeylanica, flore albo, variegato. Herb.Hert. (c).

Pee-tumba HM 9:87:46
Pêe-túmba
BL Add MS 5032:82-83.
J.Commelin in HM 9:87 Lysimachia Virginiana.
C.Commelin 1696:28 Euphrasiae affinis indica echioides Herm.Cat.668 [Hermann 1687:668,669 tab.,670].
Plukenet 1696:142 = C.Commelin 1696.
Petiver 1699(1698):330,no.36 Nella mulle Mal. = Plukenet 1696.
Ray 1704 vol.3:660 Asperifolia Indica, foliis ex adverso binis, surculis floriferis è foliorum alis exeuntibus & extantibus, vasculis seminalibus bicapsularibus erectis.
J.Burman 1737:88-89 Digitali similis, planta bicapsularis, flore tubuloso, galea trifida, & labio integro, obtuso constante.
Linnaeus 1747:8-9,no.21 Justicia foliis lanceolato-linearibus obtusis sessilibus, racemis adscendenti secundis, bracteis setaceis.
Linnaeus 1753:16 Justicia echioides.
Linnaeus 1762:22 = Linnaeus 1753.
N.L.Burman 1768:9 = Linnaeus 1762.
Houttuyn 1777 II vol.7:113-114 Justicia Echioides [Linnaeus 1767 vol.2:60].

Onapu HM 9:89-90:47
Onapú
BL Add MS 5032:84-85.
J.Commelin in HM 9:90 Balsamina faemina sp.
C.Commelin 1696:14 = J.Commelin in HM.
Ray 1704 vol.3:637 Balsamina foemina, Noli me tangere, dicta, Malabarica, flore tetrapetalo, nigerrimo & nitente.
Rumphius 1747 vol.5:256-258,Tab.XC Lacca Herba Coccinea (c).

Valli-onapu HM 9:91:48
Válli-onapú
BL Add MS 5032:86-87.
J.Commelin in HM 9:91 Balsamina faemina sp.
C.Commelin 1696:14 = J.Commelin in HM.
Ray 1704 vol.3:637-638 Balsamina foemina impatiens, major, latifolia, flore hexapetalo seminibus pilosis.
Linnaeus 1753:937 Impatiens latifolia.
Linnaeus 1763:1328 = Linnaeus 1753.
N.L.Burman 1768:187 = Linnaeus 1763.
Houttuyn 1779 II vol.11:148 Impatiens Latifolia [Linnaeus 1767 vol.2:586].

Tsjeria-onapu HM 9:93:49
Tsjería-onapú
BL Add MS 5032:88-89.
J.Commelin in HM 9:101 Balsamina faemina sp.
C.Commelin 1696:14 = J.Commelin in HM.
Ray 1704 vol.3:638 Balsamina foemina impatiens Malabarica, floribus minoribus tetrapetalis, siliquis longioribus & strictioribus.
Rumphius 1747 vol.5:256-258,Tab.XC Lacca Herba alba (c).

Man-onapu HM 9:95:50

Man-onapú
BL Add MS 5032:90.
J.Commelin in HM 9:101 Balsamina faemina sp.
C.Commelin 1696:14 = J.Commelin in HM.
Ray 1704 vol.3:638 Balsamina foemina minor, flore dipetalo.

Bellutta-onapu HM 9:99:51
Bellutta-onapú
BL Add MS 5032:91.
J.Commelin in HM 9:101 Balsamina faemina sp.
C.Commelin 1696:14 = J.Commelin in HM.
Ray 1704 vol.3:638 Balsamina foemina Indica, flore tripetalo, fructu parvo, brevi.

Tilo-onapu: ceu Notenga HM 9:101:52
Tiló-onapú
BL Add MS 5032:92-93.
J.Commelin in HM 9:101 Persicaria siliquosa Indica, flore rosaceo colore, fructu piloso.
C.Commelin 1696:14 = J.Commelin in HM.
Plukenet 1696:62 Balsamina foemina, angustis & elegantèr crenatis foliis, flore albo minore (?).
Ray 1704 vol.3:638 Balsamina foemina Indica, flore tetrapetalo, rosaceo colore, fructu piloso, brevi, striato.
J.Burman 1737:42 = Plukenet 1696 (=).
Linnaeus 1763:1328-1329 Impatiens Balsamina.
N.L.Burman 1768:187 = Linnaeus 1763.

Kaka-pu HM 9:103:53
Kakapú
BL Add MS 5032:94-95.
J.Commelin in HM 9:103 Asarina sive Hederua saxitilis Lobeli (cs).
C.Commelin 1696:35 = J.Commelin in HM.
Plukenet 1696:283 Pediculari congener Alpina, Hederae terrestris facie, utriculis amplis vasculum seminale claudentibus (?).
Ray 1704 vol.3:389 = J.Commelin in HM.
Linnaeus 1753:619 Torenia asiatica.
Linnaeus 1763:862 = Linnaeus 1753.
N.L.Burman 1768:131 = Linnaeus 1763.
Houttuyn 1778 II vol.9:512 = N.L.Burman 1768.

Schit-elu HM 9:105-106:54
Schit-elú
BL Add MS 5032:96-97.
J.Commelin in HM 9:106 Sesamum veterum B.Pin. [Bauhin 1623:27].
C.Commelin 1696:61 = J.Commelin in HM.
Plukenet 1696:344 = J.Commelin in HM.
J.Burman 1737:87 Digitalis Orientalis Sesamum dicta. Tournef.instit.pag.165.
Linnaeus 1737:318 Sesamum foliis ovato-oblongis integris.
Linnaeus 1747:107,no.237 = Linnaeus 1737.
Rumphius 1747 vol.5:204-208,Tab.LXXVI,Fig.1 Sesamum Indicum album (=).
Linnaeus 1753:634 Sesamum orientale.
Linnaeus 1763:883-884 = Linnaeus 1753.
N.L.Burman 1768:133 = Linnaeus 1763.
Houttuyn 1778 II vol.9:564-565 = N.L.Burman 1768.

Car-elu HM 9:107:55
Car-elú
BL Add MS 5032:98-99.
J.Commelin in HM 9:107 Sesamum Indicum, folio amplo, serrato, flore majore, semine nigricante.

Plukenet 1691:Tab.CIX,f.4 Sesamum alterum folijs trifidis Orientale semine obscuro (?).
C.Commelin 1696:61 = J.Commelin in HM.
Plukenet 1696:344 = Plukenet 1691.
Rumphius 1747 vol.5:204-208,Tab.LXXVI,Fig.1 Sesamum Indicum nigrum (=).

Cara-caniram HM 9:109-110:56
Cará-Cániram
BL Add MS 5032:100-101.
J.Commelin in HM 9:110 Crotolaria sp.
C.Commelin 1696:24 = J.Commelin in HM.
Plukenet 1696:142 Euphrasia, Alsines majori folio, flore galeato pallidè luteo, Jamaicensis, Phytogr.Tab.279.fig.6 (?).
Plukenet 1700:73 Euphrasia Alsines majori folio, flore galeato, purpurascente, Cimices sylvestres graviter olens, ex Ind.Orientali. Mucoorundee Malabarorum (?).
Ray 1704 vol.3:403 Siliquosa Indica, floribus labiatis, dipetalis, siliquis brevibus, in duas cameras membranâ intergerinâ divisis.
Linnaeus 1737:10 Justicia annua, hexangulari caule, foliis circaeae conjugatis, flore miniato. Houst.AA.
N.L.Burman 1768:9 Justicia paniculata.
Linnaeus 1771:317 Justicia gangetica.
Houttuyn 1777 II vol.7:117 Justicia Gangetica [Linnaeus 1767 vol.2:60].

Tsjanga-puspam HM 9:111:57
Tsjánga-puspaṁ
BL Add MS 5032:102.
J.Commelin in HM 9:111 Gentianella sp.
C.Commelin 1696:33 = J.Commelin in HM.
Plukenet 1696:132 Ecbolii Indici, s. Adhatodae cucullatus floribus aemula Hyssopifolia, Planta ex Insulis Fortunatis, Phytogr.Tab.280.fig.1 (?).
Ray 1704 vol.3:371 Gentianella Indica foliis trinerviis, flore coeruleo, bipartito.
Plukenet 1705:106 Gentianella Ascyroides Malabariensis, flore caeruleo (?).
Linnaeus 1771:174 Gratiola rotundifolia.
Houttuyn 1777 II vol.7:122-123 = Linnaeus 1771.

Katu-pee-tsjanga-puspam HM 9:113:58
Katú-peê-tsjánga-puspaṁ
BL Add MS 5032:103.
J.Commelin in HM 9:113 Teucrium sp.
C.Commelin 1696:39 = HM.
Plukenet 1696:167 Gentianella Utriusque Indiae impatiens foliis Agerati, Phytogr.Tab.186.fig.2 (r).
Plukenet 1700:44 Chamaedrifolia pusilla planta, flosculo oblongo monopetalo, capsulâ membranaceâ, ex Coromandel (?).
Plukenet 1700:73; ibid.1696:142 Euphrasia pratensis, Satureiae foliis scabris, è Maderaspatan. Phytogr.Tab.177.fig.6 (r).
Ray 1704 vol.3:422 Chamaedryi spuriae affinis Indica capsulis bivalvibus subrotundis.
N.L.Burman 1768:134 Ruellia antipoda Linn.sp.886 [Linnaeus 1763:886].
Houttuyn 1778 II vol.9:578-579 Ruëllia antipoda [Linnaeus 1767 vol.2:424].

Pee-tjanga-pulpam HM 9:115:59
Peê-tsjánga-puspaṁ
BL Add MS 5032:104.
J.Commelin in HM 9:115 Teucrium Americanum procumbens, Veronicae aquaticae foliis subrotundis Hermanni Cat.Hort. Med.Lugd.Bat. [Hermann 1687:590,591 tab.,592].
C.Commelin 1696:51 = HM.
Plukenet 1696:167 Gentianella Utriusque Indiae impatiens foliis Agerati, Phytogr.Tab.186.fig.2 (r).
Plukenet 1700:178; ibid.1696:363 = J.Commelin in HM.
Ray 1704 vol.3:422 = J.Commelin in HM.
Plukenet 1705:83 Euphrasiae species, folio Becabungae Malabariensis, flosculorum labiis amplioribus.
Linnaeus 1753:635 Ruellia antipoda.
Linnaeus 1763:886-887 = Linnaeus 1753.
N.L.Burman 1768:135 Ruellia alternata.

Nelam-parenda HM 9:117:60
Nelám-parendá
BL Add MS 5032:105.
J.Commelin in HM 9:117 Helleborine (s).
C.Commelin 1696:46 = HM.
Ray 1704 vol.3:402 Herba Indica flore monopetalo labiato, capsula seminali è rotundo trilaterali, octaspermo.
Linnaeus 1747:149-150,no.317 Viola foliis lanceolato-linearibus integerrimis distantibus, calycibus pone aequalibus.
Linnaeus 1753:937 Viola enneasperma.
Linnaeus 1763:1327 = Linnaeus 1753.
N.L.Burman 1768:186 = Linnaeus 1763.
Houttuyn 1779 II vol.11:143 = N.L.Burman 1768.

Katu-vistna-clandi HM 9:119:61
Kátu-vístna-elándi
BL Add MS 5032:106.
C.Commelin 1696:39 = HM.
Ray 1704 vol.3:660 Herba Indica foliis oblongo-rotundis, alternatim positis, floribus tripetalis papilionaceos referentibus.
Plukenet 1705:67 Crotalaria Unifolia minor Ind.Or. hirsutis foliis, floribus ad alas purpuro-caeruleis (?).
N.L.Burman 1768:135,Tab.41,f.3 Ruellia erecta β.

Manja-kurini HM 9:121:62
Mánja-kuriní
BL Add MS 5032:107-108.
Plukenet 1691:Tab.CLXXI,fig.4 Curini (fortè) prima species, s. Carim Curini, Hort.Malab.Part.2.fig.2. à Maderaspatan. (c).
C.Commelin 1696:43 = HM.
Plukenet 1696:126 = Plukenet 1691.
Ray 1704 vol.3:402 Herba fruticosa Indica, foliis ad genicula quaternis, spicis squamosis in petiolis oblongis, flore monopetalo labio tripartito, capsulis oblongis.
Linnaeus 1762:21 Justicia infundibuliformis.
N.L.Burman 1768:7 = Linnaeus 1762.
Houttuyn 1775 II vol.4:41-42,Pl.XVIII,Fig.1 = N.L.Burman 1768.

Tali-pullu HM 9:123:63
Táli-pullú
BL Add MS 5032:109.
J.Commelin in HM 9:113 Phalangium Virginianum (cs).
Plukenet 1691:Tab.XXVII,f.4 Ephemerū Phalangoides Maderaspatens. minimum folijs perangustis perfoliatum (cs).
C.Commelin 1696:53 = J.Commelin in HM.
Plukenet 1696:135 = Plukenet 1691.
Ray 1704 vol.3:564 Ephemeri seu Phalangii Virginiani species, floribus sparsis.
Plukenet 1705:73 Ephemeron Phalangoides tripetalon, gramineis foliis, Maderaspatanum (?).
Linnaeus 1762:412 Tradescantia malabarica.
N.L.Burman 1768:17 Commelina nudiflora Linn.sp.61 [Linnaeus 1762:61].
N.L.Burman 1768:79 = Linnaeus 1762.

Houttuyn 1777 II vol.8:330 = N.L.Burman 1768:79.

Upu-dali HM 9:125:64
Upu-dalí
BL Add MS 5032:110-111.
C.Commelin 1696:69 = HM.
Ray 1704 vol.3:549-550 Bicapsularis Indica, Solani foliis & flore, Convolvuli instar repens.
Linnaeus 1747:106,no.234 Ruellia foliis ovatis integerrimis, floribus solitariis sessilibus, caule procumbente.
Linnaeus 1753:635 Ruellia ringens.
Linnaeus 1763:886 = Linnaeus 1753.
N.L.Burman 1768:134 = Linnaeus 1763.
Houttuyn 1778 II vol.9:577-578,Pl.LIX,Fig.2 Ruëllia ringens [Linnaeus 1767 vol.2:424].

Soneri ila HM 9:127:65
Soneri-ilá
BL Add MS 5032:112.
J.Commelin in HM 9:127 Pulmonaria folio maculato Indica similis, floribus tripetalis, rosaceo saturis.
C.Commelin 1696:57 = J.Commelin in HM.
Ray 1704 vol.3:659 Pulmonariae maculosae foliis, floribus tripetalis rosacei coloris, in summo petiolo Paralyseos modo dispositis.

Kalu-polapen HM 9:129:66
Kalú-polapeń
BL Add MS 5032:113.
J.Commelin in HM 9:129 Plantula sedi minoris affinis folio piloso, flore monopetalo difformi Malabarica.
C.Commelin 1696:60-61 = J.Commelin in HM.
Plukenet 1696:165 Genista minima Aethiopica, foliolis Thymi confertis, hirsutie pubescentibus, Phytogr.Tab.297.fig.6 (?).
Plukenet 1696:236 Lysimachiae spicatae purpureae affinis Thymi folio, flosculis in cacumine plurimis quasi in nodis junctis. Phytogr.Tab.43 [f.6] (cs).
Plukenet 1700:73; ibid.1696:142 Euphrasia pratensis, Satureiae foliis scabris, è Maderaspatan. Phytogr.Tab.177.fig.5 (?).
Ray 1704 vol.3:502 Alsine spuria Malabarica, flore monopetalo, tetrapetaloide, difformi.

Kodatsjeri HM 9:131:67
Kodátsjari
BL Add MS 5032:114-115.
J.Commelin in HM 9:131 Portulaca foliis similis planta, flore albo galeato & labiato, semine oblongo, rotundo, russo fusco colore.
C.Commelin 1696:56 = J.Commelin in HM.
Ray 1704 vol.3:659 = J.Commelin in HM.
Plukenet 1705:145-146 Mercurialis (fortè) s. Phyllon marificum, Telephii foliis Malabaricum.
Plukenet 1705:199 Telephii legitimi quodammodo similis, Herba Maderaspatana, galeato & labiato, parvo flore candicante, in foliorum alis (?).

Corosinam HM 9:133:68
Corósinam
BL Add MS 5032:116.
C.Commelin 1696:23 = HM.
Plukenet 1700:147; ibid.1696: Pediculari congener Alpina, Hederae terrestris facie, utriculis amplis vasculum seminale claudentibus (cs).
Ray 1704 vol.3:388 Campanula Indica anomala, foliis in caule ex adverso binis, capsulis oblongis striatis.

Plukenet 1705:166 Pediculari congener Malabarica, folio oblongo-angustis, crassis ac pilosis, floribus amplis, calyce magno bullato, villis obsito exceptis (?).

Pul-colli HM 9:135:69
Pul-colli
BL Add MS 5032:117-118.
C.Commelin 1696:57 = HM.
Ray 1704 vol.3:249 Monospermos Indica gymnospermos, flore monopetalo difformiter è tubo angusto oblongo in 4 folia expanso.
Plukenet 1705:83 Euphrasiae affinis, oblongis cuspidatis foliis, subtus villosis, promissis flosculorum labiis tridentatis.
Linnaeus 1753:16 Justicia nasuta.
Linnaeus 1762:23 = Linnaeus 1753.
N.L.Burman 1768:9 = Linnaeus 1763.
Houttuyn 1777 II vol.7:115 Justicia Nasuta [Linnaeus 1767 vol.2:60].

Nelipu HM 9:137:70
Nelipú
BL Add MS 5032:119-120.
J.Commelin in HM 9:137 Planta aquatica aphyllos repens, flore caeruleo difformi.
C.Commelin 1696:54 = J.Commelin in HM.
Ray 1704 vol.3:404-405 Aphyllos aquatica Indica volubilis, flore dipetalo, difformi, calcari donato.
Plukenet 1705:73 Ephemeron Phalangoides, aphyllon, cauliculis convolvulaceis, ex Ind.Or. (?).
Linnaeus 1747:9,no.23 Utricularia scapo nudo: squamis alternis vagis subulatis.
Linnaeus 1753:18 Utricularia caerulea.
Linnaeus 1762:26-27 = Linnaeus 1753.
N.L.Burman 1768:11 = Linnaeus 1762.
Houttuyn 1777 II vol.7:137 Utricularia Coerulea [Linnaeus 1767 vol.2:62].

Kotsjiletti-pullu HM 9:139:71
Kotsjilétti-pullú
BL Add MS 5032:121-122.
J.Commelin in HM 9:139 Gladiolus palustre (s).
C.Commelin 1696:34 Gramen florens capitulo squamoso Raji H.1318 [Ray 1688 vol.2:1318].
Plukenet 1696:170 Gladiolo lacustri accedens Malabarica, è capitulo botryoide florifera.
J.Burman 1737:109 Gramen Indicum, capitulis oblongis, floridis, aureis, squammatis. Jupicai Bontii. Mus.Zeyl.pag.41 [Hermann 1717:41].
Linnaeus 1747:14-15,no.35 Xyris foliis gladiatis. Gron.virg.11.
Linnaeus 1753:42 Xyris indica.
Linnaeus 1762:62 = Linnaeus 1753.
N.L.Burman 1768:18 = Linnaeus 1762.
Houttuyn 1782 II vol.13:42-43 = N.L.Burman 1768.

Min-angani HM 9:141:72
Min-angáni
BL Add MS 5032:123.
J.Commelin in HM 9:141 Sideriti folio, caule piloso, capitulo globoso flore monopetalo, quinque partito, candido.
C.Commelin 1696:61 = J.Commelin in HM.
Ray 1704 vol.3:550 = J.Commelin in HM.
Plukenet 1705:15 Anomala Indiana, Nepetae foliis, staechadis Arabicae capitulis, ex sinu foliorum prodeuntibus (?).
Linnaeus 1762:326 Gomphrena hispida.
N.L.Burman 1768:73 = Linnaeus 1762.

Houttuyn 1777 II vol.7:802 = N.L.Burman 1768.

Tsjeru-uren HM 9:143:73
Tsjéru-úren
BL Add MS 5032:124.
J.Commelin in HM 9:143 Planta Sideritis folio insipido, flore pentapetalo, rosacei coloris, fructu rotundo in quinque partes diviso, semine triangulato.
Plukenet 1691:Tab.LXXIV,f.8 Althaea Americana frutescens Melochiae foliis angustioribus (?).
C.Commelin 1696:61 = J.Commelin in HM.
Plukenet 1696:17 = Plukenet 1691.
Plukenet 1696:25 Althaea peregrina longiore Betonicae folio, floribus albis perexiguis capitulis arctè conglomeratis. Phytogr.Tab.44.fig.5 (?).
Plukenet 1700:10; ibid.1696: Althaeae Ulmifoliae species pumila, è Maderaspatan. (?).
Ray 1704 vol.3:550 = J.Commelin in HM.
Linnaeus 1747:111-112,no.246 Melochia floribus capitatis sessilibus, capsulis subrotundis.
Linnaeus 1753:675 Melochia corchorifolia.
Linnaeus 1763:944 = Linnaeus 1753.
N.L.Burman 1768:143 = Linnaeus 1763.
Houttuyn 1775 II vol.5:367 = N.L.Burman 1768.

Iribeli HM 9:145-146:74
Iribéli
BL Add MS 5032:125-126.
C.Commelin 1696:37 = HM.
Ray 1704 vol.3:661 = HM.

Perim-munja HM 9:147:75
Perim-múnja
BL Add MS 5032:127-128.
J.Commelin in HM 9:147 Lamium (r).
C.Commelin 1696:52 = HM.
Ray 1704 vol.3:655 Lamii foliis Indica flore fructúque vidua.

Tardavel HM 9:149:76
Tardável
BL Add MS 5032:129-130.
C.Commelin 1696:65 = HM.
Ray 1704 vol.3:422 Tetrapetala Indica flore fructui dispermo insidente, foliis adversis, caulibus pilosis.
Linnaeus 1762:148-149 Spermacoce hispida (r).
N.L.Burman 1768:33 = Linnaeus 1762.

Entada HM 9:151:77
Entadá
BL Add MS 5032:131-132.
C.Commelin 1696:27 = HM.
Ray 1704 vol.3:660 Herba Indica capreolata, Orobi foliis bicompositis, spicis longis, angustis, flosculis densè stipatis.
Plukenet 1705:32 Arbuscula Maderaspatana, Coluteae foliis, Saururi fructù, spicata (?).
Linnaeus 1747:98,no.219 Mimosa foliis duplicato-pinnatis, cirrho terminatis.
Linnaeus 1753:518 Mimosa Entada.
Linnaeus 1763:1502 = Linnaeus 1753.
N.L.Burman 1768:223 = Linnaeus 1763.
Houttuyn 1776 II vol.6:445-447 Mimosa Entada [Linnaeus 1767 vol.2:676].

Pola-tsjira HM 9:153:78
Polá-tsjírá
BL Add MS 5032:133.
J.Commelin in HM 9:153 Sideriti folio Indica, palustris, caule fistuloso, flore pentapetalo subviridi, semine minutiβimo, ex albo ruffescente.
C.Commelin 1696:62 = J.Commelin in HM.
Ray 1704 vol.3:528 Herba verticillata aquatica, foliis integris serratis flosculis pentapetalis, capsulis 5 suturis striatis.
Plukenet 1705:190 Scrophulariae affinis, Sideritidis folio Malabarica, caule fistuloso striato, floribus parvis, plurimis ad nodos cauli arctè adhaerentibus (?).
Rumphius 1750 vol.6:39-40,Tab.XVI,Fig.1 Herba admirationis (c).

Pee-tardavel HM 9:153(2):79
Peê-tardável
BL Add MS 5032:134.
C.Commelin 1696:51 = HM.
Plukenet 1705:188 Scabiosis affinis Planta Malabarica, foliis crispis.

Kalu-tali HM 9:157:80
Kátu-tálij
BL Add MS 5032:135.
C.Commelin 1696:39 = HM.
Ray 1704 vol.3:404 Dipetalos Indica sterilis, foliorum angulo altero latius inferiùs quam alter exporrecto.

Ene-pael HM 9:159:81
Ené-páel
BL Add MS 5032:136.
J.Commelin in HM 9:81 Rubia (r).
C.Commelin 1696:27 = HM.
Ray 1704 vol.3:660 Herba stellata Indica quatuor ad singula genicula gemmis rotundiolis.
Linnaeus 1771:175 Rotala verticillaris.
Houttuyn 1777 II vol.7:204-205 = Linnaeus 1771.

Nelam-mari HM 9:161:82
Nélam-múrí
BL Add MS 5032:137.
J.Commelin in HM 9:161 Planta bifolia humirepa, Arifolio, flosculis aureis, villosis, articulatis.
C.Commelin 1696:54 = J.Commelin in HM.
Plukenet 1700:140; ibid.1696:270 Onobrychis Maderaspatana diphyllos siliculis clypeatis hirsutis minor. Phytogr.Tab.246. fig.6. Ponauverrepoondoo Malabarorum (?).
Petiver 1702(1701):856,no.129 Ponau verre poondoo Malab. Onobrychis Indiae utriusq; & Guineensis (?).
Ray 1704 vol.3:404 Planta bifolia humirepa Indica, floribus in propriis surculis dipetalis, globulis seminalibus monospermis.
J.Burman 1737:114,Tab.50.Fig.1 Hedysarum bifolium, foliolis ovatis, siliculis asperis, geminis, inarticulatis.
Linnaeus 1747:134-135,no.291 Hedysarum foliis binatis petiolatis: floralibus sessilibus.
Linnaeus 1753:747 Hedysarum diphyllum.
Linnaeus 1763:1053 = Linnaeus 1753.
N.L.Burman 1768:165 = Linnaeus 1763.
Houttuyn 1779 II vol.10:234-236,Pl.LXV,Fig.2 = N.L.Burman 1768.

Mallam-tsjulli HM 9:163:83
Mallam-tsjulli
BL Add MS 5032:138.
C.Commelin 1696:42 = HM.
Plukenet 1700:140; ibid.1696:270 Onobrychis Maderaspatana

diphyllos major siliculis clypeatis asperis. Phytogr.Tab.102. fig.1 (?).
Ray 1704 vol.3:660 Herba bifolia Indica foliis obtusioribus ad exortum pediculi duobus parvis oblongo-angustis foliolis appositis (sic).
J.Burman 1737:114 Hedysarum bifolium, siliquis articulatis, echinatis. Mus.Zeyl.pag.59 [Hermann 1717:59].

Beli-tsjira HM 9:165:84
Béli-tsjíra
BL Add MS 5032:139.
C.Commelin 1696:15 = HM.
Plukenet 1696:372-373 Tithymalus Indicus annuus dulcis, floribus albis, cauliculis viridantibus. PBP [Sherard 1689:381]. Phytogr.Tab.113.fig.2 (?).
Ray 1704 vol.3:654 Herba Indica foliis oblongis, acutis, ex adverso binis, floribus in foliorum alis pluribus cyathiformibus, in 4 divisis, fructu vidua.

Tsjeria-manga-nari HM 9:165:85
Tsjería-mánga-nári
BL Add MS 5032:140.
J.Commelin in HM 9:165 Alsina spuria, seu Veronica Indica, flore caeruleo, Chamaedri folio.
C.Commelin 1696:3 = J.Commelin in HM.
Plukenet 1700:44 Chamaedrifolia pusilla planta, flosculo oblongo monopetalo, capsulâ membranaceâ, ex Coromandel (r).
Ray 1704 vol.3:422 Alsine spuria Indica Chamaedryfolia, flore coeruleo tripartito.
Linnaeus 1753:17,1200 Gratiola virginiana.
Linnaeus 1762:25 = Linnaeus 1753.
N.L.Burman 1768:11 = Linnaeus 1762.
Houttuyn 1777 II vol.7:123-124 Gratiola Virginica [Linnaeus 1767 vol.2:61].

Tsjeria-narinampuli HM 9:167-168:86
Tsjeria-narinámpúli
Bl Add MS 5032:141-142.
J.Commelin in HM 9:168 Solani affinis Malabarica, flore & baccis rubescentibus.
C.Commelin 1696:64 = J.Commelin in HM.
Petiver 1704(1703):1460,no.5 Triopteris Malabarica Epimedij folio.
Ray 1704 vol.3:357 = J.Commelin in HM.
Rumphius 1750 vol.6:62,Tab.XXVI,Fig.2 Halicacabus baccifer (c).

Bahel-tsjulli HM 9:169-170:87
Bahél-tsjúlli
BL Add MS 5032:143-144.
J.Commelin in HM 9:170 Digitalis affinis Indica, Blattariae folio, flore rubicundo.
C.Commelin 1696:26 = J.Commelin in HM.
Plukenet 1700:64 Digitalis Virginiana, Lysimachiae siliquosae foliis glabris, floribus amplis pallidis, propè summitatem caulis (?).
Linnaeus 1767:90-91 Columnea longifolia.
N.L.Burman 1768:133-134 Sesamum javanicum.
Houttuyn 1778 II vol.9:586 Columnea longifolia [Linnaeus 1767 vol.2:427].

VOLUME 10

Sjasmin HM 10:1-2:1

Sjasmín
BL Add MS 5032:145-146.
J.Commelin in HM 10:2 Blattaria Indica Urticae folio piloso, flore amplo miniato.
C.Commelin 1696:16 = J.Commelin in HM.
Plukenet 1696:16 Alcea Jamaicensis fruticosa, floribus purpureis, oblongo, crenato, folio acuminato, molli & incano. Phytogr.Tab.255 (?).
Ray 1704 vol.3:523 = J.Commelin in HM.
Linnaeus 1763:958-959 Pentapetes phoenicea.
N.L.Burman 1768:144 = Linnaeus 1763 (?).
Houttuyn 1779 II vol.10:34-35 = N.L.Burman 1768.

Uren HM 10:3-4:2
Uren
BL Add MS 5032:147-148.
J.Commelin in HM 10:4 Alcea Malabarica folio vitis, flore minore, rosaceo colore, fructu lappaceo, semine rostrato.
C.Commelin 1696:3 = J.Commelin in HM.
Plukenet 1696:25 Althaea Brasiliana fructu hispido pentacocco (c).
Ray 1704 vol.3:522 = J.Commelin in HM.
J.Burman 1737:9 Alcea Indica floribus roseis parvis, fructibus parvis, quinquepartitis, hispidis, lappaceis. Mus.Zeyl.pag.8 [Hermann 1717:8].
Linnaeus 1747:117-118,no.257 Urena foliis sinuato-multifidis villosis (?).
Rumphius 1750 vol.6:59-60,Tab.XXV,Fig.2 Lappago Amboinica (c); J.Burman ibid.:60 (=).
N.L.Burman 1768:149 Urena lobata Linn.sp.974 [Linnaeus 1763:974].

Belutta-itti-canni HM 10:7:3
Velútta-jtti-Cánni
BL Add MS 5032:149-150.
C.Commelin 1696:68 = HM.
Ray 1704 vol.3:651 Parasitica Indica, foliis minoribus & angustioribus, floribus pariter se aperientibus.

Valli-itti-canni HM 10:5:4
Válli-jtti-Cánni
BL Add MS 5032:151-152.
C.Commelin 1696:68 = HM.
Petiver 1702(1701):943,no.167 Chedde meel cheddee Malab. Kanni-Viscum Eremitanum floribus spicatis (?).
Ray 1704 vol.3:651 Parasitica Indica, floribus oblongis, in medio dehiscentibus, tandémque in 5 foliola angusta se aperientibus.
Plukenet 1705:209 Viscum arboreum Indicum, crassiori folio, floribus purpureis, longo collo donatis, plurimis simùl in Surculis binatim adnascentibus.

Kanneli-itti-kanni HM 10:9:5
Kannelí-jtti-Cánni
BL Add MS 5032:153-154.
C.Commelin 1696:38 = HM.
Plukenet 1700:25 Arbor Indica baccifera Ligustri foliis densioribus & firmioribus, fructu ovali. Pungalu Malabarorum (?).
Petiver 1702(1700):704-705,no.59 Causha or Causha-chedde Malab. Baccifera Madrasp. Visci arborei foliis latioribus Mus.Petiv.38 (?).
Petiver 1702(1701):944,no.170 Cacian-cheddee Malab. Baccifera Madraspat. Visci arborei foliis latioribus Mus.Petiv.38 (?).
Ray 1704 vol.3:650-651 Parasitica Indica seu Viscus arbori Kanneli se affigens, ac uniens, flore è tubo oblongo in quatuor folia expanso, bacca oblongo-rotunda obscure rubente.

Rumphius 1747 vol.5:32-34,Tab.XX Pulassarium (s).

Manga-nari HM 10:11-12:6
Mánga-nari
BL Add MS 5032:155-156.
J.Commelin in HM 10:12 Veronica Indica aquatica, maxima, odorata, Teucri-folio, flore purpurascente.
C.Commelin 1696:68 = J.Commelin in HM.
Ray 1704 vol.3:421-422 = J.Commelin in HM.

Ana-schovadi HM 10:13-14:7
Ána-schovadí
BL Add MS 5032:157-158.
C.Commelin 1696:5 = HM.
Plukenet 1696:132-133 Echinophorae Indicae affinis, semine & floribus in capsulis, seu potiùs capitulis laevibus in caulium cymis prodeuntibus.
Linnaeus 1737:390 Elephantopus foliis integris serratis.
Linnaeus 1753:814 Elephantopus scaber.
Linnaeus 1763:1313 = Linnaeus 1753.
N.L.Burman 1768:185 = Linnaeus 1763.
Houttuyn 1776 II vol.6:150-151 = N.L.Burman 1768.

Tumba-codiveli HM 10:15:8
Túmba-codivéli
BL Add MS 5032:159-160.
C.Commelin 1696:68 = HM.
Plukenet 1696:132 Echinophora latifolia, Nerii floribus, Malabarica spicata minor.
Ray 1704 vol.3:550 Herba Indica sequenti similis florum petalis angustioribus candidis.
Linnaeus 1737:53 Plumbago foliis petiolatis.
Linnaeus 1747:30-31,no.73 = Linnaeus 1737.
Linnaeus 1762:215 Plumbago zeylanica.
N.L.Burman 1768:43 = Linnaeus 1762.
Houttuyn 1777 II vol.7:522 = N.L.Burman 1768.

Schetti-codiveli HM 10:17-18:9
Schétti-codivéli
BL Add MS 5032:161-162.
C.Commelin 1696:60 = HM.
Plukenet 1696:132 Echinophora latifolia, spicata, major, Nerii floribus Malabariensis.
Ray 1704 vol.3:550-551 Herba Indica flore è tubo oblongo in 5 folia expanso, calyce hispido, vasculo seminali monospermo.
Linnaeus 1737:53 Plumbago foliis petiolatis α.
Linnaeus 1747:30-31,no.73 = Linnaeus 1737.
Linnaeus 1762:215 Plumbago rosea.
N.L.Burman 1768:43 = Linnaeus 1762.
Houttuyn 1777 II vol.7:522-523 Plumbago Rosea [Linnaeus 1767 vol.2:254].

Schada-veli-kelangu HM 10:19-20:10
Scháda-véli-kelángú
BL Add MS 5032:163-164.
J.Commelin in HM 10:20 Asperagus aculeatus maximus sermentosus Ceylanicus Hermans in Horto Academico Lugd.Bat. [Hermann 1687:62,63 tab.,64].
C.Commelin 1696:10 = J.Commelin in HM.
Petiver 1702(1700):718-719,no.87 Tanne mutanea-tunga Malab. Corruda Zeylanica Paeoniae radicibus.
Ray 1704 vol.3:359 = J.Commelin in HM.
J.Burman 1737:37 = J.Commelin in HM.
Linnaeus 1762:450 Asparagus sarmentosus.
N.L.Burman 1768:82 = Linnaeus 1762.

Houttuyn 1777 II vol.8:344 = N.L.Burman 1768.

Coluppa HM 10:21-22:11
Colúppa
BL Add MS 5032:165-166.
J.Commelin in HM 10:22 Persicariae folio repens Malabarica, flore globoso, albescente.
C.Commelin 1696:53 = J.Commelin in HM.
Plukenet 1696:27 Amaranthoides Maderaspatanum foliis angustis ex adverso binis, floribus ad nodos verticillatis. Phytogr.Tab.132.fig.6 (?).
Petiver 1702(1701):1019,no.215 Punangunne Malab. Perexil Madraspatana, foliis oppositis angustioribus Polygoni H.Un.22. Act.Phil.no.244.p.323.22 [Petiver 1699(1698):323-324,no.22] (s).
Ray 1704 vol.3:118 = J.Commelin in HM.
Linnaeus 1747:49,no.116 Gomphrena caule repente, foliis lanceolatis, capitulis sessilibus axillaribus.
Linnaeus 1753:225 Gomphrena sessilis.
Linnaeus 1762:300 Illecebrum sessilis.
N.L.Burman 1768:66 = Linnaeus 1762.
Houttuyn 1777 II vol.7:716-717 Illecebrum sessile [Linnaeus 1767 vol.2:188].

Nir-valli-pullu HM 10:23:12
Nir-válli-pullú
BL Add MS 5032:167.
J.Commelin in HM 10:23 Potamogeton foliis bullatis.
C.Commelin 1696:47 = HM.
Plukenet 1696:135 Ephemerum phalangoides, Maderaspatanum, minimum, secundum caulem quasi ex utriculis floridum. Phytogr.Tab.174.fig.3 (r).
Plukenet 1700:155 Potamogeiton Indicum, foliis variis, partim integris, partim pennatis. Yellapashee Malabarorum (i.e.) Planta aquatica.
Ray 1704 vol.3:565 Ephemerum Virginianum ramosum procumbens tripetalon, foliis rigidis Herman.Parad.Bat. [Hermann 1698:145].

Nir-pulli HM 10:25:13
Nir-pullú
BL Add MS 5032:168.
J.Commelin in HM 10:25 Ephemerum Malabaricum flore tripetaloïde.
Plukenet 1691:Tab.CLXXIV,fig.3 Ephemerum phalangoides, Maderaspatanum, minimum, secundum caulem quasi ex utriculis floridum (?).
C.Commelin 1696:27 = J.Commelin in HM.
Plukenet 1696:135 = Plukenet 1691.
Hermann 1698:144 Ephemervm Malabaricum aquaticum procumbens tricoccum ad singulos foliorum sinus floridum.
Ray 1704 vol.3:567 Ephemerum Malabaricum flore tripetalo, in foliorum alis sessili.
Plukenet 1705:73 Ephemeron Phalangoides, longioribus angustis foliis, floribus ex inflatis, & pubescentibus foliorum vaginulis, veluti ex utriculis pronascentibus; Mella-caula Malabarorum (?).
Linnaeus 1753:42 Commelina axillaris.
Linnaeus 1762:61-62 = Linnaeus 1753.
N.L.Burman 1768:17 = Linnaeus 1762.
Houttuyn 1777 II vol.8:331 Tradescentia Axillaris Mant.321 [Linnaeus 1771:321].

Brami HM 10:27:14
Bramí

BL Add MS 5032:169.
J.Commelin in HM 10:27 Glaux Indica Portulacae folio, flore majore diluto caeruleo, albicante colore.
Plukenet 1691:Tab.CCXVI,fig.3 Portulacae similis Planta Ind. Orient. (?).
C.Commelin 1696:33 = J.Commelin in HM.
Plukenet 1696:304 = Plukenet 1691.
Plukenet 1700:9 Alsine Portulacae aquaticae folio oblongo, flore longo pedunculo insidente ex Prom.Comorin (?).
Ray 1704 vol.3:551 = J.Commelin in HM.
J.Burman 1737:197 Portulaca minima aizoides, Hingheda dicta. Mus.Zeyl.pag.62 [Hermann 1717:62] (?).

Kirganeli HM 10:29:15
Kirgáneli
BL Add MS 5032:170-171.
J.Commelin in HM 10:31 Vitis Ideae Affinis, flore hexapetalo ex albicante.
Plukenet 1691:Tab.CLXXXIII,fig.5 Fruticulus folijs brevioribus, subrotundis, & densius stipatis (?).
C.Commelin 1696:69 = J.Commelin in HM.
Plukenet 1696:159 = Plukenet 1691.
Petiver 1702(1701):1012,no.182 Keela nellee Malab. Nirouri Salawaccensis minor, Abrus folio (?).
J.Burman 1737:230-231,Tab.93,Fig.2 Urinaria Indica, erecta, vulgaris. Herb.Hart.
Linnaeus 1737:440 Phyllanthus foliis alternis alternatim pinnatis, floribus dependentibus ex alis foliolorum (?).
Linnaeus 1747:157,no.331 Phyllanthus foliis pinnatis floriferis, floribus pedunculatis, caule herbaceo erecto.
Rumphius 1750 vol.6:41-43,Tab.XXVII,Fig.1 Herba moeroris alba (c).
Linnaeus 1753:981-982 Phyllanthus Niruri.
Linnaeus 1763:1392-1393 = Linnaeus 1753.
N.L.Burman 1768:196 = Linnaeus 1763.

Tsjeru-kirganeli HM 10:31:16
Tsjéru-kirgáneli
BL Add MS 5032:172.
J.Commelin in HM 10:31 Vitis Ideae Affinis, hexapetalo ex albicante.
Plukenet 1691:Tab.CLXXXIII,fig.6 Fruticulus brevioribus, folijs & angustis (?).
C.Commelin 1696:69 = J.Commelin in HM.
Plukenet 1696:159 = Plukenet 1691.
Petiver 1704(1703):1458,no.41 Nirouri Madraspat. Mimosae foliis (?).
J.Burman 1737:231 Urinaria Indica, supina, cauliculis rubentibus. Mus.Zeyl.4 & 16 [Hermann 1717:4,16].
Linnaeus 1747:157-158,no.332 Phyllanthus foliis pinnatis floriferis, floribus sessilibus, caule herbaceo procumbente.
Rumphius 1750 vol.6:41-43,Tab.XVII,Fig.2 Herba moeroris ruber (c).
Linnaeus 1753:982 Phyllanthus Urinaria.
Linnaeus 1763:1393 = Linnaeus 1753.
N.L.Burman 1768:196 = Linnaeus 1763.

Manja-adeka-manjen HM 10:33:17
Mánja-adecá-manjén
BL Add MS 5032:173.
J.Commelin in HM 10:33 Conyza Indica, Teucri folio, flore flavo odorato.
Plukenet 1691:Tab.LXXXVIII,f.3 Eupatoria Valerianoides, flore niveo, Teucrij folijs cum pediculis Americana (?).
C.Commelin 1696:22 = J.Commelin in HM.

Plukenet 1696:56 = J.Commelin in HM.
Plukenet 1696:141 = Plukenet 1691.
Ray 1704 vol.3:155 = J.Commelin in HM.

Kurundoti HM 10:35:18
Kúrúndóti
BL Add MS 5032:174.
J.Commelin in HM 10:35 Cystus humilis Indica, ciceri folio, flore flavo.
C.Commelin 1696:20 = J.Commelin in HM.
Plukenet 1700:50 Cistus humilis Chamaedryos crispatis foliis, Promont.Bon.Spei (?).
Rumphius 1750 vol.6:44-47,Tab.XIX Silagurium rotundum (s).
N.L.Burman 1768:146 Sida retusa Linn.sp.961 [Linnaeus 1763:961].

Nelam-pullu HM 10:37:19
Nelám-púllú
BL Add MS 5032:175.
J.Commelin in HM 10:37 Phalangii affinis Indica, flore caeruleo, pallido.
C.Commelin 1696:53 = J.Commelin in HM.
Plukenet 1696:135 Ephemerum Phalangoides, Maderaspatense, minimum, foliis perangustis, perfoliatum. Phytogr.Tab.27.fig.4 (cs).
Hermann 1698:145 Ephemerum Malabaricum pumilum erectum gramineis foliis.
Ray 1704 vol.3:564 Phalangio Virginiano Tradescanti affinis, flore coeruleo pallido, per ramulos sparso.

Araka-puda HM 10:39:20
Acára-púdá
BL Add MS 5032:176.
J.Commelin in HM 10:39 Alsine Myriophylli folio, flore carneo.
C.Commelin 1696:3 = J.Commelin in HM.
Plukenet 1700:163; ibid.1696:323 Ros-Solis Lusitanicus maximus, foliis Asphodeli minoris, Tournefortij. Phytogr.Tab.117.fig.2 (cs).
Ray 1704 vol.3:501 = J.Commelin in HM.
Plukenet 1705:185 Rorella Malabarica, pennatis angustioribus foliis, folioso caule, floribus pentapetalis carneis.
J.Burman 1737:207,Tab.94,Fig.1 Ros Solis ramosus, caule folioso.
Linnaeus 1747:52,no.121 Drosera caule ramoso folioso, foliis linearibus.
Linnaeus 1753:282 Drosera indica.
Linnaeus 1762:403 = Linnaeus 1753.
N.L.Burman 1768:78 = Linnaeus 1762.
Houttuyn 1777 II vol.8:294 Drosera Indica [Linnaeus 1767 vol.2:225].

Karinta-kali HM 10:41:21
Karínta-káli
BL Add MS 5032:177.
J.Commelin in HM 10:41 Violae folio Malabarica baccifera.
C.Commelin 1696:69 = J.Commelin in HM.
Plukenet 1696:235 Lycium subrotundis foliis Malabaricum spinis brevibus, crassis recurvis, ex uno versu conjugatis, armatum. Phytogr.Tab.305.fig.5 (r).
Plukenet 1696:309 Pyrolae affinis Malabarica.
Ray 1704 vol.3:358 Violae folio, Centaurii minoris flore candido, baccifera dipyrena.
Plukenet 1705:179 Pyrolae affinis Malabarica, foliis subrotundis baccas ferens.
Linnaeus 1762:245 Psychotria herbacea.

N.L.Burman 1768:52 = Linnaeus 1762.
Houttuyn 1775 II vol.4:205-206 Psychotria herbacea [Linnaeus 1767 vol.2:165].

Nela-naregam HM 10:43:22
Nela-naregam
BL Add MS 5032:178-179.
J.Commelin in HM 10:43 Frutex Malabaricus trifoliatus, flore pentapetalo candidiβimo, fructu triangulato.
C.Commelin 1696:31-32 = J.Commelin in HM.
Plukenet 1696:195 Jasminum Malabaricum triphyllon, pediculis alatis, floribus ex oblongo tubulo, calathum quinquepartitum explicatis, colore puniceo, Phytogr.Tab.303.fig.3 (?).
Ray 1704 vol.3:527 Trifolium Indicum fruticescens, folii pediculo alato, flore pentapetalo candidissimo, fructu triangulato.

Schanganam-pullu HM 10:45:23
Schanganám-púllu
BL Add MS 5032:180-181.
J.Commelin in HM 10:45 Rubia Indica sylvestris, foliis levis, flore albicante.
Plukenet 1691:Tab.CXXX,fig.6 Alsine Holostea villosa folijs caulem ambientibus multiflora è Maderaspatan. (cs).
C.Commelin 1696:59 = J.Commelin in HM.
Plukenet 1696:22 = Plukenet 1691.
Ray 1704 vol.3:425 Alsine spuria Indica, foliis longis angustis, florum pediculis capsulis rotundis.

Kaipa-tsjira HM 10:47:24
Kaipá-Tsjíra
BL Add MS 5032:182.
J.Commelin in HM 10:47 Rubia sylvestris sp.
C.Commelin 1696:38 = HM.
Plukenet 1696:21 Alsine Indica quadrifolia & quinquefolia flosculis longiori pediculo insidentibus. Phytogr.Tab.130.fig.5 (?).
Petiver 1702(1701):1008,no.194 Terai Malab. Spergula Salawaccensis foliis inequalibus latioribus & subrotundis, alis foliorum florescens (?).
Ray 1704 vol.3:503 Alsine ramosa Indica, Rubiae facie, floribus candidis.

Tsjeru-talu-dama HM 10:49:25
Tsjéru-tálú-dáma
BL Add MS 5032:183.
J.Commelin in HM 10:49 Veronicae similis Malabarica, flore purpureo.
C.Commelin 1696:68 = J.Commelin in HM.
Plukenet 1700:164 Rubia Mariana Alsines majoris folio, ad caulem binato, flore purpuro-rubente (?).
Ray 1704 vol.3:265 Rubeola latifolia Indica, foliis in caulibus ex adverso binis, flosculis purpuro-coeruleis.

Tsjeru-jonganam-pullu HM 10:51:26
Tsjérú-tsjónganámpúllú
BL Add MS 5032:184.
J.Commelin in HM 10:51 Rubiae similis Indica, flore albicante.
C.Commelin 1696:59 = J.Commelin in HM.
Plukenet 1696:22 Alsine holostea villosa foliis caulem ambientibus multiflora è Maderaspatan. Phytogr.Tab.130.fig.6 (cs).
Ray 1704 vol.3:502 Alsine Indica, foliis oblongis cuspidatis.
N.L.Burman 1768:38,Tab.15,f.1 Oldenlandia paniculata Linn.sp. app.1667 [Linnaeus 1763:1667] (?).

Niruri vel Nirpulla HM 10:53:27

Nirúri
BL Add MS 5032:185-186.
J.Commelin in HM 10:55 Anagallis (c).
C.Commelin 1696:47 = HM.
Plukenet 1696:159 Fruticulus capsularis hexapetalos, Casiae Poetarum foliis, è Maderaspatan, Phytogr.Tab.183.fig.4 (?).

Tsjeru-vallel HM 10:55:28
Tsjérú-vallél
BL Add MS 5032:187.
J.Commelin in HM 10:55 Anagallis (c).
C.Commelin 1696:67 = HM.

Scheru-bula HM 10:57:29
Schérú-búla
BL Add MS 5032:188-189.
J.Commelin in HM 10:57 Herniaria Africana major Parkinson.
C.Commelin 1696:35 = J.Commelin in HM.
Plukenet 1696:27 Amaranthus Indicus verticillatus albus Origani foliis lanugine incanis Phytogr.Tab.75.fig.8.
Ray 1704 vol.3:132 = J.Commelin in HM.
J.Burman 1737:60,Tab.26,Fig.1 Chenopodium incanum, racemosum, folio majore, minori opposito.
Linnaeus 1747:43-44,no.104 Achyranthes caule erecto, spicis ovatis lateralibus, calycibus lanatis.
Linnaeus 1753:204 Achyranthes lanata.
Linnaeus 1762:296 = Linnaeus 1753.
N.L.Burman 1768:64 = Linnaeus 1762.
Houttuyn 1777 II vol.7:708-709 Illecebrum Lanatum Veg.XIII. Gen.290.p.205.

Bula HM 10:59:30
Búla
BL Add MS 5032:190-191.
C.Commelin 1696:16 = HM.
Plukenet 1696:27 Amaranthoides Indicum verticillatum Parietariae hirsutis foliis aculeatum. Phytogr.Tab.133.fig.3 (?).
Ray 1704 vol.3:424 Alsine-folia Indica, foliis alternis flosculis tetrapetalis minimis in foliorum alis, seminibus in singulis capsulis binis.

Nela-tsjira HM 10:61:31
Nelátsjira
BL Add MS 5032:192.
J.Commelin in HM 10:61 Sedi folio Indica, flore tetrapetalo, flavo colore.
C.Commelin 1696:61 = J.Commelin in HM.
Plukenet 1700:155 Portulaca fortè Indiae Orientalis teretifolia, flore flavo tetrapetalo.
Ray 1704 vol.3:424 = J.Commelin in HM.
N.L.Burman 1768:38,Tab.15,f.2 Oldenlandia repens.

Muriguti HM 10:63:32
Múrigúti
BL Add MS 5032:193-194.
J.Commelin in HM 10:63 Alsinae affinis planta Indica, flore candido.
C.Commelin 1696:3 = J.Commelin in HM.
Plukenet 1700:11 Amaranthus verticillatus minimus exiguis Bliti foliis pallidè viridibus ex Ind.Or. (?).
Ray 1704 vol.3:424 Alsine folio Indica, foliis binis adversis, flosculis tetrapetalis pluribus simul in foliorum alis.
Rumphius 1750 vol.6:29,Tab.XII,Fig.2 Herba memoriae (c).
Linnaeus 1753:101-102 Hedyotis Auricularia.
Linnaeus 1762:147 = Linnaeus 1753.

N.L.Burman 1768:33 = Linnaeus 1762.
Houttuyn 1777 II vol.7:264-265 Hedyotis Auricularia [Linnaeus 1767 vol.2:115].

Caicotten-pala HM 10:65:33
Caicottén-pála
BL Add MS 5032:195.
J.Commelin in HM 10:65 Veronicae similis Indica, albicante flore.
C.Commelin 1696:68 = J.Commelin in HM.
Plukenet 1696:373 Tithymalus Botryoides minor Americanus foliis hirsutis (?).
Ray 1704 vol.3:423 = J.Commelin in HM.

Naru-nundi HM 10:67-68:34
Nárú-níndi
BL Add MS 5032:196-197.
J.Commelin in HM 10:68 Anagallis (s).
C.Commelin 1696:46 = HM.
Plukenet 1696:37 Apocynum scandens angusto Rorismarini folio ex Insulis Fortunatis. Phytogr.Tab.261.fig.2 (?).
Petiver 1702(1700):593-594,no.46 Periploca Malabarica fol. angustissimo. Nanna-ree-chedde Malab. Segunda-pala Gent.
Ray 1704 vol.3:500 Graminis leucanthemi foliis & facie Indica, flosculis pentapetalis in surculis è foliorum alis egressis.
N.L.Burman 1768:70 Periploca tenuifolia Linn.sp.310 [Linnaeus 1762:310].
Houttuyn 1777 II vol.7:732-733 Ceropegia Tenuifolia.

Parpadagam HM 10:69:35
Parpadagám
BL Add MS 5032:198.
J.Commelin in HM 10:69 Spergula Indica major flore candido.
C.Commelin 1696:64 = J.Commelin in HM.
Plukenet 1696:21 Alsine holostea, s. Gramen Leucanthemum Ind: Orient. Phytogr.Tab.130.fig.3 (?).
Petiver 1699(1698):325-326,no.25 Puccapoonda Mal. Samolus Madraspatana Gram. Leucanthemi foliis (s).
Ray 1704 vol.3:424 Alsine spuria Indica, foliis angustis acutis, flore tetrapetalo.
J.Burman 1737:228 Veronica Zeylanica, folio conjugato Rorismarini, flore albescente. Herb.Hart.
N.L.Burman 1768:37,Tab.14,f.1 Oldenlandia tenuifolia.

Kara-tsjira HM 10:71:36
Cará-tsjira
BL Add MS 5032:199.
J.Commelin in HM 10:71 Portulaca Sylvestris.
C.Commelin 1696:56 = J.Commelin in HM.
Plukenet 1696:303 Portulaca angustifolia s. sylvestris CBP. [Bauhin 1623:288](?).
J.Burman 1737:11 Alkekengi Indicum, minimum, fructu virescente. Tournef.Inst.pag.151 (?).
Linnaeus 1737:207 Portulaca foliis cuneiformibus verticillatis sessilibus, floribus sessilibus.
Rumphius 1747 vol.5:268 Portulaca Indica (-); J.Burman ibid.: 268 (=).

Wadapu HM 10:73-74:37
Wádapú
BL Add MS 5032:200-201.
J.Commelin in HM 10:74 Amarantho affinis Indica Orientalis, floribus glomeratis, Ocymoidis folio. Breynii Cent. [Breyne 1678:109-111,tab.].
J.Commelin 1689:20 = J.Commelin in HM.

C.Commelin 1696:4 Amarantho affinis, seu Amaranthoides major indica ocymoides folio, & facie, flore globoso purpureo Breyn:P:2.13 [Breyne 1689:13].
J.Commelin 1697:85-86,Fig.45 = J.Commelin in HM.
Hermann 1698:14 Amaranthoides Indicum foliis Ocymastri capitulis purpureis.
Plukenet 1700:11; ibid.1696:26 Amaranthoides Indicum monospermum, foliis Ocymastri capitulis purpureis. PBP.309 [Sherard 1689:309-310].
J.Burman 1737:15-16 Amaranthoides Lychnidis folio, capitulis purpureis. Tournef.inst.pag.654.
Linnaeus 1737:86 Gomphrena caule recto, foliis ovato-lanceolatis, capitulis solitariis pedunculatis diphyllis.
Linnaeus 1747:48-49,no.115 = Linnaeus 1737.
Rumphius 1747 vol.5:289-290 Flos globosus (=).
Linnaeus 1771:348 = N.L.Burman 1768.
N.L.Burman 1768:72 Gomphraena globosa Linn.sp.326 [Linnaeus 1762:326 Gomphrena globosa].

Belutta-adeca-manjen HM 10:75:38
Belútta-adecá-manjén
BL Add MS 5032:202-203.
J.Commelin in HM 10:77 Amarantho affinis.
Plukenet 1691:Tab.CCXLI,fig.1 Anonymus Maderaspatan. Senecionis Hieracij folio similis, spicâ foliaceâ & Hormini adinstàr squamatâ (?).
C.Commelin 1696:4 = J.Commelin in HM.
Plukenet 1696:33 = Plukenet 1691.
Ray 1704 vol.3:128 Amaranto affinis spicata Indica, flosculis pentapetalis in spicis squamosis.
Plukenet 1705:12 Amaranthus spicatus albus, Persicariae foliis Maderaspatanus.
N.L.Burman 1768:64 Celosia argentea Linn.sp.96 [Linnaeus 1762:296].
Linnaeus 1771: Celosia margaritacea.
Houttuyn 1777 II vol.7:701-702 = N.L.Burman 1768.

Tsjeria-belutta-adeka-manjen HM 10:77:39
Tsjera-beluttae-adeca-manjen
BL Add MS 5032:204-205.
J.Commelin in HM 10:77 Amarantho affinis.
C.Commelin 1696:4 = J.Commelin in HM.
Ray 1704 vol.3:128 Amaranto affinis Indica, foliis angustioribus, spicis minoribus.
Plukenet 1705:12 Amaranthus spicatus angustis & longioribus foliis, capitulis dilutè purpurascentibus è Maderaspatan.
N.L.Burman 1768:64 Celosia argentea Linn.sp.96 [Linnaeus 1762:296].
Linnaeus 1771:344 = N.L.Burman 1768.
Houttuyn 1777 II vol.7:701-702 Celosia Argentea Syst.Nat.XII. Gen.286.p.187 [Linnaeus 1767 vol.2:187].

Vallia-manga-nari HM 10:79-80:40
Vallia-manga-nari
BL Add MS 5032:206-207.
J.Commelin in HM 10:80 Chrysanthemum Indicum, urticae folio, flore luteo, petalis bifidis.
C.Commelin 1696:20 = J.Commelin in HM.
Plukenet 1696:99 Chrysanthemum Maderaspatanum, latifolium, Scabiosae capitulis parvis, Phytogr.Tab.159.fig.4 (cs).
Plukenet 1705:56 Chrysanthemum Malabaricum Scrophulariae foliis octopetalon.
N.L.Burman 1768:184 Verbesina biflora Linn.sp.1272 [Linnaeus 1763:1272].
Linnaeus 1771:475 = N.L.Burman 1768.

Houttuyn 1779 II vol.10:818 Verbesina biflora [Linnaeus 1767 vol.2:568].

Cajenneam HM 10:81-82:41
Cajenneam
BL Add MS 5032:208-209.
J.Commelin in HM 10:83 Bellis major vel Chrysanthemum sp.
C.Commelin 1696:20 = J.Commelin in HM.
N.L.Burman 1768:184 Verbesina prostata Linn.sp.1272 [Linnaeus 1763:1272].

Pee-cajenneam HM 10:83:42
Pee-cajóni
BL Add MS 5032:210-211.
J.Commelin in HM 10:83 Bellis major vel Chrysanthemum sp.
C.Commelin 1696:20 = J.Commelin in HM.
Plukenet 1705:56 Chrysanthemum Malabaricum, Bellidis majoris folio, flore octopetalo.
N.L.Burman 1768:184 Verbesina calendulacea Linn.sp.1272 [Linnaeus 1763:1272-1273].
Linnaeus 1771:477 = N.L.Burman 1768.
Houttuyn 1779 II vol.10:819 = N.L.Burman 1768.

Adaca-manjen HM 10:85-86:43
Adacá-manjén
BL Add MS 5032:212-213.
J.Commelin in HM 10:86 Planta Indica alato caule, folio crenato, piloso & viscoso, flore glomerato, purpureo.
C.Commelin 1696:55 = J.Commelin in HM.
Plukenet 1696:335 Scabiosa major crispatis foliis alato caule Malabariensis.
Petiver 1699(1698):322,no.18 = J.Commelin in HM.
Ray 1704 vol.3:241 Corymbiferi affinis, caule alato, floribus in spicis seu capitulis squamosis, seminibus insidentibus.
J.Burman 1737:220-221,Tab.94,Fig.3 Sphaeranthos purpurea, alata, serrata.
Royen 1740:145 Sphaeranthus.
Linnaeus 1747:147,no.312 = Royen 1740.
Linnaeus 1753:927 Sphaeranthus indicus.
Linnaeus 1763:1314 = Linnaeus 1753.
N.L.Burman 1768:185 = Linnaeus 1763.
Houttuyn 1779 II vol.11:104-106 = N.L.Burman 1768.

Tsjetii-pu HM 10:87:44
Tsjetti-pú
BL Add MS 5032:214-215.
J.Commelin in HM 10:87 Matricaria Indica, latiore folio, flore pleno.
C.Commelin 1696:44 = J.Commelin in HM.
Plukenet 1696:243 Matricaria Japonica maxima flore roseo, s. suaverubente pleno elegantissimo Breyn.Prodr.2.66 [Breyne 1689:66-67] (?).
J.Burman 1737:153-154 Matricaria flore pleno, magno. Mus.Zeyl.pag.33 [Hermann 1717:33].
Linnaeus 1747:198,no.421 Matricaria Sinensis, flore monstroso. Vaill.act.1720.p.368.
Rumphius 1747 vol.5:259-261,Tab.XCI,Fig.1 Matricaria Sinensis (=).
Linnaeus 1763:1253 Chrysanthemum indicum.
N.L.Burman 1768:182 = Linnaeus 1763.

Katu-tsjetti-pu HM 10:89:45
Katú-tsjetti-pú
BL Add MS 5032:216-217.
J.Commelin in HM 10:89 Ambrosia Malabarica, Artemisiae folio odoratiβimo, floribus flavis.

C.Commelin 1696:4 = J.Commelin in HM.
Plukenet 1696:27 Ambrosia (forsàn) è China Anguriae foliis accedens. Phytogr.Tab.10.fig.6 (?).
Petiver 1702(1701):1020,no.221 Masia pattree Malab. Arthemisia Orientalis vulgaris facie (?).
Ray 1704 vol.3:109 = J.Commelin in HM.
Rumphius 1747 vol.5:261-262,Tab.XCI,Fig.2 Artemisia latifolia (?c).
N.L.Burman 1768:177 Artemisia vulgaris Linn.sp.1188 [Linnaeus 1763:1188-1189].

Codagen HM 10:91:46
Codagén
BL Add MS 5032:218-219.
J.Commelin in HM 10:91 Cotyledon sp.
C.Commelin 1696:21 = HM.
Plukenet 1696:314 Ranunculo adfinis Umbelliferis accedens Chelidonii minoris folio Zeylanica minor. Phytogr.Tab.106.fig.5.
Hermann 1698:238-239,tab.102 Valerianella Zeylanica palustris repens, Hederae terrestris folio, ad radicem florida.
Petiver 1704(1703):1450-1451,no.4 Asarina Malabarica fol.serrato.
Ray 1704 vol.3:635 Cotyledon aquatica acris Indica, pediculo ad marginem folii sito.
Boerhaave 1720 vol.1:71 Hydrocotyle; Zeylanica; asari folio. T.328.
J.Burman 1737:122 = Boerhaave 1720.
Linnaeus 1737:88 Hydrocotyle foliis reniformibus crenatis.
Linnaeus 1747:49-50,no.118 = Linnaeus 1737.
Rumphius 1747 vol.5:455-456,Tab.CLXIX,Fig.1 Pes Equinus (c).
Linnaeus 1753:234 Hydrocotyle asiatica.
Linnaeus 1762:339 = Linnaeus 1753.
N.L.Burman 1768:74 = Linnaeus 1762.
Houttuyn 1777 II vol.8:14 = N.L.Burman 1768.

Ana-coluppa HM 10:93:47
Ána-colúppa
BL Add MS 5032:220-221.
J.Commelin in HM 10:93 Ranunculo affinis planta Indica, floribus purpureis.
C.Commelin 1696:58 = J.Commelin in HM.
Plukenet 1700:11; ibid.1696:27 Amaranthoides humile Curassavicum foliis Cepaeae lucidis, capitulis albis. PBP.310 [Sherard 1689:310].
Petiver 1702(1700):712-713,no.78 Poordele Malab. (c).
Ray 1704 vol.3:316-317 Ranunculi facie Indica spicata, corymbiferis affinis, flosculis tetrapetalis.
Linnaeus 1747:188-189,no.399 Verbena spicis capitato-conicis solitariis, foliis serratis, caule repente.
N.L.Burman 1768:12,Tab.6,f.1 Verbena nodiflora Linn.sp.28 [Linnaeus 1762:28].

Bena-patsja HM 10:95-96:48
Béna-patsjá
BL Add MS 5032:222-223.
J.Commelin in HM 10:96 Heliotropium Indicum, latiore & rotundiore folio.
C.Commelin 1696:35 = J.Commelin in HM.
Plukenet 1696:182 Heliotropium Americanum caeruleum foliis Hormini, Dodart.Memoir. Phytogr.Tab.245.fig.4.
Ray 1704 vol.3:272 = J.Commelin in HM.
J.Burman 1737:120 Heliotropium Zeylanicum, majus, Hormini folio, flore albo, Aetsaethya Zeylonensibus. Hert.Catal.semin.
N.L.Burman 1768:40 Heliotropium indicum Linn.sp.187 [Linnaeus 1762:187].

Nelam-pata HM 10:97:49
Nelampala
BL Add MS 5032:224.
J.Commelin in HM 10:97 Jacobaea sp.
C.Commelin 1696:36 = J.Commelin in HM.
Plukenet 1696:194 Jacobaea Aethiopica procumbens, latioribus sinuatis foliis, floribus ad caulium nodos prodeuntibus (?).
Ray 1704 vol.3:180 Jacobaeae foliis Indica, flore nudo, an Corymbifera?
N.L.Burman 1768:177 Artemisia maderaspatana Linn.sp.1190 [Linnaeus 1763:1190].
Houttuyn 1779 II vol.10:589 = N.L.Burman 1768

Nanschera-canschabu HM 10:99:50
Nánschera-canschábú
BL Add MS 5032:225-226.
J.Commelin in HM 10:99 Veronica Indica Teucrii folio.
C.Commelin 1696:68 = J.Commelin in HM.
Plukenet 1696:384 = J.Commelin in HM.
Plukenet 1700:44 Chamaedrifolia pusilla planta, flosculo oblongo monopetalo, capsulâ membranaceâ, ex Coromandel (cs).
Ray 1704 vol.3:421 = J.Commelin in HM.
J.Burman 1737:228 Veronica hirsuta, latifolia, Zeylanica, aquatica. Mus.Zeyl.pag.29 [Hermann 1717:29].
Rumphius 1747 vol.5:460-462,Tab.CLXX,Fig.3 Crusta Ollae II (c).

Nir-cottam-pala HM 10:101:51
Nir-cottam-pála
BL Add MS 5032:227.
J.Commelin in HM 10:101 = Tithymalus Malabaricus androsaemi folio, flore rubro.
C.Commelin 1696:66 = J.Commelin in HM.
Plukenet 1696:373 Tithymalus botryoides Malabariensis latiori folio, floribus in Surculorum cymis.
Ray 1704 vol.3:433 = J.Commelin in HM.

Cansjan-cora HM 10:103:52
Kánsjan-córa
BL Add MS 5032:228-229.
J.Commelin in HM 10:103 Perfoliatae affinis Indica, Lychnidis aemula, flore viridi diluto.
C.Commelin 1696:52 = J.Commelin in HM.
Plukenet 1696:232 Lychnis segetum rubra foliis Perfoliatae. CBP.204 [Bauhin 1623:204] (?).
Plukenet 1700:43 Centaurium minus Hypericoides, flore luteo Lini capitulis. Narrecompree Malabar. (?cs).
Plukenet 1705: Lychnidis aemula alato caule, summis ramulis perfoliata, ex Oris Malabareis. Narrecomptee Malabarorum, Mantiss.43 (?).

Tsjeru-parua HM 10:105:53
Tsjérú-párúa
BL Add MS 5032:230-231.
J.Commelin in HM 10:105 Gratiolae affinis frutescens Americana foliis Agerati seu Veronicae erectae majoris Breyne in Prodromo secunda [Breyne 1689:54].
C.Commelin 1696:34 = J.Commelin in HM.
Plukenet 1696:14 Alcea fruticosa Malabariensis, angustioribus foliis rigidiusculis; floribus amoenè rubellis, semine papposo. Phytogr.Tab.254.fig.3 (?).
Plukenet 1696:237 Lysimachiae purpureae adfinis Americana procumbens, Anonidis Vernae frutescentis, folio singulari glabro, Phytogr.Tab.98.fig.4 (r).
Plukenet 1700:10 Althaea Coromandeliana angustis praelongis foliis, semine bicorni (cs).

Ray 1704 vol.3:527-528 = J.Commelin in HM.
Rumphius 1750 vol.6:44-47 Silagurium longifolium (c); J.Burman ibid.:47 (=).
N.L.Burman 1768:147 Sida acuta.

Katu-uren HM 10:107:54
Kátú-úren
BL Add MS 5032:232-233.
J.Commelin in HM 10:107 Althaea Americana, Betonicae folio, flore luteo.
C.Commelin 1696:3 = J.Commelin in HM.
Plukenet 1696:25 Althaea Maderaspatana subrotundo folio molli & hirsuto, multipilis, sivè seminibus ad apicem crinitis. Phytogr.Tab.131.fig.2 (?).
Plukenet 1700:10; ibid.1696:26 Althaea rostrata Coromandeliensis Pimpinellae majoris folio subrotundo. Phytogr.Tab.9.fig.3 (r).
Ray 1704 vol.3:517-518 Alcea Malabarica cordato folio, flore luteo.
J.Burman 1737:149 Malvinda bicornis, Ballotes folio. Dillen.H.Elth.pag.211.Tab.171.Fig.209.
Linnaeus 1737:347 Malva foliis cordatis villosis crenatis.
Linnaeus 1747:117,no.255 Malva foliis cordatis crenatis villosis, floribus lateralibus congestis.
N.L.Burman 1768:147 Sida cordifolia Linn.sp.961 [Linnaeus 1763:961].
Houttuyn 1779 II vol.10:39-40 = N.L.Burman 1768.

Niuren HM 10:109:55
Nirúren
BL Add MS 5032:234.
C.Commelin 1696:48 = HM.
Plukenet 1696:25 Althaea peregrina longiore Betonicae folio, floribus albis perexiguis capitulis arctè conglomeratis (...) Phytogr.Tab.44.fig.5 (?).
Ray 1704 vol.3:518 Alceae Indicae species, floribus minoribus, capsulis in 5 cellulas, singulas binos seminum ordines continentes divisis.
Plukenet 1705:6 Alcea Indica, Lamii Cannabini folio tetraphyllos, capitulis ad nodos caulium glomeratis.

Naga-pu HM 10:111:56
Nága-pú
BL Add MS 5032:235-236.
Ray 1704 vol.3:518 Alceae accedens Indicae, flore roseo amplo pentapetalo, fructu globoso quandrangulo, foliis longis, angustis crenatis.
C.Commelin 1696:46 = HM.
Plukenet 1696:18 Alceae Indicae cognata Planta, Cannabinis foliis particularibus, ex oris Cormandel.
Plukenet 1705:8 Alcea Indica.
Rumphius 1747 vol.5:288-289,Tab.C,Fig.1 Flos inpius (cs); J.Burman ibid.:289 (=).

Kilckola-tsjetti HM 10:113:57
Kilcola-tsjetti
BL Add MS 5032:237-238.
C.Commelin 1696:40 = HM.
Ray 1704 vol.3:358-359 Baccifera Indica, foliis adversis, floribus tetrapetalis umbellatim dispositis.
Plukenet 1705:146 Mercurialis cognata, Cynocrambe (fortè) genus Malabariense, Lauri densioribus, & acutioribus foliis.

Ben-pala HM 10:115:58
Ben-pála
BL Add MS 5032:239-240.

J.Commelin in HM 10:115 Tithymalus (c).
C.Commelin 1696:66 = J.Commelin in HM.
Plukenet 1696:369 Tithymalus Africanus platyphyllos, hirsuto caule, medio folii nervo in spinulam abeunte, Tab.320.fig.1 (?).
Ray 1704 vol.3:359 Cneorum Indicum fructu tricocco: seu Tithymalus fruticosus Indicus foliis adversis.

Wellia-codiveli HM 10:117:59
Wellia-codivéli
BL Add MS 5032:241-242.
J.Commelin in HM 10:117 Betae folio Malabarica, semine lappaceo.
C.Commelin 1696:15 = J.Commelin in HM.
Plukenet 1696:113 Lappula Malabarica spicata, Betae foliis Antagonistis, floribus interiùs lanuginosis.
Ray 1704 vol.3:110 = J.Commelin in HM.
Plukenet 1705:130 Lappula Maderaspatana monospermos spicata, Amaranthi folio molli, lappulis tomentosis. Vellanagumyoy Malabarorum.
J.Burman 1737:47-48,Tab.18,Fig.1 Blitum scandens fructu lappaceo.
Linnaeus 1747:43,no.103 Achyranthes floribus externe lanatis interrupte spicatis.
Linnaeus 1753:204 Achyranthes lappacea.
Linnaeus 1762:295-296 = Linnaeus 1753.
N.L.Burman 1768:63 = Linnaeus 1762.

Kalengi-cansjava HM 10:119:60
Kalengi-cansjáúa
BL Add MS 5032:243-244.
J.Commelin in HM 10:121 Cannabis similis exotica Pinac. Casp.Bauh. [Bauhin 1623:320] (s).
C.Commelin 1696:17-18 = J.Commelin in HM.
Petiver 1704(1703):1453,no.14 The Male Bangue. Bange Clus.Exot.238.c.25.&290.c.54.
Plukenet 1705:49 Cannabis sterilis, s. faemina nostras è regione Sinarum.
J.Burman 1737:135-136 Ketmia Indica, Cannabinis foliis, Bangue dicta.
Linnaeus 1737:457 Cannabis foliis digitatis. Mas.
Rumphius 1747 vol.5:208-211,Tab.LXXVII Cannabis Indica (=); J.Burman ibid.:211 (=).
N.L.Burman 1768:212 Cannabis sativa Linn.sp.1457 [Linnaeus 1763:1457].

Tsjeru-cansjava HM 10:121:61
Tsyerú-cansjáúa
BL Add MS 5032:245-246.
J.Commelin in HM 10:121 Cannibis similis exotica Pinac. Casp.Bauh. [Bauhin 1623:320] (s).
C.Commelin 1696:18 = J.Commelin in HM.
Plukenet 1696:80 Cannabis Indica trifoliata, s. Banguë Indorum.
Petiver 1704(1703):1453-1454,no.15 The Female Bangue. Bangue Malabar. trifoliata.
Linnaeus 1737:457 Cannabis foliis digitatis. Femina (?).
Rumphius 1747 vol.5:208-211,Tab.LXXVII Cannabis Indica (=); J.Burman ibid.:211 (=).
N.L.Burman 1768:212 Cannabis sativa Linn.sp.1457 [Linnaeus 1763:1457].

Nari-patsja HM 10:123:62
Narí-patsjá
BL Add MS 5032:247-248.
J.Commelin in HM 10:123 Conyza Malabarica, Lamii folio, flore purpureo.

Plukenet 1691:Tab.CLXXVII,fig.2 Eupatoria Conyzoid. Maderaspatana folijs glabris summo caule ramosior. (?).
C.Commelin 1696:22 = J.Commelin in HM.
Plukenet 1696:141 = Plukenet 1691.
Ray 1704 vol.3:155 = J.Commelin in HM.
Rumphius 1750 vol.6:34-35,Tab.XIV,Fig.1 Olus scrofinum (-); J.Burman ibid.:35 (c).
Rumphius 1750 vol.6:56-58 Conyza odorata (cs).
N.L.Burman 1768:179 Conyza odorata Linn.sp.1208β [Linnaeus 1763:1208β].

Pu-tumba HM 10:125:63
Pú-túmba
BL Add MS 5032:249.
C.Commelin 1696:57 = HM.
Ray 1704 vol.3:208 Centaurium Malabaricum majus, folio integro crenato minore, flore coeruleo diluto.
Plukenet 1705:134 Lychnis coronaria Ocymi folio singulari, floribus dilutè caeruleis, è Maderaspatan.
N.L.Burman 1768:183-184 Verbesina Lavenia Linn.sp.1271 [Linnaeus 1763:1271-1272].
Linnaeus 1771:475 = N.L.Burman 1768.
Houttuyn 1779 II vol.10:817-818 = N.L.Burman 1768.

Puam-curundala HM 10:127:64
Púam-cúrúndala
BL Add MS 5032:250-251.
J.Commelin in HM 10:127 Conyzae affinis Malabarica, Aquilegii folio, flore purpureo.
C.Commelin 1696:22 = J.Commelin in HM.
Plukenet 1696:72 Eupatorium Aquilegiae foliis, Malabaricum, flore purpureo.
Ray 1704 vol.3:155-156 = J.Commelin in HM.

Manam-podam HM 10:129:65
Manaṁ-podaṁ
BL Add MS 5032:252-253.
C.Commelin 1696:43 = HM.
Ray 1704 vol.3:660 Anomala Indica flosculis tetrapetalis in spicis squamosis deorsum reflexis, in summis caulibus & ramulis creberrimis.

Katu-mailosina HM 10:131:66
Kátú-mailósina
BL Add MS 5032:254.
C.Commelin 1696:39 = HM.
Ray 1704 vol.3:527 Herba Indica, foliis ad genicula pluribus, Floribus pentapetalis, in summis caulibus spicatim congestis, capsulis tripartitis polyspermis.
Plukenet 1705:176 Polygonum sericeum ramosius Indicum, foliolis Gallii instàr caulem cingentibus, Mantiss.154.

Pee-coipa HM 10:133:67
Peê-coipa
BL Add MS 5032:255.
C.Commelin 1696:51 = HM.
Ray 1704 vol.3:526 Herba Indica foliis binis adversis, floribus pentapetalis in capitula congestis, capsulis monospermis.
Plukenet 1705:176 Polygonum sericeum ramosius Indicum, foliolis Gallii instàr caulem cingentibus, Mantiss.154 (?).

Muel-schevy HM 10:135:68
Múél-scheví
BL Add MS 5032:256-257.
J.Commelin in HM 10:135 Planta Indica Erucae folio, caule ambiente, flore piloso.

Plukenet 1691:Tab.CCXXVII,fig.3 Sonchus laevis Maderaspatanus (?).
C.Commelin 1696:55 = J.Commelin in HM.
Plukenet 1696:354 = Plukenet 1691.
Ray 1704 vol.3:137 Sonchus Madraspatanus Lampsanae foliis.
Rumphius 1747 vol.5:297-298,Tab.CIII,Fig.1 Sonchus Amboinicus (c).
N.L.Burman 1768:175 Cacalia sonchifolia Linn.sp.1169 [Linnaeus 1763:1169].
Linnaeus 1771:463 = N.L.Burman 1768.
Houttuyn 1779 II vol.10:546 = N.L.Burman 1768.

Nela-vaga HM 10:137:69
Néla-vága
BL Add MS 5032:258.
J.Commelin in HM 10:137 Althaea sp.
C.Commelin 1696:4 = J.Commelin in HM.
Plukenet 1696:16 Alcea Malabarica, Epimedii folio, molli, & incano.
Ray 1704 vol.3:523 Alcea Indica foliis serratis, capsulis pentagonis, pentaspermis.

Inota-inodien ceu Moetoe HM 10:139:70
Inotá-inodień
BL Add MS 5032:259-260.
J.Commelin in HM 10:141 Solanum vesicarium Indicum minimum Pauli Hermanni, in Catal.Hort.Med.Lugd.Batav. [Hermann 1687:569-570,571 tab.] (c).
C.Commelin 1696:64 = J.Commelin in HM.
Plukenet 1696:352 Solanum Vesicarium Americanum caule quadrangulo, foliis Stramoniae H.Leyd.569 [Hermann 1687:569].
Petiver 1702(1701):855,no.127 Aumacarun calunga Malab. Alkakengi foliis mollibus fructu Asparagi (s).
Ray 1704 vol.3:356 Solanum vesicarium Indicum, vesiculâ acuminatâ, baccâ viridi, subflavâ.
Rumphius 1750 vol.6:61,Tab.XXIV,Fig.2 Halicacabus peregrinus (cs).
N.L.Burman 1768:55 Physalis pubescens Linn.sp.262 [Linnaeus 1762:262].

Pee-inota-inodien HM 10:141:71
Peê-inotá-inodień
BL Add MS 5032:261-262.
J.Commelin in HM 10:141 Solanum vesicarium Indicum minimum Pauli Hermanni, in Catal.Hort.Med.Lugd.Batav. [Hermann 1687:569-570,571 tab.] (c).
C.Commelin 1696:64 = J.Commelin in HM 10.
Plukenet 1696:352 = J.Commelin in HM (?).
Petiver 1702(1701):855,no.127 Aumacarun calunga Malab. Alkakengi foliis mollibus fructu Asparagi (s).
Ray 1704 vol.3:356 = HM.
Linnaeus 1737:62-63 Physalis annua ramosissima, pedunculis fructiferis petiolo longioribus.
Rumphius 1750 vol.6:61,Tab.XXIV,Fig.2 Halicacabus peregrinus (cs).
Linnaeus 1753:183-184 Physalis minima.
Linnaeus 1762:263 = Linnaeus 1753.
N.L.Burman 1768:54 = Linnaeus 1762.
Houttuyn 1777 II vol.7:663 = N.L.Burman 1768.

Caca-mulla HM 10:143-144:72
Cáca-múllú
BL Add MS 5032:263-264.
J.Commelin in HM 10:144 Solani folio Malabarica, flore monopetalo, flavo diluto colore, fructu quadrangulato.

C.Commelin 1696:64 = J.Commelin in HM.
Plukenet 1696:353 = J.Commelin in HM.
Ray 1704 vol.3:356-357 = J.Commelin in HM.
Linnaeus 1763:892 Pedalium Murex.
N.L.Burman 1768:139 = Linnaeus 1763.
Houttuyn 1778 II vol.9:592 = N.L.Burman 1768.

Nelen-tsjunda HM 10:145:73
Nélen-tsjúnda
BL Add MS 5032:265-266.
J.Commelin in HM 10:145 Solanum Officinarum acinis puniceis. Casp.Bauh. in Pinac.Varior. [Bauhin 1623:166].
C.Commelin 1696:63 Solanum officinarum Dale Pharm:270.
Plukenet 1696:349 Solanum lanuginosum hortensi, s. vulgari simile, baccis aureis, D.Morgano Raii Hist.Pl.672 [Ray 1686 vol.1:672-673].
Linnaeus 1737:60 Solanum caule inermi annuo, foliis ovatis angulatis.
Rumphius 1750 vol.6:62,Tab.XXVI,Fig.2 Halicacabus baccifer (c).
N.L.Burman 1768:55 Solanum nigrum Linn.sp.266 [Linnaeus 1762:266].

Nila-barudena HM 10:147:74
Nila-barúdena
BL Add MS 5032:267-268.
J.Commelin in HM 10:147 Solanum pomiferum, fructu oblongo, purpureo [Bauhin 1623:167].
Plukenet 1691:Tab.CCXXVI,fig.2 Solanum pomiferum, fructu incurvo. CBP [Bauhin 1623:167].
C.Commelin 1696:63 = J.Commelin in HM.
Plukenet 1696:350 = Plukenet 1691.
J.Burman 1737:157 Melongena fructu oblongo. Tournef.inst. pag.151.
Linnaeus 1747:38,no.93 Solanum calycibus aculeatis, foliis ovatis integerrimis tomentosis. Hort.cliff.61.n.14 [Linnaeus 1737:61-62].
Rumphius 1747 vol.5:238-240,Tab.LXXXV Trongum hortense (?=); J.Burman ibid.:240 (=).
N.L.Burman 1768:56 Solanum Melongena Linn.sp.266 [Linnaeus 1762:266].

Andi-malleri HM 10:149-150:75
Andí-malleri
BL Add MS 5032:269-270.
J.Commelin in HM 10:150 Solanum Mexicanum, flore magno Casp.Bauh. in Pinac. [Bauhin 1623:168].
C.Commelin 1696:44 Mirabilis peruviana flore flavo Herm.Cat. 428 [Hermann 1687:428].
Plukenet 1696:252 Mirabilis Peruviana, Park.Th.
Linnaeus 1737:53-54 Mirabilis.
Rumphius 1747 vol.5:253-255,Tab.LXXXIX Mirabilis (r).

Naga-dante HM 10:151:76
Nága-danti
BL Add MS 5032:271-272.
J.Commelin in HM 10:151 Ricinus Indicus, semine spadiceo rubro minore.
C.Commelin 1696:58 = J.Commelin in HM.
Plukenet 1696:320 Ricinus minor Indicus folio magis cuspidato, pilis rigidioribus utrinque aspero, semine spadiceo rubro.

Cottam HM 10:153:77
Cottám
BL Add MS 5032:273-274.

J.Commelin in HM 10:153 Nepetae seu Menthae Cattariae affinis Indica candido flore.
C.Commelin 1696:47 = J.Commelin in HM.
Plukenet 1696:268 Ocimum Melissophyllum Gangeticum odoratissimum, seminibus nigris Amaranthi nitentibus (?).
Boerhaave 1720 vol.1:170 Ocymum; Zeylanicum; perenne; frutescens; folio calaminthae nonnihil simil. Par.Bat.pr. [Sherard 1689:358].
J.Burman 1737:174-175,Tab.80,Fig.1 Ocymum Zeylanicum perenne, odoratissimum, latifolium.
Linnaeus 1737:315 Ocimum foliis lanceolato-ovatis, radice perenni.
Linnaeus 1747:102,no.228 = Linnaeus 1737.
Rumphius 1747 vol.5:291-292,Tab.CI Majana utraque (=).
Linnaeus 1753:597 Ocimum frutescens.
Linnaeus 1763:832 = Linnaeus 1753.
N.L.Burman 1768:129 = Linnaeus 1763.
Houttuyn 1775 II vol.5:290 Mentha Perilloides Syst.Veg.XIII.

Cadelari HM 10:155:78
Cadelári
BL Add MS 5032:275-276.
J.Commelin in HM 10:155 Verbena Indica Bontii.
C.Commelin 1696:68 = J.Commelin in HM.
Petiver 1699(1698):313-315,no.2 Naiureevee Malab. Amaranthus spicatus Dictamni Cretici folio Maderaspatensis Plukenet Phytographia, Tab.10.Fig.4 [Plukenet 1691:Tab.X,f.4] (?).
Plukenet 1700:11 Amaranthus spicatus Coromandeliens. (...) Najureevee Malabarorum.
Ray 1704 vol.3:287-288 = J.Commelin in HM.
J.Burman 1737:16-17,Tab.5,Fig.3 Amaranthus spicatus, Zeylanicus, foliis obtusis, Amaranto Siculo Boccone similis.
Rumphius 1750 vol.6:26-29,Tab.XII,Fig.1 Auris canina mas (c).
N.L.Burman 1768:63 Achyranthes indica aspera Linn.sp.295 [Linnaeus 1762:295 Achyranthes aspera β indica].
Linnaeus 1771:344 Achyranthes aspera.
Houttuyn 1777 II vol.7:696-698 = N.L.Burman 1768.

Scheru-cadelari HM 10:157:79
Scherú-cadelári
BL Add MS 5032:277-278.
J.Commelin in HM 10:157 Veronicae similis spicata Indica repens.
C.Commelin 1696:68 = J.Commelin in HM.
Rumphius 1750 vol.6:26-29,Tab.XI Auris canina femina (c).
N.L.Burman 1768:64 Achyranthes prostata Linn.sp.296 [Linnaeus 1762:296].

Belutta-modela-muccu HM 10:159:80
Belútta-módela-múccú
BL Add MS 5032:279-280.
J.Commelin in HM 10:159 Lysimachia Indica Salicis oblongo folio, flore albo spicato.
Plukenet 1691:Tab.CCX,fig.7 Persicaria Maderaspatana longiore folio hirsuto (?).
C.Commelin 1696:41 = J.Commelin in HM.
Plukenet 1696:288 = Plukenet 1691.
Ray 1704 vol.3:119 Lysimachia rectiùs Persicaria monospermos spicata Indica, foliis oblongo-angustis hirsutis.

Cupameni HM 10:161:81
Cúpaméni
BL Add MS 5032:281-282.
J.Commelin in HM 10:161 Urticae folio, flore albicante spicato, cui passim thecula bicornea insidet, foliis in summitate caulem ambientibus.

C.Commelin 1696:44 Mercurialis zeylanica tricoccos cum acetabulis, Kupamenia Zeylanensibus Herm.Cat.App.686 [Hermann 1687:686,687 tab.,688].
Plukenet 1696:248 Mercurialis tricoccos Hermophroditica, s. ad foliorum juncturas, ex foliolis cristatis julifera simùl & fructum ferens, D.Banister. Phytogr.Tab.99.fig.4 (?cs).
Petiver 1704(1703):1454,no.17 Mercurialis Madraspat. acetabulis & folijs majoribus, serratis (?).
Ray 1704 vol.3:107 = J.Commelin in HM.
J.Burman 1737:203-204 Ricinokarpos Indica, glabra, fructus in acetabulis gerens.
Linnaeus 1747:161-162,no.341 Acalypha involucris faemineis cordatis subcrenatis, foliis ovatis petiolo brevioribus.
Linnaeus 1753:1003 Acalypha indica.
Linnaeus 1763:1424 = Linnaeus 1753.
N.L.Burman 1768:202 = Linnaeus 1763.
Houttuyn 1779 II vol.11:285-286 = N.L.Burman 1768.

Pee-cupameni HM 10:163:82
Pee-cúpaméni
BL Add MS 5032:283-284.
J.Commelin in HM 10:163 Urticae folio Indica, flore tetrapetalo ramoso, capsula triquetri.
Plukenet 1691:Tab.CCV,fig.4 Mercurialis Maderaspatensis tricoccos, acetabulis destituta (?).
C.Commelin 1696:69 = J.Commelin in HM.
Plukenet 1696:248 = Plukenet 1691.
Ray 1704 vol.3:107 = J.Commelin in HM.
J.Burman 1737:205 Ricinokarpos Indica, glabra, Mercurialis folio.
Linnaeus 1747:158,no.334 Tragia foliis ovatis.
Rumphius 1750 vol.6:49,Tab.XX,Fig.2 Urtica mortua (-); J.Burman ibid.:48 (c).
Linnaeus 1753:980-981 Tragia Mercurialis.
Linnaeus 1763:1391 = Linnaeus 1753.
N.L.Burman 1768:195 = Linnaeus 1763.
Houttuyn 1776 II vol.6:222-223 Tragia Mercurialis [Linnaeus 1767 vol.2:619].

Welia-cupameni HM 10:165:83
Wélia-cúpaméni
BL Add MS 5032:285-286.
C.Commelin 1696:70 = HM.
Plukenet 1696:25 Althaea (fortè) vulgari similis, Planta Jamaicensis flosculis exiguis, ramulis tenuibus fasciculatim appensis. Phytogr.Tab.259 (?).
Plukenet 1700:129 Mercurialis Coromandeliensis, tricoccos, major, Urticae facie, cum acetabulis (?).
Plukenet 1700:163; ibid.1696:321 Ricinus (fortè) Althaeae folio Jamaicensis, floribus exiguis Phytogr.Tab.259.fig.5 (?).
Petiver 1704(1703):1454,no.19 Mercurialis Madraspat. fol. auctiore, caule piloso (?).
Ray 1704 vol.3:107 Urticae-folia Indica spicata, spicis longis rectis, flosculis & seminibus crebrioribus.
J.Burman 1737:204 Ricinokarpos Indica, hirsuta, foliis Urticae vulgaris, rotundioribus & minoribus.
Linnaeus 1747:161-162,no.341β Acalypha involucris faemineis cordatis subcrenatis, foliis ovatis petiolo brevioribus.
Linnaeus 1753:1003 Acalypha indica β.
Linnaeus 1763:1424β = Linnaeus 1753.
N.L.Burman 1768:202β = Linnaeus 1763.
Houttuyn 1779 II vol.11:285-286 = N.L.Burman 1768.

Perim-tolassi HM 10:167:84
Perím-tolassí
BL Add MS 5032:287-288.

J.Commelin in HM 10:167 Scrophularia Indica, radice fibrosa, floribus caeruleo albicantibus.
C.Commelin 1696:60 = J.Commelin in HM.
Ray 1704 vol.3:312 Verticillata Indica, floribus spicatis coeruleis, galeâ quadripartitâ. An Ocimi species?

Nala-tirtava HM 10:169:85
Nála-tirtáva
BL Add MS 5032:289-290.
J.Commelin in HM 10:173 Scrophularia Indica radice fibrosa (cs).
C.Commelin 1696:46 = HM.
Plukenet 1696:158 Ocimum odoratum vulgari simile, laxiore spicâ Malabaricum.
Ray 1704 vol.3:312-313 Sideritidis folio Verticillata Indica, galeâ floris quadripartitâ. Ocimum Indicum spurium Sideritidis folio.
Rumphius 1747 vol.5:263-264,Tab.XCII Basilicum Indicum (c).

Cattu-tirtava HM 10:171:86
Káttú-tirtáva
BL Add MS 5032:291-292.
J.Commelin in HM 10:173 Scrophularia Indica radice fibrosa (cs).
C.Commelin 1696:19 = HM.
Plukenet 1696:268 Ocimum maximum perenne utriusque Indiae foliis Atriplicis, ingrati odoris graveolens. Phytogr.Tab.308 (?).
Ray 1704 vol.3:313 Verticillata Indica, floribus spicatis, praecedentis similibus, albicantibus, foliis serratis ad Lamium accedentibus. An Ocimi species?
Rumphius 1747 vol.5:292-293,Tab.CII,Fig.1 Melissa lotoria (c).
N.L.Burman 1768:129 Ocimum gratissimum Linn.sp.832 [Linnaeus 1763:832].

Soladi-tirtava HM 10:173:87
Soladí-tirtáva
BL Add MS 5032:293-294.
J.Commelin in HM 10:173 Scrophularia Indica radice fibrosa (cs).
C.Commelin 1696:63 = HM.
Plukenet 1696:268 Ocimum odoratissimum calycibus ad verticillas deorsùm tendentibus, absque petiolis è Maderaspatan. Phytogr.Tab.208.fig.5.
Petiver 1699(1698):327,no.29a Carentulee Mal. Mentha Madraspatana cauliculis rubentibus hirsutis (?).
Ray 1704 vol.3:289 = Petiver 1699(1698) (cs).
Ray 1704 vol.3:313 Verticillata Indica foliis Sideritidis latifoliae, floribus praecedentium similibus, Ocimi Indici species.
Rumphius 1747 vol.5:265,Tab.XCII,Fig.2 Basilicum agreste (c).

Tsjadaen HM 10:175:88
Tsjadaeń
BL Add MS 5032:295-296.
J.Commelin in HM 10:175 Cardiaca Asiatica Nepetae folio, floribus brevibus, purpureis, pallidis.
C.Commelin 1696:18 = J.Commelin in HM.
Plukenet 1696:81 = J.Commelin in HM.
Petiver 1702(1700):707,no.65 Oatepemarutte Malab. Prassium Madraspatan. folio latissimo. Mus.Petiv.671 (?).
Rumphius 1747 vol.5:294-295,Tab.CII,Fig.2 Marrubium album Amboinicum (c).

Tsjeria-manga-nari HM 10:177:89
Kátú-tsjería-mánga-nári
BL Add MS 5032:297.

C.Commelin 1696:67 = HM.
Plukenet 1700:44 Chamaedrifolia pusilla planta, flosculo oblongo monopetalo, capsulâ membranaceâ, ex Coromandel (r).
Ray 1704 vol.3:389-390 Monopetala Indica, foliis in caule alternis profundè crenatis, flore è tubo oblongo in quatuor folia difforma expanso.

Katu-kurka HM 10:179:90
Kátú-kúrka
BL Add MS 5032:298.
J.Commelin in HM 10:179 Nepeta Indica rotundiore folio.
C.Commelin 1696:47 = J.Commelin in HM.
Plukenet 1700:128 Melissa Canarina multifido folio spicata, odorem Camphorae spirans, penetrantissimum (?cs).
Plukenet 1700:192; ibid.1696:401 Melissa fortè Canarina triphyllos odorem Camphorae spirans penetrantissimum Phytogr. Tab.325.fig.5 (?cs).
Plukenet 1705:15 Anomala Indiana, Nepetae foliis, staechadis Arabicae capitulis, ex sinu foliorum prodeuntibus (?).
Linnaeus 1753:571-572 Nepeta indica.
Linnaeus 1763:799 = Linnaeus 1753.
N.L.Burman 1768:126 = Linnaeus 1763.
Houttuyn 1778 II vol.9:310 = N.L.Burman 1768 (?).

Tumba HM 10:181:91
Túmba
BL Add MS 5032:299-300.
J.Commelin in HM 10:181 Nepeta Indica, sideriti folio, floribus spicatis.
C.Commelin 1696:47 = J.Commelin in HM.
Plukenet 1696:81 Cardiaca aquatica, praelongis, & perangustis foliis, Maderaspatana.
Petiver 1699(1698):323,no.21 Pea-tumba Medde Malab. = J. Commelin in HM.
Rumphius 1750 vol.6:39-40,Tab.XVI,Fig.1 Herba admirationis (s); J.Burman ibid.:40 (c).
N.L.Burman 1768:127 Leonurus indicus Linn.sp.817 [Linnaeus 1763:817].

Kattu-tumba HM 10:183:92
Kátú-túmba
BL Add MS 5032:301.
J.Commelin in HM 10:183 Nepeta Indica silvestris, flore purpureo, spicato.
C.Commelin 1696:47 = J.Commelin in HM.
Plukenet 1700:139; ibid.1696:268 Ocimum Maderaspatanum frutescens gratissimi odoris flore parvo, cauliculis villosis. Phytogr.Tab.208.fig.4 (?).
Ray 1704 vol.3:296 = J.Commelin in HM.

Carim-tumba HM 10:185:93
Carím-túmba
BL Add MS 5032:302-303.
J.Commelin in HM 10:185 Nepeta Malabarica, folio latiore, flore caeruleo ex albido.
C.Commelin 1696:47 = J.Commelin in HM.
Ray 1704 vol.3:403 = J.Commelin in HM.
Plukenet 1705:144 Mentastrum coronatum, oblongis foliis Malabaricum.
Rumphius 1750 vol.6:41,Tab.XVI,Fig.2 Majana foetida (c).
Rumphius 1750 vol.6:151,Tab.LXVIII,Fig.1 Menthastrum Amboinicum (-); J.Burman ibid.:151 (c).
Linnaeus 1771:566 Nepeta malabarica.

Tsjeru-tardavel HM 10:187:94

Tsjérú-tardável
BL Add MS 5032:304.
Ray 1704 vol.3:403 Urticae-folia Indica floribus spicatis dipetalis, ex albo & rosaceo colore variis.
C.Commelin 1696:67 = HM.
Plukenet 1696:111 Clinopodium fistulosum, pumilum, Ind. Occident summo caule floridum, Phytogr.Tab.164.fig.4 (?).
Ray 1704 vol.3:312 Verticillata Indica, floribus spicatis variis ex albo & rosaceo, foliis per margines aequalibus.
Rumphius 1750 vol.6:53-54,Tab.XXIII,Fig.1 Moretiana (s).
N.L.Burman 1768: Justicia procumbens Linn.sp.22 [Linnaeus 1762:22].

VOLUME 11

Kapa-tsjakka HM 11:1-6:1-2
Kapá-tsjakka
J.Commelin in HM 11:6 Carduus Brasilianus foliis Aloës Casp.Bauh. in Pinac. [Bauhin 1623:384].
J.Commelin 1689:23 Ananas, Pisoni, Acostae. I.Bauh.
C.Commelin 1696:5 Ananas Acost.284.
Linnaeus 1737:129-130 Bromelia foliis spinosis, fructibus coalitis caulem cingentibus.
Linnaeus 1753:285 Bromelia Ananas.
Linnaeus 1762:408 = Linnaeus 1753.
N.L.Burman 1768:79 = Linnaeus 1762.
Houttuyn 1777 II vol.8:313-316 Bromelia Ananas Syst.Nat.XII.Gen.391.p.232 [Linnaeus 1767 vol.2:232].

Kadanaku vel Catevala HM 11:7:3
Kadanákú aut catevála
J.Commelin in HM 11:7 = J.Commelin 1689.
J.Commelin 1689:14 Aloe vulgaris B.Pin. [Bauhin 1623:286].
C.Commelin 1696:3 = J.Commelin 1689.
Linnaeus 1737:130-131 Aloe foliis spinosis confertis dentatis vaginantibus planis maculatis.
Rumphius 1747 vol.5:271-272 Sempervivum majus Indicum (-); J.Burman ibid.:272 (=).
Linnaeus 1753:319-320 Aloe perfoliata var. vera.
Linnaeus 1762:458-459 = Linnaeus 1753.
N.L.Burman 1768:83 Aloe vera.
Houttuyn 1777 II vol.8:357-361 = N.L.Burman 1768.

Elettari HM 11:9-11:4-6
Eléttari
J.Commelin in HM 11:11 Cardamomum.
C.Commelin 1696:18 Cardamomum minus Bont:126.
Plukenet 1696:81 Cardamomum Indianum, fructu rotundo minore.
Boerhaave 1720 vol.2:128 Cardamomum; minus; Bontii. Raj.H.1204 [Ray 1688 vol.2:1204].
J.Burman 1737:54 Cardamomum Zeylanicum, sylvestre, aquaticum, acre, sapore Calami aromatici. Mus.Zeyl.pag.96 [Hermann 1717:46-47] (c).
Linnaeus 1747:2,no.4 Amomum scapo bracteis alternis laxis caule breviore.
Linnaeus 1753:1 Amomum Cardamom.
Linnaeus 1762:2 = Linnaeus 1753.
N.L.Burman 1768:2 = Linnaeus 1762.
Houttuyn 1777 II vol.7:14-16 Amomum Cardamomum [Linnaeus 1767 vol.2:50].
Houttuyn 1777 II vol.7:17 Amomum Grana Paradisi [Linnaeus 1767 vol.2:50].

Kua HM 11:13-14:7
Kúa
J.Commelin in HM 11:14 Zerumbeth Garzii ab Horto Pauli Hermanni [Hermann 1687:636,tab.637,638-640] (c).
J.Commelin 1689:371 Zedoaria longa B.Pin. [Bauhin 1623:35-36].
C.Commelin 1696:71 Zerumbet officinarum Dale Pharm.366 & Variorum.
Plukenet 1696:397 = J.Commelin 1689.
Ray 1704 vol.3:645 = J.Commelin 1689.
J.Burman 1737:234 Zingiber latifolium, sylvestre. H.L.Bat.pag. 636 [= Hermann 1687:636,tab.637,638-640].
Linnaeus 1737:3 Amomum scapo nudo, spica oblonga obtusa.
Linnaeus 1747:1,no.2 = Linnaeus 1737.

Tsjana-kua HM 11:15-16:8
Tsjána-kúa
J.Commelin in HM 11:16 Costus Arabicus Dioscord. Matthiolus in Diosc.lib.1.cap.15.
J.Commelin 1689:71 Zingiber latifolium sylvestre Zezumbet Garz. ab Hort.
C.Commelin 1696:71 Zingiber latifolium sylvestre Walinghuru Zeylonensibus Herm.Cat.636 [Hermann 1687:636,637 tab., 638-640].
Plukenet 1696:397 Zingiberis effigie Costus Arabicus, & Syriacus, Adversarior.
Plukenet 1700:12; ibid.1696:28 Amomum genuinum Ponae Park.Th.1566 (s).
Ray 1704 vol.3:645 = J.Commelin in HM.
Plukenet 1705:206 = J.Commelin in HM.
J.Burman 1737:78-79 Costus Indicus, Violae Martiae odore. Mus.Zeyl.pag.58 [Hermann 1717:58].
Linnaeus 1737:2 Costus.
Linnaeus 1747:2-3,no.5 = Linnaeus 1737.
Linnaeus 1762:2 Costus arabicus.
N.L.Burman 1768:2 = Linnaeus 1762.
Houttuyn 1777 II vol.7:19-21 Costus Arabicus Syst.Nat.XII. Gen.3.p.50 [Linnaeus 1767 vol.2:50] (?).

Malan-kua HM 11:17-18:9
Malankúa
J.Commelin in HM 11:18 Colchicum Ceylanicum flore Violae, odore & colore ephemero Hermannus in prodrom. Parad. Batav. [Sherard 1689:324].
J.Commelin 1689:371 Zedoaria rotunda B.Pin. [Bauhin 1623: 36].
C.Commelin 1696:71 = J.Commelin 1689.
Plukenet 1696:397 = J.Commelin 1689.
Ray 1704 vol.3:648 = J.Commelin in HM.
J.Burman 1737:67-68 = J.Commelin in HM.
Linnaeus 1762:3 Kaempferia rotunda.
N.L.Burman 1768:3 = Linnaeus 1762.
Houttuyn 1777 II vol.7:39 Kampferia rotunda [Linnaeus 1767 vol.2:51].

Manja-kua HM 11:19:10
Mánja-kúa
J.Commelin in HM 11:19 Curcuma radice rotunda [Sherard 1689:330].
C.Commelin 1696:25 = J.Commelin in HM.
Plukenet 1696:397 Zingiberis facie, Radix intùs flava rotunda.
Ray 1704 vol.3:649 = J.Commelin in HM.
J.Burman 1737:84 = J.Commelin in HM.
Linnaeus 1747:3,no.6 Curcuma foliis ovatis utrinque acuminatis: nervis lateralibus paucissimis. Roy.lugdb.12 [Royen 1740: 12].
Linnaeus 1753:2 Curcuma rotunda.

Linnaeus 1762:3 = Linnaeus 1753.
N.L.Burman 1768:2 = Linnaeus 1762.
Houttuyn 1777 II vol.7:29-33 Curcuma rotunda Syst.Nat.XII. Gen.6.p.50 [Linnaeus 1767 vol.2:50].

Manjella kua HM 11:21:11
Manjélla-kóua
J.Commelin in HM 11:21 Cucuma vera.
C.Commelin 1696:25 Curcuma radice longa Herm:Cat:208 [Hermann 1687:208,209 tab.,210-211].
Plukenet 1696:397 Zingiberis facie, Radix longa intùs flava, Malaice Cunhet Linschot Part.3.Ind.Or.cap.14.
Ray 1704 vol.3:649 = J.Commelin in HM.
J.Burman 1737:83 = C.Commelin 1696.
Linnaeus 1747:3,no.7 Curcuma foliis lanceolatis utrinque acuminatis: nervis lateralibus numerosissimis. Roy.lugdb.12 [Royen 1740:12].
Linnaeus 1762:3 Curcuma longa.
N.L.Burman 1768:3 = Linnaeus 1762.
Houttuyn 1777 II vol.7:33-36 Curcuma longa [Linnaeus 1767 vol.2:51].

Inschi vel Inschi kua HM 11:21-23:12
Ínschi
J.Commelin in HM 11:23 Inschi vel Zinziber.
J.Commelin 1689:371 Zinzibera B.Pin. [Bauhin 1623:35].
C.Commelin 1696:71 = Ray 1688.
Plukenet 1696:397 Zingiber angustiori folio, faemina, Utriusque Indiae alumna.
Ray 1704 vol.3:645-646; ibid.1688 vol.2:1314-1315 Zinziber.
J.Burman 1737:235 = Plukenet 1696.
Linnaeus 1737:3 Amomum scapo nudo, spica ovata.
Linnaeus 1747:1-2,no.3 = Linnaeus 1737.
Rumphius 1747 vol.5:156-160,Tab.LXVI,Fig.1 Zingiber majus (-); J.Burman ibid.:160 (=).
Linnaeus 1762:1 Amomum Zingiber.
N.L.Burman 1768:1 = Linnaeus 1762.
Houttuyn 1777 II vol.7:10-13 Amomum Zingiber Syst.Nat.XII. Gen.2.p.50 [Linnaeus 1767 vol.2:50].

Katou inschi kua HM 11:27:13
Káttú-ínschi
J.Commelin in HM 11:27 Zinziber sylvestris.
C.Commelin 1696:71 = J.Commelin in HM.
Plukenet 1696:397 Zingiber CBP.35 [Bauhin 1623:35].
Ray 1704 vol.3:646 Zingiber sylvestre mas Pison.Mantiss.Arom.
J.Burman 1737:234 Zingiber sylvestre, flavum, Kaluwaala Zeylonensibus. Mus.Zeyl.pag.51 [Hermann 1717:51].
Linnaeus 1762:1-2 Amomum Zerumbet.
N.L.Burman 1768:1 = Linnaeus 1762.
Houttuyn 1777 II vol.7:13-14 Amomum Zerumbeth [Linnaeus 1767 vol.2:50].

Mala inschi kua HM 11:29:14
Mála-íntschi-kúa
J.Commelin in HM 11:29 Arum Polyphyllum, Dracunculus & Serpentaria dictum Ceylanicum, caule glabro, viridi-diluto, maculis albis notato, majus & elatius [J.Commelin 1689:38].
C.Commelin 1696:71 Zingiber Montanus.
Plukenet 1700:37; ibid.1696:81 Cardamomum Indianum majus, fructu longiore (r).
Ray 1704 vol.3:646 = HM.
Linnaeus 1737:488 Cannoides (?).

Parua-kelanga HM 11:31:15

Párúa-kelángú
J.Commelin in HM 11:31 Plantagini affinis Indica, floribus sub purpureis, radice glandulosa.
C.Commelin 1696:55 = J.Commelin in HM.
Plukenet 1696:51 Arum Ambonense pumilum. Hort.Beaum.9 [Kiggelaer 1690:9] (?).
Ray 1704 vol.3:435 = J.Commelin in HM.
Linnaeus 1771:227 Saururus ? natans.
Houttuyn 1777 II vol.8:429 = Linnaeus 1771.

Nir-tsjenbu HM 11:33:16
Nir-tsjémbú
J.Commelin in HM 11:33 Arisarum Indicum Aquaticum cum pistillo rubro, pediculo globulo adhaerente.
C.Commelin 1696:9 = J.Commelin in HM.
Plukenet 1700:27 Arum Americanum foliorum auriculis magnis subrotundis ex Insulis Caribeis (?).
Ray 1704 vol.3:581 = J.Commelin in HM.

Nelenschena minor HM 11:33-34:17
Nelenschéna minor
J.Commelin in HM 11:34 Arisarum Indicum petraeum humile, radice fibrosa, pistillo rubro adhaerente granulis purpurascentibus.
J.Commelin 1689:36 Arum humile Arisarum dictum latifolium, Ceylonicum pistillo longissimo miniato colore.
C.Commelin 1696:9 = J.Commelin 1689.
Hermann 1698:78-80,tab.14 Arum trilobato folio humilius & minus Zeylanicum. Panualla Zeylan.
Ray 1704 vol.3:581 = J.Commelin in HM.
J.Burman 1737:34 Arum humile, Ceylanicum, latifolium, pistillo coccineo. H.Amst.part.1.Fig.51.pag.97 [J.Commelin 1697: 97-98,Fig.51].

Schena HM 11:35-36:18
Schéna
J.Commelin in HM 11:36 Arum Polyphyllum, Dracunculus & Serpentaria dictum, Ceylanicum, caule glabro, viridi diluto, maculis albicantibus notato, majus & elatius [J.Commelin 1689:38].
J.Commelin 1689:38 Arum polyphyllum, Dracunculus, & Serpentaria dictum Ceylonicum. Caule glabro viridi-diluto, maculis albicantibus notatus, majus & elatius.
C.Commelin 1696:9 = J.Commelin 1689.
Plukenet 1696:52 = J.Commelin 1689.
Hermann 1698:90-91 Dracontium Zeylanicum ramoso folio caule ex viridi & albo variegato laevi.
Ray 1704 vol.3:583 Arum humile, Arisarum dictum Ceylanicum, pistillo longissimo, miniato colore Commel. Cat.Amst. [J.Commelin 1689:36].
J.Burman 1737:90-91 Dracunculus Zeylanicus, polyphyllus, caule laevi, ex viridi & albo variegato. Tournef.instit.pag.161.

Mulenschena HM 11:37:19
Múlenschéna
J.Commelin in HM 11:37 Arum Polyphyllum, Dracunculus & Serpentaria dictum, Ceylanicum, caule aspero, maculis viridisfuscis, & viridi-dilutis albicantibus pulchre notato [J.Commelin 1689:38].
J.Commelin 1689:38 Arum polyphyllum, Dracunculus, & Serpentaria, dictum Ceylonicum. Caule aspero cum maculis viridi-fuscis, & viridi-dilutis albicantibus pulchre notatum, majus & elatius.
C.Commelin 1696:9 = J.Commelin 1689.
Plukenet 1696:52 Arum polyphyllum, Dracunculus & Serpentar-

ia dictum, caule aspero ex Insula Zeylon. BPB [Sherard 1689:315].
J.Commelin 1689:99-100,Fig.52 Arum polyphyllum Ceylanicum, caule scabro, viridi diluto, maculis albicantibus, notato (?).
Hermann 1698:89 Dracontium Zeylanicum ramoso folio caule ex viridi & flavo variegato aspero.
Ray 1704 vol.3:583 = J.Commelin in HM.
J.Burman 1737:90 Dracunculus Zeylanicus, polyphyllus, caule aspero, maculis viridifuscis, viridi-dilutis, & albicantibus pulchre notato. Tournef.instit.pag.160.
Rumphius 1747 vol.5:324-326,Tab.CXII Tacca sativa (-); J.Burman ibid.:326 (=).

Nelenschena major HM 11:39:20
Nelenschéna major
J.Commelin in HM 11:39 Arum minus Ceylanicum Sagittariae folio, Herman.Prodrom.Paradis.Batav. [Sherard 1689:315] (?).
C.Commelin 1696:9 = J.Commelin in HM (=).
Ray 1704 vol.3:583-584 = J.Commelin in HM (=).
J.Burman 1737:34 = J.Commelin in HM (=).
Linnaeus 1737:435 Arum acaule, foliis sagittatis triangulis acutis: angulis extrorsum flexis acutis.
Linnaeus 1747:154-155,no.325 Arum acaule, foliis subhastatis.
Rumphius 1747 vol.5:310-312,Tab.CVII Arum latifolium (-); J.Burman ibid.:312 (=).
Linnaeus 1753:966 Arum divaricatum.
Linnaeus 1763:1369 = Linnaeus 1753.
N.L.Burman 1768:193 = Linnaeus 1763.
Houttuyn 1779 II vol.11:188 Arum Divaricatum [Linnaeus 1767 vol.2:603].

Katu-schena HM 11:41:21
Kátú-schéna
J.Commelin in HM 11:41 Arum Malabaricum minus polyphyllum, radice globosa.
C.Commelin 1696:9 = J.Commelin in HM.
Plukenet 1696:52 Arum polyphyllum, Dracunculus & Serpentaria dictum, caule non maculato, minus, & humilius. H.Leyd. [Hermann 1687:60].
Ray 1704 vol.3:585 Ari foliis Malabarica minor polyphyllos, radice rotunda, flore fructúque carens.
Rumphius 1747 vol.5:329-331,Tab.CXV Tacca montana (-); J. Burman ibid.:331 (c).

Weli ila HM 11:43-44:22
Wéli-ilá
J.Commelin in HM 11:44 Arum palustre Malabaricum, folio Nimphaeae, radice arundinacea.
J.Commelin 1689:37 Arum Maximum Ceylonicum. Radice crassa, longa, rotunda; geniculata, Colocasii folio.
C.Commelin 1696:9 = J.Commelin 1689.
Plukenet 1696:51 Arum Colocasiae dictae similis, radice arundinaceâ.
Hermann 1698:73-74,tab.13 Arum Maximum, Macrorrizon Zeylanicum (r).
Ray 1704 vol.3:585-586 Aro congener palustris Malabarica, folio Nymphaeae, radice arundinacea.
J.Burman 1737:68 = J.Commelin 1689.
Linnaeus 1737:435 Arum acaule, foliis peltatis ovatis: basi semibifidis: margine integerrimis (?).
Rumphius 1747 vol.5:308-310,Tab.CVI Arum Indicum sativum (-); J.Burman ibid.310 (v).

Karin-pola HM 11:45:23
Karín-póla
J.Commelin in HM 11:45 Arum sp.
C.Commelin 1696:10 = J.Commelin in HM.
Ray 1704 vol.3:586 Aro congener palustris Malabarica, folio Nymphaeae.
Rumphius 1747 vol.5:312-313,Tab.CVIII Arum aquaticum (-); J.Burman ibid.:313 (?c).
Linnaeus 1753:967 Arum ovatum.
Linnaeus 1763:1371 = Linnaeus 1753.
N.L.Burman 1768:193 = Linnaeus 1763.
Houttuyn 1779 II vol.11:195 = N.L.Burman 1768.

Pongati HM 11:47:24
Pongati
C.Commelin 1696:55 = HM.
Plukenet 1700:148 Persicariae similis. Planta aquatica, Ind. Orientalis, floribus summo caule in capitulum oblongum glomeratis; capsulis foliaceis; Nonducallacree Malabarorum (?).
Ray 1704 vol.3:504 Salicaria aquatica Indica floribus spicatis pentapetalis, fructu bicapsulari.

Kurka HM 11:49:25
Kúrka
C.Commelin 1696:40 = HM.
Ray 1704 vol.3:655 Julala Indica radicibus glandulosis esculentis, ad Saxifragam auream accedens, flore, fructúque carens.
Plukenet 1705:129 Lamii rubri similis Malabarica, radice glandulosa eduli.
J.Burman 1737:138 = Plukenet 1705.

Ambel HM 11:51-52:26
Ámbel
J.Commelin in HM 11:52 Nymphaea Indica, flore candido, folio in ambitu serrato.
C.Commelin 1696:48 = J.Commelin in HM.
Plukenet 1696:267 Nymphaea Indica crenata, flore pleno candido.
Ray 1704 vol.3:630-631 = J.Commelin in HM.
J.Burman 1737:173 Nymphaea Indica, tuberosa, foliis per marginem crenatis; flore incarnato: seu Lotus Aegyptia Theophr. Mus.Zeyl.pag.19 [Hermann 1717:19].
Linnaeus 1747:87,no.194 Nymphaea foliis cordatis dentatis.
Linnaeus 1753:511 Nymphaea Lotus.
Linnaeus 1762:729 = Linnaeus 1753.
N.L.Burman 1768:119 = Linnaeus 1762.
Houttuyn 1778 II vol.9:113-117 = N.L.Burman 1768.

Areca ambel HM 11:52: no figure

J.Commelin in HM 11:52 Nymphaea sp.
C.Commelin 1696:9 = HM.
Ray 1704 vol.3:631 Nymphaea Indica, flore candidiore, petalis majoribus.

Cit-ambel HM 11:53:27
Citámbel
J.Commelin in HM 11:53 Nymphaea Malabarica minor, folio serrato.
C.Commelin 1696:48 = J.Commelin in HM.
Petiver 1702(1701):935-936,no.143 Alle-poo Malab. Nymphea Eremitana minor (?).
Ray 1704 vol.3:631 Nymphaea Malabarica minor, florum petalis longis acutis, foliis serratis.

Nedel ambel HM 11:55-56:28

Nédel ámbel
J.Commelin in HM 11:56 Nymphaeae minoris affinis Indica, flore albo, piloso.
C.Commelin 1696:48 = J.Commelin in HM.
Plukenet 1696:267 Nymphaea Indica subrotundo folio minor, flore albo fimbriato.
Ray 1704 vol.3:631 Nymphaea aquatica minor Indica, floribus albis pentapetalis, filamentis intus densis obsitis.
Linnaeus 1747:29-30,no.72 Menyanthes foliis cordatis subcrenatis, corollis interne pilosis.
Linnaeus 1753:145 Menyanthes indica.
Linnaeus 1762:207 = Linnaeus 1753.
N.L.Burman 1768:42 = Linnaeus 1762.

Tsjeroea cit ambel HM 11:57:29
Tsjeroea-citámbel
J.Commelin in HM 11:57 Nymphaea Indica minima, flore flavo.
C.Commelin 1696:48 = J.Commelin in HM.
Plukenet 1696:266 = J.Commelin in HM.
Ray 1704 vol.3:631-632 = J.Commelin in HM.

Tamara HM 11:59-60:30
Tamara
J.Commelin in HM 11:60 Nymphaeae affinis Malabarica, flore amplo, rosaceo albicante colore.
C.Commelin 1696:48 = J.Commelin in HM.
Plukenet 1696:267 Nymphaea glandifera Indiae paludibus gaudens foliis umbilicatis amplis, pediculis spinosis flore rosea purpureo (...) Phytogr.Tab.207.fig.5.
Ray 1704 vol.3:632 = J.Commelin in HM.
J.Burman 1737:173-174 Nymphaea alba, indica, maxima, flore albo, fabifera. Mus.Zeyl.pag.66 [Hermann 1717:66].
Linnaeus 1737:203 Nymphaea foliis undique integris.
Linnaeus 1747:86,no.193 = Linnaeus 1737.
Linnaeus 1753:511 Nymphaea Nelumbo.
Linnaeus 1762:730 = Linnaeus 1753.
N.L.Burman 1768:119 = Linnaeus 1762.
Houttuyn 1778 II vol.9:117-120 Nymphaea Nelumbo [Linnaeu 1767 vol.2:361].

Bem tamara HM 11:61:31
Bem-támara
J.Commelin in HM 11:61 Nymphaeae affinis Malabarica, folio & flore amplo, colore candido.
C.Commelin 1696:48 = J.Commelin in HM.
Plukenet 1696:267 Nymphaea glandifera Indiae paludibus flore albo, inter Delineationes nostras.
Ray 1704 vol.3:632 = J.Commelin in HM.
J.Burman 1737:173-174 Nymphaea alba, Indica, maxima, flore albo, fabifera. Mus.Zeyl.pag.66 [Hermann 1717:66].
Linnaeus 1737:203 Nymphaea foliis undique integris α.
Linnaeus 1747:86,no.193 α = Linnaeus 1737.
Linnaeus 1753:511 Nymphaea Nelumbo.
Linnaeus 1762:730 = Linnaeus 1753.
N.L.Burman 1768:119 = Linnaeus 1762.
Houttuyn 1778 II vol.9:117-120 Nymphaea Nelumbo [Linnaeus 1767 vol.2:361].

Kodda pail HM 11:63-64:32
Kódda-paíl
J.Commelin in HM 11:64 Sedum Indicum palustre, foliis latissimis, crispis, floribus albicantibus pilosis.
C.Commelin 1696:60 = J.Commelin in HM.
Plukenet 1700:137 Nymphaea Indica, latioribus albicantibus foliis, margine crispatis; Aucashdammaree Malabarorum.
Petiver 1702(1701):1018,no.213 Aucashdammaree Malab. Stratiotes Madraspatana Sedi vulgaris folio (?).
Ray 1704 vol.3:632-633 = J.Commelin in HM.
Linnaeus 1753:963 Pistia Stratiotes.
Linnaeus 1763:1365 = Linnaeus 1753.
N.L.Burman 1768:191 = Linnaeus 1763.
Houttuyn 1779 II vol.11:177-179 Pistia Stratiotes Syst.Nat.XII. Gen.1023 [Linnaeus 1767 vol.2:601].

Panover tsjeraua HM 11:65:33
Panovér-tsjeraúá
J.Commelin in HM 11:65 Hedera palustri.
C.Commelin 1696:50 = HM.
Plukenet 1696:374 Tribulus aquaticus foliis serratis Malabaricus.
Ray 1704 vol.3:634 Anomala aquatica repens, foliis subinde in Rosae formam compositis.
Linnaeus 1737:483 Trapa petiolis foliorum natantium ventricosis α.
Linnaeus 1753:120-121 Trapa natans.
Linnaeus 1762:175-176 = Linnaeus 1753.
N.L.Burman 1768:39 = Linnaeus 1762.
Houttuyn 1777 II vol.7:352-356 = N.L.Burman 1768.

Naru kila HM 11:67:34
Narú kilá
J.Commelin in HM 11:67 Cannae seu Arundinis Indicae affinis, flore globoso.
C.Commelin 1696:17 = J.Commelin in HM.
Plukenet 1696:299 Plantaginis Stellatae foliis Indiana florum spicâ ex utriculo caulis erumpente Phytogr.Tab.215.fig.4 (?).
Ray 1704 vol.3:573 Cannae Indicae foliis, floribus Juncorum modo è sinubus pediculorum foliorum erumpentibus.
Linnaeus 1753:288 Pontederia ovata.
Linnaeus 1762:412 = Linnaeus 1753.
N.L.Burman 1768:79 = Linnaeus 1762.
Houttuyn 1777 II vol.8:333 Pontederia ovata Syst.Nat.XII.Gen. 395.p.233 [Linnaeus 1767 vol.2:233].

Bela pola HM 11:69-70:35
Béla-póla
J.Commelin in HM 11:70 Gladiolus Indicus palustris latifolius, flore albicante.
C.Commelin 1696:33 = J.Commelin in HM.
Ray 1704 vol.3:560-561 = J.Commelin in HM.
Plukenet 1705:106 = J.Commelin in HM.
Rumphius 1750 vol.6:113-114,Tab.L,Fig.3 Angraecum terrestre alterum (-); J.Burman ibid.:114 (c).

Ela pola HM 11:71:36
Éla-póla
J.Commelin in HM 11:71 Gladioli affinis latifolia, flore campanulato.
C.Commelin 1696:33 = J.Commelin in HM.
Ray 1704 vol.3:561 = J.Commelin in HM.

Belam canda schularmani HM 11:73-74:37
Bésam-cánda-schúlármani
J.Commelin in HM 11:74 Gladioli affinis Malabarica, flore flavo, maculis rubris intersperso.
C.Commelin 1696:33 = J.Commelin in HM.
Ray 1704 vol.3:557-558 Flos Tigridis Indicus ramosus, floribus flavis, rubris maculis variis.
Plukenet 1705:193 Sisynrichium Malabaricum, foliis longissimis striatis, radice glandulosa; floribus flavis, maculis rubris eleganter notatis.

Linnaeus 1753:36 Ixia chinensis.
Linnaeus 1762:52 = Linnaeus 1753.
N.L.Burman 1768:16 = Linnaeus 1762.
Houttuyn 1780 II vol.12:21-22 Ixia Chinensis Syst.Nat. [Linnaeus 1767 vol.2:75].

Belutta pola taly HM 11:75-76:38
Bellútta-póla-tály
J.Commelin in HM 11:76 Tolabo Ceylanensibus, seu Lilii Ceylanici umbelliferi & bulbiferi. Herm.Cat.Hort.Acad.Lugd. Batav. [Hermann 1687:682,683 tab.,684] (cs).
C.Commelin 1696:41 = J.Commelin in HM.
Plukenet 1696:219 Lilio-Narcissus maximus Zeylanicus floribus albis umbellatis.
Petiver 1704(1702):1062,no.240 Narra Villan calunga. Malab. Liliasphodelus Malabaricus angustifolius.
J.Burman 1737:142-143 = Plukenet 1696.
Linnaeus 1747:53,no.127 Crinum foliis carinatis.
N.L.Burman 1768:81 Crinum asiaticum Linn.sp.419 [Linnaeus 1762:419].
Linnaeus 1771:362-363 = N.L.Burman 1768.
Houttuyn 1780 II vol.12:155-159, Pl.LXXXI,Fig.2 = N.L.Burman 1768.

Sjovanna pola tali HM 11:77-78:39
Sjovanna-póla-táli
J.Commelin in HM 11:78 Lilio Narcissus maximus.
C.Commelin 1696:41 = J.Commelin in HM.
Petiver 1704(1702):1062,no.240,1 Liliasphodelus Malabaricus latifolius.
Ray 1704 vol.3:553-554 Lilio Narcissus Indicus maximus.
Linnaeus 1753:291-292 Crinum latifolium.
Linnaeus 1762:419 = Linnaeus 1753.
N.L.Burman 1768:81 = Linnaeus 1762.
Houttuyn 1780 II vol.12:153-154 = N.L.Burman 1768.

Catulli pola HM 11:79:40
Catúlli-póla
J.Commelin in HM 11:79 Narcissus Ceylanicus, flore albo hexagono odorato Hermanni in Catalog.Hort.Academ.Lugd. Batav. [Hermann 1687:691-692,693 tab.].
C.Commelin 1696:46 = J.Commelin in HM.
Plukenet 1696:220 Lilio-Narcissus Indicus flore albo hexagono.
J.Burman 1737:142 = Plukenet 1696.
Linnaeus 1747:53,no.126 Pancratium spatha uniflora, petalis reflexis.
Linnaeus 1753:290 Pancratium zeylanicum.
Linnaeus 1762:417-418 = Linnaeus 1753.
N.L.Burman 1768:80 = Linnaeus 1762.
Houttuyn 1780 II vol.12:143-145 Pancratium Zeylanicum Syst. Nat.XII.Gen.400 [Linnaeus 1767 vol.2:235].

Katsjula kelengu HM 11:81:41'
Katsjúla-keléngú
J.Commelin in HM 11:81 Planta bifoliata, radice tuberosa, flore tripetalo difformi Malabariensis.
C.Commelin 1696:54-55 = J.Commelin in HM.
Ray 1704 vol.3:592-593 Bifolium aquaticum Indicum tuberosum, flore tripetalo difformi.
Plukenet 1705:15 Anomala quaedam Ind.Or. akaulon, gemino ex radice latiore folio, floribus inter folia (?).
Plukenet 1705:127 = J.Commelin in HM.
J.Burman 1737:33-34,Tab.13,Fig.1 Aro-Orchis tuberosa, platyphyllos.

Linnaeus 1737:2 Kaempferia.
Linnaeus 1747:3-4,no.8 Kaempferia foliis ovatis sessilibus.
Rumphius 1747 vol.5:173-174,Tab.LXIX,Fig.2 Soncorus (-); J.Burman ibid.:174 (c).
Linnaeus 1762:3 Kaempferia Galanga.
N.L.Burman 1768:3 = Linnaeus 1762.
Houttuyn 1777 II vol.7:37-39 Kaempferia Galanga Syst.Nat. XII.Gen.7.p.51 [Linnaeus 1767 vol.2:51].

Katu kapel seu Cadenaco HM 11:83:42
Kátú-kapél
J.Commelin in HM 11:83 Asphodeli Indicae affinis, floribus hexapetalis, spicatis.
C.Commelin 1696:10 = J.Commelin in HM.
Plukenet 1700:29 = J.Commelin in HM.
Petiver 1702(1701):1017,no.209 Maurel cheddee Malab. Nucleifera Salawaccensis Pisiformis, Yuccae folio Mus.Petiv. 661.
Ray 1704 vol.3:563 = Petiver 1702(1701).
Linnaeus 1762:456 Aletris Hyacinthoides (?).
N.L.Burman 1768:83 = Linnaeus 1762 (=).
Houttuyn 1780 II vol.12:409-411 = N.L.Burman 1768 (?).

Katu bala HM 11:85-87:43
Kátú-bála
J.Commelin in HM 11:87 Arundo Indica latifolia, Bauh. in Pinac. [Bauhin 1623:19].
C.Commelin 1696:10 = J.Commelin in HM.
Plukenet 1696:79 Canna Indica latifolia, humilior, flore saturatis rubente, Hort.Leyd.114 [Hermann 1687:113-114].
J.Burman 1737:53 Cannacorus latifolius vulgaris. Tournef.inst. pag.367.
Linnaeus 1737:1 Canna spathulis bifloris.
Linnaeus 1747:1,no.1 Canna foliis ovatis utrinque acuminatis nervosis. Roy.lugdb.11 [Royen 1740:11].
Linnaeus 1762:1 Canna indica.
N.L.Burman 1768:1 = Linnaeus 1762.

Carim gola HM 11:91-92:44
Carím-góla
C.Commelin 1696:18 = HM.
Plukenet 1696:299 Plantaginis Stellatae foliis Indiana florum spicâ ex utriculo caulis erumpente Phytogr.Tab.215.fig.4.
Ray 1704 vol.3:661 Hexapetala tricapsularis Indica, foliis nervosis cordiformibus acuminatis, floribus è sinu pediculi foliorum, Juncorum modo, erumpentibus.
Linnaeus 1747:54,no.129 Pontederia floribus umbellatis.
Linnaeus 1753:288 Pontederia hastata.
Linnaeus 1762:412 = Linnaeus 1753.
N.L.Burman 1768:80 Pontederia vaginalis.
Linnaeus 1771:222-223 = N.L.Burman 1768.
Houttuyn 1777 II vol.8:334-335 = N.L.Burman 1768.

Culi tamara HM 11:93-94:45
Cúli-támara
J.Commelin in HM 11:94 Sagitta Indica major, folio obtuso, floribus minoribus albicantibus.
C.Commelin 1696:59 = J.Commelin in HM.
Plukenet 1696:326 Sagittariae foliis, Planta glomerato fructu monopyrene, Coriandri ferè figurâ. Phytogr.Tab.220.fig.7.
Petiver 1699(1698):331-332,no.39 Neer Culuttree Mal. = J.Commelin in HM.
Petiver 1702(1701):936,no.144 Cooltee yella Malab. Sagittaria Malabarica major folio obtuso pubescente.
Ray 1704 vol.3:327 = J.Commelin in HM.

Linnaeus 1753:993 Sagittaria obtusifolia.
Linnaeus 1763:1410 = Linnaeus 1753.
N.L.Burman 1768:201 = Linnaeus 1763.
Houttuyn 1779 II vol.11:269-270 = N.L.Burman 1768.

Ottel ambel HM 11:95:46
Óttel-ámbel
J.Commelin in HM 11:95 Sagittae affinis Malabariensis, latissimo folio, floribus ex albo trifoliatis.
C.Commelin 1696:59 = J.Commelin in HM.
Plukenet 1696:326 = J.Commelin in HM.
Ray 1704 vol.3:633-634 Sagittae foliis aut Plantaginis aquaticae nervosis, floribus tripetalis, in singulis caulibus singulis.
Linnaeus 1747:99,no.223 Stratiotes foliis cordatis.
Linnaeus 1762:754 Stratiotes Alismoides.
N.L.Burman 1768:124 = Linnaeus 1762.
Houttuyn 1778 II vol.9:182 Stratiotes Alismoides [Linnaeus 1767 vol.2:373].

Tsjem cumulu HM 11:97:47
Tsjem-cúmúlú
C.Commelin 1696:67 = HM.
Ray 1704 vol.3:404 Aphyllos Indica, flore dipetalo difformi campanam emittente.
Linnaeus 1753:632 Aeginetia indica.
Linnaeus 1763:883 Orobanche Aeginetia.
N.L.Burman 1768:133 = Linnaeus 1763.
Houttuyn 1778 II vol.9:561-562 = N.L.Burman 1768.

Va embu HM 11:99:48
Vaémbú
J.Commelin in HM 11:99 Acorus verus Asiaticus radice tenuiore, vel Calamus Aromaticus Garzia ab Horto [Hermann 1687:9].
C.Commelin 1696:1 = J.Commelin in HM.
Petiver 1702(1701):943-944,no.168 Va sumboo Malab. Calamus Aromaticus Orientalis folio & radice tenuiore.
J.Burman 1737:6 = J.Commelin in HM.
Linnaeus 1747:55,no.132 = J.Commelin in HM.
Linnaeus 1753:324 Acorus Calamus var. verus.
Linnaeus 1762:462-463 = Linnaeus 1753.
N.L.Burman 1768:84 Acorus verus.
Houttuyn 1777 II vol.8:377-380 Acorus Calamus var. verus Syst.Nat.XII.Gen.430.p.249 [Linnaeus 1767 vol.2:249].

Pal modecca HM 11:101-102:49
Pal-modécca
J.Commelin in HM 11:102 Flos paβionis spurius Malabaricus, flore majore, fructu minore.
C.Commelin 1696:29 = J.Commelin in HM.
Plukenet 1700:147; ibid.1696:283 Passiflora spuria Bryonoides Malabarensis foliis variè scissis, fructu diverso (r).
Ray 1704 vol.3:374 Convolvulus Indicus, folio in lacinias aliquot diviso, floribus amplissimis in communi pediculo pluribus.
Linnaeus 1753:156 Convolvulus paniculatus.
Linnaeus 1762:223 = Linnaeus 1753.
N.L.Burman 1768:45 = Linnaeus 1762.
Houttuyn 1777 II vol.7:554 = N.L.Burman 1768.

Munda valli HM 11:103:50
Múnda-válli
J.Commelin in HM 11:103 Convolvulus Malabaricus flore amplo viridi diluto colore.
C.Commelin 1696:22 = J.Commelin in HM.
Ray 1704 vol.3:375 = J.Commelin in HM.
Plukenet 1705:63 Convolvulus Maderaspatensis, foliis amplissimis (?).

Linnaeus 1753:161 Ipomoea alba.
Linnaeus 1762:228-229 Ipomoea bona nox.
N.L.Burman 1768:49 = Linnaeus 1762.
Houttuyn 1777 II vol.7:570 = N.L.Burman 1768.

Kattu kelengu HM 11:105:51
Kattú-keléngú
J.Commelin in HM 11:105 Convolvuli affinis Malabarici, floribus pilosis, variegatis.
C.Commelin 1696:22 = J.Commelin in HM.
Plukenet 1696:113 Convolvulus arboreus, rotundifolius, Ind. Orientalis.
Plukenet 1700:54 Convolvulus Indicus amplissimo subrotundo folio caule alato, Turbith s. Tiguar Garciae quibusdaeni putatis (...) Shevada Malabarorum (?).
Petiver 1702(1701):850-851,no.113 Tagada Gent. Shevada Malab. Turbith Orientalis folio cordato (?).
Ray 1704 vol.3:376 Convolvulus Malabaricus, floribus pilosis, stellam habentibus, variegatis.
Linnaeus 1737:67 Convolvulus foliis cordatis, caule perenni villoso.
Royen 1740:429 Convolvulus foliis cordatis acuminatis, caule arboreo scandente.
Linnaeus 1753:155 Convolvulus malabaricus.
Linnaeus 1762:221 = Linnaeus 1753.
N.L.Burman 1768:44 = Linnaeus 1762.
Houttuyn 1777 II vol.7:545-546 Convolvulus Malabaricus [Linnaeus 1767 vol.2:156].

Ballel HM 11:107-108:52
Ballél
J.Commelin in HM 11:108 Convolvulus aquaticus folio longiore, floribus candidis.
C.Commelin 1696:21 = J.Commelin in HM.
Ray 1704 vol.3:376 = J.Commelin in HM.
Linnaeus 1753:158-159 Convolvulus reptans.
Linnaeus 1762:225 Convolvulus repens.
N.L.Burman 1768:47 = Linnaeus 1762.
Houttuyn 1777 II vol.7:560-561 Convolvulus repens [Linnaeus 1767 vol.2:157].

Tiru tali HM 11:109-110:53
Tírú-táli
J.Commelin in HM 11:110 Convolvulus Malabaricus floribus ex albo purpurascentibus.
C.Commelin 1696:22 = J.Commelin in HM.
Plukenet 1700:54 Convolvulus s. Scammonea Indica latifolia, Cullecaucanai Malabarorum (?).
C.Commelin 1701:101-102,Fig.51 Convolvulus Canariensis sempervirens, foliis mollibus et incanis, floribus ex albo purpurascentibus (s).
Ray 1704 vol.3:376-377 = J.Commelin in HM.
Rumphius 1747 vol.5:435,Tab.CLIX,Fig.2 Convolvulus riparius (-); J.Burman ibid.:434 (?c).

Ben-tiru-tali HM 11:111:54
Ben-tírú-táli
J.Commelin in HM 11:111 Convolvulus Malabaricus, folio longiori, flore candido.
C.Commelin 1696:22 = J.Commelin in HM.
Ray 1704 vol.3:377 = J.Commelin in HM.

Tala-neli HM 11:113:55
Tálá-néli
J.Commelin in HM 11:113 Convolvulus Malabaricus, angusto folio, flore variegato.

C.Commelin 1696:22 = J.Commelin in HM.
Ray 1704 vol.3:377 = J.Commelin in HM.
J.Burman 1737:73 Convolvulus Zeylanicus, folio sagittato. Mus.Zeyl.pag.64 [Hermann 1717:64].
Linnaeus 1747:211-212,no.475 = J.Burman 1737.
Linnaeus 1753:156 Convolvulus Medium.
Linnaeus 1762:218 = Linnaeus 1753.
N.L.Burman 1768:43 = Linnaeus 1762.
Houttuyn 1777 II vol.7:537 = N.L.Burman 1768.

Adamboe HM 11:115:56
Adámboe
J.Commelin in HM 11:115 Convolvulus Malabaricus flore maximo purpurascente.
C.Commelin 1696:22 = J.Commelin in HM.
Plukenet 1696:115 Convolvulo Cinamomeo similis, si non idem ex Americâ (?).
Ray 1704 vol.3:377 = J.Commelin in HM.
Linnaeus 1753:160 Ipomoea campanulata.
Linnaeus 1762:228 = Linnaeus 1753.
N.L.Burman 1768:49 = Linnaeus 1762.
Houttuyn 1777 II vol.7:570-571 Ipomoia Campanulata [Linnaeus 1767 vol.2:158].

Schovanna adamboe HM 11:117:57
Schovánna adámbú
J.Commelin in HM 11:117 Convolvulus Maritimus Ceylanicus folio crasso, bifido, seu cordiformi. Herman.Cat.Hort.Lugd. Bat. [Hermann 1687:174,175 tab.,176-177].
C.Commelin 1696:22 = J.Commelin in HM.
Plukenet 1696:114 Convolvulus maritimus, s. Soldanella è Maderaspatan. folio in summitate sinuato, Phytogr.Tab.24. fig.4.
J.Burman 1737:71-72 = J.Commelin in HM.
Linnaeus 1747:31-32,no.75 Convolvulus foliis bilobis.
Linnaeus 1753:159 Convolvulus Pes caprae.
Linnaeus 1762:226 = Linnaeus 1753.
N.L.Burman 1768:48 = Linnaeus 1762.
Houttuyn 1777 II vol.7:563-564 = N.L.Burman 1768.

Bel-adamboe HM 11:119:58
Bel-adámbu-walli
J.Commelin in HM 11:119 Convolvulus Malabaricus folio rotundiore, crasso, flore candido.
C.Commelin 1696:22 = J.Commelin in HM.
Petiver 1699(1698):327,no.28 Collarunan coodee Mal. Soldanella Madraspatana major (c).
Petiver 1702(1700):699,no.49 Soldanella Pearmeedoorica geniculis radicosis. Cullecaunacau Malab. (?).
Ray 1704 vol.3:378 = J.Commelin in HM.
Rumphius 1747 vol.5:419-420,Tab.CLV,Fig.1 Olus vagum (-); J.Burman ibid.:420 (?=).

Pulli-schovadi HM 11:121:59
Púllí-schovadí
J.Commelin in HM 11:121 Convolvulus Ceylanicus Villosus pentaphyllus & heptaphyllus minor, Cat.Hort.Acad.Lugd.Batav. [Hermann 1687:184,187 tab.].
C.Commelin 1696:22 = J.Commelin in HM.
Plukenet 1696:115-116 Convolvulus Maderaspatanus, villosus, quinquifidis foliis, capitulis Scabiosae quodammodò mentientibus, Phytogr.Tab.166.fig.6.
J.Burman 1737:70-71 Convolvulus Zeylanicus, hirsutus, foliis Pedis Tigridis in modum, seu quinque profundas lacinias divisis. Mus.Zeyl.pag.12 [Hermann 1717:12].
Linnaeus 1737:496 Convolvulus foliis palmatis glabris, caule hispido.
Linnaeus 1747:33,no.78 Ipomoea foliis palmatis, floribus aggregatis.
Linnaeus 1753:162 Ipomoea Pes tigridis.
Linnaeus 1762:230 = Linnaeus 1753.
N.L.Burman 1768:50 = Linnaeus 1762.
Houttuyn 1777 II vol.7:576-577 = N.L.Burman 1768.

Tsjuria-cranti HM 11:123:60
Tsjúria-cránti
J.Commelin in HM 11:123 Convolvulus exoticus annuus folio Myriophylli seu Millefolii Aquatici, flore sanguineo Morison in Hist.Oxoniens.
C.Commelin 1696:21 = J.Commelin in HM.
Plukenet 1696:117 = J.Commelin in HM.
J.Burman 1737:197-198 Quamoclit foliis tenuiter incisis, & pinnatis. Tournef.inst.pag.116.
Linnaeus 1737:66 Ipomoea foliis linearibus pinnatis, floribus solitariis.
Linnaeus 1747:32-33,no.77 = Linnaeus 1737.
Linnaeus 1762:227 Ipomoea Quamoclit.
N.L.Burman 1768:48-49 = Linnaeus 1762.
Houttuyn 1777 II vol.7:566-567 Ipomoia Quamoclit Syst.Nat. XII.Gen.215.p.158 [Linnaeus 1767 vol.2:158].

Samudra-stjogam HM 11:125:61
Samúdra-tsjógam
J.Commelin in HM 11:125 Convolvulus maximus Malabariensis, folio villoso, floribus purpurascentibus.
C.Commelin 1696:22 = J.Commelin in HM.
Plukenet 1696:113 Convolvulus maximus Malabariensis villosis foliis subrotundis, floribus purpureis.
Ray 1704 vol.3:377 = J.Commelin in HM.

Cattu-valli HM 11:127:62
Bátta-válli
C.Commelin 1696:19 = HM.
Plukenet 1696:181 Hedera monphyllos Convolvuli foliis Virginiana. Phytogr.Tab.36.fig.2 (?).
Ray 1704 vol.3:378 Convolvulus anadendras, foliis crassis subrotundis, flore fructúque carens.
Plukenet 1705:62 Cocculi Orientalis Convolvulaceus Frutex, foliis subrotundis glabris, è Maderaspatan. (?).

Mareta-inali HM 11:129:63
Máreta-ináli
C.Commelin 1696:43 = HM.
Ray 1704 vol.3:378 Convolvulus anadendras, foliis oblongorotundis angustis, ad nodos pluribus, flore carens.
Plukenet 1705:101 Frutex Sinensis baccifer Convolvulaceus, Cisti foliis pubescentibus, ad ramulorum nodos confertis.
Plukenet 1705:115 Hedera baccifera Maderaspatana, foliis plurimis ad genicula caulium prodeuntibus.

Vistnu-clandi HM 11:131-132:64
Vístnú-ilándi
J.Commelin in HM 11:132 Convolvulus Indicus minor, Alsine folio, flore rubicundo, purpureo.
C.Commelin 1696:22 = J.Commelin in HM.
Plukenet 1696:116 Convolvuli minimi species, Bisnagarica, hirsuto Alsines folio, Phytogr.Tab.9.fig.1.
Petiver 1699(1698):321,no.17 Calovee Malab. Convolvulus Madraspatanus flore auriculato, calycibus majoribus (cs).
Petiver 1702(1701):1007,no.180 Visne crantee Malab. Convolvulus Bisnagaricus Myosotidis folio (?).

Ray 1704 vol.3:379 = J.Commelin in HM.
Linnaeus 1737:68 Convolvulus foliis ovatis obtusis, caule filiformi.
Linnaeus 1747:32,no.76 Convolvulus foliis subovatis obtusis petiolatis pilosis, caule diffuso pedunculis trifloris.
Linnaeus 1753:157 Convolvulus alsinoides.
Linnaeus 1762:392 Evolvulus alsinoides.
N.L.Burman 1768:77 = Linnaeus 1762.
Houttuyn 1777 II vol.8:255 = N.L.Burman 1768.

Sendera-clandi HM 11:133:65
Séndera-clándi
J.Commelin in HM 11:133 Convolvulus Indicus minor, folio anguloso, flore ex albo flavescente.
C.Commelin 1696:22 = J.Commelin in HM.
Plukenet 1696:117 Convolvulus Indicus barbatus minor, foliorum apicibus lunulatis, Phytogr.Tab.276.fig.6 (?).
Ray 1704 vol.3:379 Convolvulus Indicus minor, folio angusto oblongo, summitate velut abscissa, flore ex albo flavescente.
Linnaeus 1753:157 Convolvulus tridentatus.
Linnaeus 1762:392 Evolvulus tridentatus.
N.L.Burman 1768:77,Tab.16,f.3 = Linnaeus 1762.
Houttuyn 1777 II vol.8:256 = N.L.Burman 1768.

VOLUME 12

Angeli maravara HM 12:1-4:1
Ansjeli Maravara
J.Commelin in HM 12:4 Orchis Abortiva Aizoides Malabariensis, flore odoratissimo variegato, intus aviculam repraesentante.
C.Commelin 1696:49 = J.Commelin in HM.
Ray 1704 vol.3:588-589 Malum arboris Angeli dictae.
Rumphius 1750 vol.6:95-98,Tab.XLII Angraecum scriptum (-); J.Burman ibid.:98 (c).
Linnaeus 1753:953 Epidendrum retusum.
Linnaeus 1763:1351 = Linnaeus 1753.
N.L.Burman 1768:190 = Linnaeus 1763.
Houttuyn 1779 II vol.11:166 Epidendrum Retusum [Linnaeus 1767 vol.2:596].

Biti maram maravara HM 12:5:2
Biti maram Maravára
J.Commelin in HM 12:5 Orchis abortiva Aizoides Malabariensis altera, flore odorato, sanguineo colore, intus aviculam purpuream referente.
C.Commelin 1696:49 = J.Commelin in HM.
Ray 1704 vol.3:589 = J.Commelin in HM.
Linnaeus 1753:953 Epidendrum retusum β.

Ponnampou-marvara HM 12:7-8:3
Ponnámpu Maravára
J.Commelin in HM 12:8 Orchis Aizodes abortiva Malabariensis, foliis obtusis, flore aureo odorato.
C.Commelin 1696:49 = J.Commelin in HM.
Petiver 1702(1701):1016,no.204 Madel velladee Malab. Orchides epidendron Salawaccensis Visci folio. Mus.Pet.662.
Ray 1704 vol.3:589-590 = J.Commelin in HM.
Rumphius 1750 vol.6:99-100,Tab.XLIV,Fig.1 Angraecum album minus (-); J.Burman ibid.:100 (=).
Linnaeus 1753:952 Epidendrum spathulatum.
Linnaeus 1763:1348 = Linnaeus 1753.
N.L.Burman 1768:188-189 = Linnaeus 1763.
Houttuyn 1779 II vol.11:160 Epidendrum Spatulatum [Linnaeus 1767 vol.2:595].

Thalia maravara HM 12:9:4
Thalia Maravára
C.Commelin 1696:66 = HM.
Ray 1704 vol.3:590 Orchis abortiva aizoides, spica florem brevi, floribus minoribus, lineis rubris udantibus radiatis.
Rumphius 1750 vol.6:110-112,Tab.L,Fig.2 Herba supplex minor (-); J.Burman ibid.:110 (c).
Linnaeus 1763:1348 Epidendrum furvum.
N.L.Burman 1768:189 = Linnaeus 1763.
Houttuyn 1779 II vol.11:160-161 Epidendrum Furvum [Linnaeus 1767 vol.2:595].

Tsjerou-mau-maravara HM 12:11:5
Mau Tsierou maravára
C.Commelin 1696:67 = HM.
Plukenet 1696:87 Caryophyllata Monomotapensis nervosis Bupleuri foliis, intùs cavis, flore caeruleo, cauliculis secundùm longitudinem alatis, Phytogr.Tab.275.fig.1 (?).
Ray 1704 vol.3:590 Anomala Indica, Orchidis aut Elleborines flore pulcherrimo versicolore, in spicis ex adverso foliorum sitis, foliis angustis carinatis.
Rumphius 1750 vol.6:107 (Auct), Tab.XLIX,Fig.1 Angraecum saxatile (-); J.Burman ibid.:107 (=).
Linnaeus 1753:952 Epidendrum tenuifolium.
Linnaeus 1763:1348 = Linnaeus 1753.
N.L.Burman 1768:188 = Linnaeus 1763.
Houttuyn 1779 II vol.11:160 Epidendrum tenuifolium [Linnaeus 1767 vol.2:595].

Kolli-tsjerou-mau-maravara HM 12:13:6
Kolly Tsierou mau Maravára
C.Commelin 1696:40 = HM.
Ray 1704 vol.3:590 = HM.

Anantali-maravara HM 12:15-16:7
Anantály Maravára
J.Commelin in HM 12:16 Orchis abortiva latifolia Malabarica, Clitoride, flore luteo, piloso.
C.Commelin 1696:49 = J.Commelin in HM.
Ray 1704 vol.3:590-591 Orchis abortiva seu Elleborine duplici cauliculo, altero florifero, altero folioso, clitoridis flore luteo piloso Commelin.
Plukenet 1705: Eriocarpos Malabarica, capsula seminali cum seminibus, gossipio candidante infarctâ (s).
Linnaeus 1753:952 Epidendrum ovatum.
Linnaeus 1763:1349 = Linnaeus 1753.
N.L.Burman 1768:189 = Linnaeus 1763.
Houttuyn 1779 II vol.11:162-163 Epidendrum Ovatum [Linnaeus 1767 vol.2:596].

Kansjiram-maravara HM 12:17-18:8
Kanspram Marabara
J.Commelin in HM 12:18 Aloë affinis Indica supra arbores crescens.
C.Commelin 1696:3 = J.Commelin in HM.
Ray 1704 vol.3:572 = J.Commelin in HM.
Linnaeus 1753:953 Epidendrum aloifolium.
Linnaeus 1763:1350 = Linnaeus 1753.
N.L.Burman 1768:189 = Linnaeus 1763.
Houttuyn 1779 II vol.11:164 Epidendrum Aloifolium [Linnaeus 1767 vol.2:596].

Maravara tsjembo HM 12:19-20:9
Máravara Tsjémbu
J.Commelin in HM 12:20 Colocasia vel Arum Aegyptiacum (r).
C.Commelin 1696:43 = HM.

Plukenet 1696:326 Sagittae (fortè) cognata Planta, ex Maderaspatan. (?).
Ray 1704 vol.3:652 Parasitica anomala, Colocasiae foliis, floribus parvis pyriformibus, in caulibus aphyllis.

Patitsjivi maravara HM 12:21:10
Patitsjevi Maravara
J.Commelin in HM 12:21 Phillitidi affinis Indica, foliis Ari.
C.Commelin 1696:54 = J.Commelin in HM.
Ray 1704 vol.3:55-56 Hemionitis foliis Ari.
N.L.Burman 1768:231 Asplenium arifolium.

Panna-kelengo-maravara HM 12:23-24:11
Panna-kelengo Maravara
J.Commelin in HM 12:24 Polypodium Indicum foliis latissimis.
C.Commelin 1696:55 = J.Commelin in HM.
Ray 1704 vol.3:64-65 Hemionitis pinnata Indica Polypodii foliis latissimis.
J.Burman 1737:195-196 Polypodium exoticum, folio Quercus C.B.Pin.pag.359 [Bauhin 1623:359].
Linnaeus 1747:181,no.382 Polypodium foliis sterilibus simplicibus brevioribus obtusis sinuatis, fructificantibus alternatim pinnatis: pinnis lanceolatis.
Rumphius 1750 vol.6:78-81 Polypodium Indicum pilosum (-); J.Burman ibid.:81 (c).
Linnaeus 1753:1087 Polypodium quercifolium.
Linnaeus 1763:1547 = Linnaeus 1753.
N.L.Burman 1768:231 = Linnaeus 1763.
Houttuyn 1783 II vol.14:161-162,Pl.XCVIII,Fig.2 = N.L.Burman 1768.

Welli-panna-kelengu-maravara HM 12:25:12-13
Wellipanna-kelengu, Weli pánna Kelengoû maravara
J.Commelin in HM 12:25 Filix non ramosa, foliis integris, non serratis maxima Indiae Orientalis Breynius, Centur.I.Cap.96 [Breyne 1678:189] (?).
C.Commelin 1696:70 = HM.
J.Burman 1737:99-100 = J.Commelin in HM 12 (r).
N.L.Burman 1768:233 Polypodium dissimile Linn.sp.1549 [Linnaeus 1763:1549].

Tama-pouel-paatsja-maravara. vel Eneadi-kourengo HM 12:27-29:14
Táma Poulpaalsja maravara
J.Commelin in HM 12:29 Selago Indiae Orientalis; sive Plegmaria admirabilis Ceylanicae Breinius, Centur.I.Cap.92 [Breyne 1678:180-183,tabs.].
C.Commelin 1696:61 = J.Commelin in HM.
Ray 1704 vol.3:655; ibid.1688 vol.2:1852 = J.Commelin in HM.
J.Burman 1737:211 = J.Commelin in HM.
Linnaeus 1747:183,no.386 Lycopodium erectum dichotomum, foliis cruciatis, spicis gracilioribus. Dill.musc.450.t.61.f.5.
Rumphius 1750 vol.6:91-93,Tab.XLI,Fig.1 Equisetum Amboinicum, seu arboreum squamatum (-); J.Burman ibid.:93 (=).
Linnaeus 1753:1101 Lycopodium Phlegmaria.
Linnaeus 1763:1564 = Linnaeus 1753.
N.L.Burman 1768:237 = Linnaeus 1763.

Para-panna-maravara HM 12:31:15
Pára-pánna Maravara
C.Commelin 1696:51 = HM.
Plukenet 1705:92 Filix ramosa Ind.Or. pinnulis subrotundis, costae nascentibus, ad nervos aversâ parte areolis, pulverulenta; summis alis mucronatis.
Rumphius 1750 vol.6:67-69,Tab.XXIX Filix esculenta, sive femina (-); J.Burman ibid.:69 (c).

Kal-panna-maravara HM 12:33:16
Kal-panna-maravara
C.Commelin 1696:38 = HM.
Plukenet 1696:150 Filix Saxatilis Rutae murariae foliis Americana, s. Adianthum album folio Filicis ex Insula Jamaicensi. Phytogr.Tab.282.fig.1 (?).

Kari-welli-panna-paravara HM 12:35:17
Kári-béli-pánna maravara
C.Commelin 1696:39 = HM.
Plukenet 1696:151 Filix mas vulgari similis, pinnulis amplioribus planis, nec crenatis, Virginiana, Phytogr.Tab.179.fig.2 (?).
Petiver 1704(1703):1452,no.11 Filix Madraspat. pyramidalis circa nervum maculata (?).
Ray 1704 vol.3:80 Filix non ramosa Indica, pinnulis obtusis non crenatis.
Plukenet 1705:92 Filix Indica non ramosa, pinnulis subrotundis, costae adnatis, in aversa parte duplici serie punctorum insignitis, summis alis mucronatis.
Linnaeus 1753:1090 Polypodium parasiticum.
Linnaeus 1763:1551 = Linnaeus 1753.
N.L.Burman 1768:234 = Linnaeus 1763.
Houttuyn 1783 II vol.14:180 = N.L.Burman 1768.

Nella-panna-maravara HM 12:37:18
Nella pánna maravara
C.Commelin 1696:47 = HM.
Ray 1704 vol.3:72-73 Filix Indica, pediculo foliis inde à latiore basi in longissimum mucronem productis, & singulari modo dentatis, vestito.
Plukenet 1705:94 Filicifolia Phyllitis cauliferae, similis, trifoliata, ex Insula Johanna, Mantiss.81.pl.6 [Plukenet 1700:81].
N.L.Burman 1768:236 Trichomanes adianthoides Linn.sp.1561 [Linnaeus 1763:1561].

Panna-mara-maravara HM 12:39:19
Panná Mara-maravara
C.Commelin 1696:50 = HM.
Ray 1704 vol.3:71 Filix Indica, folio pinnato, pinnis vetustioribus circa margines serratis, extremo in longissimum mucronem sinuatum procurrente.

Ellettadi-maravara HM 12:41:20-21
Elittadi maravara, Elettadi-maravara
C.Commelin 1696:27 = HM.
N.L.Burman 1768:231 Polypodium laciniatum.
Houttuyn 1783 II vol.14:162 = N.L.Burman 1768.

Theka-maravara HM 12:43:22
Teka-marabára
C.Commelin 1696:66 = HM.
Ray 1704 vol.3:652 Parasitica Indica repens è bulbo seu tubere simplici unicum folium producens, flore fructúque carens.

Tsjerou-tecka-maravara HM 12:45:23
Tsjerou Tecka maravara
C.Commelin 1696:67 = HM.
Ray 1704 vol.3:652 Theca Maravara minor.

Wellia-theka-maravara vel Mau-maravara HM 12:47-48:24
Wellia-Theka-maravara ofte Mau-marabára
C.Commelin 1696:70 = HM.
Ray 1704 vol.3:653-654 Parasitica Indica radice tuberosa, foliis nervosis, spica longa gracili squamosa, flore carens.

Kathou-theka-maravara HM 12:49:25

Katou-Theka-marabára
C.Commelin 1696:39 = HM.
Ray 1704 vol.3:653 Parasitica Indica bulbifera, foliis angustis, sterilis.

Katou-kayda-maravara HM 12:51-52:26
Kátou kaidà maravara
C.Commelin 1696:39 = HM.
Ray 1704 vol.3:591 Orchis abortiva Indica bulbifera, foliis longis trinervibus, floribus venis & maculis variorum colorum pulcherrime pictis.
Plukenet 1705:74 Eriocarpos Malabarica, Epidendri species hexapetalos, gladioli foliis, caule arundinaceo, capsulâ lanâ candidissimâ intus referti (?).

Basaala-poulou-maravara HM 12:53-54:27
Basaála-poulou marabara
C.Commelin 1696:14 = HM.
Ray 1704 vol.3:593 Elleborine similis Indica, flosculis monopetalis, miniato-rubris, minimis.
Plukenet 1705:115 Helleborine Malabariensis, Calceoli Mariae foliis, floribus parvis rubris, in spicam longiorem digestis (?).
Rumphius 1750 vol.6:118-119,Tab.LIV,Fig.2-3 Orchis Amboinica minor (-); J.Burman ibid.:119 (cs).

Katou-ponnam-maravara HM 12:55:28
Katou Ponnam-maravara
C.Commelin 1696:39 = HM.
Ray 1704 vol.3:593 Maravara sylvestris aurea.

Maretta-mala-maravara HM 12:57:29
Máretla-mála maravara
C.Commelin 1696:43 = HM.
Ray 1704 vol.3:651 Parasitica Indica repens, foliis crassis pyriformibus semine in siliquis seu julis longis angustis incluso.
Linnaeus 1747:180,no.378 Acrostichum frondibus integerrimus glabris petiolatis: sterilibus subrotundis, fertilibus linearibus. Heiligt.acrost.9.n.1.f.2.
Linnaeus 1753:1067 Acrostichum heterophyllum.
Linnaeus 1763:1523 = Linnaeus 1753.
N.L.Burman 1768:228-229 = Linnaeus 1763.
Houttuyn 1783 II vol.14:69-70 = N.L.Burman 1768.

Valli-vara kody-maravara HM 12:59:30
Válli vara Kody maravara
Ray 1704 vol.3:652 Hedera arborea venenata, Parasitica Indica foliis longis, angustis, julis seu siliquis longissimis teretibus, semine pulverulento.

Arana-panna HM 12:61:31
Arana-pánna
C.Commelin 1696:6 = HM.
Ray 1704 vol.3:71 Filix major Indica pinnata, pinnis praelongis, circa margines angulis rotundis excisis.
J.Burman 1737:98,Tab.44,Fig.1 Filix Zeylanica, denticulata, non ramosa. Mus.Zeyl.pag.36 [Hermann 1717:36].

Valli-panna HM 12:63:32
Válli-pánna
C.Commelin 1696:68 = HM.
Plukenet 1700:83; ibid.1696:156 Filix scandens perpulchra Brasiliana, Breyn.Cent.I.185 [Breyne 1678:185] (?).
Ray 1704 vol.3:65 Hemionitis pennata Indica scandens, trifolia, foliis venosis, tenuissimè crenatis, bifariam persaepe divisis, & in angulos ad basin excurrentibus.

Plukenet 1705:95 Filix scandens Phaseolôdes triphyllos, lobis Sisari foliorum aemulis, ut plurimùm bisectis è Maderaspatan.
Linnaeus 1747:178-179,no.375 Ophioglossum caule flexuoso angulato, petiolis diphyllis, foliolis trifido-palmatis.
Linnaeus 1753:1063 Ophioglossum flexuosum.
Linnaeus 1763:1519 = Linnaeus 1753.
N.L.Burman 1768:227 = Linnaeus 1763.
Houttuyn 1783 II vol.14:48-49 = N.L.Burman 1768.

Tsjeru-valli-panna vel Warapoli HM 12:65:33
Tsjéru-válli-pánna warapoly
C.Commelin 1696:67,70 = HM.
Plukenet 1700:83; ibid.1696:156 Filix scandens perpulchra Brasiliana, Breyn.Cent.I.185 [Breyne 1678:185] (?).
Ray 1704 vol.3:65 Hemionitis pinnata Indica scandens, foliis oblongis, angustis acutis, marginibus profundè dentatis.
Plukenet 1705:95 Filix Clematitis Malabariensis, majoribus pinnis auriculatis, per ambitum laciniarum adinstar profundé dentatis.
Linnaeus 1747:178,no.374 Ophioglossum caule flexuoso, frondibus oppositis pinnatis, foliis utrinque spiciferis. Hort.cliff.473 [Linnaeus 1737:473].
Linnaeus 1753:1063 Ophioglossum scandens.
Linnaeus 1763:1518-1519 = Linnaeus 1753.
N.L.Burman 1768:227 = Linnaeus 1763.
Houttuyn 1783 II vol.14:46-48,Pl.XCIV,Fig.2 = N.L.Burman 1768.

Tsjeru-valli-panna. Altera HM 12:67:34
Tsjeŕia Valli-pánna
C.Commelin 1696:67 = HM.
Plukenet 1700:83; ibid.1696:156 Filix scandens perpulchra Brasiliana, Breyn.Cent.I.185 [Breyne 1678:185] (?).
Plukenet 1705:95 Filix clematitis Maderaspatana, pinnulis Pistolochiae Virginianae foliorum aemulis, rachi medio huc & illuc undulatim intorto (?).
Rumphius 1750 vol.6:76,Tab.XXXII,Fig.2 Adianthum volubile minus (-); J.Burman ibid.:75 (c).

Panna-valli HM 12:69:35
Pánna-Válli
C.Commelin 1696:50 = HM.
Plukenet 1700:80; ibid.1696:153 Filix Jamaicensis pinnatis foliorum apicibus obtusis, & subrotundis, Phytogr.Tab.286.fig.4 (?).
Petiver 1704(1703):1451-1452,no.8 Phyllitis ramosa Malabarica marginibus albis.
N.L.Burman 1768:234 Polypodium palustre β.

Tsjudan-tsjera HM 12:71:36
Tsjadaen-tsjira
C.Commelin 1696:68 = HM.
Ray 1704 vol.3:265 Herba stellata Indica erecta, foliis dissectis, flosculis è tubo oblongo in quatuor segmenta expansis.
J.Burman 1737:121-122,Tab.55,Fig.1 Hottonia flore solitario ex foliorum alis proveniente.
Linnaeus 1762:208 Hottonia indica.
N.L.Burman 1768:42 = Linnaeus 1762.

Puem-peda HM 12:71:37
Puem̄-pedà
C.Commelin 1696:57 = HM.
Ray 1704 vol.3:39 Muscus capillaris, pediculis purpureo-rubescentibus, capitulis reflexis cuspidatis.

Plukenet 1705:148 Muscus aureus capillaris minor Malabaricus, apiculis cernuis.

Motta-pullu HM 12:72:38
Mótta-pullù
C.Commelin 1696:45 = HM.
Plukenet 1696:200 Juncellus Maderaspatensis capitulis compactioribus, acumine bifido longiori, Phytogr.Tab.197.fig.7 [Plukenet 1691:Tab.CXCVII,fig.7] (?).
Ray 1704 vol.3:630 Juncello omnium minimo nostrati perquam similis si non eadem.
Plukenet 1705:124 Juncellus Malabaricus tricephalos, capitulis compactis, summo caule unico foliolo cinctis.
J.Burman 1737:109-110,Tab.47,Fig.2 Gramen pusillum, Junci capitulis minimis, ad basin foliolis binis auctis (r).
N.L.Burman 1768:22 Scirpus capillaris Linn.sp.73 [Linnaeus 1762:73] (?).

Avenka HM 12:72:39
Auenka
C.Commelin 1696:10 = HM.
N.L.Burman 1768:235 Adianthum lunulatum.
Linnaeus 1771:181 Scirpus squarrosus.
Houttuyn 1782 II vol.13:117 = Linnaeus 1771.
NB The engraving has been numbered 40.

Bellan-patsja HM 12:73:40
Bellan-patsja
C.Commelin 1696:15 = HM.
Petiver 1704(1703):1451,no.5 Lycopodium Malabaric. folijs crispis.
Plukenet 1705:149 Muscus clavatus erectus crispatis foliolis, Spongiolae imitamentum, ex Chinâ.
J.Burman 1737:145 Lycopodium Zeylanicum, erectum, foliis crassioribus, & magis compressis (?).
Linnaeus 1747:183-184,no.387 Lycopodium erectum ramosissimum, spicis ovatis sessilibus nutantibus.
Rumphius 1750 vol.6:87-88,Tab.XL,Fig.1 Cingulum terrae (-); J.Burman ibid.:88 (=).
Linnaeus 1753:1103 Lycopodium cernuum.
Linnaeus 1763:1566-1567 = Linnaeus 1753.
N.L.Burman 1768:237-238 = Linnaeus 1763.
NB The engraving has been numbered 39.

Tsjama-pullu HM 12:75:41
Tsjáma-pullù
C.Commelin 1696:67 = HM.
Ray 1704 vol.3:616 Gramini pratensi paniculato majori simile, si non idem.
Plukenet 1705:107 Gramen pratense Malabaricum.
J.Burman 1737:105 Gramen amoris, minus. Mus.Zeyl.pag.25 [Hermann 1717:25] (?).
Linnaeus 1753:58 Panicum patens.
Linnaeus 1762:86 = Linnaeus 1753.
Linnaeus 1762:101 Poa tenella.
N.L.Burman 1768:26,Tab.10,f.2 = Linnaeus 1762:86.
Houttuyn 1782 II vol.13:191-192 = N.L.Burman 1768.

Wara-pullu HM 12:77:42
Wára-pullù
C.Commelin 1696:70 = HM.
Ray 1704 vol.3:618 Cyperus Indicus paniculis proliferis, spicis parvis, compressis squamosis.
Plukenet 1705:112 Gramen Cyperoides Malabaricum, tenui paniculâ sparsà (?).

Houttuyn 1782 II vol.13:70-71 Cyperus elatus Syst.Nat.XII [Linnaeus 1767 vol.2:81].

Kudira-pullu HM 12:97:43
Kudira-pullù
C.Commelin 1696:40 = HM.
Ray 1704 vol.3:616 Graminifolia Indica paniculâ è spicis cuspidatis plumulas duas cum apicibus ex aversa cauli parte emittentibus, composita.
Rumphius 1750 vol.6:13-14,Tab.V,Fig.1 Gramen aciculatum (-); J.Burman ibid.:14 (=).
N.L.Burman 1768:23 Scirpus corymbosus Linn.sp.76 [Linnaeus 1762:76].
Houttuyn 1782 II vol.13:116 = N.L.Burman 1768.

Tereta-pullu HM 12:81:44
Téreta-pullù
C.Commelin 1696:66 = HM.
Petiver 1704(1702):1260,no.289 Ponne varaga pille Malab. Sesamum granulosum bicorne Madraspatanum Mus.Petiver 570 (?).
Ray 1704 vol.3:605 Gramen Dactylon Indicum, geminâ duntaxat spicâ in summo caule.
Plukenet 1705:111 Gramen paniceum distachyophoren spicâ binis granorum ordinibus uno versu constante; Mantiss.94 [Plukenet 1700:94 Ponnevaragupille Malabarorum].

Tsjama-pullu HM 12:83:45
Tsjama pullù
C.Commelin 1696:67 = HM.
Ray 1704 vol.3:616-617 Gramen Indicum paniculâ, è ramulis alternatim singulis ad intervalla nascentibus, & spicas plures alternatim emittentibus compositâ.
Plukenet 1705:109 Gramen arundinaceâ paniculâ, ex Ind.Or. (?).
Linnaeus 1753:69 Poa malabarica.

Kerpa HM 12:85:46
Kérpa
C.Commelin 1696:39 = HM.
Petiver 1704(1702):1256,no.268 Naunel pu Malab. Arundo Madraspat. panicula sericea albissima (?).
Ray 1704 vol.3:615 Arundo Indica nostrati (cs).
N.L.Burman 1768:24 Panicum alopecuroideum Linn.sp.82 [Linnaeus 1762:82].
Linnaeus 1771:183 Saccharum spontaneum.
Houttuyn 1782 II vol.13:139-140 = Linnaeus 1771.

Beli-caraga HM 12:87:47
Béli Caragà
C.Commelin 1696:15 = HM.
Ray 1704 vol.3:617 Graminis Indica cauliferi sterilis species.
Plukenet 1705:109 Gramen arundinaceum minus Mederaspatanum, radice repente (?).

Kaden-pullu HM 12:89:48
Kadeñ-pullù
C.Commelin 1696:38 = HM.
Ray 1704 vol.3:618-619 Gramen cyperoides Indicum paniculatis è foliorum sinubus egressis, locustis raris compositis.
Plukenet 1705:113 Gramen Cyperoides Maderaspatanum, Junci paniculâ sparsâ.
Linnaeus 1753:51 Scirpus lithospermus.
Linnaeus 1762:65 Schoenus lithospermus.
N.L.Burman 1768:19 = Linnaeus 1762.
Houttuyn 1782 II vol.13:545-547 = N.L.Burman 1768.

Tagadi HM 12:91:49
Tagadi
C.Commelin 1696:65 = HM.
Ray 1704 vol.3:604 Gramen spicatum arundinaceum Indicum, spicâ squamosâ.
Linnaeus 1753:1049 Ischaemum muticum.
Linnaeus 1763:1487 = Linnaeus 1753.
N.L.Burman 1768:221 = Linnaeus 1763.
Houttuyn 1782 II vol.13:600-601 = N.L.Burman 1768.

Pota-pullu HM 12:93:50
Pótta-pullū
C.Commelin 1696:56 = HM.
Ray 1704 vol.3:618 Cyperus Indicus radice geniculata longa, transversum progediente, panicula è summo caule egressa, è spicis longioribus & habitioribus composita.
Plukenet 1705:113 Gramen Cyperoides Maderaspatanum, Cyperi paniculis viridifuscis, radice irrepente (?).

Kouda-pullu HM 12:95:51
Kónda-pullu
C.Commelin 1696:34 Gramen dactylon Aegyptiacum Park.Theat.1.179.
Petiver 1704(1702):1256,no.264 Poo pillee Malab. Gramen Dactylon Madraspat. spicis pilosis Mus.Petiv.60 (?).
Ray 1704 vol.3:605 Gramen dactylon Indicum, spicis pluribus surrectis, aristatis.

Mottenga HM 12:97:52
Motténga
C.Commelin 1696:45 = HM.
Petiver 1704(1702):1257,no.273 Mucutang Corea Malab. Cyperus Jamaicensis fere tricephalus (?).
Ray 1704 vol.3:619 Cyperus Indicus radicibus glandulosis, spicis parvis sessilibus in summitate caulis congestis.
Plukenet 1705:111 Gramen cyperoides Americanum, spicis speciosis plurimis compressis, summo caule absq; pediculis adnatis.
Rumphius 1750 vol.6:8-9,Tab.III,Fig.2 Gramen capitatum (-); J.Burman ibid.:8 (=).
N.L.Burman 1768:19 Schoenus tuberosus.
Houttuyn 1782 II vol.13:53-54 Schoenus Niveus = N.L.Burman 1768.

Pee-mottenga HM 12:99:53
Pee Motténga
C.Commelin 1696:51 = HM.
Petiver 1704(1702):1254,no.253 Chenduppee Corea Malab. Cyperus Madraspatanus Allij capitulo (?).
Ray 1704 vol.3:619 Gramini maritimo cyperoidi J.B. (s).
Plukenet 1705:113 Gramen Cyperoides Maderaspatanum, capitulo grandiore simplici, globoso, candicante, ad summum caulem medio foliorum sessili.
Linnaeus 1753:52 Scirpus glomeratus.
Linnaeus 1762:64-65 Schoenus coloratus.
N.L.Burman 1768:18-19 = Linnaeus 1762.
Houttuyn 1782 II vol.13:54-55 = N.L.Burman 1768.

Mulen-pullu HM 12:101:54
Muleṅ-pullu
C.Commelin 1696:45 = HM.
Petiver 1704(1702):1256,no.265 Shanee Coree Malab. Cyperus Filicinus medius, panicula comosa e Madraspatan Mus.Petiver 593 (?).
Plukenet 1705:111 Gramen cyperoides Maderaspatanum minus, paniculâ compactiore, summo caule trium foliorum in sinu sessili.

Ira HM 12:103:55
Ira
C.Commelin 1696:37 = HM.
Ray 1704 vol.3:618 Cyperus Indicus, radice fibrosa, paniculâ è spicis magis confertis & habitioribus compositâ.
Plukenet 1705:111 Gramen Cyperoides Maderaspatanum, paniculâ magnâ insignitèr squammata.

Ira ceu Balari HM 12:105:56
Ira eu Balurī
C.Commelin 1696:37 = HM.
Ray 1704 vol.3:618 Cyperus Indicus paniculâ proliferâ, è spicis rarioribus & compressioribus compositâ.
Plukenet 1705:112 Gramen Cyperoides majus Malabaricum, summo caule ex unâ panicula plures proferens.
Linnaeus 1753:45 Cyperus Iria.
Linnaeus 1762:67 = Linnaeus 1753.
N.L.Burman 1768:20 = Linnaeus 1762.
Houttuyn 1782 II vol.13:69-70 = N.L.Burman 1768.

Kodi-pullu HM 12:107:57
Kodì-pullù
C.Commelin 1696:40 = HM.
Petiver 1704(1702):1251-1252,no.250 Comachee pillee Malab. Juncus odoratus (cs).
Ray 1704 vol.3:617 Gramen paniculatum Indicum paniculâ convolutâ, è locustis oblongis squamosis compositâ.

Beera-kuida HM 12:109:58
Beêra-kaidā
C.Commelin 1696:14 = HM.
Ray 1704 vol.3:619 Gramen cyperoides Indicum, paniculis è foliorum sinubus egressis, pluribus acerosis capitulis compositis.
Plukenet 1705:112 Gramen Cyperoides Ind.Or. latiore folio striato, summo caule paniculis paleaceis.

Mella-pana-kelangu HM 12:111:59
Néla pána kelangu
C.Commelin 1696:44 = HM.
Ray 1704 vol.3:617 Gramen Indicum, è foliorum summitate radices agens, & novam plantam producens.

Katou-tsjolam HM 12:113:60
Kálu-Tsjólam
C.Commelin 1696:39 = HM.
Ray 1704 vol.3:617 Gramen paniculatum Indicum, granis in calycibus foliosis, rotundis nigris.
Linnaeus 1753:991 Zizania terrestris.
Linnaeus 1763:1408 = Linnaeus 1753.
N.L.Burman 1768:200 = Linnaeus 1763.
Houttuyn 1782 II vol.13:551-552 = N.L.Burman 1768.

Kuren-pullu HM 12:115:61
Kureṅ-pullù
C.Commelin 1696:40 = HM.
Ray 1704 vol.3:617 Gramen Indicum caulibus pilosis paniculatum, locustis conicis, singulis singula semina continentibus.
Plukenet 1705:146 Milium Indicum hirsuto caule, semine aureo.

Tsjeria-kuren-pullu HM 12:117:62
Tsjería kureṅ pullū

C.Commelin 1696:67 = HM.
Plukenet 1696:177 Gramen geniculatum brevifolium crispum, spicâ purpuro-sericeâ Maderaspatanum, Phytogr.Tab.119.fig.1 [Plukenet 1691:Tab.CXIX,f.1] (?).
Petiver 1704(1702):1255,no.258 Mautangee pilloo Malab. Alopecuros Malabarica folijs undulatis spica praetenue.
Ray 1704 vol.3:616 Canna Indica gracilis, geniculata, spicâ in vertice longâ plumosâ.
Linnaeus 1753:54 Saccharum spicatum.
Linnaeus 1762:79 = Linnaeus 1753.
N.L.Burman 1768:23,Tab.9,f.3 = Linnaeus 1762.
Houttuyn 1782 II vol.13:148-149 = N.L.Burman 1768.

Kol-pullu HM 12:119:63
Kol-pullù
C.Commelin 1696:40 = HM.
Ray 1704 vol.3:618 Gramen cyperoides Indicum, spicis habitioribus, in pediculis oblongis è caulium vertice pluribus simul egressis.
Plukenet 1705:112 Gramen Cyperoides Ind.Or. spicis ex communi centro plurimis, Alopecuro accedentibus, summo caule paniculatis.

Tsjeru-tsjurel HM 12:121:64
Tsjéru-tsjúrel
C.Commelin 1696:67 = HM.
Plukenet 1696:276 Palma (fortè) s. Phaenico-Scorpiuros, aut Heliotropium Palmites spinosum Polygonati angustis foliis Maderaspatana. Phytogr.Tab.106.fig.1.&2. [Plukenet 1691:Tab. CVI,f.1,2] (c).
Petiver 1699(1698):326-327,no.27 Perrepan Chedde Mal. Rottang Malabaricus minor.
Ray 1704 vol.3:616 = HM.
Rumphius 1747 vol.5:97-101,Tab.LI Palmijuncus Calapparius (-); J.Burman ibid.:101 (r).
Linnaeus 1762:463 Calamus Rotang.
N.L.Burman 1768:84 = Linnaeus 1762.
Houttuyn 1775 II vol.4:445-448 = N.L.Burman 1768.

Katu-tsjurel HM 12:123:65
Kátu-tsjúrel
C.Commelin 1696:39 = HM.
Ray 1704 vol.3:616 = HM.
Plukenet 1705:40 Arundo nucifera spinosa, foliis plurimis, circà caulem ambientibus, fructu nigricante.
Rumphius 1747 vol.5:97-101,Tab.LI Palmijuncus Calapparius (-); J.Burman ibid.:101 (r).
Linnaeus 1762:463 Calamus Rotang.

Perim-tsjurel HM 12:125:66
Perim-tsjúrel
C.Commelin 1696:52 = HM.
Ray 1704 vol.3:616 = HM.
Rumphius 1747 vol.5:97-101,Tab.LI Palmijuncus Calapparius (-); J.Burman ibid.:101 (r).

Tsjolap-pullu HM 12:127:67
Tsjolap-pullu
C.Commelin 1696:67 = HM.
Ray 1704 vol.3:619 Gramen cyperoides Indicum, paniculis pluribus è foliorum alis egressis, spicis parvis oblongis angustis, compositis.
Plukenet 1705:112 Gramen cyperoides Indianum, foliis variè adumbratis, paniculis Graminis pratensis minoris Nostratium, ex alis foliorum prodeuntibus.

Rumphius 1750 vol.6:20-22,Tab.VIII Carex Amboinica (-); J.Burman ibid.:22 (c).
N.L.Burman 1768:19 Schoenus paniculatus.

Tsjeru-kotsjiletti-pullu HM 12:129:68
Tsjéru Cotsjilétti-pullū
C.Commelin 1696:67 = HM.
Ray 1704 vol.3:604 Graminifolia aquatica, Indica, cauliculis nudis capitula rotunda gestantibus.
Plukenet 1705:188 Scabiosa graminifolia, s. Statice minima argentea, foliis & cauliculis hirsutie pubescentibus, ex Provincia Floridanâ (r).
Linnaeus 1747:20-21,no.50 Eriocaulon culmo sexangulari, foliis setaceis.
Linnaeus 1753:87 Eriocaulon setaceum.
Linnaeus 1762:129 = Linnaeus 1753.
N.L.Burman 1768:31 = Linnaeus 1762.
Houttuyn 1782 II vol.13:451-452 = N.L.Burman 1768.

Kavara-pullu HM 12:131:69
Cavára-pullū
C.Commelin 1696:34 Gramen dactylon Aegyptiacum B:Pin:7 [Bauhin 1623:7].
Plukenet 1696:175 Gramen dactylon Americanum cruciatum Barbadensibus nostratibus Dutch-Grass dictum (?).
Ray 1704 vol.3:605 Gramen crucis, sive Niem-el-Salib Alpini J.B.
J.Burman 1737:106 Gramen cruciatum Zeylanicum, an Gramen Miliaceum latifolium? Mus.Zeyl.pag.6 [Hermann 1717:6].
Rumphius 1750 vol.6:9-10,Tab.IV Gramen vaccinum, aliaque Gramina (-); J.Burman ibid.:10 (cs).
Linnaeus 1762:106 Cynosurus indicus.
N.L.Burman 1768:28-29 = Linnaeus 1762.
Houttuyn 1782 II vol.13:284-286,Pl.XCI,Fig.3 = N.L.Burman 1768.

Catri-conda HM 12:133:70
Cafrī-Condā
C.Commelin 1696:19 = HM.
Plukenet 1696:250 Milium arundinaceum semine Lithospermi facie maximo durissimo, H.Leyd. [Hermann 1687:426] (?).
Petiver 1704(1702):1264-1265,no.313 Nelle monnee Malab. Arundo Lithospermos Ger.emac.
J.Burman 1737:138 Lachryma Jobi Zeylanica, omnium maxima.
Linnaeus 1737:437-438 Coix seminibus ovatis.
Linnaeus 1747:156-157,no.330 = Linnaeus 1737.
N.L.Burman 1768:194 Coix Lacryma Jobi β Linn.sp.1378 [Linnaeus 1763:1378].
Linnaeus 1771:494 = N.L.Burman 1768.
Houttuyn 1782 II vol.13:511-513 Coix Lacryma β Syst.Nat.XII. Gen.1043 [Linnaeus 1767 vol.2:615].

Tsjeli HM 12:135:71
Tsjélli
C.Commelin 1696:67 = HM.
Plukenet 1696:200 Juncus glomeratus Maderaspatensis acumine longo geniculato, Phytogr.Tab.197.fig.6 [Plukenet 1691:Tab. CXCVII,fig.6] (?).
Petiver 1704(1702):1259-1260,no.287 Souta Cora Malab. Juncus Malabar. major, super capitulum tantum geniculatus.
Ray 1704 vol.3:629 Juncus Indicus geniculatus, paniculâ compactâ, è spicis squamosis compositâ.
Linnaeus 1753:47 Scirpus articulatus.
Linnaeus 1762:70 = Linnaeus 1753.
N.L.Burman 1768:21 = Linnaeus 1762.
Houttuyn 1782 II vol.13:86 Scirpus articulatus [Linnaeus 1767 vol.2:82].

Ramacciam HM. 12:137:72
Ramácciam
C.Commelin 1696:58 = HM.
Plukenet 1696:200 Juncus acutus maritimus, caule triquetro rigido, mucrone pungente, Phytogr.Tab.40.fig.1 [Plukenet 1691:Tab.XL,f.1] (?).
Ray 1704 vol.3:604 Gramen Indicum spicatum foliis confertis concavis, floribus spicatis plumosis.
N.L.Burman 1768:219 Andropogon schoenanthus β Linn.sp. 1481 [Linnaeus 1763:1481].

Nain-canna HM 12:139:73
Naiñ Cánna
C.Commelin 1696:46 = HM.
Petiver 1704(1702):1260,no.291 Corki pullu Malab. Donax Madraspat. vulgaris facie (?).
Ray 1704 vol.3:615 Canna seu Arundo Indica, paniculis latè diffusis in summis caulibus, è pluribus spicis squamosis compositis.
Plukenet 1705:39 Arundo Maderaspatana, panicula maximè sparsâ.

Tiri-panna HM 12:141:74
Tiri-pánna
C.Commelin 1696:66 = HM.
Ray 1704 vol.3:604 Graminifolia Indica flore & fructu destituta; Gramen Indicum flore & fructu carens, foliis crassis, pulverulenta lanugine obductis.
Linnaeus 1747:180,no.380 Acrostichum frondibus linearilanceolatis acutis, caule scandente. Heiligt.acrost.9.n.2.
Linnaeus 1753:1067 Acrostichum lanceolatum.
Linnaeus 1763:1523 = Linnaeus 1753.
N.L.Burman 1768:228 = Linnaeus 1763.
Houttuyn 1783 II vol.14:68 = N.L.Burman 1768.

Ily-mullu HM 12:143:75
Ilỹ mullũ
C.Commelin 1696:37 = HM.
Petiver 1704(1702):1265,no.315 Ravana Mese Malab. Spartum Avenaceum pumile Malabaricum.
Ray 1704 vol.3:614 Arundo Indica foliosa, aculeata, panicula in summo è spicis pluribus compositâ.
Rumphius 1750 vol.6:6,Tab.II,Fig.2 Cyperus litoreus (-); J.Burman ibid.:6 (c).
Linnaeus 1767:34 Stipa spinifex.
N.L.Burman 1768:29-30 Stipa littorea.

Linnaeus 1771:300-301 Spinifex squarrosus.
Houttuyn 1782 II vol.13:561-564,Pl.XCII,Fig.2 Spinifex squarrosus Syst.Nat.Veg.XIII.Gen.1333.p.757.

Velutta-modela-muccu HM 12:145:76
Velutta-mudela-muceu
C.Commelin 1696:68 = HM.
Ray 1704 vol.3:117-118 Persicaria Indica mitis, non maculosa, angustifolia, alba.
Linnaeus 1762:518 Polygonatum barbatum (descr.).

Schovanna-modela-muccu HM 12:147:77
Schovánna-múdela-múccu
C.Commelin 1696:60 = HM.
Plukenet 1696:288 Persicaria Maderaspatana longiore folio hirsuta Phytogr.Tab.210.fig.7 [Plukenet 1691:Tab.CCX,fig.7] (?).
Ray 1704 vol.3:118 Persicaria Indica mitis non maculosa, angustifolia, rubra.
Linnaeus 1762:518 Polygonum barbatum (tab.).
N.L.Burman 1768:89 = Linnaeus 1762.
Houttuyn 1777 II vol.8:473-474 Polygonum Orientale [Linnaeus 1767 vol.2:275].

Tsjtti-pullu HM 12:149:78
Tsjétti-pullũ
C.Commelin 1696:67 = HM.
Petiver 1704(1702):1261,no.295 Cavaree Codee Malab. Dactylon Malabar. maximum esculentum.
Ray 1704 vol.3:606 Gramen dactyloides ramosum, spicis habitioribus, fructu esculento.
Rumphius 1747 vol.5:203 (Auct), Tab.LXXVI,Fig.2 Panicum Gramineum, seu Naatsjoni (=).
Linnaeus 1762:106-107 Cynosurus coracanus.
N.L.Burman 1768:29 = Linnaeus 1762.
Houttuyn 1782 II vol.13:281-283 = N.L.Burman 1768.

Tenna HM 12:151:79
Teñna
C.Commelin 1696:66 = HM.
Plukenet 1696:174 Gramen Paniceum spicâ asperâ. CBP.8 [Bauhin 1623:8].
Petiver 1704(1702):1265,no.314 Tenne Malab. Panicum Malab. esculentum, spica aristata e plurimis capitulis dense stipata.
N.L.Burman 1768:220 Holcus spicatus Linn.sp.1483 [Linnaeus 1763:1483-1484].

Appendices

APPENDIX 1

Genealogical table of the Van Reede family

The dedications of the volumes 4-10 of Hortus Malabaricus are given in italics

```
Antoni Utenhove                                                                          Gerard van Reede
? - 1625                                                                                 ? - 1612
lord of Rijnestein                                                                       lord of Nederhorst
x                                                                                        x
Agnes van Renesse                                                                        Machteld Peunis van Diest
van Baer                                                                                 ? - 1615
? - 1613
                    Godard van Reede                                                     Ernst van Reede
                    1588 - 1648                                                          1588 - 1640
                    lord of Nederhorst                                                   lord of De Vuursche and Drakenstein
                    x                                                                    x
Charles Utenhove    Catharina Utenhove                                                   Elisabeth Utenhove
? - 1641            ? - 1656                                                             ? - 1637
lord of Rijnestein
x                   Margaretha van Reede  Gerard van Reede    xx  Agnes van Reede   Maria van Reede    Gerard van Reede          Machteld van Reede
Alexandrina van Tuyll   ? - 1669          1624 - 1670             ? - 1692           ? - 1662           ? - 1669                  x
van Serooskerken    x                     lord of Nederhorst      x                  x                  lord of De Vuursche      Bartholomeus van Panhuiz
? - 1640            Gerard van Reede                              René van Tuyll     Diederik van Baer  and Drakenstein          ? - 1676
|                   ? - 1669                                      van Serooskerken   ? - 1683           x                        lord of Voorn
Hendrik Utenhove    lord of De Vuursche                           ? - 1652           volume 7 (1688)   Margaretha van Reede
1630 - 1715         and Drakenstein                               lord of Rijnhuizen                    ? - 1669
lord of Amelisweerd
volume 10 (1690)
                                Anna Elisabeth van Reede  x  Hendrik Jacob van Tuyll  Elisabeth van Tuyll  Frederik Hendrik      Antoni Carel van Panhuize
                                ? - 1682                     van Serooskerken         van Serooskerken    van Reede              1657 - 1714
                                                             1647 - 1692              ? - 1689            x                      lord of Te Vliet
                                                             lord of Zuilen           x                   Johanna Schade
                                                             volume 4 (1683)          Paul de la Baye
                                                                                      ? - 1699
                                               Reinoud Gerard van Tuyll               baron of Theil                             Bartholomeus Cornelis
                                               van Serooskerken                                                                  van Panhuizen
                                               1677-1729                                                                         1696 - 1716
                                               lord of Zuilen, Mijdrecht,                                                        lord of Te Vliet
                                               etc.
```

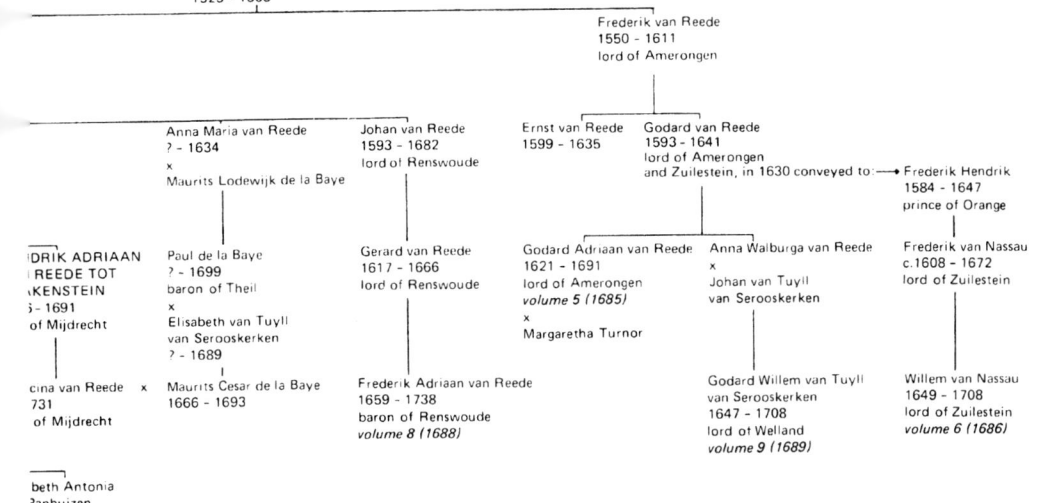

Appendix 1

```
                            Goert van Reede
                            1516 - 1585
                            lord of Saasveld
                            x
                            Geertruid van Nijenrode
                            1525 - 1605
                                            |
                                    Frederik van Reede
                                    1550 - 1611
                                    lord of Amerongen
```

	Anna Maria van Reede	Johan van Reede	Ernst van Reede	Godard van Reede		Frederik Hendrik
	? - 1634	1593 - 1682	1599 - 1635	1593 - 1641		1584 - 1647
	x	lord of Renswoude		lord of Amerongen		prince of Orange
	Maurits Lodewijk de la Baye			and Zuilestein, in 1630 conveyed to: →		

iDRIK ADRIAAN	Paul de la Baye	Gerard van Reede		Godard Adriaan van Reede	Anna Walburga van Reede	Frederik van Nassau
REEDE TOT	? - 1699	1617 - 1666		1621 - 1691	x	c.1608 - 1672
KENSTEIN	baron of Theil	lord of Renswoude		lord of Amerongen	Johan van Tuyll	lord of Zuilestein
1 - 1691	x			*volume 5 (1685)*	van Serooskerken	
of Mijdrecht	Elisabeth van Tuyll			x		
	van Serooskerken			Margaretha Turnor		
	? - 1689					

cina van Reede x	Maurits Cesar de la Baye	Frederik Adriaan van Reede		Godard Willem van Tuyll	Willem van Nassau
731	1666 - 1693	1659 - 1738		van Serooskerken	1649 - 1708
of Mijdrecht		baron of Renswoude		1647 - 1708	lord of Zuilestein
		volume 8 (1688)		lord of Welland	*volume 6 (1686)*
				volume 9 (1689)	

beth Antonia
Panhuizen
? - ?
 of Te Vliet

aume Jalabert
764

tte Le Roux
   |
ander Le Roux Jalabert
 - 1803

APPENDIX 2

Translation of the commission for Hendrik Adriaan van Reede tot Drakenstein as commander of Malabar, dated 26 September 1669, at Batavia.
VOC 893:711-713.

Commission for Commander Hendrik van Rheede.

Joan Maetsuijcker, Governor-General, together with the councillors on behalf of the General Chartered Dutch East India Company in India, salute all those that shall see or hear this read, and give notice that in the resolution of 6 September last we have thought fit to appoint Mr. Lucas van der Dussen, at present commander of Cochin and dependencies on the coast of Malabar, as director of the Company's important trade in the Persian Empire. At the same time it has been established in the said resolution that Cochin and dependencies shall be raised to a separate commandment that is not dependent on the Ceylonese Government, under which it fell up to this day. In order to fill this vacancy again, a qualified, able, and experienced person is necessarily required. Therefore, since we are pleased with the good qualities of the Honourable sergeant-major Hendrick van Rede, who, having militated many years in Ceylon as well as on the aforesaid coast of Malabar and having filled several distinguished functions, has served us to our great satisfaction and content, we have thought fit to nominate, appoint, and authorize him, as we herewith do, as absolute commander of the Company's towns, fortresses, trade, and further administration on the said coast of Malabar, charging and commissioning him to go by the earliest opportunity from Ceylon, or wherever he should be, to Cochin. Upon his arrival there he shall take over from Mr. Lucas van der Dussen, as is proper, all the Company's effects, books, documents, etc. Further, under the title of commander he shall hold supreme authority and command over the administration and the subordinates of the Company, pursue and attend to the trade, administer, with the recommendation of the councillors, law and justice, in both criminal and civil cases, and further do everything that has to be done and befits a good and faithful commander. We therefore ordain and order all commanders, merchants, masters, under merchants, assistants, lower officials, soldiers, sailors, and in general all other people under our authority, both those who are already stationed at Cochin and on the coast of Malabar and those who should arrive there later, no one excepted, to acknowledge, respect, and obey the said Mr. Hendrick van Rheede as such as has been said, and furthermore to assist him by word and deed to the best of their ability, in accordance with the oath by which everyone is bound to the General Chartered Dutch East India Company as we consider it to be proper for its prosperity. Given at Batavia, at the castle, on 26 September 1669. Signed: Joan Maetsuijcker, who has affixed, in the open space underneath, the Company's stamp in red sealing-wax. Subscription: signed by the order of the aforesaid
Honourable gentlemen,
Joan van Riebeeck, Secretary.

APPENDIX 3

Translation of the contract of Hendrik Adriaan van Reede tot Drakenstein, Johannes Munnicks, and Jan Commelin with Pieter van Someren, Hendrik Boom, and the widow of Dirk Boom, dated 7 January 1681 OS (= 17 January 1681), at Utrecht.
GAU, Notarial Records U080a006 (protocols W. Zwaardekroon):241-241v.

Today, the 7th of January 1681 Old Style, appeared as above the Right Honourable Mr. Henrick van Rhede tot Draeckesteijn, Lord of Mijdrecht, etc., appearing as a member of the Nobility and the Equestrian Order of the Country of Utrecht at the meeting of the Right Honourable Gentlemen the States of this country, and Mr. Johannes Munnix, professor of medicine in the University of this City, in case of need having confidence in and standing surety proportionally for Mr. Johannes Commelijn, Councillor in the City Council of Amsterdam, on the one hand, and on the other hand Mr. Pieter van Someren, bookseller at Amsterdam, on his own behalf and having confidence in and standing surety proportionally for Mr. Hendrick and the widow of Mr. Diderick Boom, also booksellers in that city, all four of them being Dutch subjects. They declared that they had agreed with each other by a contract that the above-mentioned gentleman, the first appearant, as the principal author, and the said Professor Munnix as well as Mr. Commelijn shall deliver every year to the said Van Someren cum suis the descriptions and drawings of two volumes of the book or work called Hortus Malabaricus (of which two volumes have already been printed and published), to be printed by the said Van Someren cum suis in the same way and the same format as the two aforesaid volumes already published. On the other hand, the said Van Someren cum suis, as he promises herewith, shall complete and publish the said volumes, properly printed and properly constituted, each time within one year from the date of the exact delivery of the said descriptions and drawings by or on behalf of the aforesaid first three gentlemen. Furthermore this shall be done and accomplished by one party as well as the other, year by year, until the aforesaid work is complete. All this shall be done at the expense of and be chargeable to the aforesaid Van Someren cum suis. Van Someren, in the aforesaid capacity, promises that he cum suis will uphold this. Moreover, as each volume is printed, they shall deliver to the aforesaid first three gentlemen, without receiving any payment therefor: ten copies to the said Mr. Van Rhede tot Draeckesteyn, twelve to the said professor, and six to the said Mr. Commelijn, thus together twenty-eight unbound copies. Furthermore, the condition has been made that this contract shall only apply during the lifetime of the aforesaid Van Someren and Hendrick Boom. If one of them should die, the longest-lived shall retain the option to perform this contract. On the other hand, the said Mr. Van Rhede tot Drakesteyn engages that after his death his heirs shall be obliged to deliver the aforesaid descriptions and drawings from time to time, until the work is complete. For the fulfilment of this, the appearants severally stand surety with their persons and property. They submit it to the judgment and execution of the Supreme Courts of Holland and Utrecht, the Courts of Utrecht, and any other judges. They request proper subscription thereof, which has been done herewith.

Thus done and ratified at Utrecht, at the house of the said Mr. Rhede, in the presence of Johannes van Steenbergen and Engelbert Bouwman, who have been requested to act as witnesses.

Johannes Munnicks	H. v. Reede
Pieter van Someren	
Engelbertus Bouman	Johannes van Steenbergen
Zwaerdecroon, Notary Public 1681	

APPENDIX 4

Translation of the instruction of Hendrik Adriaan van Reede tot Drakenstein to the Western Quarters, dated 5 March 1691, at Cochin.
VOC 1478:1382-1385.

To all the Governors, Directors, and Commanders of Ceylon, Coromandel, Bengal, Surat, Persia, Malabar, and Cape of Good Hope,

Gentlemen and friends, etc.

I have requested some of you orally to send annually by the homeward-bound ships from Ceylon to the fatherland all kinds of seeds, bulbs, or roots of the trees, plants, herbs, flowers, etc. which each of you is able to collect in his district for a whole year. I did so because on leaving Europe I had received a large order from distinguished gentlemen to send annually such a consignment from the Indies, not only to satisfy the curiosity of many wise men and amateurs, but also in order that the world might enjoy and be served by the medicinal virtues which the above-mentioned wise men might discover and find out. Since I could only be in one place at a time, I was not very well able to execute this order without calling in the assistance of some of you and informing you of the order imposed upon me. But considering that during my presence here, in the Indies, this recommendation not only was renewed annually, but it was also made with an ever-increasing earnestness and insistence, because in the fatherland good results have already been got from these seeds, and also because I do not as yet consider myself capable of complying with this request (and this order) without your assistance and aid, I request you jointly and individually to cooperate and to cause all kinds of seeds, without exception, to be gathered, to pack them in two parts with the registers of their names, and to send them every year via Ceylon to the fatherland, with the exception of the governor at the Cape of Good Hope, who can do this directly. The cases are to be addressed to Mr. Pieter van Dam, solicitor and councillor of the above-mentioned Company in Amsterdam, one half to be forwarded and handed to His Majesty of Great Britain and the other part to Mr. Jan Crommelin [sic], for the account of the Academic garden of Amsterdam.

But since apart from the renewals of these orders I have also got the report that some of the seeds received in the fatherland were found to become contaminated and rotten, which is believed to have been caused by the fact that they were too old, or packed in a moist condition, I must not omit to inform you of this and to warn you accordingly that the young, hard, and bone-dry seeds are least prone to this, to which your good care can greatly contribute and which will give me, and those who have given me these orders, reason to thank you. You should also know that by the name of seeds we do not mean seeds of garden plants with are tame and known in Europe, but exotic, wild, and unesteemed seeds which occur and can be found without exception in the forests. As we have said above, they must be packed in two parts, so that for greater security they may be sent by two different ships. It would be desirable that they are packed in wooden, turned boxes, which can be put together in two general boxes or cases, because it has been found in the fatherland that the seeds which had been sent by us in this way in boxes from Bengal, all arrived in very good condition, but that, on the contrary, the seeds which were sent in small bags or cloths were usually received in decayed condition. By the present I commend you together to the protection of the Lord, and I remain, signed, H. van Reede, aside, Chocim [sic], the 5th of March 1691.

APPENDIX 5

Translation of the last will of Hendrik Adriaan van Reede tot Drakenstein, dated 17 September 1691, at Cochin.
VOC 1505:420-422.

By the content of this instrument it is notified to all concerned that today, the 17th of September of the year 1691, I, Hendrik Swaerdekroon, secretary in the Commission representing the meeting of the XVII, and I, Johannes Scholten, provisional secretary of the commandment of the Coast of Malabar in India, both in the service of the General Chartered Dutch East India Company, and the witnesses to be mentioned hereinafter, have come and appeared personally before the Right Honourable Mr. Hendrik Adriaen van Rheede tot Drakenstijn, Lord of Mijdrecht, etc., head of the above-mentioned Commission, today present on the said Coast of Malabar, within the town of Cochin, still unmarried and well known to both of us, sound in body, moving at his free will, and to all outward appearances in full possession and use of his mental faculties, reason, and memory.

He declared -taking into account the general infirmity and fragility of man's life on earth, fleeting like a shadow, and the common debt of nature, viz. death, to which he is subject with the greatest certainty, although the time and the hour of it are uncertain- that he therefore intended and meant not to depart this life without having ordained and disposed finally of the temporal property which Almighty God has bestowed on him. He openly declared that he does this of his own free will and his own disposition, without anyone inciting or misleading him. First commending his immortal soul to the merciful and charitable hands of Almighty God, his Creator and Saviour, and his dead body to the earth, with a decent burial, he revokes, cancels, and annuls herewith any such testaments, codicils, and similar legacies as he, testator, should have executed before the present date. He does not wish or desire that such documents or any of them shall in any point be followed or performed in any way whatsoever, and, disposing anew, he, testator, declares that he bequeaths, wills, and secures to Agnes van Reede tot Drakenstijn, Lady of Nederhorst, his only sister, a sum of three hundred guilders annually as long as she shall live, unconditionally. In the same way to various persons mentioned in a separate memorandum such legacies and sums of money as are specified therein. He has executed and confirmed this memorandum with his own signature before me, Hendrik Swaerdekroon, secretary in the aforesaid commission, and handed it to the heiress to be mentioned hereinafter. He wishes and desires that the said memorandum shall have such force as if it were expressed herein verbatim. But should it happen that through travel, mislaying, or displacement, or similar accidents, the heiress cited above and mentioned by name below were so unfortunate as to lose the aforementioned memorandum or to have it filched from her, the testator explicitly desires that no one, whoever this might be, judging from the presupposed supposition or other cause that he might be mentioned or interested in the said memorandum, shall have power to demand it from her or cause her any difficulty. In that case the memorandum shall be regarded as cancelled and as if it had never been issued or published. The testator in that case is content that the heiress is aware of what legacies and legatees are mentioned therein. He remains quite easy about this and does not leave to anyone the right to summon her about this or to implicate her, no matter whether the memorandum is still in her power or hands or not. In the present case all the aforesaid legacies shall have to be paid within one year after testator's decease by his heiress to be mentioned hereinafter, and that freely, without any deduction of a falcidian portion or any other defalcations known in the law. Furthermore, he, testator, declares that he institutes and presents, as he institutes and presents herewith, as his only, universal, and entire heiress Francina van Reede, whose godfather he, testator, was and whom he brought up and always considered as his adopted daughter and still does. After the decease of the testator she shall therefore as heiress acquire the full possession and ownership of the property left by him, testator, both personal and real property, consisting of manors, landed estate, houses, gardens, orchards, debentures, annuity bonds, cash, gold, silver, jewelry, gems, account of the salary earned in the service of the Company, and so on, such and in such manner as the testator may have possessed, administered, and owned or acquired during his lifetime. As regards the above-mentioned salary earned, he especially requests and authorizes the Right Honourable Gentlemen, directors of the General Dutch East India Company at the meeting of the aforesaid XVII, to be executors of this testament.

He, testator, has declared that all this is his testament and his last will, and he ordains that it shall be completely valid, either as testament, codicil, or any other legacies, as will be most appropriate, even if there are any omissions or all the solemnities could not be performed or maintained perfectly herein according to law. He therefore implores all judges and courts if this should be necessary to have his last will executed and performed in this way. When cognizance has been taken of this, he further requests that one or more instruments of this be furnished in the proper form and also that the originals be registered twice, viz. one at the office of the aforesaid Commission and the other at the office of the Commandment of this Coast.

Thus done and executed at the town of Cochin, on the day and in the month and year as shown above, in the presence of Gerrit van Thol, first clerk of the aforesaid Commission, and Louis Taijspil, first clerk of the commandment of Malabar, requested to be present as reliable and sworn witnesses, who have signed the originals of this last will, along with the testator and us, secretaries. Each of the originals consists of two separate sheets, which have been fastened together by the seal of the Commission and thus are deposited at the two above-mentioned offices.

Subscriptions and signatures:
Hk. Swaerdecroon, secretary of the Commission, and
J. Scholten, provisional secretary of the Commandment.

APPENDIX 6

Translation of the report of the funeral of Hendrik Adriaan van Reede tot Drakenstein at Surat on 3 January 1692.
VOC 1529:404v-408v, resolution of 3 January 1692 of the provisional directors and the council of Surat.

Funeral procession held at Surat, on the 3rd of January 1692, on the occasion of the interment of the body of the Right Honourable Hendrick Adriaan van Reede tot Drakesteijn, Lord of Meijdregt, member of the Equestrian Order and delegated from said order into the ordinary deputies of the Honourable Gentlemen the States of the Country of Utrecht, commissioner on behalf of the General United Dutch Company, representing in that capacity the Right Honourable Gentlemen, directors at the meeting of the XVII, who died blissfully after a prolonged, lingering illness on 15 December 1691, between 3 and 4 o'clock in the afternoon on the ship Dreghterland, sailing to Surat off Bombay. Insofar as this was possible in connection with the spot, the body was properly embalmed and later, on the 24th of December, brought by the same ship to the royal roadstead and on the 26th by the sloop Nassouw in front of the Company's garden and further put ashore, with the black flags and pennons flying and with the small guns firing, and taken into the hall downstairs on the place of the Company's garden, surrounded by the new Kanarese and hung with crape inside. There a watch was kept until, as said, on the 3rd of January 1692, at 9 o'clock in the morning, the funeral procession began from there in the following order; it was completed in the evening.

First of all, the four gold-imprinted prince's flags and pennon of the Company, flying from five silver-coated staffs, borne by five natives.

Fifty peons or native soldiers, commanded by the native minda or ruler.

Twenty small prince's flags, each according to the country's custom, flying from a heavy rocket as used in war by the natives, borne by 20 natives.

Four native runners with plumes on the turbans, according to the country's custom.

The canopy hung with crape, borne by the Kanarese or native bearers, and by their side four chobdars or native ushers with silver staffs, on horseback according to the country's custom, on either side two behind each other.

Two small field-guns one behind the other, each of them drawn by eight natives, and two gunners with mourning-bands and sashes, bearing the linstocks upside down under their arms, near the muzzle of each gun.

A company of soldiers under the Hon. Lieutenant Cs. van Pothuijsen, who was on horseback on account of his indisposition, 36 men strong, including eight grenadiers, who marched together with their rifles upside down in 8 files of 4 men each; the partisan and the halberds were provided with mourning-ribbons and the standard had such ribbons wound around it, the drums were covered with crape, and further the officers and the drummers were provided with mourning-bands and sashes.

Two small field-guns, each of them drawn by 6 natives, and for the rest as above.

A company of soldiers, 35 men strong, under the Sergeant Alexander Wast, marching in eight files of 4 men each, and for the rest as above.

Three trumpeters, to wit two in mourning with banderoles and black tassels and one in the middle with a tunic of red cloth, with the arms of the Company embroidered in silver thread on the chest, and with the Company's banderoles on the silver trumpet; they all blew through sordinos and they were all provided with mourning-bands and veils.

A Company with a gold-imprinted prince's flag and the pennon at the top, flying from a silver-coated staff, borne by Jan van Leeuwen.

The first standard, borne by the bookkeeper Abraham van Helsdingen.

The horse equipped for tournaments, with a golden harness, white and black plumes on its head and hung with trappings of red cloth, profusely embroidered with gold, and led in this way by the first mate of the ship Dreghterland, called Jacob Coensz., and the first assistant Lt. Limbeek.

The large blazon borne by the bookkeeper Mode and the first assistant Js. Sloffen.

The four main quarters borne one behind the other by Cornelis Jansz. Croon, master, Joris Harting, bookkeeper on the ship Dreghterland, François Mol, assistant in the Commission, and Ludolff Tervile, under surgeon at this office.

A horse hung with crape, with the arms fastened thereon and led by the provisional assistants Js. de Meester and Samson Jansz.

The spurs borne on the black staff by the dispenser Joan Uldrix.

The gloves borne as above by the bookkeeper and warehouse master Mr. Aij. Erhart.

The gilt foil in the sheath with the hilt upward, borne by the bookkeeper Mr. Dirk Debuson.

The helmet with black and white plumes, on a black velvet cushion, borne by the under merchant Mr. Isaak Seloven.

The coat of mail borne by the provisional under merchant Gt. van Thol, first clerk in the Commission.

A led horse hung with crape and with the arms fastened thereon, led by the assistants Pr. Uldrix and Hendrik Dijkman d'Jonge.

The second standard borne by the first assistant Pr. van Breen, first clerk at the office and secretary of Justice at this place.

The mourning-horse, hung with black velvet, with the arms fastened thereon, and black and white plumes on its head, led by Sijbrant de Jong, first mate on the ship the Maes, and the first assistant Jacob Schuijnderman.

The sword in the sheath, with the point upwards, borne by the chief of equipage Jan Roeloffsz.

The first assistant Meijndert Vos, having a presentable stature or figure, in full armour or cuirass, with the regimental staff in his hand.

The body on a hearse drawn by 4 ordinary draught animals, all of them hung with crape and with the arms fastened thereon, the coffin covered with a new black velvet pall and with eight quarters fastened thereon; the four corners of the pall being lifted by 4 men, and the hearse further being followed by 12 bearers, to wit

the four pall-bearers
Gillis van Markel, master of the ship Dregterlant
Cors van der Mast, ditto of the ship the Maes
Cornelis van Oosterhoff, under merchant
Mattheus de Haan, ditto
the 12 bearers
Dirk Westerhuijse, upper surgeon on the Dreghterlant
Jan Wassenaar, sick-visitor on the Maes
Adriaan Christiaansz. Wijnants, ditto at this place

Jacob Groenrijs,	first assistant
Jan van Nes,	assistant
Jacob van Hoorn,	ditto ditto
David van Orleij,	,, ,,
Jan van Ommen,	,, ,,
Jan Martensz.,	,, ,,
Jan Martin Rason,	,, ,,
Jacob Sterksel,	,, ,,
Anthonij van Helsdingen.	

All the aforesaid ornament-bearers, pall-bearers, and bearers as well as all further persons were in mourning and provided with mourning-bands and sashes; they were further followed by Mr. Adriaan van Ommen, in a funeral coach, representing the kinsfolk of His Honour, of laudable memory.

And further Mr. Pr. van Helsdingen, the Abyssinian Ambassador Godsja Moraat, the English free merchant Mr. George Boucher, Mr. Swaardecroon, secretary, and other members of the Commission, the Jewish merchants, the further qualified persons, and members of the Council of Surat, together with the other Dutchmen stationed and residing here as well as many of the most distinguished Armenian merchants, all of them in coaches or on horseback. All the qualified persons and the members of the Council of Surat were also in mourning and provided with mourning-bands and sashes. The other friends, both foreigners and Dutchmen, all wore white veils. The retinue would have been even larger if the English President Mr. Bartholomeus Hannijs and his party had been present; they had been invited the day before, but His Honour had asked to be excused.

In this way the procession marched from the Company's garden, around the town, to the cemetery, where during the last ceremonies the militia fired 3 salvoes and between each of them a gunshot with the small field-guns. Meanwhile, from the morning, the black flags and pennons were flying from the Company's office. The ships of the Company, also provided with black flags and pennon, in addition to the admiral's flag, were anchored upstream in the town. After the aforesaid salvoes and gunshots they blazed away three times in succession all round, to wit the Speelman each time ..., the Orangie ..., the Amsterdam ..., and the sloop Nassouw ..., being together ... shots*. This was done while the militia, according to the custom, drew off with drums beating and colours flying, followed by all the aforesaid gentlemen and friends in coaches, native carts, or on horseback. They marched through the town, across the large square, past the castle, back to the Company's garden, where the main dwelling was covered and hung with crape. The whole company was regaled with a repast which had been prepared in two different places. After this, all took their leave.

---

* In the manuscript the number of shots has not been filled in.

Map of Malabar

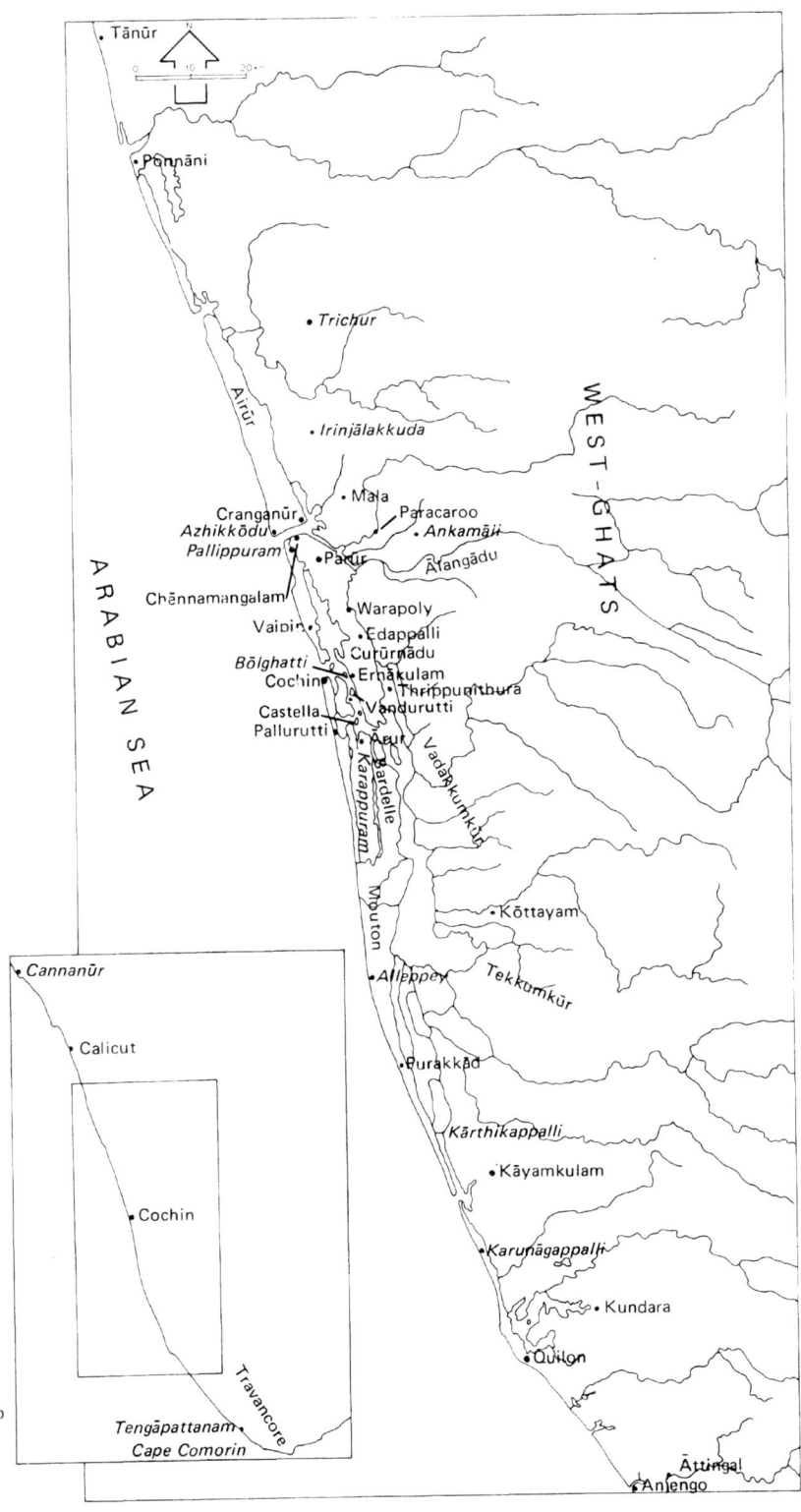

Map of Malabar (after s'Jacob 1976). The names in Roman type are finding-places of plants mentioned in Hortus Malabaricus, volumes 3-8.

Literature

Aa, A.J. van der 1852-1878. Biographisch Woordenboek der Nederlanden. 21 vols. Haarlem.
Aalbers, J. 1916. Rijcklof van Goens, commissaris en veldoverste der Oost-Indische Compagnie, en zijn arbeidsveld, 1653/54 en 1657/58. Groningen.
Ablaing van Giessenburg, W.J. d' 1859. De Ridderschap van de Veluwe, of Geschiedenis der Veluwsche Jonkers. 's-Gravenhage.
Acosta, Cr. 1578. Tractado De las Drogas y medicinas de las Indias Orientales. Burgos.
Adamson, R.S. & T.M.Salter 1951. Flora of the Cape Peninsula. Cape Town, Johannesburg.
[Albinus, B.S.] 1771. Pars Bibliothecae sive Catalogus Librorum Medicorum . . . Bernardus Siegfriedus Albinus . . . D.Lun. Oct. & seqq. 1771. Lugduni Batavorum.
Album Studiosorum Academiae Lugduno Batavae MDLXXV-MDCCCLXXV accedunt nomina curatorum et professorum per eadem secula. Hagae Comitum. 1875.
Album Studiosorum Academiae Rheno-Traiectinae MDCXXXVI-MDCCCLXXXVI. Ultraiecti. 1886.
Aldini, T. 1625. Exactissima Descriptio Rariorvm Quarundam Plantarvm, Quę continentur Romę in Horto Farnesiano. Romae.
Alpini, P. 1640 (J.Veslingi ed.). De Plantis Aegypti Liber. Patavii.
Andel, M.A. van (ed.) 1931. Bontius tropische geneeskunde. Bontius on tropical medicine. Amstelodami. - Opuscula Selecta Neerlandicorum de Arte Medica 10.
Andreas, Ch.H. 1953. Hortus Muntingiorum. Geschiedenis van de Groningse hortus in de zeventiende eeuw. Groningen, Djakarta. - Scripta Academica Groningana.
Arasaratnam, S. 1958. Dutch power in Ceylon 1658-1687. Amsterdam.
Baldaeus, Ph. 1672. Naauwkeurige Beschryvinge van Malabar en Choromandel, Derzelver aangrenzende Ryken, En het machtige Eyland Ceylon. Amsterdam.
Bauhin, C. 1620. Prodromos Theatri Botanici. Frankfurt.
Bauhin, C. 1623. Pinax Theatri Botanici. Basileae.
Bauhin, J. & J.H.Cherler 1650-1651. Historia Plantarum Universalis. 3 vols. Ebroduni.
Beek, J.R.J. van 1983. Pieter Hertog (1695-1728). - Type-script, Biohistorical Institute, Utrecht University.
Berendes, J. 1902. Des Pedanios Dioskurides aus Anazarbos Arzneimittellehre in fünf Büchern. Stuttgart.
Bijl, M. van der 1975. Utrechts weerstand tegen de oorlogspolitiek tijdens de Spaanse Successieoorlog. De rol van de heer van Welland van 1672 tot 1708, in: H.L.Ph.Leeuwenberg & L.van Tongerloo (eds), Van Standen tot Staten 600 Jaar Staten van Utrecht 1375-1975, pp. 135-199.
Bloys van Treslong Prins, P.C. & J.Belonje 1928-1931. Genealogische en Heraldische Gedenkwaardigheden in en uit de Kerken der Provincie Noord-Holland. 5 vols. Utrecht.
Bodaeus a Stapel, J. 1644. Theophrasti Eresii De Historia Plantarvm Libri Decem Graecè & Latinè. Amstelaedami.
Boeree, Th.A. 1943. De kroniek van het geslacht Backx. Wageningen.
Boerhaave, H. 1710. Index Plantarum, Quae In Horto Academico Batavo Reperiuntur. [Lugduni Batavorum].

Boerhaave, H. 1720. Index Alter Plantarum Quae In Horto Academico Lugduno-Batavo Aluntur. 2 vols. Lugduni Batavorum.
Bontius, J. 1942. De Medicina Indorum. Lugduni Batavi.
Bontius, J. 1658 (G.Piso ed.). Historiae Naturalis & Medicae Indiae Orientalis Libri Sex. Amstelaedami.
Botter-Weissleder, R. c.1950. Leben und Verdienste des Botanikers Paul Hermann. (1646-1695). - Type-script, thesis Halle University.
Boxer C.R. 1957. The Dutch in Brazil 1624-1654. Oxford.
Breyne, J. 1678. Exoticarum aliarumque minus cognitarum Plantarum Centuria Prima. Gedani.
Breyne, J. 1680. Prodromus Fasciculi Rariorum Plantarum Anno M.DC.LXXIX. in Hortis Celeberrimis Hollandiae . . . observatarum. Gedani.
Breyne, J. 1689. Prodromus Fasciculi Rariorum Plantarum Secundus, Exhibens Catalogum Plantarum Rariorum, Anno M.DC.LXXXIIX. in Hortis Celeberrimis Hollandiae observatarum. Gedani.
Breyne, J.Ph. 1739. Jacobi Breynii, Gedanensis, Prodromi Fasciculi Rariorum Plantarum Primus et Secundus. Gedani.
Bridel, Ph.C. 1857. Biographie de Laurent Garcin. Le Conservateur Suisse ou Recueil complet des Etrennes Helvétiennes (ed. 2 Lausanne), 13:69-76.
Brinkman, J. 1980. Surinaamse planten in Nederland in de zeventiende eeuw. - Type-script, Biohistorical Institute, Utrecht University.
Briquet, J. 1940. Biographies des botanistes à Genève de 1500 à 1931. Genève.
Brosterhuysen, J. 1647. Catalogvs Plantarvm Horti Medici Illvstris Scholae Avriacae Quae est Bredae. Bredae.
Bruijn, J.R., F.S.Gaastra, I.Schöffer & E.S.van Eyck van Heslinga 1979. Dutch-Asiatic shipping in the 17th and 18th centuries. Vols. 2 and 3. The Hague. - Rijks Geschiedkundige Publicatiën, Grote Serie, 166, 167.
Bunt, A.W. van de 1949. Een Utrechtenaar in de 17de eeuw. Jhr. Godard Willem van Tuyll van Serooskerken. Historia 14: 6-12.
Burman, J. 1737. Thesaurus Zeylanicus, exhibens Plantas In Insula Zeylana Nascentes. Amstelaedami.
Burman, J. 1769. Flora Malabarica, sive Index in omnes Tomos Horti Malabarici. Amstelaedami.
Burman, N.L. 1768. Flora Indica. Lugduni Batavorum.
Camerarius, J. 1588. Hortvs Medicvs Et Philosophicvs. Francofurti ad Moenum.
Chick, H.G.J. 1939. A Chronicle of the Carmelites in Persia. 2 vols. London.
Chijs, J.A. van der (ed.) 1891, see Dagh-Register . . . (1663).
Chijs, J.A. van der (ed.) 1893, see Dagh-Register . . . (1664).
Chijs, J.A. van der (ed.) 1903, see Dagh-Register . . . (1676).
Chijs, J.A. van der (ed.) 1904, see Dagh-Register . . . (1677).
Churchill, W.A. 1967 (ed.3). Watermarks in paper in Holland, England, France, etc., in the XVII and XVIII centuries and their interconnection. Amsterdam.
Clusius, C. 1601. Rariorvm Plantarvm Historia. Antverpiae.

Clusius, C. 1605. Exoticorvm Libri Decem: Quibus Animalium, Plantarum, Aromatum, aliorumque peregrinorum Fructuum historiae describuntur. Ex Offininâ Plantinianâ Raphelengii.

Commelin, C. 1696. Flora Malabarica sive Horti Malabarici Catalogus Exhibens Omnium ejusdem Plantarum nomina. Lugduni Batavorum.

Commelin, C. 1701. Horti Medici Amstelaedamensis Rariorum Tam Africanarum, quàm Utriusque Indiae, aliarumque Peregrinarum Plantarum. Pars Altera. Amstelaedami.

Commelin, C. 1703. Praeludia Botanica Ad Publicas Plantarum exoticarum demonstrationes. Lugduni Batavorum.

Commelin, J. 1676. Nederlandtze Hesperides. Amsterdam.

Commelin, J. 1683. Catalogus Plantarum Indigenarum Hollandiae. Amstelodami.

Commelin, J. 1689. Catalogus Plantarum Horti Medici Amstelodamensis. Pars Prior. Amstelodami.

Commelin, J. 1697. Horti Medici Amstelodamensis Rariorum Tam Orientalis, quàm Occidentalis Indiae, aliarumque Peregrinarum Plantarum. Amstelodami.

Coolhaas, W.Ph. 1960-1985. Generale Missiven van Gouverneurs-Generaal en Raden aan Heren XVII der Verenigde Oostindische Compagnie. 8 vols. 's-Gravenhage. - Rijks Geschiedkundige Publicatiën, Grote Serie, 104, 112, 125, 134, 150, 159, 164, 193.

Cordus, V. 1561 (C.Gesner ed.). Historiae Stirpium Libri IV. Argentorati.

Cornelisz, H. 1661. Catalogus Plantarum Horti Publici Amstelodamensis. Amstelodami.

Cornut, J.Ph. 1635. Canadensivm Plantarvm, aliarúmque nondum editarum Historia. Parisiis.

Dagh-Register gehouden int Casteel Batavia vant passerende daer ter plaetse als over geheel Nederlandts-India Anno 1624-1682 (J.A.van der Chijs, H.T.Colenbrander, F.de Haan, J.E.Heeres, J.de Hullu eds). 1887-1931. 31 vols. Batavia, 's-Gravenhage.

Dalgado, D.G.C. 1896. Vires Plantarum Malabaricum Ou Virtudes das Plantas do Malabar extrahidas do "Hortus Indicus Malabaricus" de Henrique van Rheede. Goa.

Dandy, J.E. 1958. The Sloane Herbarium. London.

De Ficalho, Conde 1891-1895. Coloquios dos Simples e Drogas da India por Garcia da Orta. 2 vols. Lisboa.

Dennstedt, A.W. 1818. Schlüssel zum Hortus Malabaricus, oder dreifaches Register zu diesem Werke. Weimar. - Also published in: Fortsetzung des allgemeinen Teutschen Garten-Magazins 3: 23-43, 76-87.

The Dictionary of National Biography from the Earliest Times to 1900 (L.Stephen & S.Lee eds). 1921-1922. 22 vols. London.

Dictionary of Scientific Biography (Ch.C.Gillispie ed.). 1970-1978. 15 vols. New York.

Dictionnaire de Biographie française. 1933-1985. 16 vols. Paris.

Dillwyn, L.W. 1839. A review of the references to the Hortus Malabaricus of Henry van Rheede van Draakestein. Swansey.

Dioscorides, see Berendes 1902.

Dodonaeus, R. 1616. Stirpivm Historiae Pemptades Sex Sive Libri XXX. Antverpiae.

Eeghen, I.H. van 1960-1978. De Amsterdamse boekhandel 1680-1725. 5 vols. Amsterdam.

Einarson, B. & G.K.K.Link 1926. Theophrastus De Causis Plantarum in three volumes. 3 vols. London, Cambridge (Massachusetts).

Elias, J.E. 1903-1905. De Vroedschap van Amsterdam 1578-1795. 2 vols. Haarlem. - Facsimile ed. 1963. Amsterdam.

Encyclopaedie van Nederlandsch-Indië. 1917-1939 (ed.2). 8 vols. 's-Gravenhage, Leiden.

Eustachio di S. Maria 1719. Istoria della Vita, Virtù, Doni, e Fatti Illustri del Ven. Monsignor Fr. Gioseppe di S.Maria de' Sebastiani Dell'Ordine de' Carmelitani Scalzi, Delegato e Visitatore Apostolico all'Indie Orientali. Roma.

Evers, G.A. 1933. De schuilkerk der Remonstrantsche gemeente in de Rietsteeg en hare bezitting op 't Heilig-Leven. Jaarboekje van 'Oud-Utrecht' 1933:97-114.

Ferrari, G.B. 1633. De Florvm Cvltvra Libri IV. Romae.

Flacourt, E. de 1661. Histoire de la grande isle Madagascar. Troyes, Paris.

Florijn, P.J. 1984. Christiaan Kleynhoff († 1777) achttiende eeuwse arts-botanicus op Java. Onderzoek naar de botanische en medische activiteiten op Java (Indonesië) rond 1750. 2 vols. - Type-script, Biohistorical Institute, Utrecht University.

Florijn, P.J. 1985a. Christiaan Kleynhoff, een Culemborgs oud-Indië-ganger. De Drie steden. Regionaal-historisch tijdschrift voor Tiel, Buren en Culemborg 6:3-7.

Florijn, P.J. 1985b. Geschiedenis van de eerste hortus medicus in Indië. Tijdschrift voor de Geschiedenis der Geneeskunde, Natuurwetenschappen, Wiskunde en Techniek 8:209-221.

Fournier, M. 1978. H.A. van Reede tot Drakestein en de Hortus Malabaricus. Spiegel Historiael 13(9):571-578.

Fournier, M. 1980. Hortus Malabaricus of Hendrik Adriaan van Reede tot Drakestein, in: K.S.Manilal (ed.), Botany and History of Hortus Malabaricus, pp. 6-24.

Giuseppe di S. Maria 1666. Prima speditione all'Indie Orientali. Roma.

Giuseppe di S. Maria 1672. Seconda Speditione All'Indie Orientali di Monsignor Sebastiani Fr. Givseppe di S. Maria dell'Ordine de' Carmelitani Scalzi prima Vescovo di Hierapoli, hoggi de Bisignano, e Barone di Santa Sofia. Roma.

Gola, G. 1947. L'Orto Botanico. Quattro secoli di attività (1545-1945). Padova.

Gommans, J. 1984. Malo Mori Quam Foedari. Een onderzoek naar het ontstaan van de commissie van Reede tot Drakestein en haar verrichtingen op het eiland Ceylon (1684-1691). - Type-script, Nijmegen University.

Goor, J.van 1978. Jan Kompenie as Schoolmaster, Dutch Education in Ceylon 1690-1795. Utrecht.

Govindankutty, A. 1983. Some observations on seventeenth century Malayalam. Indo-Iranian Journal 25:241-273.

Greshoff, M. (ed.). 1902. Rumphius Gedenkboek 1702-1902. Haarlem.

Greshoff, M. 1906. De grafstede van H.A. van Rheeden te Surat. Eigen Haard 1906:612-613.

Grimm, H.N. 1677. Laboratorium Chymicum, Gehouden op het voortreffelycke Eylandt Ceylon, Soo in 't Animalische, Vegetabilische, als Mineralische Ryck. Batavia. - Facsimile ed. by F.A.H.Peeters 1982. Tilburg.

Gronovius, J.F. 1739. Flora Virginica Exhibens Plantas Quas V.C. Johannes Clayton In Virginia Observavit atque collegit . . . Pars Prima. Lugduni Batavorum.

Gulick, F.W. van 1960. Nederlandse kastelen en landhuizen. Den Haag.

Gunn, M. & L.E.Codd 1981. Botanical Exploration of Southern Africa. An illustrated history of early botanical literature on the Cape flora. Biographical accounts of the leading plant collectors and their activities in southern Africa from the days of the East India Company until modern times. Cape Town.

Gunn, M. & E. du Plessis 1978. The Flora Capensis of Jakob and Johann Philipp Breyne. Johannesburg.

Haan, F. de 1903. Uit oude notarispapieren II. Andreas Cleyer. Tijdschrift voor Indische Taal-, Land- en Volkenkunde 46:423-464.

Haan, F. de (ed.) 1907, see Dagh-Register . . . (1678).

Haan, F. de 1910-1912. Priangan. De Preanger-Regentschappen onder het Nederlandsch Bestuur tot 1811. 4 vols. Batavia, 's-Gravenhage.

Haan, F. de (ed.) 1919, see Dagh-Register . . . (1681).

Haan, F. de 1922-1923. Oud Batavia. 3 vols. Batavia.

Hamilton, F. 1822-1835. A commentary on *Hortus Malabaricus*. Transactions of the Linnean Society of London 13(1822):474-560, 14(1824):171-312, 15(1825):78-152, 17(1835):147-252.

Harmsen, Th.W. 1978. De Beknopte Lant-Meet-Konst, beschrijving van het leven en werk van de Dordtse landmeter Mattheus van Nispen (circa 1628-1717). Delft.

Hasskarl, J.K. 1861. Horti Malabarici Clavis nova. Flora oder allgemeine botanische Zeitung 44:401-408, 481-488, 545-552, 577-

584, 609-616, 641-648, 705-712, 737-745.

Hasskarl, J.K. 1862. Nachträge und Verbesserungen zu "Horti Malabarici Clavis nova". Flora oder allgemeine botanische Zeitung 45:41-48, 73-80, 121-128, 153-160, 187-192. - Also published separately in Regensburg in 1862.

Hasskarl, J.K. 1867. Horti Malabarici Rheedeani Clavis locupletissima. Abhandlungen der Königlichen Leopoldina Carol. d.A. 34:1-34. - Also published separately in Dresden in 1867.

Havart, D. 1693. Op- en Ondergang van Coromandel. Amsterdam.

Heeres, J.E. & F.W.Stapel (eds) 1907-1955. Corpus diplomaticum Neerlando-Indicum. 6 vols. 's-Gravenhage.

Heller, J.L. 1959. Index auctorum et librorum a Linnaeo (*Species Plantarum*, 1753) citatorum, in: C.Linnaeus 1753, Species Plantarum, vol.2 Facsimile ed. 1957-1959. Ray Society, London, pp.3-60.

Heniger, J. 1969. Der wissenschaftliche Nachlass von Paul Hermann. Wissenschaftliche Zeitschrift der Martin-Luther-Universität Halle-Wittenberg 18:527-560.

Heniger, J. 1971. Some botanical activities of Herman Boerhaave, professor of botany and director of the botanic garden at Leiden. Janus, Revue Internationale de l'Histoire des Sciences, de la Médecine, de la Pharmacie et de la Technique 58:1-78.

Heniger, J. 1973. De eerste Nederlandse wetenschappelijke reis naar Oost-Indië, 1599-1601. Leids Jaarboekje 1973:27-49.

Heniger, J. 1980. Van Reede's Preface to Volume III of Hortus Malabaricus and its Historical and Political Significance, in: K.S.Manilal (ed.), Botany and History of Hortus Malabaricus, pp.35-69.

Hermann, P. 1687. Horti Academici Lugduno-Batavi Catalogus exhibens Plantarum omnium Nomina, quibus ab anno MDCLXXXI ad annum MDCLXXXVI Hortus fuit instructus. Lugduni Batavorum.

Hermann, P. 1698. Paradisus Batavus, Continens Plus centum Plantas affabrè aere incisas & Descriptionibus illustras. Lugduni Batavorum.

Hermann, P. 1711. Catalogus Musei Indici Continens varia Exotica, tum Animalia, tum Vegetabilia, Nativam Figuram servantia, Singula in Liquore Balsamico asservata. Lugduni Batavorum.

Hermann, P. 1717. Musaeum Zeylanicum, sive Catalogus Plantarum, In Zeylana sponte nascentium, observatarum & descriptarum. Lugduni Batavorum.

Hernandez, F. 1651 (N.A.Recchus ed.). Rervm Medicarvm Novae Hispaniae Thesavrvs. Romae.

Herport, A. 1669. Eine kurze Ost-Indianische Reisz-Beschreibung. Bern.

Honoré Naber, S.P. L' (ed.) 1930. Albrecht Herport. Reise nach Java, Formosa, Vorder-Indien und Ceylon 1659-1668. Neu herausgegeben nach der zu Bern im Verlag von Georg Sonnleitner im Jahre 1669 erschienenen Original-Ausgabe. Haag. - Reisebeschreibungen von deutschen Beambten und Kriegsleuten im Dienst der niederländischen West- und Ost-Indischen Kompagnien 1602-1797, vol.5.

Horden Jz., P. 1952. Een kleine geschiedenis van Het Land van Vianen. s.l.

Hort, A. 1916. Theophrastus Enquiry into Plants and minor works on odours and weather signs. 2 vols. London, New York.

Hotton, P. 1702. Excerpta ex literis D. Petri Hotton Med. & Botan. Profess. in Acad. Lugduno Batava, ad Editorem de Acmella & ejus faculate lithontriptica. Philosophical Transactions 22:760-762.

Houttuyn, M. 1761-1785. Natuurlyke Historie of Uitvoerige Beschryving der Dieren, Planten en Mineraalen, Volgens het Samenstel van den Heer Linnaeus. 37 vols. Amsterdam.

Hullu, J. de (ed.) 1904, een Dagh-Register . . . (1656-1657).

Hulshof, A. 1941. H.A. van Reede tot Drakestein, journaal van zijn verblijf aan de Kaap. Bijdragen en Mededeelingen van het Historisch Genootschap (Utrecht) 62:1-245.

Hulshof, A. 1942. Compagnie's dienaren aan de Kaap in 1685. Bijdragen en Mededeelingen Historisch Genootschap (Utrecht) 63:347-369.

Hunger, F.W.T. 1925. Jan of Johannes Commelijn (Johannes Commelinus) 1629-1692. Nederlandsch Kruidkundig Archief 1924:187-202.

Hunger, F.W.T. 1927-1942. Charles de l'Escluse (Carolus Clusius) Nederlandsch Kruidkundige 1526-1609. 2 vols. 's-Gravenhage.

Hutchinson, J. 1946. A botanist in Southern Africa. London.

Jacob, H.K. s' 1976. De Nederlanders in Kerala 1663-1701. De memories en instructies betreffende het commandement Malabar van de Verenigde Oost-Indische Compagnie. 's-Gravenhage. - Rijks Geschiedkundige Publicatiën, Kleine Serie, 43.

Japikse, N. 1927-1937. Correspondentie van Willem III en van Hans Willem Bentinck, eersten graaf van Portland. 5 vols. 's-Gravenhage. - Rijks Geschiedkundige Publicatiën, Kleine Serie, 23-24, 26-28.

Jeurissen, M. & M.Fournier 1970. Een bio-bibliographie van de botanici Jan Commelin (1629-1692) en Caspar Commelin (1668-1731). - Type-script, Nijmegen.

Johnston, M.C. 1970. Still no herbarium records for Hortus Malabaricus. Taxon 19:655.

Jonge van Ellemeet, B.M. de 1939. De Utrechtsche Claaskerk. Jaarboekje van 'Oud-Utrecht' 1939:89-111.

Jurriaanse, M.W. 1942. The lady of Trincomalee. Fact and fiction about Francina van Reede. The Ceylon Daily News, April 15, 16, 1942.

K . . . 1956. Saken van cleynder importantie. "Huwelijksreis". Maandblad van 'Oud-Utrecht' 29:101-102.

Kaempfer, E. 1712. Amoenitatum Exoticarum Politico-Physico-Medicarum Fasciculi V. Lemgoviae.

Kaempfer, E. 1906 (1727). The history of Japan Together with a Description of the Kingdom of Siam 1690-1692. 3 vols. Glasgow.

Kalff, S. 1905. De Maecenas van Malabar. Elsevier's Geïllustreerd Maandschrift 15:241-257, 312-322.

Karsten, M.C. 1951. The old Company's Garden at the Cape and its superintendents. Involving a Historical Account of Early Cape Botany. Cape Town.

Karstens, W.K.H. & H.Kleibrink 1982. De Leidse Hortus een botanische erfenis. Zwolle.

Kasbeer, T. 1965. Early Herbarium Specimen Volumes Are Acquired. UCLA Librarian 1965:80.

Kern, H. & H.Terpstra 1955-1957 (ed.2). Itinerario, Voyage ofte Schipvaert van Jan Huygen van Linschoten naer Oost ofte Portugaels Indien 1579-1592. 3 vols. 's-Gravenhage. - Werken Linschoten-Vereeniging 57, 58, 60.

Ketner, F. (ed.) 1936. Album Promotorum, qui inde ab anno MDCXXXVI° usque ad annum MDCCCXV^{um} in Academia Rheno-Trajectina gradum doctoratus adepti sunt. Traiecti ad Rhenum.

Keuning, J. 1938-1951. De Tweede Schipvaart der Nederlanders naar Oost-Indië onder Jacob Cornelisz. van Neck en Wybrant van Warwijck 1598-1600. 5 vols. 's-Gravenhage. - Werken Linschoten-Vereeniging 42, 44, 46, 48, 50.

Kiggelaer, F. 1690. Horti Beaumontiani Exoticarum Plantarum Catalogus, Exhibens Plantarum minus cognitarum & rariorum nomina, quibus idem Hortus Anno Domini MDCLXXXX instructus fuit. Hagae-Comitis.

Kooiman, H.N. & H.J.Venema 1942. Jan Moninckx en de illustraties van Commelyn's Hortus Medicus, in: Gedenkboek J. Valckenier Suringar. Wageningen, pp.260-263.

Kuijlen, J. 1976. Onderzoek naar de tuinen en oranjeriëen van Simon van Beaumont. - Type-script, Nijmegen University.

Kuijlen, J. 1982. De Danziger botanicus en koopman Jacob Breyne (1637-1697) en zijn betekenis voor de Hollandse plantkunde. Tijdschrift voor de Geschiedenis der Geneeskunde, Natuurwetenschappen, Wiskunde en Techniek 6:116-118.

Kuijlen, J. 1983. Bibliografie, in: J.Kuijlen et al., Paradisus Batavus, pp.67-192.

Kuijlen, J., C.S.Oldenburger-Ebbers & D.O.Wijnands 1983. Paradisus Batavus Bibliografie van plantencatalogi van onderwijstuinen, particuliere tuinen en kwekerscollecties in de Noordelijke

en Zuidelijke Nederlanden (1550-1839). Wageningen.
Lasègue, A. 1845. Musée botanique de M. Benjamin Delessert. Paris, Leipzig.
Legré, L. 1904. La botanique en Provence au XVIe siècle. Les deux Bauhin, Jean-Henri Cherler et Valerand Dourez. Marseille.
Leiden 1659 – Res Curiosae & Exoticae, Quae in Ambulacro Horti Academiae Leydensis curiositatem amantibus offeruntur. Anno 1659. s.l. – Copy in British Library 1882. c.2 (198).
Leiden 1659 – Verscheyden Rarietyten, Inde Galderije des Universiteyts Kruyt-Hoff, tot Leyden. Anno 1659. s.l. – Dutch version of the preceding one; copy in State Archive Gent (Belgium), Council of Flanders F.51.
Leiden 1670 – Res Curiosae et Exoticae, In Ambulacro Horti Academici Lugduno-Batavi conspicuae. s.l., s.d. – Copy in British Library 728. c.38.
Lequin, F. 1982. Het personeel van de Verenigde Oost-Indische Compagnie in Azië in de achttiende eeuw, meer in het bijzonder in de vestiging Bengalen. 2 vols. Leiden.
Lewin, L. 1889. Ueber Areca Catechu, Chavica Betle und das Betelkauen. Stuttgart.
Lindeboom, G.A. 1968. Herman Boerhaave. The Man and his Work. London.
Lindeboom, G.A. 1974. Florentius Schuyl (1619-1669) en zijn betekenis voor het Cartesianisme in de geneeskunde. Den Haag.
Linnaeus, C. 1737. Hortus Cliffortianus Plantas exhibens Quas In Hortis tam Vivis quam Siccis, Hartecampi in Hollandia, Coluit . . . Georgius Clifford. Amstelaedami. – Facsimile ed. 1968. Lehre.
Linnaeus, C. 1737. Viridarium Cliffortianum, In quo exhibentur Plantae omnes, quas Vivas aluit Hortus Hartecampensis Annis 1735. 1736. 1737. Amstelaedami.
Linnaeus, C. 1738. Classes Plantarum Seu Systemata Plantarum. Lugduni Batavorum.
Linnaeus, C. 1747. Flora Zeylanica Sistens Plantas Indicas Zeylonae Insulae. Holmiae.
Linnaeus, C. 1748. Hortus Upsaliensis, Exhibens Plantas Exoticas, Horto Upsaliensis Academiae a sese illatas, Ab anno 1742, in annum 1748. Stockholmiae.
Linnaeus, C. 1753. Species Plantarum, exhibentes Plantas rite cognitas, ad Genera relatas. 2 vols. Holmiae. – Facsimile ed. 1957-1959. Ray Society London.
Linnaeus, C. 1758-1759 (ed.10). Systema Naturae Per Regna Tria Naturae. 2 vols. Holmiae.
Linnaeus, C. 1762-1763 (ed.2). Species Plantarum, Exhibentes Plantas rite cognitas, ad Genera relatas. 2 vols. Holmiae.
Linnaeus, C. 1766-1768 (ed.12). Systema Naturae Per Regna Tria Naturae. 2 vols. Holmiae.
Linnaeus, C. 1767. Mantissa Plantarum. Holmiae.
Linnaeus, C. 1771. Mantissa Plantarum Altera. Holmiae.
Linnaeus, C. 1774 (ed.13 by J.A.Murray). Systema Vegetabilium. Gottingae, Gothae.
Linschoten, J.H. van 1596, see Kern & Terpstra 1955-1957.
Lobelius, M. 1581. Plantarvm Sev Stirpivm Icones. Antverpiae.
Lobelius, M. 1605. Stirpium Adversaria Nova. Londini.
Lotsy, J.P. 1902. Over de in Nederland aanwezige botanische handschriften van Rumphius, in: Rumphius Gedenkboek 1702-1902, pp.46-58.
Lourteig, A. 1966. L'Herbier de Paul Hermann, base du Thesaurus zeylanicus de Johan Burman. Taxon 15:23-33.
Majumdar, N.C. & D.N.Guha Bakshi 1979. A few Linnaean specific names typified by the illustrations in Rheede's Hortus Indicus Malabaricus. Taxon 28:353-354.
Manilal, K.S., C.R.Suresh & V.V.Sivarajan 1977. A reinvestigation of the plants described in Rheede's "Hortus Malabaricus" – an introductory report. Taxon 26:549-550.
Manilal, K.S. (ed.) 1980. Botany and History of Hortus Malabaricus. New Delhi, Bombay, Calcutta.
Manilal, K.S. 1980. The Epigraphy of the Malayalam Certificates in Hortus Malabaricus, in: K.S.Manilal (ed.), Botany and History of Hortus Malabaricus, pp.113-120.

Marcgraf, G. 1648, see Piso 1648.
Maris, A.J. 1956. Repertorium op de Stichtse Leenprotocollen uit het Landsheerlijke tijdvak. I. De Nederstichtse leenacten (1394-1581). 's-Gravenhage.
Martinoli, M. 1963. L'Orto Botanico di Pisa. Rivista Agricoltura 1963(7):1-10.
Matthioli, P.A. 1583. Commentarii in VI. libros Pedacij Dioscoridis Anazarbei de Medica materia. Venetijs.
Meier-Lemgo, K. 1933. Engelbert Kämpfer: 1651-1716 Seltsames Asien (Amoenitates exoticae). Detmold.
Meier-Lemgo, K. 1952. Das Stammbuch Engelbert Kämpfers. Mitteilungen aus der lippischen Geschichte und Landeskunde 21:142-200.
Meier-Lemgo, K. 1965. Die Briefe Engelbert Kaempfers. Abhandlungen der Akademie der Wissenschaften und der Literatur (Mainz), mathematisch-naturwissenschaftliche Klasse 1965:267-314.
Meier-Lemgo, K. 1968. Die Reisetagebücher Engelbert Kaempfers. Wiesbaden.
Meinsma, K.O. 1896. Spinoza en zijn kring. Historisch-kritische studiën over Hollandsche vrijgeesten. 's-Gravenhage.
Memoir written in the year 1677 A.D. by Hendrik Adriaan van Rheede . . . for his successor. Madras, 1911. – Selections from the records of the Madras Government. Dutch records no.14.
Mentzel, Chr. 1685. Observatio XIII. De S. Thomae Christianis Indiae Or. pedibus strumosis. Miscellanea Curiosa sive Ephemeridum Medico-Physicarum Caesario-Leopoldina Academiae Naturae Curiosorum 2(3):52-53.
Meyer, E.H.F. 1854-1857. Geschichte der Botanik. 4 vols. Königsberg.
Möbius, M. 1937. Geschichte der Botanik. Von den ersten Anfängen bis zur Gegenwart. Jena. – Ed.2 1968. Stuttgart.
Moes, E.W. & K.Sluyterman 1912-1915. Nederlandsche kasteelen en hun historie. 3 vols. Amsterdam.
Molhuysen, P.C. 1913-1924. Bronnen tot de geschiedenis der Leidsche Universiteit. 7 vols. 's-Gravenhage.
Monti, C. 1742. Jacobi Zanonii Rariorum Stirpium Historia ex parte olim edita. Bononiae.
Mulder, A.W.J. & D.F.Slothouwer 1949. Het kasteel Amerongen en zijn bewoners. Maastricht.
Muller, S., R.Fruin, J.G.C.Joosting & W.J.Hora Siccama van de Harkstede (eds) 1915. Catalogus van het Archief der Staten van Utrecht, 1375-1813. Utrecht.
Muntschick, W. 1984. Ein Manuskript von Georg Meister, dem Kunst- und Lustgärtner, in der British Library. Medizinhistorisches Journal 19:225-232.
Neck, J.C. van, see Keuning 1938-1951.
Nieuhof, J. 1682. Gedenkwaerdige Zee- en Lant-reize door de voornaemste Landschappen van West en Oost Indien. Amsterdam.
Nieuw Nederlandsch Biografisch Woordenboek. 1911-1937. 10 vols. Leiden.
Nuys, W. van 1978. Andries Cleyer, leven en werk. – Type-script, Biohistorical Institute, Utrecht University.
Oliver, E.G.H. 1980a. Some observations on two early Cape florilegia. Bothalia 13:115-125.
Oliver, E.G.H. 1980b. Book-review: The Flora Capensis of Jakob. and Johann Philipp Breyne, ed. M.Gunn & E. du Plessis. Bothalia 13:259.
Ooststroom, S.J. van 1937. Hermann's collection of Ceylon plants in the Rijksherbarium (National Herbarium) at Leyden. Blumea Suppl. 1:193-209.
Opuscula Selecta Neerlandicorum de Arte Medica, see Van Andel 1931.
Orta, G. da 1563, see De Ficalho 1891-1895.
Panikkar, K.M. 1931. Malabar and the Dutch. Being the history of the fall of the Nayar power in Malabar. Bombay.
Parkinson, J. 1629. Paradisi In Sole Paradisus Terrestris. or A Garden of all sorts of pleasant flowers. London.
Pasti Jr., G. 1950. Consul Sherard: Amateur botanist and patron

of learning, 1659-1728. – Thesis, University of Illinois.
Peeters, F.A.H. 1982, see Grimm 1677.
Petiver, J. 1699. An Account of some Indian Plants, &c. with their Names, Descriptions and Vertues; Communicated in a Letter from Mr. James Petiver, Apothecary and Fellow of the Royal Society; to Mr. Samuel Brown, Surgeon at Fort St. George. Philosophical Transactions 20(1698):313-335.
Petiver, J. 1701-1704. An Account of parts of a Collection of Curious Plants and Drugs, lately given to the Royal Society by the East India Company. Philosophical Transactions 22:579-594 (May and June 1700), 699-721 (November and December 1700), 843-861 (May and June 1701), 933-946 (September 1701), 1007-1022 (November and December 1701); 23:1055-1065 (January and February 1702), 1251-1265 (November and December 1702), 1450-1460 (September and October 1703).
Piso, G. 1658. De Indiae Utriusque Re Naturali et Medica Libri Quatuordecim. Amstelaedami.
Piso, G. & G.Marcgraf 1648. Historia Natvralis Brasilae. Lvgdvn. Batavorvm, Amstelodami.
Pliny, see Rackham et al. 1938-1962.
Plukenet, L. 1691-1692. Phytogrphia, Sive Stirpium Illustriorum & minùs cognitarum Icones. 3 pts. Londini.
Plukenet, L. 1696. Almagestum Botanicum sive Phytographiae Pluc'netianae Onomasticon Methodo Syntheticâ digestum. Londini.
Plukenet, L. 1700. Almagesti Botanici Mantissa. Londini.
Plukenet, L. 1705. Amaltheum Botanicum. Londini.
Pool-Stofkoper, E. van der 1984. Een reconstructie van de Hortus Medicus Amstelodamensis 1682-1800. – Type-script, Institute of History of Art, University of Amsterdam.
Rackham, H., W.H.S.Jones & D.E.Eichholz 1938-1962. Pliny Natural History with an English translation in ten vols. 10 vols. London.
Raven, Ch.E. 1942. John Ray Naturalist. His life and works. Cambridge.
Raven-Hart, R. 1971. Cape Good Hope 1652-1702. The first fifty years of Dutch colonisation as seen by callers. 2 vols. Cape Town.
Ray, J. 1682. Methodus Plantarum Nova, Brevitatis & Perspicuitatis causa Synoptice in Tabulis Exhibita. Londini.
Ray, J. 1686-1704. Historia Plantarum. 3 vols. Londini.
Ray, J. 1703. Methodus Plantarum Emendata. Londini, Amstelaedami.
Recchus, N.A. 1651, see Hernandez 1651.
Reede tot Drakenstein, H.A. van 1677, see Memoir.
Reede tot Drakenstein, H.A. van 1678-1693. Hortus Indicus Malabaricus, Continens Regni Malabarici apud Indos celeberrimi omnis generis Plantas rariores. 12 vols. Amstelodami.
Reede tot Drakenstein, H.A. van (A.van Poot transl.) 1689. Malabaarse Kruidhof, Vervattende het raarste slag van allerlei soort van Planten Die in het Koningrijk van Malabaar worden gevonden. 2 vols. Amsterdam. – Ed. 2 1720. 's-Gravenhage.
Reede tot Drakenstein, H.A. van (J.Hill ed.) 1774. Horti Malabarici Pars Prima, De Variis Generis Arboribus Et Fruticibus Siliquosis. London.
Regius, H. 1650. Hortus Academicus Ultrajectinus. Ultrajecti.
Reynolds, G.W. 1950. The Aloes of South Africa. Johannesburg.
Rientjes, A.E. & J.G.Böcker 1947. Het kerspel Jutphaas. s.l.
Rochefort, Ch. de 1681 (ed.1 1658). Histoire naturelle et morale des Iles Antilles de l'Amerique. Rotterdam.
Rookmaker, L.C. 1976. An early engraving of the black rhinoceros (*Diceros bicornis* (L.)) made by Jan Wandelaar. Biological Journal of the Linnean Society 8:87-90.
Rouweler, M. 1982. Het verband tussen een aantal brieven van Mr. Joan Huydecoper van Maarseveen (1625-1704), en de herkomst van, in Zuid-Afrika gevonden, aquarellen van Kaapse planten uit de 17e eeuw. – Type-script, Biohistorical Institute, Utrecht University.
Royen, A. van 1740. Florae Leydensis Prodromus, exhibens Plantas quae in Horto Academico Lugduno-Batavo aluntur. Lugduni Batavorum.
Rumphius, G.E. 1741-1750. Herbarium Amboinense. 6 vols. Amstelaedami, Hagae Comitis, Ultrajecti.
Rumphius, G.E. 1755. Herbarii Amboinensis Auctuarium. Amstelaedami.
Rumphius Gedenkboek, see Greshoff 1902.
Saccardo, P.A. 1895-1901. La botanica in Italia. 2 vols. Venezia.
– Memorie del Reale Istituto Veneto di Scienze, Lettere ed Arti 25(4) and 26(6).
Sachs, J. 1875. Geschichte der Botanik vom 16. Jahrhundert bis 1860. München. – Facsimile ed. 1966. New York, Hildesheim.
Scannel, M.J.P. 1979. A 17th century *Hortus Siccus* made in Leyden, the property of Thomas Molyneux, at DBN. Irish Naturalist's Journal 19(9):320-321.
Schenk, M.G. & J.B.Th.Spaan 1967 (ed.2). Drakensteyn en zijn bewoners. Baarn.
Schoute, D. 1929. De geneeskunde in den dienst der Oost-Indische Compagnie in Nederlandsch-Indië. Amsterdam.
Schouten, W. 1708 (ed.1 1676). Gedenk-waardige Reysen Naar Oost-Indiën. Amsterdam.
Schuyl, F. 1668. Catalogus Plantarum Horti Academici Lugduno-Batavi Quibus is instructus erat Anno MDCLXVIII . . . Lugd. Batav.
Sherard, W. 1689. Pauli Hermani Paradisi Batavi Prodromus Sive Plantarum Exoticarum in Batavorum Hortis observatarum. Index, in: J.Pitton de Tournefort, Schola Botanica, pp.301-386,(4).
Sirks, M.J. 1915. Indisch Natuuronderzoek. [Amsterdam].
Smit, P. 1969. Paul Hermann (1646-1695). Ein Vertreter der niederländische Botanik des 17. Jahrhunderts. Wissenschaftliche Beiträge der Martin-Luther-Universität Halle-Wittenberg 1969(2):69-88.
Snippendael, J. 1646. Horti Amstelodamensis Alphabetico ordine exhibens eas, quibus is instructus fuit, atq, quibus auctior factus est, stirpes. Amstelodami.
Soulsby, B.H. 1933 (ed.2). A Catalogue of the Works of Linnaeus (and publications more immediately relating thereto) preserved in the libraries of the British Museum (Bloomsbury) and the British Museum (Natural History) (South Kensington). London.
Spigelius, A. 1633. Isagoges in rem herbariam Libri duo. Lvgdvni Batavorvm.
Sprengel, K. 1817-1818. Geschichte der Botanik. Neu bearbeitet. 2 vols. Altenburg, Leipzig.
Stafleu, F.A. 1967-1971. Taxonomic literature. A selective guide to botanical publications with dates, commentaries and types. Utrecht, Zug.
Stafleu, F.A. & R.S.Cowan 1976-1983 (ed.2). Taxonomic literature. A selective guide to botanical publications and collections with dates, commentaries and types. 4 vols. Utrecht, Antwerpen-The Hague, Boston.
Stapel, F.W. 1936. Cornelis Janszoon Speelman. 's-Gravenhage.
Stapel, F.W. (ed.) 1938-1940. Geschiedenis van Nederlandsch Indië. 5 vols. Amsterdam.
Stevens, P.F. 1980. The Correct Names for the Three Elements in the Protologue of *Calophyllum calaba* L., in: K.S.Manilal (ed.), Botany and History of Hortus Malabaricus, pp.168-176.
Sweertius, E. 1612. Florilegivm. 2 vols. Francofurti ad Moenum.
Swillens, P. 1925. Schilders en beeldhouwers in Oud-Utrecht. Jaarboekje van Oud-Utrecht 1925:50-71.
Tachard, G. 1686. Voyage de Siam, des Pères Jésuites envoyez par le Roy aux Indes et à La Chine. Paris.
Taets van Amerongen, M.J.L. 1914. Hooge en vrije Heerlijkheid van Renswoude en Emmickhuysen. 's-Gravenhage.
Theophrastos. Historia Plantarum, see Hort 1916.
Theophrastos. De Causis Plantarum, see Einarson & Link 1926.
Thieme, U. & F.Becker 1907-1950. Allgemeines Lexikon der bildenden Künstler von der Antike bis zur Gegenwart. 37 vols. Leipzig.

Tonkelaar, I. den 1983. De correspondentie tussen Pieter Hotton (1648-1709) en Hans Sloane (1660-1753). – Type-script, Biohistorical Institute, Utrecht University.

Tournefort, J.Pitton de 1689. Schola Botanica Sive Catalogus Plantarum . . . in Horto Regio Parisiensi. Amstelaedami.

Tournefort, J.Pitton de 1700. Institutiones Rei Herbariae Editio Altera, Gallica Longe Auctior. 3 vols. Parisiis.

Valentijn, F. 1724-1726. Oud en Nieuw Oost-Indien. 8 vols. Dordrecht, Amsterdam.

Vallot, A. 1665. Hortus Regius Parisiensis. Parisiis.

Veendorp, H. & L.G.M.Baas Becking 1938. 1587-1937 Hortus Academicus Lugduno Batavus. The development of the gardens of Leyden University. Harlemi.

Veslingi, J. 1640, see Alpini 1640.

Veth, P.J. 1887. Hendrik Adriaan van Reede tot Drakestein. De Gids 51(3):423-475, 51(4):113-161.

Vincenzo Maria di S. Caterina da Siena 1672. Il Viaggio All'Indie Orientali. Roma.

Vliet, M. van 1961. Het Hoogheemraadschap van de Lekdijk Bovendams. Assen.

Vorstius, A. 1633. Catalogvs Plantarvm Horti Academici Lvgdvno-Batavo, Quibus is instructus erat Anno MDCXXXIII, in: A.Spigelius 1633:225-272.

Vos, F.H. de 1903. Genealogische en heraldische aanteekeningen aangaande Hollandsche Familiën op Ceylon. De Navorscher 53:415-421, 692-699.

Vrijer, M.J.A. de 1917. Henricus Regius een "Cartesiaansch" hoogleeraar aan de Utrechtsche Hoogeschool. 's-Gravenhage.

Waller, F.G. & W.R.Juynboll 1938. Biographisch Woordenboek van Noord Nederlandsche Graveurs. 's-Gravenhage.

Warmington, E.H. 1974 (ed.2). The commerce between the Roman empire and India. London.

Waterhouse, G. 1979 (ed.2). Simon van der Stel's journal of his expedition to Namaqualand, 1685-6. Cape Town, Pretoria.

White, A. & B.L.Sloane 1937 (ed.2). The Stapelieae. 3 vols. Pasadena.

Wijnaendts van Resandt, W. 1944. De Gezaghebbers der Oost-Indische Compagnie op hare Buiten-Comptoiren in Azië. Amsterdam.

Wijnands, D.O. 1983. The botany of the Commelins. Rotterdam.

Winckler, E. 1854. Geschichte der Botanik. Frankfurt a.M.

Wit, H.C.D. de (ed.) 1959. Rumphius Memorial Volume. Baarn.

Wittert van Hoogland, E.B.F.F. 1909-1912. Bijdragen tot de Geschiedenis der Utrechtsche Ridderhofsteden en Heerlijkheden. 2 vols. 's-Gravenhage.

Wurzbach, A. von 1906-1911. Niederländisches Künstler-Lexikon auf Grund archivalischer Forschungen bearbeitet. 3 vols. Wien, Leipzig.

Zanoni, G. 1675. Istoria Botanica. Bologna.

Index of persons

Aarssen van Sommelsdijk, Cornelis van 161, 162, 169, 170
Abcoude van Meerten, Ernst van 5, 6
Abramsen, Cornelis 53, 55
Achates 53
Achudem, Itti, see Itti Achudem
Acosta, Cristobal 12, 95, 154, 165
Adair, Patrick 176
Adrichem, Dirk van 49
Ainikkur Nambidi 85
Albinus, Bernard Siegfried 125, 138
Aldini 156, 165
Alexander VII 38
Almeloveen, see Theodorus Janssonius van Almeloveen
Almonde, Jan van 22, 26
Alpini, Prospero 150, 157, 162, 165
Anguillara, Luigi 139
Apati, Nicolaus 103
Apollo 53
Appelman, Gonsalez 97, 102, 104
Apu Botto 40, 43, 100, 145-147, 149
Aristotle 140
Artafa, Saladin 38, 49, 111, 122, 151

Baar, Diederik van 58, 59, 66, 103, 266-267
Bacherus, Johannes 65, 67, 69, 78
Baldaeus, Philippus 14, 16-18, 22, 26, 165
Barentsz., Dirk 85
Barrelier, Jacques 107, 108, 120, 122, 123
Bauhin, Caspar 140, 141, 156-159, 162, 163, 165, 166
Bauhin, Jean 140, 150, 156
Bax van Herentals, Joan 9, 10, 19, 23, 28, 31, 35, 39, 47, 52, 55-57, 67, 70-73, 82, 83, 153, 160, 164
Baye, Maurits Cesar de la 58, 65, 78, 85, 88, 91, 266-267
Baye, Maurits Lodewijk de la 4-6, 266-267
Baye, Paul de la 58, 66, 88, 266-267

Beatrix, queen of The Netherlands 6
Beaumont, Simon van 161, 169, 172, 173, 176
Becker, Everard 90
Beet, Jan Hendriksz. de 71, 72, 76, 82-84
Bentinck, Hans Willem 173
Beverningk, Hieronimus 8, 10, 65, 71, 76, 155, 163, 166, 171-173, 176
Block, Agnes 71, 83, 172, 173
Bloock, Cornelis van der 50
Bobart, Jacob 63
Bodaeus a Stapel 8
Boddens, Pieter 56
Boerhaave, Herman 77, 84, 153, 174
Bontius, Jacobus vii, 12, 13, 95, 154, 165, 167
Boom, Dirk (Theodorus), widow of 62, 268
Boom, Hendrik 62, 268
Booth, Everard 89
Borghorst, Jacob 31, 70, 82
Borsselen, van 98
Bort, Balthasar 55
Botto, see Apu Botto, Ranga Botto
Boucher, George 271
Bouwman, Engelbert 268
Brandenburg, van 97, 98, 104
Breen, Pr. van 271
Brent, Isabella 41
Breyne, Jacob 63, 67, 70, 71, 83, 153, 155, 160, 162, 165, 171-173, 176
Breyne, Johann Philipp 71
Broussard, Carel Hendrick 91
Broussard, Cornelia Lydia 91
Broussard, Daniel 89
Broussard, Fransina Margriet 91
Browne, Samuel viii, 174, 176
Bruynink, Joan Jacob 85
Burman, Johannes 84, 164, 174-176
Burman, Nicolaas Laurens 172, 175, 176

Caesalpino, Andrea 140
Camerarius 165
Camphuijs, Joan 54, 56
Campo, Alexander da 39, 40, 45, 49
Carneiro, Emanuel (Manuel) 34, 43, 45, 100, 148, 149
Caron, François 35, 48
Casearius, Johannes 5, 34, 41, 42-47, 49, 52, 53, 100, 101, 143, 146, 147, 149-151, 156
Casier, Cornelis 49
Casier, Johannes 49
Cero, Johan 42
Chahestachan 77
Cherler, Jean 140
Choisy, François-Timoléon de 10
Christiaansz., Adriaan 271
Claudius, Hendrik 52, 53, 55, 67, 72, 74-76, 83, 84
Cleyer, Andries vii, 11-13, 29, 31, 37, 38, 42, 50, 52-55, 63, 70, 72, 75, 83, 84, 154, 161, 169, 171, 172
Cleyer, Anna Elisabeth 50
Cleyer, Cornelis 55
Clifford, George 174
Clusius, Carolus 8, 12, 140, 150, 153, 154, 165
Coelestinus of St. Liduina 38, 106-108, 111, 120, 122
Coensz., Jacob 270
Commelin, Caspar 66, 71, 77, 160, 163, 173-175
Commelin, Caspar (sr.) 160
Commelin, Isaac 160
Commelin, Jan viii, 9, 30, 59, 62-65, 67, 71, 75-78, 84, 101, 102, 119, 125, 128, 130, 138, 140, 141, 150, 154, 155, 157-175, 268, 269
Compton, Henry 173
Cordus, V. 165
Cornut 165
Cortesi, Vittoria 38
Coste, Pieter van der 75
Couchetez, Pierre 75, 84
Coulster, Ludolph van 26
Croon, Cornelis Jansz. 271
Crudop, Hendrik 56, 70, 82

Dam, Pieter van 52, 61, 66, 101, 269
Debuson, Dirk 271
Desmarets, Daniel 71, 83, 159
Dielen, Isaac van 80, 81, 86
Diest, van 97, 98, 104
Diest, Machteld Peunis van 266-267
Dijkman d'Jonge, Hendrik 271
Dioscorides 139-141, 165

Dodonaeus, Rembertus 153, 165
Donep, Christiaan Herman van 34, 42, 43, 45, 53, 59, 100, 149-151
Doorslagh, G. van den 91
Dussen, Lucas van der 23, 27, 28, 30, 31, 47, 268
Duyn, Adam van der 50
Dyck, Jan van 61, 62, 99

Elsevier, N. 76, 85, 161, 169
Emden, Margaretha Willemsdr. van 50
Erhart, Aij. 271
Erorma, prince of Cochin 48
Erpecum, Jan van 161
Essem, Cornelis 15

Fagel, Caspar 65, 71, 72, 75, 83, 161, 163, 164, 172, 173
Farnese 165
Ferrari 165
Fijbeecq, Joachim 31, 42
Flacourt, de 35, 48
Flacourt, E. de 154
Flines, Mrs De 83
Foglia, Pietro 38, see also Matthew of St. Joseph
Foglia, Scipio 38
Frederik Hendrik, prince of Orange 58, 266-267
Frederik Willem, elector of Brandenburg 66

Gaasbeek van Abcoude, Johanna van 104
Gaesbeeck, van 98
Gangadhara Laksmi, rãni of Cochin 14, 19
Garcin, Laurent 175, 176
Gaymans, Antoni 154, 159
Gesel, Antoni van 174
Giacinto di S. Vincenzo 39
Giuseppe di S. Maria 19, 38-40, 49, 106, 118
Gōda Varma, rãja of Cochin 15, 18, 48, 80
Godsja Moraat 271
Godske, Isbrand 14, 21-23, 26, 27, 47, 70, 82
Godtke, Daniel 56
Goens Jr., Rijklof van 10, 30, 44, 48, 50, 57, 64, 69, 71, 72, 74, 75, 82, 83, 85, 145, 155
Goens Sr., Rijklof van viii, 7, 9,

Index of persons

10, 13-15, 18, 19, 22, 23, 26-33, 35-37, 39, 42, 44, 46-52, 54, 55, 57, 63, 64, 67, 77, 145, 150
Goetjens, Abraham 30
Goetkint, Antoni 49
Goetkint, Antoni Jacobsz. 42, 45, 49, 99, 100, 125
Goetkint, Pieter 49
Gool, Jacob van 38, 40, 171
Gool, Pieter van 38, *see also* Coelestinus of St. Liduina
Goor, van 97
Gottifredo di S. Andrea 39
Grimm, Herman 52-55, 68
Groenrijs, Jacob 271
Gunst, Pieter Stevensz. van 98, 104

Haan, Mattheus de 271
Hammius, G.J. 92
Hannijs, Bartholomeus 271
Hardenbroek, Jan Louis van 90
Harting, Joris 271
Hasencamp, Herman 34, 43, 47
Havart viii
Heeck, Gijsbert 7, 8, 10-13
Heermans, Cornelis 83
Hees, H. van 91
Helsdingen, Abraham van 270
Helsdingen, Anthonij van 271
Helsdingen, Pr. van 271
Henninius, Henricus Christianus 102
Hermann, Paul vii, viii, 30, 31, 53, 61-63, 65-67, 70, 75-78, 82, 85, 105, 138, 144-146, 151, 153-156, 159, 160, 162-166, 171-176
Hermans, Dr. 61
Hermansz. 61
Hernandez 154, 165
Herport, Albrecht viii, 14, 18, 19, 26
Hertog, Jan 76, 84
Hertog, Pieter viii, 84, 174, 175
Hertog, Willem 76, 84
Hill, John 97, 104
Hinlopen, Aletta 55, 56, 67, 85
Hinlopen, Cornelia Isabella 56
Hoefijser, Pieter Martsz. 5
Hoepels, Antoni 85
Hoey, Pieter de 85
Hondius, Petronella 160
Hoorn, Jacob van 271
Hotton, Pieter 62, 63, 77, 166, 172, 174
Houttuyn, Martinus 175, 176
Hudde, Joan (Johannes) 61, 66, 67, 101
Huisman, Martin 45, 48, 50
Hurt, Antoni 55
Hustaert, Jacob 21, 26
Huydecoper van Maarsseveen, Jan Elias 65, 68
Huydecoper van Maarsseveen, Joan ix, 9, 10, 29, 31, 52, 55-57, 61, 63, 64, 65, 67, 68, 70-73, 75-77, 82-85, 87, 88, 90, 101, 160, 161, 163, 164, 169
Huydecoper van Maarsseveen Jr., Joan 82
Huygens Jr., Constantijn 87, 90

Itti Achudem 42, 43, 50, 100, 146-148

Jager, Herbert de 171, 172
Jalabert, Abraham André 92
Jalabert, Emilia 92
Jalabert, Guillaume 91, 92, 266-267
Jalabert, Magdalena 92
Jalabert, Margarita 92
Jalabert, Susanna 92
Jans, Susanna 49
Janssonius van Almeloveen, Theodorus 63, 67, 102, 150, 151
Jansz., Samson 271
Jong, Sijbrant de 271
Jussieu, Antoine de 122

Kadensky, Barbara Margaretha 50
Kadensky, Simon 42, 50, 53, 146
Kaempfer, E. vii, 78, 80, 86
Kale, de 98
Ketel 98
Kiggelaer, Frans 160
Kis, Nicolaus 103
Kleynhoff, Christiaan 175, 176

Lamotius, Isaac 63, 71, 161, 162, 169
Leersum, baron of 102, *see also* Willem van Nassau
Leeuwen, Jan van 270
Leeuwenhoek, Antoni van 5, 59
Le Roux, Alexander 90
Le Roux, Gillette Jeanne 90, 266-267
Le Roux Jalabert, Alexander 90, 92, 266-267
Limbeek, Lt. 271
Linnaeus, Carolus vii, viii, 140, 154, 166, 171, 172, 174-176
Linschoten, Jan Huygen van 12, 14, 18, 26, 154, 165
Lobelius 8, 153, 165
Lobs, Jacob 37, 38, 45, 46, 49-52
Lords XVII viii, 7, 10, 19, 26-31
Louis XIV 59

Maas, Jan Pietersz. 85
Maatsuiker, Joan 44, 51, 55, 56, 61, 100, 268
Mans, Raphael du 78
Manteau, Samuel 56
Marcello 39
Marcgraf vii, 154, 165
Markel, Gillis van 271
Martensz., Jan 271
Mary II 173
Mast, Cors van der 271
Matthaeus à S. Joseph, *see* Matthew of St. Joseph

Matthew of St. Joseph viii, ix, 5, 19, 25, 38-50, 53, 100, 105-108, 111, 118, 119, 122, 123, 125, 143-146, 149, 151, 156, 166, 171
Matthioli 120, 165
Mazius, Marcus 34, 42, 44, 47, 49, 50
Meerseveen, Mr. 82
Meester, Js. de 271
Meister, Georg 53, 75, 84
Mentzel, Christian 53
Meysner, Paulus 37, 38, 42
Michael di S. Eliseo 39, 106
Minnes, Pieter 43
Mode 271
Mogul 31
Mol, François 271
Moninckx, Jan 64
Monti, Gaetano (Cajetanus) Lorenzo 48, 106-108, 111, 118, 119, 123
Monti, Giuseppe 124
Morison, Robert 63, 153
Mulart 98
Munck, Alexander de 85
Munnicks, Johannes 62, 63, 101, 102, 150, 151, 268
Munting, Abraham 62, 63
Murray, J.A. 175

Nassau, Frederik van 58, 266-267
Nassau, Willem van 58, 66, 102, 266-267
Neck, J. van 165
Neck van Monnikendam, Jan 56
Nes, Jan van 271
Nienrode, Jan van 97
Nieuhof, Joan 7-10, 14, 16, 21-27, 166
Nijenrode, van 98
Nijenrode, Ernst van 104
Nijenrode, Geertruid van 266-267
Noetavile-Virola 101, *see also* Vīra Kērala Varma

Olatche 26
Oldenland, Hendrik Bernard vii, 75, 76, 84, 161
Ommen, Adriaan van 271
Ommen, Jan van 271
Oosterhoff, Cornelis van 271
Oosterwijck, Joan van 55, 56
Orleans, Duke of 59
Orleij, David van 271
Orta, Garcia da 12, 95, 120, 154, 163-165
Outgaerden, Hendrik Otto van 175, 176

Padbrugge, Robert 29-31
Paludanus 154
Panhuis, Pieter van 91
Panhuis tot Voorn, Cornelia Agnes van 91
Panhuizen, Bartholomeus van 266-267

Panhuizen, Bartholomeus Cornelis van 90, 91, 266-267
Panhuizen, Elisabeth Antonia van 90, 91, 266-267
Panhuizen, Henriette Adriana van 90, 91
Panhuizen, Machteld Louise van 89, 90
Panhuizen tot Voorn, Antoni Carel van 26, 65, 88, 89, 91, 266-267
Panhuizen van Voorn, Elisabeth Catharina van 91
Parkinson 165
Paviljoen, Antoni 31
Perimbala 36, 37, 45
Pernis, W. 91
Petiver, James 174-176
Pijl, Laurens 21, 22, 26, 28, 31, 63-67, 76, 77, 85, 161, 162, 164
Piso, Willem vii, 12, 154, 165
Pliny 139, 140
Plukenet, Leonard 74, 160, 173, 174, 176
Poel, Adriaan van der 56
Poot, Abraham van 63, 67, 97, 103, 125, 151
Pothuijsen, Cs. van 270
Pryon, Albert 175, 176

Radermacher 175
Radja Sinha 77
Rāma Varma, prince of Cochin 35, 48
Rāma Varma, rāja of Cochin 14, 80
Rāma Varma, rāja of Travancore 19, 20
Ranga Botto 40, 43, 100, 145-147, 149
Ranst, Constantin 52, 54, 55, 57
Rason, Jan Martin 271
Ray, John 63, 140, 160, 166-175
Reaal, Pieter 91
Recchus 154, 165
Reede, van, family 3-6, 57, 266-267
Reede, Anna Maria van 6, 266-267
Reede, Anna Walburga van 266-267
Reede, Ernst van 3, 5, 6, 266-267
Reede, Francina van 26, 42, 47, 49, 58, 65, 81, 86-91, 266-267, 270
Reede, Frederik Hendrik van 90, 266-267
Reede, Godard van 66
Reede, Goert van 266-267
Reede, Lucia van 6
Reede, Margaretha van 266-267
Reede tot Amerongen, Ernst van 5, 266-267
Reede van Amerongen, Frederik van 5, 98, 266-267
Reede van Amerongen, Godard Adriaan van 3, 4, 59, 66, 102, 266-267

Index of persons

Reede tot Drakenstein, Agnes van 5, 57, 88, 91, 266-267, 269
Reede tot Drakenstein, Anthonie van 5
Reede tot Drakenstein, Frederik van 5
Reede tot Drakenstein, Geertruid van 5, 6
Reede van Drakenstein, Gerard (Gerrit) van 3-6, 65, 266-267
Reede tot Drakenstein, Hendrik van 5, 6
Reede tot Drakenstein, Hendrik Adriaan van vii-ix, 3-78, 80-91, 97, 98, 100-106, 111, 118, 119, 123, 129, 130, 134, 135, 141, 143-147, 150, 151, 153, 155-169, 171-176, 266-270
Reede tot Drakenstein, Jan van 5
Reede tot Drakenstein, Karel van 5, 59, 66
Reede tot Drakenstein, Machteld van 5, 57, 266-267
Reede tot Drakenstein, Margaretha van 5, 57
Reede tot Drakenstein, Maria van 5, 59, 66, 266-267
Reede van Nederhorst, Anna Elisabeth van 58, 266-267
Reede van Nederhorst, Gerard van 3, 4, 104, 266-267
Reede van Nederhorst, Godard van 3-6, 266-267
Reede van Renswoude, Frederik Adriaan van 5, 59, 66, 103, 266-267
Reede van Renswoude, Gerard van 4, 266-267
Reede van Renswoude, Johan van 3, 266-267
Reets, Gabriel 68
Reets, Sandrina 65, 67, 87
Regius, Henricus 41
Renesse, van 97, 98
Renesse van Baar, Agnes van 266-267
Renesse van Baar, Johan van 104
Rensen, Catharina van 55
Reuters, Petronella 5
Rhee, Thomas van 76, 77, 84, 91
Rhijne, Willem ten 12, 13, 52, 54-56, 59, 63, 70, 84, 101, 134, 145, 149, 150, 161, 169, 171, 172
Richter 175
Riebeek, Jan van 7, 8, 70, 73, 268
Rochefort, de 154, 165
Roeloffsz., Jan 271
Royen, Adriaan van 174
Rumphius, Georg Everhard vii, 29, 54, 55, 95, 174-176
Ruysch, Frederik 160, 161

Saint-Martin, Isaac de 9, 10, 15, 27, 30, 51, 52, 64-69, 76, 84, 85, 88, 90, 161, 169
Sanen, Johannes van 76
Schade, Johanna 90, 266-267
Schade, Herman 65
Schimmelpenn(ing) 98
Scholten, Johannes 269, 270
Schouten, Wouter viii, 7, 8, 10, 14, 16, 17
Schuijnderman, Jacob 271
Schuyl, Florens 54, 153, 154
Seloven, Isaak 271
Servaes, Jacob 65
Sevenhuizen, Frans van 171, 176
Sherard, William 166, 173, 176
Sibelius van Goor, Caspar 129, 138
Sipkens, see Francina van Reede 87
Sleeswijk, Paulus 56
Sloane, Hans 82, 174, 176
Sloffen, Js. 271
Solman, Isaac 65, 67
Solms, van 98, 104
Someren, Johannes van 61, 62, 99
Someren, Pieter van 62, 268
Speelman, Cornelis 51-55, 64, 67
Spiegel, Elbert 5
Spinola 4
Spinoza, Baruch de 41, 49
Splinter, Gerrit (I) 49
Splinter, Gerrit (III) 50
Splinter, Marcelis (I) 50
Splinter, Marcelis (II) 42, 45, 50, 102, 128, 129
St. John the Baptist 118
Staden, Jan van 59
Steenbergen, Johannes van 268
Stel, Simon van der 63, 67, 69-77, 82-85, 161, 169
Stel, Willem Adriaan van der 71, 84
Sterksel, Jacob 271
Stomphius, Johannes Frederik 85
Stoopendael, Bastiaan 97, 100, 104, 161
Strick Berts, Jacob Frederik 31, 42
St. Thomas 120
Sweertius 8
Syen, Arnold viii, 30, 59, 61-63, 100, 101, 134, 141, 145, 153-161, 164-166, 169, 171-173, 176

Tachard, Gui 74, 75, 84
Taijspil, Louis 270
Tervile, Ludolff 271
Tetterode, Karel van 9, 70, 73
Theophrastos 139, 140
Thol, Gerrit van 270, 271
Thunberg, Carl Per 175
Tournefort, Joseph Pitton de 63
Trap, Gottschalk 50
Turnor, Margaretha 266-267
Tuyll van Serooskerken, Alexandrina van 266-267
Tuyll van Serooskerken, Elisabeth van 66, 266-267
Tuyll van Serooskerken, Godard Willem van 58, 59, 66, 88, 91, 103, 266-267
Tuyll van Serooskerken, Hendrik Jacob van 58, 59, 65, 66, 88, 89, 91, 102, 266-267
Tuyll van Serooskerken, Johan van 266-267
Tuyll van Serooskerken, René van 4, 57, 266-267
Tuyll van Serooskerken, Reinoud Gerard van 66, 89, 266-267
Tuyll van Serooskerken, Trajectina van 58, 90

Uijttenbogaert, Isabella 65
Uldrix, Joan 271
Uldrix, Pr. 271
Utenhove 98
Utenhove, Antoni 4, 266-267
Utenhove, Catharina 266-267
Utenhove, Charles 5, 266-267
Utenhove, Elisabeth 3, 5, 266-267
Utenhove, Hendrik (van) 59, 103, 266-267
Uytter, Burghart 36, 37, 42, 45, 48, 147

Valckenier, Gillis 52, 61, 66, 101
Valentijn viii
Valerius of St. Joseph 39, 106, 119
Valkenburg, Cornelis 15
Vallot, A. 156, 165
Vauquet, Susanne 92
Veelen, van 97, 98, 104
Veslingi 165
Vinaique Pandito 34, 40, 43, 100, 145-147, 149
Vincenzo Maria di S. Caterina da Siena 38, 49, 166
Vinck, Belia 160
Vīra Kêrala Varma, prince of Cochin 14
Vīra Kêrala Varma, rāja of Cochin 14, 18, 19, 26, 36, 37, 45, 64, 80, 85, 147
Visser, Laurens 70, 82
Vlaer, E. 91
Volger, Willem 47, 51
Voorn, Mr. Van 90
Vos, Meijndert 271
Vosburg, Gelmer 31, 32, 34, 37, 43-48, 51, 52, 85
Vosch, Joriphaes 22, 26, 89, 91
Vries, Simon de 49

Wassenaar, Jan 271
Wast, Alexander 270
Weert, Cornelis van 91
Weert, Johannes Fransz. van 89
Westerhuijse, Dirk 271
Wichelman, Magnus 76, 85, 161, 169
Wickevoort, F. van 169
William III of Orange 44, 50, 57-59, 63, 65, 71, 75, 77, 153, 155, 159, 160, 164, 172, 173
Wisdorpius, Johannes 43, 47
Wissel, Digna van 160
Wissel, Johannes van 160
Withoos, Alida 162
Witsen, Nicolaas 104
Witte, Willem de 30

Zanoni, Giacomo 39, 40, 106-108, 111, 118, 119, 123, 166
Zwaardekroon, Hendrik 48, 65, 67, 69, 78, 80, 85, 87, 89-91, 269, 270
Zwaardekroon, W. 67, 268

Index of geographical names

Aachen 59, 84
Abbenhoesen 49
Abessinia, ambassador of 82, 271, *see also* Godsja Moraat
Africa vii, 62, 73, 74
  East 38
  South 11, 63, *see also* Cape of Good Hope
Aicotta, *see* Azhikkōdu
Ālangādu 19, 26, 85, 119
  prince of 147
Aldea 16
Alur 78, 85
Ambon vii, 29
  governor of, *see* Antoni Hurt, Robert Padbrugge
Ameliswerd, house 59
  lord of, *see* Hendrik van Utenhove
America vii, 62, 75, 140, 154, 161, 162, 172, 174
Amerongen 58, 66
  castle 4, 6, 59
  house ix, 59, 98
  lord of, *see* Frederik, Godard, and Godard Adriaan van Reede van Amerongen
  seigniory 59
Amersfoort 31
Amsterdam ix, 5, 7, 12, 28, 41, 42, 44, 49-51, 57, 62-65, 67, 76, 88, 91, 97, 104, 125, 134, 138, 150, 151, 160-164, 170, 173, 174, 269
  admiralty 3, 5, 6, 66
  alderman, *see* Willem Adriaan van der Stel
  Athenaeum library 165
  botanical garden, *see* Hortus Medicus
  burgomaster, *see* Joan Hudde, Joan Huydecoper van Maarsseveen, Gillis Valckenier
  Chamber of viii, 26, 31, 51, 65, 82, 83
  chemist's shop 8, 11
  classis of 34, 41, 49
  council of 10, 63, 66, 160, 161, 268
  councillor of, *see* Jan Commelin, Joan Hudde, Joan Huydecoper van Maarsseveen, Joan van Oosterwijck, Gillis Valckenier
  Hortus Medicus 12, 22, 63, 64, 71, 72, 75-78, 83, 84, 104, 138, 150, 154, 160-166, 168, 169, 172-175, 269
  Municipal Archive ix
  Municipal Garden, *see* Hortus Medicus
  Nieuwe Plantage 63
  Old Church 3, 50
  Remonstrant Church 5, 6
  university library 165
Anchi Kaimal, *see* Curūrnādu
Angimal (Ankamāli), church 38
Antilles 154, 165
Antwerp 42, 49, 104
Arabian Sea 13
Arnhem 66
Asia vii, 7, 10-14, 16, 29, 41, 51, 58, 61, 62, 64, 65, 70, 75, 77, 78, 88, 95, 130, 140, 143, 145, 150, 154, 156-158, 160, 167, 171, 172, 174
  Eastern Quarters 51
  Western Quarters viii, 10, 23, 27, 52, 76-78, 80, 269
  Western Quarters, commissioner-general, *see* Hendrik Adriaan van Reede tot Drakenstein
Asia Minor 139
Āttingal 20
Azhikkōdu 35

Bagueur, palace 35
Bantam 154
Barbados 164
Bārssalūr 13
Basra, *see* Bassora
Bassora 38-41, 111, 149, 151
Batavia viii, 7, 10-14, 21-23, 26, 27, 29, 31, 34, 35, 37-40, 42, 44-47, 50-55, 59, 61, 63, 65, 67, 69, 70, 72, 75-78, 81-85, 88, 95, 101, 120, 130, 134, 135, 145, 149, 151, 154, 159, 161, 171, 268
  castle 11, 29
  church council 49
  Council of Justice 51
  laboratory 29, 37, 53
  Leper House 54
  medical shop 13, 29, 37, 42, 53, 70, 76
  printing office 53, 55
  sergeant-major, *see* Isaac de Saint-Martin
  surgeon's shop 11-13
Bender Abbas 39, 78
Bengal 31, 40, 42, 56, 64, 69, 77, 78, 161, 173, 175, 269
  director of, *see* Constantin Ranst, Willem Volger
  king of, *see* Chahestachan
Bergrivier 75
Berlin 59
Bijapūr 13
Bisignano, bishop of, *see* Giuseppe di S. Maria
Black Notley 172
Bōlghatti island 15
Bologna 40
  botanical garden 123, 124
  university 123, 124
Bombay 81, 87, 270
Brandenburg, ambassador in 3, 102
  elector of, *see* Frederik Willem
Brazil 5, 55, 154, 159, 162, 165
Breda, botanical garden 12

C, *see also* K
Cailpatnam 30
Caleture 29
Calicut 13, 35, 39, 95, 119
  university viii
  Zamorin of 13, 15, 17, 18, 28, 33, 35, 36, 45, 47
Canara 13, 33, 36, 39, 67, 119
Cannanūr 14, 15, 18, 33, 35, 48
  chief of, *see* Gelmer Vosburg Kolathiri of 13
Cape Comorin 13, 20, 21, 27, 95
Cape of Good Hope viii, 7-12, 29, 44, 50, 52, 54-56, 63, 65, 67, 69-77, 82, 84, 85, 88, 153, 161, 171, 173, 174
  castle 69
  commander of, *see* Jacob Borghorst, Jan van Riebeek, Simon van der Stel
  commissioner of 72
  Company's garden 7, 8, 70, 72-77, 83
  council of 31, 56, 82-84
  governor of, *see* Joan Bax van Herentals, Hendrik Crudop, Isbrand Godske, Simon and Willem Adriaan van der Stel
  hospital 69
  school 69
Capua 38
Carembaly 22
Ceylon vii, viii, 9, 10, 13, 22-25, 27, 29-32, 34-38, 40, 42, 44, 46-57, 61-64, 67, 69-72, 76-78, 82, 84-86, 88, 89, 91, 118, 140, 144-146, 153-156, 159, 161-165, 171-176, 268, 269
  captain (first), *see* Hendrik Adriaan van Reede tot Drakenstein
  church council of 42
  commander (first), *see* Laurens Pijl
  council of 22, 26, 27, 30, 31, 47, 49, 50, 88, 91, 159
  councillor of, *see* Hendrik Adriaan van Reede tot Drakenstein
  governor of 88, 159, 269, *see also* Rijklof van Goens Jr. and Sr., Laurens Pijl, Thomas van Rhee
  king of 85, *see* Radja Sinha
  laboratory 54
  North 7, 10, 22, 78, 85
  sergeant-major, *see* Joan Bax van Herentals, Hendrik Adriaan van Reede tot Drakenstein
  South 7, 76, 77, 84
Chathiath 40
Chennanmangalam 45
China 26
Città di Castello, bishop of, *see* Giuseppe di S. Maria
Cochin 5, 12, 13, 17-19, 22, 23, 31-36, 38, 40, 43-45, 47-51, 53, 61, 66, 76, 78, 80, 81, 87, 97, 100
  commander of 268
  kingdom 13, 15, 16, 18, 19, 36, 37, 43, 44
  prince of 36, *see also* Erorma, Rāma Varma, Vīra Kērala Varma
  rāja of 13, 21, 33, 35-37, 45, 48, 64, 101, 147, 151, *see*

also Gōda Varma, Rāma Varma, Vīra Kērala Varma rāni of 32, *see also* Gangadhara Laksmi
regedore maior of 48, *see also* Olatche, Hendrik Adriaan van Reede tot Drakenstein
Cochin (city) 13-16, 18, 27, 30, 33, 35-37, 39-42, 48, 95, 105, 119, 134, 144-147, 149, 151, 172, 268, 270
church council of 42, 49, 87, 90
clergyman, *see* Johannes Casearius, Marcus Mazius
council of 16, 18
hospital 38, 42
laboratory 37, 38, 41-43
medical shop 42
secretary, *see* Christiaan Herman van Donep
Codda Carapalli (Coddacarappalli) 50
Coladda (Collada), house 43, 50, 146
Cologne 47
Colombo 10, 13, 22, 26, 27, 29, 30, 42, 46, 47, 50, 70, 76-78, 85, 87, 88, 91, 174
hospital 28
secretary 47
Copenhagen 66
Coromandel viii, 29, 40, 42, 64, 70, 78, 85, 87, 118, 161, 174, 175
council of 31
governor of 269, *see also* Antoni Paviljoen
Cranganore River 35, 45
Cranganūr 23, 25, 33, 35, 47, 48, 85
prince of 147
rāja of 15, 35
Cranganūr (city) 14, 15, 18, 35, 48
church 17
Curaçao 161-163, 170
director of, *see* Jan van Erpecum
Curūrnādu 15, 36, 85
Cyprus 118

Damao 118
Danzig 70, 71, 171
Decima 54
Delft 5, 59
Denmark, ambassador 3
De Schuur 69
Deventer 138
Devil's Peak 73
Diu 38, 118
Drakenstein (Netherlands), house 4, 6
lord of 64, *see also* Ernst van Reede, Gerard van Reede van Drakenstein
Drakenstein (South Africa) 75, 76

East, Middle 40, 41

Near 40, 41, 139
Emmerich 4
England 14, 23, 153, 172
ambassador 3
Ernakulam 15, 40
Europe vii, 8, 23, 29, 30, 33, 38, 39, 54, 63, 67, 70, 73-77, 139, 140, 149, 158, 165, 174, 269

False Bay 69
France 35, 75, 123, 153
Franeker 54
Fulham Palace, gardens 173

Galle 76, 77, 85, 88
church council of 49
clergyman, *see* Philippus Baldaeus
commander of, *see* Joan Bax van Herentals
Gamron, *see* Bender Abbas
Ganges 77
Geervliet, clergyman, *see* Philippus Baldaeus
Gelderland 59
Germany 47, 59, 84, 153
Ghats, mountains 13
Western 13, 119
Ginkel, lord of, *see* Godard van Reede
Goa 12-14, 19, 26, 32-34, 38, 39, 95, 118, 149, 154, 159
Gouda 67, 150
Goudestein, house 63, 71, 160
Great Britain, king of, *see* William III of Orange
Groningen 62, 87
botanical garden 62
Guarapes 5
Gujarat 38

Haarlem 63, 83, 160
Hague, The 66, 97, 160, 171, 172
General State Archive viii, ix
Hampton Court, royal gardens 173
Harderwijk, university 67
Hartekamp 174
Hasselt 59
Heemstede 174
Heer-Hugowaard 4
's-Hertogenbosch 59
Hieropolis, bishop of, *see* Giuseppe di S. Maria
Holland 82, 160, 163
Court of 66, 268
North 4
States of 170, 176
Honselaarsdijk, garden 172, 173
menagerie 71
palace 71, 164
Horstermeer 4
Hottentots Holland 69, 70
Houtbaai 73, 74

India vii, 7, 10, 13, 14, 19-24, 26, 27, 35, 38-41, 67, 69, 87, 90, 118, 120, 140, 162, 171, 268

commissioner-general of, *see* Rijklof van Goens Jr.
council of 11, 26, 30-32, 44-52, 55-57, 64, 66, 67, 81, 82, 85, 90, 91, 151, 159
councillor of, *see* Rijklof van Goens Sr., Laurens Pijl, Constantin Ranst, Cornelis Speelman
director-general of, *see* Rijklof van Goens Sr.
extraordinary councillor of, *see* Joan Bax van Herentals, Rijklof van Goens Jr., Robert Padbrugge, Constantin Ranst, Hendrik Adriaan van Reede tot Drakenstein, Isaac de Saint-Martin
governor-general of, *see* Joan Camphuijs, Rijklof van Goens Sr., Joan Maatsuiker, Hendrik Zwaardekroon
High Government of 10, 11, 23, 27-32, 37, 38, 46, 50, 51, 59, 61, 64, 67, 69, 77, 83, 134
Portuguese 154
Indian Archipelago 12, 95, 174
Indies 39
East 7, 154, 165
West 153, 162
Iraq 38
Irinjālakkuda 48
Isfahan 39
Italy 38, 40, 106, 123

Jaffanapatnam, Jaffna 13, 22, 26, 29, 34, 78, 89-91
chief of, *see* Joriphaes Vosch
clergyman, *see* Philippus Baldaeus
commander of, *see* Laurens Pijl, Hendrik Zwaardekroon
school 42
seminary 78, 85
Japan 11, 54, 75, 78, 84
chief of, *see* Joan Camphuijs, Constantin Ranst
Shogun of 54, 55
Java 11, 12, 72, 156, 161, 175
East 51
Jedo 54, 56
Johannesburg, Brenthurst Library 71

K, *see also* C
Karapurram (Carrapuram, Carrappuram) 43, 50
Kārthi-kappalli, rāja of 19, 26
Karunāgappalli 22
rāja of 27
Kassel 50
Kāyamkulam 21-23, 25, 33
Kerala, *see* Malabar
Klapmuts 69
Konkan 149
Koperberg 70

Kōttayam 42, 47
school 34, 42, 147, 149
Kuravilanādu, kattanār of, *see* Alexander da Campo
Kyūshū 56

Lebanon 38, 118
Lebanon Mountain 111
Leeuwenberg 72
Leeuwenhorst, house 71, 164
Leiden 12, 30, 31, 41, 54, 63, 66, 84, 145, 153, 156, 160, 164, 165, 171-173
Ambulacrum 12, 83, 154, 159
botanical garden, *see* university garden
curators of 63, 134, 153, 164, *see also* Hieronimus Beverningk
Rijksherbarium 104
university 12, 41, 61, 83, 138, 144, 153
university garden 8, 12, 13, 31, 63, 64, 76, 77, 83, 138, 153, 154, 159, 160, 163, 165, 172, 174
university library 77, 154
Leipzig 31
Lek, river 89
Lekdijk Bovendams, dike-reeve of 59
Lequas 54
Lion-Hill 8
Loenen 71, 172
Lokhorst 164
London 66, 97, 156
bishop of, *see* Henry Compton
British Library ix, 59, 125
British Museum 125
Royal Society of 174
Loochoo islands 56
Lopik 89, 91
Lopikerkapel 26, 89, 91
church 90
Verwershoef 91
Lübeck 50

Maarssen 71, 82
Maarsseveen 63, 71, 82
Maarsseveen, lord of 101, *see also* Joan Huydecoper van Maarsseveen
Madagascar 26, 71, 154
Madras 174
Madurai 21, 25, 26, 28, 30, 31
Neik of 21, 22, 27-31
Pulle of 28
Teuver of 21
Malabar vii, viii, 4, 5, 7, 9, 10, 12-49, 51-54, 59, 61, 63, 64, 69, 72, 75, 77, 80, 81, 83, 85-88, 91, 95, 105, 118, 119, 123, 130, 134-136, 140, 143-151, 153, 156, 158, 161-169, 171-176, 268-270
commander of vii, viii, 3, 9, 10, 18, 21-23, 25-27, 29-31, 44-48, 50-52, 80, 82, 85, 86,

Index of geographical names

90, 268, 269, *see also* Isaac van Dielen, Lucas van der Dussen, Adam van der Duyn, Isbrand Godske, Martin Huisman, Jacob Hustaert, Jacob Lobs, Hendrik Adriaan van Reede tot Drakenstein, Gelmer Vosburg, Magnus Wichelman, Hendrik Zwaardekroon
- commissioner of 48, 50, 67, *see also* Martin Huisman, Hendrik Zwaardekroon
- council of 10, 18, 23, 26, 31, 32, 47-51, 85, 90, 151
- council of justice 32
- councillor of 16, 18, 23, *see also* Joan Bax van Herentals
- flora of 13, 29
- natural history 17, 20, 25, 26
- princess of 18
- South 22, 23, 27

Malacca 11
- governor of 47, 55, *see also* Balthasar Bort, Gelmer Vosburg

Malay, lord of 66, *see also* Paul de la Baye
Manapar 30
Mandurty (Noordwijk) 35
Mangalīer, river 13
Mangati 119
Manike Magalan 48
Mannar 22, 26
Marcianise 38
Mataram, kingdom 51, 52
Maturé 76, 85
Mauritius 63, 71, 161, 162
Mediterranean area 14
Mexico 154
Middelburg 7
Mijdrecht 57
- house 57, 65, 88, 89
- lord of, *see* Hendrik Adriaan van Reede tot Drakenstein, Reinoud Gerard van Tuyll van Serooskerken
Milan 39
Morenbril, *see* Carembaly
Mount Carmel 38, 41, 106
Mouton 39, 43, 45, 146
- lords of 36
- Union 37, 48
Mozambique 38, 118
Münster, Treaty of 3

Nāgappattinam 27, 78, 84, 85, 118
Nalur 85
Namaqualand 72, 74, 75, 83, 84
Naples 38
- university 38
Nederhorst, colony 4-6
- house 4, 6, 59
- lady of, *see* Agnes van Reede tot Drakenstein
- lord of, *see* Gerard and Godard van Reede van Nederhorst
Nederhorst-den-Berg 4, 6
Neerdijk, lord of 101, *see also* Joan Huydecoper van Maarsseveen
Negombo 50, 76
Netherlands vii, 3, 5-12, 14, 22, 26, 29-31, 33, 39-41, 44, 47, 51, 52, 55-57, 59, 61-64, 70, 71, 76-78, 82, 84, 85, 88, 91, 98, 134, 136, 145, 150, 151, 153-156, 158, 160, 163-165, 171, 172, 174
- queen of, *see* Beatrix
New York 4, 5
Nigtevecht, lord of 91
Noordwijk (Mandurty) 35, 47
Noordwijk (Netherlands) 71
Noordwijkerhout 164, 172
Noudepont sur Varne, lord of 66, *see also* Paul de la Baye

Oostende 4
Orange, prince of, *see* Frederik Hendrik, William III
Oranienburg, governor of 66, *see also* Karel van Reede tot Drakenstein
Ougli 51, 56, 77, 85
Oxford 63

Padova, *see* Padua
Padua 31
- botanical garden 139
Pagodinso 22
Palestine 38, 118
Paliacatte 78
Pallippuram 33, 35, 48
Pallurutti, church 45
Papeneiland 15
Paradijs, Het 69, 73
Paramaribo, botanical garden 161
Paris 66, 75, 156
- Institut de France 63, 164
- Jardin Royal (Jardin des Plantes) 156, 165
- Musée d'Histoire Naturelle ix, 107
Parūr 15, 85
- prince of 19, 147
Nāyars of 36
Pernambuco 55
Persia 11, 31, 33, 39, 40, 45, 78, 118
- mission 38, 39
- director of 85, 269
- directorate of 82
Pisa, botanical garden 139
- museum of natural history 139
- university 139
Poland 59
Pondecail 30
Ponnāni 35
Porka 39
Portland, earl of, *see* Hans Willem Bentinck
Portugal 34, 39, 40
Pretoria, Botanical Research Institute 83
Purakkād 18, 32
- rāja of 15, 33

Quilon 13, 14, 19-27, 31-33, 84
- chief of 21-23, *see also* Joan Nieuhof, Hendrik Adriaan van Reede tot Drakenstein
- fortress 14, 35, 48
- Portuguese governor 23
- princess of 33
- queen of 27
- St. Thomas Church 23
Quilon de China 24

Renswoude, house 4, 6, 59
- lord of, *see* Frederik Adriaan, Gerard, and Johan van Reede van Renswoude
Republic, Dutch, *see* Netherlands
Rhedervaart 4
Rijnestein, lord of 104, 266-267, *see also* Johan van Renesse van Baar, Antoni Utenhove, Charles Utenhove, Hendrik Utenhove
Rijnhuizen, house 4
- lord of 266-267, *see also* René van Tuyll van Serooskerken
Rijnsburg 41
Rome 33, 34, 39, 40, 44, 106, 165
Rondebosje 69, 71-73
Roukiou 54
Rustenburg, outer garden 71-73, 76, 83, 84
Ryu-Kyu islands 56

Saasveld, lord of, *see* Goert van Reede
Saxony, ambassador in 102
Scandinavia 59
Schavan, lord of 66, *see also* Paul de la Baye
Serra, Church of the 38-40, 119
Siam 75
- French embassy to 10, 84
Sindi 48
Smyrna 41
Spain 3, 59, 123
St. Germain 75
Staten Island 4
Steenberg 83
Stellenbosch 69, 76
Surat viii, 33, 38-40, 45, 47, 67, 81, 86-88, 91, 118, 162, 270
- Company's garden 87
- council of 49, 81, 86, 91, 270, 271
- director of 47, 49, 51, 86, 90, 91, 269, 270, *see also* Dirk van Adrichem, Willem Volger, Gelmer Vosburg, Hendrik Zwaardekroon

Surinam 84, 161, 162, 169
- director of 56, *see also* Joan van Oosterwijck
- governor of 161
- Society of 162, 170
Sweden 175
- ambassador in 3
Syria 118
- mission 38

Table Bay 69
Table Mountain 8, 73, 74
Table Valley 73
Taprobana (Ceylon) 25
Tatta 48
Tekkumkūr 32, 39, 47
- kingdom 34
- rāja of 21, 33, 34, 47, 147, 149
Tēngāpattanam 20-22, 33, 35
Ternate, governor of 31, 51, *see also* Jacob Lobs, Robert Padbrugge
- sultanate 64
Texel 55, 57, 65, 66, 69, 75
Teylingen, Oud-, estate 164
Theil, baron of 66, 266-267, *see also* Paul de la Baye
Tonkin, chief of 55, *see also* Constantin Ranst
Travancore 25, 26, 32, 78
- rāja of 13, 19, 21, 22, *see also* Rāma Varma
Trichur 36
Trincomalee 87
Turkey 41, 139
Tuticorin 21, 22, 25-31, 35, 78, 80, 82, 84, 85, 87, 91
- chief of 21, 26, 28, 31, 67, 85, *see also* Joan Nieuhof, Laurens Pijl, Thomas van Rhee
- pearl-fishery 13, 28

Uppsala, university garden 175
Utrecht ix, 3-5, 26, 42, 49, 50, 57-59, 62-67, 87, 88, 90-92, 102, 150, 268
- Audit Office 57, 66
- botanical garden 12, 63
- Court of 66, 88, 91, 268
- Equestrian Order ix, 3, 5, 57-59, 65, 66, 89, 98, 103, 268, 270
- Orphans Court 5, 6
- States of ix, 3, 4, 57-59, 65, 87, 102, 270
- university 5, 41, 63, 67, 268

Vadakkumkūr 39
- prince of 147
Vaipin 25, 35, 37, 39
- Company's garden 37, 48
Vecht, river 63, 71, 83
Veluwe, Equestrian Order of 59
Vengurla 13, 19, 31, 33, 38, 39
Vianen 90
- burgomaster 90
- Chamber of Justice 92
- dike-reeve 92

Dutch Reformed Church 92
Feudal Court 92
Walloon clergyman 90, *see also* Guillaume Jalabert
Vijverhof, house 71
Vliet, lady of 91, 266-267, *see also* Elisabeth Antonia van Panhuizen, Francina van Reede
  lord of 90, 91, 266-267, *see also* Antoni Carel van Panhuizen tot Voorn, Bartholomeus van Panhuizen
Vliet, Te, house 26, 89-91
Vreeland, baron of 102, *see also* Hendrik Jacob van Tuyll van Serooskerken
  seigniory 91
Vuursche, De 4, 6
  country-seat 6
  lord of 266-267, *see also* Ernst van Reede, Gerard van Reede
  seigniory 6

Warapoly 40, 45, 49
Warmond 164, 171
Waveren, lord of, *see* Johannes Hudde
Welland, lord of, *see* Godard Willem van Tuyll van Serooskerken
Western Quarters, *see* Asia, Western Quarters
Wittenberg 31

Zevenhoven 89, 91, 92
  seigniory 91
Zuiderhout, country-seat 63, 160
Zuilen, castle 57, 58, 66
  house 98
  lord of 58, 59, 65, 66, 102, 266-267, *see also* Hendrik Jacob van Tuyll van Serooskerken, Reinoud Gerard van Tuyll van Serooskerken
Zuilestein 66
  house 58
  lord of, *see* Frederik and Willem van Nassau, Ernst van Nijenrode, Godard van Reede van Amerongen

Index of plant names

abele 76
Acacia 82
Acacia Tinctoria Hermans 165
Acmella 63, 77, 85
Acorus Asiaticus radice tenuiore 159
Acorus Calamus L. var. *verus* L. 155, 164
Acorus verus Asiaticus radice tenuiore 164
Adenanthera pavonina L. 162
Aehaela 155
Aeschynomene (?) sp. 155
Aeschynomene aspera L. 155
Aeschynomene grandiflora L. 156
Aeschynomene mitis secunda 159
Agaty 156, 157
Alcea 157, 158
Alcea Bengalensis 78
Alcea Malabarensis 158
Alcea Malabarensis, Abutili folio 157
Alcea Malabarensis Pentaphylla (etc.) 157
alder 72, 73
almond tree, wild 8, 71
Aloe 17, 118, 124
Aloe perfoliata L. var. *vera* 162
Aloe vera (L.) Burm. f. 162
Aloe vulgaris 162
Alou 167
Alu 157, 158
Amarantho affinis Indiae Orientalis (etc.) 162
Amaranthus major Zeilanicus spinosus (etc.) 159
Amaranthus spinosus L. 155
Ambalam 159
Ambapaja 100
Ambel 169
Ameri 155
amfiun 17
Amomum Zingiber L. 119, 163
Amorphophallus paeoniifolius (Dennst.) Nicolson 163
Ampana 169
Anakokke 162
ananas 170
Ananas comosus (L.) Merr. 162
Ana-parua 158
Apocynaceae 169
Apocynum 128, 173
apple 72
apricot 8
Araceae 163

Arbor Acaciae foliis Malabarica (etc.) 156
Arbor Lanigera (etc.) 167
Arbor Sancti Thomae (etc.) 155
Arbor siliquosa Malabarica 157
Arbor Triste 17
Arbores bacciferae 167, 168
Arbores Pomiferae 167
Arbores pruniferae 167
Arbores siliquosae 168
Archam Indiae Orientalis arbor 111
Arealu 143
areca 14, 33, 119, 122
  growers 37
  orchard 37, 43
  palm 37
  trade 37, 48
Areca catechu L. 37, 119
Aria-bepou 135
Arkasond 119
Arna Vareca 119
Artocarpus integrifolia L. 119
Arum 151
Arum bifolium Arabicum maculatum 145
Arum humile Arisarum dictum (etc.) 163
Arum Maximum Ceylonicum 163
Arum polyphyllum, Dracunculus (etc.) 163
Arundo indica 13
Arundo Indica arborea maxima, cortice spinosa 155
Asclepias gigantea L. 164
ash-tree 73
asparagus 8
Asparagus sarmentosus L. 163
Asperagus aculeatus maximus (etc.) 163
Astragalus Indicus spicatus (etc.) 164
Asymbolae 122
Atty-alu 143
Avanacu 158
Awari 155

Bakeli tree 118
Bala 156
Bamboo 12, 13, 24, 154
Bambusa arundinacea (Retz.) Willd. 12
banana 11, 14, 17, 62, 155, 156, 158

Barleria Prionitis L. 163, 165
Basella 162
Basella rubra L. 162
batata 11, 17
Batatta sylvestre 163
Bauhinia scandens L. 165
Bauhinia variegata L. 120, 155
bay-tree 72
bean, red 162
beech 71, 73, 76
beet 8
Bella modagam 167
Beloere 162
Belutta pola taly 164
Belutti-areli 162
Bem-curini 136
Ben Kalesjam 167
Bentheka 167
betel 11, 37, 38, 48
Betel-nut Palm 119, 156
Bignonia indica L. 155
Biophytum sensitivum (L.) DC. 12
Blitum monospermum Indicum (etc.) 159
Borassus flabellifer L. 26
borborri, borreborry 17
Brabeium stellatifolium L. 8
Bromelia Ananas L. 162
Bryonia Ceilanica foliis profunde laciniatis 163
Bryonia laciniosa L. 163
bulbs 141
  Cape 10, 71, 75, 83, 161
  Ceylon 76
  Malabar 44
Bupariti 157, 158

cabbage 8, 11, 72, 122
  white 8
Caesalpinia Sappan L. 165
Caicotten-pala 157
Calamus Aromaticus 164
Calli 158
Calophyllum calaba L. 159, 164
Cambogia Gutta L. 155
Campanulaceae 166
camphor-tree 71, 75
  Sumatran 161
Canella ex qua Cinnamomum 155
Caniram 154, 159
Canna indica L. 164
Canschena-pou 159
Capo-molago 164

Capraria biflora L. 163
Capsicum frutescens L. 164
carambola 17, 122
Carambu 158
cardamom 14, 17, 33
Cardamomo 39
Caretti 164
Carim-curini 136
Carimpana 26
Cari-villandi 168
Carrion-Flower 8
carrot 8, 11
carrot-salad 8
Carua 155
Cassia 156
Cassia Cinamomea 159
Cassia fistula L. 108, 155
Cat-ambalam 157
Cattu-gasturi 158
Catulli pola 162, 164
Caunga 119, 156
Cavalam 155
Cerbera Manghas L. 118, 155
Champacam 158
cherry 72
chervil 8
chestnut 76
Chovanna-mandaru 120, 155
cinnamon 22, 33
  Ceylon 61
  trees 44, 71, 72, 76, 82, 155
  wild, distillation of 38, 42
  wild Malabar 61
Clitorius Ternatea L. 162
clove 17, 71
coconut 11, 14, 17, 23, 82
  palm 17, 33
  tree 16, 20, 37, 50, 156
Codaga-pala 158
Codda-panna 167
Coddam-pulli 155
Coletta-veetla 163, 165
colocynth apple 29, 31
Compositae 166
Coniferae 122
Conna 155, 159
convolvuli 128
Convolvulus Ceylanicus Villosus (etc.) 164
Convolvulus Maritimus Ceylanicus (etc.) 164
Convolvulus Pes caprae L. 164
copal, Indian 14
corn-salad 8
Costus arabicus L. 163

cress 8
Crinum asiaticum L. 164
Crinum zeylanicum (L.) L. 164
Crocus Aladar 108
Crotolaria sp. 162
Cruciferae 166
cryptogams 166, 168
cucumber 11, 122
Cucumis sp. 162
Cucumis puniceis Zeylanicus minor (etc.) 170
Cucurbitaceae 168
Cudupariti 157
Curcuma longa L. 163
Curcuma Officinarum 163
Curcuma radice longa 108, 163, 164
Curini 158
currant-bush 73
Curuta-pala 156
Cycas circinalis L. 164
Cyperaceae 169
Cystus Indicus quinque nervis (etc.) 165
Cytisus cajan L. 162

Danti 118, 124
date-palm, Persian 78
Datura 17, 138
 fruits 136
dicotyledons 166, 169
Dioscorea sp. 163, 168
durian 11

Ela-calli 161
elder 73
Elengi 158
elm 76
Epidendrum furvum L. 11
epiphytes 166
Ericu 164
Eryngium Ceilanicum febrifugum (etc.) 165
Erythroxylum japonicum non spinosum (etc.) 170
Eugenia malaccensis L. 155
Euphorbia antiquorum L. 161
Euphorbia neriifolia L. 161
Euphorbia Tirucalli L. 161

ferns 141, 166, 168, 169
fico grande 39
Ficus sp. 157, 167
Ficus bengalensis 78
Ficus Malabarensis 157
fockie fockie 11
foodplants 8, 12
fruit-trees 8, 72
 European 72
 Indian 24
Frutices bacciferi 168
Frutices bacciferi Scandentes 168
Frutices floribus Papilionaceis 168
Frutices qui cum herbis (etc.) 168
Fula Mestica 162
Fulfel 111
fungi 141

Galega Aegyptiaca siliquis articulatis 157
Galegae affinis Malabarica arborescens (etc.) 157
Genista 162
Ghoraka 155
giacca 119
Giassoan 108
Giasson 108
ginger 14, 17, 119
Gloriosa superba L. 162-164, 168
Glycine Abrus L. 162
Gomphrena globosa L. 162
Gossympium arboreum L. 158
Gramineae 169
grapefruit 11
grasses 72, 141, 169
Guilandina Bonducella L. 164

Haemanthus coccineus Jacq. 8
hazelnut tree 71
heather 72, 74
Hedera 168
Hedyotis Auricularia L. 119
Hedysarum diphyllum L. 165
Helicteres isora L. 119
Herba viva Christoph. à Costa 13
Herbae anomalae 169
Herbae flore monopetalo 168
Herbae flore monopetalo vasculiferae 169
Herbae flore papilionaceo (etc.) 169
Herbae flore pentapetaloide anomalae 169
Herbae Pomiferae 168, 169
Herbae radice bulbosa 169
Hibiscus sp. 168
Hibiscus Abelmoschus L. 158
Hibiscus cannabinus L. 161
Hibiscus mutabilis L. 162
Hibiscus populneus L. 158
Hibiscus Rosa sinensis L. 158
Hibiscus tiliaceus L. 158
Hina paretti 97, 162
horse-chestnut 76
Hummatu 136, 158

Iambos major 155
Iambos Sylvatica, fructu cerasi magnitudine 155, 156
Igera muri 119, 120
Ily 154, 155
Indigo 78, 156
Indigofera hirsuta L. 164
Indigofera tinctoria L. 155
Inota-inodien 163
Inschi 119, 163
Ipomoea Pes tigridis L. 164
isāramūri 120
Isora-murri 119, 129
Itty-alu 143

Jack Fruit Tree 119
jagerboom 26
Jaka 119
jambo 122
Jusala 163

Kada-kandel 168
Kadanaku 162

Kaempferia rotunda L. 163
Kaginri 118, 124
Kaida 157, 158
Kaida-taddi 158
Kaida tsjerria 158
kaita cakka 120
Kalesjam 167
Kanna Ghoraka 155
Kapa-tsjakka 162
Kareta-tsjori-valli 135
Karetta amelpodi 102
Kasjavo maram 136
Kata Tartavè 119
Kathou-theka-maravara 35
Kathukarohiti 165
katjang 17
Katou belluta amelpodi 102
Katou-indel 167
Katou-kadali 165
Katou-kalesjam 167
Katou-naregam 135
Katou-theka 167
Katou-tsjeroe 134
Kattu-tagera 164
Kattu tirtava 119
Katu bala 164
Katu-conna 129
Katu-uren 163
kayita cakka 120
Kelengu 163
Kiridiwael 156
kitsery 17
Kniphofia uvaria (L.) Hook f. 8
Knowltonia vesicatoria (L.f.) Sims 84
Konni 162
Kua 162, 164
Kurka 163
Kurudu 155

lac-tree 17
laurus 119
leguminosae 128, 169
lemon 24, 72, 73
letter wood 22
lettuce 11
Lilium superbum Ceylanicum 163
lime 11, 73, 76
Limonia 167
long grass 8
Lonicera 168

mace 17
Mail-anschi 158
Malacca schambu 136
Malan-kua 163
Mal-naregam 165
Malus limonia 167
Malus limonia pumila (etc.) 165
Malvaceae 158, 168
Mandaru 120, 157-159
Mandragora 111
Mandsjadi 162
Mangas fructu venenato C. Bauhin 155
Mangas Tree 119
mango 11, 17, 24, 82
Manjapumeram 158
Manjella Kua 164
māntāram 120

marjoram 8
Marotti 154, 159
Melastoma aspera L. 165
Mendoni 162-164, 168
Milium indicum sive Sorghum rubrum 13
milkwood tree 8, 22
Mimosa non spinosa major (etc.) 155
mimosacea 122
Minari 163
Modagam 167
Momordica Charantia L. 162-164
monocotyledons 166, 169
mosses 141, 166
Moul-elavou 134
Mountain Ebony 119
Mudela-nila-hummatu 136
Mu-kelengu 163
Mulenschena 163
Mullen-belleri 162
Murigati 119
Muri Kuti 119
myrabolans 31
Myrsine africana L. 84

Naga-dante 111
Naga-mu-valli 165
Nai 118, 124
Nalla-mulla 162
Nandi-ervatam 164
Narcissus Ceylanicus, flore albo (etc.) 164
Naregam 167
Narinam-poulli 161
Nati-schambu 155, 156
Natsjatam 168
Nehoemeka 163
Nelam-mari 165
Nelenschena minor 163
Neli-pouli 125
Nelta 161
Nerium Indicum (etc.) 162
Nerium indicum Mill. 162, 163
Nerium Oleander L. 162, 163
Nianghala 162
Niir-pongelion 168
Nila-hummatu 136
Nilicamaram 156, 159
Nosi 158
Nuayhas 155
Nyctanthes multiflora Burm. f. 162

oak 71, 76
 European 73
Ocimum gratissimum L. 119
Odallam 119, 155, 157, 159
Oepata 136
olive 8
 olive-trees 76
onion 8, 17
opium 17, 21, 22, 25, 33
orange 8, 11, 72
orchids 169
Oxyacantha 158

Paeru 135
Pala 158

Index of plant names

Palega-pajaneli 155
palm 78, 122, 155, 156, 158, 167
Palma cujus fructus sessilis Faufel dicitur 156
Palma humilis longis latisque folijs C.B.P. 156
Palma Indica coccifera angulosa (etc.) 156
palmyra palms 20, 26
Pana 158
pana, Canara 119
  Guzerat 119
Pancratium zeylanicum L. 162, 164
Pandi-pavel 162, 163
Panel 157, 158
Panja 167
Papajamaram 100, 153
papaya 11, 153, 156
Papilionaceae 166
Pariti 156-158
Pariti seu Tali-pariti 158
parsley 8
Parthenocissus 168
Pavel 163
peach 8, 72
pear 72, 73
peas 8
Pee-inota-inodien 163
Pee-tsjanga-pulpam 163
pepper 13, 14, 17, 22, 24, 33, 37, 44
Peralu 143
Perin-kaida-taddi 158
Perin-toddali 164
Persica 157
Persicae similis angusti folia (etc.) 157
phanas, see pana
Physalis minima L. 163
Physalis pubescens L. 163
pinang 11
pine 76
pine-apple 11, 17, 162
pineapple tree, wild 8
Pinxevi 111
plane-tree 76
Planta bifolia humirepa (etc.) 165
Podocarpus latifolius 8
Poerinsii 164

Poinciana pulcherrima L. 162
Polygala Indica frutescens (etc.) 155
pomegranates 72
Pomiferae 122
Pongam seu Minari 163
poplar 76
Pulli 158, 159
Pulli-schovadi 164
pumpkin 11
Pungam 163

quercus 119

rais de deos 11
Red-Hot Poker 8
radish 8
Rambora 111
reed 7
  Indian 12
Rhamnus Jujuba L. 164
rice 14
  black 17
  white 17
Ringhini 108
Root-tree 17
Rosa Brasiliensis 162
rotangs 169
Ruellia antipoda L. 163
runner-bean 8, 11

Sabsanta 39
sage 8
Sagou 54
salsaparilla 29
Sambac Lesmin Arabicum 162
Samstravadi 134
sandalwood 14
Sapindus trifoliata L. 164
Schada-veli-kelangu 163
Schadida-calli 161
Schambu 158, 159
Schanga-cuspi 162
Schem-pariti 158
Schena 163
Schetti 158
Schorigenam 158
Schovanna adamboe 164
Schulli 158
Schunda 158

Schunda-pana 136
Scrophulariaceae 166
Sensitive Trees 17
Sesban 157
Sida acuta Burm. f. 163
Sida asiatica L. 162
Sidracalli 161
Siliquosae 122
Solanum vesicarium Indicum minimum 163, 164
Sorghum rubrum, see *Milium indicum*
Sorghum saccharatum (L.) Moench. 12
sorrel 8
Spilanthes paniculata Wall. ex DC 63
Spinosae 122
Stapelia variegata L. 8
Sterculia Balangas L. 155
Stipsdanti 111
St. Thomas Flower 119
sugar cane 11
Sweet flag 155

Tabernaemontana alternifolia L. 156
Tagera 158
Tolabo 164
taloets 82
Tamara 169
tamarind 17
Tandale-cotti 162
tarragon 8
tea-shrub 75
teak 14
Telabo 155
Tenga 97, 125, 155, 156
Teregam 167
Teucrium Americanum procumbens (etc.) 170
Thalia maravara 119
Theka 167
Thora-paërou 162
Tinda-parua 135, 158
Tiru-calli 161
Todda-panna 54, 164, 167
Todda-vaddi 118
Toppi 111

Tota Vari 118, 120
tottã vati 120
Touch-me-not 12
Trachyandra divaricata (Jacq.) Kunth 84
Tsetti-mandarum 162
Tsiapangam 165
Tsjaka-maram 119
Tsjakela 167
Tsjana-kua 163
Tsjela 167
Tsjerou-panna 164
Tsjerou-theka 167
Tsjeru-pana 163
Tsjovanna-aleri 162
Tsjurel 169
Tulip, African 8
Turbith 118, 124

Umbelliferae 166
Unaghas 155

Va embu 155, 164
Valli-kara 168
Vicia Benghalensis 78
vine 72
Vitis 168
vomit nut 31

Wadapu 162
Waelambillu 164
Walgambu 155
walnut-tree 76
waterlemon 8
Wel-ila 163
Wonder-tree 17

Zambal 108
Zantedeschia aethiopica (L.) Spreng. 84
Zedoaria longa 163
Zedoaria rotunda 163
Zenzaro 39
Zerumbeth 163, 164
Zingiber 119
Zingiberaceae 162
Zinziber 163
Zinziber latifolium sylvestre 163
Ziziphus Indica argentea (etc.) 164

Index of animal names

Aianda Polagen (snake) 39

Basij 77, 85
bat 18
bird 12, 39, 74, 136, 138
    Malabar 14, 18
    of prey 136
boar 22

camel 18, 21
canary 71
civet-cat 82
cobra 78
cow 7, 74
crocodile 39

deer 22, 74
dog, wild 73, 84

duck, tame 18

elephant 21, 28, 44
    trade 78
elk 74

falcon 77
fishes 12, 18, 24

Goa cod-fish 17
goat 73, 136
    billy-goat 74
goose 18
    Cape 71

hen 7
heron 18
horse 21

jackal 18
jackdaw 71

leopard 73
lion 73, 138
Lycaon pictus 84

mallard 18

Naghaja 154
Naja 154
Negombo devils 82

owl 136
ox 73

parakeet 18, 71, 82
parrot 18, 71

partridge 18
peacock 18
Perimpambo (snake) 39
pig 7

rhinoceros 71, 83
roe 74

sheep 7, 73
singing-bird 18
siskin 71
snake 12, 18, 39, 71, 136, 138, 154

turtle dove 18

wolve 73

Index of ship names

Adrichem 76, 82
Africa 83
Amersfoort 82
Amsterdam 65, 271
Azië 54, 56

Bantam 65, 67, 76, 82, 85, 91
Blauwe Hul(c)k 51, 55

Capelle 78
Carbares 39
Cortgene 54

Den Helder 75, 84

Dregterland 78, 81, 85-88, 270, 271

Eemnes 82
Eenhoorn 85

Geldria 78
Gouda 82

Ipensteyn 82

Kalf 10

Land van Schouwen 55, 67

Maas 85, 87, 271

Nassouw 270, 271
Nieuwpoort 10

Oranje 7, 271

Pampus 84
Pouleron 46, 47
Purmer 65, 67, 76, 78, 82, 85, 88

Ridderschap 82
Roemerswaal 159

Silversteyn 47, 51
Sparendam 42
Speelman 271
Sumatra 82

Verenigde Provinciën 10
Vrije Zee 55, 56, 83

Waalstroom 84
Wapen van Goes 159
Waterland 85